COLLECTED MATHEMATICAL PAPERS

OF

HENRY J. S. SMITH

Henry J S Smith

THE COLLECTED

MATHEMATICAL PAPERS

OF

HENRY JOHN STEPHEN SMITH

M.A., F.R.S.

LATE SAVILIAN PROFESSOR OF GEOMETRY IN THE UNIVERSITY OF OXFORD

EDITED BY

J. W. L. GLAISHER, Sc.D., F.R.S.

FELLOW OF TRINITY COLLEGE, CAMBRIDGE

WITH A MATHEMATICAL INTRODUCTION BY THE EDITOR, BIOGRAPHICAL SKETCHES AND A PORTRAIT

IN TWO VOLUMES

VOLUME I

CHELSEA PUBLISHING COMPANY

BRONX, NEW YORK

First Published, In Two Volumes, 1894
Reprinted, In Two Volumes, 1965
Reprinted, In Two Volumes, 1979
Library Of Congress Catalog Card No. 65-11859
International Standard Book Number 0-8284-0187-X

Note: This Work Is Printed On "Long-life" Acid-free Paper. All Of The Paper In Volume II And Most Of The Paper In Volume I Is A Cream-color Acid-free Paper. However, Several Of The "Signatures" In Volume I Are Printed On A Paper Bluer In Color, Also Acid-free. This In No Way Affects The Longevity Of The Book Nor Its Readability.

Printed In The United States Of America.

CONTENTS OF VOLUME I

LIST OF PAPERS.

VOLUME II

APPENDIX.

BIOGRAPHICAL SKETCH

THE short record of Henry Smith's life, which I have compiled at the request of his sister, is chiefly based upon a Memoir by herself, which I was anxious to give in its entirety, but for which she has desired me to substitute my own words. I have to thank Professor Irving for a letter containing his recollections of Henry Smith at Balliol in his first years as Fellow and Lecturer. The Memoir and Letters from which I have worked are unfortunately most defective during the years of my own absence from England, 1871–1883; and I must ask my readers to bear in mind under what disadvantages I have attempted to perform a sacred duty. Happily, Henry Smith's character, about which there is really no difference of opinion, exhibits an unbroken continuity of growth. As a boy he seemed to have something of the mature wisdom of a man; and to the day of his death he retained the simplicity and high spirits of a boy. My own estimate of him, based on the close intimacy of more than twenty years, represents, I hope and believe, what his friends thought and would wish said. To those who did not know him, it will perhaps appear that my judgment has been influenced by friendship. Those who knew him will notice points I have missed or excellences I have slurred, and will condemn my inadequacy.

The original plan of this Memoir assumed that it would be supplemented by the publication of a large number of Henry Smith's letters. This was over-ruled in Oxford while I was in Australia, and cannot now be reverted to. Of these letters a few only have been published. Neither, unfortunately, have I leisure or strength to recast the Memoir altogether. I have done what I can in that direction, and can only hope that the difficulties under which I have worked may be borne in mind.

CHARLES H. PEARSON*.

MELBOURNE, 1888.
LONDON, 1893.

HENRY JOHN STEPHEN SMITH was the second son of John Smith, a distinguished though short-lived Irish barrister, who, after graduating at Trinity College, Dublin, was law pupil, at the Temple, of Serjeant Henry John Stephen, the learned editor of Blackstone's Commentaries. Mr. John Smith went back to practise in Dublin, and in 1818 married Mary Murphy. By this marriage he had four children—a daughter, who died of consumption in 1834 at the age of fifteen; a son, Charles, who died also of consumption in 1843, being then a cadet at Addiscombe; a daughter who still lives; and the subject of this memoir, who was named after his father's old tutor, and who was born on the 2nd of November, 1826.

Mr. Smith died in 1828, of abscess of the liver, and his widow was left for a time in very straitened circumstances. Fortunately, after delays which seemed interminable, the Courts affirmed the validity of a bequest of £10,000, which had been made to Mr. Smith by his cousin the Marchioness of Ormond, and which her husband disputed. With this money, and with that produced by the sale of a house which Mr. Smith had just built, the widow had wherewithal to provide adequately for her family; and partly to escape from sorrowful associations, partly to secure for her children that good education which it had been their father's earnest wish they should receive, Mrs. Smith resolved within six months of her husband's death to pass over into England. The family wandered successively to the Isle of Man, 1829; to Harborne near Birmingham, 1829–30; to Leamington, 1830–31; and then to Ryde in the Isle of Wight, where nine or ten years were spent.

Probably no widow left in charge of a young family could have been better fitted to train them for eminence in after life than was Mrs. Smith. A tall

---

* Dr. Pearson's death has deprived this 'Sketch' of the benefit of the author's final revision.

distinguished-looking lady, who retained the traces of great beauty to the day of her death, and who united a certain stateliness of manner and reserve of temperament to Irish ease and kindly charm of manner, she was also considerably more than an accomplished and clever woman, for she possessed powers and learning such as are rare even among able and learned men. Henry Smith inherited genius on both sides. He was a sickly child, and he was also short-sighted from a very early age, perhaps partly from being allowed to read too much when he was quite young. In 1831, when he was only four years old, he was able to follow English and French lessons. He also picked up an old Greek grammar of his mother's, rendered additionally formidable by contractions, and learned the alphabet, the nouns, the adjectives and the pronouns for his own pleasure. 'His practice,' says his sister, 'was to lay himself at full length on his stomach on the floor with the book he wished to study under his chin to suit his sight. When he was between seven and eight I remember Prideaux' *Connection* being for long an absorbing study.' As soon as his mother found out how marked his taste for language was, she took him into her own hands for the classics, and for the next six or seven years he owed all his chief training to her. In 1838 the pupil had got so far that his mother thought it desirable to call in other aid. She was fortunate enough to meet with a highly trained tutor in the person of Mr. R. Wheler Bush, who has put his recollections of Henry Smith on record in the following terms:

'In the years 1838–39 Henry Smith, then a boy of eleven years of age, read with me for about nine months at Ryde, in the Isle of Wight. He had been previously taught by his widowed mother—a remarkably clever and highly educated woman. After reading with Henry Smith I had a large experience of boys during a head-mastership of more than thirty-three years, but I have often remarked that the brilliant talents of Henry Smith prevented me from ever being really astonished at the abilities of any subsequent pupil. His power of memory, quickness of perception, indefatigable diligence, and intuitive grasp of whatever he studied were very remarkable at that early age. What he got through during those few months, and the way in which he got through it, have never ceased to surprise me. From a record which I have before me I see that during that short time he read all Thucydides, Sophocles, and Sallust, twelve books of Tacitus, the greater part of Horace, Juvenal, Persius, and several plays of Æschylus and Euripides. I see also that he got up six books of Euclid, and algebra to simple equations; that he read a considerable quantity of Hebrew; and that, among other things, he learnt all the Odes of Horace by heart. I could scarcely understand at the time how he contrived at his early age to translate so well and so accurately the most difficult speeches of Thucydides, without note or comment to guide him. He was a deeply interesting boy, singularly modest, lovable, and affectionate.' (*Times*, Feb. 12, 1883.)

Scarcely less valuable for the boy's development were the abundant leisure that he enjoyed, and the comparative isolation. His lessons never occupied more than five hours a day, and the obligatory 'constitutional' was only of an hour. During the rest of the time the brothers and sisters were turned out to play by themselves. Their story books were limited to *Robinson Crusoe*, *Evenings at Home*, *Sandford and Merton*, and Miss Edgeworth's *Frank*; their toys consisted of hoops and tops, and one or two dissected games. They grew up like the young Brontës, in a world of their own, improvising plays from *Robinson Crusoe* or combats from Homer. In one of these fights Henry had his finger badly hurt by an arrow from the bow of Achilles, his elder brother, and the surgeon's aid had to be called in. These amusements could not occupy their whole leisure. In idle hours the children became diligent students of animal and insect life, learning much about the habits of bees and ants and spiders and wood-lice and garden moths. They were directed in these pursuits by two books, *Insect Architecture* and *Insect Transformations*, from the Library of Entertaining and Useful Knowledge, and assisted in them by two neighbours, a lady who was something of a botanist and a conchologist, and a Mr. Jacques, who had some knowledge of chemistry. That the interest they took in these matters was more than cursory seems proved by the fact that they supplied Dr. Blomfield, who was engaged on a Flora of the Isle of Wight, with several new homes of rare plants.

In 1839 Mr. Bush was called away to a head-mastership. It proved difficult to supply his place, though an excellent mathematical master was found at Newport, who came over twice a week and carried his pupils through the advanced parts of Arithmetic, elementary Algebra, and Euclid. Henry continued to be a very docile pupil, sometimes asking, when he received an order which displeased him, whether he was 'forced' to obey it, but never demurring if he understood that obedience was required. In 1840, however, he lost his chief fellow-student, through his elder brother going to Addiscombe, and Mrs. Smith decided on moving to Oxford, where it was certain that better teaching could be found than in the Isle of Wight. The Oxford of those days was comparatively a small place. Resident professors, married tutors, and married fellows were almost or quite unknown, while the Heads of Houses, then the governing body of the University, formed a little society of their own. Consequently, the widow lived in comparative solitude, though even so she could not avoid hearing something of the war of opinion that was beginning: of the angry opposition provoked by Tract 90, of Newman's sermons, of the coalition of Evangelicals and Liberals against Puseyites, and now and again of the few Liberals who stood outside the strife of the Churches. Meanwhile she had been exceptionally fortunate in the tutor she secured for her son. The Rev. Henry Highton, Fellow of Queen's and then Curate of St. Ebbe's, was a sound though not a brilliant scholar, and a really good mathematician, far above the average of Oxford in those days. No one could be better fitted to develope Henry Smith's varied capacities, and in Mr. Highton's class-room Henry, for the first time, was able to measure himself with boys of his own age. In the summer of 1841 Mr. Highton received the offer of a Mastership at Rugby, which at that time was chiefly valuable when a boarding-house was attached to it. Mr. Highton accepted the offer, which allowed of his marrying, and proposed that he should take Henry Smith with him as his first boarder. Mrs. Smith agreed, and Henry was thus launched into school life under the most famous teacher of the day, Dr. Arnold.

I have always regarded it as singularly fortunate for Henry Smith that he was at Rugby in its best days, and that he was not there long enough to acquire that part of its tone which was not generally popular. Whether that sweet buoyant nature, with its supreme sense of proportion, and lively humour, could ever have been really spoiled, made pedantic or harsh, is perhaps more than doubtful; but I cannot doubt that the years of travel on the Continent, which two chances, that seemed unkindly, substituted for school and Oxford life, were really of the greatest use to the sufferer. He carried on his studies abroad less methodically, but quite as profitably, as he could have done at home; he learned French, German, and Italian, and he gained some acquaintance with foreign ideas and methods. Meanwhile his first years at Rugby were certainly profitable to him. It was a rule of the school that no one should be in the Sixth Form until he was sixteen, and in deference to this rule

Henry Smith was kept for a year doing work below his capacity in the Upper Fifth and the Twenty Form, though by a curious anomaly he was allowed to act as præpostor in Mr. Highton's house, where he was the senior boy. His Report for the first half-year, which was spent in the Fifth Form, has been preserved ; and might have been written of him at almost any time :—'*Classics* : In extent and variety of knowledge he is certainly the best in the Form, and he is particularly fortunate in combining accurate and literal construing with an excellent choice of English words. His composition also (though it has sometimes been careless) is spirited and clever. *Mathematics*: Very good. *Modern Languages*: He has made great progress in German, and is getting on very well. He has been late for morning prayers oftener than I like; and I should wish him to get rid of a few trifling irregularities, such as occasional inattention at lesson and inaccuracy in saying his lines. G. E. L. COTTON, Master of the Fifth Form.' When in the Twenty he came under Mr. (afterwards Professor) Bonamy Price, confessedly the ablest teacher on the very able staff of which Rugby then boasted, and probably never surpassed as a teacher of classics. In the Midsummer examination of that year, Henry Smith passed into the Sixth, and was accordingly entitled to bid the Doctor good-bye. A few days later he received a letter from Mr. Highton (June 12) :—'You hardly supposed that when you bid Dr. Arnold "good-bye" on Friday it was for the last time. He was taken to his rest at six this morning. . . . You may imagine how the loss is felt here. It is almost as if a common parent were taken away. I felt it so quite myself.'

The true education of a boy at a public school is even more in the playground than in the class-room. What Henry Smith was in this regard has not come down to me. Going to Rugby just before he left it, I remember to have heard 'Highton Smith,' as he was popularly called, spoken of with vague reverence for his great ability, but in no other way. Nevertheless there are indications from reports and letters that he was abundantly capable of healthy enjoyment, and not merely what the Rugbeans used to call a 'swat' or bookworm. Mr. Highton twice reports of him in his first half-year that he was not working as hard as he ought ; and his sister says that 'he came home for his first holidays "astonishing" us by the buoyancy of his spirits and even more by a propensity for "grub," unknown to the ascetic days of his childhood. By one who learned so easily as he did, a little idleness was easily made up for.' In the examination of June 1843 he obtained a Junior Scholarship, being ineligible for the University Scholarship because he had not been three years in the school. In July 1843 the new head master, Dr. Tait, wrote to Mrs. Smith to say :—'There is no young man in the Sixth Form from whose abilities I am led to expect more than from him; and . . . . . I have formed a very high opinion of his character and conduct generally.'

Rugby however was not to keep him. In September 1843 his elder brother Charles died of rapid consumption, and the uncle, who was also guardian and adviser, declined under these circumstances to consent to Henry's remaining any longer at a school in a bleak part of England. The boy bore the blow to his ambition with his unvarying sweetness, and wintered with his family at Nice, while he spent the following summer by the Lake of Lucerne. These were months of steady reading, though his books were few, and he was even unprovided with a Greek lexicon. In the autumn of 1844 he went back to his old friend, Mr. Highton, for a month, that he might be 'coached' for the Balliol scholarship. He won it easily, and, as he was not to go into residence till Easter, went back to join his family at Rome. His journey was a series of disasters. He missed the mail-boat at Dover, had his pocket picked at Paris, and, even after pledging his books, could only muster funds enough to carry him in the roughest way to Rome. This misadventure involved a journey of seventy hours, on the outside of a diligence during a severe frost, to Marseilles; and, after a third-class passage from that port to Civita Vecchia, he arrived in Rome with both his feet frost-bitten, and was laid up for a long time. Presently came an attack of small-pox. 'All the same,' writes his sister, 'the winter was a time of intense enjoyment, and a gathering and growing time.' By Easter he was well enough to go to Oxford, and spend his first term there.

When the Long Vacation came Henry Smith rejoined his mother and sister in Italy. Unfortunately they arranged to spend the summer at Frascati in the Alban Hills, and Henry soon became languid and ailing, and at last ill enough to need a doctor. The doctor who came, an Italian physician of eminence, declared after three weeks that his patient was undoubtedly consumptive, and ordered him to the sea at Naples by way of the Pontine Marshes. Even in their alarm the family were discreet enough to substitute the hill route for the deadly road along the plains; but this involved a four days' journey, during which the sufferer became delirious, and, when Naples was at length reached, the English doctors had all left the city. One however was to be found at Castellamare, and he, when he was called in, declared that the disease was nothing but long-neglected malaria, which an ordinary Italian doctor should have recognised. It was now thought right to revert to the use of strong tonics. Severe inflammatory attacks subsequently came on. These have been, since then, attributed to the presence of gall-stones, which may have possibly laid the foundation of his latest illness. Moreover, with spring (1845), the malaria itself returned, and it became necessary to leave the South. He himself at a later time described his illness to me as a sort of euthanasia, in which he seemed to be gliding painlessly out of life. The sister who helped to nurse him remembers that he was too weak even to put up his glass that he might look at an eruption of Mount Vesuvius. Still he was able to enjoy being read aloud to, and his mother used to read to him incessantly from English newspapers and standard authors, but especially from the Latin and Greek classics, while on Sundays he would honour the day by forbearing to correct or even to shudder at a false quantity. The move from Naples was to Wiesbaden, and there the waters restored him to comparative health. It was thought better however that he should not return to England, and accordingly the next winter (1845–6) was spent at Paris, where Henry Smith attended several of the courses at the College of France or the Sorbonne, and derived especial advantages from the lectures of Arago and of Milne Edwards. By this time his strength was thoroughly re-established, and, though he went a second time to Wiesbaden in the summer of 1847, he had already resumed work at Oxford (Easter, 1847), and never afterwards needed to suspend it. His health, as I remember it for more than twenty years of unbroken intimacy, during which I was constantly seeing him, was always good, though never what could be called robust. He suffered especially as a young man from weak eyes; and he had to be a little careful of himself in diet and exercise; but he was rarely depressed, and he habitually worked beyond what most men could have endured without breaking down. There was one attack of low fever in 1856, the result of course of the earlier Roman fever, in consequence of which he was ordered to ride, and the obligation to take horse exercise was undoubtedly very good for him, and contributed a great deal to his enjoyment of life.

The Oxford into which Henry Smith was now thrown had almost recovered from the strong ferment which ended in Newman's going over to the Church of Rome. The leaders of the High Church party had either followed their captain, like Christie and Bowles, or had satisfied themselves, like Mark Pattison, that

Protestantism does not admit of a divided allegiance. There were still High Church cliques among the undergraduates, such as Newman has sketched with caustic subtlety in *Loss and Gain*, which discussed Church matters from the Anglo-Catholic point of view, but they rarely got beyond a mild dilettanteism. Even this was not always treated with a proper tolerance. I remember a debating society of young Churchmen, which so irritated the Protestantism or the common sense of a rather sporting College by carrying a resolution that 'St. Augustine's interference with the British Church was uncatholic and uncalled for,' that at its next meeting the orators were dispersed by the agency of hot pepper thrown into the room, and saluted with a baptism little short of total immersion as they left the quad. Of the fast men of that day, it need only be said that they have been inimitably limned for good and bad in *Tom Brown at Oxford*. Outside these two sets, which have perhaps attracted more attention than they deserve, and also outside the common and obscure men, were the abler young men of the University, Conservative or Liberal in their politics, as temperament or training determined, but mostly with a wholesome share of English secularism, and neither High Church, except in rare instances, nor aggressively Protestant, nor to any appreciable extent Freethinkers. Lord Salisbury, Lord Kimberley, Lord Brabourne, Sir M. E. Grant Duff among politicians; Goldwin Smith, Sellar, Grant, Sandars, Poste, and Conington among men of letters or scholars; Spottiswoode and Rolleston among men of science; Chitty among judges; Sandford and Ducane among officials, were some of the Oxford men of Henry Smith's day, and with most of these he was more or less intimate at some time, while Grant Duff and Conington were among his dearest friends. Whatever time or thought men of this type could spare from work for the schools, was divided between politics and literature; and Henry Smith's University letters are a singularly faithful reflex of the spirit of the period. They are more mature and temperate than perhaps any one but himself could have written, but they show the enthusiasm for intellectual eminence which is the salt of Oxford life; and the admiration evinced for Mill, and the praise, however qualified, of Robert Chambers, are evidence that the writer was already to be numbered among the few on whom Carlyle had no hold.

Of Henry Smith's Oxford career it may briefly be noted that in 1848 he won the Ireland University Scholarship, the blue ribbon of classical scholars; was a double first-class in the Lent Term of 1849; was elected Fellow of Balliol in

November 1849; and gained the Senior Mathematical Scholarship in 1851. He was unable through absence to stand for the Hertford Scholarship, which falls to the best Latin scholar of the year; or for the Junior Mathematical; and he was beaten for the Senior Mathematical Scholarship in 1850, the first time that he stood for it, by Mr. Ashpitel of Brasenose; the single defeat of the kind, I believe, which Henry Smith sustained.

Balliol College, to which Henry Smith belonged, was far away the best in the University during the time of his residence, and for some years afterwards. A variety of circumstances had contributed to build up its pre-eminence. The first cause was the far-sighted integrity of the old Master, Dr. Jenkyns, who was almost singular among the Heads of his day in regarding it as the first duty of a College to promote intellectual distinction, and who waged an incessant war with privilege, abolishing gentlemen-commoners and throwing open close endowments as far as he legally could. Dr. Jenkyns could not have done much single-handed, but he gradually found or created men, often no doubt abler than himself, who were glad to carry on his work in the same spirit; and the late Master of Balliol, Mr. Jowett, then one of the tutors, was undoubtedly the soul of the College during the whole time of Henry Smith's connection with it. At the time of Henry Smith's election, the College wanted a mathematical lecturer. There is no doubt, I think, that he was chosen in the well-warranted expectation that he would consent to reside and lecture. In this way began his own lifelong union with Oxford, for until then he had been a mere bird of passage. Having once decided to accept the office thrust upon him, he gave himself up heart and soul to doing his work well.

It was a common story in Oxford at that time that Henry Smith, being uncertain after he had taken his degree whether he should devote himself to classics or mathematics, had solved the doubt by tossing up a halfpenny. His sister remembers how he actually expressed a wish that some one would do this for him. He was, in fact, the last man on earth to have committed any important decisions to chance; and he has himself told me that his choice was partly determined by the fact that having at that time weak sight he found he could do more work in thinking out problems than in any other way without using his eyes. The decisive reason was of course a pre-eminent genius for mathematics—the born aptitude that is itself fate—and the cause why the determination was made at that particular time may have been this offer of a lectureship. Nevertheless the Oxford tradition is so far valuable as it testifies to the general belief that Henry Smith could have made his mark in any study he embraced. 'I do not know,' Professor Conington once said to me, 'what Henry Smith may be at the subjects of which he professes to know something; but I never go to him about a matter of scholarship, in a line where he professes to know nothing, without learning more from him than I can get from any one else.' Once it seemed as if he would be attracted into chemistry. The College demanded of him that he should give chemical lectures (1853), and Henry Smith accordingly became a pupil under Professor Story-Maskelyne, who then occupied a laboratory under the Ashmolean Museum and gave instruction in chemical analysis. Here H. Smith showed that in delicate manipulation and in accuracy of work he possessed a sort of instinctive faculty. A lifelong friendship grew out of the hours spent in this way; although the demands of a new and engrossing science on his time were too great to permit his sacrificing to chemistry the many other important subjects and duties that filled up his life. Even then, however, I remember, his idea was to seek numerical relations connecting the atomic weights of the elements and some mathematical basis for their various properties*, so that we might anticipate experiments by the operations of the mind—an ambition which was very interesting to Alexander von Humboldt, when Sir M. E. (then Mr.) Grant Duff told him of it. Ultimately Henry Smith of course found that science is too jealous a mistress to admit of a divided allegiance; and, though his reading was always wide and various and singularly well digested, he practically devoted himself to mathematics, and as I understand to two or three great subjects with which his name will always be associated, the Theory of Numbers, the Theory of Elliptic Functions, and certain new processes of Geometry.

One point for which the generations younger than Henry Smith and John Conington will always remember them gratefully was the way in which they mingled in undergraduate society. The distinctions of academical rank were at that time rather jealously marked in Oxford. If the tutors and fellows were

* His conviction that such a numerical and mathematical basis underlay the phenomena of chemistry was even stronger in the case of crystals. At my suggestion he undertook the discussion of the principles involved in the parallelism of zone-axes and face-normals in a crystal system with rational indices; the results of which were given by him to the London Mathematical Society in vol. viii of its Proceedings.—N. S. M.

not, like the Heads of Colleges, in inaccessible isolation, they were scarcely better known to us through the formal breakfast parties which were submitted to on both sides as a very irksome duty. Here and there a man, like the admirable Charles Marriott of Oriel, made it a duty to invite young men, in order that they might feel at liberty to go to him if they needed religious advice, which was never obtruded; but men of this stamp were never, as far as I remember, more than genial hosts. Henry Smith and Conington, men of the most opposite temperament though devoted to one another, threw themselves with such unaffected simplicity into our interests and occupations, that we all came to regard them as personal friends, and to talk as freely before them as before one another. Looking back I can see that their position was that of very helpful and sympathetic seniors at a children's party, and I can conceive how Smith's playful sense of fun and Conington's grave humour must often have been tried by the obligation to treat our criticisms of men and things or our forecasts of the future seriously. That the intercourse was begun and carried on on their part from a conscientious desire—due partly to Arnold's teaching—to convey a serious interest into everyday life, I at least cannot doubt. As one who profited by the association let me record that meetings of this kind were not only the most pleasurable part of a chequered Oxford life to many of us, but unquestionably did more to stimulate thought and form character than the more formal influences of the chapel and the lecture-room.

A letter from one of Henry Smith's old pupils, Professor Irving, of Melbourne, will complete the description of this part of his career, and speaks with authority on some matters which I only knew from report.

Melbourne,
1st September, 1888.

MY DEAR PEARSON,

It is by no means an easy task to carry back the memory nearly forty years, and recall scattered reminiscences of one from whom you have been altogether parted, with whom you have not even kept up communication by letter for more than thirty years.

Yet such in the old Oxford days was my affection for, and so highly have I ever honoured him whom I was then proud to call my friend, that I must accede to your request and do what little I can towards your presentment to the world of Henry Smith, Scholar and Mathematician.

My first introduction to him was, I think, in my Freshman's Term at Oxford, Michaelmas 1849, the term in which he gained his Fellowship at Balliol; but our intimacy really dated from the beginning of 1850, when he became our Mathematical Tutor. From the first there had been this link between us, that I had inherited the little third-floor rooms in the inner quad of Balliol, which had belonged to Scholar Smith, as he was called in those days to distinguish him from sundry others of the same patronymic.

I remained his pupil in Mathematics throughout my undergraduate time, and was able to do him the credit of winning the Junior Mathematical University Scholarship in 1850. Of his power as a Class Teacher I cannot speak, as I was almost alone with him: but my individual experience was that as a Teacher he was all a learner could desire, most patient with all one's difficulties, most clear and full in his explanations. But he was too kind to me. He sympathized strongly with my disappointment when the authorities refused me permission to compete for the Ireland in the spring of 1850, and knowing how keen my desire was to win that distinction in the following years, he did not force me to work at Mathematics as I ought to have done, and so I failed to do justice to his teaching, and attain his own class in the Final Honour Schools.

Through Smith I made another valued friendship, that of John Conington, Professor of Latin, and might have gained access to the intellectual circle in which they moved, and of which they were such brilliant ornaments. But I must honestly confess that my own work for the Schools was sufficient mental exercise for me, and that I sought my relaxation, not in other spheres of thought, but on the river.

Yet in our frequent intercourse there was quite opportunity enough for me to learn and to appreciate the manysidedness of Henry Smith's mind. All of my generation were prepared to look up to and to admire one who had done so brilliantly as he had in Oxford: and when you came to know him personally you could not look upon that splendid forehead of his without assurance of the powerful intellect it betokened: you could not converse with him without realising that he was one of whom it might be said that 'omne scibile novit.' And what he knew, he knew, not as so much stored up learning to be brought forth as required, but he had made it all his own, he had thought as well as read.

Still with all his vast erudition, and his great intellectual power, he was the humblest, the gentlest of men. Ignorance, even if pretentious, was not to him something to be crushed with sledgehammer Johnsonian blows, but a thing to be pitied and kindly enlightened.

In fact were I asked to select his peculiar moral characteristic, I should say he was the most gracious man I ever knew.

Reading over these lines, I recognize with regret how very imperfectly are therein expressed the love and admiration I felt for him, how feebly they serve to set him forth.

Yet inadequate though they are, they may help somewhat. If I cannot do better—

> 'His saltem accumulem donis, et fungar inani
> Munere.'

I am, my dear Pearson,

Yours ever sincerely,

M. H. IRVING.

In 1855 some of the abler Oxford and Cambridge men determined to publish a yearly volume of Oxford and Cambridge Essays, and Oxford led the way, T. C. Sandars, I believe, being the guiding spirit and editor. Henry Smith contributed an essay on the Plurality of Worlds to this publication. He took for his theme Whewell's then unacknowledged book on that subject, and Sir David Brewster's fiery answer, 'More worlds than one, the Creed of the Philosopher and the Hope of the Christian.' The subject was a fascinating one, and Henry Smith was in many respects admirably calculated to do it justice. He wrote as simply and clearly as Herschell or Tyndall, and he was skilful in dissecting arguments of every kind with faultless impartiality, so as to reduce them to their absolute value or no-value. A single passage will serve as an illustration of his method. Whewell had argued in the spirit of Lucretius that there was, so to speak, a law of waste traceable in the Divine economy, and supported this position by facts from the origin of species. 'If we found,' Henry Smith remarks, 'that Jupiter's four seasons differed so slightly from one another that they hardly deserved the name, and that they could not be conceived to be of any use to his hypothetical inhabitants, we should be reminded by those "rudimentary seasons" of the osteological facts on which the essayist dwells so much, of the rudimentary fingers in the hoof of a horse, or the rudimentary paws with which a snake is said to be endowed. But the one thing we should not be prepared to find would be a wasted, imperfect, uninhabitable planet. We should know of no facts in Zoology with which to compare such an occurrence. The crust of our earth is filled with the remains of departed life, but we find not a vestige of imperfect attempts, of forms moulded after the vertebrate type, and yet incapable of animation.' It will be remembered that both essayist and critic wrote in the days before Darwin, and that Henry Smith had never made any special study of comparative anatomy. With this allowance it must, I think, be recognised that Whewell's conception of a general law producing a single successful result and failing in every other case was substantially and hopelessly wrong, and that his critic's conception of 'a law uniformly asserted in a multitude of individual cases, and uniformly productive of variously perfect results,' was a singularly correct expression of all that science was then able to teach.

There are passages of characteristic irony scattered through the essay. We are told that in one of Plutarch's Dialogues, 'the lunar world is connected with the future destiny of the human soul, after a manner which, we conceive,

Sir David Brewster would allow to be highly creditable for a heathen, and on the whole corroborative of his own opinion.' Of the literary form of Whewell's essay it is said: 'In the dialogue at the beginning of the essay the earlier letters of the alphabet, who appear as the objectors, conduct themselves so much like simpletons that one wonders at their being thought worthy of so long an interview with the enlightened Z.' Of theological arguments intruded into the domain of science it is observed: 'We cannot imagine a more painful spectacle of human presumption than that which would be afforded by a man who should sit down to arrange "in a satisfactory way" a scheme for the extension of Divine mercy to some distant planet, and who, when he found "great difficulty in conceiving" such an extension of the Divine attribute, instead of desisting from his vain attempt, should go a step further still, and infer that no such scheme can exist because he fails to discover a *modus operandi* for it.'

Still, though the article on the 'Plurality of Worlds' was read with pleasure and spoken of with esteem, its success was not so unquestionable as to tempt its author into larger literary work; and his only other contributions of any length to English prose are, I believe, a review of Mr. Freeman's *Federal Government*, which he wrote for the *National Review* in 1864, and the Memoir of Professor Conington, which was prefixed to the volumes of his works in 1871, and which is a very perfect example of skill in recording a quiet life so as to invest it with interest. It would have been a great misfortune for science if a man capable of enlarging its boundaries had wasted his powers upon mere criticism or exposition; yet, considering Henry Smith's unambitious temperament, which made him careless of personal fame, and his invariable readiness to oblige friends, I cannot doubt that he might have been seduced into Quarterly Reviewing or some other form of ephemeral literature if he had possessed some of the minor qualifications of a journalist. The fault of his argumentative writing is a disposition to hold the balance and to avoid summing up; and it is in keeping with this quality that his style, though it has the subflavour of irony and the point inseparable from lucid concentration, is not epigrammatic or what would be called strong. The writer's tenderness of disposition had something to do with this characteristic. He who as a boy of fifteen had stopped himself in a caustic criticism in a private letter because the subject of it was 'somebody's bairn' carried the same thought for others into his words and dealings through life. Reserve on matters that lay near the heart was another modifying influence. To many men, the scholar who identified himself with every movement for religious or intellectual or political freedom in the University was still more or less a sphinx because he never propounded his convictions as a topic for conversation. The world that reads is rather like the world that listens to a platform orator. It likes its instructors to be positive, even where they cannot be certain, and to have its conclusions presented to it in the form of short and simple aphorisms, which it may swallow and retain without trouble. Henry Smith could not have attained to this ideal, and it is matter of some satisfaction that he did not aspire to it.

In 1857 Mrs. Smith died. The attachment of mother and son to one another had been deeper than is common, and the course of their lives had drawn them nearer together than can often be the case. It was now arranged that Miss Smith should keep house with her brother in Oxford, the two spending the term together, and each being allowed complete liberty of movement during the vacations. I cannot doubt that this arrangement contributed very much to Henry Smith's happiness. He was eminently domestic and hospitable, and having the cares of household life taken off his hands, and being supplemented by one who was almost another self, was able to fill his house with friends, who were certain of an Irish welcome, however unseasonably they might arrive to ask for a dinner or a bed. He was also able under his own roof to gratify his passion for pets—Persian cats of distinction and two aristocratic dogs—to which there are frequent allusions in his letters. During the vacations he often visited the Continent, going once to Sweden and Norway; more than once to North Italy; to Spain with Grant Duff in 1864; and to Greece in 1872 with Mr. and Mrs. Grant Duff. From time to time he paid visits to an old friend of the family, Miss Theodora Price, who had lived with his mother during the whole time of her widowhood, and who, on Mrs. Smith's death, established herself at Tunbridge Wells. For some years, too, Henry Smith was a prominent figure at the various meetings of the British Association in England, Scotland and Ireland. It will be seen that his life was in no sense that of a recluse; and it may be added that he entered with zest into every form of social enjoyment in Oxford, from croquet parties and picnics to dinners. That the irregular, desultory life, with its frequent breaks, suited his health is probable; and, as he possessed a rare power of utilising stray hours so as never to intermit work altogether, even when his distractions were most numerous, it seemed possible that he might be among the singular few who have combined residence in an English University with unswerving devotion to the claims of abstract research.

Fortune appeared to favour this anticipation. In 1860 Mr. Baden Powell, the Savilian Professor of Geometry, died, and Henry Smith became a candidate for the vacant post. Some years before, he had told a friend that to occupy this chair was the great ambition of his life. He said that the two Savilian Professorships were the most honourable offices in the University: they were open to the whole world of Mathematicians, and had usually been held by distinguished men. His wish was now to be gratified in the pleasantest way. Those who would naturally have been his rivals, the other Oxford mathematicians, were the first to draw back in his favour and sign a common testimonial to his pre-eminent claims*. I well remember the generous warmth with which two of the senior professors, Mr. Walker and Mr. Bartholomew Price, but especially the latter, expressed themselves to me at the time about Smith's undoubted genius and the chances that he would one day leave a name to be remembered beside those of Newton and Laplace. The electors for the chair chose him, as I have understood, without hesitation, taking the view that as no other Oxford man was a candidate, and as Henry Smith was pre-eminently qualified, it was needless to scrutinise the testimonials of outsiders. He himself was a little troubled by a doubt whether the claims of an older man, Dr. Boole, of Queen's College, Cork, ought not to have received further consideration. That Henry Smith justified his electors by the magnificent work he did later on, is beyond question. He was also a very successful teacher, having what must be considered large classes in a University where mathematics have, at least in recent times, attracted comparatively few students. Passages in his letters prove how keenly he

---

* I can express no opinion worth having on this subject; but I see from a notice in the *Academy* (Feb. 17, 1883), written evidently by a personal friend, that much of the work given to the world in later years had been produced before he was thirty-five. 'He (Professor Smith) communicated at different times a good many notes and papers to the Mathematical Society, especially during his Presidency in 1874–76; and we believe that all the results he gave he had had in his possession for fifteen years.' His work on the Theory of Elliptic Functions and the Introduction to Professor Clifford's *Remains* belong however to the last seven years of his life.

promoted the well-doing of his pupils, and what his views were about the reforms desirable in mathematical teaching.

The present volumes show what splendid contributions Henry Smith made to science during the short twenty-nine years of his speculative activity. Nevertheless, it must, I think, be admitted that his unrivalled powers were often employed upon work that scores of able men might have been found to do efficiently, and which his friends should not have asked him nor he have consented to undertake. From 1850 to 1870 he was Lecturer at Balliol, not being able to afford to give up his Fellowship, and having scruples about retaining it if he did not teach, as the number of Fellowships was limited and the stipend of a Lecturer was too small by itself to remunerate any one for the work. It must be borne in mind that during part of this time he was also Savilian Professor, and during the whole of it he was constantly doing other College and University work*, assisting backward men, or taking part in examinations, or serving on University Boards and Committees. In 1873 he freed himself from the worst of this drudgery, the College Lectureship, by accepting a flattering and generous offer from Corpus Christi College of a Fellowship upon that foundation†. Not long afterwards he obtained the Keepership of the University Museum, left vacant by the death of Professor Phillips. The office gave him a pleasant house, a small stipend, and not very uncongenial duties, half as master, half as servant,

---

* The Master and Fellows of Balliol College, for instance, once asked him to give a course of lectures on the Schoolmen; and he complied.

† A friend at Balliol writes :—'We knew perhaps better than others how necessary this relief was to Henry Smith, and we rejoiced that it had come to him; we knew likewise the perfect loyalty towards his old College which prompted his resignation. Nevertheless, it was a grievous thing to us that he should be obliged to leave our body. Never had we felt so bitterly the difference between a poor foundation and a rich one. Henry Smith, as Steward of Common Room, was our chairman on social occasions, especially at our annual "gaudy" on St. Catharine's Day. The last speech he made in this capacity was immediately before his migration to Corpus. He assumed a playful tone, and tried to amuse us by various quaint comparisons into forgetting our loss, but he was quite unable to subdue his own emotion, and he was weeping himself before he had made us laugh. This was the only time that I ever knew him break down. Though ceasing to be a Fellow, he continued to give us the benefit of his presence and counsel at our College meetings. By the next St. Catharine's Day the keen and constant interest which he took in our affairs had somewhat reconciled us to the change, and this feeling was warmly expressed by the Master in proposing Henry Smith's health. I remember the Master's concluding words, which struck me at the time as a note of warning, and which have now a sadder significance: "I will only venture to express the hope that he will not suffer himself to be numbered among those men of varied powers and charming manners who have given up to society and business what was meant for science and posterity."'

which sate lightly upon one who was genial and full of instinctive tact. Nevertheless, it cannot be said that his work was sensibly lightened for any long time. Partly, he was himself to blame. He had a speculative element in his nature, and had invested so much money in mines—almost always, I am afraid, unremunerative—that it became important now and again to eke out his regular income. I remonstrated with him very strongly, when he added the duties of Mathematical Examiner at the University of London to his other heavy work (1870), for he seemed to be breaking down at the time he undertook it, and I felt sure that whatever he did for the day's need was so much taken from more enduring labours. It seemed however as if the world was in a conspiracy to force duties of every kind upon one whose talent was so flexible and whom men of all opinions agreed to welcome as a coadjutor. He was for years a member of the Royal Commission on Scientific Education, having been appointed in 1870, and he drafted a large portion of its report. In 1877 he became a member of the Oxford University Commission under Lord Salisbury's Act; and in the same year he agreed to be chairman of the new Meteorological Office, the governing body of which was practically nominated by the Royal Society. This latter work was specially congenial, and the associates were so considerate and able as to give a charm to toil; and Henry Smith enjoyed the fortnightly visit to London, and the temporary rest from the turmoil of Oxford business. Still, when all is said, it can hardly be doubted that the labours of all these various offices meant a partial interruption of nobler toil and may have hastened a premature death.

It may perhaps be said, and not without some truth, that those who knew of the condition of his health should have refrained from heaping work upon him and should even have compelled him to take a long term of real rest. But in fact these demands came on him from several and distinct quarters, and what might seem to each person or group of persons making the demand a light and congenial undertaking for the always gracious counsellor of 'golden speech,' became, when added to the aggregate of such undertakings, a serious, perhaps even a fatal burden. The truth is that his presence was always welcome on Boards and Committees; for he possessed the rare gift of suggesting some middle course which would often effect an agreement between persons who had been advocating opposite points of view, and of so bringing about a welcome end to a weary discussion.

It will be a melancholy satisfaction to some of the friends with whom Henry Smith went on holiday expeditions, to reflect that these intervals of rest were not only periods of unmixed enjoyment to him, but probably helped to prolong his life. The letters he wrote from Greece in 1871 will not bear reproduction, but are full of the pleasure he experienced. 'We certainly had a most perfect voyage.' 'Except Pylos, I don't think we missed seeing anything we could have seen. Grant Duff, in a place he has not seen and wants to see, is quite perfect; and we shall now work morning and night, till we have done Athens well.' 'All that is in Athens we have done to the greatest perfection,' is the comment in another letter. If only excursions of this kind could have been more frequent! The last I find commemorated is a visit to Rome in 1879, when Henry Smith represented the Meteorological Council at the International Meteorological Congress.

In 1878 Mr. Gathorne Hardy, who then represented the University of Oxford, was raised to the Peerage; and the Oxford Liberals determined to bring forward Henry Smith for the vacant seat, in the hope that his great personal popularity and unrivalled academical reputation might win over many votes from moderate Conservatives. Moreover Henry Smith was not emphatically opposed to the Jingo or war policy of the Beaconsfield Ministry; the test by which Conservatives especially weighed politicians in that particular year. He did not expect success, and he hardly desired it, but he would not shrink from a fight if he was asked to stand forward as the representative of a principle. I am told his friends were sanguine of success for a time. Friends are bound to be; but no sane looker-on could have anticipated any other result than that which actually took place, that the Conservative candidate would be elected by an overwhelming majority. I have never felt that, in this particular instance, the rejection of an eminently good and wise man was unconditionally to be regretted. Personally, I have no sympathy with the doctrine that scholars are out of place in Parliament—a doctrine which would have excluded Macaulay, Gladstone, Cornewall Lewis, Goschen, Fawcett, Grant Duff, Morley, and Bryce among office-bearers of recognised ability, as well as Grote and Mill and a host of others who have added distinction to the House of Commons although they never attained to office. I am convinced that Henry Smith would have been as popular in the House of Commons as he was everywhere else, would always have been listened to when he spoke, and would have spoken with effect. Still I cannot persuade myself that his magnificent powers could have been adequately employed in debating or administering: and I am certain that dozens of inferior men would have played their part as usefully in St. Stephen's; while there was no one but himself in England so peculiarly fitted to increase knowledge in one very difficult and abstruse department of enquiry.

If however Henry Smith had ever gone into Parliament, he would have been something more than a representative of learning or even of academic Liberalism. He had a genuine sympathy with the poor of his own land; and his last public appearance anywhere was in the Oxford Town Hall to support a resolution by Mr. Arch in favour of giving the franchise to the agricultural labourer. The speech then delivered will bear reproduction for its own merits, and is a good specimen of the speaker's style; that of a man thinking aloud in simple words, yet with an instinctive perception of rhetorical effect.

Professor Henry Smith said it was as a Liberal that he would say a few words with reference to the resolution before them. In the course of his life he had found himself sometimes on the extreme left of the Liberal party, sometimes verging towards its right—for every party had a right and left—and pretty often about in the middle of it. He was bound to say that his belief was, in the first place, that the whole of the Liberal party, right, left, and middle, was unanimous in thinking that the National Agricultural Labourers' Union had rendered a great service to the United Kingdom. He further believed that the whole of the Liberal party rejoiced to think that the great benefit which that Union had conferred upon the agricultural labourers of this country would remain for ever associated with the name of Mr. Joseph Arch. He would endeavour to support the resolution by an argument different a little from those they had heard. The extension of the household franchise to the counties was inevitable; whether they liked it or not, it was a thing which must be done. He believed there were but few men in this country—sensible men, men who looked at what was around them, and who listened to what was said—but felt that it was inevitable. He was one of those who, when it was clear that a thing must be done, believed that the sooner it was done the better; and if it were for that reason only, he would heartily support the resolution. But, in addition to that, he did believe that the extension of the franchise to the great classes who now were excluded from it would, as had been well put before them already, exercise a beneficial influence upon the future course of their legislation. It might be true that some persons might ask what would the agricultural labourer and the rural artisan do with the franchise when they got it? They would do like other people. He feared they would do some mischief, for he knew no class of his countrymen among whom there were not some who did mischief with any right that was entrusted to them, but he firmly believed on the whole they would exercise the franchise for good. If he were told they would exercise the franchise for selfish objects, for objects peculiar to their own class, he would say let it be so; but if so, what would they say to the other classes who already possessed political powers? Could any one of them send a representative who could say that his hands were free from selfish legislation? If they must count—for, alas, they must—upon some strain of selfishness in their common nature, at least let them take care that their representation was not one-sided, but that at any rate each class had a fair means given it of defending itself from others. It was for these reasons that he for one most heartily supported the resolution. He hoped to see the household franchise extended as rapidly as possible to the counties, and he was not one of those who shrank from a still greater consequence which would come when that great measure of enfranchisement should be followed by an equally sweeping measure of redistribution of seats.

From that platform Henry Smith went home to die. Overwork and sedentary work had gradually undermined his constitution. When I last saw him in 1879 he still looked well, and in some respects, from having filled out a little, less delicate than as a young man, but I noticed that he was less capable of sustained effort. In 1881 premonitory signs of a break-up of the constitution showed themselves, and were unhappily not heeded as they should have been. First he suffered from his digestion, and had to put himself under Sir H. Thompson's care; and then a stoppage in one of the veins of his leg confined him for many months to the sofa, and made all but occasional carriage exercise impossible. He seemed to be tiding over this illness, when a rush of University work threw him back again into the condition of an invalid.

When he spoke at the Town Hall meeting he was suffering from a cold. The exposure and excitement were followed by congestion of the liver, which was the more dangerous after the severe attacks which had followed on the Roman fever, from which he suffered in 1845. On the morning of Thursday (February 8) there seemed to be a change for the better, but at noon the worst symptoms returned; and Sir William Gull, who had been telegraphed for and who arrived about eight o'clock, held out little hope. About four o'clock next morning (Friday, February 9, 1883), the patient's state was declared desperate, and three hours later he passed painlessly away.

He was buried (writes a friend who was present) at St. Sepulchre's Cemetery in Oxford on Tuesday, February 13. So great a concourse of undergraduates as well of senior members of the University and friends and strangers from a distance has rarely been seen on an occasion of the kind in an English cause and the men he assailed. Henry Smith's tenderness of feeling interfered with his command of literary form. He had a feminine instinct for avoiding whatever would give pain, and never allowed his buoyant spirits to betray him into a word that might seem harsh, or his inimitable persiflage to pass the boundary line into sarcasm. Those who heard him talk were conscious of wit that played round every subject with a perpetual sparkle, and that left a delicate aroma behind; but no one ever knew it employed as a weapon of offence. Reading over his private letters I find the same kindliness, the nearest approach to personal satire being perhaps the description of a heavy dinner, 'with four pièces de résistance, not including X and Z,' two rather overwhelming talkers. Therefore if Henry Smith had ever written on any of the subjects on which he felt strongly—and he was an ardent Liberal on every University question and on almost every political topic of interest—I cannot doubt that he would have adopted a style of earnest simplicity, and would have trusted for effect to argument, enhanced at most by a restrained eloquence. Bearing in mind that he was confessedly one of the most brilliant talkers of his day, so that every obituary notice dwelt lingeringly upon this trait, and considering how easily the playful but keen humour might have been transformed into caustic satire, I can only wonder at the mixture of kindliness with strong self-discipline that prevented even an occasional lapse. Both in this matter and in his judgments of men and things, a singularly fine character gave law to the intellect. He was clear and just in expression because he was accurate and truthful in thought; he was irreproachable in speech, because he never allowed himself to cherish an ill-natured thought.

Any one who has been often in the society of brilliant talkers can hardly have failed to notice how little of the best conversation is of a kind to bear record or is practically remembered. Dr. Johnson was singular in attracting an unrivalled biographer, who took notes unblushingly, and was skilful enough in literary form to polish up what he took; and Sydney Smith's fertility was so great that some of his mirth has survived him: but of George Selwyn and Luttrell, of Fox and Canning, of Macaulay and Bagehot, we know disappointingly little. Two or three trifling instances may serve to show what Henry Smith's manner was. He was once winding up a mathematical lecture by explaining a new solution of an old problem. 'It is the peculiar beauty of this method, gentlemen,' he concluded, 'and one which endears it to the really scientific mind, that under no circumstances can it be of the smallest possible utility.' He was obliged to pass through France in 1870, when fortune had just turned against the French armies, and the cry of treachery was raised everywhere. A guard noticed the tall Englishman with a blonde beard and spectacles, and instantly denounced him as a German spy. A suspicious crowd collected in a moment around the carriage. 'Gentlemen,' said Henry Smith with an amused smile, 'I speak French very badly, but not I hope with a German accent.' The proof and the speaker's impressive serenity carried conviction, and the crowd melted away. 'You take tea in the morning,' was the remark with which he once greeted a friend, 'If I did that I should be awake all day.' A friend mentioned to him the enigmatical motto of Marischal College. 'They say; what say they; let them say.' 'Ah,' said Henry Smith, 'it expresses the three stages of an undergraduate's career. In his first year he is reverent, and accepts everything he is told as inspired: "they say;" in his second year he is sceptical and asks "what say they?" and "let them say" expresses the contemptuous attitude of his third year.' At a time when English society was perhaps extravagantly fluttered by Lord Beaconsfield's apparent success at the Berlin Conference, Henry Smith reduced the event to something like its proper proportions. 'Dizzy,' he said, 'has taken John Bull to Cremorne, and the old gentleman is rather pleased to have been there.' On the news that a distinguished friend, who was also markedly pessimist by temperament, had been appointed to a high post in India: 'How fortunate!' was the remark; 'it will give him another world to despair of.' He summed up X, a brilliant writer but inconsecutive thinker, in the criticism, 'X is never right and never wrong; he is never to the point.'

It is sometimes said of loveable men, that they diffuse their affections so evenly as to be incapable of strong personal attachments. With Henry Smith to be a friend once was to be a friend for life. The masterly biographical sketch which he wrote as an introduction to Conington's Miscellaneous Writings will give a measure of one friendship that lasted from school days till it was interrupted by death. Professor Rolleston, who could hardly ever speak of him without some epithet such as 'the golden-mouthed,' confided his family when he died to Henry Smith's care, and the trust was accepted and discharged with exemplary fidelity. Probably no other great student was ever so ready as he always was to put aside books and papers when a friend entered the house.

Yet his nature, genial and hospitable in the extreme, was not what is called an effusive one. He has noticed in the life of Professor Conington, that that great scholar, after he underwent a spiritual conversion, used to speak of his experiences unreservedly though in the simplest language. On this as on every subject of delicacy, Henry Smith was absolutely reticent. He would discuss religious topics if they were started as matter of interest, but I never knew him talk of his own faith, and I should be slow to believe that he ever did. My impression is that he accepted Christianity not only as a habit and a conviction, but as a rule of life, and in fact his character can scarcely be explained, except by ranking him with those who feel that they are 'ever in the great Task-master's eye.' I think he regarded much popular theology as irrational, and much fashionable doubt as a mere winnowing of chaff. Some of the weightiest words I ever heard from him were on religion. Beyond this I can say and surmise nothing, and his letters are not more unguarded than his speech was*.

The one question as to Henry Smith's character that appears to be still undecided is, whether his inaptitude for self-assertion, his scorn of personal ambition, his severe acceptance of duty in whatever shape it came to him, are to be regarded as blemishes or excellences. The distinguished friend who wrote about him in the *Spectator*† has shown in thoughtful and wise words how much there is admirable in the 'philosophic life'—'life of exemplary moderation, far removed from even a suspicion of worldliness and vanity.' 'Great moral gifts,' as the writer goes on to say, 'can be found when occasion demands them; talents grow on every tree. But the serenity of heart which enables its possessor to wear the gifts of genius with sobriety, and to use them nobly and well, without seeking to expend them in the purchase of fame, or wealth, or of advancement, is a quality which modern society little cultivates and seldom sees.' It may seem to those who ponder this temperate and lofty apology, that it is a sufficient answer to the regrets I have freely expressed in this sketch over genius that was often lavish of itself on work for the moment's need or work of ordinary compass. Let me say for myself and for those who think with me, that we never desired wealth for Henry Smith except in such measure as might free him from sordid

* He on more than one occasion spoke to me on these subjects. His position was perhaps most simply expressed in a conversation in the course of which I remember his saying that the essential features of the Christianity held to-day were held in the time of Justin Martyr.—N. S. M.

† The article is printed at p. xlvi.

necessities, or fame except as a recognition of what he might achieve; and that the world's opinion or the state's honours could not have raised him in the estimation of those to whom he was already above all men. Our feeling about him was essentially what Newton expressed when he said that 'if Cotes had lived, we should have known something.' It is very possible that those who saw how much of his time Henry Smith gave to Examinations, and Boards, and Commissions, and who unconsciously estimated the range of human effort by their own measure, did injustice to the special capacity I have noted in him for carrying on consecutive work in stray moments. It has been said by one who can speak with authority on such a subject that Henry Smith was 'the author of mental achievements in the most abstract and complicated of the sciences, which will rank as scarcely second to any in the century.' On this matter the collected works will be conclusive evidence. What, however, surviving friends will and must feel is that genius is as rigidly bound to husband its powers as mere capacity, and that nothing can be spared from the supreme work of life without loss *. Prove to us that Henry Smith's work was indeed scarcely second to any in the century, and we are constrained to assume that, had his energies been more severely economised, it must have been second to none. Certainly the ordering of these things is not in our hands. He who gave the perfect intellect gave also the fine temperament, the tenderness that shrank from disobliging, the modesty that esteemed no duty undignified, the absolute disregard of self. To us who knew him, let me repeat, the man was always greater than any possible work he might do, though we set no limits to its possibilities; and to us the ever-green wound of his loss is partially compensated by the remembrance of an ideal character. What we grieve for is that generations that did not know him as we have known him will try him by the only standard possible, that of his completed work, and will give him less than the measure of his real capacity, though they can never refuse to number him among the great names of the century.

---

* I may add, however, to this that, frequently as I urged on H. Smith to turn a deaf ear to some of the too many supplicants for his time, and to give up some of his less important occupations, his answer to me always was that he did 'get all that he could out of himself' as it was; that in truth his greatest work could only be done now and then, and could not be reeled off the mind indefinitely. Much interval was necessary to him.—N. S. M.

RECOLLECTIONS OF PROFESSOR SMITH.

I. BY PROFESSOR JOWETT *.

MY recollections of Professor H. J. S. Smith extend over about 40 years. I first heard his name mentioned in the year 1843, by the late Archbishop of Canterbury, then recently elected Head Master of Rugby, who told me that there was a boy in the School quite deserving of a place by the side of Conington and Walrond, who were the great names of Rugby in those days. He was going to try for the Balliol Scholarship. At the end of the year, on that occasion, he beat all the other candidates, of whom one was the late Sir Alexander Grant, elected Scholar at the same time with him. I remember him in the *viva voce* part of the examination, a youth of eighteen, rather overgrown and stiff, as youths of eighteen are apt to be, construing Latin and Greek authors in a pious and evangelical tone of voice, which provoked a smile in the Examiners, but with never-failing accuracy. The old Master, as we used to call him, took up his English Essay and showed it to me, saying, in his emphatic way, 'There's mind in that.' The subject given for Latin Hexameters at that examination was the Pelasgi, of whom he did not forget to mention in his verses that 'they worshipped nameless gods.' Meeting Arthur Stanley on the Woodstock Road the day after the election, he congratulated me on our having chosen a youth whose fame had preceded him at Oxford.

He more than justified the promise which he had given. Though not a poet or creative genius, he was, I think, possessed of greater natural abilities than any one else whom I have known at Oxford. He had the clearest and most lucid mind, and a natural experience of the world and of human character hardly ever to be found in one so young. He took up all subjects at the right end; he knew whereabouts the truth lay even when he was imperfectly acquainted with the facts. And he was the most amiable and good-natured of

* Professor Jowett's death has deprived these 'Recollections' of the author's final revision.

young men. I might apply to him the words in which Plato describes the youthful Athenian Mathematician, Theætetus, where he says: 'In all my acquaintance, which is very large, I never knew any one who was his equal in natural gifts. He had a quickness of apprehension which was almost unrivalled, and he was exceedingly gentle. There was a union of qualities in him which I have never seen in any other, and should scarcely have thought possible, for quick wits have generally quick tempers . . . but he moved surely and smoothly and successfully in the paths of knowledge and enquiry. He flowed on silently like a river of oil. At his age it was wonderful. He was also surprisingly liberal about money, though his fortune was only moderate' (Theætetus, 144).

The facility with which the youthful Scholar of Balliol picked up all sorts of knowledge was equally wonderful. During the first two years after his election to the Scholarship he was struck down by a serious and almost fatal illness, and did not come up to Oxford until what is usually the third year of residence had commenced. In the interval, residing in Italy he acquired a considerable knowledge of Roman inscriptions and antiquities, and also of modern languages. Within the year and a half which remained of his Undergraduate course he obtained the Ireland Classical Scholarship, and a double first. Two years later, the Senior Mathematical Scholarship was awarded to him. Similar honours have only been gained by one other person—the late Very Rev. G. H. S. Johnson, Dean of Wells, an eminent man, but little known, who, from ill-health, was unable to do justice to his great natural talents.

After he had taken his degree, he at one time thought of going to the Bar, for which he was very well suited—he would have risen rapidly to the high places of the profession. But the feebleness of his constitution when a young man led him to abandon this intention, and he soon settled down in the post of Mathematical Lecturer at Balliol College. The other tutors were the late Rev. E. C. Woollcombe, Dr. Lake, the present Dean of Durham, and myself. The Bishop of London (Temple) who preceded him in the office had just left us. In those days he was almost equally a lover of Classics and Mathematics. There was a time when he was quite divided in his allegiance between them, and used to say, in his free and easy way, that he 'must toss up a shilling to decide.' Even in the last years of his life he was in the habit of taking with him Greek books to read during the Vacation. In conversation he left the impression of being a well-read scholar, and a real critic, who was never led away by ingenious conjectures or uncertain fancies. For some time he was intending to edit the Timæus of Plato for the Clarendon Press, but he never had leisure to carry out this project. He finally determined, and probably he was right, to make Mathematics the chief work of his life.

The Mathematician is more cut off by his pursuits from his fellow-men than the student of any other branch of knowledge. He has interests which are locked up in his own breast, pleasures and also pains which he cannot communicate to others; the better part of him is moving about in a world of numbers and figures which have no connection with ordinary life (cp. Plato, Theætetus, 184 D). His study is apt to become a passion with him and affects his character. I am sure that this was true of Professor Henry Smith. It was the smaller part of him which we knew or could appreciate. His mathematical speculations could have been shared by a very few, not more than two or three, of his contemporaries at Oxford. Yet he did not withdraw himself from business or society. He was not the silent philosopher who is lost in reverie, or who, while acknowledged to be a mathematical genius, is pointed at by mankind as a poor and eccentric mortal. He was a thorough man of the world and greatly liked by everybody. He was very manly in his bearing, and quite free from shyness and nervousness in any company. He had a kind greeting for servants, and felt a real kindness for them—they were devoted to him. His manner of behaviour towards all sorts and conditions of men might be described as exhibiting a singular 'urbanity.' He was decidedly good-looking, and there was a certain intellectual distinction in his features and expression. It is necessary to combine these various aspects of him if we would duly estimate him. He was everywhere, and known to everyone, the life and soul of a social gathering. But he was also a thorough student, and an omnivorous reader, passing several hours of the day in abstruse Mathematics, but nevertheless acquainted with all new books, and on a level with every recent scientific enquiry.

He went on teaching at Balliol College as Mathematical Tutor for about thirteen years; at the end of that time he was appointed Professor of Geometry; he then combined the duties of Tutor and Professor. While only a Tutor of Balliol he had hardly any pupils worthy of him. The College, having at that time no Mathematical Scholarships, had seldom any good Mathematical students (those who were being usually men who read for double honours). His duties were, for the most part, confined to the preparing and examining men for Responsions. But he never thought it beneath him to take pains with any one, and he was an admirable teacher. He used to have his pupils on a Sunday afternoon to be examined by him, and would tell them that 'it was lawful on the Sabbath Day to pull an ass out of the ditch.' The better men were of opinion that they learned more from him in a few minutes than from another in a whole hour. He was constitutionally apt to be late and irregular in lecture, and on occasions of business as well as at a dinner party was often the last to arrive, but every one was very willing to wait for him. The circumstances of the University hardly admitted of his raising up a School of Mathematical pupils, but he was the life and centre of the study while he was with us.

He was very desirous to promote the interests of Natural Science in Oxford, and was in favour of some measure which would have made the knowledge of a portion of some one of the Natural Sciences the condition of obtaining a degree. The teachers of these sciences had long been fighting a battle against the older traditions of the University; they had now become the study of a few, but he clearly saw that they could never truly flourish until an interest in them was more generally diffused and they had a congenial atmosphere. But he was also the best friend that the older studies then had in the University, for he could speak with authority, and he was firmly convinced that in education Science should not supersede Literature. He deplored equally the want of literary culture which he observed in many scientific men, and the gross ignorance of the most general facts of Science which prevails in the world at large, especially at English Universities and Public Schools. In a similar spirit he was anxious to encourage at Oxford the study of Medicine and also of Engineering, thinking that they would supply a missing link between the Physical Sciences and the older studies of the University.

A considerable portion of his time was devoted to College and University business. Though he transferred his name to C. C. C. about ten years before his death and nominally ceased to be a member of Balliol College, he continued to show the same earnest interest in its concerns which he had always done. He took the same part in its Examinations and College Meetings—the only difference being that he no longer received the stipend of a Fellowship from it. There was never any one more affectionately regarded by the Fellows, or whose opinion had greater weight with them. He had the art not only of doing business well, but of making it pleasant, often with a slight jest or play of words smoothing away difficulties. I do not remember his ever having had a quarrel or difference with any one in the University. It will be easily understood that such a man was well adapted to keep men together and to carry things forward. At the Hebdomadal Council, where he usually appeared rather late in the day, he gave life and animation to every discussion. He seemed to say things in a better way than anybody else, and in an argument there was no one who was a match for him. When a new measure had been put into form by the Council he was very often selected to carry it through Convocation, his popularity and his manner of speaking having great weight in that assembly; and it was whispered that 'the Council relied for the success of their measures too much upon Henry Smith's oratory.' Though well aware that the order and discipline of the University must be maintained, he was always a very earnest supporter of freedom, and a great enemy to the imposition of useless restrictions upon Undergraduates. He was indulgent to the failings of young men, and felt a humane pity for persons who had lost their character. He was one of whom it might be said that 'he would have stood by a friend, not only in adversity, but in disgrace.' Two occasions on which he distinguished himself were long remembered by those who heard him,—once in the Common Room, more than thirty years ago, when some of the elder members of the College sought to impose a new-fangled test upon the undergraduates instead of the time-honoured Thirty-Nine Articles. He pleaded earnestly for the retention of the latter, alleging that 'old chains were smoother and easier to the wearer of them.' The other occasion was in Congregation, about twelve years ago [1880], when he introduced a measure granting privileges to Colonial Universities, and drew a sketch of the growth of the London University, and of the mistaken policy of Oxford and Cambridge in their opposition to it.

He was not an orator, but a very good speaker, who had the faculty of thinking when on his legs, never faltering for a word, able to strike out, right and left, good-humoured and telling blows. His speeches were clear and luminous, and they also had the merit of keeping up the attention. Above all, he had tact. He said what he ought to have said, and abstained from giving needless offence. As a writer, he never attained to considerable eminence. He was the author, when quite a young man, of a very clever review in the Oxford Essays of Sir David Brewster's 'More Worlds than One.' In this paper the fallibility, both of men of science and of theologians, was impartially exposed, and I remember that Bishop Temple remarked, after reading it, that 'the author could do many things well, but that he would write better than he did anything else.' The prophecy was not destined to be fulfilled. His mind was drawn in another direction, and he had not the poetical gifts which seem to be indispensable in a great writer, whether of poetry or prose.

He was wanting in initiative. Though a very able supporter of the plans of others, he rarely, if ever, took the first step in introducing a measure himself. He was easy-going, not burnt up with a fiery zeal for change, but satisfied in general with the world as it is, and really, I think, somewhat deficient in practical originality. He was contented to be a follower rather than a leader in most of our University contests. When he was brought forward a few years ago, rather against his will, as a Candidate for the representation of the University in Parliament, he was, I believe, absolutely indifferent to the result. I never knew a man, possessing so much ability, with so little ambition. Hence he interfered with no one, and as no suspicions were entertained of him, everybody was willing to do justice to his great merits. Nor had he any sympathy with new opinions in politics or theology. In politics he would have professed himself a Liberal, but he was not an advanced one. He was willing to talk about them, and his views were always worth hearing, because they were not strained out of newspapers, but the result of his own reflections. He was an acute political economist, a disciple of Mill and Ricardo, not much interested in the wider field which is sometimes claimed for the science. Nor was he at all disposed to undervalue the influence of theology. He was well acquainted with the results of German criticism on the Scriptures, but they seemed to make no difference to him. When he first came up to the University he was an Evangelical, and, for a while, retained his old belief. Indeed, some years after, on the occasion of a high-church sermon at St. Mary's, he would say, with indignation, 'that was not the sort of religion in which he had been brought up.' But, in time, the old clothes of his youth naturally fell off—he had out-grown them, and there remained a blameless character, a singular kindness and generosity, a love of justice and fairness, and a sense of religion which was wrapped in impenetrable silence—it was one of the subjects of which he least desired to talk.

He was very reserved. Like many other persons who pour themselves out freely in conversation, there was the appearance of *abandon*, but there were many subjects about which he rarely, if ever, spoke. One of these was himself. He was probably the confidant of many, for no man could give better advice in a difficulty, or was more willing to assist others. There was a feeling that he could be absolutely trusted, and even if a foolish thing were said to him, that he would not repeat it. His insight into human character was said, by one of his friends, to be 'terrible,' but it was never used by him except for some kind purpose. He could see through the vanity and folly of a friend, and yet retain a never-changing affection for him. Of his own life, he seldom or never spoke; he was not an egotist, and his own sayings or doings did not seem to interest him afterwards.

It is difficult to give an idea of his conversation. It was gay rather than serious, full of life and chaff, arising naturally out of the circumstances of the hour. If a stranger had come across him in a railway train, or had been his companion on a voyage, he would probably have found that this unknown person was one of the most agreeable men he had ever met. It was a great pleasure to have a tête-à-tête with him, for he was not one of those who required a company in which to show off. I have often decoyed him into my room for the sake of having a chat with him, and when once there, he was very willing to stay, for he was one of those who like to have a talk out and did not hurry away when the clock struck. In society he was ready to talk to every one, and every one was ready to talk to him. He had the art of setting people at their ease. He would at times break out into fits of laughter and joviality, which showed that the original Irish nature was not extinguished, but only kept under by him. Stories were repeated of his performances at Meetings of the British Association, which must greatly have enlivened that sedate assembly. He was certainly a wit, but his good sayings were of too delicate a fibre to be transplanted. To use Boswell's expression, his *bons mots* 'would not carry.' But they were the delight and admiration of those who heard them at the time. They possessed also one of the highest qualities of wit and humour, spontaneity. They were made on the spur of the moment with reference to something which was said or done at the time. And this very quality tended to impair their effect on those who were not present when they were first uttered, and did not know the occasion which had given rise to them. A light irony seemed to be always playing about his mind. It was the form under which he inclined to regard all human things, for he was very unimpassioned. An old school-friend would sometimes be the target at which he aimed. The great scholar, Professor Conington, a man so unlike himself, that their mutual friends wondered what could be the tie which united them, was often the butt of his humour. But the slight humiliation to which he was subjected was more than made up to him by the constancy and faithful attachment of his friend, who afterwards collected his literary remains and wrote his life. He was a little provoking to some others, especially to those who were too earnest or of too pushing a temper. He knew how, in Aristotle's language, 'to overcome seriousness by laughter,' or in other words, to make such persons appear slightly ridiculous. An enthusiastic friend might have thought him deficient in sympathy, but he was really always kind and considerate.

I hardly venture to repeat some of his good sayings, lest, detached from their surroundings, they should seem not to justify the high opinion which has been expressed of his conversational gifts. They are not of course of the quality of the best sayings of Charles Lamb or Sydney Smith, yet they are such as might have been said by them. The reader is requested to bear in mind their impromptu or occasional character. He who made them could have made many such every day of his life, and never aspired to be a wit, but only to amuse himself and his companions. At any rate they may serve to remind his friends of pleasant hours which they passed with him, never to return.

A friend told him of a rather ponderous jest made by Sir George Lewis, who, when Minister of War, once proclaimed in the House of Commons in a loud voice that he had ordered experiments to be tried respecting the comparative effect of 'short and long bores.' To this heavy piece of artillery Henry Smith instantly replied by asking whether he was not aware that 'smooth bores' were the most deadly of all.—Another friend said to him: 'What a wonderful man Ruskin is, but he has a bee in his bonnet.' 'Yes,' replied Henry Smith, 'a whole hive of them; but how pleasant it is to hear the humming!'—The Lectures of a certain College Tutor were reported to be 'cut and dried.' 'Yes,' said Henry Smith, '*dried* by the Tutor and *cut* by the men.'—A dispute arose at an Oxford dinner-table as to the comparative prestige of Bishops and Judges. The argument, as might be expected at a party of Laymen, went in favour of the latter. 'No,' said Henry Smith; 'for a Judge can only say, "Hang you," but a Bishop can say, "D—n you."'—The next is of a higher class of wit. Speaking of an eminent scientific man to whom he gave considerable praise, he said: 'Yet he sometimes forgets that he is only the editor and not the author of Nature.'

The two remaining ones are autobiographical. He once said to a friend: 'C., I was kept in bed by illness when quite young for six weeks; I then began to study mathematics, and I wish I had remained there ever since.'—Speaking to a newly elected Fellow of a College, he advised him, in the low whisper which we all remember, to write a little and to save a little, adding: 'I have done neither.'

These slight jests may perhaps be thought disappointing: it is probable that they are marred in the telling. They were the bubbles which were always rising to the surface of his mind, and though but poorly reported, may help to give to those who did not know him personally a faint idea of the charm of his character and conversation.

Though not rich, he was extremely liberal. He never seemed to think either about gaining or spending. He used to say that not enough money was to be had in Oxford to make it worth while to take trouble about it. Yet a certain love of speculation which was latent in his nature once led him into an unfortunate venture, from which he extricated himself by taking the affairs of a Company into his own hands, and at a considerable loss. For his services as a College Tutor he received a very moderate remuneration, but, having enough for his wants, he never seemed to desire that it should be increased. He did not wish to impose on the College a burden which it could ill afford to bear.

I have endeavoured, in a few pages, to give a sketch of one with whom I was in daily intercourse during thirty-five years of his life, and who I think may be regarded, without exaggeration, as one of the most remarkable persons of his time. Yet he lived and died almost unknown to the world at large. I have sometimes asked myself what was the reason of this contrast between his reputation and his real merits. It has been said that 'the world knows nothing of its greatest men,' but this familiar line, whether true or not, is not the whole account of the matter in his case. The explanation is partly to be sought in his own character. He had no ambition, he had not a strong will, and he had never made himself known to the public. He was once reproached by a friend for 'giving up to society what was meant for mankind,' and the reproof, as far as it applied to his life at Oxford, was not without foundation. He was not the author of any considerable work. His Mathematical writings, on which his fame chiefly rests, await the judgment of time. Though he managed, in great part, the affairs, not only of the University, but of several other great institutions such as Winchester and Rugby Schools, University College, Bristol, the University Commission, the Meteorological Office, the Oxford Museum, of which he was Keeper, and the Ashmolean Society, of which he was the Secretary, he could hardly be said to have left his mark upon any of them, however valuable his services have been to those institutions. To understand his superiority over his contemporaries, it was necessary to have lived with him and known him, to have heard him lecture, to have been with him at a College Meeting, to have enjoyed his society at a dinner-party, or on an excursion of pleasure. He never offended you, never disappointed you, he was never tired or out of humour. His greatness was shown in the peaceful continuity of a private life, not in great actions, or on striking occasions.

B. JOWETT.

II. RECOLLECTIONS BY LORD BOWEN *.

GREAT statesmen, successful generals, famous authors, distinguished men of science, eminent theologians—all those who have been raised by industry, talent, or the caprice of fortune, to prominence in a profession—become by degrees actors on whose movements our attention rests, and whose familiar figures are part of the spectacle of life. The public they have interested during their time bids them, when they die, a kindly and sympathetic farewell, retraces their career, counts up their successes, and assesses their general apparent value. Professor Henry Smith, whose loss this week casts a shadow both over Oxford and through many circles of educated men and women, belonged to none of these categories. To by far the greater number of Englishmen, his name is probably unknown. Some will vaguely recollect it as that of a candidate put forward unsuccessfully a few years ago by Oxford Liberals for the representation of the University. Many even of those who are aware that a man in the fulness of his powers is just dead, whose brilliant intellectual attainments have probably not been surpassed by any other of their English contemporaries, may, nevertheless, be surprised at regret so widely felt and so loudly expressed over the loss of one who wrote no great books, patented no great invention, amassed no fortune, made no famous speeches, and led no conspicuous movement, political or social. Measured by the popular measure of publicity and fame, Professor Henry Smith would hardly seem, to most of us, to have been one of the great men of the time. Yet it would be difficult among the world's celebrities to find one who in gifts and nature was his superior. Generally speaking, there is a rough justice in the sentence passed upon intellectual men who achieve no definite worldly success. We surmise, and often with truth, that some weak spot somewhere in their powers has been the cause of their failure to acquire those sublunary distinctions and rewards which coarser and more practical people manage to secure. To the case of Professor Smith, this kind of criticism would be inapplicable, for he possessed both the qualities and the character which might have made him famous in many active walks of life. His mental attainments were of the highest order. A finished classical

* [Reprinted (by permission) from *The Spectator*, Feb. 17, 1883.]

scholar, a mathematician, in some respects of European distinction, a considerable metaphysician, a trained master of most branches of knowledge, literary, economic, and scientific, an adequate linguist, and a man of sound judgment, perfect temper, and wise aptitude for affairs, he combined with his other special excellences a delicate gaiety of spirit, a brilliant conversational power, which made him one of the most accomplished and attractive ornaments of any educated company in which he moved. To what eminence in public or professional life accomplishments so varied might not have led him, it is difficult to feel sure, if only he had ever plunged into the stream of competition or adventure. But some delicate touch of indifference to worldly success mingled itself with his genius, and he remained to the last content with playing, and with playing well, whatever part fortune brought to him to play. Incessantly occupied in the discharge of duties both of a public and a private kind, that thickened round him as years went by, he was satisfied with what had fallen to his share in the lottery of life, and neither solicited nor ostentatiously avoided anything beyond. The 'note' of personal ambition seemed absent from his composition. And so it happens that the great public which takes its knowledge of men from newspapers and books, from debates in Parliament and the records of our Law Courts, hardly knew—if, indeed, it knew at all—Professor Henry Smith.

As the personal 'note' was wanting in Smith, so, on the other hand, the intellectual or academic 'note' was one which he possessed in, perhaps, its most attractive form. Vanity and self-seeking, every form of mental intemperance and extravagance, seemed to have no place in anything that he ever said or did. The last, the rarest triumph of education, is when it destroys the desire of self-assertion in a man of genius, and substitutes in its place the crowning flower of perfect moderation and equanimity. The greatest of Greek philosophers, in the greatest of moral treatises, has elaborated a theory that virtue consists in a golden mean, and in the avoidance of dangerous extremes; but when driven into a corner for a standard by which the mean is to be measured, the illustrious moralist has no better compass to furnish for our guidance than this,—that the golden mean in each case must be that which is defined by the reason of some thoroughly temperate man. The result of Henry Smith's genius and culture combined seemed to make him the very man required by a philosopher for his human measuring-rod. A University life sometimes spoils and sometimes perfects natural capacities, but it usually leaves its mark upon them, whether it be for good or evil. Nobody could doubt but that Henry Smith, as he issued from the Academic mould, was a natural genius, with an impress of his University stamped distinctly upon him; and Oxford has, perhaps, never had a more happy specimen to produce of her best influence than the late Savilian Professor of Geometry.

Smith came from Rugby to the University as a remarkable boy, and won the blue ribbon in all the great intellectual competitions of his undergraduate days. He became in due course a Fellow of Balliol, and joined a Common Room which consisted of a small group of very distinguished men. The present* Master of Balliol was already conspicuous in the society of Balliol Fellows, as the most successful and most energetic tutor of the first of the Oxford Colleges of the period. Among the rest were names of academic fame—Mr. Lake, the present Dean of Durham; Riddell, an accomplished hero even among Shrewsbury scholars, whose beautiful character and refinement of mind were prematurely lost to the University by an early death; Archdeacon Palmer, not the least distinguished of a trio of brothers with all of whom Oxford had reason to be content; Lonsdale, Wall, Woollcombe, Walrond, and a few years later, Newman and Green. These were the days when Oxford, always passing through some phase or other, was entering on a new situation. The Tractarian movement had subsided, but the University was not at rest. A reforming Parliamentary Commission was troubling the waters. The old system of close Scholarships and Fellowships was slowly giving way, and like the rotten boroughs of a past political period, the close preserves of the Colleges were being either extinguished, or thrown open to public competition. But Oxford was still Conservative at heart. Leaders of the old school and their followers held the University pulpits, dominated Congregation, monopolized the best preferments, resisted to the best of their powers all local change, and were ready on provocation to ostracize unorthodox reformers for being, like Socrates, the corrupters of youth. Married Fellows were as yet unknown; it had not yet become necessary to build whole suburbs of semi-detached villas to receive the feminine colonists of the future. But there was a stir and an agitation throughout the Academic world which the sense of changes, present and to come, had produced. University politics and polemics were, as always, of absorbing interest. Mansel and Goldwin Smith tilted against each other in debate before an admiring and competent academic audience. Oxford was, in fact, at war,—a war, it is true, polite, polished, and courteous.

* The late Professor Jowett.

Into this atmosphere, charged as it was with considerable personal electricity, Henry Smith was thenceforward absorbed; for nearly thirty years, no more attractive, brilliant, or genial figure was to be found in the perturbed society of the University. Some happy combination of judgment and temper made him acceptable even to those with whose opinions he had nothing in common. He succeeded in being a politician, without wearing the obnoxious colours of a partisan. He had the great art of never pressing a victory home, and of bearing defeat with pleasant equanimity. His business powers, his modesty, his wisdom, and his entire freedom from egotism and dogmatic presumption, a delicate gaiety that never flagged, wit that sparkled without wounding, and which rose incessantly to real brilliancy, made him not merely an effective personage in the Oxford world, but universally acceptable in any society, whatever the shade of its opinions. His finished persiflage, his pleasant epigrams, will long be remembered, though the brightest conversation is often the most evanescent, and the finesse of wit, like a musical laugh, disappears with the occasion, and cannot be reproduced upon paper or in print. As by degrees his attainments were recognized, both in England and abroad, his influence at Oxford naturally deepened; but neither within nor without the University did he grasp at opportunities for notoriety. Such power and authority as he possessed he held without an effort, without solicitation, apparently without any personal satisfaction in them. In offices of friendship he was constant; in such public or civic duties as came in his way, assiduous; no good or benevolent work ever needed a helping hand, but his was at its service, without ostentation, and without any expectation of personal advantage. He was a good speaker, without being a rhetorician; his death, indeed, last week was hastened by a chill caught or increased while he was addressing a gathering of agricultural labourers.

A life like Henry Smith's, of exemplary moderation, far removed from even a suspicion of worldliness and vanity, is seldom found in these days in combination with intellectual powers and practical ability on so considerable a scale. There are, no doubt, many nooks and corners in which at times may be seen flowering 'the wise indifference of the wise.' Students, divines, men of science or of letters, not seldom seem content to retire from the world, as if they had measured the true value of the things we most of us eagerly compete for, and were perfectly satisfied, of deliberate choice, to remain spectators of the fever of mankind. Some physical inaptitude, or some constitutional tendency, not unfrequently lies at the bottom of this apparently philosophic temper. Patient self-possession, and a sober estimate of the world and of what it can give, are rarely found in a man who lives in constant contact with other men and their affairs, who shares in the interests of his generation, occupies himself with its business, and whose genius seems to bring high honour and success almost within his reach. Professor Henry Smith was not buried away from his fellow-creatures in literature, or study, or contemplation; he was no recluse or invalid, but a man of the world, active, competent, social, *only*—not ambitious. Personal serenity of such a type is rather a classical than a modern virtue; perhaps an age different to our own may yet regard it as one of the highest forms, not merely of intellectual, but of civic excellence. It is the characteristic of recent civilization, that in almost all its aspects it seems based upon a theory of personal competition. The prominent figures on every stage are the result of a struggle, not for existence, but for success. It is a contest which all seem satisfied to recognize as one of the conditions of ordinary life; which constitutes the essence of our politics, of our commerce, of our political economy, of our laws of property themselves. In the general race to possess more than the average share of wealth, power, fame, it is, perhaps, a wholesome lesson to turn for a short breathing-time to the uneventful example of the life of a man of genius, who was fitted for most distinctions, if he had cared to seek them; but who was unaffected by the universal fever, possessed his soul in perfect patience, and remained to the last content to discharge all the duties which Providence allotted to him, without affectation, and with that composure of soul to which great gifts are not always allied. The secret of the philosophic temperament, exhibited in this its most manly shape, is one which is not easy to explore; but when the phenomenon is seen, its charm attracts us the more in proportion to its rarity. Essayists and moralists for the last two thousand years have preached it, and inculcated it; some have gone so far as to boast of its acquisition,—its praise, certainly, is among all the prophets. Probably it is the product neither of Nature, nor of education singly, but of a happy, and of an admirable combination of the two. Among the many friends, acquaintances, admirers, whose thoughts have in the last few days been saddened or sobered by the unexpected death of a brilliant man of genius, there are none who will not readily accord to Professor Henry Smith the tribute of unaffected respect for what without extravagance may be termed his extraordinary powers of mind, his gentle and Laelian wisdom, and the sweetness of character which never made an enemy, lost a friend, or sought a personal advantage for itself. But besides this and beyond this, it may not be out of place, before a personality in many ways so complete fades into indistinctness, and a life ceases to be familiar to us which must hereafter be treasured rather in the memory of his contemporaries and friends than in the history of his time, to recognize in the Professor Oxford has lost that special type of wholesome and of manly virtue the growth of which is not much favoured by the rush and turmoil of these times. Great mental gifts can be found, when occasion demands them; talents grow on every tree. But the serenity of heart which enables its possessor to wear the gifts of genius with sobriety, and to use them nobly and well, without seeking to expend them in the purchase of fame, or wealth, or of advancement, is a quality which modern society little cultivates, and seldom sees.

III. RECOLLECTIONS BY MR. J. L. STRACHAN-DAVIDSON.

THE death of Henry Smith will be felt as the greatest loss which could have befallen Oxford. In him the University possessed a student whose knowledge and genius were honoured throughout Europe. Of those amongst whom he lived few indeed could follow him to the height of his scientific speculations. Most of us did not know enough to understand where and how he was working in the field of Mathematical Science. By us he is lamented as the wisest counsellor of the University, and as the delightful companion who gave life and charm to its society. Though his activity extended far beyond the limits of the University, he was very constant to Oxford. Since he took his degree he did not miss a single Term's residence. Re-elected time after time to the Hebdomadal Council, his assistance was called for whenever any serious business required sound judgment or delicate handling. His advice was generally followed, and if not, the neglect of it was almost always regretted in the sequel.

In Henry Smith were united to a rare degree knowledge of business and knowledge of men. He seemed most thoroughly in his element when swaying and guiding his fellows. To every matter which he took in hand, he seemed to come with a fresh mind, throwing off all the multitude of concerns which beset him, and unburdened by care or anxiety. Then under the cover of his easy playful manner it would soon become manifest that he had grasped all the true points at issue, and was ready with a firm and wise decision. He always looked facts in the face, and strove hard to distinguish the difficult from the impossible. To the more ardent spirits among his followers it was sometimes a matter of disappointment that he would not lead them to assaults which he saw to be fruitless. Though he could fight hard when the moment for fighting came, no one was more averse to multiply occasions of controversy. He saw things without passion and without prejudice, and laboured quietly and steadily for all that could advance the studies and promote the efficiency of the University.

In this spirit he accepted the thankless task of serving on the University Commission. It is the inevitable fate of such a body that their work is attacked at once by the criticisms of those who think that it has gone too far, and of those who would have had it go further. Henry Smith knew well that it was impossible to satisfy either the one party or the other. But it was a source of keen satisfaction to him to notice that when those who joined in complaining of the Commission came to propose alternative schemes they found that these divided them more than that to which they had objected. In the same way he was much gratified that it was only a minute point in the Commissioners' arrangements which was finally contested by the University. He claimed, and with justice, that when the Proctorial appointment of Examiners was the only portion of the old constitution which was defended to the last, it was pretty clear that the more important changes were acknowledged to be wise and necessary.

It may be interesting to note what Henry Smith thought of these greater changes. He fully appreciated the charms of the old system of celibate Fellowships, and never for a moment cherished the illusion that the new seven years' tenure could ever have the value and dignity of the old. But he felt that the old system could be practically worked only so long as the majority of Fellows were willing to take Orders and retire to a College Living in middle life. When it became evident that the University must either renounce the service of its most efficient members or be content to be served by laymen, he recognized that, at whatever sacrifice, a career must be provided into which a man could enter as his profession for life.

Another important question often present to his mind was the effect of the College system on the life and teaching of Oxford. He felt the difficulties as keenly as many who urged radical changes; but he felt likewise that it was worth making an effort to preserve this distinctive feature of the English Universities by transforming it to suit the new conditions. When in conversation he summed up the work of the Commission, it was, 'we have given a fresh lease of life to the College system.' He was not very sanguine that this system could be permanent, but he was convinced that it ought to have another chance, and that the best chance was secured to it by the reforms which he and his colleagues had effected.

It is difficult to speak of the charm of his life and conversation. The light touch and happy play of mind with which he enlivened the most serious business, and softened the jarring of controversy, was a source of real power, and procured a ready acceptance for the wisdom of his practical suggestions. In social life the same qualities shone forth at every moment. It seems hardly credible to those who knew him best that a deep-seated disease had been sapping his life for years. His temper was always unruffled, his spirits always gay and easy, and his sympathy always ready. In the midst of a mass of business which would have absorbed any ordinary man he could always find time to attend to the interests and concerns of his friends. To cheer a sick friend with the sunshine of his presence, to be the protector of the children of those who were taken away, to lend a ready ear to the perplexed and a helping hand to those who had committed themselves by any foolish action—all such kindnesses seemed so easy and natural to him, that men claimed and accepted his benefits almost as a matter of course. He seemed to be good not in obedience to any external law nor as the result of any internal struggle, but because goodness was the simple outcome of his nature.

His wit and gaiety were the delight of all who listened to him. It was not so much that he was a sayer of good things to be remembered and repeated—though of these too there was no lack—but the really characteristic feature of his talk was that its interest never flagged; a certain flavour of freshness and originality pervaded it, and revealed itself even in his commonest remarks. To walk or ride with him was to enjoy a conversation in which not a sentence was commonplace. There was always some new light, some refinement or subtlety of thought or expression which gave a charm to the most ordinary topics. This effect was due mainly to the keen and delicate temper of his mind, but partly also to the wonderful breadth of his knowledge and his interests. He knew the literature of Greece and Rome as if he had made their study the work of his life, whereas it was really the amusement of his leisure hours. He had the sincerest love for the classical writings and the most profound belief in their value. His retentive memory and delicate taste made his conversation on these topics a storehouse of interesting and instructive criticism.

Though the resources of his own mind filled to overflowing every moment he could snatch for quiet study, yet he never shut himself up or held aloof from his fellows. There was absolutely nothing stern or forbidding about him. He seemed to take in the society of his friends the same pleasure which his presence imparted to them. In every relation of life there was in him the perfect ease and grace which flows naturally from complete and sufficing strength. The sweetness of his character and the perfect cordiality of his nature seemed to offer all the rare gifts of his genius to minister to the happiness of his friends. His death leaves dark what was a ray of sunlight in the lives of many.

He was entirely free from superstition. He held deliberately that the questions whose solution is hidden from man, and above all the prospect of death, should never be allowed to cast a shadow over the life and work of the present hour. He believed that it became a man to live at his best and to labour at his best during every day allotted to him, even as though an endless succession of such days were in store. It was permitted to Henry Smith to give a bright example of his theory. Till within a week of his death he was teaching from his chair, attending to all the varied work of government and management for which he was responsible, and living a bright and happy life which shed cheerfulness and comfort on all around him. He always maintained, that there is no such thing as a necessary man, and that every place left vacant can be adequately filled. Of all that Henry Smith taught, this doctrine is the one which it seems most difficult to realize at this moment.

[February, 1883.]

NOTE BY MR. ALFRED ROBINSON.

I HAVE been asked to give some account of the contested election in which Professor Smith was a candidate for one of the University seats in the House of Commons; and I do this with much pleasure, because, although he was defeated, the amount and kind of support which he received in the contest show how much he was valued by men of all parties in Oxford, and how unique was his position in the University.

In the spring of the year 1878 it became known that Mr. Gathorne Hardy, who was then one of the University representatives, was about to be summoned to the House of Lords. The rejection of Mr. Gladstone in 1865, and the defeat of Sir Roundell Palmer by Sir John Mowbray in 1868, had proved that no one but a Conservative could win in a contest conducted upon the lines of political party. But it was thought by many persons that the Members for the University ought to be chosen upon academical rather than upon political grounds, and ought to represent learning, science, and education, without special reference to party interests.

Professor Henry Smith was brought forward as a man most eminently qualified to represent the University in this sense. The fact that he was a Liberal in politics of course was not disguised. He was indeed at this time not fully in sympathy with some of the Liberal leaders. The Eastern question then filled the political foreground, and Professor Smith, while disapproving of the general policy of the Conservative Government, thought that Lord Salisbury ought to be supported in maintaining against Russia in her dealings with Turkey, the rights of the neutral Powers and the general interests of Europe. Perhaps, also, Professor Smith was too critical, and too fond of looking at questions from every point of view to have ever made a first-rate party man. But still he belonged undeniably to the Liberal party, and he was not a man to be led by any waywardness, or by any love of fads or crotchets, into a position of political isolation. So his election by the University would no doubt have been a transfer of a seat from the Government to the Opposition side of the House, and this was the aspect in which the contest presented itself to the great majority of the voters. It was not, however, on a contest of this kind that Professor Smith's chief supporters wished to enter. In their view the special representation of the University in Parliament was useless if the University Members were to be party men of the ordinary type, without special qualifications for dealing with the questions with which the University was specially concerned, and their main object in bringing forward Professor Smith was that this view should be put before the constituency and the country

With the arrangements for his own candidature Professor Smith had very little to do. An old custom of the University, which had been observed by Mr. Gladstone throughout his long tenure of his seat, precluded a candidate from issuing any address, or from making any speech to the electors. At an early stage in the proceedings Professor Smith was invited to stand, and he agreed to allow himself to be nominated, but he took no part in the initiation of his own candidature; he was never present at the meetings of his committee; and his supporters defrayed the expenses of the contest, declining a request which he made that he might at least be permitted to contribute to the subscription list.

Professor Smith's Committee was formed in the month of April, 1878, and consisted of two sections.

(1) The London Committee, the Chairman of which was Mr. Mountague Bernard; and which had for its Vice-Chairmen the then Marquis of Tavistock, Mr. Goschen, Mr. Knatchbull Hugessen (afterwards Lord Brabourne), Mr. Dodson

(now Lord Monk Bretton), Dean Stanley, the Dean of Canterbury, the Dean of Durham, and the late Sir Benjamin Brodie; and for its Secretaries, Mr. Buller of All Souls, Mr. Ilbert, late Bursar and Fellow of Balliol, Mr. Pope, formerly Fellow of Lincoln, Mr. Robertson, Fellow of Corpus, Mr. A. L. Smith, then Fellow of Trinity.

(2) The Oxford Committee, having for its Chairman the Dean of Christ Church (Dr. Liddell); for its Vice-Chairmen, the then President of Corpus and Archdeacon Palmer; and for its Secretaries, Mr. Crowder, Bursar of C. C. C., Professor Green, Mr. Jackson, Fellow and now Rector of Exeter, Mr. Monro, Fellow and now Provost of Oriel, Mr. Papillon, Fellow of New College, Mr. Salwey, Student of Christ Church.

It is not likely that any of these persons were under the illusion that their cause was going to win. The Conservative feeling of the constituency was soon found to be so strong that under no circumstances could any one but a Conservative have been elected. And at this time the excitement of the two political parties with reference to the Eastern Question much increased the difficulties with which Professor Smith's Committee had to contend. On the one hand, a large section of the constituency saw in him only an opponent of the Government which was patriotically defending British interests against Russia. And on the other, some well-known Liberals considered that he was too lukewarm in his censure of the Party and of the Ministry which they associated with the notorious atrocities in Bulgaria.

Opposition of the former kind, which insisted that the representative of the University of Oxford must be a supporter of a Conservative government, it was impossible to disarm. But an effort was made to conciliate the critics and opponents who belonged to the Liberal ranks. Professor Smith was requested by his Committee to put forward some definite statement of his views on the Eastern Question, and in the following letter to a member of his Oxford Committee he complied with this request.

April 25, 1878.

DEAR ——,

I am well aware that the custom of the University imposes a great measure of reserve upon any candidate for the honour of representing it in Parliament. But I do not think that I shall be departing from a tradition, which I am most anxious to see maintained, if I venture to write a few lines to you in explanation of the views which I entertain with regard to the Eastern Question.

There has been much in the foreign policy of the Government during the last two years of which I cannot approve. I think that they should have recognised, at a far earlier period than they did, that the condition of the Christian Provinces of Turkey had become unendurable, and that the maintenance of the *status quo* was no longer possible. A grave, but long foreseen, emergency had arisen; and this country should have been prepared with a well-considered policy to meet it. Instead of this, the Ministry seem to me to have drifted with the stream of events, until at last they find themselves in a position in which it is immeasurably more difficult, than it would have been twelve months ago, to assert the right of the neutral powers to have a decisive voice in the settlement of a question affecting such vast European interests.

Looking at the most recent events, I have to express a general concurrence with the main tenor of Lord Salisbury's Despatch; and I have observed, with great satisfaction, that it has been received with cordial approval by the Liberal Press on the Continent. Interpreting that document, as I think I am justified in doing, by the light thrown on it by Lord Salisbury's own conduct at the Conference of Constantinople, I do not perceive in it any intention to restrict the liberties to be granted to the Christian subjects of the Porte; but I regard it as a protest in favour of the recognition of international obligations, and against any attempt on the part of Russia to dispose of the Eastern Question in her own way.

For the reasons which I have stated, I feel that I could not enter Parliament, except upon the condition of preserving the right to form while there an independent judgment with regard to the future action of the Government in these important matters. If there is a war party in England, I have no sympathy with it: but I am not for peace at any price; and, if any of the great interests of the country should be endangered, I should hope to see all Englishmen, without distinction of party, united in defending them.

Believe me to remain,

Very faithfully yours,

HENRY J. S. SMITH.

P.S.—You are at liberty to make any use which you may think fit of this reply to your letter.

This letter removed some of the misapprehensions as to Professor Smith's position, and probably produced some effect upon the canvass. Promises of support were received from some of the Liberals who had originally stood aloof, including one from Mr. Gladstone, which arrived a few days before the opening of the Poll. His example, however, was not imitated by all his followers, a few of whom, more Gladstonian than their chief, remained neutral to the last.

May 13th was fixed for the nomination. On that day the two candidates were proposed to the House of Convocation in Latin speeches—Professor Henry Smith by the Dean of Ch. Ch., Mr. J. G. Talbot by the President of St. John's. The Dean dwelt upon the scientific and literary qualifications of Professor Smith, his ability in business and in debate, and his suavity and fairness of judgment, which conciliated the regard of all. He recommended him to the electors as a man whom the Ministry of the day had entrusted with the most weighty affairs, and as one 'unice idoneum qui ipse academicus academicos suos in Parliamento repræsentet.' The President of St. John's, in nominating Mr. Talbot, made some kindly remarks on the undesirableness of withdrawing Professor Smith from the Professorial duties and from the sciences which he adorned.

Immediately after the nomination the Poll opened, and under the Act governing University elections it was not to be closed until the 17th, unless either of the candidates were withdrawn in the meantime.

The electors could vote either in person or by voting papers. At the time when the voting began Professor Smith's Committee had received less than a thousand promises from a constituency numbering more than four thousand members, and the last hope of the most sanguine of his supporters had disappeared. It was, however, thought best that all the votes should be recorded, in order that the amount and kind of support with which his candidature had been received might be accurately measured and generally known; so the polling was continued daily for five days in all. At the close Mr. J. G. Talbot was declared to be elected, the numbers being—Talbot, 2687 ; Smith, 989.

Defeated by a majority of more than two to one, Professor Smith's Committee might to some extent console themselves with the thought that they had conducted the contest with great economy. The expenses amounted to about £420, the chief item being the bills for advertising the lists of supporters in the chief London papers. This sum was less than half of what was believed to have been spent on the Liberal side in each of the two preceding Oxford elections.

Still more consolatory was the analysis of the Poll Book, which was published soon after the result of the election was declared. This proved that the majority of the electors and the working staff of the University had been ranged under opposite banners. The following table shows how certain sections of the constituency had voted :—

	SMITH.	TALBOT.	Abstained from voting.
Heads of Colleges, including two acting Heads......	10	10	3
Professors, Readers, and University Teachers.........	28	11	6
Tutors and Lecturers of Colleges and Halls	91	39	13
Fellows of Colleges, resident and non-resident	159	82	53
*Residents	152	117	44

* Residents—Members of Congregation qualified by residence—i. e. all electors who were in residence Oct. 1876—Oct. 1877, including the parochial clergy in Oxford and others not engaged in University work.

Whether this table points to any practical conclusion or not may be doubted. That the Members for the University should be chosen by those who are identified with it as the place of their work or residence in the present, and not merely of their education in the past, may seem reasonable, but no change could possibly be made which would reduce the constituency to less than one-tenth of its former number; and perhaps the special representation of the Universities in Parliament is more likely in the future to be abolished than to be reformed.

But whatever inferences of this kind might be drawn from the result of the election, the Poll Book was unequivocal in its recognition of Professor Smith's personal qualifications and eminence. And even the numbers set forth in the foregoing table, emphatic as they are, do not fully express the estimation in which he was held by that portion of the constituency in the midst of which he had lived, and with which he had been officially connected. For among those who abstained from voting there were some who remained neutral, in spite of their high appreciation of his claims, because they thought that a seat in Parliament would be incompatible with his Oxford work; and there were others who on ordinary occasions would have been ranked among his warmest supporters, but were unable at this time, when the foreign policy of the country filled the political horizon, to vote for a man who was variously criticised as going too far, or not far enough, in support of, or in opposition to, the Government of the day.

But, in spite of these abstentions, the preponderance of opinion in the working staff of the University was clearly marked, and was most significant.

It may be confidently said that no other man could have enlisted at this time among his supporters in a Parliamentary contest so many of the men who were identified by their position or occupation with Oxford; and it may be doubted whether in any of the controversies, political and academic, which have divided the University at various times in its history, so many of its resident graduates have ever enrolled themselves upon one side.

INTRODUCTION

TO THE

COLLECTED MATHEMATICAL PAPERS

OF

HENRY J. S. SMITH.

THE present volumes contain all the mathematical papers published by the late Professor H. J. S. Smith in his lifetime, as well as those which were in the press or which had been written out for printing. The reader is therefore in possession of all that he had already published, or had wholly or partially prepared for publication at the time of his death.

The arrangement of the papers is strictly chronological, the order being that of the date of reading or publication. The only partial exception is the Report on the Theory of Numbers, which is printed as a whole, although several other papers which follow it were published in the six years during which it was in progress. It is possible therefore by merely glancing over the titles of the papers to trace the course of Professor Smith's mathematical studies and tastes. The first two papers, written in 1851 and 1852, show that his mind was then occupied by Geometry. Within three years he published his first paper on the Theory of Numbers, consisting of a characteristic proof of Fermat's theorem that every prime number of the form $4n+1$ is the sum of two squares. From this time until 1867 the printed papers relate almost exclusively to the Theory of Numbers. Then follow a number of geometrical papers. In the last years of his life he was occupied principally with the subject of Elliptic Functions. It will be seen therefore that his work falls into three distinct groups: (i) Geometry, (ii) Theory of Numbers, (iii) Elliptic Functions. From the fact that the two earliest papers relate to Geometry we may infer that this was the subject which originally proved most attractive to him. The first of these papers was written in the year in which he obtained the Senior Mathematical Scholarship, and only a little more than a year after his election to a Fellowship at Balliol. It would seem that about 1853 he commenced the study of the Higher Arithmetic, a subject which engaged his almost undivided attention for many years, and which was never afterwards quite absent from his thoughts. The short notes which bear the dates of 1854 and 1857 show the tendency of his mind at this time: and in 1859 the first part of his Report on the Theory of Numbers was contributed to the British Association. The subsequent instalments appeared in the annual volumes of the Association for 1860, 1861, 1862, 1863, and 1865. These reports, which contain in a very condensed form the result of an immense amount of research, are models of clear exposition and systematic arrangement. Besides the accounts there given of the work of others, many of the paragraphs contain results of his own. These original contributions are not, however, noted as such, and they can only be detected by those who are already well acquainted with the details of the subjects to which they belong.

During the preparation of this Report he carried out elaborate researches of his own in several important branches of the Higher Arithmetic. The principal investigations undertaken at this time, which were completed for publication, relate to systems of indeterminate linear equations and congruences and to the orders and genera of ternary quadratic forms containing more than three indeterminates. These memoirs appeared in the *Philosophical Transactions* for 1861 and 1867. He also contributed several shorter papers to the Proceedings of the same Society, which indicate much more extended investigation in the same field: one especially (No. xviii, vol. i.) consists merely of a brief statement of results which were obtained by means of a very long and delicate analysis.

A considerable part of the last instalment of the Report is concerned with arithmetical formulæ derived from Elliptic Functions, and it seems likely that it was in this way that he was first attracted to this Theory; for his first published paper on the subject (No. xvi, vol. i.) bears the date 1866. The remaining papers included in the first volume relate to Geometry, principally homographic figures.

In the second volume (1869–1883) there is more Elliptic Functions and less Theory of Numbers: but the sequence of the papers no longer affords an indication of the author's train of thought: for, in the later years of his life, he was frequently compelled, by various circumstances, to leave the subject upon which he was engaged, in order to prepare for publication theorems and demonstrations that formed part of the many unfinished investigations stored up in his note-books.

The first paper in the second volume was a prize memoir for which, conjointly with another memoir, the Steiner prize of the Berlin Academy was awarded. The subject was announced in 1866, and the memoirs were to be sent in, each designated by a motto, before March 1, 1868 *. Four were received, and the prize of six hundred thalers was divided between Professor Smith and Dr. Hermann Kortum, of Bonn, the two memoirs being regarded as of equal merit. The report on the memoirs received, which was laid before the Academy by Professor Kummer on July 2, 1868, contained the following remarks relating to that sent in by Professor Smith :

'Die vierte, in französischer Sprache abgefasste Preisschrift mit dem Motto : "Haud facilem esse viam voluit" führt den Titel : "Mémoire sur quelques problèmes cubiques et biquadratiques," und ist in drei Abschnitte eingetheilt. Der erste Abschnitt beschäftigt sich mit der Theorie des Imaginären in der Geometrie, der zweite enthält verschiedene Methoden, die gemeinsamen Punkte zweier durch ihre Elemente gegebener Kegelschnitte mittels des Lineals, des Cirkels und eines festen Kegelschnitts zu construiren, in dem dritten Abschnitte endlich löst der Verfasser ausser einigen andern sogenannten kubischen und biquadratischen geometrischen Aufgaben namentlich das speciell in der Preisfrage hervorgehobene, die Curven vierten Grades betreffende Problem. Die ganze Arbeit zeichnet sich durch übersichtliche und systematische Behandlung des Stoffes aus. Der Verfasser macht bei seinen Constructionen, wie es in der Preisfrage verlangt wird, nur von den einfachsten erforderlichen und ausreichenden Hilfsmitteln Gebrauch, aber bei den Constructions-Methoden selbst hat er mehr auf gedankliche als auf praktische Einfachheit, mehr auf die vollständige Darlegung aller Gesichtspunkte als auf die Ausführung aller einzelnen Operationen sein Bestreben gerichtet. Dadurch ist es ihm gelungen, im zweiten Abschnitte das an sich dürftige und trockene Material in gediegener und interessanter Weise zu verarbeiten und im dritten Abschnitte die specielle dort behandelte Frage mit allgemeineren zu verknüpfen. Fast überall lässt die Arbeit zum Vortheil für ihren wissenschaftlichen Werth deutlich erkennen, dass der Verfasser zu seinen umfassenderen Untersuchungen durch algebraische Betrachtungen gelangt ist, deren genauer Zusammenhang mit dem Gegenstande der Preisfrage schon in deren Formulirung enthalten ist †.'

The origin of the long memoir on the Theta and Omega Functions—the last paper but one in the second volume—was as follows. At the end of 1873, or the

* The announcement of the subject was made in the following terms : 'Für diejenigen geometrischen Probleme, deren algebraische Lösung von Gleichungen von höherem als dem zweiten Grade abhängt, fehlt es noch an der Feststellung der zur constructiven Lösung derselben erforderlichen und ausreichenden fundamentalen Hilfsmittel, so wie an den Methoden zur systematischen Benutzung dieser Hilfsmittel.

'Indem die Akademie die Frage, die sie stellt, auf die Probleme beschränkt, welche auf kubische Gleichungen führen, wünscht sie, dass wenigstens an einer Anzahl von speciellen Beispielen gezeigt werde, wie diese Lücke in dem Gebiete der constructiven Geometrie ausgefüllt werden könne. Namentlich verlangt sie die vollständige Lösung des folgenden Problems :

'Wenn dreizehn Punkte in der Ebene gegeben sind, so sollen durch geometrische Construction diejenigen drei Punkte bestimmt werden, welche mit den gegebenen zusammen ein System von sechzehn Durchschnittspunkten zweier Curven vierten Grades bilden.

'Bei der Lösung sind die Fälle zu berücksichtigen, in welchen einige der dreizehn Punkte imaginär und demgemäss nicht als individuelle Punkte, sondern als Durchschnittspunkte vorgelegter Curven gegeben sind. Gewünscht wird ferner, dass sämmtliche geometrische Constructionen durch die entsprechenden algebraischen Operationen erläutert werden.'

† *Monatsberichte* for 1868, p. 420.

beginning of 1874, when I was passing through the press the Tables of the Theta Functions which I had calculated in connexion with a Committee of the British Association, I asked Professor Smith, who was a member of the Committee, if he would contribute an Introduction to the volume. He replied that he did not see his way to writing anything appropriate to the tables themselves, but that he 'could say something with respect to the constants at the head of the pages.' These constants were K, K', E, J, J', &c., the numerical values of which were given for every minute of the modular angle. The memoir grew in extent, and was subject to frequent interruptions; in fact a number of other papers were written and published during its progress. These papers were generally called into existence by special circumstances unconnected with the memoir, but a few of them, and especially the Notes on the Transformation of Elliptic Functions (Nos. xli, xlii, vol. ii.), which immediately precede it in the volume, arose directly out of it. The first two of these Notes were given to me in the summer and autumn of 1882 for the *Messenger of Mathematics*, and appeared in the numbers from August to November of that year. The remaining Notes were printed after his death from a draft manuscript which he had shown to me, and explained in some detail, in October, 1882. The memoir itself, with the Notes that were connected with it, formed the principal new work upon which he was engaged from the time of its commencement until his death: most of the other papers, published in the interval, containing results which were mainly derived from his earlier investigations. It was left incomplete: Arts. 1–31 (pp. 415–484) had been passed for press: Arts. 32–48 (pp. 485–535) had been revised, and Arts. 49–73 (pp. 535–585) were in type in quarto pages and had been partially corrected. The succeeding Articles up to Art. 88 inclusive were in type in octavo slip, and had been partially corrected in this form *. The last two Articles (89 and 90) are printed from a manuscript found among his papers, and which he had marked as following on after Art. 88. I believe that no more was written, even in draft. The figures which occur in the Memoir had not been drawn.

The object of Professor Smith's first paper on Elliptic Functions (No. xvi,

* The whole of the Memoir was originally set up all in octavo slip, and remained in this form for a long time, during which it was greatly altered and extended. It was reset in quarto pages during 1881 and 1882, and passed for press in this form. It had been intended that it should appear as an Introduction, but it was finally decided that it should follow the tables with the title 'Memoir on the Theta and Omega Functions.'

vol. i.) was, as stated in the first paragraph, to enunciate and demonstrate a fundamental theorem, the nature of which had been indicated in a letter, written in 1845, from Jacobi to M. Hermite, in which he mentioned that he used it as the starting-point in his Königsberg lectures. Jacobi died in 1851, and as the theorem referred to had never been published, Professor Smith reproduced it, in 1866, in this paper. Guided by Jacobi's suggestion, he multiplied together four general Theta series and expressed the product as the sum of four terms, each of which was the product of four Theta series with different arguments. From this theorem he derived, as indicated by Jacobi, all the principal results of Elliptic Functions, either as particular cases or as simple corollaries. In 1881 the first volume of the Collected Works of Jacobi was issued, and his Königsberg lectures on Elliptic Functions were there printed for the first time. By comparing them with Professor Smith's paper it will be seen that, although the theorem itself is of course essentially the same, still there are differences in the mode in which it is presented which enhance the interest of the latter. Professor Smith treated the question with great generality, and with absolute precision, and this short paper is very characteristic of his style of work.

At the meeting of the London Mathematical Society on January 8, 1879, Professor Cayley communicated to the Society the theorem

$$k^2 k'^2 \operatorname{sn}\alpha \operatorname{sn}\beta \operatorname{sn}\gamma \operatorname{sn}\delta - k^2 \operatorname{cn}\alpha \operatorname{cn}\beta \operatorname{cn}\gamma \operatorname{cn}\delta + \operatorname{dn}\alpha \operatorname{dn}\beta \operatorname{dn}\gamma \operatorname{dn}\delta - k'^2 = 0,$$

where α, β, γ, δ are any quantities whose sum is zero. Professor Smith, who was present at the meeting, remarked that it was a special case of a theorem relating to the multiplication of four Theta functions, and at the next meeting in February he communicated to the Society the general Theta Function formulæ which dominate all results of this class. This paper (No. xxxviii, vol. ii.), which is supplementary to that of 1866, was written out from notes which he had had by him since that date.

The paper on the conditions of perpendicularity in a parallelepipedal system (No. xxxii, vol. ii.) was written in response to a request from his friend Professor Maskelyne, who was seeking for a general treatment of a problem which, in the particular case of its application to crystallography and the distribution of molecules in a crystal, was of paramount importance.

The circumstances connected with the publication of the memoir which concludes the second volume require a more extended notice. In February, 1882, he was surprised to see in the *Comptes Rendus* that the subject proposed by the French Academy for the Grand Prix des Sciences Mathématiques was the decomposition of a number into five squares *. His feelings in the matter are shown by the following extracts from letters to myself. In the first, dated Oxford, February 17, 1882, he wrote—'The Paris Academy have set for their Grand Prix for this year the theory of the decomposition of numbers into five squares, referring to a note of Eisenstein, *Crelle*, vol. xxxv, in which he gives without demonstration the formulæ for the case in which the number to be decomposed has no square divisor. In the Royal Society's Proceedings, vol. xvi, pp. 207, 208, I have given the complete theorems, not only for five, but also for seven squares: and though I have not given my demonstrations, I have (in the paper beginning at p. 197) described the general theory from which these theorems are corollaries with some fulness of detail. Ought I to do anything in the matter? My first impression is that I ought to write to Hermite, and call his attention to it. A line or two of advice would really oblige me, as I am somewhat troubled and a little annoyed;' and in the second, of date February 22, he proceeded, 'You see I take your advice entirely upon the point that he ought to be written to. The worst of it is that it would take me a year, and a hundred pages, to work out the demonstrations of the paper in the Royal Society's Proceedings.'

The following reply was received from M. Hermite:

MON CHER MONSIEUR,

Aucun des membres de la commission qui a proposé pour sujet du prix des sciences mathématiques en 1882 la démonstration des théorèmes d'Eisenstein sur la décomposition des nombres en cinq carrés n'avait connaissance de vos travaux contenant depuis bien des années cette démonstration et dont j'ai pour la première fois connaissance par votre lettre. L'embarras n'est point pour vous, mais pour le rapporteur des mémoires envoyés au concours, et si j'étais ce rapporteur je n'hésiterais pas un moment à faire d'abord l'aveu complet de l'ignorance où il s'est trouvé de vos publications, et ensuite à proclamer hautement que vous aviez donné la solution de la question proposée. Une circonstance pourrait ôter tout embarras et rendre sa tâche facile autant qu'agréable. S'il avait en effet à rendre compte d'un mémoire adressé par vous-même dans lequel vous rappelleriez vos anciennes recherches en les complétant, vous voyez que justice vous serait rendue en même temps que les intentions de l'Académie

---

* The subject of the prize for 1882 had also been announced a year previously, but the notice had then escaped his attention. The following are the terms of the announcement:

Grand Prix des Sciences Mathématiques. (Prix du Budget.) Question proposée pour l'année 1882. L'Académie propose pour sujet du prix la *Théorie de la décomposition des nombres entiers en une somme de cinq carrés*, en appelant particulièrement l'attention des concurrents sur les résultats extrêmement remarquables énoncés sans démonstration par Eisenstein dans une Note écrite en langue française au Tome 35 du *Journal de Mathématiques de Crelle* (p. 368, année 1847).

'Le prix consistera en une médaille de la valeur de trois mille francs.

'Les Mémoires devront être remis au Secrétariat avant le 1er juin 1882; ils porteront une épigraphe ou dévise répétée dans un billet cacheté qui contiendra le nom et l'adresse de l'auteur. Ce pli ne sera ouvert que si la pièce à laquelle il appartient est couronnée.' (*Comptes Rendus*, vol. xcii. p. 622, March 14, 1881, and vol. xciv. p. 330, Feb. 6, 1882.)

seraient remplies puisqu'on lui annoncerait la solution complète de la question proposée. Jusqu'ici je n'ai pas eu connaissance qu'aucune pièce ait été envoyée, ce qui s'explique par la direction du courant mathématique qui ne se porte plus maintenant vers l'arithmétique. Vous êtes seul en Angleterre à marcher dans la voie ouverte par Eisenstein. M. Kronecker est seul en Allemagne ; et chez nous M. Poincaré, qui a jeté en avant quelques idées heureuses sur ce qu'il appelle les invariants arithmétiques, semble maintenant ne plus songer qu'aux fonctions Fuchsiennes et aux équations différentielles. Vous jugerez s'il vous convient de répondre à l'appel de l'Académie à ceux qui aiment l'Arithmétique ; en tout cas soyez assuré que la commission aura par moi connaissance de vos travaux si elle a se prononcer et à faire un rapport à l'Académie sur des mémoires soumis à son examen . . . Je vous renouvelle, mon cher Monsieur, l'expression de ma plus haute estime et de mes sentiments bien sincèrement dévoués.

CH. HERMITE.

PARIS, 26 Février, 1882.

In consequence of an accident when riding, Professor Smith had been confined to his sofa for some weeks ; but, as far as his strength permitted, he had been working steadily at subjects connected with the memoir on the Theta and Omega subjects, which he was very reluctant to lay aside. Nevertheless, he thought it his duty to accede to the suggestion of M. Hermite, and bring his demonstrations before the Academy in the form of a memoir sent in for the *concours*. For a while he divided his spare time between Elliptic Functions and the work connected with the prize subject, but in April he wrote : 'I fear I cannot let you have the Transformation papers before the end of June. As I foresaw, getting the quadratic forms of n indeterminates into my mind again, putting my proofs into a rigorous form, and writing them out, will take up every moment till the end of May (the paper has to be in Paris by June 1). My sole reason for taking this trouble is that sooner or later I should have had to do it unless I was to allow my demonstrations to perish.'

Professor Smith died on February 9, 1883, and it was not till nearly two months after his death (at the meeting of the Academy on April 2) that the report of the Commission was announced, two prizes being awarded, one to Professor Smith and one to M. Minkowski, of Königsberg. The following is the text of the report :

Grand Prix des Sciences Mathématiques (Prix du Budget).

(Commissaires : MM. Hermite, Bonnet, Bertrand, Bouquet ; Jordan, rapporteur.)

L'Académie avait proposé pour sujet de prix la 'Théorie de la décomposition des nombres entiers en une somme de cinq carrés,' en appelant particulièrement l'attention des concurrents sur les résultats extrêmement remarquables énoncés sans démonstration par Eisenstein dans une Note écrite en langue française au tome 35 du *Journal de Mathématiques de Crelle*, p. 868, année 1847.

Ce problème semble assez restreint au premier abord; mais on avait lieu de penser que les théorèmes obtenus par cet illustre géomètre s'étaient offerts à lui comme conséquences dernières d'une longue série de recherches, où devaient se trouver combinées les notions d'*ordre* et de *genre*, établies par Gauss pour les formes binaires, et transportées par Eisenstein dans le domaine des formes ternaires, celle de la *densité*, qu'il avait introduite pour la première fois, enfin les méthodes infinitésimales de Dirichlet. L'Académie était donc fondée à espérer que ce voyage de découvertes imposé aux concurrents à travers une des régions les plus intéressantes et les moins explorées de l'Arithmétique produirait des résultats féconds pour la Science. Cette attente n'a pas été trompée.

Trois Mémoires ont été transmis au Concours; ils portent les épigraphes suivantes:

No. 1. Quotque quibusque modis possint in quinque resolvi
Quadratos numeri, pagina nostra docet.

No. 2. Felix qui potuit rerum cognoscere causas!

No. 3. Rien n'est beau que le vrai; le vrai seul est aimable.

Le Mémoire No. 2 montre chez son auteur des connaissances étendues et renferme plusieurs résultats intéressants; mais la question posée par l'Académie ne s'y trouve même pas abordée. La Commission a donc principalement concentré son étude sur les deux autres Mémoires. Tous deux sont des œuvres considérables, où se trouvent exposés d'une manière magistrale plusieurs des points fondamentaux de la théorie des formes quadratiques. Les formules relatives à la décomposition en cinq carrés n'y figurent que comme conséquences très particulières des principes généraux.

Il est d'ailleurs aisé de discerner dans ces deux Mémoires, à travers les différences d'exposition, une singulière identité dans la filiation des idées, au point qu'il serait difficile de signaler dans l'un d'eux une notion ou un théorème important qu'on ne retrouvât pas dans l'autre, et que, pour éviter les redites et faire mieux ressortir les nuances qui les séparent, nous devions les analyser simultanément.

L'auteur du Mémoire No. 1 montre tout d'abord qu'à une forme quadratique quelconque on peut associer une série de formes adjointes *; la valeur numérique du plus grand commun diviseur des coefficients de ces diverses formes et leur ordre de parité servent de base à une distribution en ordres de même déterminant.

L'auteur du Mémoire No. 3 ne parle pas de ces formes adjointes, si ce n'est de la première, que Gauss avait déjà définie; mais il considère la série de leurs coefficients, ce qui lui donne un résultat identique au précédent. La marche suivie dans les deux Mémoires est d'ailleurs la même et consiste à transformer la forme proposée en une autre équivalente, telle que son résidu par rapport à un module donné soit ramené à une expression canonique.

Cette expression canonique contient encore des coefficients indéterminés dont la valeur dépendra de la manière de conduire les calculs; mais de quelque façon que l'on opère, en partant d'une forme donnée, certaines combinaisons de ces coefficients conserveront toujours un caractère quadratique déterminé par rapport aux nombres premiers qui divisent le déterminant et par rapport aux nombres 4 et 8. L'ensemble de ces caractères, invariables pour toutes les formes d'une même classe, définira le genre.

Ainsi que Gauss l'avait déjà signalé pour les formes binaires, en insistant tout particulièrement sur ces circonstances, qui sont pour l'Arithmétique du plus haut intérêt, toutes les combinaisons de caractères ne sont pas admissibles. Les deux auteurs indiquent d'une façon précise les conditions que doit remplir une semblable combinaison pour correspondre à un genre réellement existant.

Ils passent ensuite à la recherche du nombre des solutions des congruences du second degré à plusieurs inconnues. Cette question se lie intimement aux précédentes. La méthode élégante fondée sur l'emploi de la résolvante de Lagrange, par laquelle elle est traitée dans le Mémoire No. 3, mérite

* Ces formes avaient déjà été considérées par M. Darboux dans le *Journal de Liouville*.

une mention particulière. L'Auteur énonce ensuite cette proposition, dont il est facile de rétablir la démonstration: *Deux classes de formes qui appartiennent au même genre sont congrues par rapport à un module quelconque.*

Cette nouvelle définition du genre, déjà formulée d'ailleurs par M. Poincaré, a l'avantage de s'étendre immédiatement aux formes d'ordre supérieur au second.

Les deux auteurs s'occupent ensuite de la représentation des nombres par une forme quadratique à n variables. Ils montrent, en généralisant une méthode de Gauss, que cette recherche revient à celle de la représentation d'une forme quadratique à $n-1$ variables. Abordant ensuite ce dernier problème, ils font voir comment l'ordre et le genre de la forme représentée peuvent se déduire de l'ordre et du genre de la forme qui la représente. Les résultats précédents leur permettent de ramener la recherche de la densité des représentations d'un nombre donné par l'ensemble des formes d'un même genre à celle de la densité d'un genre donné.

L'application des méthodes de Dirichlet a fourni la solution de ce problème à l'auteur du Mémoire No. 1 pour les formes quaternaires; à celui du Mémoire No. 3 pour les formes à un nombre quelconque de variables dont toutes les adjointes sont des formes impaires. Mais chacun d'eux, pressé par le temps, n'a donné la démonstration de ses résultats qu'autant qu'il était nécessaire pour résoudre le problème posé par l'Académe. Tous les deux le ramènent à la sommation d'une série infinie,

$$\sum\left(\frac{M}{m}\right)\frac{1}{m^2},$$

fort analogue à celle que Dirichlet avait rencontrée dans son célèbre Mémoire sur les applications du Calcul infinitésimal à la Théorie des nombres.

L'auteur de Mémoire No. 3 s'arrête à ce point; celui du Mémoire No. 1 donne sans démonstration le résultat de la sommation, d'où découlent immédiatement les théorèmes d'Eisenstein.

De même que nous n'avons pu séparer ces deux beaux Mémoires dans la courte analyse qui précède, nous ne saurions les présenter l'un sans l'autre aux suffrages de l'Académie. Tous deux en sont également dignes. Ils font faire un pas considérable à l'Arithmétique, en fixant d'une manière définitive la théorie de l'ordre et du genre dans les formes quadratiques. Le talent déployé par les auteurs nous est d'ailleurs garant qu'ils sauront mener à terme les questions difficiles qu'ils ont dû traiter un peu hâtivement à la fin de leur travail.

Dans l'impossibilité où elle se trouve de mettre l'un d'eux au second rang, la Commission à l'unanimité émet le vœu que l'Académie accorde à chacun d'eux la totalité du prix, si elle le juge possible. Nous devons faire observer, en terminant, que le Mémoire No. 3 est écrit en allemand, contrairement à l'une des conditions du programme. L'auteur s'en excuse dans sa Préface, en disant que le temps lui a manqué pour faire la traduction de son Mémoire. Nous n'avons pas pensé qu'il y eût lieu de repousser *a priori*, pour une irrégularité de forme, un travail de cette importance. Mais, tout en l'accueillant, à titre exceptionnel, l'Académie devra faire toutes réserves pour l'application des règles ordinaires aux concours à venir.

L'Académie adopte les propositions de la Commission et décide qu'elle décernera deux prix de même valeur aux auteurs des Mémoires inscrits sous les Nos. 1 et 3.

Conformément au Règlement, M. le Président procède à l'ouverture des plis cachetés qui accompagnent ces Mémoires et proclame pour le No. 1 le nom de M. J. S. Smith, professeur à l'Université d'Oxford, et pour le No. 3 nom de M. Hermann Minkowski, étudiant de Mathématiques à l'Université de Königsberg.

It will be seen that in this report, which has been reproduced in its entirety, no mention is made of Professor Smith's previous publications, nor is there even a reference to his having completed Eisenstein's formulæ for five squares, and given the corresponding formulæ for seven squares, more than fifteen years before: in fact, the report shows that the writer regarded Professor Smith's memoir as perfectly new work called into existence by the prize competition. Under these circumstances Miss Smith, as the representative of her brother, wrote to M. Hermite recalling his attention to the expression in his letter of February 26, 1882, 'En tout cas soyez assuré que la commission aura par moi connaissance de vos travaux si elle a se prononcer et à faire un rapport à l'Académie sur des mémoires soumis à son examen,' and expressing the hope that he would give the explanation that had become necessary. M. Hermite replied that the omission of which she complained was an error which was due to absolutely involuntary forgetfulness ('ce tort ne consiste que dans un oubli, qui a été absolument involontaire'); but he made no further statement of any kind. The award of the prize gave rise however to a good deal of comment in the Paris newspapers. The Academy was blamed for having been unaware of work published by the Royal Society in 1868, and it was pointed out that the award was necessarily unsatisfactory, in spite of Professor Smith himself having sent in a memoir, as any other competitor might have availed himself of the indications contained in his published writings. The striking identity between the first and third memoirs, which is emphasized in the report, gave rise to the statement, which appeared in the newspapers, that this had actually taken place. In consequence of these criticisms M. Bertrand made certain explanations at the meeting of the Academy on April 16, 1883. The proceedings commenced with the reading of an appreciative obituary notice of Professor Smith by M. Camille Jordan, in which special reference was made to his arithmetical researches. The account then proceeds:

M. Bertrand demande à l'Académie la permission d'ajouter quelques mots à la lecture qu'elle vient d'entendre.

'La Commission chargée de proposer le sujet du prix de Mathématiques avait demandé aux concurrents l'étude d'un théorème énoncé, il y a près de quarante ans, par l'illustre géomètre Eisenstein, enlevé à la science avant d'en avoir publié la démonstration.

'Un seul Mémoire depuis la mort d'Eisenstein avait été consacré à cette difficile question: il était de M. Smith et, comme celui d'Eisenstein, contenait l'énoncé seulement des résultats principaux. Si le concours proposé par l'Académie n'était pas venu reporter l'attention de M. Smith vers ces recherches déjà anciennes, il n'aurait, de même qu'Eisenstein, légué sur ce sujet aux géomètres qu'un énigme difficile à déchiffrer.

'Sur les trois Mémoires présentés au concours, le premier a été écarté comme insuffisant.

'Le deuxième suivait exactement la marche tracée par M. Smith et donnait la démonstration de ses énoncés; celui des Commissaires qui a accepté la tâche d'en faire l'examen a pu, sur ces indices, deviner le nom de l'auteur. Peu importait, d'ailleurs, que le Mémoire fût de M. Smith ou inspiré par le travail depuis longtemps livré au public par le savant professeur d'Oxford : il méritait incontestablement le prix.

'Un troisième Mémoire résolvait la question ; il était difficile que deux géomètres assez habiles pour parcourir ce terrain élevé, mais un peu étroit, ne s'y rencontrâssent pas sur plus d'un point. Les méthodes avaient de l'analogie, mais chaque Mémoire portait la marque d'un esprit original et distingué ; tous deux étaient excellents et il semblait impossible de donner à l'un d'eux le second rang.

'Les deux Mémoires seront publiés, et l'Académie se félicitera d'avoir donné à leurs savants auteurs, l'un à la fin, l'autre au début de sa carrière, l'occasion de montrer les ressources d'un esprit ingénieux et la preuve, inscrite à chaque page, d'une science étendue et profonde.'

These official remarks, which are supplementary to the report of the Commission, render justice to M. Minkowski, and offer a carefully framed defence of the Academy, but without admitting that the subject was proposed in ignorance of Professor Smith's work, or that the reporter was not aware of the existence of the paper of 1867 until after the publication of the report. In a historical statement relating to the subject and award of the prize, drawn up a fortnight after the publication of the report, and in reply to adverse criticisms, a full avowal of all the circumstances might have been looked for. It is right to say that M. Camille Jordan, the reporter, was not a member of the Academy when the subject was announced, and that it was only at the last moment that he was charged with the duty of reporting upon the three memoirs.

It is much to be regretted that it should have been necessary to devote so much space to the matters connected with this memoir. A very brief notice would have sufficed if M. Hermite had communicated the existence of the paper of 1867 to the other members of the Commission, or if after the award he had given a brief account of the facts, or caused such an account to be given. But the only statement made was that of M. Bertrand, and it therefore became impossible to avoid details and quotations, as Professor Smith would not have been willing to send in a memoir for the competition except under the special circumstances of the case and in response to M. Hermite's suggestion.

An Appendix at the end of the second volume contains four writings which, though not of the same original character as the papers themselves, necessarily find a place in a collected edition of Professor Smith's mathematical works. The last of the four is a portion of the Introduction to the collected edition of Clifford's Mathematical Papers, which was written in the summer of 1881. Only so much of this Introduction has been included as could be of interest to a reader who had not the book itself before him. A reference should be here added to a review by Professor Smith of Campbell and Garnett's *Life of Professor*

Clerk Maxwell which appeared in the *Academy* for January, 1883 (vol. xxiii, pp. 19, 35). This review, being almost wholly biographical, is not reprinted.

He contributed verbally to the meetings of the London Mathematical Society and British Association a number of papers, which unfortunately were never written out. The following is a list of the titles of these papers :

London Mathematical Society.

1. Construction of the last point of intersection of a cubic curve by a curve of a superior order. March 26, 1868 (vol. ii, p. 61).
2. Geometrical note on the concomitants of a binary cubic. March 26, 1868 (vol. ii, p. 61).
3. Theory of certain systems of conics which present themselves in connexion with cubic curves. May 28, 1868 (vol. ii, p. 67).
4. On a problem in kinematics, and focal properties of skew surfaces. April 14, 1870 (vol. iii, p. 99).
5. On elliptic integrals. December 8, 1870 (vol. iii, p. 195).
6. On skew cubics. March 9, 1871 (vol. iii, p. 224).
7. On the partition of geometrical curves. February 10, 1876 (vol. vii, p. 90).
8. On the aspects of circles on a plane or on a sphere. April 13, 1876 (vol. vii, p. 172).
9. On some elliptic function properties. January 11, 1877 (vol. viii, p. 139).
10. On Eisenstein's Theorem. June 14, 1877 (vol. viii, p. 289).
11. Note relating to the theory of the division of the circle. April 11, 1878 (vol. ix, p. 102).
12. On a correction in Sohncke's tables. January 9, 1879 (vol. x, p. 44).
13. Upon a modular equation. January 9, and February 13, 1879 (vol. x, pp. 42 and 75).
14. Two geometrical notes relating to surfaces of the second order. March 13, 1879 (vol. x, p. 104).
15. Two geometrical notes. June 12, 1879 (vol. x, p. 167).
16. Geometrical notes (3). February 12, 1880 (vol. xi, p. 50).

British Association.

1.	On a property of surfaces of the second order	1866, p. 6.	Nottingham.
2.	On the large prime numbers calculated by Mr. Barrett Davis	1866, p. 6.	,,
3.	On a construction for the ninth cubic point	1868, p. 10.	Norwich.
4.	On geometrical constructions involving imaginary data	1868, p. 10.	,,
5.	On a property of the Hessian of a cubic surface	1868, p. 10.	,,
6.	On the circular transformation of Möbius	1872, p. 24.	Brighton.
7.	On modular equations	1873, p. 24.	Bradford.
8.	On singular solutions	1875, p. 21.	Bristol.
9.	On the effect of quadric transformation on the singular points of a curve	1875, p. 21.	,,
10.	On the modular curves	1878, p. 463.	Dublin.
11.	On quadric transformation	1878, p. 465.	,,
12.	On inverse figures in geometry	1880, p. 476.	Swansea.
13.	On a mathematical solution of a logical problem	1880, p. 476.	,,
14.	On the distribution of circles on a sphere	1880, p. 476.	,,

15.	Note on the skew surfaces of the third order	1880, p. 482.	Swansea.
16.	On a kind of periodicity presented by some elliptic functions	1880, p. 482.	,,
17.	On the differential equations satisfied by the modular equations	1881, p. 535.	York.
18.	On the equation of the multiplier in the theory of elliptic transformation	1881, p. 538.	,,
19.	On the linear relation between two quadratic surds . . .	1881, p. 538.	,,
20.	On a property of a small geodesic triangle on any surface .	1881, p. 548.	,,

I have omitted a title from this list whenever I knew that the paper in question was published elsewhere. Thus a paper 'On Continued Fractions' which was communicated to the British Association in 1875 was afterwards published in the *Messenger*, and forms No. xxviii. of the present reprint. It is probable that the contents of a few others are included in the published papers. No doubt all the results upon which these communications were founded are contained in his note-books.

With respect to the character of Professor Smith's mathematical writings a very noticeable feature is the arithmetical spirit that runs through the whole of his work. The years of study which produced the Report upon the Theory of Numbers exercised a lasting influence upon his mode of thought; and his familiarity with the ideas and methods of the Higher Arithmetic continually shows itself in his treatment of Geometry and Elliptic Functions. In the latter subject the arithmetical tendency of his mind is especially evident in the point of view from which the theory of Transformation is always regarded. Another characteristic feature of his work is its completeness, both as regards attention to details and accuracy of demonstration. He had a very strong dislike to careless or slovenly work of any kind, and thought that it was nowhere so much out of place as in Pure Mathematics. He was ready enough to pass over the ground boldly and rapidly, without regard to ambiguities or details, when he was seeking after new theorems, or merely endeavouring to decide upon the truth of generalizations or guesses; but he was of opinion that a mathematician should refrain from publication until he had established his results by perfectly rigorous demonstration. He had no sympathy with those who were contented to give imperfect demonstrations, or to regard results as proved merely because they had satisfied themselves of their truth. No task is more irksome to a mathematician than that of working out in detail all the various particular cases of a theorem, when the novelty of the investigation by which it was discovered has long since worn off. The general result, too, of such examinations is to produce modifications and limitations which at the same time add to the cumbrousness of the demonstrations and detract from the simplicity of the theorems themselves. But he held that any slurring-over of difficulties or ambiguities was utterly repugnant to the nature of the subject, and that a mathematician was bound to spare no amount of labour that was requisite in order to give to his results the highest degree of precision of which they were susceptible. The comparatively slow rate of progress of the memoir on the Theta and Omega Functions was no doubt primarily due to the many other claims upon his time, but it was also attributable, in no slight degree, to the extreme care taken to avoid ambiguities of every kind, and to the attention bestowed upon the systematic examination of all the special cases of the general theorems. His natural love of precision in thought and expression was no doubt strengthened by his early study of the writings of Gauss, for whom he always felt the most unbounded admiration. The following notes, which he wrote for Mr. Tucker*, on the occasion of the celebration of the centenary of Gauss's birth, find a fitting place here, as they show, in his own words, not only his deep reverence for the great master of the Higher Arithmetic, but also the extreme importance that he attached to perfection of form in the presentation of mathematical results.

If we except the great name of Newton (and the exception is one which Gauss himself would have been delighted to make) it is probable that no mathematician of any age or country has ever surpassed Gauss in the combination of an abundant fertility of invention with an absolute rigorousness in demonstration, which the ancient Greeks themselves might have envied. It may be admitted, without any disparagement to the eminence of such great mathematicians as Euler and Cauchy, that they were so overwhelmed with the exuberant wealth of their own creations, and so fascinated by the interest attaching to the results at which they arrived, that they did not greatly care to expend their time in arranging their ideas in a strictly logical order, or even in establishing by irrefragable proof propositions which they instinctively felt, and could almost see, to be true. With Gauss the case was otherwise. It may seem paradoxical, but it is probably nevertheless true, that it is precisely the effort after a logical perfection of form which has rendered the writings of Gauss open to the charge of obscurity and unnecessary difficulty. The fact is that there is neither obscurity nor difficulty in his writings, as long as we read them in the submissive spirit in which an intelligent schoolboy is made to read his Euclid. Every assertion that is made is fully proved, and the assertions succeed one another in a perfectly just analogical order; there is nothing so far of which we can complain. But when we have finished the perusal, we soon begin to feel that our work is but begun, that we are still standing on the threshold of the temple, and that there is a secret which lies behind the veil and is as yet concealed from us. No vestige appears of the process by which the result itself was obtained, perhaps not even a trace of the considerations which suggested the successive steps of the demonstration. Gauss says more than once that, for brevity, he only gives the synthesis, and suppresses the analysis of his propositions. '*Pauca sed matura*' were the words with which he delighted to describe the character which he endeavoured to impress upon his mathematical writings. If, on the other hand, we turn to a memoir of Euler's there is a sort of free and luxuriant gracefulness about the whole performance, which tells of

* 'Carl Friedrich Gauss,' by R. Tucker. *Nature*, vol. xv, p. 537 (April 19, 1877).

the quiet pleasure which Euler must have taken in each step of his work; but we are conscious nevertheless that we are at an immense distance from the severe grandeur of design which is characteristic of all Gauss's greater efforts. The preceding criticism, if just, ought not to appear wholly trivial; for though it is quite true that in any mathematical work the substance is immeasurably more important than the form, yet it cannot be doubted that many mathematical memoirs of our own time suffer greatly (if we may dare to say so) from a certain slovenliness in the mode of presentation; and that (whatever may be the value of their contents) they are stamped with a character of slightness and perishableness, which contrasts strongly with the adamantine solidity and clear hard modelling, which (we may be sure) will keep the writings of Gauss from being forgotten long after the chief results and methods contained in them have been incorporated in treatises more easily read, and have come to form a part of the common patrimony of all working mathematicians. And we must never forget (what in an age so fertile of new mathematical conceptions as our own, we are only too apt to forget) that it is the business of mathematical science not only to discover new truths and new methods, but also to establish them, at whatever cost of time and labour, upon a basis of irrefragable reasoning.

The *μαθηματικὸς πιθανολογῶν* has no more right to be listened to now than he had in the days of Aristotle; but it must be owned that since the invention of the 'royal roads' of analysis, defective modes of reasoning and of proof have had a chance of obtaining currency which they never had before. It is not the greatest, but it is perhaps not the least, of Gauss's claims to the admiration of mathematicians, that, while fully penetrated with a sense of the vastness of the science, he exacted the utmost rigorousness in every part of it, never passed over a difficulty as if it did not exist, and never accepted a theorem as true beyond the limits within which it could actually be demonstrated.

These words certainly express the ideal which Professor Smith had always in his mind, and which has governed the character of his own work.

In passing the papers through the press I have corrected all the misprints and errors that I detected, but no other alterations of any kind have been made in the text. I have added notes only in those cases where they seemed to be absolutely necessary. All additions, references, or notes which are not in the original papers are enclosed in square brackets [ ]. The correction of misprints or slips often involved matters of some delicacy, and occasioned frequent delays. There were also other difficulties connected with the papers that were printed from manuscript. The sheets containing the concluding portion of the Report on the Theory of Numbers were passed through the press (during my absence abroad) by Professor Cayley, by whom the index to the Report was made. Professor Cayley also kindly undertook the revision of the uncorrected portion of the Memoir on the Theta and Omega Functions.

Professor Smith did not leave many separate mathematical manuscripts, most of his work being contained in note-books. These books, about forty in number, cover the whole period of his mathematical career. Some contain his early notes when making his first studies in Geometry, or reading the memoirs upon which the Report on the Theory of Numbers was founded; others relate to his University lectures; and rather more than a dozen are devoted to the records of original work, a very large portion of which has never been published. I have repeatedly examined the note-books relating to the subjects with which I was most familiar in hopes of being able to make extracts that could have been included in the present volumes. But in this I have been unsuccessful, for Professor Smith entered in these books not only the finished theorems which he had demonstrated, but also results which he had arrived at by rough explorations and inductions, as well as mere guesses sometimes; and it is certain that he would have published nothing himself from these books without submitting it to the most careful examination and working out the demonstrations afresh. Under these circumstances it was decided, but with great reluctance, to confine the present work to the published writings, and make no attempt to give an account of the varied contents of the note-books. The editing of any considerable portion of the unpublished work would be a matter of great difficulty, requiring much time and research, but it would not be so serious an undertaking to prepare separately for publication some of the investigations which he has left upon special subjects. In particular, it is very desirable that his researches relating to the decomposition of numbers into seven squares should be published; and it would probably be found that the editing of this application of the general formulæ has been greatly simplified by his own treatment (in the prize memoir) of the corresponding work on the five-square problem.

The principal subjects upon which he lectured in the University were Modern Geometry, Analytical Geometry, Theory of Numbers, Calculus of Variations, and Differential Equations. With the exception of the Theory of Numbers, his lecture-notes on these subjects are very fragmentary; but full and accurate transcripts of the lectures themselves as delivered were kindly supplied by Mr. Lazarus Fletcher, F.R.S., Mr. Thiselton Dyer, F.R.S., Mr. H. T. Gerrans, Mr. Walter Larden, the late Mr. Arthur Buchheim, and other pupils. As no other teaching on Modern Geometry was given in an English University, and as his lectures on this subject exercised great influence upon the direction of mathematical studies in Oxford, it was considered very important that they should be published. The editorship was undertaken by Mr. H. M. Taylor, Fellow of Trinity College, Cambridge, who after a careful comparison of the lectures as delivered in different years wrote out for press a fair copy of what might be regarded as the standard form of the course. It was, however, finally decided to abandon the publication, partly because the same ground was more systematically covered by foreign treatises, and partly because the extent of the lectures was so limited (owing to the fact that students did not specialize in the subject) that the volume would be scarcely adequate to form an independent treatise on so important a branch of Mathematics. It may be mentioned that the courses delivered in various years differed very much from one another, and it would appear as if their nature and extent had to some degree depended upon the audience.

It is well known that Professor Smith intended to write an Introduction to the Theory of Numbers, and regret was frequently expressed to him that the work was still unpublished. Among his note-books there are several in which the elements of the subjects are very clearly and succinctly explained in methodically arranged paragraphs, and it cannot be doubted that these are successive editions of the commencement of such a work, in which he was striving after greater perfection. Other note-books contain carefully written articles which may have been intended as chapters in such a work. The completed portion of the treatise, however, is so small, only reaching to quadratic forms, that the idea at first entertained of publishing it separately as a fragment was ultimately given up.

I hope that it will not be thought out of place for me to include in this Introduction a few reminiscences of my own with respect to Professor Smith, as he appeared to me, and to attempt a sketch, however slight, of his personality. In the eleven years that have elapsed since his death many of those to whom his presence was so familiar have passed away, and a new generation of mathematicians has arisen to whom he is but a name, so that the time seems to have already come when it is allowable to place on record matters which were once of common knowledge or might have seemed too trivial for mention in print.

I first saw Professor Smith at the British Association meeting at Nottingham in 1866, when he was one of the secretaries of Section A (Mathematics and Physics). I can perfectly remember his attitude and manner both when as secretary he read the papers of others, and when standing by the blackboard, he explained, so simply and gracefully, the nature of his own communications to the section. His tall handsome figure, his commanding presence, and the charm of his manners, stand out clearly before me, as I watched him then; and in no essential respect was there any change in him between the first time I saw him and the last.

At this meeting he spoke upon the average frequency of prime numbers, and I then for the first time heard of Legendre's approximate formula $\frac{x}{\log x - 1{\cdot}08366}$ for the number of primes inferior to x, a result which interested me intensely, although I little thought that it was subsequently to occupy so much of my own time *.

I was introduced to him in the committee-room of the section by my father, and although I was not eighteen years of age, he welcomed me with as much cordiality as if I had been a fellow-mathematician of equal standing with himself. I was a shy and retiring schoolboy, but, in spite of the respect with which his knowledge inspired me, his kind and friendly manner at once placed me at my ease. I mention so particularly this experience of my own because it was very characteristic of his gentle and considerate nature. I am sure that no one was ever treated by him with less courtesy or attention on account of youth or junior standing: on the contrary, I believe that in such cases he instinctively and unconsciously showed even more consideration. I may perhaps mention that on this occasion he gave me the first separate reprint of a mathematical paper which I ever possessed: it was not a paper of his own, but one which had been given to him, and seeing me interested in it he told me I might have it, as he could procure another copy from the author.

I did not see him again till the meeting of the British Association at Brighton in 1872, the year after that in which I took my degree at Cambridge. At that meeting he spoke upon the circular transformation of Möbius. I was then able for the first time to appreciate his wonderful power as an expositor of abstruse mathematics. His winning manners and graceful delivery charmed me as before, but I was even more struck with the skill with which he succeeded in giving, in the simplest language, a correct idea of complicated theories to those to whom they were entirely new.

---

* My memory is quite distinct that this account was given as a 'Report on the Theory of Numbers,' and that he briefly explained to the section the nature of the subjects dealt with in the report. This impression is confirmed by the fact that among the sectional papers there is no title under which Legendre's formula could have been introduced; for the paper 'On the large prime numbers calculated by Mr. Barrett Davis' (which I also distinctly remember), was given on a different day and in a different room. (Mr. Barrett Davis had communicated a manuscript list of large prime numbers, and Professor Smith, in laying them before the section, merely called attention to the fact that, as in the case of the smaller primes, they were sometimes clustered thickly together and sometimes widely separated.) It would therefore appear, almost with certainty, that Professor Smith had intended to write a seventh part of the report, which should relate to the frequency of primes and other asymptotic formulæ in the Theory of Numbers. The early note-books written while the report was in preparation contain references to Legendre's law, and a résumé of Lejeune Dirichlet's memoir on asymptotic formulæ in the Berlin Abhandlungen for 1849. Professor Plücker was present at this meeting, and exhibited some models of complexes to the section on the same morning as that on which Professor Smith spoke upon the subject of his Report.

All the papers of which he gave any verbal account in public after this date were communicated either to the London Mathematical Society or to Section A of the British Association, and I believe I was present on every such occasion; for I was a very regular attendant at the meetings of the Mathematical Society; and was one of the secretaries of Section A from 1871 to 1880 inclusive. I was also present at the meetings in 1881 and 1882. Several of these papers, like the one I have just referred to, related to subjects with which I was quite unfamiliar; but I never failed to derive some benefit from his explanations or to feel a deeper interest in the theories of Pure Mathematics in consequence of what he had said. In general I do not readily gain an insight into new mathematical methods merely from verbal explanations, but his papers had a wholly exceptional effect upon me in this respect. He had the gift of fixing the complete attention of his audience, and imparting valuable knowledge, no matter how remote or technical the subject. Those who were present at the reading of any paper of his will know that there is no exaggeration in this. Some mathematicians of our day have regarded the reading of a technical paper before a society as a mere formal preliminary to printing, which exists only as a survival from the past: but the beautiful mathematical expositions by which Professor Smith could gently lead on his audience into the remote intricacies of a difficult subject, prove that it is possible even in Pure Mathematics to convey a true idea of highly technical researches without being technical at all. He always began at a point from which an ordinary mathematical listener could take up the thread, and, laying down the main lines of his subject in a series of simple and clear sentences, following each other in logical order, succeeded, apparently with the greatest ease, in placing his audience in possession of sufficient general knowledge to enable them to grasp the nature and scope of the new work that he was bringing before them. He spoke slowly, with a marked emphasis and a measured and almost rhythmical utterance, which were very distinctive and attractive. His language was always peculiarly felicitous, both in formal expositions and in private conversation; and the elegance of his style may be fairly judged by the papers printed in the Appendix, which, I think, those who knew him could scarcely read without fancying that they heard in them the cadence of his voice. Although dignified in words, manner, and bearing, he was utterly free from any trace of formality: and indeed no small part of the charm of his character was due to the way in which natural dignity was modified by sweetness of disposition and gaiety of heart. Even when explaining the most abstract theories with the severest logical accuracy, his liveliness and wit would frequently peep out unexpectedly in parenthetical remarks. He was always in touch with his surroundings, but never more perfectly so than when addressing a mathematical audience, for his modesty and unselfishness rendered it impossible for him ever to weary others by allowing himself to be carried away by the interest which he felt himself in the researches he was explaining. On no single occasion was he ever dull or tedious, and his papers were always looked forward to with pleasure at the Mathematical Society and by the habitués of the mathematical Saturdays at the British Association meetings. The power to render advanced researches intelligible and interesting to a mixed audience is a rare gift; and the only other brilliant expositor of mathematics whom I have ever heard was Clifford, whose style however differed widely in almost every respect from that of Professor Smith. Clifford spoke very rapidly and fluently, in cleverly-worded sentences that were often startling or paradoxical. The art with which he could invest familiar things with a new interest, or connect them with novel ideas, and the facility with which this was done, apparently on the spur of the moment, were truly surprising, but it seemed to me that the effect produced was greatly dependent upon the exact words which he used and upon his mode of delivery. In Professor Smith's expositions there was never anything paradoxical or artificial. The explanations which he gave were perfectly matter-of-fact, his power being shown in the skill with which he held the sustained attention of his hearers as he proceeded from step to step.

It should be mentioned that very few of his papers were produced quite spontaneously. Mr. Tucker, the secretary of the Mathematical Society, was always anxious to have several communications announced for each meeting, and if he had not received enough titles would write to those who were likely to have papers in progress or suitable matter for verbal communication to the Society. Professor Smith always responded willingly to such appeals, and would mention subjects upon which 'he could say something, if required.' In the same way, at the meetings of the British Association at which he was present, I always asked him for papers, and he would give me a list of subjects which he could bring before the section, sometimes offering me a choice and letting me select those which I preferred. In making his verbal communications he generally placed one of his quarto note-books on the table, open at the place, and occasionally referred to it as he proceeded with his explanation. These quarto note-books in their greyish covers were well-known objects to all who attended mathematical meetings between

1873 and 1883. After laying before the Mathematical Society the results of some researches of his own, probably carried out years before, great pressure would be brought to bear to induce him to write out an account which would be suitable for publication. This he did whenever he could find the time, but unfortunately many of his most interesting communications remained unwritten when death removed him. The communications to Section A were never intended to be published in the volumes of the Association. One he wrote out for me for the *Messenger* (No. xxviii, vol. ii.), and others which I had specially asked for had been promised to me for the same journal.

The address which he delivered before the Mathematical Society on retiring from his two years of office as President in 1876 possesses so much mathematical interest that I felt justified in including it among the papers (No. xxxi, vol. ii.). I think it would be admitted without question that this was by far the most remarkable presidential address, both in substance and in mode of delivery, which has been made to the Society.

I have been thus particular in trying to describe the characteristic features of his method of exposition, partly because for some years before his death there was no more conspicuous personal figure in English Mathematics, and partly because in the severe style of the papers themselves there is no trace of the bright and winning gaiety of manner with which their first introduction to a mathematical audience was so often adorned.

I should despair of the possibility of myself conveying any adequate impression of Professor Smith's position in University and general society, but fortunately I am saved from the anxiety of any such attempt by the excellent article in the *Spectator* from the accomplished pen of the late Lord Bowen, which is reprinted on pp. xlvi–li. This tribute of affectionate appreciation, in which Professor Smith's character is delineated with perfect justice and delicacy, enables the reader to form a true idea of the unique place which he held in the larger world in which he moved, while his special claims as a mathematician were unknown to all except a few experts. His general attainments were so great and varied, and his personal and social qualities so brilliant, that his mathematical powers were completely overshadowed by other more conspicuous gifts. In an article in the *Times*, published the day after his death, it was truly said: 'It is probable that of the thousands of Englishmen who knew Henry Smith, scarcely one in a hundred ever thought of him as a mathematician at all. . . . He was a classical scholar of wide knowledge and exquisite taste, and there were few who talked to him on English, French, German, or Italian literature, who were not struck by his extensive knowledge, his capacious memory, and his sound and critical judgment.'

It always seemed to me very strange that it should have been possible for him to have held so distinguished a position in the foremost rank of mathematicians without his eminence, or his devotion to the subject, becoming more widely recognized among his friends and colleagues. His official post in the University was that of Professor of Geometry, and it was of course well known that he was an accomplished mathematician of high reputation. But I am sure that very few even of his intimate friends were aware that in his own subjects he stood alone in England, and that his papers upon the Higher Arithmetic held a place among the most important productions of the century in abstract science. Even fewer still had any idea of the extent to which his heart and mind were engrossed by his mathematical researches. This want of recognition (if it may be so called) was no doubt partly due to his disinclination to speak of his own work except occasionally to those whom he knew to be interested in it, and his non-mathematical friends may be pardoned for not discovering an enthusiasm which showed itself so little; in fact it cannot be doubted that he would have been spared much of the voluntary work which he so unselfishly undertook at the solicitation of others, if the depth of his devotion to his own subject had been generally known. But I think a truer explanation is to be found in the fact that, as his whole time and powers were apparently given up to other occupations, such as University work of all kinds and Royal Commissions, it could scarcely be supposed that he would be much more than a distinguished amateur in so exacting a science. There was nothing that suggested the specialist in his actions or conversation; and it is indeed truly remarkable that, in the midst of so many varied pursuits all requiring constant care and attention, he should have been able to carry out original work which can compare in extent and profundity with the researches of the ablest mathematicians, who have concentrated their whole lives upon their special subjects. Except in vacations he seemed to have no time for mathematical investigation, and the amount that he accomplished was always a mystery to me until I learned that after a hard day's work, closing perhaps with a dinner party at which his lively wit and brilliant conversation had made him seem the gayest and the brightest of the circle, he would quietly settle down to work in his own room for some hours before going to bed. What he then wrote related probably to matters that had been more or less in his mind all day, and to which at intervals he had actively turned his thoughts, making a few stray notes perhaps on slips of paper. The last thing of all at night he would enter the results of the day's work or thoughts in his note book. Most of his mathematical work he did in his head, by sheer mental effort, and he scarcely ever committed an investigation to paper in any detail except when writing it out for publication. The notes which he made while thinking out a subject were often written on scraps of paper or backs of envelopes, which were destroyed as soon as he had made a definite advance which would allow of an entry in his notes. The fact that he used pen and paper so little, relying on his brain as it were, increased the mental strain of his mathematical production, so that, as a rule, when struggling with difficulties or exploring new fields, he did not work for long at a time. After an hour or two he would leave the subject as it were to grow of itself in the background and permeate his mind, while he was actively employed on something less exciting*. I may here mention that the high standard of completeness which he exacted from himself in his published writings, and which has been referred to on p. lxxiv, added considerably to the effort with which his finished work was produced. The logical sequence of propositions, the absolute sufficiency of definitions, and the rigour of demonstrations, were all matters that exactly suited the quality of his mind; but his mode of working did not readily adapt itself to the laborious classification of the separate cases of a general theorem, or other details requiring merely industry and attention.

As his attention was not specially directed to mathematics until after his degree, he was in fact as regards its higher branches a self-made mathematician. It was during the long period of isolated study in which he familiarized himself with every formula in the greatest of the abstract Theories that his powers were developed and that his interest in mathematics grew into the almost passionate attachment of later years. Led on by pure fascination, under no pressure, but without either assistance or encouragement, he slowly and surely mastered everything that had been accomplished, and gained such an insight into the principles of the subject, and such a command over its methods, as could

* In an article in the *Fortnightly Review* for May, 1883, I wrote: 'His victories were won by the hardest of intellectual conflicts, in which for the time his whole heart and soul and powers were entirely and absolutely absorbed. It was in his wide interests and sympathies, the pleasure of intercourse with others, and the love of all that was good and cultivated, that he found relief from these severe mental efforts. Had he not been gifted with a disposition that gave him the keenest sympathy with every human interest, that attracted him to society and endeared him to his friends, that gave him, in fact, his other noble life—the life the world knew—his fierce devotion to the subject he loved would have ended his days long since.'

only have resulted from so long and complete a self-devotion. But one unfortunate result of his comparative isolation was that he allowed too much of his own work to accumulate in manuscript, and that, the 'note' of personal ambition (as Lord Bowen described it) being wanting in his character, and no external stimulus prompting him, he remained indifferent to the advantages of early publication, and was too little sensible of the difficulties that would stand in the way of preparing for the press any work which has been too long on hand. Thus, when he was forty years of age, besides the Report, he had published only one important memoir, although he was in possession of an immense amount of original work relating to Quadratic Forms, Geometry, and Elliptic Functions.

The foundation of the London Mathematical Society in 1865 was an event which exercised a marked influence upon the subsequent course of all his work. He was a fairly regular attendant at the council and ordinary meetings, and there met other mathematicians who appreciated his unique knowledge, and urged him to bring papers before the Society. Wherever he was known he was a *persona grata*, but nowhere more so than at the Society's rooms in Albemarle Street. During his presidency he communicated nine papers to the Society (besides the Address), only four of which however were written out. As time went on, and engagements and duties thickened upon him, he became more and more uneasy about the mass of work that lay unfinished in manuscript. In declining to undertake a fresh piece of work he wrote: 'I have twenty papers embedded in my note-books. I extricated and published seven last year.' He found it impossible to obtain the amount of consecutive leisure that was requisite to complete long and difficult investigations; and he was continually distracted between the fascination of new work and the desire to publish portions of the old. He would often say, 'I must bind my sheaves;' and only a few days before his death he said to his sister, 'My mind is teeming with new ideas *.'

His power of reading rapidly the mathematical writings of others, seizing the principles and grasping the methods as if by intuition, always struck me as very remarkable. Up to the last, and in spite of the scanty allowance of time that

* Three months before his death, after the meeting at Cambridge for a memorial to the late Professor F. M. Balfour, referring to the opinion expressed by one of the speakers that a man's original ideas came to him before he was thirty, he said to me that in his own case he was certain that not only had his power of seeing and understanding increased without interruption all through his life, but that his thoughts and ideas and invention had undergone a corresponding progression and development.

he could devote to Mathematics, he continued to read new mathematical literature with the same ardour, and he never allowed the pursuit of his own work to prevent him from keeping abreast of what was being done by others, not only in his own departments of study, but also in other branches of the exact sciences.

I cannot refrain from devoting one brief paragraph to recording his admirable style and perfect taste when addressing a mixed scientific audience. Of this I recollect three remarkable instances: the first when he proposed the late Professor Tyndall as President of the British Association for the meeting at Belfast in 1874; the second when, in reply to Lord Grimthorpe, he spoke on the endowment of research, at a special meeting of the Royal Astronomical Society, in 1881; and the third at the Balfour Memorial meeting, which has just been alluded to. On the second occasion especially, I think that none who heard the speech are likely to forget the power and brilliance of the speaker.

He spoke so lightly, and often with such whimsical disparagement of his own attainments and performances, that even those who were conversant with the nature of his published writings and the varied character of his pursuits were frequently surprised to find how well acquainted he was with matters and subjects which would not have been thought likely to be of special interest to him. This was the case also in Mathematics; and I can remember my own astonishment when, long after I knew him well, I accidentally discovered how familiar he was with every page of Jacobi's *Fundamenta Nova*. From the way in which the subject of Elliptic Functions was treated in his writings, I had not suspected that the *Fundamenta Nova* would have possessed so much attraction for him.

The following extracts from a letter addressed to the late Mr. Todhunter (in acknowledgment of some reprints of his papers) seem to be of sufficient interest to deserve preservation:

I have been also reading, and with great interest too, your 'Conflict of Studies.' I am afraid I am a shade less conservative than yourself. I have been led to entertain a somewhat higher impression of the value of experimental science, at least when the pupil is made to experiment himself. I am perhaps a little more willing than you are to consider favourably attempts to improve Euclid, though I have a great dread of the Association for the Improvement of Geometrical Teaching. Further (as I am a professor, and as there is nothing like leather), I am for having more professors, with more work, and more pay. But I so heartily agree with much, or rather with most of your book, that I should not have troubled you with this letter, if it were not that I cannot wholly subscribe to your estimate of the present state of Mathematics. All that we have, one may say, comes to us from Cambridge; for Dublin has not of late quite kept up the promise she once gave. Further, I do not think that we have anything to blush for in a comparison with France; but France is at the lowest ebb, is conscious that she is so, and is making great efforts to recover her lost place in Science.

Again, in Mixed Mathematics, I do not know whom we need fear: Adams, Stokes, Maxwell, Tait, Thomson will do to put against any list, even though it may contain Helmholtz and Clausius.

But in Pure Mathematics I must say that I think we are beaten out of sight by Germany; and I have always felt that the *Quarterly Journal* is a miserable spectacle, as compared with Crelle, or even Clebsch and Neumann. Cayley and Sylvester have had the lion's share of the modern Algebra (but even in Algebra the whole of the modern theory of equations, substitutions, &c., is French and German). But what has England done in Pure Geometry, in the Theory of Numbers, in the Integral Calculus? What a trifle the symbolic methods, which have been developed in England, are compared with such work as that of Riemann and Weierstrass!

But it is with the younger, or at least the less-known, people that I feel the difference most. Our English papers are so often quite free from anything really new, whereas a German takes care to know what is known before he begins to work, and besides generally takes care to work at some really important problem, and not at some trifling expression for the co-ordinates of the focus.

If I had room, I should vent my spleen (or perhaps my envy) by saying that I attribute the mischief to the business of problem-making: ninety per cent. of the good problems in Pure Mathematics that I see, are, if I mistake not, mere fragments of some great theory, of which the candidate is supposed to be ignorant.

In the last paragraph but one he refers to the want of a sufficiently important object in the papers of many English mathematicians. This was a subject which was often in his mind, and I have heard him more than once express his regret that so many writers, instead of attacking recognized difficulties or those parts of their subject where real advances might be expected, should be content to occupy themselves with developments of a comparatively trifling character. In connexion with the reference to Cambridge problems, I may mention that on one occasion, when I was telling him about a proposal to abolish the order of merit in the Mathematical Tripos, he said that in his opinion a system which was successful in extracting a great amount of hard work from the students should not (in spite of many drawbacks) be lightly abandoned.

My own friendship with Professor Smith arose in connexion with the interest I felt in some of the subjects in which he was an accomplished master, but it was not until he began to write the Introduction to the Theta Tables for me that I became intimate with him. The progress of this work naturally brought us into closer and more frequent contact. I used to meet him at the Mathematical and Astronomical Societies, often walking with him to the Athenæum Club at the close of the meetings, and we had long mathematical conversations at Cambridge when he came to the dinners of the Ad Eundem Club. When the memoir on the Theta Functions in its final form was passing through the press, we both read the proof-sheets, and at the same time he was sending me the Notes on Elliptic Transformation for the *Messenger*: I also had occasion to consult him on several mathematical and other questions;

and all these causes combined to produce a rapid interchange of correspondence during the last two years of his life.

It was not until I became really intimate with him that I had any idea of the intensity and earnestness of his devotion to Mathematics. Even among mathematicians he referred so gaily and with so light a heart to his own studies and pursuits that I have been almost startled to find, when alone with him, how engrossed he really was with mathematical researches, and how completely they possessed his mind and affections. He derived intense pleasure both from working at Mathematics and from the contemplation of its truths and processes; and although he was undoubtedly anxious in the latter part of his life that what he had accomplished should not perish in his note-books, he seemed quite indifferent to the amount of recognition that was accorded to his published writings by his contemporaries*: in fact, the only word of impatience that, so far as I know, ever escaped him with reference to the slight attention that had been paid to his best work, was the sentence quoted in the private letter to myself on p. lxvi.

The last paper of which he gave a verbal account had for its title 'On a property of a small geodesic triangle on any surface' (p. lxxiii), and was communicated to the Meeting of the British Association at York in 1881. The object of this note was to point out that if a, b, c are the sides of a small geodesic triangle, then the correction to be applied to the formula $a^2 = b^2 + c^2 - 2bc \cos A$ is $-\frac{4}{3}(\text{Area})^2 \times$ curvature.

I have no word to express the admiration and affection with which I regarded him myself. As regards his qualities and abilities, if I had not known him as I did it would have seemed to me incredible that such varied gifts and powers could be combined in the same person. All the assistance that I have ever received with respect to the direction of my own work, or the manner of conducting research, came from him, and I have never ceased to miss his advice and help: and more and more with each succeeding year. It will be long indeed before his place in Mathematics can be held by another; but in the lives of those who were personally indebted to him the void can never be filled.

* In communicating a paper to the Mathematical Society he once had occasion to refer to some results contained in one of his memoirs in the Philosophical Transactions, and he playfully apologized for having 'to quote from a paper which he had no reason to think that any one had ever looked at.' His indifference to personal prominence or display of any kind was frequently shown at the meetings of the British Association, for whenever there was any pressure upon the limited time of the section, he always waived his own claims in favour of those of others.

It is always somewhat hazardous to quote from private letters (except for the sake of facts), as they so often give to strangers a very different impression from that conveyed to those who knew the writer personally. Still I am tempted to close this Introduction with a few extracts from letters which, though too trivial perhaps to deserve publication on their own account, are yet not without a certain interest in connexion with the published papers. All the extracts are from letters written to myself during the last two or three years of his life; and most of them have been selected because they relate to the progress of the Introduction (or Memoir) on the Theta and Omega Functions and the Notes on Elliptic Transformation, with which he was occupied to the very last.

Oxford, 2 November, 1880.

I enclose the penultimate copy of the four θ-functions. The Society is reprinting its early numbers, and I have ordered fifty separate copies. There is an erratum in the note on p. 9, viz. it should be, I think, $\beta = \nu' - \nu$ not $\beta = \nu - \nu'$. This I have altered in the reprint.

The trodden worm will turn; and I feel sure that even Cayley will admit any defender of suffixes to all the privileges which appertain to the status of a worm. I therefore, speaking as a worm, declare that I do not in the least care for *suffixes*, but that any one who does not admit that a *double* notation is, for certain purposes, imperatively required by the circumstances of the case, is not fit to be an annulated animal at all, but only a mere zoophyte. I will, seriously, quite as willingly write $\vartheta\left(\begin{matrix}\mu \\ \nu\end{matrix}, x\right)$, or $\vartheta(\mu, \nu; x)$, as $\vartheta_{\mu,\nu}(x)$; indeed to me it is a mere printer's question. But if I am told that $\vartheta_1, \vartheta_2, \vartheta_3, \vartheta_4$ (however convenient as abbreviations), or again Θ, H, Θ_1, H_1, are as handy for use in general formulæ applying to all the four θ-functions, I am disposed to dissent. The Germans, I perceive, are great lovers of suffixes; and I confess that when I try to do without them, I soon want another alphabet.

Of course you are most welcome to do what you please with my paper: it will be much honoured by any use you may make of it. The 'Logic,' such as it is, you should have had long since, but that I sit seven hours a day, day after day, with our Commission. . . . It is my birthday and I am feeling very old.

The paper referred to is No. xvi (vol. i). I had accused him of exulting in the number and complication of the suffixes, and had said that the criticism of Professor Cayley (who disliked suffixes and avoided their use as much as possible) would be, 'Too many suffixes!' I was in the habit of giving the principal theorem of the paper in my lectures, and had asked for the separate copies, as the formulæ were unsuited for writing on the blackboard. I had also said that when, in printing my 'Lectures,' I came to the Theta Functions I wished to reproduce the whole of the paper just as it stood as a separate chapter. The 'Logic' was the paper whose title appears as No. 13 on p. lxxii. It had been promised for the *Messenger*.

Oxford, 5 June, 1881.

Best thanks; only I have not time to express them. [He then refers in detail to some misprints.] It is very kind of you to take the trouble you have done about a wretched little paper, of which the only interest, if any, is that it applies Liouville's theorem to a question of convergence.... When I sent you the manuscript of my paper I had almost asked you to print (at the end of it) Riemann's proof of Abel's 'little theorem.' There would then have been a good tail to a poor little thing; because Riemann's proof is a model of what such a proof should be. (I notice Todhunter in his 'Laplace's Functions' refers a little contemptuously to the 'little theorem'—this designation is mine, not his;—and in this he is quite wrong, as I think the trigonometrical series at once shows.)

The paper referred to in this letter is No. xl (vol. ii.), 'On some discontinuous series considered by Riemann.'

Ryde, 15 July, 1881.

Alas! I am not yet at the Elliptic Functions. For three weeks I was tied to my sofa in Oxford by a sprained thigh: and during that time I was exposed to continual interruptions, as in addition to the usual Oxford business at the end of term I had become (just at that time) executor of my dear friend Rolleston's will, and guardian of his children... Finding I could not be quiet enough to work at my Introduction to your Tables, I took up a very different bit of work, the Introduction to Clifford's Collected Mathematical Works. This is three-parts done and must be finished next week early: indeed it would have been done long ago except that thinking about it takes me into space of many dimensions, &c. I grudge the time I am giving to it because I can say nothing on the one hand fit for mathematicians to read, nor on the other fit for non-mathematicians. So I have to maunder a good deal, which is neither acceptable to me nor suitable to my ideas of the right way of honouring Clifford's memory. I long to be at the Elliptic Functions, I can tell you.

If you think the *Messenger* would like a note of three pages on one or two points in Riemann's 'Hypotheses which lie at the basis of Geometry' (viz. on the only two results which he announces in formulæ), the said *Messenger* would be most welcome.

On another occasion, he said that this Introduction was inferior to the similar work which he had done in connexion with the writings of the late Professor Conington, and that it 'savoured of the sick couch on which it had been written.'

Oxford, 12 December, 1881.

I have stolen a few hours for the Elliptic Functions, chiefly to try and get my hand in again for work immediately after Christmas. (Till then I am liable to many interruptions.) I must rewrite the transformations of the second order; I fear that nine of them must be given, viz. the nine which give different transformations of the *elliptic* functions.

Oxford, 7 February, 1882.

I have not seen you for a long time, and am afraid I am not likely to see you very soon. I am a close prisoner to my sofa with an inflamed vein in my thigh (gouty phlebitis, they call it). I hope I am beginning to get slowly better, but it will be a good bit of time before I am able to move about again. I have had to rewrite 'Transformations of the Second Order,' Art. 33, and while about it I have also made many changes in Art. 31 ('Linear Transformations of Elliptic Functions'). All this I could send to the printer if it were any good as yet for me to do so. I am not allowed to work very much, and I find I can only do rather easy things. But I think I am up to doing what remains to be done with the Memoir. It is horrible to me to think you should take any more trouble over the thing. And so I hope you will do no more than look through the proof-sheets very hastily indeed, if indeed you do as much as that.

I have been preparing a little 'paperlet' to show (1) that the coefficients a, b of the general elliptic transformation

$$y = \frac{x}{M} \frac{1 + a_1 x^2 + a_2 x^4 + \dots}{1 + b_1 x^2 + b_2 x^4 + \dots}$$

are rational in k^2 and λ^2, not only in u and v as they appear in Jacobi and Cayley; (2) that when λ^2 is an equal root of the modular equation, they are not rational (in general) in k^2 and λ^2 (viz. in this case Cayley's system of equations at the beginning of his memoir admits of more than one solution); (3) giving a new (slightly new) process for determining them which shows that in all cases (even of equal roots) they are rational in k^2 and λ^2 and $\frac{1}{M}$ (in the equal-root case, $\frac{1}{M}$ need not be rational in k^2 and λ^2). I had thought of bringing this to the Mathematical Society on Thursday; but finding such a journey out of my reach I am thinking of inflicting it on you for the *Messenger*. By the way, I will try and finish the little fragment of Logic for you. My difficulty is that I cannot get upstairs to my study and no one can find my papers for me.

Oxford, 22 February, 1882.

I am putting several interesting *little* things together in the 'Notes on Transformation' which I am writing for you.

Another extract from this letter has been given on p. lxvi.

Oxford, 9 March, 1882.

I am sorry to say that to-morrow I shall not be able to be at the Astronomical Society. I shall however probably venture up to London in order to go to the Meteorological Office, under a solemn promise to my doctors to be carried up and down stairs and do nothing else. So you see I am getting on, and if I am only patient I may soon hope to be about again.

I have just finished going over the revise of sheets 2, 3, 4; and am sending them to you at Cambridge. I am ashamed to have kept them so long. I find a few errata of my own, but none (I hope) to give much trouble to the printer. I am putting together several 'Notes on Transformation' for you. The paper is getting rather larger than I expected, because I have found two or three new (to me) little things while lying on my sofa.

Several of his friends were desirous that he should be nominated as the President of the British Association. The following is an extract from a reply to a letter of mine on this subject:

Oxford, 14 March, 1882.

I can tell you in a very few words what I feel about the Presidency of the British Association; indeed I do not know any one more likely to understand my feelings with regard to the matter than yourself. I should esteem the office a most horrid nuisance; at the same time I know my duty better to the British Association, to the University here, and to myself, than to refuse it if it were offered to me. For the honour (which I know to be a great one) I cannot bring myself to care (perhaps this is owing to a temporary weariness of the world, induced by lying on a sofa); but on the other hand I have a great horror of the indolence which induces one to refuse a position because the duties of it are irksome; and I think Dante was quite right to put the man in hell 'che il gran rifiuto fè' (I forget who he was, and what he declined). What makes me say that the position would be an unmixed nuisance, is that I have (by this time), in the University and out of it, had my full share of the sort of work which calls my mind away from the subjects which interest me most, and I am very anxious (before the evening closes in) to concentrate myself as much as I can. If I had to be President of the British Association, the best work of a year would have to be given to my address, and that is much more than I can afford. It would certainly be a sad interruption to my plans of work, and I should have a perpetual sense of unreality about it.

He afterwards said in conversation that the only scientific topics of general public interest upon which he could usefully discourse in a Presidential address were the motion of the atmosphere, the law of storms, &c.

Folkestone, 13 April, 1882.

I have been here for a fortnight, and can now limp about enough for purposes of business. I hope to meet you on Friday, and to have a few words on Mathematics with you then.

A quotation from a letter written a few days afterwards has been given on p. lxvii. The prize memoir was completed and sent off by the end of May.

Oxford, 30 July, 1882.

Have you returned from the United States? and, if so, when and where can we have a conference? I have been absolutely idle for thirty days at Royat in Auvergne, and have returned, a good deal better, I hope; but I am totally demoralized, and I feel as if I was too sleepy ever to do anything like a day's work again. However, I am now your slave, till I have accomplished my engagements with you (Introduction—that was—and *Messenger*). But my mental forces are in complete disarray, and you will have to use the whip severely to rally them.

Do you see that Lindemann has covered himself with immortal renown by proving the transcendentality of π? Of course, nine-tenths of the discovery is really Hermite's: but then Lindemann has the immense glory of having seen that Hermite's method could be applied to prove the transcendentality of π, when Hermite himself despaired of it. I have never examined Hermite's method closely, but taking his results for granted, Lindemann's reasoning *seems* all right. It is difficult not to envy, as well as admire, people who do such beautiful things: Lindemann's name is sure of a place in every history of mathematics hereafter *.

* Nine years before (May 31, 1873) he had written to me: 'I am much pleased in particular with the way in which you call attention to the question of arithmetical irrationality. So far as π is concerned, I do not believe that any one has ever proved even so much as that π cannot be the root of an affected quadratic equation. And I always maintain that, until geometers have done this, they should not treat the problem of the rectification of the complete circumference as a *demonstrated* impossibility. Perhaps, however, the proof of the quadratic equation theorem may be obtained by Lambert's method. But this I have never tried.' Dr. Lindemann was the guest of Professor Smith (when I was so too) at the Oxford Commemoration in 1876.

Ryde, 20 August, 1882.

I have four of my notes nearly ready for you, and hope to finish them before I leave. They will make about eighty of my little pages; will this fill a number for you?

I have been led in Note II to your question about convergence of series like sin am u in powers of u. The only one giving any trouble is $\frac{u}{\text{sin am } u}$; here the radius of convergence is the analytical modulus of K or iK' or $K \pm iK'$, whichever of these four is least; and the question is to find the values of k^2 for which each of these is least.

I have put headings to the Memoir, but have not sent it off, having been absorbed, so far as I had time, in my 'Notes.' But I will send it before I leave.

On August 10 I had gone through the first seven sheets of the Memoir with him at the Athenæum Club.

Ryde, 23 August, 1882.

Can you let me have a figure in the *Messenger*? Here it is*.... It is one of the modular curves of order 4; it divides the plane (as you see) into five regions. The *least* possible 'quarter periods' of sin am u are, if k^2 lies in 1, 2, ..., 5 (i. e. if the extremity of the vector k^2 lies in 1, 2, ..., 5), 1 . $K, \frac{1}{2}iK'$; 2 . $\frac{1}{2}iK', K$; 3 . $\frac{1}{2}iK', K \pm \frac{1}{2}iK'$; 4 . $\frac{1}{2}iK', K \pm iK'$; 5 . $K \pm iK', \frac{1}{2}iK'$, the $\pm$ sign being taken according as k^2 is below or above the axis. The absolutely least period is put first; of course K and K' are the rectilinear integrals, and least refers to absolute magnitude, *i. e.* to analytical modulus. Of course also there is a general theory relating to transformation to which this proposition belongs (it is in fact the theory of a problem which Jacobi touches on in the *Fundamenta Nova*, saying it is very difficult).

A more complete account of these results is given on pp. 411–413 of vol. ii.

The first portion of the manuscript of the 'Notes' was given to me at the meeting of the British Association at Southampton on August 29.

Margate, 8 September, 1882.

I return the proof. I am heartily ashamed of the state it is in.... My excuse is that I pressed myself a little too much to deliver the manuscript to you at Southampton. I am very glad I did so, however, for I think that it would have taken a longer time if I had tried to revise it thoroughly in manuscript, even allowing for the time it will take the printer to go through it.

I think I have now made it hang together in an intelligible way. I confess that till I wrote out the pages, which I sent you from Spottiswoode's, I had imagined that, when the modular equation has equal roots, the multiplier might be a root (square or cube) of a rational function of k^2 and λ^2. But I found that what really happens is that the multiplier (when the roots are equal) still continues to be a rational function of k^2 and λ^2, but is a function of k^2 and λ^2 with irrational coefficients, viz. the coefficients contain an imaginary quadratic surd such as $\sqrt{-m}$, where m is a whole number; whereas in all other cases the coefficients are rational numbers. I had said nothing to contradict this; but some of my

---

* The figure represented a symmetrical closed curve, consisting of four loops, each of which included the next smaller one, and having three double points on the axis of x. The region '1' was the interior of the smallest loop, the region '2' the space interior to the next loop but exterior to the smallest, the region '3' the space interior to the next and exterior to the second loop, and similarly for the region '4'; the region '5' being the space exterior to the whole curve.

present alterations are made with a view to lead up to it. More of them however are made simply to make the meaning, and connexion, clearer. It all lies close to what is known, but I think it is full enough of new *little* things to make it fit for the *Messenger*.

I send all that I have received from you so that it ends abruptly. I ought to have before me this portion when I revise the remainder. Correcting this has taken me two and a half days of (for me) hard work. I return at once to Note II: but would you not prefer to follow up Note I with something else, and let Note II take its chance by and bye? Notes I and II together would carry you nearly to the end of a third number; and this would be dreary for your subscribers.

Please send a card to say you have received this and have not gone mad with indignation at the state of the proof.

London, 16 September, 1882.

I enclose the revise. Of course I need not see another revise, and I should think you need not, as Metcalfe might well be trusted to make the corrections. On Monday morning you shall have the manuscript of the remainder of Note I. Of course I could not resist the temptation of re-scribbling it.

Enough of Note II to fill up the September number, and more, shall, if I can possibly manage it, be in your hands on Monday morning also.

London, 17 September, 1882.

I enclose the remainder of Note I, rewritten and made as tidy as I can.

As for Note II, a great part of it is nearly ready, but none of it quite. I will send you, very soon indeed, as much as you are likely to want—and more. I am sorry to tell you it will make more than a number and a half. Now this is intolerable, and I must divide it, for I will not take up three numbers running (even if you would let me, which, for your credit as an editor, I hope you would not). I think I can manage to divide it, though some of the beginning part is written solely with a view to the end. Till I get the September number of the *Messenger* safe in your hands I don't look at the Memoir: alas!

Oxford, 24 September, 1882.

Here is some more copy for Metcalfe. It will take him a good bit on into the October number. But now the worst of it is, that a lot more of Note II remains—I think twenty-five slips at least—and this is after my cutting off all about the absolutely least periods (with the curly cue curves), which I now propose to make into Note III (when you have got over the surfeit occasioned by Notes I and II). So that you see Note II, if allowed to run on, will take up nearly all, perhaps quite all, the October number. I cannot divide it into two Notes, because it really has a unity of its own, and the arithmetic of Arts. 2 and 3 (especially Art. 3) would be unmeaning (in a note on Elliptic Functions) without what follows. But there can be no objection to your dividing it in print, with a 'To be continued.' And this I should advise you to do. But I put myself wholly in your hands, and will do what you please. I think I could let you have the rest of Note II very soon. The *Messenger* must have as many lives as a cat, if it survives my Notes. Still I am prepared to maintain that the stuff in them is reasonably good, though by trying to be complete and exact I have become diffuse.

Brockham, 29 September, 1882.

I enclose the rest of Note II. There is not quite so much of it as it looks. Still I think it will run on pretty far into the October number. I have (as I said) left out the parts that would require a diagram or two.

If the London Mathematical Society are in want of food at their first meeting, I could give them an account of these omitted portions, which are to be Note III (when you allow such a thing to appear). This would also give me an opportunity of saying briefly what Note II comes to 'when it comes to be fired.'

The next thing that I shall do is to send you the revise of the Memoir, and to this I shall now stick till it is done; I shall begin at it this very evening.

If Metcalfe could, without putting himself out, send me the whole of Note II together, it would save time. But I have treated him abominably about Note I, and only hope that Note II will come out decently straight; there really are some things in it worth a moment's attention.

Brockham, 4 October, 1882.

I return the proofs. This time they are very clear, and Mr. Metcalfe will not be able to reproach me. A couple of references to Gauss and to my own Report have to be inserted.

I send with the proof a little fragment which comes in after the end of the August number, and before the beginning of the slips now sent to me. I also return the copy for the October number. All I shall want will be a revise (in pages) of the September number, and that will enable me to correct the part that comes out in October. I do hope, and I think it likely, that I shall not run quite to the end of the October number. I think I shall be more easily forgiven by your public, if they see that I really have come to an end, and that someone else is going ahead. . . .

All this interests me very much, because it turns on the theory of 'reduction' as applied to doubly periodic functions, and seems to me to excuse the amount of space I have made you give to it in the *Messenger*. I have still to make out whatever I can about the course of the curves P; but I fear this will not be much. I shall try (whenever Note III comes into existence) to put all this stuff into it. So Note III will want figures. My *hexagon*, curiously enough, had already been considered by Dirichlet; not, of course, in relation to Elliptic Functions, but in proving Gauss' famous theorems about the minimum value of a ternary definite quadratic form.

The concluding paragraph relates to the limits of convergence of a series for arg sn x, about which I had consulted him. He took great interest in the question, and several letters were entirely devoted to it.

The portion that was written of Note III, referred to in the last three letters, appears as No. X of the 'Notes' (vol. ii, pp. 408–414).

Abergavenny, 7 October, 1882.

I enclose the revise; I see there is one page over, to run on into the October number . . . Your remark as to the complexity of the result in the case of the value of the series for arg sn x, has made me begin to doubt whether I am really right in saying that one of the branches of the three-forked curve of discontinuity does really enter the circle of convergence. If it does the nature of things is a fool; if it does not, I am a fool; the latter hypothesis seems to me the more probable, and I gladly embrace it. Besides, I begin to see dimly a weak point in my demonstration. If only the curve can be coaxed into staying outside the circle the result will be the simplest possible, viz. that the series, when convergent, always gives the least value possible.

It was on the 12th of October, that I went over all the manuscript of the 'Notes' at the Athenæum Club with him (p. lxiv).

Oxford, 31 October, 1882.

At last I return the revise. I dare say it is full of blunders of mine, and is peppered over with printer's errata, but I cannot find any more than I have marked.

If the alteration of the note on p. 88 and the rearrangement on p. 89 are troublesome, it would not be ruinous if they were left alone. *Item*, on p. 96 the signs in lines 2 and 3 are not very wrong as they stand, and might be left as they are; I have now made them correspond exactly with the 'elementary matrices of Art. 3'; as they stand, they do not.

I am very sorry to have kept you waiting so long. I comfort myself by thinking that the number has not been expected with great impatience by any one.

Oxford, 7 November, 1882.

I am almost sorry you took the trouble of sending me a revise. When I returned the proof I had intended to tell you that you might print it straight off. The small world that reads the *Messenger* will give an audible sigh of relief when they come to p. 99, and find there is no more of me. However, you will have, within a year or so, to print Note III, and some figures with it. That done I absolve you from all further Notes on Elliptic Functions, and if I ever write them, I will inflict them on the London Mathematical Society, or the *Quarterly Journal*, or on the new Scandinavian journal, or on Sylvester's Journal, or on any one but you.

London, 30 December, 1882.

Do you happen to have a copy of the sheet pp. 423–431 that you could send me? I have two, but the printer could hardly make them out. I mean now to do nothing but proofs for a long time. I have the two sheets which follow those now printed off practically ready, and there is nothing to cause delay for a long time to come.

Oxford, 20 January, 1883.

I enclose four more sheets of the Memoir: the rest (as far as set up in 4to) will follow immediately. I am sorry to say that the first two of the sheets I send have had to undergo great alterations. This will not happen with any of the remaining sheets. I should be very glad to have, as soon as you can, the manuscript which is in your hands set up. For the next three or four months I can give a great deal of time to this work, and hope (D.V.) to bring it to a close.

Oxford, 1 February, 1883.

Best thanks for your letter. I cannot be at the R. A. S. to-morrow. . . . I have returned to the printers four sheets of the Memoir for revise—but this includes the sheet which really has to be set up again, and made, I should think, into two. I find the stuff (now that I have quite forgotten it) more intelligible and hanging together better than I supposed. I find many little slips of mine and some of the printers', but very few great blunders so far. It takes an enormous amount of time to go through it. I must write to you before the week is over about *figures*, and about completing, or rather shutting up, the whole thing: there are now 136 pp.; I *think* it will run to about 170, or a little over. Please regard this letter as not needing any answer. I shall see you on the ninth.

These were the last words I was to hear from him. The 9th of February was the anniversary meeting of the Royal Astronomical Society, and I entered the Society's rooms expecting to meet him, and go over some of the sheets of the Memoir in the way that had become habitual to us; but Mr. Stone, who had just arrived from Oxford, told me that he had died at seven o'clock that morning.

J. W. L. GLAISHER.

COLLECTED MATHEMATICAL PAPERS

OF

HENRY J. S. SMITH

I.

ON SOME OF THE METHODS AT PRESENT IN USE IN PURE GEOMETRY.

[Transactions of the Ashmolean Society, Vol. II, No. xxv. Read December 1, 1851.]

THE principles of the Analytical Geometry introduced by Descartes effected a change in the nature of the science, the importance of which it is impossible to over-estimate. This change consisted in two things principally; first, in the creation of a wholly new system of Geometry by the side of the old synthetical methods; and secondly, in the complete, though gradual, metamorphosis which these methods themselves underwent. For a long time, it is true, this second effect did not manifest itself. No one, for example, would say that the Pure Geometry of our great English mathematicians, of Newton, or Maclaurin, or Matthew Stewart, exhibits so essential a difference from the Geometry of the ancients, as that which strikes us in the works of far less original writers in the present day. But the change, though long delayed, appears now to be complete, and the geometrical methods, by long contact with analysis, seem to have acquired much of its spirit, and of its peculiar power and facility, and this without losing the intuitiveness proper to themselves.

This has been the work of the last sixty or seventy years; but it is historically interesting to observe that two of Descartes' contemporaries had anticipated the change, and had introduced methods into Pure Geometry wholly unlike its ancient resources. The theorems due to Desargues and Pascal are still primary in the geometrical theory of the conic sections, and the methods by which those results were obtained, so far as it is possible to judge, appear to have partaken fully of the generality of the results themselves. In particular, had the great work of Pascal upon Conic Sections been published, there can be no doubt that geometrical theories, which we now owe to the writers of the last half century, would have been in the possession of the world for a much longer period. Unfortunately all that now remains of Pascal's Conic Sections is comprised in a letter of Leibnitz to Pascal's executor, giving an account of the work, which had been submitted to him for his inspection, and earnestly recommending its immediate publication; and in a fragment of three or four pages in length, entitled 'Essais pour les coniques,' and written by Pascal when he was only sixteen years old. But even this fragment, though printed and circulated by Pascal himself, was lost for nearly a hundred years after his death, and the magnificent theorem contained in it remained fruitless till comparatively recent times. Nor has Desargues been more fortunate. His works are completely lost, and we are left to form our opinion of him from the isolated expressions of Pascal and Descartes, and from the virulent attacks of his enemies, of whom he appears to have had a great many. Had it not been for the ill-fortune of their works, Pascal and Desargues would have been the founders of modern Geometry; as it is, to Monge, before all others, this honour justly belongs. Himself a great master of Analytical Geometry—how little, for example, has been added to the general analytical theory of surfaces since his time—he yet seems to have been the first who fully felt of how great an extension the old Geometry was capable. He not only enriched it with a method and a body of doctrine, which has rendered thoroughly rational the relation of the science to the arts depending on it, but he also infused into Geometry two qualities which had seemed peculiar to analysis—method in its processes and generality in its results. What gave, and what still gives, analysis so immense an advantage over Geometry, is, that when we have once expressed analytically a definition, say, or a theorem, the known laws of the combination of symbols enable us to transform that expression in a thousand different ways, and whenever we can interpret any such transformation we have a new theorem. To take an instance, perhaps too simple, if we write $x^2+y^2=a^2$ we express merely the common definition of the circle; if we write $y^2=a^2-x^2$ we express a theorem, deduced from the definition, and deduced, too, by changing the place of a single letter in an equation. If we add to this copious power of transformation, first, the generality of analysis, that is, its power of expressing theorems, essentially the same, however they may differ accidentally, in one and the same formula, and, secondly, the facility with which relations dependent on the consideration of infinity may be algebraically expressed and transformed, we shall perhaps have stated the three principal prerogatives of analysis over the old methods of Pure Geometry. Now it would be hazardous to say that rational Geometry can at present compete with analysis in any one of these respects, or even that it can ever hope to do so, but it is not too much to say that it is in possession of extensive and powerful methods for the transformation of theorems; that it has attained to a generality of expression unknown before; and, finally, that it has learned to employ all the resources of the theory of infinitesimals, and to employ them with facility and elegance for the demonstration of theorems that seemed formerly to require the aid of analysis. Nay, further, whereas since the time of Monge Analytical Geometry has received three principal improvements, the introduction of symmetry, the method of triliteral or quadriliteral coordinates, and the method of tangential coordinates, it is a fact that for the two last we are indebted, in the first instance, to the rapid development of Pure Geometry, and to the efforts successfully made by analytical writers to reconquer the ground that seemed for the moment lost to their favourite branch of the science.

Among the principal peculiarities by which the present Geometry distinguishes itself from the old we must reckon the theory of transversals, the different theories of the transformation of figures, and the frequent use of the geometrical method of infinitesimals. The object of the present paper is very briefly to characterise the first two of these theories, and to illustrate the last by a few examples of its application to the theory of geodesic lines.

But before doing so it will be well to allude to a question, very obscure in itself, but which puts the spirit of the new Geometry in a clear light, I mean the theory of imaginary quantities and imaginary figures. The analytical writers on Geometry have adopted one of two courses in this matter. They have either attempted to construct imaginary magnitudes geometrically, or else they have asserted the impossibility of constructing them, and have thence inferred the impossibility of getting any good at all out of them. With respect to the attempted constructions of imaginaries, it cannot be denied that they have been of great use; but they have been of use not to Geometry, which they have conducted to no new results, but to analysis, which they have enriched by an important interpretation of symbolical expressions. In fact, the persons by whom these constructions were introduced were much more familiar with analysis than Geometry, and hence they were guided much rather by ideas of analytical than of geometrical continuity. If, for example, in the equation $y^2 = a^2 - x^2$ we assign to x values beyond the limits $\pm a$, and proceed to construct the imaginary ordinates in a plane perpendicular to the plane of reference, we preserve it is true the algebraic continuity of the function, and we give an admissible interpretation of the symbol $\sqrt{-1}$; but what geometer can persuade himself that the equilateral hyperbola thus obtained stands in any but the most arbitrary relation to the circle? The continuity of the circle requires that every line in its plane, without exception, should cut it twice, and that every point in its plane, without exception, should be the intersection of a pair of tangents to it; does the equilateral hyperbola enable us to realise or to interpret either of these properties in the cases in which they become unmeaning? Again, it is an obvious remark, though apparently either not made or not attended to, that through every imaginary point there passes one, and but one, real line, and that on every imaginary line there exists one, and but one, real point. Can the principle of perpendicularity furnish any explanation of this fact? Apparently not; and yet this is only one out of many cases that might be mentioned. Analytically considered, the theory is faultless; but geometrically, it introduces discontinuity, it is inadequate to explain the phenomena, and (what is still worse in the eyes of a geometer) it is barren of results. Nor should we forget that till the constructions in question can be extended to loci in space, their use can never become general even in Plane Geometry, since it is frequently requisite to consider plane curves as sections of surfaces. For these reasons all attempts to construct imaginaries have been wholly abandoned in Pure Geometry; but, by asserting once for all the principle of continuity, as universally applicable to all the properties of figured space, geometers have succeeded, if not in explaining the nature of imaginaries, yet, at least, in deriving from them great advantages. They consider it a consequence of the law of continuity, that if we once demonstrate a property for any figure in any one of its general states, and if we then suppose the figure to change its form, subject of course to the conditions in accordance with which it was first traced, the property we have proved, though it may become unmeaning, can never become untrue, even if every point and every line, by means of which it was originally proved, should wholly disappear. In this way geometers are enabled not only to present theorems, in appearance the most dissimilar, as really identical (which in a scientific point of view is of immense importance), but also to make one easy demonstration serve where many dissimilar ones were before required.

The practice of demonstrating real properties by means of imaginary ones was first introduced by Monge, who employed it, though tacitly and without enunciating the general principle, in demonstrating what we should now call the properties of poles and polars in surfaces of the second order. The principle was first broadly stated, I believe, by Poncelet, in his 'Traité des Propriétés Projectives.' It was somewhat unfavourably criticised by the Commission which was appointed by the Institute to report on that work, and which consisted of Cauchy, Arago, and Poisson. It seems however to have more than held its ground, and to be frequently appealed to by all the most eminent writers.

A single example of its application must suffice. It has been shown very recently that if we draw two tangents to a conic from a point without it, and inscribe a circle touching the two tangents and the curve, the arc of the curve intercepted between the two tangents will be divided at its point of contact with the circle into two parts such that their difference shall be geometrically rectifiable. This theorem is of great importance: it enables us to construct with extreme simplicity the principal formulae for the addition and multiplication of elliptic functions of the two first species, and its demonstration depends ultimately on two general properties of confocal conics, viz., that if from any point on a conic A we draw two tangents to a confocal B, the angle between these tangents is bisected by the tangent and normal to A; and secondly, that if from two points on A we draw pairs of tangents to B, these four lines shall be tangents to one and the same circle. Now it is a known theorem, due to Quetelet, that if two spheres be inscribed in a cone, so as to touch a plane section, the two points of contact will be the two foci of the section. It is also known that if a surface of the second order be inscribed in another surface of the same order, the two sections determined on the two surfaces by any plane will have double contact, the chord of contact being the intersection of the plane of the sections and the plane of contact. We may therefore consider the foci of a conic as two evanescent sections of a sphere, that is, as two evanescent circles, having double contact with the conic. If we now observe that an evanescent circle degenerates into a pair of imaginary lines, we shall perceive that a system of confocal conics may be regarded as all inscribed in one and the same imaginary quadrilateral. That is to say, they may be regarded as possessing all properties incident to a system inscribed in a real quadrilateral, and therefore as forming a system of curves in tangential involution, and from this general property the two properties required to be proved are at once deducible, though since the demonstration at this point ceases to be imaginary the subsequent steps are here omitted.

Did time permit it might be shown how from the same imaginary property of confocal conics all the known theorems respecting such a system might be easily deduced, the geometrical theory of involution enabling us to transform the imaginary relation in a multitude of ways, and with a facility that renders the process almost mechanical. Nay, more, with merely verbal, or at least very simple alterations, the same proofs would apply to spherical as well as to plane conics; so that, for example, we might extend the constructions before alluded to for elliptic functions of the second order, so as to include those of the third also. If any one will give himself the pains to examine any one of those theories of modern Geometry in which frequent use is made of imaginaries—for example, Poncelet's theory of homological figures, or Chasles' construction for the semi-axes of an ellipsoid—he will rise perhaps with no clearer idea of imaginary magnitude than when he began, but he will probably be satisfied that this is one of those cases to which Dr. Woodhouse's remark applies with all its force, that a method which leads to true results must have its logic.

In passing to speak of the theory of transversals, it is hardly necessary to observe that only a very few of its most characteristic features can be alluded to here. Under the term Theory of Transversals I mean to include (somewhat improperly, though not without precedent) all those methods for the investigation of the properties of curves and surfaces which rest upon metrical rather than descriptive relations, and which operate on the figure as it stands, and not on figures derived by transformation from it. Several of the isolated theories belonging to this head appear to bear a special character, from the simple fact of their having been first developed as parts of the theory of Conic Sections. This is the case, for example, with the principles of harmonic and anharmonic section, and especially of involution. But it is a mistake to suppose that this special character is more than accidental. In fact, the harmonic properties of all geometric curves have been known since the days of Cotes, and have in recent times acquired great interest from the discoveries of Poncelet, and from the still imperfect theory of polar curves. But the importance of the theory of harmonic section cannot be fully understood till we reflect that of all conceivable metrical relations among points on the same right line, not one can by possibility belong to the sphere of linear Geometry, except it belong to the class of harmonic properties. With the ruler alone we cannot bisect lines or draw parallels, still less take proportionals, but we can always determine harmonic means and construct harmonic progressions. This observation alone would suffice to show that the harmonic relation, far from being special in its application to curves of the second order, must meet us at every turn in the science of space, and such in fact we find to be the case. For example, let a geometric curve be traced on a sheet of paper, and let it be required to assign its tangent at a given point, not singular, with the ruler alone. Let P be the given point; through it draw four transversals, which will cut the curve in $n-1$ points apiece. On each of these transversals take the harmonic centre of its $n-1$ points with respect to P, and consider the four points thus obtained as determining a conic passing through P. Pascal's theorem will now assign the tangent of this conic at P, and this tangent will be the tangent required. We may add that were the curve of the nth class, instead of the nth order, *i. e.* were it such that only n tangents could be drawn to it from a given point, and if we supposed that we were given, or could construct, the tangents passing through any point in the plane, we might, by a reciprocal construction, determine with the ruler alone the point of contact on any tangent, supposed not to be a double tangent to the curve. This solution of the general problem of tangents, being purely linear, is so far simpler than any that has ever been deduced from analysis, and it is worth while to ask why, in this particular case, Geometry possesses an advantage over analysis. The answer plainly is, that the analytical method introduces elements foreign to the real question, videlicet, a pair of right lines termed axes, and standing in a purely arbitrary relation to the curve and its tangent at the point P. When therefore we proceed to construct the tangent by means of its relation to these axes (by constructing its intercepts, for example), we lose sight for the moment of its immediate connection with the curve, and substitute for that immediate connection a mediate, and therefore a less simple relation. It will perhaps be found that similar observations apply to almost all those cases in which the Cartesian analysis is outstripped either by the present Geometry or by the newer methods of triliteral and indeterminate coordinates.

Similarly, by properly generalising the definition of involution, we can obtain without any trouble at all very many general properties of systems of curves and surfaces, some of which would not be without interest, and would admit of a multitude of corollaries.

Thus, if four surfaces have their complete intersection common, the anharmonic ratio of their four tangent planes is constant for every point on the line of intersection, and reciprocally, if any number of surfaces of the nth order be completely circumscribed by one and the same developable, the anharmonic ratio of the points determined by four of the lines of contact on any edge of the developable is constant for all the edges of the developable; so that, in particular, if the lines of contact be lines of curvature on three of the surfaces, they are lines of curvature upon all of them.

But, after all, it is in the method of transversals, strictly so called, that we find the principal resources of geometrical inquiry respecting general loci. That method is comprised entirely in the general theorem known by the name of Carnot, and in the means that have been devised for transforming and developing the equation supplied by it. Its importance in Pure Geometry is so unquestionable that a few remarks upon the place which it holds in the theory of curves, and on the geometrical considerations by which it may be established, cannot be out of place here. What renders at all possible a purely geometrical theory of curves, is precisely the introduction of considerations involving imaginary points and lines into the principles of the science. It is impossible to define geometrically a plane curve of the nth order, except by saying that it is a curve such that every right line in its plane must of necessity cut it in neither more nor less than n points, and of course this definition is inadmissible so long as imaginaries are excluded from consideration. But this definition once admitted, it is found possible to construct a purely geometrical theory of curves. If, for instance, we wish to find the locus of a point subject to certain conditions, these conditions give us the means of determining how many points of the locus can lie on one line, and, since the continuity of the locus implies that all lines in its plane cut it in the same number of points, we can in general determine the order of any proposed locus; and as soon as this is done, we are in a condition to inquire still further into its properties, to construct it by points, to determine its general form, &c. We will take one or two very simple examples of this *a priori* determination of the order of a locus. The first shall be Cotes' theorem already alluded to, namely, that if we take a fixed point P in the plane of a geometrical curve, and draw transversals through it, and then take the harmonic centres with respect to P of the n points in which every transversal cuts the curve, the locus of these centres for every position of the transversal shall be a right line. Now it is clear that the distance of the harmonic centre from the fixed point, being a *symmetrical* function of the distances of the n points of intersection from the same point, can have but one value for one position of the transversal, and therefore but one point of the locus can lie on each transversal through P. Unless therefore P be itself upon the locus, either as an ordinary or as a singular point, the locus is a right line. But that P is not on the locus, may readily be verified in the case before us, so that the theorem is proved. As a second instance, we will take the following from solid geometry. If a body be anyhow in motion, it is required to prove that the tangents to the trajectories of all the points in it, which lie upon a given line, form at any instant a hyperbolic paraboloid. To show this, we observe that no two tangents can lie in the same plane (except in a particular case, when all of them lie in one plane), and that consequently the complete section of the surface, formed by any plane containing the given line, consists of that given line and of one tangent only; so that the locus is a surface of the second order. That it is a paraboloid now follows from the fact that all the tangents are manifestly parallel to one and the same plane.

It should be added, that precisely in the same way in which we define a curve of the nth *order* by the number of its points which lie on a line, we define a curve of the nth *class* by the number of tangents (real or imaginary) that can be drawn to it from a given point, and this definition enables us to find envelopes, just as the former enabled us to find loci: an obvious example will suffice. 'One side of a constant angle passes through a fixed point, and the vertex lies on a fixed line, the other side will envelope a parabola.' For here the fixed line is itself one of the tangents, and therefore from each point upon it two, and only two, tangents can be drawn to the envelope—the envelope is therefore a conic section—and since the tangent line can in one position remove to an infinite distance, it is a parabola.

We see then that when a locus is investigated geometrically, the most general and the most important question we can ask respecting it is, in how many points it can be cut by a right line, and the preceding examples may serve to show how this question may in many cases be answered, that is, how the loci occurring in particular problems may be brought under the general definition of geometric curves. The next step is, from the purely descriptive relation asserted in the definition, to deduce an equally general metrical one; and this is exactly what is effected by Carnot's theorem. Carnot gives, in the 'Géométrie de Position,' two demonstrations of his theorem, one analytical and one geometrical. The former is very simple and elegant, but the latter is unsatisfactory; and though there can be no real objection to rest a general geometrical theorem on an analytical proof, it was still a problem of some interest to show that geometry could dispense with this assistance. To do this, M. Poncelet first showed that Carnot's theorem passed into Newton's by perspective, and then succeeded in demonstrating the latter by considerations analogous to those which we have been just employing for the theorem of

Cotes. Newton's theorem is, that if we take a fixed axis of abscissas, and draw ordinates to it, the ratio of the continued product of the abscissas to the continued product of the ordinates is constant, at whatever point of the axis of abscissas the ordinate is drawn. To prove this, we need only take upon each ordinate a line proportional to the value of the ratio for that ordinate, there is then no difficulty in establishing, first, that the locus of the extremities of these lines is a right line, and secondly, that it is a right line parallel to the axis of abscissas. And this is evidently equivalent to the required proof.

Carnot's theorem, notwithstanding its simplicity, is not very easily enunciated. If we take a triangle ABC in the plane of a curve of the nth order, and if its sides taken in order be cut in the $3n$ points, $p_1 p_2 \ldots p_n$, $q_1 q_2 \ldots q_n$, $r_1 r_2 \ldots r_n$; and if we denote by (Aq) the continued product of $Aq_1 Aq_2 \ldots . Aq_n$, we shall have, by the theorem,

$$(Aq) . (Br) . (Cp) = (Ar) . (Bp) . (Cq).$$

We see that this equation establishes a relation between the $3n$ points, which must of necessity subsist in order that they may all lie on one and the same curve of the nth order, and that when all these points are given, except one, we are able to determine that remaining one. If, for example, all the branches of a curve be completely described excepting one, and if two points upon the remaining branch be given, the theorem enables us immediately to describe it, or at least to determine as many points as we please upon it. But the principal applications of the theorem depend mainly upon the facility with which the fundamental equation may be modified and transformed. These modifications are rendered possible, first by the absolutely arbitrary position of the triangle in the plane of the curve, and secondly, by the facility with which evanescent segments may be eliminated from the equation, by introducing new transversals, and combining the equations supplied by them with the original equation. Thus, if we take one of the vertices of the triangle of transversals as A upon the curve, Carnot's equation will contain an evanescent segment on either side; and if we introduce a new transversal, passing through the extremities of the evanescent segments, and cutting the third side in a point t, the equation connecting the segments determined by this line on the sides of the triangle will immediately eliminate the evanescent segments, and give an equation for determining the point t, that is, for determining the tangent at A. By continuing the same process we might with the utmost facility determine the position and magnitude of the circle of curvature at A, and with a little more trouble might extend the investigation to the case of singular as well as ordinary points. But our present limits preclude the possibility of our pursuing this subject further. We will only add that it is possible so to transform Carnot's equation as to render the relation given by it capable of linear construction, and that in this way an immense number of descriptive properties of geometric curves may be obtained. For instance, we might demonstrate in this way Cotes' theorem, or the linear construction before given for the tangent at any point, which was exhibited as a consequence of the harmonic property of curves. But we will confine ourselves to merely stating one very general property due to Poncelet, 'If the points determined on the curve by a sufficient number of transversals be given, it is possible to determine the intersections of the curve with any other transversal by means of curves depending on the intersections of right lines only.'

We must dismiss with a still more imperfect notice the theory of the transformation of figures; and this, not because the subject is less interesting, but because, both in this country and on the continent, it has attracted so much more attention than any other part of pure geometry, that it would be no easy task to give a summary view of the whole system, while at the same time it would be hard to present anything with respect to particular applications that should have the interest of novelty. A few general remarks is all that will be attempted here. Figures may be transformed in two ways; either directly, that is, into others of the same kind, or inversely into reciprocal ones. In the first case, to every point in the original figure a point corresponds in the derived, and a line to every line. In the second case, this relation is inverted, and a point of one figure corresponds to a line of the other, and *vice versa.* It follows from this, that to the points of a curve line of the nth order there will correspond in the first case the points of a curve of the same order, but in the second case we shall have as the correlatives of the points of a curve in the primitive figure the tangent lines of a curve, no longer now of the nth order, but of the nth class. Consequently, the descriptive properties of a figure derived directly will be precisely the same as those of its primitive, but the descriptive properties of a figure derived inversely will be reciprocal to those of its primitive. This will frequently enable us to extend a descriptive relation from a particular to a general state of a figure, and from a descriptive relation of one figure to deduce another belonging to a different figure. The use of such processes in discovering new theorems, or in establishing a connection between ones already known, is too obvious to be dwelt on. We also see that any method of transformation, which satisfies the single condition, that to the points of a right line there should correspond the points of a right line in the first case, and a system of right lines passing through a point in the second, will enable us to generalise or transform any purely descriptive property. And it is possible to invent an unlimited number of methods of transformation, which shall comply with this restriction, and which, so far, possess no advantage one over another. But by a proper selection of the methods to be employed it has been found that we are enabled to transform not only all descriptive relations, but very many metrical ones also; in fact, all such as can be enunciated in a sufficiently general form. The two methods of projection and of reciprocal polars, for both of which geometry is mainly indebted to M. Poncelet, besides including in themselves almost all the methods for transformation that had previously been proposed, possess this last property in a pre-eminent degree. What is singular is, that though the principles upon which the two methods rest are so widely different, exactly the same class of metrical theorems which can be brought under the first are capable of being also transformed by the second. All harmonic properties, and consequently the whole 'geométrie de la règle,' all anharmonic and involutional relations, and, besides, all the general theorems of the theory of transversals, can be operated on by either of the two methods. Wherever it is applicable, the method of projection will enable us to make the proof of a general theorem depend upon its simplest cases, and on the other hand to explain and follow the modifications which a general principle undergoes in its application to particular instances, while the method of reciprocal polars unveils the singular duality which pervades so large a portion of the science of space, and which now finds its analytical expression in the method of tangential coordinates, but which at the time of M. Poncelet's invention had hardly been observed at all. It was of course known that to any triangle upon the surface of the sphere there always corresponded a second triangle, the angles and sides of which answered to the sides and angles of the first triangle, and were connected with them by an uniform and simple metrical relation. But this remark had never been generalised so as to extend to all spherical figures, still less had it been perceived that the property in question was so far from being confined to figures on a sphere, that it was only a particular case of a general property of a far more extensive class of figures. Now, of course, it is well ascertained, that it is impossible to assert any theorem respecting a figure on a sphere without, at the same time, asserting a different property of a different figure on the same sphere, and this whether the theorem be metrical or descriptive. Even any proposition respecting the rectification of a spherical curve gives us at once a theorem respecting the quadrature of the supplementary curve, so that if we could find all spherical quadratures we could rectify all spherical curves, and *vice versa*. We can even in this way obtain transformations of definite integrals; for example, of elliptic functions of the third order. For if we express first the area of a spherical conic, and then the length of the arc of its supplemental conic, we shall obtain two elliptic functions of the third order, with different moduli and parameters, and the supplementary relation of the two figures will at once establish an equation between these two integrals. The duality, then, of spherical figures is absolute, and extends to every conceivable case; but as soon as we pass to plane figures, or to figures in space, the case is different, and it is only in certain definite, though still very extensive classes of properties, that we find the principle manifesting itself, though this perhaps may be partly owing to the imperfection of our means of investigation. For it is certain that in many particular cases it is a matter of considerable difficulty to discover the reciprocal relations between theorems, even where it can be shown to exist. Take, for instance, the two theorems, 'A tangent to the interior of two similarly placed and concentric conics cuts off a constant area from the exterior conic;' and again, 'The sum of two tangents to an ellipse, which intersect on a confocal ellipse, diminished by the arc intercepted between them, is constant.' No one would have suspected, at first sight, that these two theorems are supplementary to one another, in exactly the same sense in which the word is understood in Spherical Trigonometry. But we should find, that if we were to imagine the two figures to become infinitely small, and to be placed upon a sphere, they would become supplementary, and the properties specified would follow the one from the other. In the same way many properties of the asymptotes of an hyperbola might be shown to be supplementary to the focal properties of an ellipse. From the constancy of the sum of the radii vectores in the ellipse, we might deduce the constancy of the triangle contained by the asymptotes and any tangent to the hyperbola; and from the equality of the angles made by a tangent to the ellipse with the focal radii vectores, we might infer the known theorem, that the intercept determined by the asymptotes on any tangent to a hyperbola is bisected at the point of contact. These instances may serve to show how cautious we should be in inferring that theorems, which seem to give rise to no reciprocal property, are really incapable of assuming this double character. It is therefore quite conceivable that future discoveries in Geometry may render the application of the principle of duality to the general properties of space quite as universal as it already is in the case of the sphere. But in the present state of the science it would be hard to name a case in which the existence of duality can be proved, and in which, nevertheless, it cannot be brought to light by the method of reciprocal polars. This method, therefore, still extends to the full extent of our present knowledge. It is now equalled, but it has not yet been surpassed in this respect, by the analytical method of tangential coordinates.

In forming the polar reciprocal of any proposed plane figure, we replace every point by its polar, and every line by its pole, with respect to an auxiliary conic taken in the plane of the figure. M. Poncelet has himself observed, that it would be unfair to argue against the generality of his method on account of its reposing (as it thus is made to do) on a particular property of curves of the second order. In fact, provided we once assure ourselves that the transformation we are employing is capable of being applied to any proposed figure, it will seldom signify whether the principles upon which the transformation rests be general or not; the only object is to obtain such a transformation as will enable us to transform the greatest number of metrical relations possible. It has been said, that any projective property may be transformed into a reciprocal one; but other relations, not projective, can nevertheless be made to yield reciprocal properties, by employing a circle or a parabola as the transforming conic. These particular applications, though of inferior interest with respect to general Geometry, are of great importance in the case of curves and surfaces of the second order, and in some physical applications of pure geometry —for example, in Professor Mac Cullagh's theory of apsidals, and his demonstration of Fresnel's construction of the wave surface.

Poncelet's account of Reciprocal Polars is to be found in a memoir in the third volume of Crelle's Journal, but he has devoted a separate work (the Traité des Propriétés Projectives) to the use of central projection, that is to say, of perspective in geometry. Few works, perhaps, could be named more calculated to awaken a taste for Pure Geometry than this admirable treatise: though the greater part of it is occupied with applications to the theory of conic sections, the reader feels all along that the methods developed in it are perfectly general, and that it only needs the genius of the author to apply them with equal success in almost any investigation. It is so natural an idea to simplify a diagram by forming a perspective representation of it, that it is surprising it should not have been introduced long before into geometry, especially when we remember for how long a time the conic sections were studied only on the cone.

Perhaps no better proof can be given of the rationality of the projective method than that which is supplied by the fact, that it is to it mainly that we owe the introduction of triliteral coordinates, a modification of the conception of Descartes, which possesses, it is true, many undeniable advantages over the purely geometrical perspective of Poncelet, but which cannot be properly understood in its relation to space till it is regarded, if we may so express ourselves, as a translation of perspective into the language of analysis. In fact, when we express the equation of a curve in triliterals, we are merely putting it into a form in which it becomes common to all possible perspective representations of the curve we are considering; we, as it were, divest the curve of all its non-projective properties, for the purpose of exhibiting in a more palpable and explicit form those essential ones which still continue to characterise it.

We can immediately determine whether any proposed metrical relation be projective or not, by merely examining whether it leads or does not lead to a relation involving the angles that are at any point subtended by the lines of the figure, and not involving the projecting lines. Applying this criterion, we should find, for example, that Carnot's theorem is projective, or, on the other hand, that it is impossible for a quadrature or a rectification to be so. But it would not be easy to lay down any general formula for determining *a priori* what properties are projective and what are not. Such a determination, though practically of little use, would theoretically be of the greatest value. However, the criterion we have given enables us now to see what was before observed, that every projective property can be made to yield a reciprocal one. For let the auxiliary conic be a circle, we shall have a relation between the angles contained by the rays drawn from its centre to the points of the figure, and since this same relation will subsist between the angles formed by the polars of those points, the new figure will possess a property reciprocal to and derived from that of the old.

Lastly, it may be remarked that the principle of perspective seems well calculated to form the basis of classification for geometric curves of any given order; a splendid example of this is given by Newton's famous theorem, that all curves of the third order may be generated by the shadows of five of them. This surely is the first step towards a purely natural classification of these curves, and it is much to be regretted that Plücker in his enumeration of curves of the third order, which is the most complete that has yet appeared, appears to have paid so little attention to the fundamental distinction between curves that can and that cannot be cut from the same cone, for there can be no doubt that it would have enabled him to construct his classification of the 219 varieties on a principle at once more general and more simple than that which he has adopted.

We come last of all to the method of infinitesimals. Ever since the invention of the Differential Calculus many of its resources have been at the command of Pure Geometry; and yet, in the first instance, the effect of its introduction upon that part of mathematical science was anything but favourable to it. For as soon as Newton and his immediate successors had passed away, the geometrical methods ceased (with rare exceptions) to be cultivated; the attention of mathematicians was so engrossed by the brilliant successes of the new calculus, that the era which is the most remarkable of all in the history of analysis was almost wholly unproductive in Pure Geometry. But the school of Monge, who delighted in finding geometrical solutions for all kinds of problems, soon attempted to rival the Differential Calculus on its own ground. And in particular cases they not only obtained very simple proofs of known theorems, but succeeded in discovering new properties, to which analysis might not have guided them so easily. Two striking instances of this might be mentioned, both from the works of Ch. Dupin. One is his celebrated theorem, that three series of surfaces which cut one another orthogonally cut one another in their lines of curvature. This he demonstrated by direct and purely geometrical considerations; and yet, perhaps, his proof is not the most simple that might be given. The other is the proposition which is the most general yet obtained in Dioptrics, 'That if a system of rays possess the property of being all normal to some one and the same surface, they will still continue to possess it after any number of refractions or reflexions at surfaces of absolutely arbitrary form and position.' Malus had succeeded in showing that if a system of rays emanate from a point they will form after a first reflexion two series of developables intersecting orthogonally, i.e. that they will all be normal to the same surface. He then proceeded to inquire whether they would continue to possess this property after a second reflexion, and, deceived by a slight error in his analysis, he concluded that they would not. But Dupin showed, that if the surface normal to the incident rays were imagined to envelope a system of spheres of variable radius, but having their centres on the reflecting surface, those spheres would determine a second envelope behind the reflecting surface, and that every reflected ray would be normal to this the second sheet of the complete envelope of the spheres, that is, that all the reflected rays would be normal to one and the same surface. In the case of refraction, we have only to substitute for the single sphere a pair of concentric spheres, having their radii in the ratio of the refractive index to unity*.

Many other instances, some of them of even greater interest, might be given; but instead of doing so we will conclude this Paper with demonstrations, simpler perhaps than those usually given, of some of the principal theorems relating to geodesic lines; first, upon surfaces in general, and then upon the ellipsoid.

We will set out with the assumption, which is easily justified, that a plane drawn parallel to the tangent plane at any point P of a curved surface, and at an infinitely small distance from it, cuts the surface in a curve which for all points indefinitely near the point of contact assumes the form and properties of an evanescent conic section, having its centre upon the normal at the point P. If we now consider a point Q situated on the circumference of the conic, it is plain that the normal to the conic at Q will be the orthogonal projection on its plane of the normal to *the surface* at the same point Q. We can therefore find the angle which the normal at Q makes with the normal section PQ, and this angle, divided by the arc PQ, is equal to the reciprocal of the radius of torsion of the geodesic line PQ, since the normal section is the osculating plane of that curve at P, and the normal to the surface at Q is its principal normal at that point. Transforming the expression thus obtained for the radius of torsion, we obtain the still simpler one,

$$\frac{1}{T^2} = \frac{1}{\rho_1 \rho_2} - \frac{1}{R_1 R_2};$$

where R_1, R_2 denote the principal radii of curvature, while ρ_1, ρ_2 are the radii of curvature of the normal sections tangent, and perpendicular to, the geodesic line PQ. Several consequences may be deduced from this formula, a few of which will be mentioned here.

First, if a geodesic line be tangent to a line of curvature, its torsion is invariably suspended at the point of contact. If therefore a line of curvature become a geodesic line it must at the same time become plane, since every point upon it will be a point of suspended torsion.

Secondly, if two geodesic lines intersect at right angles, their torsions at the point of intersection are equal.

Thirdly, it is possible to trace upon a given surface lines of maximum

* For an analytical proof of this theorem see the first part of Sir William Hamilton's Essay on Systems of Rays, or Prof. Minding in Poggendorff, 1847, p. 268.

geodesic torsion, i.e. curves such that at any point upon them the geodesic tangent to the curve will have greater torsion than any geodesic line passing through the point. Two of these curves will pass through every point on the surface. They will be orthogonal trajectories upon one another, and will intersect the lines of curvature at angles of 45°.

Fourthly, if a point be found on a curve surface such that it is a point of suspended torsion on every geodesic line passing through it, it must be an umbilic.

Consequently, if a surface have all its geodesic lines plane it must be umbilical at every point. But Monge has shown that the sphere is the only surface possessing this property; therefore the sphere is the only surface all whose geodesic lines are plane curves.

Let us now consider any curve S traced on a given surface A. If P be a point on S, and if we project S orthogonally on the tangent plane at P, the projected curve will pass through P, and will possess at that point a definite curvature. This curvature we shall term the tangential curvature of the curve S at the point P; it is obviously equivalent to the curvature of S multiplied by the cosine of the inclination of its osculating plane upon the tangent plane. If S be a geodesic line, its tangential curvature will be zero, and its tangential projection will be inflected at P, so that any small arc in the immediate vicinity of P may be regarded as rectilinear. It hence appears, that when S is not a geodesic line its tangential curvature is equal to the angle between two consecutive geodesic tangents, divided by the arc intercepted between the points of contact; or again, it is equal to an evanescent geodesic chord, divided by the square of the sagitta bisecting it perpendicularly. Now if A' be a second surface such that A can be developed on it without disruption or duplication, the minimum property of geodesic lines shows us that every geodesic line on A will determine a geodesic line on A', and that consequently the tangential curvatures of corresponding curves are equal at corresponding points on the two surfaces. In particular, we see that if A be a surface developable on a plane, the tangential curvature of S is precisely the curvature of the plane curve into which S is transformed, when A is developed on a plane. Or, when A is any surface whatever, if we imagine it to be circumscribed by a developable along S, since the tangential curvature will continue the same, whether we regard S as traced on the developable or on A, we may define the tangential curvature of S as the curvature of the plane curve into which S is transformed by the complanation of the developable circumscribing A along S.

Gauss, in his celebrated memoir 'Disquisitiones generales circa superficies curvas,' has introduced one or two expressions into Geometry which it is convenient to preserve. If we take a finite area upon a curved surface, bounded by any closed contour whatever, and if through any fixed point we draw parallels to the normals to the surface along the given contour, they will intercept a spherical area on a concentric sphere of radius unity, which is termed the 'spherical value' of the given curved area. The spherical value divided by the true value is the 'integral curvature' of the area; and if instead of a finite area we take an evanescent one, including a given point P, the limiting value of the integral curvature is the curvature of the surface at P. By the 'integral curvature' of a finite arc of a plane curve we understand (it is hardly necessary to observe) the angle between the extreme normals divided by the arc, and the angle itself may be called the 'circular value' of the arc.

The theorems which we shall now endeavour to establish geometrically are the following:

I. The curvature of a surface at any point is equal to the product of the reciprocals of the radii of curvature. (For simplicity, we consider only surfaces doubly concave, but the demonstrations, *mutatis mutandis*, will apply to surfaces having their curvatures of opposite signs.)

II. The spherical curve, which is supplementary on the auxiliary sphere, to the spherical value of any proposed area, is equivalent to the integral of the angle of tangential curvature extended over the whole contour of the area.

III. Corresponding areas on surfaces developable upon one another have equal spherical values, and consequently equal 'integral curvatures.'

I. If we take an evanescent rectangle dS contained by four lines of curvature, it is plain that if $\delta\phi$, $\delta\phi'$ denote the angles subtended by two adjacent sides at their respective centres of curvature, we shall have the equation $dS = RR'\,\delta\phi\,\delta\phi'$. But if $d\Omega$ be the spherical value of dS, we also have $d\Omega = \delta\phi\,\delta\phi'$. Therefore $\dfrac{d\Omega}{dS} = \dfrac{1}{RR'}$, and the truth of the result is independent of the peculiar form we have assigned to the element dS. For, whatever form we assign to that evanescent element, we can always imagine it made up of an infinite number of such rectangles, for every one of which the product $\dfrac{1}{RR'}$ will retain the same value within an infinitesimal, so that if $d\sigma$, $d\sigma'$ etc. be the little rectangles, $d\omega$, $d\omega'$ etc. their spherical values, we shall always find

$$dS = \Sigma d\sigma, \qquad d\Omega = \Sigma d\omega;$$

and, in the limit, when dS is itself rendered evanescent,

$$\frac{1}{RR'} = \frac{d\omega}{d\sigma} = \frac{d\omega'}{d\sigma'} = \ldots = \frac{d\Omega}{dS}, \text{ as before.}$$

II. Let S represent the contour of the given area, and Σ the contour of the corresponding spherical area. The spherical curve supplementary to Σ, which we will term Σ', is the envelope of great circles having their poles on Σ, and, consequently, having their planes parallel to the tangent planes of the given surface along S. The consecutive intersections of the planes of the great circles determine the sides of the cone subtended by Σ' at the centre of the sphere, and, in like manner, the tangent planes of the surface determine the arêtes of the developable circumscribing the surface along S. Therefore the sides of the cone are respectively parallel to the arêtes of the developable, so that Σ', which is obviously equal to the sum of the angles subtended by its elements at the centre of the sphere, is equal to the sum of the angles contained by the consecutive arêtes of the developable. Observing that every arête intersects S once, and once only, we see that, after complanation, the sum of the consecutive angles will be precisely equal to the angle contained by the two extreme arêtes, that is, to the angle contained by the two extreme normals, or, finally, to the integral of the angle of tangential curvature extended over S.

A particular case of this theorem deserves special attention.

Let S, instead of a continuous curve, form a polygon composed of geodesic lines. The theorem will evidently still subsist, only the quantity we have designated as the integral of the angle of tangential curvature will be simply replaced by the sum of the external angles of the polygon. We have therefore this theorem given by Gauss:

'The excess of the angles of any geodesic polygon, above the sum of the angles of a plane polygon of the same number of sides, is equal to the spherical value of the area of the polygon.'

This property will serve, in its turn, to establish another of Gauss' propositions: 'If from any point O two geodesic lines OP, OP' be drawn containing an evanescent angle ω, and if $OP = \rho$, PP' perpendicular to $OP = P$, the quantity P will satisfy the differential equation of the second order,

$$\frac{d^2P}{d\rho^2} + \frac{P}{RR'} = 0.\text{'}$$

Let Ω denote the spherical value of the triangle OPP'. Then, since

$dP = d\rho \cos OP'P$, we find $OP'P = \frac{\pi}{2} - \frac{dP}{d\rho}$; and therefore the preceding theorem gives at once $\frac{dP}{d\rho} + \Omega - \omega = 0$; or, differentiating, $\frac{d^2P}{d\rho^2} + \frac{d\Omega}{d\rho} = 0$. But, by theorem I, $\frac{d\Omega}{d\rho} = \frac{P}{RR'}$, and, substituting, $\frac{d^2P}{d\rho^2} + \frac{P}{RR'} = 0$.

We may also observe that this demonstration supplies us with a first and second integral of Gauss' equation. In fact, we have not only

$$\frac{dP}{d\rho} + \Omega - \omega = 0, \quad \text{but also } P = \rho\omega - \int_0^\rho \Omega d\rho.$$

The first arbitrary constant is ω, the second has been put equal to zero in the inferior limit of the definite integral in the expression of P.

In this form the equation suggests some interesting remarks, which our limits compel us to omit.

III. The integral of the angle of tangential curvature, extended over any arc of a curve, is constant for all developments of the surface on which the curve is traced. For since each element of the integral is an angle contained by two consecutive geodesic lines, it is apparent that neither the number nor the magnitude of the elements will be affected by the transformation, and consequently the sum will remain unaltered. If the arc become a closed contour, it will follow that not only this integral of contingence, but also the quantity supplementary to it, that is, the spherical value of the area, will remain constant for all developments of the surface.

In particular, if we consider an evanescent area we shall find $\frac{1}{RR'} = \frac{d\Omega}{dS}$. But since $d\Omega$ and dS are both constant, it follows that $\frac{1}{RR'}$ will be so too. This gives us the theorem which Gauss has demonstrated by a singularly beautiful analysis:

'If two surfaces be developable one upon another, the product of the principal radii of curvature is the same for any two corresponding points.'

The propositions we have been considering are of great importance in the theory of surfaces. The properties of geodesic lines, or more generally those properties of a surface which remain unchanged so long as the geodesic distances of its points remain unaltered, are doubtless as yet but very imperfectly known, notwithstanding the attention bestowed on them since the publication of Gauss' memoir. The subject is one of great difficulty, as those who have tried it well know, but it is at the same time of great interest, as it is certain that any general results obtained here would find frequent and useful application. Recently, too, the properties discovered by Mr. Roberts on the geodesic lines of the ellipsoid have attracted increased attention to the general question, and have themselves furnished fresh examples of the resources of Pure Geometry; for though Mr. Roberts' results were obtained in the first instance analytically, the geometrical proof of them since given is so direct that one is almost surprised that they were not discovered sooner.

As it is possible to exhibit this proof in a very simple form, and one which shows clearly its connection with the general theory of surfaces, we will allow it to find a place here. We have to show that the sum of two geodesic lines drawn from the umbilics of an ellipsoid to any point on a line of curvature, including the two umbilics, is constant; or, which comes to the same thing, that the angles made by the two geodesic radii vectores with the line of curvature are equal. Now it is a well-known property of confocal surfaces that the cones which envelope them from any point in space are themselves confocal, and consequently orthogonal, so that from whatever point in space two confocal surfaces be viewed their apparent contours will intersect at right angles. If we compare this property with the theory given by Monge, of the 'Surface of Centres' of any given surface, we shall perceive that any two confocal surfaces may be regarded as forming the two sheets of the surface of centres of some one and the same transcendental surface, and that every geodesic line existing on either confocal surface, and touching their common intersection, will be the cuspidal line of a developable circumscribing the second confocal. This proposition is the geometrical expression of Liouville's equation, $\mu^2 \cos^2 i + \nu^2 \sin^2 i = \rho^2$, or Joachimsthal's, $PD = \text{const.}$ As a particular case, we observe that every tangent to an umbilical geodesic line on an ellipsoid will pass through the focal hyperbola, so that the two tangents to the two geodesic radii vectores of any point on a line of curvature will be sides of the cone which from that point envelopes the focal hyperbola. But since these two sides lie in a principal plane of the cone they will make equal angles with the principal axes of the cone, that is, with the tangent and normal to the line of curvature.

The same principles would serve to demonstrate the other theorems discovered by Mr. Roberts; but the proof of this one may be enough to show, that though any property of space is doubtless discoverable by analysis, yet it sometimes happens that in particular cases it is more convenient to lay aside for a moment our analytical formulae, and consider the questions that arise in some more special but less artificial manner. This is especially requisite when the subject of inquiry is one about which little is as yet known. For here we cannot be sure *a priori* that the analysis of coordinates is the most natural method, and therefore the surest to lead to results. In such cases we find ourselves obliged to adopt, though it be but tentatively, a more direct study of the circumstances of the case in preference to a method which is apt at times to conceal from us in a singular manner the real grounds of the results with which it supplies us. In the words of M. Poinsot, 'Rien ne nous dispense d'étudier les choses en elles-mêmes, et de nous bien rendre compte des idées qui font l'objet de nos spéculations. Si le calcul seul peut quelquefois nous offrir une vérité nouvelle, il faut songer que cette vérité étant indépendante des méthodes ou des artifices qui ont pu nous y conduire, il existe certainement quelque démonstration simple qui pourrait la porter à l'évidence; ce qui doit être le grand objet et le dernier résultat de la science mathématique.'

[The following abstract of the preceding paper was published in the Proceedings of the Ashmolean Society, Vol. II. pp. 305, 306.]

The object of this paper was to show how, during the course of the last fifty years, the geometrical methods had acquired that generality and facility which had been, at an earlier period, regarded as exclusively characteristic of analysis. This rapid development of the resources of Pure Geometry was illustrated partly by general observations on the nature of some of its principal theories, and partly by a series of more particular examples.

The use of imaginary magnitudes in Geometry was especially dwelt on, and it was shown how by their aid we may sometimes comprehend in one and the same statement theorems at first sight widely different, and exhibit them as expressions of some one and the same general principle, assuming a different form, under different accidental circumstances. Other illustrations were taken from the application of the theory of transversals to the investigation of the properties of algebraic curves, and, among other conclusions, a linear solution of the direct and inverse problem of tangents (analogous to that given by M. Poncelet) was deduced from the harmonic properties of such curves.

The various methods for the transformation of figures, whether into other of the same kind, or into reciprocal ones, were also alluded to; and the law of Geometric Duality, which manifests itself in these transformations, was commented on, and an attempt made to fix the limits within which it is applicable.

In order similarly to exemplify the use of infinitesimals in Pure Geometry, some applications were made of this method to the theory of curved surfaces, and outline demonstrations of some of the results of Gauss on geodesic lines, and the mutual developability of surfaces, were given without the aid of coordinate Geometry.

Lastly, a proof was proposed of Mr. Roberts' theorems respecting the geodesic lines of ellipsoid, in which those results were exhibited as immediate corollaries from two well-known theorems of Pure Geometry, due to Monge and Jacobi.

II.

ON SOME GEOMETRICAL CONSTRUCTIONS.

[Cambridge and Dublin Mathematical Journal, vol. vii. pp. 118–126; May, 1852.]

If a geometrical curve be completely traced on a sheet of paper, the principles of the Theory of Transversals enable us to assign its tangent line and radius of curvature at any point, without supposing its equation known, and without employing any operations excluded from the sphere of elementary geometry.

The construction given by M. Chasles for this purpose is the following. Let m be the point on the curve, M any point assumed in the plane, mp, mq, two transversals, and MP, MQ, two parallels to them. Let also p, q, P, Q, denote the continued products of the segments on mp, mq, MP, MQ, respectively, excepting the evanescent segments on mp, mq. Then if we take on mp, mq, two lines respectively proportional to $\frac{P}{p}$, $\frac{Q}{q}$, the line joining their extremities shall be parallel to the tangent at m. To find the osculating circle, let t denote the continued product of the segments on the tangent at m, excepting the two evanescent segments, T the continued product of the segments on MT drawn parallel to mt; then, if on any transversal mp we take mc equal to $\frac{T}{t}\cdot\frac{p}{P}$, the point c shall lie on the osculating circle. For the diameter of this circle we have the expression $\frac{T}{t}\cdot\frac{n}{N}$; n, N denoting products of segments on the normal and on a parallel to it.

If the point be a double point, the preceding constructions fail; but by slightly modifying them, we may determine the two tangents and two radii of curvature, if the point be nodal: or, if it be conjugate, we can assign the elements of an ellipse, whose imaginary asymptotes shall be the imaginary tangents in question; that is to say, an ellipse concentric, similar, and similarly placed, with the evanescent conic formed by the conjugate point. In this case the two radii of curvature are in general imaginary and therefore cannot be constructed: but, since they are conjugate imaginary magnitudes, any rational symmetrical functions of the two (for example, the rectangle under them, or their harmonic or arithmetic mean), may readily be determined. The process to be employed is as follows. Take three transversals mp, mq, mr, and three parallels to them MP, MQ, MR, and let MR cut mp, mq in A, B, and the two tangents in θ_1, θ_2. We shall have

$$A\theta_1 . A\theta_2 = (mA)^2 \cdot \frac{R}{r} \cdot \frac{p}{P}, \qquad B\theta_1 . B\theta_2 = (mB)^2 \cdot \frac{R}{r} \cdot \frac{q}{Q}.$$

Now if the point m be conjugate, the products $A\theta_1$, $A\theta_2$, $B\theta_1$, $B\theta_2$, are essentially positive; and if it be nodal we can ensure their being so, by taking the three transversals mp, mq, mr, in one and the same pair of vertically opposite angles. Hence, if we put

$$(mA)^2 \frac{R}{r} \cdot \frac{p}{P} = a^2, \qquad (mB)^2 \frac{R}{r} \cdot \frac{q}{Q} = b^2,$$

the lines a and b can always be constructed. Therefore to determine θ_1, θ_2, describe two circles round A and B, with radii a and b, respectively. The radical axis of these circles will intersect MR at o the middle point of θ_1, θ_2; and any circle of the system orthogonal to the two circles (A) and (B), (*i.e.* any circle having its centre on the radical axis and its radius equal to the tangential distance of its centre from either of those two circles), will intersect MR in θ_1, θ_2. If (A) and (B) intersect in real points, their radical axis is instantly found; but in this case θ_1 and θ_2 are always imaginary. Let s_1, s_2 be the points in which the radical axis is cut by any one of the orthogonal circles; on mr take mo' a mean proportional between os_1 and os_2 (*i.e.* equal to the tangential distance of o from any one of the orthogonal circles): the ellipse having its centre at m, and mo, mo' for semi-conjugate diameters, will have the two imaginary tangents for its asymptotes. If (A) and (B) touch, the points θ_1, θ_2 coincide in o, and the double point becomes a cusp, having mo for its tangent. Lastly, if (A) and (B) intersect in imaginary points, the radical axis, though not immediately given, can always be determined by the ruler alone, and in this case, θ_1, θ_2 being always real, the tangents $m\theta_1$, $m\theta_2$ can be directly constructed.

The direction of the tangents once ascertained, the radii of curvature may be immediately found. In fact, if we denote by c_1, c_2 the chords intercepted on mp by the two circles, and by θ_1, θ_2 the two points in which the tangents are cut by MP, parallel to mp, we have

$$c_1 = \frac{\theta_1\theta_2}{m\theta_1}\cdot\frac{T_1}{t_1}\cdot\frac{p}{P}, \qquad c_2 = \frac{\theta_1\theta_2}{m\theta_2}\cdot\frac{T_2}{t_2}\cdot\frac{p}{P};$$

and by making mp coincide successively with the two normals, we get the values of the two diameters of curvature. Or we may first determine one, and then obtain the second by the proportion, which is easily demonstrated,

$$R_1 : R_2 :: \frac{T_1}{t_1} : \frac{T_2}{t_2}.$$

If the point be triple the determination of the directions of its tangents, which in analysis depends on the solution of a cubic equation, is not in general possible by the intersections of right lines and circles. The problem in its simplest form is this: Given three points on a right line ABC, and given the products $A\theta_1.A\theta_2.A\theta_3$, $B\theta_1.B\theta_2.B\theta_3$, $C\theta_1.C\theta_2.C\theta_3$, find θ_1, θ_2, θ_3. But whatever the order of the point, if the direction of its tangents be once known, the construction of its radii of curvature is very easy. If, for example, the order of the point be r, the chord determined on mp, by the circle tangent to $m\theta_1$, is readily seen to be given by the equation

$$c_1 = \frac{T_1}{t_1}\frac{p}{P}\frac{\theta_1\theta_2.\theta_1\theta_3\ldots\theta_1\theta_r}{(m\theta_1)^{r-1}};$$

which chord is therefore imaginary for an imaginary tangent, as it ought to be.

Returning to the case of double points, we see from the formula

$$c_1 = \frac{\theta_1\theta_2}{m\theta_1}\cdot\frac{T_1}{t_1}\frac{p}{P},$$

that if the two tangents coincide, the osculating circles become simultaneously evanescent, except a fourth segment on the tangent become evanescent also, that is, except the tangent cut the curve in four coincident points at m. In this case the point m is not cuspidal, but is a point of osculation, and possesses two radii of curvature, for which we proceed to give a graphical construction. If D_1, D_2 be the two diameters, we find readily enough $D_1 D_2 = \frac{T}{t}\cdot\frac{n}{N}$; but the theorem of Newton's, which has hitherto guided us, is perhaps insufficient immediately to furnish a second relation. Such a relation, however, may be obtained by the following considerations. It is well known that the polar conic of a point of inflexion breaks up into two lines: one of these is the tangent at the point of inflexion, the other will be found to be the locus of the harmonic centres of the $n-1$ points in which the curve is cut by a transversal through the point. Exactly in the same way the polar curve of the third order at a point m of the nature here considered, resolves itself into the tangent line and into a conic section; this conic touches the tangent at m, and is the locus of harmonic centres of the $n-2$ points in which the curve is cut by a transversal through m; consequently its curvature at m, multiplied by $n-2$, is precisely the sum of the curvatures sought. Now the diameter of curvature in a conic is to the chord intercepted on the normal, as the rectangle under the segments of a parallel to the tangent is to the rectangle under the segments of the normal chord. Hence, if mp_1 be any radius vector of the conic, p_1p_2 a chord parallel to the tangent at m, the radius of curvature is known as soon as the point p_2 has been constructed. To effect this, take any circle tangent to the conic at m; this circle and the conic being homological, their axis of homology may be first found, and then the line homologous to p_1p_2; this will give the point homologous to p_2, and therefore p_2 itself; in fact, the circle once described, p_2 may be found by the ruler alone. We now know the rectangle under the two radii of curvature, and the harmonic mean between them: the radii may therefore themselves be found by a simple and well-known construction.

It may be observed that the theory of polar curves leads to a construction for the tangent of a curve line, which is different from M. Chasles', and in fact linear. Through the given point P draw four transversals; each of these will cut the curve in $n-1$ points. Take the harmonic centre of each of these four groups with respect to P, and consider the four points thus obtained as determining a conic section passing through P. Pascal's theorem will then determine the tangent to this conic at P; that is, the tangent required. It is unnecessary to give the reciprocal construction, which enables us, when a curve of the nth class is given tangentially, to determine with the ruler alone the point of contact on any one of its tangents, supposed not to be a double tangent. It should however be added, that a method for the linear solution of these two problems has been long since given in a different and less explicit form by M. Poncelet, in his excellent memoirs on the Analysis of Transversals.

The radius of curvature of any point is, of course, by its nature, incapable of linear construction; but if we imagine ourselves to have constructed the normal at any point, and to have determined on it the centre of curvature of the given curve or of any one of its superior or inferior polar curves; and if, in addition, a line parallel to the normal be given, in order that the point at infinity on the normal may be known; we can find linearly the centre of curvature of every single curve of the polar system continued as far upwards as we please. This is a consequence of the following theorem: The distances of the centres of curvature of the successive polar curves from their common tangent form a harmonic progression, commencing at infinity and having the given point for its point of evanescence.

All the preceding methods admit of an easy application to the theory of surfaces. For example, to determine the tangent plane at m, we must draw three transversals mp, mq, mr, not in one plane, and then proceed as the case of plane curves. If the surface be of the nth order, its tangent plane will determine on it a curve of the same order; the point m being a double point in that curve nodal, cuspidal, or conjugate, according as the contact is hyperbolic, parabolic, or elliptic. Taking the last case, we must determine two conjugate semidiameters of an ellipse concentric, similar, and similarly placed with the evanescent conic in the tangent plane, and therefore with the indicatrix of the point m: the directions and magnitudes of the semiaxes of this ellipse may now be deduced by a construction of extreme simplicity (vide Note XXV. on M. Chasles' *History of Geometry*), and therefore the ratio of the principal curvatures, and the traces of the principal normal sections on the tangent plane are known. If now we construct the radius of curvature in either of these normal sections, the square of one semi-axis of the indicatrix is found, and therefore that curve may be regarded as completely determined. It hence appears, that to find the tangent plane and the indicatrix of any point, it is requisite to draw sixteen transversals; not that so many are absolutely essential, but the trouble is rather increased than lessened by taking fewer.

From their connexion with the present subject the following geometrical demonstrations of Meunier's and Euler's theorems on curvature may find a place here. If we take a point P on a curve line, and if we consider an evanescent chord drawn parallel to the tangent at P as an infinitesimal of the first order, this chord will be bisected by the normal at P; that is to say, it will be divided into two segments whose difference will be infinitesimal of the second order. Moreover, if we take any sagitta perpendicular to the chord, and intersecting it in a point distant only by an infinitesimal of the second order from its centre, it is readily seen that the square of either segment of the chord, divided by the sagitta, may be taken to represent the diameter of curvature of the evanescent arc. Hence, if we take two plane sections of a surface intersecting in an evanescent chord, the radii of curvature of the evanescent arcs are inversely as any two sagittæ perpendicular to the chord, and bisecting it approximately. If, now, one of the sections be a normal one, we may take for the sagitta in that section the intercept on the normal to the surface. Consequently the triangle formed by joining the extremities of the two sagittæ will be right-angled; and therefore the radius of the oblique section is equal to the orthogonal projection of the normal radius upon the plane of the oblique section, which is Meunier's theorem.

It follows also from what has been said, that if we draw a plane parallel to a tangent plane, and distant from it by an infinitesimal of the second order, the curve surface will, in general, determine upon this plane an evanescent hyperbolic or elliptic oval; and that the chords of the oval, being themselves infinitesimals of the first order, will be bisected within an infinitesimal of the second order at the point at which the normal meets the plane, and which we will call C. We may therefore consider the oval as a central curve having its centre at C: it only remains to shew that it is a conic section. This may be done as follows: Every transversal passing through C and lying in the plane of the oval, will cut the surface in two points belonging to the oval, and in $n-2$ points whose distance from C is infinitely great compared with that of the two first points. Now, if for a moment we consider the diameters of the oval to be finite, the remaining $n-2$ points will lie at infinity, and therefore an infinitely magnified representation of the section we are considering would consist of a finite central conic, replacing the oval, and of the line at infinity $n-2$ times repeated, replacing the $n-2$ branches which lie at a finite distance from C. Since, then, the radii of curvature of the normal sections vary as the squares of the diameters of the evanescent oval, they vary as the squares of the central radii vectores of a conic section.

If there be a double line upon the surface we can construct the two tangent planes at any point m by taking two plane sections passing through m and constructing the tangents of the double points. Each of these tangent planes will cut the surface in a curve having a triple point at m; but as the direction of one of the three tangents is known *a priori*, being the intersection of the tangent planes, the directions of the remaining two may be found by the construction used for double points; consequently the directions of the tangents to the principal sections on each sheet of the surface are known, and the principal radii of curvature may be determined by the construction before given for finding either radius of curvature at a double point. The two indicatrices at the point m may therefore be considered as ascertained in magnitude and position.

If the osculating plane and radius of curvature of the double line itself be required, they may be obtained very simply by a method to be given below.

If the singular line be of the rth order ($r > 2$), a little consideration will shew that though we cannot determine the tangent planes by any elementary construction, yet, if we assume these planes as known, the indicatrices upon each sheet may be found as easily as if the point were not singular. This is the more remarkable, since the expressions given by analysis for the principal radii of curvature at such points appear to be of great complexity.

Let us now take a point m on a curve surface where two sheets of the surface meet and have a common tangent plane. This tangent plane will intersect the surface in a curve having a quadruple point at m; but the directions of the four tangents may always be ascertained by a quadratic construction. For at such a point the polar surface of the third order will resolve itself into the tangent plane and into a surface of the second order. And it may be shewn that the two generatrices (real or imaginary) of that surface which lie in the tangent plane are in involution with the two pair of asymptotes of the two indicatrices; that is, with the four tangents before mentioned. Now these two generatrices may be determined by means of the theory of homological figures, since that theory enables us to assign a pair of semi-conjugate diameters of a section of the surface of the second order parallel to the tangent plane at m, whence the directions of the asymptotes of that section become known, and therefore the two generatrices required. The problem now will be: Given four points in a line $PQRS$, and the four products $P\theta_1.P\theta_2.P\theta_3.P\theta_4$, &c., determine $\theta_1, \theta_2, \theta_3, \theta_4$, a pair of points G_1, G_2 being also given which form an involution with the two pairs θ_1, θ_2 and θ_3, θ_4. It is plain that, A being any point whatever, any symmetrical function of the distances $A\theta_1$, $A\theta_2$, $A\theta_3$, $A\theta_4$, may be constructed. Hence H_1, H_2, the harmonic centres of the four points $\theta_1, \theta_2, \theta_3, \theta_4$ with respect to G_1, G_2, are known. But H_1, H_2 form a pair of points in involution with the two pair sought: and therefore H_1, H_2 together with G_1, G_2 completely determine the involution. Therefore the centre and foci of the system are known, and consequently θ_1, θ_2 and θ_3, θ_4 may be now quadratically determined. Points of the nature here considered may exist isolated on a curve surface; but if there be a continuous series of them, we shall have a line along which two sheets of the surface envelope one another (not a cuspidal line, for any transversal plane will determine a section having not cusps, but points of osculation at its intersections with the singular line), and at any point on such a line the two generatrices before mentioned will be found to coincide: and consequently the surface of the second order will degenerate into a cone. The side of this cone, existing in the tangent plane at m, may be determined by proceeding as in the general case:

and then, instead of a pair of points in involution with θ_1, θ_2, θ_3, θ_4, we shall have one focus of that involution given. The second focus may next be constructed (being the harmonic centre of θ_1, θ_2, θ_3, θ_4 with respect to the given focus), and the problem becomes quadratic as before.

Lastly, let there be a point on a curve surface having a tangent cone of the second order. It will be possible to determine three conjugate diameters of that cone. For, take any two planes passing through the vertex, and having constructed the tangent lines of the double points in those planes, take the harmonic conjugates of the line of intersection of the two planes with respect to each pair of tangents. This will give the plane conjugate to the line of intersection; and by taking any two lines harmonically conjugate with respect to the two tangent lines existing in that plane, we shall obtain the directions of the three semi-diameters required. Likewise, their ratios, or rather the ratios of their squares, may be found, since the two sides of the cone in each conjugate plane may be constructed. Hence, we may deduce the directions and the ratios of the squares of the principal semiaxes of the cone. But this determination, involving the solution of a cubic equation, requires the construction of a conic, and is consequently not within the limits of elementary geometry. (Vide the Note on M. Chasles' *History*, already quoted, and a paper by Mr. Townsend in this *Journal.*)

If a curve of double curvature be given in space the principles of the theory of transversals are not immediately applicable: but if we regard it as the intersection of two geometrical surfaces completely given, we may immediately find its tangent, osculating plane, and radius of curvature. The tangent at m is of course determined by the intersection of the two tangent planes; and if we take the two normal sections containing that tangent, and, having constructed their radii of curvature, let fall a perpendicular from m on the line joining the two centres, this perpendicular will represent in magnitude and direction the radius of curvature of the given curve. This (it will be seen) follows at once from Meunier's theorem, or from that known as Hachette's.

III.

DE COMPOSITIONE NUMERORUM PRIMORUM FORMAE $4\lambda+1$ EX DUOBUS QUADRATIS.

[Crelle's Journal, vol. L. pp. 91, 92; 1855.]

SIT
$$q_1+\cfrac{1}{q_2+\cfrac{1}{q_3+\ddots+\cfrac{1}{q_n}}}$$
fractio continua, cujus numerator, qui determinanti
$$\begin{vmatrix} q_1, & 1, & 0, & 0, & . & . & . & 0 \\ -1, & q_2, & 1, & 0, & . & . & . & 0 \\ 0, & -1, & q_3, & 1, & . & . & . & 0 \\ 0, & 0, & -1, & q_4, & . & . & . & 0 \\ . & . & . & . & . & . & . & . \\ . & . & . & . & . & . & . & 1 \\ 0, & 0, & 0, & 0, & . & & -1, & q_n \end{vmatrix}$$
aequalis est, per hujusmodi formulam $(q_1 q_2 q_3 \cdots q_{n-1} q_n)$ exprimatur. Erit ergo
$$[q_1 q_2 \cdots q_{i-1} q_i] = [q_i q_{i-1} \cdots q_2 q_1]$$
et
$$[q_1 \cdots q_n] = [q_1 q_2 \cdots q_i] \cdot [q_{i+1} \cdots q_n] + [q_1 q_2 \cdots q_{i-1}] \cdot [q_{i+2} \cdots q_n];$$
quae aequationes pendent ab illa forma determinantali, ambae autem L. Eulero debentur.

Itaque, si quantitatum q *par* sumatur numerus, ipsaeque ita serie symmetrica disponantur, ut binae inter se aequales fiant, elucet, quantitatem

$[q_1 q_2 \dots q_i q_i \dots q_2 q_1]$ summam fore duorum quadratorum inter se primorum; fit enim $[q_1 q_2 \dots q_i q_i \dots q_2 q_1] = [q_1 q_2 \dots q_i]^2 + [q_1 q_2 \dots q_{i-1}]^2 \dots$.

Contra in numero quotientium *impari*, erit

$$[q_1 \dots q_{i-1} q_i q_{i-1} \dots q_2 q_1] = (q_1 \dots q_{i-1}) \cdot \{[q_1 \dots q_i] + [q_1 \dots q_{i-2}]\},$$

unde colligis, numerum $[q_1 \dots q_i \dots q_1]$ primum esse non posse, nec duplicem numeri primi; si quidem casus excipis, in quibus, aut i unitati aequatur, aut i binario, q unitati.

Sit p numerus integer datus; $\mu_1, \mu_2, \dots \mu_s$ series numerorum, qui ad p primi sunt, ipsiusque p dimidio minores.

Formentur fractiones continuae $\frac{p}{\mu_1}, \frac{p}{\mu_2}, \dots \frac{p}{\mu_s}$; quae omnes ita terminentur, ut is quotiens qui in extremo loco ponatur unitatem superet. Hinc patet, quanta fuerit numerorum $\mu_1, \mu_2, \dots \mu_s$ multitudo, tantum fore numerum determinantium $[q_1 \dots q_n]$, qui dato numero p aequales erunt, neque praeter illos ullum dare ejusdem formae determinantem, cujus et primus et extremus quotiens unitate major sit, quique numero p aequalis esse possit.

Jam vero, quum duo determinantes $[q_1 \dots q_n]$ et $[q_n \dots q_1]$ aequales sint, quumque ipsum q_n unitate majus sit, apparet $[q_n \dots q_1]$ ex una aliqua fractionum $\frac{p}{\mu}$ oriri. Unde sequitur, data quavis fractione $\frac{p}{\mu}$, inveniri posse aliam in eadem serie, quae quotientes eosdem, ordine inverso, repraesentet.

Sit p primus, formae $4\lambda+1$; ut numerus determinantium ipsi p aequalium par existat. Quum ipse p unus e determinantium serie fiat, unus certo alius inveniri poterit in quo quotientium ordo invertendo non mutatur. Cum sit ergo

$$p = [q_1 q_2 \dots q_i q_i \dots q_2 q_1]$$

erit denique

$$p = [q_1 q_2 \dots q_i]^2 + [q_1 q_2 \dots q_{i-1}]^2.$$

Quam theorematis Fermatiani demonstrationem maxime elementarem esse patet, quum pendeat a conversione fractionum vulgarium in fractiones continuas.

Singulos autem formae $1+x^2$ divisores ex duobus quadratis conflari, eodem modo demonstrare in promptu est. Sit enim

$$\mu\nu = 1 + x^2,$$

apparet fore

$$\mu = [q_1 q_2 \dots q_i q_i \dots q_2 q_1]$$
$$\nu = [q_2 q_3 \dots q_i q_i \dots q_3 q_2]$$
$$x = [q_1 q_2 \dots q_i q_i \dots q_2].$$

OXFORD, *Maio* 1854.

IV.

ON THE HISTORY OF THE RESEARCHES OF MATHEMATICIANS ON THE SUBJECT OF THE SERIES OF PRIME NUMBERS.

[Proceedings of the Ashmolean Society, Vol. III. No. xxxv. pp. 128–131. Read March 2, 1857.]

IT is probable that the Pythagorean school was acquainted with the definition and nature of prime numbers; nevertheless the arithmetical books of the elements of Euclid contain the oldest extant investigations respecting them; and, in particular, the celebrated, yet simple, demonstration that the number of the primes is infinite. To Eratosthenes of Alexandria, who is for so many other reasons entitled to a place in the history of the sciences, is attributed the invention of the method by which the primes may successively be determined in order of magnitude. It is termed, after him, 'the sieve of Eratosthenes;' and is essentially a method of exclusion, by which all composite numbers are successively erased from the series of natural numbers, and the primes alone are left remaining. It requires only one kind of arithmetical operation; that is to say, the formation of the successive multiples of given numbers, or, in other words, addition only. Indeed it may be said to require no arithmetical operation whatever; for if the natural series of numbers be represented by points set off at equal distances along a line, by using a geometrical compass we can determine without calculation the multiples of any given number. And it was in fact by a mechanical contrivance of this nature, that M. Burckhardt calculated his table of the least divisors of the first three millions of numbers. But simple as this process is, the questions to which it gives rise are among the most obscure of the theory of numbers.

Adopting (with a slight variation in its meaning) an expression introduced by M. Polignac, we may call the series of numbers left unerased, after the erasure of the multiples of any given primes, the *diatomic* series of those primes. Thus the diatomic of 2.3.5 is 1, 7, 11, 13, 17, 19, 23, 29 : and it is unnecessary to continue the series further, because if a denote any one of the eight numbers we have written down, the remaining terms of the series are included in the eight arithmetical progressions $30m+a$. In general, if Ω denote the product of any given primes, in forming their diatomic we need only attend to the terms less than Ω. A few of the properties of this finite series are very easily seen. In the first place, the number of the diatomic terms is that of the numbers less than Ω and prime to it, or $\Pi.(P-1)$, if P be any one of the given primes and Π denote a continued product. Secondly, the diatomic terms are distributed symmetrically; that is to say, if a be a diatomic term, $\Omega - a$ is one too. The sum of the diatomic terms is therefore $\frac{1}{2}\Omega\Pi(P-1)$ or $\frac{1}{2}\Pi P(P-1)$. It is not difficult similarly to form expressions giving the sums of any positive and integral powers of the diatomic numbers; or series giving expressions convergent up to a certain point, for their product or the powers of their reciprocals. It is therefore possible to form an equation of which the coefficients are functions of the given primes, and the roots are the diatomic numbers. But it does not appear that this equation throws much light on the nature of its roots.

A remark of greater interest is due to Legendre. Let us denote by (r) the greatest whole number not surpassing a given positive numerical quantity r; let $u_1\, u_2\, u_3 \ldots$ be the diatomic terms of any given primes, and let a be a numerical quantity inferior to u_{k+1}, but not inferior to u_k; then

$$k = (a) - \Sigma\left(\frac{a}{P}\right) + \Sigma\left(\frac{a}{P_1 P_2}\right) - \Sigma\left(\frac{a}{P_1 P_2 P_3}\right) + \ldots.$$

The series is to be continued till it stops of itself, and the signs of summation Σ extend to every possible combination of the given primes $P_1\, P_2 \ldots$ taken one by one, two by two, &c. The principle on which the demonstration of this equation (and of many resembling it, which occur in the theory of numbers) is founded, may be termed the principle of *cross classification*, and may be enunciated thus. If $\sigma_1\, \sigma_2 \ldots$ be a system of cross-classifying classes, and if (σ_1) denote the number of things in the class σ_1, $(\sigma_1\, \sigma_2)$ the number of things common to the two classes σ_1 and σ_2, $(\sigma_1\, \sigma_2\, \sigma_3)$ the number of things common to the three classes $\sigma_1, \sigma_2, \sigma_3$: then $\Sigma(\sigma) - \Sigma(\sigma_1\, \sigma_2) + \Sigma(\sigma_1\, \sigma_2\, \sigma_3) - \ldots$ will express the whole number of things present in the classes $\sigma_1, \sigma_2 \ldots$.

Legendre's formula, it is readily seen, assigns the index k of any given diatomic term u_k. Conversely we can express u_k in terms of k. Let us denote the function

$$\Sigma\left(\frac{a}{P}\right)-\Sigma\left(\frac{a}{P_1 P_2}\right)+\Sigma\left(\frac{a}{P_1 P_2 P_3}\right)-\ldots \text{ by } \phi(a):$$

and let us form the series of terms

$$\phi(k)=a_1,\quad \phi(k+a_1)=a_2,\quad \phi(k+a_2)=a_3\ldots$$

till we arrive at last (as we shall certainly do) at two consecutive terms equal to one another, say a_n and a_{n+1}; we may then stop, for we should find

$$a_{n+1}=a_{n+2}=a_{n+3}=\ldots.$$

The expression for u_k will then be

$$u_k=k+a_n,\quad \text{or}\quad u_k=k+\phi(k+\phi(k+\phi(k+\ldots))).$$

It must be confessed that this result is one which, for diatomics derived from a numerous group of primes, would involve far too much labour to be of any use. But it is of some slight theoretical interest. For if the given primes $P_1 P_2 \ldots P_x$ be the x first primes in order of magnitude, it is clear that their 2d, 3d, ... diatomic terms will be $P_{x+1} P_{x+2}\ldots$, and that the first diatomic term which will not be a prime is P^2_{x+1}. Given therefore the first x primes, the formula prescribes a direct method for the calculation of the primes intermediate between P_x and P^2_{x+1}. Thus let the given primes be 2, 3, 5, and let it be required to determine the next prime after 5, we find

$$\phi(2)=1,\ \phi(2+1)=2,\ \phi(2+2)=3,\ \phi(2+3)=4,\ \phi(2+4)=5,\ \phi(2+5)=5.$$

We have therefore

$$u_2=2+\phi(2+\phi(2+\phi(2+\ldots)))=2+5=7.$$

It may be added, that the calculation of the functions ϕ does not absolutely require the knowledge of the primes $P_1\ldots P_x$; only the arithmetical operations which would be requisite for determining their value would involve more trouble than the determination of the primes $P_1\ldots P_x$.

V.

REPORT ON THE THEORY OF NUMBERS.

PART I.

[Report of the British Association for 1859, pp. 228-267.]

---

1. THE 'Disquisitiones Arithmeticae' of Karl Friedrich Gauss (Lipsiae, 1801 {ed. 1. 1798}*) and the 'Théorie des Nombres' of Adrien Marie Legendre (Paris, 1830, ed. 3) are still the classical works on the Theory of Numbers. Nevertheless, the actual state of this part of mathematical analysis is but imperfectly represented in those celebrated treatises. The arithmetical memoirs of Gauss himself, subsequent to the publication of the 'Disquisitiones Arithmeticae;' those of Cauchy, Jacobi, Lejeune Dirichlet, Eisenstein, Poinsot, and, among still living mathematicians, of MM. Kummer, Kronecker, and Hermite, have served to simplify as well as to extend the science. From the labours of these and other eminent writers, the Theory of Numbers has acquired a great and increasing claim to the attention of mathematicians. It is equally remarkable for the number and importance of its results, for the precision and rigorousness of its demonstrations, for the variety of its methods, for the intimate relations between truths apparently isolated which it sometimes discloses, and for the numerous applications of which it is susceptible in other parts of analysis. 'The higher arithmetic,' observes Gauss†, confessedly the great master of the science, 'presents us with an inexhaustible store of interesting truths,—of truths, too, which are not isolated, but stand in a close internal connexion, and between which, as our knowledge increases, we are continually discovering new and sometimes wholly

---

* The additions enclosed in { } are taken from manuscript notes in the author's interleaved copy; they are all in his own handwriting.

† Preface to Eisenstein's 'Mathematische Abhandlungen,' Berlin, 1847.

unexpected ties. A great part of its theories derives an additional charm from the peculiarity that important propositions, with the impress of simplicity upon them, are often easily discoverable by induction, and yet are of so profound a character that we cannot find their demonstration till after many vain attempts; and even then, when we do succeed, it is often by some tedious and artificial process, while the simpler methods may long remain concealed.'

2. It is the object of the present report to exhibit an outline of the results of these later investigations, and to trace (so far as is possible) their connexion with one another and with earlier researches. An attempt will also occasionally be made to point out the *lacunae* which still exist in the arithmetical theories that come before us; and to indicate those regions of inquiry in which there seems most hope of accessions to our present knowledge. In order, however, to render this report intelligible to persons who have not occupied themselves specially with the Theory of Numbers, it will be occasionally necessary to introduce a brief and summary indication of principles and results which are to be found in the works of Gauss and Legendre. It is hardly necessary to add that we must confine ourselves to what we may term the great highways of the science; and that we must wholly pass by many outlying researches of great interest and importance, as we propose rather to exhibit in a clear light the most fundamental and indispensable theories, than to embarrass the treatment of a subject, already sufficiently complex, with a multitude of details, which, however important in themselves, are not essential to the comprehension of the whole.

3. There are two principal branches of the higher arithmetic:—the Theory of Congruences, and the Theory of Homogeneous Forms. The first of these theories relates to the solution of indeterminate equations, of the form

$$a_n x^n + a_{n-1} x^{n-1} + \ldots + a_1 x + a_0 = Py,$$

in which $a_n a_{n-1} \ldots a_1 a_0$ and P are given integral numbers, and x and y are numbers which it is required to determine. The second relates to the solution of indeterminate equations of the form

$$F(x_1 x_2 \ldots x_m) = M,$$

in which M denotes a given integral number, and F a homogeneous function of any order with integral coefficients. In this general point of view, these two theories are hardly more distinct from one another than are in algebra the two theories to which they respectively correspond,—the Theory of Equations, and that of Homogeneous Functions; and it might, at first sight, appear as if there was not sufficient foundation for the distinction. But, in the present state of our knowledge, the methods applicable to, and the researches suggested by these two problems, are sufficiently distinct to justify their separation from one another. We shall therefore classify the researches we have to consider here under these two heads; those miscellaneous investigations, which do not properly come under either of them, we shall place in a third division by themselves.

(A) *Theory of Congruences.*

4. *Definition of a Congruence.*—If the difference between A and B be divisible by a number P, A is said to be *congruous to B for the modulus P*; so that, in particular, if A be divisible by P, A is *congruous to zero for the modulus P*. The symbolic expressions of these congruences are respectively

$$A \equiv B, \text{ mod } P,$$
$$A \equiv 0, \text{ mod } P.$$

Thus $7 \equiv 2$, mod 5; $13 \equiv -3$, mod 8.

It will be seen that the definition of a congruence involves only one of the most elementary arithmetical conceptions,—that of the divisibility of one number by another. But it expresses that conception in a form so suggestive of analogies with other parts of analysis, so easily available in calculation, and so fertile in new results, that its introduction into arithmetic (by Gauss) has proved a most important contribution to the progress of the science. It will be at once evident, from the definition, that congruences possess many of the properties of equations. Thus, congruences in which the modulus is the same may be added to one another; a congruence may be multiplied by any number; each side of it may be raised to any power whatever, and even may be divided by any number prime to the modulus.

5. *Solution of a Congruence.*—If $\phi(x)$ denote a rational and integral function of x with integral coefficients (we shall, throughout this report, attach this meaning to the functional symbols F, f, ϕ, &c., except when the contrary is expressly stated); the congruence $\phi(x) \equiv 0$, mod P, is said to be solved, when all the integral values of x are assigned which make the left-hand number of the congruence divisible by P; *i.e.* which satisfy the indeterminate equation $\phi(x) = Py$. It is evident that if $x = a$ be a solution of the congruence $\phi(x) \equiv 0$, every number included in the formula $x = a + \mu P$ is also a solution of the congruence. But the solutions included in that formula are all congruous to one another and to a. It is proper, therefore, to consider all these congruous solutions as identical, and in speaking of the number of solutions of a congruence

to understand the number of sets of incongruous solutions of which it is susceptible. To assign, by a direct method, all the solutions of which a proposed congruence is capable, is the general problem which, in the Theory of Numbers, corresponds to the problem of the solution of numerical equations in ordinary algebra. But the solution of the arithmetical problem is attended with even greater difficulties than that of the algebraical one; and the attention of geometers has been turned with more success to the improvement of the indirect or tentative methods of solution, and to the discovery of criteria of possibility or impossibility for congruential formulae, than to their direct solution. It is to be observed that, by virtue of a remark already made, the *tentative* solution of a congruence involves no theoretical difficulty. For if $x=a$ be a solution, every number included in the formula $x=a+\mu P$ is also a solution, and among these numbers there is always one, and only one, comprised within the limits 0 and $P-1$ inclusively. By substituting, therefore, for x all numbers in succession less than the modulus, and rejecting those which do not satisfy the congruence, we shall obtain its complete solution. But the interminable labour attending this operation, notwithstanding all the abbreviations in it suggested by the Calculus of Finite Differences, renders its application impossible, except when the modulus is a low number.

6. *Systems of Residues.*—The set of numbers $0, 1, 2 \dots P-1$ (or any set of P numbers respectively congruous for the modulus P to those numbers) is termed a *complete system of residues for the modulus P*. By *a system of residues prime to P*, we are to understand a complete system, from which every residue has been omitted which has any common divisor with P. Thus 1, 5, 7, 11, or 1, 5, -5, -1, are the terms of a system of residues prime to 12. The word Residue is employed instead of Remainder, because the word Remainder would suggest the idea of a positive number less than the modulus or divisor; whereas it is frequently convenient to consider residues differing from those positive remainders by any multiples of the modulus whatever.

7. *Linear Congruences.*—The general form of a linear congruence is

$$ax+b \equiv 0, \text{ mod } P;$$

a, b, and P denoting given numbers, and x a number to be determined.

The theory of these congruences may be considered to be complete, both as regards the determination of the solutions or *roots* themselves and of their number. If a be prime to the modulus, there is always one solution, and one only; if a have a common divisor with the modulus which does not also divide b, the congruence is irresoluble; if δ be the greatest common divisor of a

and P, and if δ also divide b, the congruence has δ solutions. In every case when the congruence is resoluble, the direct determination of its roots may be made to depend on the solution of a congruence of the form $ax \equiv 1$, mod P, in which a is prime to P. This congruence coincides with the indeterminate equation $ax = 1 + Py$, methods for the solution of which were known to the ancient Indian geometers *, and have been given in Europe by Bachet de Meziriac †, Euler ‡, and Lagrange §. The methods of these writers ultimately depend on the conversion of a vulgar fraction into a continued fraction, and in one form or another have passed into every book on algebra. Nor would it have been proper to allude to them here, were it not that they serve to supply us with a clear conception of what we have a right to expect in the solution of an arithmetical problem. In such problems, we cannot expect to express the *quaesita* as (discontinuous) analytical functions of the *data*. Such expressions may indeed, in many cases, be obtained (by the use of the roots of unity or by other methods); but the results of the kind which have hitherto been given, though sometimes of use in calculation, may be said, with few exceptions, to conceal rather than to express the real connexion between the numbers required and the numbers given. The arithmetical solution of a problem should consist in prescribing a finite number of purely arithmetical operations (exempt from all tentative processes), by which all the numbers satisfying the conditions of the problem, and those only, are obtained. It is clear that this description exactly applies to the methods on which the solution of linear congruences depends; but, unfortunately, the higher arithmetic presents but few examples of solutions of equal perfection.

8. Besides the older methods for the solution of the equation $ax = 1 + Py$, others have, in very recent times, been suggested. Of these the following may serve as examples:—

A. In the equation $ax = 1 + Py$, or the congruence $ax \equiv 1$, mod P, form

---

* See the Arithmetic of Bhascara, cap. xii, and the Algebra of Brahmegupta, cap. i, in Mr. Colebrooke's translation, London, 1817.

† Problèmes plaisans et délectables, qui se font par les nombres. Seconde édition. Par Claude Gaspar Bachet, Sieur de Meziriac, Lyon, 1624. (See Props. xv to xxv.)

‡ Comment. Acad. Petropol. tom. vii. p. 46, or in the Collection of Euler's Arithmetical Memoirs (L. Euleri Commentationes Arithmeticae Collectae, Petropoli, 1849), vol. i. p. 2; and in his Elements of Algebra, part ii. cap. 1.

§ Sur la Résolution des Problèmes Indéterminés du seconde degré. Hist. de l' Acad. de Berlin, 1767, p. 165. (See Arts. 7, 8, and 29 of the Memoir.) Also in the Additions to Euler's Algebra, sects. i and iii. (Lyon, an. III.)

the residues of the successive powers of a for the modulus P. If a be prime to P, we shall at last arrive at a power which has $+1$ for its remainder or residue. The residue of the power immediately inferior to this power is the value of x in the congruence $ax \equiv 1$, mod P. This solution is evidently an application of Fermat's Theorem*.

B. Let there be P points $A_1, A_2, \ldots A_p$, arranged at equal distances on the circumference of a circle. Join A_1 to A_{a+1}, A_{a+1} to $A_{2a+1} \ldots$ and so on continually. It can be proved that if a be prime to P, we shall not return again to A_1, until we have passed through every one of the P points, and have formed a polygon of P sides. Let $X_1, X_2, \ldots X_p$ be the vertices of this polygon, taken in order, and let $A_2 = X_{m+1}$; then $x \equiv m$ is the value of x in the congruence $ax \equiv 1$, mod P†.

C. Let an origin and a pair of axes be assumed in a plane, and let all the points be constructed whose coordinates are integral multiples of the linear unit; call these points unit points. Join the origin to the point (a, P). If a be prime to P, no unit point can lie on the joining line, but on each side of the joining line there will be a point lying nearer to it than any other. Let $(\xi_1\, \eta_1)$, $(\xi_2\, \eta_2)$ be the coordinates of these points, and let $\xi_1 : \eta_1 < \xi_2 : \eta_2$; then ξ_1, η_1, and ξ_2, η_2 are the least positive numbers satisfying the equations

$$a\eta_1 - P\xi_1 = 1, \quad a\eta_2 - P\xi_2 = -1.$$

The late M. Crelle, of Berlin, in the 45th volume of his Journal (p. 299), has given a very useful table, containing the least positive numbers x_1 and x_2 which satisfy the equation $a_1 x_1 - a_2 x_2 = 1$, for all values of a_1 up to 120, and for all values of a_2 prime to a_1 and less than it.

9. *Systems of Linear Congruences.*—The theory of these systems is left imperfect in the work of Gauss (see 'Disq. Arith.' art. 37); but, by the aid of a few subsidiary propositions relating to determinants, we may, in every case, obtain directly all possible solutions of any proposed system; and (what is frequently of more importance) we can decide *a priori* whether a given system of linear congruences be resoluble or not, and if it be resoluble we can assign the

---

* Binet, sur la Résolution des équations du premier degré en Nombres entiers. (Journal de l'École Polytechnique, cahier xx. p. 289.)

Libri, Mémoires de Mathématique et Physique (Florence, 1829), pp. 65–67.

Poinsot, Réflexions sur les Principes Fondamentals de la Théorie des Nombres (Paris, 1845), cap. iii. nos. 19 and 20. For another solution by M. Binet, see Comptes Rendus, xiii. p. 349. See also Cauchy, Comptes Rendus, xii. p. 813. {Exer. d'Anal. et de Phys. Math., vol. ii. p. 1.}

† Poinsot, Reflexions, &c., cap. iii. nos. 17 and 18.

number of its solutions. The following theorems by which the determination of the number of solutions is, in every case, effected, will sufficiently indicate the nature of these investigations.

Let the proposed system of congruences be represented by

$$\begin{array}{l}(1, 1)\, x_1 + (1, 2)\, x_2 + (1, 3)\, x_3 + \ldots + (1, n)\, x_n \equiv u_1, \\ (2, 1)\, x_1 + (2, 2)\, x_2 + (2, 3)\, x_3 + \ldots + (2, n)\, x_n \equiv u_2, \\ \cdot \quad \cdot \quad \cdot \quad \cdot \quad \cdot \quad \cdot \quad \cdot \quad \cdot \quad \cdot \quad \cdot \\ (n, 1)\, x_1 + (n, 2)\, x_2 + (n, 3)\, x_3 + \ldots + (n, n)\, x_n \equiv u_n ; \end{array} \qquad (A)$$

let the modulus be q, and the determinant $\Sigma \pm (1, 1)\,(2, 2) \ldots (n, n) = D$. If the determinant be prime to the modulus, these congruences will always admit of one, and only one, system of solutions, namely, that supplied by the system of congruences

$$Dx_r \equiv \sum_{k=1}^{k=n} \frac{dD}{d(k, r)}\, u_k.$$

But if D be not prime to q, let $q = p_1^{m_1} . p_2^{m_2} \ldots$ where p_1, p_2, &c. denote different primes. In order that the proposed system should be resoluble for the modulus q, it must be separately resoluble for each of the modules $p_1^{m_1}$, $p_2^{m_2}$, &c.; and, conversely, if it be resoluble for each of those modules, and admit P_1 solutions when taken with respect to the modulus $p_1^{m_1}$, P_2 solutions when taken with respect to the modulus $p_2^{m_2}$, and so on, it will be also resoluble for the modulus q, and will admit $P_1 \times P_2 \times P_3 \ldots$ solutions for that modulus. It is, therefore, only necessary to assign the number of solutions of the congruences (A), for a modulus p^m which is the power of a prime. Let I_n be the index of the highest power of p which divides D; and similarly, let I_r denote the index of the highest power of p which divides all the minors of D which are of order r; then if $I_n - I_{n-1} \leqq m$, the system (A) (if resoluble at all) admits of p^{I_n} solutions; but if $I_n > m + I_{n-1}$, it will always be possible, in the series of differences

$$I_n - I_{n-1}, \quad I_{n-1} - I_{n-2}, \ldots,$$

to assign a pair of consecutive terms $I_{r+1} - I_r$, $I_r - I_{r-1}$, satisfying the inequalities

$$I_{r+1} - I_r > m \geqq I_r - I_{r-1};$$

and then the number of solutions (supposing always that the congruences are resoluble) is expressed by the formula $p^{I_r + (n-r)m}$.

The analogy of this theory with the corresponding algebraic theory of systems of linear equations is in particular cases very striking. For example, we have in Algebra the theorem:

'The system of n linear equations

$$(1, 1)\, x_1 + (1, 2)\, x_2 + (1, 3)\, x_3 + \dots + (1, n)\, x_n = 0,$$
$$(2, 1)\, x_1 + (2, 2)\, x_2 + (2, 3)\, x_1 + \dots + (2, n)\, x_n = 0,$$
$$\cdot \quad \cdot \quad \cdot \quad \cdot \quad \cdot \quad \cdot \quad \cdot \quad \cdot \quad \cdot \quad \cdot \quad \cdot \quad \cdot \quad \cdot \quad \cdot$$
$$(n, 1)\, x_1 + (n, 2)\, x_2 + (n, 3)\, x_3 + \dots + (n, n)\, x_n = 0,$$

implies either that $D = \Sigma \pm (1, 1)\,(2, 2) \dots (n, n) = 0$, or else that $x_1, x_2, \dots x_n$ are separately equal to zero.'

In the Theory of Numbers we have the corresponding theorem:

'If n linear and homogeneous functions of an equal number of indeterminates be congruous to zero for a prime modulus, either the determinant of the system is congruous to zero for that modulus, or else every one of the indeterminates is separately congruous to zero.'

10. *Fermat's Theorem.*—The theory of congruences of the higher orders is so essentially connected with Fermat's Theorem, that it will be proper before proceeding further to introduce a few considerations relating to that celebrated proposition.

It may be considered from two different (though closely connected) points of view, each of which has proved equally fertile in consequences. First, it may be regarded as asserting that, if p be a prime number, and x any number prime to p, the remainder left by the power x^{p-1} when divided by p is unity. It is thus the fundamental proposition in the arithmetical theory of the residues of powers, or, which is the same thing, of binomial congruences. Or, secondly, it may be regarded as asserting that the congruence $x^{p-1} \equiv 1$, mod p, has precisely $p - 1$ roots; and that these roots are the terms of a system of residues prime to p. It is in this latter point of view that the theorem is the basis of the general theory of congruences.

We may observe that the demonstrations of Fermat's Theorem point to this twofold aspect.

The proof, which is found in most English treatises of Algebra (it is the first of those given by Euler*), and which depends on the property of the binomial or multinomial coefficient, would naturally lead us to regard the Theorem in the first point of view. The same may be said of Euler's second

* Comment. Acad. Petropol., vol. viii. p. 141, or Comment. Arith., vol. i. p. 21. This is the first demonstration of the Theorem discovered, since the time of Fermat. The memoir containing it was presented to the Academy of St. Petersburg, Aug. 2, 1736.

demonstration*, which consists in showing that the index of the lowest power of x in the series $1, x, x^2, x^3$, &c., which leaves unity for its remainder when divided by p, is either $p-1$, or some submultiple of $p-1$; or again of the demonstration of MM. Dirichlet †, Binet ‡, and Poinsot §, which depends on the observation that the terms of a system of residues prime to any modulus, being multiplied by any residue prime to the modulus, still form a system of residues prime to the modulus.

But a remarkable proof of the theorem, in the second expression we have given to it, occurs in a memoir of Lagrange ‖. As this proof (though very elementary) has not been copied by subsequent writers, and is consequently but little known, its nature may be indicated here.

Let the product $(x+1)(x+2)(x+3)\ldots(x+p-1)$ be represented by

$$x^{p-1}+A_1 x^{p-2}+A_2 x^{p-3}+\ldots+A_{p-2}x+A_{p-1},$$

x denoting an absolutely indeterminate quantity. Writing $x+1$ for x, and multiplying by $x+1$, we obtain the identity

$$\begin{aligned}(x+1)^p+A_1(x+1)^{p-1}+A_2(x+1)^{p-2}+\ldots+A_{p-1}(x+1)\\ =(x+p)\left[x^{p-1}+A_1x^{p-2}+A_2x^{p-3}+\ldots+A_{p-2}x+A_{p-1}\right];\end{aligned}$$

whence, by equating the coefficients of like powers of x, we find

$$A_1=\frac{p(p-1)}{1.2},$$

$$2A_2=\frac{p(p-1)(p-2)}{1.2.3}+\frac{(p-1)(p-2)}{1.2}A_1,$$

$$3A_3=\frac{p(p-1)(p-2)(p-3)}{1.2.3.4}+\frac{(p-1)(p-3)}{1.2.3}A_1+\frac{(p-2)(p-3)}{1.2}A_2,$$

. . . .

$$(p-1)A_{p-1}=1+A_1+A_2+A_3+\ldots+A_{p-2}.$$

---

* Novi Commentarii Petropol., vol. vii. p. 49, or Comment. Arith., vol. i. p. 260. From the point of view in which Fermat presents his theorem, it is not improbable that the demonstration he had found of it was no other than this of Euler's. (See Fermati Opera Mathematica, Tolosae, 1679, p. 163.) It has been adopted by Gauss in the Disquisitiones, Art. 49.

† Crelle's Journal, vol. iii. p. 390.

‡ Journal de l'École Polytechnique, Cahier xx. p. 289.

§ Reflexions sur la Théorie des Nombres, p. 32. But the principle of this demonstration is employed by Gauss in a memoir published in the Comm. Soc. Gotting. vol. xvi. p. 69, to which we shall have again to refer. (See Art. 19 of this Report.)

‖ Démonstration d'un Théoreme nouveau concernant les Nombres Premiers (Nouveaux Mémoires de l'Académie Royale de Berlin, 1771, p. 125). The 'new theorem' is that known as Sir. J. Wilson's.

From these equations we successively infer the congruences $A_1 \equiv 0$, $A_2 \equiv 0$, $A_3 \equiv 0$, ... $A_{p-2} \equiv 0$, and lastly, $A_{p-1} \equiv -1$, mod p. We have, therefore, the indeterminate congruence

$$(x+1)(x+2)(x+3)\dots(x+p-1) \equiv x^{p-1}-1, \text{ mod } p,$$

which is evidently *identical*, *i.e.* it subsists for all values of x. And since, if $a_1, a_2, \dots a_{p-1}$ be the terms of any system of residues prime to p, the factors $x-a_1$, $x-a_2$, $x-a_3$, ... $x-a_{p-1}$ are one by one congruous to the factors $x+1$, $x+2$, $x+3$, ... $x+p-1$ taken in a certain order, the products

$$(x-a_1)(x-a_2)\dots(x-a_{p-1}) \text{ and } (x+1)(x+2)\dots(x+p-1)$$

are also identically congruous for the modulus p, so that we may write

$$(x-a_1)(x-a_2)\dots(x-a_{p-1}) \equiv x^{p-1}-1, \text{ mod } p.$$

This congruence exhibits in the clearest manner possible what the real nature of the function $x^{p-1}-1$ is when considered with respect to the modulus p, and explains to us why it assumes a value divisible by p, when we assign to x any integral value not divisible by p.

It will be observed that the last of the $p-1$ congruences included in the congruence

$$(x-1)(x-2)(x-3)\dots(x-p-1) \equiv x^{p-1}-1, \text{ mod } p,$$

(which is a particular case of that last written), namely, the congruence

$$1.2.3\dots p-1 \equiv -1, \text{ mod } p,$$

is the symbolic expression of Sir J. Wilson's Theorem.

11. *Lagrange's Limit of the Number of Roots of a Congruence.*—The full development of the consequences of Fermat's Theorem requires the aid of the following proposition, which was first given, in a slightly different form, by Lagrange *.

'If $F(x)$ be a function of x of n dimensions, such that $F(a) \equiv 0$, mod p, then a function of x of $n-1$ dimensions, $F_1(x)$, can always be assigned such that we shall have the identical congruence $F(x) \equiv (x-a) F_1(x)$, mod p.'

Hence we may infer that no congruence, of which the modulus is prime, can have more incongruous roots than it has dimensions; and, if a congruence have congruous roots, we obtain a definition of their multiplicity; viz., if $F(x) \equiv (x-a)^r F_1(x)$, mod p, then we may say that $F(x) \equiv 0$, mod p, has

* Nouvelle Méthode pour resoudre les Problemes Indéterminés en Nombres entiers. (See Hist. Ac. Berl. 1768, p. 192.) The case of binomial congruences of the form $x^n \equiv 1$ had already been treated by Euler. (See Nov. Comment. Petropol. vol. xviii. p. 85, or Comment. Arith. vol. i. p. 516, Art. 28 of the Memoir.)

r roots congruous to a. We may also observe that this theorem enables us at once to infer Lagrange's indeterminate congruence from the first expression of Fermat's Theorem. For since $x^{p-1}-1$ is $\equiv 0$ for the values $x \equiv 1$, $x \equiv 2$, $\ldots x \equiv p-1$, we may, by successive applications of the preceding theorem, show that

$$x^{p-1}-1 \equiv (x-1)(x-2)\ldots(x-p+1), \text{ mod } p.$$

12. *Theory of the Residues of Powers.*—The principal elementary theorems relating to the Residues of Powers are the following. They are all due to Euler *, who was the first to demonstrate Fermat's Theorem, and to develope the numerous arithmetical truths connected with it.

I. If e and f be *conjugate* divisors of $p-1$ so that $p-1=ef$; the congruence $x^f \equiv 1$, mod p, always admits of f incongruous roots. Let these roots be denoted by $a_1, a_2, \ldots a_f$. Then each of the f congruences $x^e \equiv a_r$ admits of e solutions, and the ef roots of these f congruences exhaust completely the $p-1$ residues prime to p. It appears, therefore, that if we raise the residues of p to the power e, they will divide themselves into f groups of e numbers apiece; the e numbers of each group giving, when raised to the power e, the same residue for the modulus p. The numbers $a_1, a_2, \ldots a_f$ are termed the quadratic, cubic, biquadratic, quintic, &c. residues of p, according as $e=2$, $e=3$, $e=4$, $e=5$, &c., because they are each of them congruous to an eth power (and indeed to an eth power of e different numbers), and because no other number beside them can be congruous to such a power. Thus every uneven prime has $\frac{1}{2}(p-1)$ quadratic, and as many non-quadratic residues; every prime of the form $4n+1$ has $\frac{1}{4}(p-1)$ biquadratic residues, and three times as many non-biquadratic residues, &c.

II. It is readily seen that if the same number x satisfy the two congruences $x^{f_1} \equiv 1$, and $x^{f_2} \equiv 1$, it also satisfies the congruence $x^d \equiv 1$, mod p; where d is the greatest common divisor of f_1 and f_2. If therefore f be the *lowest* index for which the number x satisfies the congruence $x^f \equiv 1$, mod p, f is a divisor

---

* Euler's memoirs on this Theory are :—

(i.) Theorematum quorundam ad numeros primos spectantium demonstratio. Comment. Arith. vol. i. p. 21.

(ii.) Theoremata circa residua ex divisione potestatum relicta. Ibid. p. 260.

(iii.) Theoremata arithmetica novo methodo demonstrata. Ibid. p. 274.

(iv.) Disquisitio accuratior circa residua ex divisione quadratorum aliarumque potestatum per numeros primos relicta. Ibid. p. 487.

(v.) Demonstrationes circa residua ex divisione potestatum per numeros primos resultantia. Ibid. p. 516.

of $p-1$; as indeed appears directly from Euler's second demonstration of Fermat's Theorem. Let $\psi(f)$ denote the number of numbers less than f and prime to it; then there are always $\psi(f)$ roots of the congruence $x^f \equiv 1$, mod p, which cannot satisfy any other congruence of lower index and similar form. These are called *primitive roots* of the congruence $x^f \equiv 1$, mod p; they are also said to *appertain to the exponent f.* If $f=p-1$, the $\psi(p-1)$ primitive roots of the congruence $x^{p-1} \equiv 1$, mod p, are termed for brevity (though the designation is somewhat improper) the *primitive roots of p.* There are therefore $\psi(p-1)$ primitive roots of p.

13. *Primitive Roots.*—The problem of the direct determination of the primitive roots of a prime number is one of the 'cruces' of the Theory of Numbers. Euler, who first observed the peculiarity of these numbers, has yet left us no rigorous proof of their existence *; though, assuming their existence, he succeeded in accurately determining their number. The defect in his demonstration was first supplied by Gauss †, who has also proposed an indirect method for finding a primitive root. This method ‡ consists in taking any residue a of p, and determining (by the successive formation of its powers) the exponent f to which it appertains. If $f=p-1$, a is itself a primitive root of p; if not, let b be a second residue of p, not contained in the *period* of a, (*i.e.* not congruous for the modulus p to any one of the numbers a^0, $a, a^2, \ldots a^{f-1}$,) and let the exponent to which b appertains be determined. This exponent cannot (as is shown by Gauss) be identical with, nor yet a divisor of, the exponent to which a appertains; but it is always possible by a comparison of the values of a and b to determine a third number, c, which shall appertain to an exponent divisible by each of the exponents to which a and b appertain. By proceeding in this way we shall evidently obtain numbers appertaining to exponents continually higher, till at last we come to a number appertaining to the exponent $p-1$; *i.e.* to a primitive root of p.

M. Poinsot § proposes the following method. If $2, q_1, q_2, \ldots$ &c. be all the prime divisors of $p-1$, raise the numbers

$$\pm 1, \ \pm 2, \ \pm 3, \ \ldots \ \pm \tfrac{1}{2}(p-1),$$

which form a system of residues prime to p, to the powers of which the

---

* See the memoir (i) of the preceding note; and Gauss's criticism on it; Disq. Arith. Art. 56.

† Disq. Arith. Art. 52–55.

‡ Ibid. Art. 73–74.

§ Reflexions sur la Théorie des Nombres, cap. iv. art. 3.

indices are 2, q_1, q_2, &c.; so as to determine all the quadratic residues of p, and its residues of the powers q_1, q_2, &c. If from the system of residues 1, 2, 3, ... $p-1$, we successively exclude these residues of squares and higher powers, we shall have $\psi(p-1)$ numbers left, which cannot be congruous to any power having an index that divides $p-1$, and which are consequently (as may easily be shown) the primitive roots of p.

This method is very symmetrical; and if the problem proposed be to find *all* the primitive roots of p, it is sufficiently direct. But it is (like many other direct methods in the Theory of Numbers) of interminable prolixity; and becomes absolutely impracticable if p be a number even of moderate size, as it requires us to form the residues of the successive powers of the numbers 1, 2, 3, ... $\frac{1}{2}(p-1)$. Of course, in performing this operation, the multiples of p are to be rejected as fast as they arise; but, notwithstanding this abbreviation, and others which a little experience will readily suggest, Gauss's method is, for any practical purpose, greatly preferable.

In a memoir by M. Oltramare in Crelle's Journal (vol. xlix. p. 161), several considerations are offered for facilitating the determination of the primitive roots of primes in numerous special cases. Some, however, of the general results of this memoir are erroneous, at least in expression, and the demonstrations of the more particular conclusions contained in it involve no new principle, but may be obtained by combining the definition of primitive roots with the criteria by which (as we shall hereafter see) we are enabled to decide on the quadratic or cubic characters of the residues of given primes. The following may serve as examples of the very interesting results which are thus obtained by M. Oltramare:—

'If a be a prime number and $2a+1$ be also a prime, 2 or a is a primitive root of $2a+1$, according as a is of the form $4n+1$ or $4n+3$.' Thus 2 is a primitive root of 83, 11 is a primitive root of 23, 83 of 167, &c.

'If a be a prime number, other than 3, and if $p=2^m a+1$, where m is >1, be also a prime, 3 is a primitive root of p, unless the congruence

$$3^{2m-1}+1 \equiv 0, \text{ mod } p,$$

be satisfied.' Thus 3 is a primitive root of 89, and of 137.

Theorems of the same character will be found in the 'Théorie des Nombres*' of M. Desmarest. By their aid M. Desmarest has constructed a table giving a primitive root for every prime less than 10,000.

* Paris, 1852. See pp. 275–279.

14. *Indices.*—If γ be a primitive root of p, the least positive residues of the $p-1$ successive powers of γ, $\quad \gamma^1, \gamma^2, \gamma^3, \dots \gamma^{p-2}, \gamma^{p-1}$,
which we may denote by $\quad \gamma_1, \gamma_2, \gamma_3, \dots \gamma_{p-2}, 1$,
are all incongruous for the modulus p. These residues, therefore, irrespective of the order in which they occur, coincide with the numbers $1, 2, 3, \dots p-1$, *i.e.* they represent the terms of a complete system of residues prime to p. If $\gamma^\kappa \equiv a$, mod p, then κ, or any number congruous to κ for the modulus $p-1$, is termed the *index** of a for the primitive root or *base* γ; and this is expressed symbolically by writing

$$\kappa \equiv \operatorname{Ind} a, \bmod (p-1), \quad \text{or} \quad \kappa \equiv \operatorname{Ind}_\gamma a, \bmod (p-1).$$

The principal properties of these indices, which it is clear are a kind of arithmetical logarithm, are as follows:—

$$(1) \operatorname{Ind}(AB) \equiv \operatorname{Ind} A + \operatorname{Ind} B, \bmod (p-1).$$

$$(2) \operatorname{Ind}(A^\sigma) \equiv \sigma \operatorname{Ind} A, \bmod (p-1).$$

$$(3) \operatorname{Ind}\left(\frac{A}{B}, \bmod p\right) \equiv \operatorname{Ind} A - \operatorname{Ind} B, \bmod (p-1).$$

[The symbol $\left(\frac{A}{B}, \bmod p\right)$ is used to denote the value of x deduced from the congruence $Bx \equiv A \bmod p$.]

$$(4) \operatorname{Ind}_\gamma A \equiv \operatorname{Ind}_\gamma \gamma' . \operatorname{Ind}_{\gamma'} A, \bmod (p-1).$$

$$(5) \text{If } A \equiv B, \bmod p, \operatorname{Ind} A \equiv \operatorname{Ind} B, \bmod p-1.$$

In these congruences A and B represent numbers prime to p, σ any integral number, and γ and γ' two different primitive roots.

The great importance of these indices in arithmetical researches has induced the Academy of Berlin to publish a volume containing tables of the numbers corresponding to given indices, and of the indices corresponding to given numbers for all primes less than 1000. This volume, the 'Canon Arithmeticus†,' was edited by C. G. J. Jacobi, and contains, besides the Tables, a preface

---

* The reader must be careful to distinguish between the *index* of a number and the *exponent* to which the number appertains. The *exponent* does not depend on the choice of the primitive root: for a given number it has but one value, a, which is such that $\frac{p-1}{a}$ is the greatest common divisor of the index and of $p-1$. The index may have any one of $\psi(a)$ different values; which of these it has depends on the particular primitive root chosen.

† Berlin, 1839.

explaining the methods which he adopted in their construction. The annexed specimen will serve to exemplify the arrangement of the Tables :—

$$p=29,$$
$$p-1=2^2.7.$$

Numeri.

I.	0	1	2	3	4	5	6	7	8	9
		10	13	14	24	8	22	17	25	18
1	6	2	20	26	28	19	16	15	5	21
2	7	12	4	11	23	27	9	3	1	

Indices.

N.	0	1	2	3	4	5	6	7	8	9
		28	11	27	22	11	10	20	5	26
1	1	23	21	2	3	17	16	7	9	15
2	12	19	6	24	4	8	13	25	14	

M. Burckhardt, to whom arithmetic is indebted for an excellent Table of the divisors of numbers from 1 to 3,036,000*, has inserted in his work, and apparently only to fill up a blank-page at the end of the first million, a table stating the number of figures in the decimal period of the fraction $\frac{1}{p}$, for every prime number p less than 2500. It is evident that the number of terms in the decimal period of $\frac{1}{p}$ is nothing else than the exponent to which 10 appertains for the modulus p. M. Burckhardt's Table, therefore, at once apprises us that out of the 365 primes inferior to 2500 (2 and 5 are not counted in this enumeration, as being divisors of 10), 10 is a primitive root of 148; because there are 148 primes p below 2500, the reciprocals of which have decimal periods consisting of $p-1$ figures. Again, for 108 of the remaining primes below 2500, the exponent to which 10 appertains is $\frac{1}{2}(p-1)$. Of these 108 primes, 73 are of the form $4n+3$, from which it may be inferred that -10 is a primitive root of those 73 numbers. M. Burckhardt's Table supplies us, therefore, with a primitive root (and that root the most convenient for the purposes of computation) of $148+73=221$ out of the 365 primes inferior to 2500. Nor is this the limit to its usefulness; for when the exponent to which 10 appertains is as high as $\frac{1}{2}(p-1)$ or $\frac{1}{3}(p-1)$ or $\frac{1}{4}(p-1)$, it is possible by methods which Jacobi has indicated to construct the Table of Indices with very little labour.

Jacobi says that had it not been for this table of Burckhardt's he should hardly have ventured on the construction of the 'Canon Arithmeticus,' on

* Paris, 1814-1817. A Table containing the exponents to which 10 appertains, for every prime less than 10,000, has since been given by M. Desmarest. (See p. 308 of his 'Théorie des Nombres.')

account of the prolixity and uncertainty of the tentative methods for the investigation of primitive roots. But, while endeavouring to avail himself of the results of M. Burckhardt's Table, for the computation of his own Tables of Indices, in other cases besides those in which that Table immediately furnishes a primitive root, he was led to the invention of a general method of procedure, which, as he says, would have enabled him to dispense with the assistance of Burckhardt's Table altogether, or to extend his Canon to any higher limit which the expense of printing would have admitted. This method is not in principle very different from Gauss's process for finding primitive roots, but the form which Jacobi has given to it possesses great advantages, for the purpose to which he has applied it. He first of all takes a number a (not quite at hap-hazard, for quadratic residues can at any rate be excluded by the law of reciprocity; see *inf.* Art. 16); and determines its *period* of residues, and the exponent α to which it appertains. Let $\alpha\alpha' = p-1$, and let the residues of $a, a^2, a^3, \ldots a^\alpha$ be entered in a Table of which the arguments are the indices $1, 2, 3, \ldots p-1$, opposite to the indices, $\alpha', 2\alpha', 3\alpha', \ldots \alpha\alpha'$, respectively. It has been shown by Gauss that there are always $\frac{\psi(p-1)}{\psi(\alpha)}$ primitive roots for which this assignment is true. A number b is then taken, not contained in the period of a, and the residues of its successive powers are formed till we come to the lowest power of it that is congruous to any power of a; so that $b^B \equiv a^A$, mod p. Let β be the exponent to which b appertains, θ the greatest common divisor of α and β, and $\lambda = \frac{\alpha\beta}{\theta}$ their least common multiple; let also $\beta\beta' = p-1$. It may be proved that $B = \frac{\beta}{\theta}$; $A = \frac{k\alpha}{\theta}$; where k is some number less than θ and prime to it, so that $\frac{\alpha}{\theta}$ is the greatest common divisor of A and α. These relations show, that when we know the numbers α, A, and B, we can immediately find θ, k, and β, without having to raise b to any power higher than b^B. We may then assign to b any index of the form $l\beta'$, where l is prime to β, and congruous to k for the modulus θ. The number of such values of l (incongruous for the modulus β) is $\frac{\psi(\beta)}{\psi(\theta)}$; and, whichever of them we take, there will be $\frac{\psi(p-1)}{\psi(\lambda)}$ primitive roots, for which b will have the index $l\beta'$, while a retains the index α'. We must next form the residues of the $\lambda - \alpha$ products included in the formula $a^x b^y$; where x has any value from 1 to α inclusive, and y any value from 1 to $B-1$. These residues are all incongruous; the indices of all of them are known;

and, together with the α powers of a already entered in the table, they exhaust all the numbers which have indices divisible by $\frac{p-1}{\lambda}$.

In practice, it will almost always happen that λ is equal to $p-1$. When this is so, nothing remains to complete the operation but to enter in the Table the residues of the numbers $a^x b^y$ opposite to the indices corresponding to them. But, if $\lambda < p-1$, we may take that residue which has $\frac{p-1}{\lambda}$ for its index, and use it to replace a in the preceding operation, while b is replaced by some other residue not yet entered in the Table. In this way we shall ultimately (and in practice very speedily) obtain a complete Table of Residues corresponding to given indices, which, of course, immediately supplies us with the inverse Table of Indices corresponding to given residues. It will be seen (as has been already observed) that the process is not dissimilar to Gauss's method for determining a number appertaining to the exponent λ when we already know two numbers a and b appertaining to the exponents α and β respectively. But it is so arranged by Jacobi that hardly a single figure is wasted, the primitive root, instead of being found by a preliminary investigation, presenting itself at the end of the operation, and being recognized by its standing opposite to the index 1.

To calculate with rapidity the residues of the powers of a number, Jacobi employs a method proposed by M. Crelle in his Journal, vol. ix. p. 30, and which is most easily explained by an example.

Let $p=11$, and let it be required to determine the residues of the powers of 3; and the residues of those powers multiplied by 7.

Column	I.	1, 2, 3, 4, 5, 6, 7, 8, 9, 10;
,,	II.	3, 6, 9, 1, 4, 7, 10, 2, 5, 8;
,,	III.	3, 9, 5, 4, 1;
,,	IV.	10, 8, 2, 6, 7.

The first column contains the numbers 1, 2, 3, ... $p-1$. The second column begins with 3 (the number the powers of which we are considering), and consists of numbers formed by successive additions of 3, multiples of 11 being rejected as fast as they arise. The third column also commences with 3, and is so formed that any number r in it is followed by the number which in column II. stands under r in column I. This column contains the residues of the powers of 3 taken in order, and stops at 3^5 because after that the same residues recur.

Lastly, column IV. begins with 10 (the number which in column II. stands under 7 in column I.), and is formed in the same way as column III. It represents the residues of 7.3, 7.3^2, &c. ....

15. *Quadratic Residues.*—It appears from the theorems cited in Art. 12, that the numbers 1, 2, 3, ... $p-1$ divide themselves into two classes of Quadratic Residues and Quadratic non-Residues, comprising $\frac{1}{2}(p-1)$ numbers each. Every quadratic residue a satisfies the congruence $x^{\frac{1}{2}(p-1)} \equiv 1$, mod p; every quadratic non-residue b satisfies, instead, the congruence $x^{\frac{1}{2}(p-1)} \equiv -1$, mod p. Again, for every quadratic residue the congruence $x^2 \equiv a$, mod p, is resoluble; for every non-quadratic residue the congruence $x^2 \equiv b$, mod p, is irresoluble. The solution of almost every problem relating to the indeterminate analysis of quadratic functions involves a congruence of the simple form $x^2 \equiv A$, mod p. It is therefore of great importance to obtain a criterion which shall enable us to determine *a priori* whether a given number is or is not a quadratic residue of a given prime. If we have a Table of Indices for the given prime, we have only to see whether the index of the given number is even or uneven; if even, it is a quadratic residue; if uneven, it is a quadratic non-residue. Or, again, we may raise the given number a (by M. Crelle's method, or any other) to the power $\frac{1}{2}(p-1)$, and see whether the residue is $+1$ or -1. It is usual to denote the positive or negative unit which is the remainder of $a^{\frac{1}{2}(p-1)}$, mod p, by the symbol $\left(\frac{a}{p}\right)$, which is known as 'Legendre's Symbol'; so that in every case

$$a^{\frac{1}{2}(p-1)} \equiv \left(\frac{a}{p}\right), \text{ mod } p, \text{ and } \left(\frac{a}{p}\right) = +1 \text{ or } = -1,$$

according as a is or is not a quadratic residue of p. It will be seen that we also have in every case the equation

$$\left(\frac{a_1}{p}\right)\left(\frac{a_2}{p}\right) = \left(\frac{a_1 a_2}{p}\right).$$

If a instead of being prime to p be divisible by p, it is convenient to attribute to $\left(\frac{a}{p}\right)$ the value zero.

16. *Legendre's Law of Reciprocity.*—The two methods alluded to for the discrimination of quadratic and non-quadratic residues, or, which is the same thing, for the determination of the value of the symbol $\left(\frac{a}{p}\right)$, are not satisfactory, —the first because it supposes a reference to a Table of Indices (*i.e.* to a

recorded solution of the problem it is proposed to solve), the second on account of its inapplicability to high numbers. A very different solution of the problem is supplied by a theorem which is known as 'Legendre's Law of Quadratic Reciprocity,' and which is, without question, the most important general truth in the science of integral numbers which has been discovered since the time of Fermat. It has been called by Gauss* 'the gem of the higher arithmetic,' and is equally remarkable whether we consider the simplicity of its enunciation, the difficulties which for a long time attended its demonstration, or the number and variety of the results which have been obtained by its means. The theorem is as follows:—

'If p and q be two *uneven* prime numbers,

$$\left(\frac{p}{q}\right) = (-1)^{\frac{1}{4}(p-1)(q-1)} \left(\frac{q}{p}\right), \text{'} \quad \text{(i);}$$

to which we must add the complementary propositions relating to the residues -1 and 2,

$$\left(\frac{-1}{p}\right) = (-1)^{\frac{1}{2}(p-1)}, \quad \text{(ii)}; \qquad \left(\frac{2}{p}\right) = (-1)^{\frac{1}{8}(p^2-1)}, \quad \text{(iii)}.$$

In (ii), p is supposed to be positive; in (i), p and q are supposed not to be simultaneously negative.

The equation

$$\left(\frac{p}{q}\right)\left(\frac{q}{p}\right) = (-1)^{\frac{1}{4}(p-1)(q-1)}$$

may be expressed in words by saying that 'if p and q be two primes, the quadratic character of p in regard to q is the same as the quadratic character of q in regard to p; except both p and q be of the form $4n+3$, in which case the two characters are opposite instead of identical.'

Gauss, who attributes the first enunciation of this theorem to Legendre, while he justly claims the first demonstration of it for himself†, appears to have considered that Euler was unacquainted with the theorem, at least in its simple

---

* {Jacobi, Crelle, vol. xix. p. 314.}

† 'Pro primo hujus elegantissimi Theorematis inventore ill. Legendre absque dubio habendus est, postquam longe antea summi geometræ Euler et Lagrange plures ejus casus speciales jam per inductionem detexerant. . . . . In ipsum theorema proprio marte incideram anno 1795, dum omnium, quæ in arithmetica sublimiori jam elaborata fuerant, penitus ignarus, et a subsidiis literariis omnino præclusus essem. Sed per integrum annum me torsit, operamque enixissimam effugit, etc.'—Comm. Soc. Gött. vol. xvi. p. 69.

form. (See Disq. Arith., Art. 151.) Nevertheless, we find in the 'Opuscula Analytica' of Euler, vol. i. p. 64, a memoir* the concluding paragraph of which contains a general and very elegant theorem, from which the Law of Reciprocity is immediately deducible, and which is, *vice versâ*, deducible from that law. But Euler (*loc. cit.*) expressly observes that the theorem is undemonstrated; and this would seem to be the only place in which he mentions it in connexion with the theory of the Residues of Powers; though in other researches he has frequently developed results which are consequences of the theorem, and which relate to the linear forms of the divisors of quadratic formulae. But here also his conclusions repose on induction only; though in one memoir he seems to have imagined (for his language is not very precise) that he had obtained a satisfactory demonstration. The theorem, in a form precisely equivalent to that in which we have cited it, was first given by Legendre, in a Memoir contained in the 'Histoire de l'Académie des Sciences' for 1785. (See pp. 516, 517.) But the demonstration with which he has accompanied it is invalid for several reasons. (See Gauss, Disq. Arith., Arts. 151, 296, 297, and the Additamenta.)

[*Addition*†. Legendre's investigation of the law of reciprocity (as presented in the 'Théorie des Nombres,' vol. i. p. 230, or in the 'Essai,' ed. 2, p. 198) is invalid only because it assumes, without a satisfactory proof, that if a be a given prime of the form $4n+1$, a prime b of the form $4n+3$ can always be assigned, satisfying the equation $\left(\frac{a}{b}\right) = -1$. M. Kummer (in the Memoirs of the Academy of Berlin for 1859, pp. 19, 20) says that this postulate is easily deducible from the theorem demonstrated by Dirichlet, that every arithmetical progression, the terms of which have no common divisor, contains prime numbers. It would follow from this, that the demonstration of Legendre (which depends on a very elegant criterion for the resolubility or irresolubility of equations of the form $ax^2+by^2+cz^2=0$) must be regarded as rigorously exact (see, however, the 'Additamenta' to arts. 151, 296, 297 of the Disq. Arith.). In the introduction to the memoir to which we have just referred, the reader will find some valuable observations by M. Kummer on the principal investigations relating to laws of reciprocity.]

---

* Observationes circa divisionem quadratorum per numeros primos (Comment. Arith. vol. i. p. 477).

† The additions to Arts. 16, 20, 22, 24, 25, 36, 37, and 38 were published at the end of Part II. of the Report (1860).

17. *Jacobi's extension of Legendre's Symbol.*—The symbol $\left(\frac{q}{p}\right)$, the introduction of which has greatly contributed to simplify the theories of the higher arithmetic, does not appear in Legendre's Memoir of 1785. It first occurs in the 'Essai sur la Théorie des Nombres'; the first edition of which appeared at Paris in 1798, and the second in 1808.

Jacobi, in a note communicated to the Academy of Berlin in 1837*, has extended the notation of Legendre. If $P = p_1 p_2 p_3, \ldots$ where p_1, p_2, p_3 denote (equal or unequal) uneven prime numbers, Jacobi defines the symbol $\left(\frac{k}{P}\right)$ by the equation

$$\left(\frac{k}{P}\right) = \left(\frac{k}{p_1}\right)\left(\frac{k}{p_2}\right)\left(\frac{k}{p_3}\right)\ldots,$$

and observes that we then have the equations

$$\left(\frac{P}{Q}\right) = (-1)^{\frac{1}{4}(P-1)(Q-1)}\left(\frac{Q}{P}\right), \quad \text{(i)};$$

$$\left(\frac{-1}{P}\right) = (-1)^{\frac{1}{2}(P-1)}, \quad \text{(ii)}; \qquad \left(\frac{2}{P}\right) = (-1)^{\frac{1}{8}(P^2-1)}, \quad \text{(iii)};$$

P and Q denoting any two uneven numbers relatively prime, the signs of which are subject to the same restrictions as the signs of p and q in the corresponding formula of Art. 16. The theorems expressed by these formulae of Jacobi are very easily deducible from the formulae of Legendre, and will be found in the Disq. Arith. (Art. 133). To prevent misconception, however, it is proper to observe that, while Legendre's equation $\left(\frac{k}{p}\right) = 1$ is a necessary and sufficient condition for the resolubility of the congruence $x^2 \equiv k$, mod p, Jacobi's equation $\left(\frac{k}{P}\right) = 1$, where P is not a prime number, though a necessary, is not a sufficient condition for the resolubility of the corresponding congruence $x^2 \equiv k$, mod P. That congruence requires for its resolubility that the conditions

$$\left(\frac{k}{p_1}\right) = 1, \quad \left(\frac{k}{p_2}\right) = 1 \ldots$$

should separately be satisfied; $p_1, p_2, \ldots$ denoting the unequal prime factors of P.

Gauss (who had in the course of his own early researches arrived inde-

* Ueber die Kreistheilung und ihre Anwendung auf die Zahlentheorie. See the Monats-Bericht of the Berlin Academy, vol. ii. p. 127 (Oct. 16, 1857), or Crelle's Journal, vol. xxx. p. 166.

pendently at the Law of Quadratic Reciprocity), before finally abandoning the theory, succeeded in obtaining no fewer than six demonstrations of this fundamental proposition. The first two are contained in the Disq. Arith. (Arts. 125–145, and Art. 262); the third and fourth in two memoirs presented in 1808 to the Society of Göttingen (Comm. Soc. Gött. vol. xvi. p. 69, Jan. 15, and Comm. Recentiores, vol. i. Aug. 24), of which the latter bears the title 'Summatio serierum quarundam singularium.' The fifth and sixth appeared nine years later in the memoir entitled 'Theorematis Fundamentalis in doctrina de Residuis quadraticis demonstrationes et ampliationes novae' (Comm. Recent. vol. iv. p. 3, Feb. 10, 1817). The fourth of these demonstrations is probably that which is promised in the Disq. Arith., Art. 151, but which does not appear in that work, because (as it would seem) Gauss had not yet succeeded in overcoming the difficulties connected with it.

Independently of the fundamental importance of Legendre's Law of Reciprocity, these demonstrations of Gauss possess such intrinsic interest, and have contributed so much to the progress of the science, that we shall briefly review them here.

18. *Gauss's First Demonstration.*—The first demonstration (Disq. Arith., Arts. 125–145), which is presented by Gauss in a form very repulsive to any but the most laborious students, has been resumed by Lejeune Dirichlet in a memoir in Crelle's Journal (vol. xlvii. p. 139), and has been developed by him with that luminous perspicuity by which his mathematical writings are distinguished.

Let λ represent any uneven prime. The single observation that

$$\left(\frac{3}{5}\right) = -1 = \left(\frac{5}{3}\right)$$

shows that the theorem of reciprocity is true for primes inferior to 7. To establish its universal truth, it is, consequently, sufficient to show that, if true for all primes up to λ exclusively, it is also true for all primes up to λ inclusively. Let the theorem therefore be assumed to be true for all primes inferior to λ; let p be any one of those primes; and let the eight cases $[2 \times 2 \times 2 = 8]$ be considered separately, which arise from every possible combination of the hypotheses

$$(\alpha),\ \left(\frac{p}{\lambda}\right) = +1, \text{ or } = -1;\ (\beta),\ \lambda \equiv 1, \text{ or } \equiv 3, \text{ mod } 4;\ (\gamma),\ p \equiv 1, \text{ or } \equiv 3, \text{ mod } 4.$$

It has to be shown that, in each of these eight cases, the symbol $\left(\frac{\lambda}{p}\right)$ actually has the value which the Law of Reciprocity assigns to it. The nature of the proof in the four cases in which $\left(\frac{p}{\lambda}\right) = +1$ will be rendered intelligible by a single example.

Let $\left(\frac{p}{\lambda}\right) = 1$ and let $\lambda \equiv p \equiv 1$, mod 4. By virtue of the symbolic equation $\left(\frac{p}{\lambda}\right) = 1$, we can establish the congruence $x^2 \equiv p$, mod λ, or (which is the same thing) the equation $x^2 = p + \lambda y$; in which we may suppose x *even* and less than λ, y positive, less than λ and of the form $4n+3$. From this equation it appears that $\left(\frac{\lambda y}{p}\right) = 1$, and $\left(\frac{p}{y}\right) = 1$, the symbol $\left(\frac{p}{y}\right)$ being here used with the meaning Jacobi has assigned to it. But every prime divisor of y is less than λ; and, therefore, by Jacobi's formula of reciprocity (which is valid for all uneven numbers less than λ, since by hypothesis Legendre's law is valid for all primes less than λ), $\left(\frac{y}{p}\right) = \left(\frac{p}{y}\right) = 1$. But $\left(\frac{\lambda y}{p}\right) = 1 = \left(\frac{\lambda}{p}\right)\left(\frac{y}{p}\right)$; so that, finally, $\left(\frac{\lambda}{p}\right) = 1$ in conformity with Legendre's law. We have here assumed that x is prime to p; a slight modification in the proof will adapt it to the contrary supposition.

Again, the two cases in which $\left(\frac{p}{\lambda}\right) = -1$, and $\lambda \equiv 3$, mod 4, admit of similar treatment. For the equation $\left(\frac{p}{\lambda}\right) = -1$ involves also the equation $\left(\frac{-p}{\lambda}\right) = +1$, because $\lambda \equiv 3$, mod 4. We have therefore the congruence $x^2 \equiv -p$, mod λ, which will serve to replace the congruence $x^2 \equiv p$, mod λ, which presents itself in the four cases first mentioned.

But the two remaining cases, in which $\left(\frac{p}{\lambda}\right) = -1$, $\lambda \equiv 1$, mod 4, require a different mode of treatment. By a singularly profound analysis, Gauss has succeeded in showing that every prime of the form $4n+1$ is a non-quadratic residue of some prime less than itself. Assume, therefore, the existence of a prime ϖ, less than λ, and satisfying the condition $\left(\frac{\lambda}{\varpi}\right) = -1$. This condition implies that $\left(\frac{\varpi}{\lambda}\right) = -1$; for if $\left(\frac{\varpi}{\lambda}\right)$ were equal to $+1$, we should have $\left(\frac{\lambda}{\varpi}\right) = +1$, by one of the first four cases. Hence we may infer that $\left(\frac{\varpi p}{\lambda}\right) = +1$, and may establish the congruence $x^2 \equiv \varpi p$, mod λ, which, treated as in the preceding cases, will lead us to the conclusion that $\left(\frac{\lambda}{p}\right)\left(\frac{\lambda}{\varpi}\right) = 1$, *i.e.* that $\left(\frac{\lambda}{p}\right) = -1$.

19. *Gauss's Second, Third, and Fifth Demonstrations.*—The second demonstration (Disq. Arith. 262) depends on the theory of quadratic forms, and will be referred to in its proper place in this Report [see Art. 115].

The third and fifth (which are in principle very similar to one another) depend on much simpler considerations.

A *half-system of Residues for a prime modulus p* is a system of $\frac{1}{2}(p-1)$ numbers $r_1, r_2, \dots r_{\frac{1}{2}(p-1)}$, such that the $p-1$ numbers $\pm r_1, \pm r_2, \dots \pm r_{\frac{1}{2}(p-1)}$ constitute a system of residues prime to p. We might take for the numbers r_1, r_2, &c., the even numbers less than p (as Eisenstein has done: see Crelle's Journal, vol. xxviii. p. 246), but Gauss has preferred to take the numbers $1, 2, 3, \dots \frac{1}{2}(p-1)$.

Let q be any number prime to p, and let k be the number of the numbers, $qr_1, qr_2, qr_3, \dots qr_{\frac{1}{2}(p-1)}$, which are congruous, not to numbers in the series $r_1, r_2, \dots r_{\frac{1}{2}(p-1)}$, but to numbers in the series $-r_1, -r_2, \dots -r_{\frac{1}{2}(p-1)}$. It may be shown (by a method similar to that employed in Dirichlet's proof of Fermat's Theorem) that $q^{\frac{1}{2}(p-1)} \equiv (-1)^k$, mod p; so that $\left(\frac{q}{p}\right) = (-1)^k$.* Hence if q be a prime as well as p, and k' denote the number which replaces k, when p and q are interchanged in the preceding considerations, we find that

$$\left(\frac{p}{q}\right)\left(\frac{q}{p}\right) = (-1)^{k+k'}.$$

It has, therefore, to be shown that $k+k' \equiv \frac{1}{4}(p-1)(q-1)$, mod 2. The way in which this is done is different in each of the two demonstrations, and is a little complicated in each of them; but by the aid of a diagram the congruence may be demonstrated intuitively (compare Eisenstein: Crelle, xxviii. p. 246 {translated by Cayley in the Quart. Jour. of Math. vol. i. p. 186}). With a pair of axes Ox and Oy construct a system of unit-points in a plane: only let no such points be constructed on the axes themselves. If S be any geometrical figure, let (S) stand for the number of unit-points contained inside it or on its contour. On Ox and Oy respectively take $OA = \frac{1}{2}q$, $OB = \frac{1}{2}p$. Complete the parallelogram $OACB$, and draw its diagonals, OQC, AQB. It is then easily seen that

---

* {Mr. Morgan Jenkins in a paper read to the London Mathematical Society [vol. ii. p. 29, 1867] shows that $\left(\frac{Q}{P}\right) = (-1)^k$, Q and P uneven, and $\left(\frac{Q}{P}\right)$ being Jacobi's symbol.}

$$\begin{aligned} k &= (QCA) \quad - \quad (QBO), \\ k' &= (QBC) \quad - \quad (QOA), \\ k+k' &= (ABC) \quad - \quad (AOB), \\ &= (OABC) - 2(AOB), \\ &\equiv (OABC), \text{ mod } 2. \end{aligned}$$

But $(OABC) = \frac{1}{4}(p-1)(q-1)$; therefore, finally,

$$k+k' \equiv \tfrac{1}{4}(p-1)(q-1), \text{ mod } 2.$$

These demonstrations (the 1st, 3rd, and 5th) introduce no heterogeneous elements into the inquiry (the geometrical method of the present article is to be regarded only as an abbreviation of an equivalent and purely arithmetical process); they are based on the principles of the two theories with which the Law of Reciprocity is most intimately connected,—those of the residues of powers, and of quadratic congruences. The third, in particular, appears to have commended itself above the rest to Gauss's judgment*.

20. *Gauss's Fourth Demonstration.*—The fourth and sixth demonstrations, though somewhat different from one another, are both intimately connected with the theory of the division of the circle. They must, therefore, be regarded as less direct than the earlier proofs, but they have contributed even more to the methods and resources of the higher arithmetic.

The fourth depends on the formula

$$1 + r + r^4 + r^9 + \ldots + r^{(n-1)^2} = i^{\frac{1}{4}(n-1)^2}\sqrt{n} \ldots, \qquad \text{(A)}$$

in which i represents (as throughout this Report) an imaginary square root of -1; n is any uneven number, $\sqrt{n}$ its positive square root,

$$r = \cos\frac{2\pi}{n} + i \sin\frac{2\pi}{n}.$$

Let the series

$$1 + r^k + r^{4k} + r^{9k} + \ldots + r^{(n-1)^2 k} \text{ be denoted by } \psi(k, n);$$

in the particular case in which n is a prime number, it is easy to see that $\psi(k, n) = \left(\frac{k}{n}\right)\psi(1, n)$. Further, p and q denoting two prime numbers, it is

* 'Sed omnes hæ demonstrationes,' (he is speaking, apparently, of the 1st, 2nd, 4th, and 6th,) 'etiamsi respectu rigoris nihil desiderandum relinquere videantur, e principiis nimis heterogeneis derivatæ sunt; primâ forsan exceptâ, quæ tamen per ratiocinia magis laboriosa procedit, operationibusque prolixioribus premitur. Demonstrationem itaque *genuinam* hactenus haud affuisse non dubito pronunciare; esto jam penes peritos judicium, an ea, quam nuper detegere successit,' (the 3rd,) 'hoc nomine decorari mereatur.'—Comm. Soc. Gott. vol. xvi. p. 70.

found by actual multiplication of the two series $\psi(p, q)$ and $\psi(q, p)$ that

$$\psi(p, q) \times \psi(q, p) = \psi(1, pq); \text{ that is } \left(\frac{p}{q}\right)\left(\frac{q}{p}\right) = \frac{\psi(1, pq)}{\psi(1, p)\,\psi(1, q)}.$$

If we substitute for the functions ψ their values given by the equation (A), we find

$$\left(\frac{p}{q}\right)\left(\frac{q}{p}\right) = i^{\frac{1}{4}(pq-1)^2 - \frac{1}{4}(p-1)^2 - \frac{1}{4}(q-1)^2},$$

an equation which gives a relation between $\left(\frac{p}{q}\right)$ and $\left(\frac{q}{p}\right)$ coincident with that assigned in Legendre's Law of Reciprocity.

The equation (A) is not easy to demonstrate. It is not indeed difficult to show that the sum of the series on the left-hand side is $\pm\sqrt{n}$ when $n \equiv 1$, mod 4; and $\pm i\sqrt{n}$ when $n \equiv 3$, mod 4. But the determination of the ambiguous sign in these values appears to have long occupied Gauss. He has effected it in his memoir (the 'Summatio Serierum &c.') by establishing the equality

$$1 + r + r^4 + r^9 + \ldots + r^{(n-1)^2} = (r - r^{-1})(r^3 - r^{-3}) \ldots (r^{n-2} - r^{-n+2}) \ldots, \qquad \text{(B)}$$

which he obtains by writing r for x, and $n-1$ for m, in the series

$$1 - \frac{1 - x^m}{1 - x} + \frac{(1 - x^m)(1 - x^{m-1})}{(1 - x)(1 - x^2)} - \frac{(1 - x^m)(1 - x^{m-1})(1 - x^{m-2})}{(1 - x)(1 - x^2)(1 - x^3)} + \ldots.$$

This series when m is a positive integer becomes an integral algebraical function, and is proved by Gauss to be zero if m be uneven; and if m be even, to be equal to the product $(1 - x)(1 - x^2) \ldots (1 - x^{m-1})$. From this last observation, the demonstration of the formula (B) naturally flows. If n be an even number, the formula (A) becomes

$$1 + r + r^4 + r^9 + \ldots + r^{(n-1)^2} = (1 + i)\sqrt{n} \text{ or } = 0, \qquad \text{(A')}$$

according as n is evenly or unevenly even.

A very different, but a simpler demonstration of these formulae (A) and (A′), depending on the properties of the definite integrals

$$\int_{-\infty}^{+\infty} \cos x^2\, dx, \quad \int_{-\infty}^{+\infty} \sin x^2\, dx, \quad \text{or} \quad \int_{-\infty}^{+\infty} e^{ix^2}\, dx,$$

has been given by Dirichlet in his memoir, 'Application de l'Analyse Infinitésimale à la Théorie des Nombres' (Crelle, vol. xxi. p. 135).

The same formulae have also been deduced by Cauchy from the equation

$$a^{\frac{1}{2}}\left(\tfrac{1}{2} + e^{-a^2} + e^{-4a^2} + e^{-9a^2} + \ldots\right) = b^{\frac{1}{2}}\left(\tfrac{1}{2} + e^{-b^2} + e^{-4b^2} + e^{-9b^2} + \ldots\right),$$

or

$$\tfrac{1}{2} + e^{-a^2} + e^{-4a^2} + e^{-9a^2} + \ldots = \frac{\sqrt{\pi}}{a}\left(\tfrac{1}{2} + e^{-b^2} + e^{-4b^2} + e^{-9b^2} + \ldots\right),$$

in which $ab = \pi$, {a^2 and b^2 denoting real positive quantities, or imaginary quantities the real parts of which are positive; the real parts of a, b have the same sign as $a^{\frac{1}{2}} b^{\frac{1}{2}} = \pm\sqrt{\pi}$}*. This equation Cauchy obtained, as early as 1817, by the principles of his theory of reciprocal functions; but it is also deducible from known elliptic formulae. (See a note by M. Lebesgue in Liouville's Journal, vol. v. p. 186. {See also Crelle, xvii. p. 57, and the Berlin Transactions for 1835; the former Memoir contains the criticism of M. Libri's proof.}) If in it we write

$$a^2 - \frac{2i\pi}{n} \text{ for } a^2, \quad \text{and} \quad \beta^2 + \frac{ni\pi}{2} \text{ for } b^2,$$

α and β being two evanescent quantities connected by the relation $n\alpha = 2\beta$, the two series

$$n\alpha\left(\tfrac{1}{2} + e^{-\alpha^2} + e^{-4\alpha^2} + e^{-9\alpha^2} + \dots\right)$$

and

$$2\beta\left(\tfrac{1}{2} + e^{-b^2} + e^{-4b^2} + e^{-9b^2} + \dots\right)$$

become respectively

$$\psi(1, n) \times \int_0^\infty e^{-x^2}\,dx, \quad \text{and} \quad \left(1 + e^{-\frac{1}{2}ni\pi}\right) \times \int_0^\infty e^{-x^2}\,dx;$$

whence, dividing by the definite integral, and observing that

$$a = \sqrt{\frac{2\pi}{n}}\, e^{-\frac{1}{4}i\pi},$$

we obtain finally, in accordance with the formulae of Gauss,

$$\psi(1, n) = \tfrac{1}{2}\sqrt{n}\,(1+i)\left(1 + e^{-\frac{1}{2}ni\pi}\right) = \sqrt{n}\,\frac{1 + e^{-\frac{1}{2}ni\pi}}{1 + e^{-\frac{1}{2}i\pi}}\ \dagger.$$

For the case in which n is a prime number, the equality (B) has been

---

* {Put $a^2 = -i\pi\omega$; we get the formula in Lacroix, vol. ii. p. 408; and the condition that a is positive is the same as that the real part of $\sqrt{-i\omega}$ is positive. See Liouville (II.) vol. iii. p. 30 for a general formula.}

† See M. Cauchy's 'Mémoire sur la Théorie des Nombres' in the Mémoires de l'Académie de France, vol. xvii, notes ix, x, and xi. See also the Comptes Rendus for April 1840, or Liouville's Journal, vol. v. p. 154; and compare (besides the note of M. Lebesgue quoted in the text) a memoir by the same author in Liouville, vol. v. p. 42.

{Writing $a^2 = \alpha^2 - \frac{2mi\pi}{n}$, $b^2 = \beta^2 + \frac{ni\pi}{2m}$, I find $\psi(m, n) = \frac{1}{4}\sqrt{\frac{n}{m}}\,(1+i)\,\psi(-n, 4m)$. If $m = 1$, this is right. If $n = 4\nu$, since $\psi(-4\nu, 4m) = 4\psi(-\nu, m)$, we have

$$\psi(m, 4\nu) = 2\sqrt{\frac{\nu}{m}}\,(1+i)\,\psi(-\nu, m),$$

making the case when n is even depend on the case when n is uneven, and agreeing with Art. 104. note.}

established in a very simple manner by M. Cauchy* and M. Kronecker†. But, as these latter methods have not been extended to the case in which n is a composite number, they cannot be used to replace Gauss's analysis in this demonstration of the law of reciprocity.

From the formula (A) combined with the equation $\psi(k,p)=\left(\frac{k}{p}\right)\psi(1,p)$, p denoting a prime number, we may infer

$$\left(\frac{k}{p}\right)\sqrt{p}=\sum_{s=0}^{s=p-1}\cos s^2\frac{2k\pi}{p}\,; \qquad \sum_{s=0}^{s=p-1}\sin s^2\frac{2k\pi}{p}=0\,;$$

or

$$\left(\frac{k}{p}\right)\sqrt{p}=\sum_{s=0}^{s=p-1}\sin s^2\frac{2k\pi}{p}\,; \qquad \sum_{s=0}^{s=p-1}\cos s^2\frac{2k\pi}{p}=0\,;$$

according as $p\equiv 1$, or $\equiv 3$, mod 4.

These formulae serve to express the value of the symbol $\left(\frac{k}{p}\right)$ by means of a finite trigonometrical series, and are, therefore, of very great importance. Conversely, the circumstance that a trigonometrical summation should depend on the quadratic characters of integral numbers, may serve of itself to show the use of abstract arithmetical speculations in other parts of analysis.

[*Addition.* Dirichlet's demonstration of the formulae (A) and (A′) first appeared in Crelle's Journal, vol. xvii. p. 57. Some observations in this paper on a supposed proof of the same formulae by M. Libri (Crelle, vol. ix. p. 187) were inserted by M. Liouville in his Journal, vol. iii. p. 3, and gave rise to a controversy (in the Comptes Rendus, vol. x) between MM. Liouville and Libri. The concluding paragraphs of Dirichlet's paper contain the application of the formulae (A) and (A′) to the law of reciprocity (Gauss's fourth demonstration).]

21. *Gauss's Sixth Demonstration.*—This demonstration depends on an investigation of certain properties of the algebraical function

$$\xi_k=\sum_{s=0}^{s=p-2}(-1)^s x^k\gamma^s,$$

in which p is a prime number, γ a primitive root of p, k any number prime to p, and x an absolutely indeterminate symbol. These properties are as follows :—

(1) $\xi_k^2-(-1)^{\frac{1}{2}(p-1)}p$ is divisible by $\dfrac{1-x^p}{1-x}$,

(2) $\xi_k-\left(\frac{k}{p}\right)\xi_1$ is divisible by $1-x^p$,

---

* In the Mémoire sur la Théorie des Nombres, Note xi, or Liouville, vol. v. p. 161.

† Liouville, New Series, vol. i. p. 392.

(3) If $k=q$ be a prime number,

$\xi_1{}^q-\xi_q$ is divisible by q.

From (1) we may infer that $\xi_1{}^{q-1}-(-1)^{\frac{1}{4}(p-1)(q-1)}p^{\frac{1}{2}(q-1)}$ is divisible by $\dfrac{1-x^p}{1-x}$; and, by combining this inference with (1) and (2), we may conclude that

$$\xi_1(\xi_1{}^q-\xi_q)-(-1)^{\frac{1}{2}(p-1)}p\left[(-1)^{\frac{1}{4}(p-1)(q-1)}p^{\frac{1}{2}(q-1)}-\left(\frac{q}{p}\right)\right]$$

is also divisible by $\dfrac{1-x^p}{1-x}$; that is to say,

$$(-1)^{\frac{1}{2}(p-1)}p\left[(-1)^{\frac{1}{4}(p-1)(q-1)}p^{\frac{1}{2}(q-1)}-\left(\frac{q}{p}\right)\right]$$

is the remainder left in the division of the function $\xi_1(\xi_1{}^q-\xi_q)$ by $\dfrac{1-x^p}{1-x}$. But every term in that function is divisible by q; the remainder is therefore itself divisible by q. We thus obtain the congruence

$$(-1)^{\frac{1}{4}(p-1)(q-1)}p^{\frac{1}{2}(q-1)}\equiv\left(\frac{q}{p}\right), \bmod q,$$

which involves the equation

$$\left(\frac{p}{q}\right)\left(\frac{q}{p}\right)=(-1)^{\frac{1}{4}(p-1)(q-1)}.$$

Gauss has given a purely algebraical proof of the theorems (1), (2), and (3), on which this demonstration depends. The third is a simple consequence of the arithmetical property of the multinomial coefficient, already referred to in Art. 10 of this Report; to establish the first two, it is sufficient to observe that $\xi_k{}^2-(-1)^{\frac{1}{2}(p-1)}p$ and $\xi_k-\left(\frac{k}{p}\right)\xi_1$ vanish, the first, if x be any imaginary root, the second, if x be any root whatever, of the equation $x^p-1=0$. If, for example, in the function ξ_k we put $x=r=\cos\dfrac{2\pi}{p}+i\sin\dfrac{2\pi}{p}$, we obtain the function $\psi(k,p)$, which satisfies, as we have seen, the two equations $[\psi(k,p)]^2=(-1)^{\frac{1}{2}(p-1)}p$, and $\psi(k,p)=\left(\frac{k}{p}\right)\psi(1,p)$. It is, indeed, simplest to suppose $x=r$ throughout the whole demonstration, which is thus seen to depend wholly on the properties of the same trigonometrical function ψ, which presents itself in the fourth demonstration; only it will be observed that here no necessity arises for the consideration of composite values of n in the function $\psi(k,n)$; nor for the determination of the ambiguous sign in the formula (A). In this specialized form, Gauss's sixth proof has been given by Jacobi (in the 3rd edit. of Legendre's 'Théorie des

Nombres,' vol. ii. p. 391), Eisenstein (Crelle, vol. xxviii. p. 41), and Cauchy (Bulletin de Férussac, Sept. 1829, and more fully Mém. de l'Institut, vol. xviii. p. 451, note iv. of the Mémoire), quite independently of one another, but apparently without its being at the time perceived by any of those eminent geometers that they were closely following Gauss's method. (See Cauchy's Postscript at the end of the notes to his Mémoire; also a memoir by M. Lebesgue in Liouville, vol. xii. p. 457; and a foot-note by Jacobi, Crelle, vol. xxx. p. 172, with Eisenstein's reply to it, Crelle, vol. xxxv. p. 273.)

MM. Lebesgue* and Eisenstein† have even exhibited a proof essentially the same in a purely arithmetical form, from which the root of unity again disappears, and is replaced by unity itself. Eisenstein considers the sum

$$C_a = \Sigma \left(\frac{k_1}{p}\right)\left(\frac{k_2}{p}\right)\cdots\left(\frac{k_q}{p}\right),$$

in which $k_1, k_2, \ldots k_q$ denote q terms (equal or unequal) of a system of residues prime to p, the sign of summation extending to every combination of the numbers $k_1, k_2, \ldots k_q$, that satisfies the congruential condition

$$k_1 + k_2 + k_3 + \ldots + k_q \equiv a, \text{ mod } p.$$

This sum is, in fact, the coefficient of r^a in the development of the qth power of the function $\sum\limits_{k=1}^{k=p-1}\left(\frac{k}{p}\right)r^k$, {reduced by the equation $r^p - 1 = 0$; not $\frac{r^p-1}{r-1} = 0$}, which is equivalent in value to Gauss's function $\psi(1, p)$. From the equation

$$\left[\sum_{k=1}^{k=p-1}\left(\frac{k}{p}\right)r^k\right]^2 = (-1)^{\frac{1}{2}(p-1)}p,$$

it follows that

$$\left[\sum_{k=1}^{k=p-1}\left(\frac{k}{p}\right)r^k\right]^q = (-1)^{\frac{1}{4}(p-1)(q-1)}p^{\frac{1}{2}(q-1)} \times \sum_{k=1}^{k=p-1}\left(\frac{k}{p}\right)r^k;$$

whence

$$C_a = (-1)^{\frac{1}{4}(p-1)(q-1)}\left(\frac{a}{p}\right)p^{\frac{1}{2}(q-1)}.\ddagger$$

---

* See Liouville's Journal, vol. ii. p. 253, and vol. iii. p. 113. (The proof of the law of reciprocity will be found in sect. i. art. 5, and sect. iii. art. 2, of the memoir). See also the memoir referred to in the text, Liouville, vol. xii. p. 457.

† Crelle's Journal, vol. xxvii. p. 322.

‡ {This assumes that $C_0 = 0$. Adding the $p-1$ equations

$$\Sigma\, C_a r^a = (-1)^{\frac{1}{4}(p-1)(q-1)}\, p^{\frac{1}{2}(q-1)}\, \Sigma\left(\frac{k}{p}\right)r^a$$

to the equation $\Sigma_0^{p-1} C_a = 0$, we obtain $C_0 = 0$; then the equation follows from the irreducibility of $\frac{r^p-1}{r-1} = 0$.}

And again, since

$$\left[\sum_{k=1}^{k=p-1}\left(\frac{k}{p}\right)r^k\right]^q \equiv \sum_{k=1}^{k=p-1}\left(\frac{k}{p}\right)r^{kq} \equiv \left(\frac{q}{p}\right)\sum_{k=1}^{k=p-1}\left(\frac{k}{p}\right)r^k, \text{ mod } q,$$

we have the congruence $$C_a \equiv \left(\frac{a}{p}\right)\left(\frac{q}{p}\right), \text{ mod } q.$$

But these results, which, taken together, establish the law of reciprocity, are obtained by Eisenstein from his arithmetical definition of C_a, without any reference to the trigonometrical function $\psi(1, p)$. If we write that function in the form $\sum_{k=0}^{k=p-1} r^{k^2}$, instead of the form $\sum_{k=1}^{k=p-1}\left(\frac{k}{p}\right)r^k$, we obtain from its qth power the coefficient C'_a considered by M. Lebesgue. This coefficient, which is connected with C_a by the equation $C'_a = p^{q-1} + C_a$, represents the number of solutions of the congruence $x_1^2 + x_2^2 + x_3^2 + \dots + x_q^2 \equiv a$, mod q. From this definition M. Lebesgue deduces the equation $C'_a = p^{q-1} + (-1)^{\frac{1}{4}(p-1)(q-1)}\left(\frac{a}{p}\right)p^{\frac{1}{2}(q-1)}$, and the congruence $C'_a \equiv 1 + \left(\frac{a}{p}\right)\left(\frac{q}{p}\right)$, mod q, by processes which, though different from those of Eisenstein, involve, like them, the consideration of integral numbers only.

22. Other proofs of the Theorem of Reciprocity have been suggested to subsequent writers by a comparison of the different methods of Gauss. The symbol r denoting a root of the equation $\frac{x^p - 1}{x - 1} = 0$, it is very easily shown that

$$(r - r^{-1})^2 (r^2 - r^{-2})^2 \dots (r^{\frac{1}{2}(p-1)} - r^{-\frac{1}{2}(p-1)})^2 = (-1)^{\frac{1}{2}(p-1)} p. \qquad \text{(C)}$$

It is natural therefore to employ this equation to replace the equation

$$\left[\sum_{k=1}^{k=p-1}\left(\frac{k}{p}\right)r^k\right]^2 = (-1)^{\frac{1}{2}(p-1)} p,$$

which presents itself in the 4th and 6th methods of Gauss. It is also found that the product $$\prod_{k=1}^{k=\frac{1}{2}(p-1)} \frac{r^{kq} - r^{-kq}}{r^q - r^{-q}} \text{ is equal to } \left(\frac{q}{p}\right). \qquad \text{(D)}$$

This is an immediate consequence of the property of *a half-system* of Residues (see Art. 19 *supra*) on which Gauss's 3rd and 5th methods depend. From a combination of the equations (C) and (D), the law of reciprocity is immediately deducible. (See a note by M. Liouville, Compt. Rend. vol. xxiv., or Liouville's Journal, vol. xii. p. 95, and especially a memoir by Eisenstein, entitled 'Application de l'Algèbre à l'Arithmétique transcendante,' Crelle, vol. xxix. p. 177. The proof by the same author in vol. xxxv. p. 257, is the same as that in the earlier memoir, only that the properties of the circular functions, which here replace the roots of unity, are in the later memoir deduced immediately from the definition of the sine as the product of an infinite number of factors.)

[*Addition.* From a general theorem of M. Kummer's (see Arts. 43, 44 of this Report), it appears that the congruence $r^2 \equiv (-1)^{\frac{1}{2}(\lambda-1)}\lambda$, mod q, is or is not resoluble, according as $q^{\frac{1}{2}(\lambda-1)} \equiv +1$, or $\equiv -1$, mod λ,—a result which implies the theorem of quadratic reciprocity. This very simple demonstration (which is, however, only a transformation of Gauss's sixth) appears first to have occurred to M. Liouville (see a note by M. Lebesgue in the Comptes Rendus, vol. li. pp. 12, 13).]

23. *Algorithm for the Determination of the Value of the Symbol* $\left(\frac{Q}{P}\right)$.—Gauss has shown in the memoir 'Demonstrationes et ampliationes novæ,' already quoted, that, if p be a prime number, the value of the symbol $\left(\frac{Q}{p}\right)$ may be obtained by developing the vulgar fraction $\frac{Q}{p}$ in a continued fraction, and considering the evenness or unevenness of a certain function of the quotients and remainders which present themselves in the development. Jacobi has observed (see Crelle, vol. xxx. p. 173) that a much simpler rule may be obtained, by the use of his extension of Legendre's symbol to the case when p is not a prime. The following is the form in which the rule has been exhibited by Eisenstein (see Crelle, vol. xxvii. p. 319). Let p_1, p_2 be two uneven numbers prime to one another, and let us form by division the series of equations

$$p_1 = 2k_1 p_2 \quad + \epsilon_2 p_3,$$
$$p_2 = 2k_2 p_3 \quad + \epsilon_3 p_4,$$
$$\cdots\cdots\cdots\cdots$$
$$p_\mu = 2k_\mu p_{\mu+1} + \epsilon_{\mu+1},$$

in which $\epsilon_2, \epsilon_3, \ldots \epsilon_{\mu+1}$ denote positive or negative units, and $p_1, p_2, p_3, \ldots$ which are all positive and uneven, form a descending series. Let σ denote the number of the quantities $2k_{r-1}p_r + \epsilon_r p_{r+1}$, in which *both* p_r and $\epsilon_r p_{r+1}$ are of the form $4n+3$; then $\left(\frac{p_1}{p_2}\right) = (-1)^\sigma$. The demonstration of this result flows immediately from the definition of Jacobi's symbol of reciprocity.

A numerical example is added (see Disq. Arith., Art. 328) from which the reader will perceive the utility of these researches in their practical application to congruences.

Let the proposed congruence be $x^2 \equiv -286$, mod 4272943, where 4272943 is a prime number.

We have to investigate the value of the symbol $\left(\frac{-286}{p}\right)$, in which p is written for 4272943. Now

$$\left(\frac{-286}{p}\right) = \left(\frac{-1}{p}\right) \times \left(\frac{2}{p}\right) \times \left(\frac{143}{p}\right) = -\left(\frac{143}{p}\right),$$

because $\left(\frac{-1}{p}\right) = -1$, and $\left(\frac{2}{p}\right) = +1$, p being of the form $8n-1$. To find the value of $\left(\frac{143}{p}\right)$, we have

$$\begin{aligned} 143 &= 0 \times 4272943 + 143\dagger, \\ 4272943 &= 29880 \times 143 + 103\dagger, \\ 143 &= 2 \times 103 - 63, \\ 103 &= 2 \times 63 - 23, \\ 63 &= 2 \times 23 + 17, \\ 23 &= 2 \times 17 - 11, \\ 17 &= 2 \times 11 - 5\dagger, \\ 11 &= 2 \times 5 + 1. \end{aligned}$$

The obelisk (†) denotes that the equation to which it is affixed is one of those enumerated in σ. Hence

$$\left(\frac{143}{4272943}\right) = (-1)^3 = -1, \quad \text{and} \quad \left(\frac{-286}{4272943}\right) = +1,$$

or the proposed congruence is resoluble. Its roots (as determined by Gauss) are ± 1493445.

24. *Biquadratic Residues.*—Reverting to the general theory alluded to in Art. 12, we see that, when p is a prime of the form $4n+1$, the congruence $x^4 - 1 \equiv 0$, mod p, admits four incongruous solutions; these are $+1$, -1, and the two roots of the congruence $x^2+1 \equiv 0$, mod p, which we shall denote by $+f$ and $-f$, or by f and f^3, so that the four roots of $x^4 - 1 \equiv 0$ are $1, f, -1$, and f^3. Further, if k be any number prime to p, k satisfies one or other of the four congruences—

(i.) $k^{\frac{1}{4}(p-1)} \equiv 1$, mod p. (iii.) $k^{\frac{1}{4}(p-1)} \equiv -1$, mod p.

(ii.) $k^{\frac{1}{4}(p-1)} \equiv f$, mod p. (iv.) $k^{\frac{1}{4}(p-1)} \equiv f^3$, mod p.

We see therefore that the $p-1$ residues of p divide themselves into four classes, comprising each $\frac{1}{4}(p-1)$ numbers, according as they satisfy the 1st, 2nd, 3rd, or 4th of these congruences. The first class comprises those numbers α for which the congruence $x^4 \equiv a$, mod p, is resoluble; that is, the *biquadratic residues* of p; the third comprises those numbers which are quadratic, but not biquadratic, residues of p; the second and fourth classes divide equally between them the non-quadratic residues.

We owe to Gauss two memoirs* on the Theory of Biquadratic Residues, which, while themselves replete with results of great interest, are yet more remarkable for the impulse they have given to the study of arithmetic in a new direction. Gauss found by induction that a law of reciprocity (similar to that of Legendre) exists for biquadratic residues. But he also discovered that, to demonstrate or even to express this law, we must take into consideration the imaginary factors of which prime numbers of the form $4n+1$ are composed. By thus introducing the conception of imaginary quantity into arithmetic, its domain, as Gauss observes, is indefinitely extended; nor is this extension an arbitrary addition to the science, but is essential to the comprehension of many phenomena presented by real integral numbers themselves.

Gauss's first memoir (besides the elementary theorems on the subject) contains a complete investigation of the biquadratic character of the number 2 with respect to any prime $p=4n+1$. The result arrived at is that if p be resolved into the sum of an even and uneven square (a resolution which is always possible in one way, and one only), so that $p=a^2+b^2$ (where we may suppose a and b taken with such signs that $a \equiv 1$, mod 4; $b \equiv af$, mod p), 2 belongs to the first, second, third, or fourth class, according as $\frac{1}{2}b$ is of the form $4n$, $4n+1$, $4n+2$, or $4n+3$. The considerations by which this conclusion is obtained are founded (see Art. 22 of the memoir) on the theory of the division of the circle, and we shall again have occasion to refer to them. In the second memoir Gauss developes the general theory already referred to, by which the determination of the biquadratic character of any residue of p may in every case be effected. The equation $p=a^2+b^2$ shows that $p=(a+bi)(a-bi)$, or that p, being the product of two conjugate imaginary factors, is in a certain sense not a prime number. Gauss was thus led to introduce *as modulus* instead of p one of its imaginary factors: an innovation which necessitated the construction of an arithmetical theory of complex imaginary numbers of the form $A+Bi$. The

---

* Theoria Residuorum Biquadraticorum. Commentatio prima et secunda. (Gottingæ, 1828 and 1832, and in the Comm. Recent. Soc. Gott., vol. vi. p. 27 and vol. vii. p. 89.) The articles in the two memoirs are numbered continuously. The dates of presentation to the Society are April 5, 1825, and April 15, 1831.

elementary principles of this theory are contained in the memoir in question; they have also been developed by Lejeune Dirichlet with great clearness and simplicity in vol. xxiv. of Crelle's Journal (pp. 295–319, sect 1–9)*. The following is an outline of the definitions and theorems which serve to constitute this new part of arithmetic.

[*Addition.* A note of Dirichlet's, in Crelle, vol. lvii. p. 187, contains an elementary demonstration of Gauss's criterion for the biquadratic character of 2. From the equation $p = a^2 + b^2$, we have $(a+b)^2 \equiv 2ab$, mod p, and hence

$$(a+b)^{\frac{1}{2}(p-1)} \equiv 2^{\frac{1}{4}(p-1)} a^{\frac{1}{4}(p-1)} b^{\frac{1}{4}(p-1)} \equiv (2f)^{\frac{1}{4}(p-1)} a^{\frac{1}{2}(p-1)},$$

or, which is the same thing,

$$\left(\frac{a+b}{p}\right) \equiv (2f)^{\frac{1}{4}(p-1)} \left(\frac{a}{p}\right). \quad \text{(A)}$$

But $\left(\frac{a}{p}\right) = \left(\frac{p}{a}\right) = 1$, because $p \equiv b^2$, mod a; and $\left(\frac{a+b}{p}\right) = \left(\frac{p}{a+b}\right)$, or, observing that $2p = (a+b)^2 + (a-b)^2$,

$$\left(\frac{a+b}{p}\right) = \left(\frac{2}{a+b}\right) = (-1)^{\frac{1}{8}\{(a+b)^2-1\}} \equiv f^{\frac{1}{4}(p-1)+\frac{1}{2}ab},$$

since $f^2 + 1 \equiv 0$, mod p. Substituting these values in the equation (A), we find $2^{\frac{1}{4}(p-1)} \equiv f^{\frac{1}{2}ab}$, mod p, which is in fact Gauss's criterion.]

25. *Theory of Complex Numbers.*—The product of a number $a+bi$ by its conjugate $a-bi$ is called its *norm*; so that the norm of $a+bi$ is a^2+b^2; the norm of a (which is its own conjugate) is a^2. This is expressed by writing $N(a+bi) = N(a-bi) = a^2+b^2$; $N(a) = a^2$. If α and β be two complex numbers, we have evidently $N(\alpha) \times N(\beta) = N(\alpha\beta)$. There are in this theory four units, $1, i, -1, -i$, which have each of them a positive unit for their norm. The four numbers $a+bi$, $ia-b$, $-a-ib$, $-ia+b$ (which are obtained by multiplying any one of them by the four units in succession, and which consequently stand to one another in a relation similar to that of $+a$ and $-a$ in the real theory) are said to be *associated* numbers. These four associated numbers with the numbers respectively conjugate to them form a group of eight numbers (in general

* The death of this eminent geometer in the present year (May 5, 1859) is an irreparable loss to the science of arithmetic. His original investigations have probably contributed more to its advancement than those of any other writer since the time of Gauss; if, at least, we estimate results rather by their importance than by their number. He has also applied himself (in several of his memoirs) to give an elementary character to arithmetical theories which, as they appear in the work of Gauss, are tedious and obscure; and he has thus done much to *popularize* the theory of numbers among mathematicians—a service which it is impossible to appreciate too highly.

different), all of which have the same norm. These definitions are applicable whatever be the nature of the real quantities a and b. If a and b are both rational, the complex number is said to be rational; if they are both integers, $a+bi$ is a complex integral number. One complex integer α is said to be divisible by another β, when a third γ can be found such that $\alpha=\beta\gamma$. Adopting these definitions, we can show that Euclid's process for investigating the greatest common divisor of two numbers is equally applicable to complex numbers; for it may be proved that, when we divide one complex number by another, we may always so choose the quotient as to render the norm of the remainder not greater than one-half of the norm of the divisor*. If, therefore, we apply Euclid's process for finding the greatest common divisor to two complex numbers, we shall obtain remainders with norms continually less and less, thus at last arriving at a remainder equal to zero; and the last divisor will be, as in common arithmetic, the greatest common divisor of the two complex numbers. Similarly the fundamental propositions deducible in the case of ordinary integers from Euclid's theory are equally deducible from the corresponding process in the case of complex integral numbers. Thus, 'if a complex number be prime to each of two complex numbers, it is prime to their product.' 'If a complex number divide the product of two factors, and be prime to one of them, it must divide the other.' 'The equation $ax-by=1$, where a and b are complex numbers prime to one another, is always resoluble with complex numbers x and y, and admits an infinite number of solutions,' &c.

A prime complex number is one which admits no divisors besides itself, its associates, and the four units.

There are three distinct classes of primes in the complex theory:—

1. Real prime numbers of the form $4n+3$ (with their associates).
2. Those complex numbers whose norms are real primes of the form $4n+1$.
3. The number $1+i$ and its associates, the norm of which is 2.

Instead of dividing numbers into even and uneven, we must here divide them into three classes, uneven, semi-even, and even, according as they are (1) not divisible by $(1+i)$; (2) divisible by $1+i$, but not by $(1+i)^2$; (3) divisible by $(1+i)^2=2i$, or, which is the same thing, by 2.

---

* Since $\dfrac{a+bi}{c+di}=\dfrac{ac+bd}{c^2+d^2}+\dfrac{bc-ad}{c^2+d^2}i$; if p be the integral number nearest to $\dfrac{ac+bd}{c^2+d^2}$, and q that nearest to $\dfrac{bc-ad}{c^2+d^2}$, $p+qi$ is the quotient required.

Of four associated uneven numbers, there is always one, and only one, such that b is even and $a+b-1$ evenly even. This is distinguished from the others as *primary*. Thus -7 and $-5+2i$ are primary numbers. A primary number is congruous to $+1$ for the modulus $2(1+i)$; whence it appears that the product of any number of primary numbers is itself a primary number. The conjugate of a primary is also primary. In speaking of uneven numbers, unless the contrary is expressed, we shall suppose them to be primary. This definition of a primary number is that adopted by Gauss (*l.c.* Art. 36), and after him by Eisenstein, and we shall adhere to it in this Report. But Gauss has also suggested a second definition (which is for some purposes slightly more convenient), and which has been adopted by Dirichlet, who defines a primary uneven number to be one in which b is even, and $a \equiv 1$, mod 4. The object of singling out one of the four associated numbers is merely that it serves to give definiteness to many theorems. For example, the theorem that 'every real number may be expressed as the product of powers of real primes in one way, and in one only,' may be now transferred in an equally definite form to the complex theory, 'Every complex number can be expressed in one way only in the form

$$i^m (1+i)^n . A^\alpha . B^\beta . C^\gamma \ldots,$$

where m, n, α, β, γ, &c. are real integral numbers, A, B, C ... *primary* complex primes.'

If $a+bi$ be a complex number, and $N=N(a+bi)=a^2+b^2$, and if h be the greatest common divisor of a and b, it can be shown that every number is congruous, for the modulus $a+bi$, to one, and one only, of the numbers $x+iy$, where

$$x=0, 1, 2, \ldots \frac{N}{h}-1; \qquad y=0, 1, 2, \ldots h-1.$$

These numbers therefore (or any set of numbers congruous to them) form a complete system of residues for the modulus $a+bi$. The number of the numbers $x+iy$ is evidently N, so that the norm of the modulus represents the number of residues in a complete system. In particular, therefore, if the modulus $a+bi$ be a prime of the second kind, having p for its norm, the numbers $0, 1, 2, \ldots p-1$ represent a complete system of residues; and if the modulus be a prime of the first kind, as q, the numbers included in the formula $x+iy$, where x and y may have any values from 0 to $q-1$ inclusive, will represent a complete system of residues.

[*Addition.* Although the second definition has been adopted by Dirichlet in his memoir in Crelle's Journal, vol. xxiv (see p. 301), yet in the memoir

'Untersuchungen über die complexen Zahlen' (see the Berlin Memoirs for 1841), sect. 1, he has preferred to follow Gauss.]

26. *Fermat's Theorem for Complex Numbers.*—Dirichlet's proof of this theorem for ordinary integers is equally applicable to complex numbers, and leads us to the following result:—

'If p be a prime in the complex theory, and k any complex number not divisible by p, then $k^{Np-1} \equiv 1$, mod p.'

Again, the demonstration of the theorem of Lagrange (see Art. 11) is equally applicable here (see Gauss, Theor. Res. Biq., Art. 50), and therefore the general theorems mentioned in Art. 12 may be extended, *mutatis mutandis*, to the complex theory. In particular, the number of primitive roots will be $\psi[N(p)-1]$, or the number of numbers less than $N(p)-1$, and prime to it. It will follow from this that, if the modulus be an *imaginary* prime p, every primitive root of Np in the real theory will be a primitive root both of p and its conjugate. Those Tables of Indices, therefore, in the 'Canon Arithmeticus,' which refer to primes of the form $4n+1$ will continue to hold, if for the real modules we substitute either of the imaginary factors of which they are composed. For primes of the form $4n+3$ (considered as modules in the complex theory), it would be requisite to construct new tables,—a labour which no one as yet appears to have undertaken.

27. *Law of Quadratic Reciprocity for Complex Numbers.*—If p and q be any two uneven primes (not necessarily *primary*, but subject to the condition that their imaginary parts are even), and if we denote by $\left[\frac{p}{q}\right]$ the unit-residue of the power $p^{\frac{1}{2}[Nq-1]}$, mod q; so that $\left[\frac{p}{q}\right] = +1$, or $= -1$, according as p is or is not a quadratic residue of q: then a law of reciprocity exists, which is expressed by the equation $\left[\frac{p}{q}\right] = \left[\frac{q}{p}\right]$.

If p and q are both real primes, it is easily seen that either of them is a quadratic residue of the other in the complex theory, or $\left[\frac{p}{q}\right] = \left[\frac{q}{p}\right] = 1$. But, as p may or may not be a quadratic residue of q in the theory of real integers, we see that the values of the symbols $\left[\frac{p}{q}\right]$ and $\left(\frac{p}{q}\right)$ are not necessarily identical.

This theorem is only enunciated in Gauss's memoir (Art. 60), and, as he speaks of it as a special case of the corresponding theorem for biquadratic residues, it is probable that his demonstration of it was of the same nature with that which he had found of the law of biquadratic reciprocity. However, a simple proof of it, depending on Legendre's law of reciprocity, has been given by Dirichlet in Crelle's Journal*. He shows that, if q be a prime of the first kind, $\left[\frac{a+\beta i}{q}\right] = \left(\frac{a^2+\beta^2}{q}\right)$; and that, if $a+bi$ be any prime of the second kind in which b is even, $\left[\frac{a+\beta i}{a+bi}\right] = \left(\frac{a\alpha+b\beta}{a^2+b^2}\right)$. The law of reciprocity is easily deducible from these transformations. If, for example, $a+bi$, $\alpha+\beta i$, be primes of the second species in which both b and β are even, we have simultaneously

$$\left[\frac{\alpha+\beta i}{a+bi}\right] = \left(\frac{a\alpha+b\beta}{p}\right); \qquad \left[\frac{a+bi}{\alpha+\beta i}\right] = \left(\frac{a\alpha+b\beta}{\varpi}\right),$$

where $p=N(a+bi)$; $\varpi=N(\alpha+\beta i)$. But $\left(\frac{a\alpha+b\beta}{p}\right) = \left(\frac{p}{a\alpha+b\beta}\right)$, by Jacobi's formula (see Art. 17 *supra*); and $\left(\frac{a\alpha+b\beta}{\varpi}\right) = \left(\frac{\varpi}{a\alpha+b\beta}\right)$. Also

$$p\varpi = (a\alpha+b\beta)^2 + (a\beta-b\alpha)^2;$$

whence we infer $\left(\frac{p\varpi}{a\alpha+b\beta}\right) = 1$, or, which is the same thing,

$$\left(\frac{p}{a\alpha+b\beta}\right) = \left(\frac{\varpi}{a\alpha+b\beta}\right); \text{ and therefore finally, } \left[\frac{\alpha+\beta i}{a+bi}\right] = \left[\frac{a+bi}{\alpha+\beta i}\right].$$

The complementary theorems which have to be united with this formula are

$$\left[\frac{i}{\alpha+\beta i}\right] = (-1)^{\frac{1}{4}(\varpi-1)}; \qquad \left[\frac{1+i}{\alpha+\beta i}\right] = (-1)^{\frac{1}{8}\{(\alpha+\beta)^2-1\}}$$

(see Dirichlet, Crelle, vol. xxx. p. 312); and they, as well as the formula of reciprocity itself, admit of an extension similar to that which Jacobi has given to the corresponding formulae of Legendre.

28. *Reciprocity of Biquadratic Residues.*—We now come to the theorem which first suggested the introduction of complex numbers.

If p be any (complex) prime, and k be any residue not divisible by p, we denote by $\left(\frac{k}{p}\right)_4$ the power i^e of i, which satisfies the congruence $k^{\frac{1}{4}(Np-1)} \equiv i$. It

* Crelle, vol. ix. p. 379.

will be observed that when p is a prime of the second species, the quadripartite classification of the real residues of p which we thus obtain is identical with that which we obtain for Np in the real theory (see Art. 24 *supra*); for the numbers f and $-f$ being the roots of the congruence $x^2+1\equiv 0$, mod Np, satisfy the same congruence for either of the complex factors of Np, and are therefore congruous to $+i$ and $-i$, for one of those factors, and to $-i$ and $+i$ for the other. Admitting this definition of the symbol $\left(\frac{k}{p}\right)_4$, Gauss's law of biquadratic reciprocity is expressed by the equation

$$\text{(i.)}\quad \left(\frac{\alpha}{\beta}\right)_4 = (-1)^{\frac{1}{4}(A-1).\frac{1}{4}(B-1)}\left(\frac{\beta}{\alpha}\right)_4,$$

α and β denoting two primary uneven primes, and A and B being their norms.

The complementary theorems relating to the unit i and the semi-even prime $1+i$ are

$$\text{(ii.)}\quad \left(\frac{i}{a+ia'}\right)_4 = i^{-\frac{1}{2}(a-1)}; \qquad \text{(iii.)}\quad \left(\frac{1+i}{a+ia'}\right)_4 = i^{\frac{1}{4}((a+a')-(1+a')^2)},$$

in which $a+ia'$ denotes a primary uneven prime. These formulae, like those of the last article, are susceptible of the same generalization which Jacobi has applied to Legendre's symbol; and we may suppose in the first that α and β are any two primary uneven numbers, prime to one another; and in the second and third that $a+ia'$ is any primary uneven number.

If, in the formula (i.) which expresses the law of reciprocity, $\alpha=a+ia'$, $\beta=b+ib'$, it may be easily seen that the unit $(-1)^{\frac{1}{4}(A-1).\frac{1}{4}(B-1)}$ is equal to $(-1)^{\frac{1}{2}(a-1).\frac{1}{2}(b-1)}$. This gives us a second expression of the theorem. (See Eisenstein, 'Math. Abhandl.' p. 137, or Crelle, vol. xxx. p. 193.)

Further, if we observe that every primary number is either $\equiv 1$, mod 4, or else $\equiv 3+2i$, mod 4; and that $\frac{1}{4}(A-1).\frac{1}{4}(B-1)$ and $\frac{1}{2}(a-1).\frac{1}{2}(b-1)$ are even numbers, unless *both* α and β satisfy the latter congruence, we may enunciate the law of biquadratic reciprocity by saying—

'The biquadratic characters of two primary uneven prime numbers with respect to one another are identical, if either of the primes be $\equiv 1$, mod 4; but if neither of them satisfy that congruence, the two biquadratic characters are opposite.'

This theorem is only enunciated by Gauss, who never published his demonstration of it. 'Non obstante,' he observes, 'summâ huius theorematis simplicitate ipsius demonstratio inter mysteria arithmeticae sublimioris maxime recondita referenda est, ita ut, saltem ut nunc res est, per subtilissimas tantum modo investigationes enodari possit, quae limites praesentis commentationis longe transgrederentur.'—*Theor. Res. Biq.* Art. 67.

Soon after the publication of the theorem, its demonstration was obtained by Jacobi, and communicated by him to his pupils in his lectures at Königsberg in the winter of 1836–37 (see his note to the Berlin Academy, already cited in Art. 17). These lectures have unfortunately never been published; but Jacobi's demonstration, from his criticism (see *ibid.*) on the first of those given ten years later by Eisenstein, appears to have been very similar to it.

It is to Eisenstein that we are indebted for the only published proofs of the theorem in question. That great geometer (so early lost to arithmetical science—a victim, it is said, to his devotion to his favourite pursuit) has left us as many as five demonstrations of it; the two earlier based on the theory of the division of the circle; the last three, on that of the lemniscate. We proceed to explain the principles on which each of these two classes of proofs depends:—

29. *Biquadratic Residues—Researches of Eisenstein.*—It is possible, as we have seen, to obtain a proof of Legendre's law of Reciprocity by considerations relating to the function $\sum\limits_{k=1}^{k=p-1}\left(\frac{k}{p}\right)x^k$, p denoting a real prime, and x a root of the equation $\frac{x^p-1}{x-1}=0$. This function is a particular case of the well-known function (introduced by Gauss and Lagrange into the theory of the division of the circle) $F(\theta, x)=\sum\limits_{s=0}^{s=p-2}\theta^s x^{\gamma^s}$, where θ is *any* root of the equation $\frac{\theta^{p-1}-1}{\theta-1}=0$, γ a *primitive* root of the congruence $x^{p-1}\equiv 1$, mod p, and x a root of the equation $\frac{x^p-1}{x-1}=0$. In the quadratic theory we assign to θ the value -1; in the theory of Biquadratic Residues we put $\theta=i$, and are thus led to consider another particular form of the same function, viz. $F(i, x)=\sum\limits_{s=0}^{s=p-2}i^s x^{\gamma^s}$, p denoting a prime of the form $4n+1$.

30. The function $F(\theta, x)$ or $F(\theta)$ is characterized by the following general properties; which have been given by Jacobi, Cauchy, and Eisenstein. (See Jacobi, Crelle, vol. xxx. p. 166; Cauchy, Mémoire sur la Théorie des Nombres in the Mém. de l'Acad. de l'Institut de France, vol. xviii; Eisenstein, Crelle, vol. xxvii. p. 269. {Also M. Lebesgue, Liouville, vol. xix}.)

$$\text{I.}\quad F(\theta, x^k) = \theta^{-\operatorname{Ind}_\gamma k} F(\theta, x),$$

$$\text{II.}\quad F(\theta)\, F(\theta^{-1}) = \theta^{\frac{1}{2}(p-1)}\, p,$$

$$\text{III.}\quad \frac{F(\theta^{-m})\, F(\theta^{-n})}{F(\theta^{-(m+n)})} = \psi(\theta),$$

where $\psi(\theta)$ does not involve x, and is an integral function of θ with integral coefficients *. The function $\psi(\theta)$ satisfies the equation

$$\text{IV.}\quad \psi(\theta)\, \psi(\theta^{-1}) = p.$$

Lastly, let θ be a *primitive* root of $\dfrac{x^{p-1}-1}{x-1} = 0$, and in the function

$$\psi(\theta) = \frac{F(\theta^{-m})\, F(\theta^{-n})}{F(\theta^{-m-n})}$$

let γ be written for θ; then if m and n be positive and less than $p-1$,

$$\text{V.}\quad \psi(\gamma) \equiv -\frac{\Pi(m+n)}{\Pi m \,.\, \Pi n}, \bmod p;$$

Πm denoting the continued product $1 \,.\, 2 \,.\, 3 \ldots m$.

Applying these equations to the particular form of the function F which we have to consider here, we find

$$F(i)\, F(-i) = i^{\frac{1}{2}(p-1)} p, \quad \psi(i) = \frac{F(i)\, F(i)}{F(-1)}, \text{ if } \theta^{\frac{1}{4}(p-1)} = i, \text{ and } m = n = \tfrac{3}{4}(p-1).$$

$$[F(i)]^4 = p[\psi(i)]^2,$$

$$[F(-i)]^4 = p[\psi(-i)]^2, \qquad \psi(i)\, \psi(-i) = p,$$

$$\psi(\gamma^{\frac{1}{4}(p-1)}) \equiv 0, \bmod p.$$

Let $\psi(i) = a + bi = p_1$; $\psi(-i) = a - bi = p_2$, so that $p_1 p_2 = p$.

The congruence

$$\psi[\gamma^{\frac{1}{4}(p-1)}] \equiv 0, \bmod p, \quad \text{or} \quad a + b\gamma^{\frac{1}{4}(p-1)} \equiv 0, \bmod p,$$

involves also the congruence

$$a + b\gamma^{\frac{1}{4}(p-1)} \equiv 0, \bmod p_1; \quad i.e.\ \gamma^{\frac{1}{4}(p-1)} \equiv i, \bmod p_1;$$

so that $\left(\dfrac{\gamma^k}{p_1}\right)_4 = i^k$. Hence we have, putting $\gamma^s \equiv k, \bmod p$,

$$F(i) = \sum_{k=1}^{k=p-1} \left(\frac{k}{p_1}\right)_4 x^k = S,$$

$$F(-i) = \sum_{k=1}^{k=p-1} \left(\frac{k}{p_1}\right)_4^3 x^k = T.$$

* In this equation θ^{-m} and θ^{-n} are supposed not to be reciprocals.

From these formulae two cases of the law of Reciprocity are directly deducible.

a. Let q be a real prime of the form $4n+3$. Raising S to the power q, we have

$$S^q \equiv \sum_{k=1}^{k=p-1} \left(\frac{k}{p_1}\right)_4^q x^{qk} \equiv \sum_{k=1}^{k=p-1} \left(\frac{k}{p_1}\right)_4^3 x^{qk} \equiv \left(\frac{q}{p_1}\right)_4 T, \text{ mod } q, \text{ by (I.)}.$$

Multiplying by S, we find

$$S^{q+1} = (S^4)^{\frac{1}{4}(q+1)} = p^{\frac{1}{4}(q+1)}\, p_1^{\frac{1}{2}(q+1)} \equiv (-1)^{\frac{1}{4}(p-1)}\, p \left(\frac{q}{p_1}\right)_4, \text{ mod } q\,;$$

or, observing that $p_2 \equiv p_1^q$, mod q, and $p = p_1 p_2$,

$$p_1^{\frac{1}{4}(q^2-1)} \equiv (-1)^{\frac{1}{4}(p-1)} \left(\frac{q}{p_1}\right)_4, \text{ mod } q\,;$$

that is to say
$$\left(\frac{p_1}{q}\right)_4 = \left(\frac{-q}{p_1}\right)_4, \qquad \text{(A.)}$$

which is in accordance with the law of Reciprocity.

β. Again, let q be a prime of the form $4n+1$;

then
$$S^q \equiv \left(\frac{q}{p_1}\right)_4^3 S, \text{ mod } q\,; \text{ that is, } S^{q-1} \equiv \left(\frac{q}{p_1}\right)_4^3, \text{ mod } q,$$

or
$$p^{\frac{1}{4}(q-1)}\, p_1^{\frac{1}{2}(q-1)} \equiv \left(\frac{q}{p_1}\right)_4^3, \text{ mod } q\,;$$

whence, if $q = q_1 q_2$,
$$\left(\frac{p_2}{q_1}\right)_4 \left(\frac{p_1}{q_1}\right)_4^3 = \left(\frac{q}{p_1}\right)_4^3.$$

But, by changing i into $-i$,

$$\left(\frac{p_1}{q_1}\right)_4^3 = \left(\frac{p_2}{q_2}\right)_4, \quad \text{and} \quad \left(\frac{q}{p_1}\right)_4^3 = \left(\frac{q}{p_2}\right)_4,$$

so that
$$\left(\frac{p_2}{q}\right)_4 = \left(\frac{q}{p}\right)_4 \qquad \text{(B.)}$$

The symbolic equations (A.) and (B.) lead immediately to the conclusion that if a and b be any two primary uneven numbers, one, at least, of which is real, we have $\left(\frac{a}{b}\right)_4 = \left(\frac{b}{a}\right)_4$; and that if a and b be both real, the common value of these symbols is $+1$. By combining with these results the supplementary equation $\left(\frac{i}{a+ia'}\right)_4 = i^{-\frac{1}{2}(a-1)}$, in which $a+ia'$ denotes any primary uneven number, and also the self-evident equations,

$$c(a \pm bi) = (ac + bd) \pm bi(c \pm di),$$
$$a(c \pm di) = (ac + bd) \pm di(a \pm bi),$$
$$\left(\frac{a+bi}{c+di}\right)_4 \left(\frac{a-bi}{c-di}\right)_4 = 1,$$

Eisenstein* investigates a relation between the symbols $\left(\frac{a+bi}{c+di}\right)_4$ and $\left(\frac{c+di}{a+bi}\right)_4$, which, when $a+bi$ and $c+di$ are primary, coincides with that expressed by the law of reciprocity.

31. The proof in Eisenstein's second memoir† is identical in its essential character with that in the first; but he has given it a purely arithmetical form, independent of the theory of the division of the circle. Instead of the sum $S = \sum_{k=1}^{k=p-1} \left(\frac{k}{p_1}\right)_4 x^k$, in which x is a root of the equation $\frac{x^p - 1}{x - 1} = 0$, he considers the powers of the series $\sum_{k=1}^{k=p-1} \left(\frac{k}{p_1}\right)_4$, and arrives by a process purely arithmetical at the equations (A.) and (B.) of the preceding article. Thus the two forms in which he has exhibited his demonstration are precisely analogous to the two expressions which he has given to Gauss's sixth demonstration of Legendre's law (see above, Art. 21).

32. The proofs of the Law of Biquadratic Reciprocity, which are taken from the theory of elliptic functions, no less than those which we have just considered, depend in great measure on a generalization of the principles introduced by Gauss into his demonstrations of Legendre's law. Indeed, Gauss himself tells us‡ that his object in multiplying demonstrations of Legendre's law, was that he might at last discover principles equally applicable to the Biquadratic Theorem. It would be interesting to know whether the proof which he ultimately obtained of this theorem depended only on the division of the circle, or on elliptic transcendents. Jacobi appears to have believed the latter; for he expresses his opinion that his own demonstration of the Biquadratic Theorem

* See the memoir entitled 'Lois de Réciprocité,' in Crelle, vol. xxviii. pp. 53–67.

† 'Einfacher Beweiss und Verallgemeinerung des Fundamental-Theorems für die biquadratischen Reste,' in Crelle, vol. xxviii, p. 223.

‡ See the memoir, 'Theorematis Fundamentalis Demonstrationes et Ampliationes Novæ,' p. 4: 'Hoc ipsum incitamentum erat ut demonstrationibus jam cognitis circa residua quadratica alias aliasque addere tantopere studerem, spe fultus, ut ex multis methodis diversis una vel altera ad illustrandum argumentum affine aliquid conferre posset.'

was widely different from that of Gauss*; and he further conjectures that what induced Gauss to introduce complex numbers, *as modules*, into the theory of numbers, was not the study of any purely arithmetical question, but that of the elliptic functions connected with the Lemniscate Integral $\int \frac{dx}{\sqrt{(1-x^4)}}$†. This opinion of Jacobi's will not appear improbable, when we remember that in the 'Disquisitiones Arithmeticæ' (Art. 335) Gauss promises an 'amplum opus' on these transcendents; and that a casual remark of his in relation to them renders it perfectly certain (as Dirichlet has observed)‡ that he was at that early period in possession of the principle of the double periodicity of elliptic functions—thus anticipating by twenty-five years the discoveries of Abel and Jacobi. Nevertheless the close analogy we have endeavoured to point out between Gauss's sixth proof of the quadratic theorem, and the trigonometric demonstration of the biquadratic one, may perhaps incline us to the opposite opinion. Nor is the introduction of complex numbers, *as modules*, an idea unlikely to have suggested itself, when once complex numbers were admitted; though it is remarkable that Jacobi, in the first printed memoir in which complex numbers appear, and to which we shall presently refer, seems not to have thought of this extension of his theory§.

33. *Application of the Lemniscate Functions to the Biquadratic Theorem* ‖.—Let p_1 be a complex prime (real or imaginary), p its norm; and let the $p-1$ residues, prime to p_1, be divided into four groups of $\frac{1}{4}(p-1)$ terms, after the following scheme:—

$$\begin{array}{llll}
(0) & r_1, & r_2, \ldots\ldots\ldots & r_{\frac{1}{4}(p-1)}, \\
(1) & ir_1, & ir_2, \ldots\ldots\ldots & ir_{\frac{1}{4}(p-1)}, \\
(2) & -r_1, & -r_2, \ldots\ldots\ldots & -r_{\frac{1}{4}(p-1)}, \\
(3) & -ir_1, & -ir_2, \ldots\ldots\ldots & -ir_{\frac{1}{4}(p-1)},
\end{array}$$

* 'Ueber die Kreistheilung,' Crelle, vol. xxx. p. 171.

† Crelle, vol. xix. p. 314, or in the 'Monatsbericht' of the Berlin Academy for May 16, 1839.

‡ In his 'Gedächtnissrede über Karl Gustav Jacob Jacobi,' Mém. de l'Académie de Berlin, 1852. This remarkable éloge is also inserted in Crelle's Journal, vol. lii, and in a French translation in Liouville's Journal, vol. ii, 2nd series.

§ {Gauss's demonstration seems after all to have been more nearly comparable to the second and fifth of the Quadratic Theorems. See his Works, vol. ii: but I have not yet examined the paper carefully.}

‖ See Eisenstein's memoir, 'Applications de l'Algèbre à l'Arithmétique transcendante,' in Crelle's Journal, vol. xxx. p. 189, or in Eisenstein's 'Mathematische Abhandlungen,' p. 121.

so that of any four *associated* numbers one, and only one, appears in each group. Let q_1 be any residue prime to p_1; $k_1, k_2, k_3, \ldots$ the numbers of the residues

$$q_1 r_1, \quad q_1 r_2, \ldots\ldots\ldots q_1 r_{\frac{1}{4}(p-1)}$$

which belong to the groups (1), (2), (3), respectively; then

$$q_1^{\frac{1}{4}(p-1)} \equiv i^{k_1+2k_2+3k_3}, \text{ mod } p_1,$$

or

$$\left(\frac{q_1}{p_1}\right)_4 = i^{k_1+2k_2+3k_3}.$$

(See Gauss, Theor. Res. Biq., Art. 71.)

The expression on the right-hand side of this equation may now be transformed by means of the Lemniscate function ϕ, defined by the equations

$$v = \int_0^x \frac{dx}{\sqrt{(1-x^4)}}, \qquad x = \phi(v).$$

The function $\phi(v)$ is doubly periodic, the arguments of the periods being 2ω and $(1+i)\omega$, or, more simply, $(1+i)\omega$ and $(1-i)\omega$, where

$$\tfrac{1}{2}\omega = \int_0^1 \frac{dx}{\sqrt{(1-x^4)}};$$

so that we have $\phi(v+2k\omega) = \phi(v)$, k denoting any complex integer whatever. From this it appears that the relation of the Lemniscate functions to the theory of complex numbers is the same as the relation of circular functions to the arithmetic of real integers. The function $\phi(v)$ also satisfies the equation* $\phi(i^k v) = i^k \phi(v)$, whence

$$i^{k_1+2k_2+3k_3} = \frac{\Pi\phi\left(\frac{2rq_1\omega}{p_1}\right)}{\Pi\phi\left(\frac{2r\omega}{p_1}\right)} = \left(\frac{q_1}{p_1}\right)_4, \quad \ldots\ldots \quad (1)$$

the sign of multiplication Π extending to every residue r included in the group (0). Similarly, if q_1, like p_1, be a prime,

$$\left(\frac{p_1}{q_1}\right)_4 = \frac{\Pi\phi\left(\frac{2sp_1\omega}{q_1}\right)}{\Pi\phi\left(\frac{2s\omega}{q_1}\right)}, \quad \ldots\ldots \quad (2)$$

s denoting the general term of a *quarter-system* of Residues for the modulus q_1.

---

* {This, which is evident from the definition of $\phi(v)$, is also readily verified by applying the transformation $\omega = \dfrac{-i+\omega}{-1+\omega}$; we find the multiplier $\dfrac{1}{M} = i$, &c.}

By an elementary* theorem in the calculus of Elliptic Functions, $\frac{\phi(kv)}{\phi(v)}$ is for every uneven value of k a rational and fractional function of $x=\phi(v)$. If p_1 be primary, as we shall now suppose, and if we put $\alpha_r=\phi\left(\frac{2r\omega}{p_1}\right)$, we have, by the principles of that calculus,

$$\frac{\phi(p_1 v)}{\phi(v)}=\frac{\Pi(x^4-\alpha^4)}{\Pi(1-\alpha^4 x^4)}, \quad \ldots \ldots \ldots \quad (3)$$

the sign Π extending to all the different values of α; and similarly,

$$\frac{\phi(q_1 v)}{\phi(v)}=\frac{\Pi(x^4-\beta^4)}{\Pi(1-\beta^4 x^4)}, \quad \ldots \ldots \ldots \quad (4)$$

if $\beta_s=\phi\left(\frac{2s\omega}{q_1}\right)$. Combining the equations (3) and (4) with (1) and (2), we find

$$\left(\frac{q_1}{p_1}\right)_4=\frac{\Pi(\alpha^4-\beta^4)}{\Pi(1-\alpha^4\beta^4)},$$

$$\left(\frac{p_1}{q_1}\right)_4=\frac{\Pi(\beta^4-\alpha^4)}{\Pi(1-\alpha^4\beta^4)};$$

the sign of multiplication extending to the $\frac{1}{4}(p-1)(q-1)$ combinations of the values of α and β; whence, evidently,

$$\left(\frac{q_1}{p_1}\right)_4\left(\frac{p_1}{q_1}\right)_4=(-1)^{\frac{1}{4}(p-1)(q-1)}.$$

The priority of Eisenstein in this singularly beautiful investigation is indisputable.

34. In a later memoir ('Beiträge zur Theorie der Elliptischen Functionen,' Crelle, xxx. p. 185, or Math. Abhandl. p. 129), Eisenstein has put this proof into a slightly different form. He shows, by a peculiar method, that if p_1 be an *imaginary* and primary complex prime, every coefficient in $\Pi(x^4-\alpha^4)$ except the first is divisible by p_1, and that for every primary uneven value of p_1 (whether prime or not) the last coefficient is p_1, so that $(-1)^{\frac{1}{4}(p-1)}p_1=\Pi\alpha^4$. Representing therefore by p_1 an imaginary and primary prime, by q_1 any complex prime, the equation

$$\left(\frac{p_1}{q_1}\right)=\Pi\frac{\phi\left(\frac{2sp_1\omega}{q_1}\right)}{\phi\left(\frac{2s\omega}{q_1}\right)}=\Pi\frac{\beta^4-\alpha^4}{1-\alpha^4\beta^4}$$

* {It is not the ordinary theorem of multiplication, for k is complex. Doubtless equations (3) and (4) may be immediately proved by the general method of comparing the zeros and infinities of either side.}

assumes the form $\left(\frac{p_1}{q_1}\right) \equiv (-1)^{\frac{1}{4}(p-1)(q-1)} q_1^{\frac{1}{4}(p-1)}$, mod p,

or $$\left(\frac{p_1}{q_1}\right) = (-1)^{\frac{1}{4}(p-1)(q-1)} \left(\frac{q_1}{p_1}\right),$$

which establishes the law of Reciprocity for every case except that of two real primes, when the value of the symbols $\left(\frac{p_1}{q_1}\right)_4 = \left(\frac{q_1}{p_1}\right)_4 = 1$ is at once apparent from their definition and from Fermat's Theorem.

35. A third, and no less interesting application of the theory of elliptic functions to the formula of Biquadratic Reciprocity, occurs in the memoir, 'Genaue Untersuchung der Unendlichen Doppel-Producte, aus welchen die Elliptische Functionen als Quotienten zusammengesetzt sind' (Mathematische Abhandl. p. 213, or Crelle's Journal, vol. xxxv. p. 249). The function

$$F(x) = \prod_{n=-\infty}^{n=+\infty} \prod_{m=-\infty}^{m=+\infty} \left(1 - \frac{tx}{m+ni}\right),$$

which is considered in this memoir, and in which the factor $1 - \frac{tx}{0}$ is to be replaced by tx, coincides (if we disregard a constant factor) with the numerator of $\phi(v)$, when that function is expressed as the quotient of one infinitely continued product divided by another. This may be seen by comparing $F(x)$ with the expression of the *general* elliptic function $\phi(a)$ given by Abel, viz.

$$\phi(a) = a \prod_{\mu=1}^{\mu=\infty} \left(1 + \frac{a^2}{\mu^2 \varpi^2}\right) \prod_{m=1}^{m=\infty} \left(1 - \frac{a^2}{m^2 \omega^2}\right)$$

$$\times \prod_{m=1}^{m=\infty} \prod_{\mu=1}^{\mu=\infty} \left[\frac{1 + \frac{(a+m\omega)^2}{\mu^2 \varpi^2}}{1 + \frac{\left[a + (m-\frac{1}{2})\omega\right]^2}{(\mu-\frac{1}{2})^2 \varpi^2}} \cdot \frac{1 + \frac{(a-m\omega)^2}{\mu^2 \varpi^2}}{1 + \frac{\left[a - (m-\frac{1}{2})\omega\right]^2}{(\mu-\frac{1}{2})^2 \varpi^2}} \right] \times \left[\frac{1 + \frac{(m-\frac{1}{2})^2 \omega^2}{(\mu-\frac{1}{2})^2 \varpi^2}}{1 + \frac{m^2 \omega^2}{\mu^2 \varpi^2}} \right]^2 .$$

(See Abel, Œuvres, vol. i. p. 213, equat. 178.)

If we particularize this expression, by putting $\omega = \varpi$ (which changes $\phi(a)$ into the Lemniscate-function) and then write $\omega t x$ for a, we shall find that the function of x which appears in the numerator is precisely Eisenstein's function $F(x)$. This function (which is, consequently, a particular case of Jacobi's function H in his 'Fundamenta Nova') is only singly periodic; so that $F(x) = F\left(x + \frac{2\mu}{t}\right)$, if μ denote any real integer; but $F\left(x + \frac{2\mu}{t}\right)$ is equal to the product of $F(x)$ by an exponential function, if μ be an *imaginary* complex number. (Compare the formulae of sect. 61 of the 'Fundamenta Nova.') The difficulty occasioned by this imperfect periodicity of $F(x)$ Eisenstein has overcome by the introduction of the number t, which is supposed to represent a real even indeterminate integer. The formulae on which his proof depends, are

$$\text{(i)}\quad F(x+k) = e^{wt^2}F(x),$$

$$\text{(ii)}\quad F(ix) = ie^{wt^2}F(x),$$

$$\text{(iii)}\quad \frac{F(p_1 x)}{F(x)} = c^{p-1}e^{wt^2}\Pi . F\left(x+\frac{r}{p_1}\right).$$

The symbol w which depends on x, but is independent of t, is different in each of these equations: in the first, k is any complex integer; in the third, c is a numerical constant independent of x and p_1; p_1 a *primary* number prime to t; p its norm; and r the general term of the $p-1$ residues of p_1, the sign of multiplication Π extending to every value of r. These equations, the first two of which depend on the most elementary properties of the function $F(x)$ or H (see 'Fundamenta Nova,' *loc. cit.*), while the third is of a more abstruse character, Eisenstein has established by methods which are peculiar to himself, and which it would take us too far from our present subject to describe. They serve to replace the formulae

$$\phi(v) = \phi(v+2k\omega); \qquad \phi(iv) = i\phi(v);$$

$$\frac{\phi(p_1 v)}{\phi(v)} = \frac{\Pi(x^4-\alpha^4)}{\Pi(1-\alpha^4 x^4)}$$

in Eisenstein's earlier demonstration; and lead to the conclusion

$$\left(\frac{p_1}{q_1}\right) = \left(\frac{q_1}{p_1}\right)(-1)^{\frac{1}{4}(p-1)\cdot\frac{1}{4}(q-1)}e^{wt^2},$$

w still denoting some quantity independent of t. And since in this formula t may have any even value prime to p_1 and q_1, it is impossible that e^w should have any value but that of one of the fourth roots of unity, so that we have $e^{wt^2}=1$; which gives the law of Reciprocity.

36. An algorithm has been given by Eisenstein* for calculating the value of the symbol $\left(\frac{a+ia'}{b+ib'}\right)_4$ by means of the development of $\frac{a+ia'}{b+ib'}$, in a continued

* Crelle's Journal, vol. xxviii. p. 243. But the first invention of this algorithm, and of the similar one which exists in the Theory of Cubic Residues, is due to Jacobi. (See the note, 'Ueber die Kreistheilung &c.,' so often cited in this report.)

fraction. This algorithm, in a slightly simplified form, is as follows:—Let $a+ia'=p_0$, $b+ib'=p_1$, and form the series of equations

$$\begin{aligned} p_0 &= k_0 p_1 \quad + i^{\mu_1} . p_2, \\ p_1 &= k_1 p_2 \quad + i^{\mu_2} . p_3, \\ &\cdots\cdots\cdots\cdots \\ p_n &= k_n p_{n+1} + i^{\mu_{n+1}}. \end{aligned}$$

The numbers p_0 and p_1 are supposed to be uneven, and prime to one another; p_1 is primary; the quotients $k_0, k_1, k_2, \ldots k_n$ are all divisible by $1+i$, and are so chosen that the norms of $p_2, p_3, \ldots$ form a continually decreasing series (as is always possible); lastly, the units i^μ are so chosen as to render $p_2, p_3, \ldots$ primary. Let $p_s = a_s + i a_s$; let $-\frac{1}{2}(a_s - 1) \equiv \theta_s$, mod 4; and in the series $\theta_1, \theta_2, \ldots \theta_{n+1}$, let ρ be the number of sequences of uneven terms. Then $\left(\frac{p_1}{p_2}\right)_4 = i^{2\rho + \Sigma\theta\mu}$.

Example. Let it be required to determine whether the congruence

$$x^4 \equiv -3381, \text{ mod } 11981$$

be possible or impossible.

Since $11981 = 109^2 + 10^2$, and is a prime number, the resolubility of this congruence depends on that of the congruence $x^4 \equiv -3381$, mod $(-109+10i)$. We have therefore to investigate the value of the symbol $\left(\frac{-3381}{-109+10i}\right)_4$. This gives us the series of equations

$$\begin{aligned} -3381 &= (31+3i)(-109+10i) + i^3(-17+28i), \\ -109+10i &= (2+2i)(-17+28i) + i^0(-19-12i), \\ -17+28i &= -2i\,(-19-12i) + i^0(+7-10i), \\ -19-12i &= -2i\,(7-10i) + i^2(-1-2i), \\ 7-10i &= (3+5i)(-1-2i) + i. \end{aligned}$$

Here $\theta_1 = -1$, $\theta_2 = +1$, $\theta_3 = 2$, $\theta_4 = 1$, $\theta_5 = 1$; so that $\rho = 2$, $\Sigma\mu\theta = 0$, and $\left(\frac{-3381}{-109+10i}\right)_4 = 1$, or the proposed congruence is resoluble. Its four roots are ± 87, ± 2646, as may be found by any of the indirect methods for the solution of Quadratic congruences.

[*Addition.* In the algorithm given in the text, the remainders $p_2, p_3 \ldots$ are all uneven; and the computation of the value of the symbol $\left(\frac{p_0}{p_1}\right)_4$ is thus rendered independent of the formula (iii) of Art. 28. The algorithm given by Eisenstein is, however, preferable, although the rule to which it leads cannot be expressed with the same conciseness, because the continued fraction equivalent to $\frac{p_0}{p_1}$ terminates more rapidly when the remainders are the least possible, and not necessarily uneven.]

37. *Cubic Residues.* The Theory of Cubic Residues is less complex than that of Biquadratic Residues, and is at the same time so similar to it, that it will not be necessary to treat it with the same detail.

If p be a real prime of the form $3n+1$, and if $1, f, f^2$ denote the roots of the congruence $x^3-1\equiv 0$, mod p, the $p-1$ residues $k_1, k_2, \dots k_{p-1}$ of p divide themselves into three classes according as $k^{\frac{1}{3}(p-1)}\equiv 1$, or $\equiv f$, or $\equiv f^2$, mod p; the first class comprising the cubic residues, the two other classes comprising the cubic non-residues. Now it can be proved that every prime number of the form $3n+1$ may be represented by the quadratic form A^2-AB+B^2; *i.e.* it may be regarded as the product of two conjugate complex numbers of the forms $A+B\rho$, $A+B\rho^2$, where ρ and ρ^2 are the two imaginary cube roots of unity; just as the theory of biquadratic residues involves the consideration of the quadratic form A^2+B^2, and of complex numbers of the type $A+Bi$. The real integer A^2-AB+B^2 is the *norm* of the complex numbers $A+B\rho$ and $A+B\rho^2$, and expresses the number of terms in a complete system of residues for either of those modules.

The theory of these complex numbers has not been treated of in detail by any writer (see Eisenstein, Crelle, vol. xxvii. p. 290); but the methods of Gauss or Dirichlet are as applicable to them as to complex numbers involving i*. Thus it will be found that every fraction of the form $\frac{A+B\rho}{C+D\rho}$ can be developed in a finite continued fraction, having for its quotients complex integers; that Euclid's process for finding the greatest common divisor is applicable in this case also, and that the same arithmetical consequences may be deduced from it as in the case of ordinary integers. The prime numbers to be considered in this theory are—

(1) Real primes, as 2, 5, 11, 17, &c. of the form $3n+2$.

(2) Imaginary primes of the form $A+B\rho$, having for their norms real primes of the form $3n+1$.

(3) The primes $1-\rho$, $1-\rho^2$, having 3 for their norm.

The units are ± 1, $\pm\rho$, and $\pm\rho^2$.

If $A+B\rho$ be any complex number not divisible by $1-\rho$, it may be seen

* {There is a note by Gauss on this subject in vol. ii. of his Works.}

that of the three pairs of numbers, $\pm(A+B\rho)$, $\pm\rho(A+B\rho)$, $\pm\rho^2(A+B\rho)$, there is always one, and one only, which, when reduced to the form $a+b\rho$, satisfies the congruences $a\equiv\pm 1$, $b\equiv 0$, mod 3. Such a number is called a primary number. The product of two primary numbers, taken {positively or} negatively, is itself primary.

If α be any prime of this theory, and k any number not divisible by α, Fermat's Theorem is here represented by the congruence $k^{N\alpha-1}\equiv 1$, mod α.

Denoting by $\left(\frac{k}{\alpha}\right)_3$ that power ρ^s of ρ which satisfies the congruence $k^{\frac{1}{3}(N\alpha-1)}\equiv\rho^s$, the law of cubic reciprocity is contained in the formula

$$\left(\frac{\alpha}{\beta}\right)_3=\left(\frac{\beta}{\alpha}\right)_3,$$

α and β denoting any two *primary* complex primes.

The demonstration of this theorem follows quite naturally from the formulae cited in Art. 30. Applying them to this particular case, we have, if p denote a real prime of the form $3n+1$,

$$\text{(i)}\quad F(\rho).F(\rho^2)=p,$$

$$\text{(ii)}\quad [F(\rho)]^3=p\psi(\rho),$$

$$\text{(iii)}\quad \psi(\rho).\psi(\rho^2)=p,$$

$$\text{(iv)}\quad \psi(\gamma^{\frac{1}{3}(p-1)})\equiv 0, \bmod p\,;$$

from which we may infer that $\gamma^{\frac{1}{3}(p-1)}\equiv\rho$, mod $\psi(\rho)$. (Compare Art. 29.) In the equation (iii), $\psi(\rho)$ and $\psi(\rho^2)$ are primary; for from the equation $[F(\rho)]^3=p\psi(\rho)$, it appears that $\psi(\rho)\equiv -1$, mod 3. The congruence $\gamma^{\frac{1}{3}(p-1)}\equiv\rho$, mod $\psi(\rho)$, implies that $\left(\frac{\gamma^s}{\psi(\rho)}\right)=\rho^s$, whence if $\gamma^s\equiv k$, mod p,

$$F(\rho)=\sum_{k=1}^{k=p-1}\left(\frac{k}{p_1}\right)_3 x^k,$$

$$F(\rho^2)=\sum_{k=1}^{k=p-1}\left(\frac{k}{p_1}\right)_3^2 x^k;$$

where $p_1=\psi(\rho)$. By these formulae the several cases of the theorem of reciprocity may be proved, as follows* :—

---

* Eisenstein in Crelle's Journal, vol. xxvii. p. 289. But in this, as in many of his earlier researches, Eisenstein had been anticipated more than ten years by Jacobi.

First, let q be a prime of the form $3n+2$. Then

$$[F(\rho)]^q \equiv \sum_{k=1}^{k=p-1} \left(\frac{k}{p_1}\right)_3^q x^{qk}, \bmod q,$$

$$\equiv \left(\frac{q}{p_1}\right)_3 F(\rho^2), \bmod q,$$

or $$[F(\rho)]^{q+1} \equiv \left(\frac{q}{p_1}\right)_3 p, \bmod q.$$

But also $$[F(\rho)]^{q+1} = p^{\frac{1}{3}(q+1)} p_1^{\frac{1}{3}(q+1)};$$

so that $$p^{\frac{1}{3}(q-2)} p_1^{\frac{1}{3}(q+1)} \equiv \left(\frac{q}{p_1}\right)_3, \bmod q;$$

or raising each side of this congruence to the power $q-1$,

$$p_1^{\frac{1}{3}(q^2-1)} \equiv \left(\frac{q}{p_1}\right)_3 \quad \text{or} \quad \left(\frac{p_1}{q}\right)_3 = \left(\frac{q}{p_1}\right)_3.$$

Secondly, let q be a real prime of the form $3n+1$; we find

$$[F(\rho)]^q \equiv \left(\frac{q}{p_1}\right)_3^2 F(\rho), \bmod q, \quad \text{or} \quad F(\rho)^{q-1} \equiv \left(\frac{q}{p_1}\right)_3^2, \bmod q;$$

and also $$[F(\rho)]^{q-1} = p^{\frac{1}{3}(q-1)} p_1^{\frac{1}{3}(q-1)}.$$

Hence $\left(\frac{p}{q_1}\right)_3 \left(\frac{p_1}{q_1}\right)_3 = \left(\frac{q}{p_1}\right)_3^2$, where q_1 is either of the complex factors of q; or, observing that $\left(\frac{p_1}{q_1}\right)_3^2 = \left(\frac{p_2}{q_2}\right)_3$, and $\left(\frac{q}{p_1}\right)_3^2 = \left(\frac{q}{p_2}\right)_3$, we may write

$$\left(\frac{p_2}{q_1}\right)_3 \left(\frac{p_2}{q_2}\right)_3 = \left(\frac{q_1}{p_2}\right)_3 \left(\frac{q_2}{p_2}\right)_3.$$

It is clear from this, that if we denote the four symbols

$$\left(\frac{p_1}{q_1}\right)_3, \quad \left(\frac{p_1}{q_2}\right)_3, \quad \left(\frac{p_2}{q_1}\right)_3, \quad \left(\frac{p_2}{q_2}\right)_3$$

by a_1, b_1, b_2, a_2 respectively, and the reciprocal symbols by a_1', b_1', b_2', a_2', we have the equations

$$a_1 b_1 = a_1' b_1', \quad a_1 b_2 = a_1' b_2', \quad a_1 a_2 = a_1' a_2' = 1,$$
$$a_2 b_1 = a_2' b_1', \quad a_2 b_2 = a_2' b_2', \quad b_1 b_2 = b_1' b_2' = 1,$$

which imply that $a_2 = a_2'$, $b_2 = b_2'$, &c., or, since a, a', b, b', ... are cubic roots of unity,

$$\left(\frac{p_1}{q_1}\right)_3 = \left(\frac{q_1}{p_1}\right)_3.$$

If p_1 and q_1 be *conjugate* primes, the preceding proof fails; but it is easily seen that in this case also

$$\left(\frac{p_1}{p_2}\right)_3 = \left(\frac{p_2}{p_1}\right)_3 .$$

Lastly, if p and q are both of the form $3n+2$, it follows from the definition of the symbols, and from Fermat's Theorem, that

$$\left(\frac{p}{q}\right)_3 = \left(\frac{q}{p}\right)_3 .$$

The complementary theorems* relating to the unit ρ and the prime $1-\rho$ (which are not included in the preceding investigation) are

$$\left(\frac{\rho}{p_1}\right)_3 = \rho^{\frac{1}{3}(Np_1-1)} = \rho^{\alpha+\beta},$$

$$\left(\frac{1-\rho}{p_1}\right)_3 = \rho^{2\alpha};$$

where p_1 is a primary prime, and α and β are defined by the equality

$$p_1 = 2\alpha - 1 + 3\beta\rho.$$

Eisenstein has observed† that a demonstration of the law of cubic reciprocity, precisely similar to that analysed in Art. 33 of this Report, may be obtained by considering the integral $\int \frac{dx}{\sqrt{(1-x^3)}}$ and its inverse function, instead of the Lemniscate integral and Lemniscate function. He has not, however, entered into any details on this interesting subject (which is the more to be regretted, because there appears to be no published memoir treating specially of the integral $\int \frac{dx}{\sqrt{(1-x^3)}}$); although his latest proof of the Biquadratic law (see Art. 35) has been exhibited by him in such a form as to extend equally to Cubic Residues, and even to residues of the sixth power.

[*Addition.* In the definition of a primary number, for '$a \equiv \pm 1$,' read '$a \equiv -1$.' But, for the purposes of the theory of cubic residues, it is simpler to consider the two numbers $\pm(a+b\rho)$ as both alike primary (see Arts. 52 and 57).]

38. The first enunciation of the law of Cubic Reciprocity is due to Jacobi, and the demonstration of it which we have inserted in the preceding article

---

* Eisenstein, Crelle's Journal, vol. xxviii. p. 28 (the continuation of the memoir cited in the preceding note).

† In the memoir, 'Application de l'Algèbre' &c., already referred to.

is doubtless the same with that which he gave in his Königsberg Lectures. In one of his earliest memoirs ('De residuis cubicis commentatio numerosa,' Crelle, vol. ii. p. 66), which was composed after the announcement, but before the publication, of Gauss's memoirs on Biquadratic Residues, Jacobi had already arrived at two theorems relating to Cubic Residues, which involve the law of Reciprocity, and which he seems to have deduced from his formulae for the division of the circle. But, as it had not occurred to Jacobi, at the time when this memoir was written, to introduce, as modules, instead of the prime numbers themselves, the complex factors of which they are composed, the law of Cubic Reciprocity in its simplest form does not appear in the memoir.

To complete the present account of the Theory of the Residues of Powers, or of Binomial congruences, we should have in the next place to review the recent investigations of M. Kummer on complex numbers, and on the reciprocity of the residues of powers of which the index is a prime. But the consideration of these investigations, as well as of the other researches belonging to our present subject, our limits compel us to postpone to the second part of this Report.

[*Addition.* Jacobi's two theorems cannot properly be said to involve the cubic law of reciprocity. If $\left(\frac{p_1}{p_2}\right)_3 = 1$, it will follow from those theorems that $\left(\frac{p_2}{p_1}\right)_3 = 1$. But if $\left(\frac{p_1}{p_2}\right)_3 = \rho$, or ρ^2, they do not determine whether $\left(\frac{p_2}{p_1}\right)_3 = \rho$, or ρ^2. It is remarkable that these theorems, 'formâ genuinâ quâ inventa sunt,' may be obtained by applying the criteria for the resolubility or irresolubility of cubic congruences (Art. 67) to the congruence $r^3 - 3\lambda r - \lambda M \equiv 0$, mod q (Art. 43), which, by virtue of M. Kummer's theorem (Art. 44), is resoluble or irresoluble according as q is or is not a cubic residue of λ.]

VI.

REPORT ON THE THEORY OF NUMBERS.

PART II.

[Report of the British Association for 1860, pp. 120–169.]

---

39. *Residues of the Higher Powers. Researches of Jacobi.*—The principles which have sufficed for the determination of the laws of reciprocity affecting quadratic, cubic, biquadratic, and sextic residues, are found to be inadequate when we come to residues of the 5th, 7th, or higher powers. This was early observed by Jacobi, when, after his investigations of the cubic and biquadratic theorems, he turned his attention to residues of the 5th, 8th, and 12th powers*. It was evident, from a comparison of the cubic and biquadratic theories, that in the investigation of the laws of reciprocity the ordinary prime numbers of arithmetic must be replaced by certain factors of those prime numbers composed of roots of unity; and Jacobi, in the note just referred to, has indicated very clearly the nature of those factors in the case of the 5th, 8th, and 12th powers respectively. He ascertained that the two complex factors composed of 5th roots of unity into which every prime number of the form $5n+1$ is resoluble by virtue of Theorem IV. of art. 30 of this Report, are not prime numbers, *i.e.* are each capable of decomposition into the product of two similar complex numbers; so that every (real) prime number of the form $5n+1$ is to be regarded as the product of *four* conjugate complex factors; and these factors are precisely the complex primes which we have to consider in the theory of

---

* See a note communicated by him to the Berlin Academy, on May 16, 1839, in the Monatsberichte for that year, or in Crelle, vol. xix. p. 314, or Liouville, vol. viii. p. 268, in which, however, he implies that he had not as yet obtained a definitive result; nor does he seem at any subsequent period to have succeeded in completing this investigation.

quintic residues, in the place of the real primes they divide. To this we may add that primes of the forms $5n \pm 2$ continue primes in the complex theory; while those of the form $5n-1$ resolve themselves into *two* complex prime factors. Thus

$$7=7; \quad 11=(2+\alpha)(2+\alpha^2)(2+\alpha^3)(2+\alpha^4); \quad 13=13;$$

$$19=(4-3(\alpha+\alpha^4))(4-3(\alpha^2+\alpha^3)); \quad 29=(5-(\alpha+\alpha^4))(5-(\alpha^2+\alpha^3));$$

$$31=(2-\alpha)(2-\alpha^2)(2-\alpha^3)(2-\alpha^4), \text{ \&c.,}$$

where α is an imaginary 5th root of unity. Precisely similar remarks apply to the theories of residues of 8th and 12th powers,—real primes of the forms $8n+1$, $12n+1$, resolving themselves into four factors composed of 8th and 12th roots of unity respectively. By considerations similar to those previously employed by him in the case of biquadratic and cubic residues, Jacobi succeeded in demonstrating (though he has not enunciated) the formulæ of reciprocity affecting those powers for the particular case in which one of the two primes compared is a real number. But it would seem that he never obtained the law of reciprocity for the general case of any two complex primes; and indeed, for a reason which will afterwards appear, it was hardly possible that he should do so, so long as he confined himself to the consideration of those complex numbers which present themselves in the theory of the division of the circle. No less unsuccessful were the efforts of Eisenstein to obtain the formulæ relating to 8th powers, by an extension of the elliptical properties employed by him in his later proofs of the biquadratic theorem*. It does not appear that any subsequent writer has occupied himself with these special theories; while, on the other hand, the theory of complex numbers composed with roots of unity of which the exponent is any prime, has been the subject of an important series of investigations by MM. Dirichlet and Kummer, and has led the latter eminent mathematician to the discovery and demonstration of the law of reciprocity, which holds for all powers of which the exponent is a prime number not included in a certain exceptional class.

40. *Necessity for the Introduction of Ideal Primes.*—The fundamental proposition of ordinary arithmetic, that if two numbers have each of them no common divisor with a third number, their product has no common divisor with that third number, is, as we have seen, applicable to complex numbers formed

---

* See M. Kummer, 'Ueber die allgemeinen Reciprocitätsgesetze,' p. 27, in the Memoirs of the Berlin Academy for 1859.

with 3rd or 4th roots of unity, because it is demonstrable that Euclid's theory of the greatest common divisor is applicable in each of those cases. With complex numbers of higher orders this is no longer the case* ; and it is accordingly found that the arithmetical consequences of Euclid's process, which are of so much importance in the simpler cases, cease to exist in the general theory. In particular, the elementary theorem, that a number can be decomposed into prime factors in one way only, ceases to exist for complex numbers composed of 23rd† or higher roots of unity—if, at least (in the case of complex as of real numbers), we understand by a prime factor, a factor which cannot itself be decomposed into simpler factors‡. It appears, therefore, that in the higher complex theories, a number is not necessarily a prime number simply because it cannot be resolved into complex factors. But by the introduction of a new arithmetical conception —that of ideal prime factors—M. Kummer has shown that the analogy with the arithmetic of common numbers is completely restored. Some preliminary observations are, however, necessary to explain clearly in what this conception consists.

41. *Elementary Definitions relating to Complex Numbers.*—Let λ be a prime number, and α a root of the equation $\frac{\alpha^\lambda - 1}{\alpha - 1} = 0$; then any expression of the form

$$F(\alpha) = a_0 + a_1\alpha + a_2\alpha^2 + \ldots + a_{\lambda-2}\alpha^{\lambda-2}, \quad \ldots \ldots \quad \text{(A)}$$

in which $a_0, a_1, a_2, \ldots a_{\lambda-2}$ denote real integers, is called a complex integral number. To this form every rational and integral function of α can always be reduced; and it follows, from the irreducibility of the equation $\frac{\alpha^\lambda - 1}{\alpha - 1} = 0$, that the same complex number cannot be expressed in this reduced form in two different ways. The *norm* of $F(\alpha)$ is the real integer obtained by forming the product of all the $\lambda - 1$ values of $F(\alpha)$, so that

$$N.F(\alpha) = N.F(\alpha^2) = \ldots = N.F(\alpha^{\lambda-1}) = F(\alpha).F(\alpha^2).F(\alpha^3)\ldots F(\alpha^{\lambda-1}).$$

* {See Gauss, vol. ii, 'Zur Theorie der complexen Zahlen,' and Schering's Note.}

† For complex numbers composed with 5th or 7th roots of unity, the theorem still exists; for 23 and higher primes it certainly fails; whether it exists or not for 11, 13, 17, and 19, has not been definitely stated by M. Kummer (see below, Art. 50).

‡ 'Maxime dolendum videtur' (so said M. Kummer in 1844) 'quod hæc numerorum realium virtus, ut in factores primos dissolvi possint, qui pro eodem numero semper iidem sint, non eadem est numerorum complexorum, quæ si esset, tota hæc doctrina, quæ magnis adhuc difficultatibus premitur, facile absolvi et ad finem perduci posset.' (See his Dissertation in Liouville's Journal, vol. xii. p. 202.) In the following year he was already able to withdraw this expression of regret.

The operations of addition, subtraction and multiplication present no peculiarity in the case of these complex numbers; by the introduction of the norm, the division of one complex number by another is reduced to the case in which the divisor is a real integer. Thus

$$\frac{f(\alpha)}{F(\alpha)} = \frac{f(\alpha)\,F(\alpha^2)\,F(\alpha^3)\ldots F(\alpha^{\lambda-1})}{N.F(\alpha)};$$

and $f(\alpha)$ is said to be divisible by $F(\alpha)$ when every coefficient in the product $f(\alpha)\,F(\alpha^2)\,F(\alpha^3)\ldots F(\alpha^{\lambda-1})$, developed and reduced to the form (A), is divisible by $N.F(\alpha)$. When $f(\alpha)$ is not divisible by $F(\alpha)$, it is not, in general, possible to render the norm of the remainder less than the norm of the divisor; and it is owing to this circumstance that the common rule for finding the greatest common divisor is not generally applicable to complex numbers. If, in the expression (A), we consider the numbers $a_0, a_1, \ldots, a_{\lambda-2}$ as indeterminates, the norm is a certain homogeneous function of order $\lambda-1$, and of $\lambda-1$ indeterminates; so that the inquiry whether a given real number is or is not resoluble into the product of $\lambda-1$ conjugate complex factors, is identical with the inquiry whether it is or is not capable of representation by a certain homogeneous form, which is, in fact, the *resultant* of the two forms

$$a_0x^{\lambda-2} + a_1x^{\lambda-3}y + \ldots + a_{\lambda-2}y^{\lambda-2},$$

and

$$x^{\lambda-1} + x^{\lambda-2}y + x^{\lambda-3}y^2 + \ldots + y^{\lambda-1}.$$

The problem is considered in the former aspect by M. Kummer, in the latter by Dirichlet. The methods of Dirichlet appear to have been of extreme generality, and are as applicable to complex numbers, composed with the powers of a root of *any* irreducible equation having integral coefficients, as to the complex numbers which we have to consider here. Nevertheless, in the outline of this theory which we propose to give, we prefer to follow the course taken by M. Kummer: for Dirichlet's results have been indicated by him, for the most part, only in a very summary manner*; nor is it in any case difficult to assign to them their proper place in M. Kummer's theory; while, on the other hand, it would, perhaps, be impossible to express adequately, in any other form than that which M. Kummer has adopted, the numerous and important results (including the law of

* See his notes in the Monatsberichte of the Berlin Academy for 1841, Oct. 11, p. 280; 1842, April 14, p. 93; and 1846, March 30; also a Letter to M. Liouville, in Liouville's Journal, vol. v. p. 72; a note in the Comptes Rendus of the Paris Academy for 1840, vol. x. p. 286; and another in the Monatsberichte for 1847, April 15, p. 139.

reciprocity itself) contained in the elaborate series of memoirs which he has devoted to this subject*.

---

* The following is a list of M. Kummer's memoirs on complex numbers:—

1. De numeris complexis qui radicibus unitatis et numeris realibus constant, Breslau, 1844. This is an academical dissertation, addressed by the University of Breslau to that of Königsberg, on the tercentenary anniversary of the latter. It has been inserted by M. Liouville in his Journal, vol. xii. p. 185.

2. Ueber die Divisoren gewisser Formen der Zahlen, welche aus der Theorie der Kreistheilung entstehen.—Crelle, vol. xxx. p. 107.

3. Zur Theorie der complexen Zahlen, in the Monatsberichte for March 1845, or in Crelle, vol. xxxv. p. 319.

4. Ueber die Zerlegung der aus Wurzeln der Einheit gebildeten complexen Zahlen in ihre Primfactoren.—Crelle, vol. xxxv. p. 327. The date is Sept. 1846.

5. A note addressed to M. Liouville (April 28, 1847), in Liouville's Journal, vol. xii. p. 136.

6. Bestimmung der Anzahl nicht äquivalenter Klassen für die aus λten Wurzeln der Einheit gebildeten complexen Zahlen, und die idealen Factoren derselben.—Crelle, vol. xl. p. 93.

7. Zwei besondere Untersuchungen über die Classen-Anzahl, und über die Einheiten der aus λten Wurzeln der Einheit gebildeten complexen Zahlen.—Crelle, vol. xl. p. 117. (See also the Monatsberichte of the Berlin Academy for 1847, Oct. 14, p. 305.)

8. Allgemeiner Beweis des Fermat'schen Satzes, dass die Gleichung $x^\lambda + y^\lambda = z^\lambda$ unlösbar ist, für alle diejenigen Potenz-Exponenten λ, welche ungerade Primzahlen sind, und in den Zählern der ersten $\frac{1}{2}(\lambda-3)$ Bernouillischen Zahlen als Factoren nicht vorkommen.—Crelle, vol. xl. p. 131. (See also the Monatsberichte for 1847, April 15, p. 132.) This and the two preceding memoirs are dated June 1849.

9. Recherches sur les Nombres Complexes.—Liouville, vol. xvi. p. 377. This memoir contains a very full *résumé* of the whole theory, and may be read by any one acquainted with the elements of the theory of numbers.

10. A note in the Monatsberichte of the Berlin Academy for May 27, 1850, p. 154, which contains the first enunciation of the law of reciprocity.

11. Ueber die Ergänzungssätze zu den allgemeinen Reciprocitätsgesetzen.—Crelle, vol. xliv. p. 93 (Nov. 30, 1851), and vol. lvi. p. 270 (Dec. 1858).

12. A note on the irregularity of determinants, in the Berlin Monatsberichte for 1853, March 14, p. 194.

13. Ueber eine besondere Art aus complexen Einheiten gebildeter Ausdrücke.—Crelle, vol. l. p. 212 (Aug. 31, 1854).

14. Ueber die den Gaussischen Perioden der Kreistheilung entsprechenden Congruenzwurzeln.—Crelle, vol. liii. p. 142 (June 5, 1856).

15. Einige Sätze über die aus den Wurzeln der Gleichung $\alpha^\lambda = 1$ gebildeten complexen Zahlen für den Fall dass die Klassenanzahl durch λ theilbar ist, nebst Anwendung derselben auf einen weiteren Beweis des letzten Fermat'schen Lehrsatzes.—Memoirs of the Berlin Academy for 1857, p. 41. An abstract of this memoir will be found in the Monatsberichte for 1857, May 4, p. 275.

16. Theorie der idealen Primfactoren der complexen Zahlen, welche aus den Wurzeln der Gleichung

42. *Complex Units.* A complex unit is a complex number of which the norm is unity. If $\lambda = 3$, there is only a finite number [six] of units included in the formula $\pm \alpha^k$. But for all higher values of λ, the number of units is infinite. Nevertheless it is always possible to assign a system of $\mu - 1$ units (putting, for brevity, $\frac{1}{2}(\lambda - 1) = \mu$) such that *all* units are included in the formula $\pm \alpha^k u_1^{n_1} u_2^{n_2} \dots u_{\mu-1}^{n_{\mu-1}}$; in which $u_1, u_2, u_3, \dots u_{\mu-1}$ are the assigned units, and $k, n_1, n_2, \dots n_{\mu-1}$ are real (positive or negative) integral numbers. A system of units, capable of thus representing all units whatsoever, is called a fundamental system. The existence, for every value of λ, of fundamental systems of $\mu - 1$ units may be established by means of a general proposition due to Dirichlet and relating to any irreducible equation having unity for its first coefficient, and all its coefficients integral. If, in such an equation, R be the number of real, and $2I$ of imaginary roots, there always exist systems of $R + I - 1$ fundamental units, by means of which all other units can be expressed; or, in other words, the indeterminate equation 'Norm $= 1$' is always resoluble in an infinite number of ways, and all its solutions can be expressed by means of $R + I - 1$ fundamental solutions *. The demonstration of the actual existence, in every case, of these

---

$\omega^n = 1$ gebildet sind, wenn n eine zusammengesetzte Zahl ist.—Memoirs of the Berlin Academy for 1856, p. 1.

17. Ueber die allgemeinen Reciprocitätsgesetze unter den Resten und Nicht-resten der Potenzen, deren Grad eine Primzahl ist.—Memoirs of the Berlin Academy for 1859, p. 20. It was read on Feb. 18, 1858, and May 5, 1859. An abstract will be found in the Monatsberichte of the former year.

A memoir by M. Kronecker (De unitatibus complexis, Berlin, 1845; it is his inaugural dissertation on taking his doctorate) connects itself naturally with the earlier memoirs of the preceding series.

* To enunciate Dirichlet's theorem with precision, let $f(x) = 0$ be the proposed equation; let $a_1, a_2, \dots a_n$ be its roots, and $\psi(a_1), \psi(a_2), \dots \psi(a_n)$ a system of n conjugate units. If the analytical modulus of every one of the quantities $\psi(a_1), \psi(a_2), \dots \psi(a_n)$ be unity, the system of units is an isolated or *singular* system. The number of singular systems (if any such exist) is always finite, whence it is easy to infer that the units they comprise are simply roots of unity. For if $\psi(a)$ be a singular unit, its powers are evidently also singular units, and therefore cannot be all different from one another; *i.e.* $\psi(a)$ is a root of unity. If $f(x)$ be of an uneven order, there are no singular units; if $f(x)$ be of an even order, -1 is a singular unit; and if $f(x) = 0$ have any real roots, it is the only singular unit; whereas if all the roots of $f(x) = 0$ be imaginary, other singular units may in special cases exist. Thus the equation $\frac{x^\lambda - 1}{x - 1} = 0$ has $2(\lambda - 1)$ singular units included in the formula $\pm \alpha^k$. Admitting this definition of singular units, we may enunciate Dirichlet's theorem as follows:—A system of h units $[h = I + R - 1]$, $e_1(a), e_2(a), \dots e_h(a)$, composed with any root a, can always be assigned such that every unit composed with the same root can be represented (and in one way only) by the formula

$$\omega \,.\, e_1^{n_1}(a) \,.\, e_2^{n_2}(a) \,.\, e_3^{n_3}(a) \dots . \, e_h^{n_h}(a),$$

systems of fundamental units (a theorem which is, as Jacobi has said*, 'un des plus importants, mais aussi un des plus épineux de la science des nombres') is of essential importance in the theory of complex numbers, and has the same relation to that theory which the solution of the Pellian equation $x^2 - Dy^2 = 1$ has to the theory of quadratic forms of determinant D. It may be observed, however, that in the case which we have to consider here, that of the equation $\frac{\alpha^\lambda - 1}{\alpha - 1} = 0$, the existence of fundamental systems of $\mu - 1$ units has been demonstrated independently of Dirichlet's general theory by MM. Kronecker and Kummer†.

If $\lambda = 5$, $\alpha + \alpha^{-1}$ is the only fundamental unit; so that every unit is included in the formula $\pm\, \alpha^k (\alpha + \alpha^{-1})^n$.
If $\lambda = 7$, the complex units are included in the formula

$$\pm\, \alpha^k (\alpha^1 + \alpha^{-1})^{n_1} (\alpha^2 + \alpha^{-2})^{n_2}.$$

But for higher primes the actual calculation of a system of fundamental units involves great labour; and a method practically available for the purpose has not yet been given. It is remarkable that every unit can be rendered real (*i.e.* a function of the binary sums or *periods* $\alpha^1 + \alpha^{-1}$, &c.) by multiplying it by a properly assumed power of α. We shall therefore suppose, in what follows, that the units of which we speak have been thus reduced to a real form.

For all values of λ greater than 5, the number of systems of fundamental units is infinite. For if $u_1, u_2, \ldots u_{\mu-1}$ still represent a system of fundamental units, it is evident that the system $E_1, E_2, \ldots E_{\mu-1}$, defined by the equations

$$\left.\begin{aligned} E_1 &= u_1^{(1,1)} . u_2^{(1,2)} \ldots\ldots . u_{\mu-1}^{(1,\mu-1)}, \\ E_2 &= u_1^{(2,1)} . u_2^{(2,2)} \ldots\ldots . u_{\mu-1}^{(2,\mu-1)}, \\ &\cdots\cdots\cdots\cdots \\ E_{\mu-1} &= u_1^{(\mu-1,1)} . u_2^{(\mu-1,2)} \ldots\ldots . u_{\mu-1}^{(\mu-1,\mu-1)}, \end{aligned}\right\} \quad \ldots\ldots \quad (A)$$

is also a fundamental system, if the indices (1, 1), &c. be integral numbers, and if the determinant $\Sigma \pm (1, 1)\,(2, 2) \ldots (\mu - 1, \mu - 1)$ be equal to unity. And

---

where $n_1, n_2, \ldots n_h$ are positive or negative integral numbers and ω is unity, or some one of the singular units composed with α.

The principles on which the demonstration of this theorem depends are very briefly indicated in the notes presented by Dirichlet to the Berlin Academy in 1841, 1842, and 1846. {See also Liouville, vol. v. p. 72.}

* Crelle's Journal, vol. xl. p. 312.

† See Kronecker, De unitatibus complexis, pars altera; and Kummer, in Liouville's Journal, vol. xvi. p. 323.

conversely, every system of fundamental units will be represented by the equations (A), if in them we assign to the indices (1, 1), (2, 2), &c. all systems of integral values in succession consistent with the condition

$$\Sigma \pm (1, 1)(2, 2)(3, 3) \dots (\mu - 1, \mu - 1) = \pm 1;$$

so that a single system of fundamental units represents to us all possible systems.

We shall also have occasion to allude to independent systems of units. A system of $\mu - 1$ units, $u_1, u_2, \dots u_{\mu-1}$, is said to be independent when it is impossible to satisfy the equation

$$u_1^{n_1} u_2^{n_2} u_3^{n_3} \dots u_{\mu-1}^{n_{\mu-1}} = 1,$$

whatever integral values are assigned to the indices $n_1, n_2, n_3, \dots n_{\mu-1}$. The equations (A) will represent all possible systems of independent units, if we suppose that in them the indices (1, 1), (2, 2), (3, 3) ... receive all positive and negative integral values, subject only to the condition that the determinant $\Delta = \Sigma \pm (1, 1)(2, 2) \dots (\mu - 1, \mu - 1)$ does not vanish. Every system of fundamental units is also independent; but not conversely. Every unit can be represented as a product of the powers of the units of an independent system; but if the system be not also fundamental, the indices of the powers are not in general integral, but are fractions having denominators which divide Δ. Lastly, if $c_1(a), c_2(a), \dots c_{\mu-1}(a)$ be a system of independent units, the logarithmic determinant

$$\begin{vmatrix} L.c_1(a) & , L.c_2(a) & , \dots\dots, L.c_{\mu-1}(a) \\ L.c_1(a^\gamma) & , L.c_2(a^\gamma) & , \dots\dots, L.c_{\mu-1}(a^\gamma) \\ \cdot & \cdot & \cdot \\ L.c_1(a^{\gamma^{\mu-2}}) & , L.c_2(a^{\gamma^{\mu-2}}) & , \dots\dots, L.c_{\mu-1}(a^{\gamma^{\mu-2}}) \end{vmatrix}$$

in which γ denotes a primitive root of λ, is different from zero; and conversely, if the determinant be different from zero, the system of units is independent. For all systems of fundamental units, the absolute value of the logarithmic determinant is the same; for any other independent system, its value is Δ times that least value. The quantities denoted by the symbols $L.c_1(a)$, $L.c_2(a)$, &c. are the arithmetical logarithms of the real units $c_1(a)$, &c. taken positively.

43. *Gauss's Equations of the Periods.*—In Gauss's theory of the division of the circle, it is shown that if λ be a prime number, and if $ef = \lambda - 1$, the e *periods* of f roots each, that is the quantities $\eta_0, \eta_1, \eta_2, \dots \eta_{e-1}$, defined by the equations

$$\begin{aligned}
\eta_0 &= \alpha^{\gamma^0} + \alpha^{\gamma^e} + \alpha^{\gamma^{2e}} + \ldots\ldots + \alpha^{\gamma^{(f-1)e}},\\
\eta_1 &= \alpha^{\gamma^1} + \alpha^{\gamma^{e+1}} + \alpha^{\gamma^{2e+1}} + \ldots\ldots + \alpha^{\gamma^{(f-1)e+1}},\\
&\cdots\cdots\cdots\cdots\cdots\cdots\\
\eta_{e-1} &= \alpha^{\gamma^{e-1}} + \alpha^{\gamma^{2e-1}} + \alpha^{\gamma^{3e-1}} + \ldots\ldots + \alpha^{\gamma^{fe-1}}
\end{aligned}$$

(γ still denoting a primitive root of λ), are the roots of an irreducible equation of order e having integral coefficients, which we shall symbolize by

$$F(y) = y^e + A_1 y^{e-1} + A_2 y^{e-2} + \ldots + A_{e-1} y + A_e = 0$$

(see Disq. Arith., Art. 346). This equation is of the kind called Abelian; that is to say, each of the e periods is a rational function of any other, in such a manner that we may establish the equations

$$\eta_1 = \phi(\eta_0), \quad \eta_2 = \phi(\eta_1), \quad \eta_3 = \phi(\eta_2), \ldots, \eta_0 = \phi(\eta_{e-1});$$

where it is to be observed that the coefficients of the function ϕ are not in general integral. The determination of the coefficients of the equation $F(y) = 0$ may be effected, for any given prime λ, and any given divisor e of $\lambda - 1$, by methods which, however tedious, present no theoretical difficulty. Every rational and integral function of the periods can be reduced to the form

$$a_0 \eta_0 + a_1 \eta_1 + a_2 \eta_2 + \ldots + a_{e-1} \eta_{e-1}.$$

If we combine the equation

$$1 + \eta_0 + \eta_1 + \eta_2 + \ldots + \eta_{e-1} = 0$$

with the $e-2$ equations, by which $\eta_0^2, \eta_0^3, \ldots \eta_0^{e-1}$ are expressed in that linear form, we may eliminate $\eta_2, \eta_3, \ldots \eta_{e-1}$, and shall thus obtain an equation of order e, satisfied by η_0, *i.e.* the equation of the periods, or $F(y) = 0$. This is the method proposed by Gauss (Disq. Arith., Art. 346); M. Kummer, instead, forms the system of equations

$$\begin{aligned}
\eta_0^2 &= n_0 f + (0, 0)\eta_0 + (0, 1)\eta_1 + (0, 2)\eta_2 + \ldots + (0, e-1)\eta_{e-1},\\
\eta_0\eta_1 &= n_1 f + (1, 0)\eta_0 + (1, 1)\eta_1 + (1, 2)\eta_2 + \ldots + (1, e-1)\eta_{e-1},\\
\eta_0\eta_2 &= n_2 f + (2, 0)\eta_0 + (2, 1)\eta_1 + (2, 2)\eta_2 + \ldots + (2, e-1)\eta_{e-1},\\
&\cdots\cdots\cdots\cdots\cdots\cdots\\
\eta_0\eta_{e-1} &= n_{e-1} f + (e-1, 0)\eta_0 + (e-1, 1)\eta_1 + (e-1, 2)\eta_2 + \ldots + (e-1, e-1)\eta_{e-1}.
\end{aligned}$$

and eliminates $\eta_1, \eta_2, \ldots \eta_{e-1}$ from them. The symbol (k, h) represents the number of solutions of the congruence $\gamma^{ey+h} \equiv 1 + \gamma^{ex+k}$, mod λ, x and y denoting any two terms of a complete system of residues for the modulus f: n_k is zero for all values of k, excepting that $n_0 = 1$ if f be even, and $n_{\frac{1}{2}e} = 1$ if f be uneven *.

* Liouville's Journal, vol. xvi. p. 404.

The systems of equations corresponding to the particular cases $e=3$, $e=4$, have been given by Gauss, who has succeeded in expressing the values of the coefficients (k, h) in each of those cases by means of numbers depending on the representation of λ by certain simple quadratic forms; and has employed these expressions to demonstrate the criterion already mentioned in this Report for the biquadratic character of the number 2 *. A third method has been given by M. Libri †: he establishes the formula

$$\lambda N_k = \lambda^k + \eta_0(1+e\eta_0)^k + \eta_1(1+e\eta_1)^k + \ldots + \eta_{e-1}(1+e\eta_{e-1})^k,$$

in which N_k represents the number of solutions of the congruence

$$1 + x_1^e + x_2^e + \ldots + x_k^e \equiv 0, \bmod \lambda \ddagger.$$

If S_1, S_2, $S_3 \ldots$ denote the sums of the powers of the roots of the equation $F(y)=0$, this formula may be written thus,—

$$\lambda N_k = \lambda^k + S_1 + ke\,S_2 + \frac{k.k-1}{1.2}e^2 S_3 + \ldots + e^k S_{k+1},$$

or, solving for S_1, S_2, ...,

$$e^k S_{k+1} = \lambda\Big[N_k - kN_{k-1} + \frac{k(k-1)}{1.2}N_{k-2} - \ldots - (-1)^k N_1\Big] - (\lambda-1)^k.$$

From this equation, when the values of N_1, N_2, &c., have been determined, S_1, S_2, ... may be calculated, and thence by known methods the values of the coefficients of the equation $F(y)=0$. Lastly, M. Lebesgue has shown that, if we denote by σ_k the number of ways in which numbers divisible by λ can be formed by adding together k terms of the series γ^0, γ^1, ..., $\gamma^{\lambda-2}$, subject to the condition that no two powers of γ be added the indices of which are congruous for the modulus e, the function $(\lambda-1)F(y)$ assumes the form

$$\lambda[y^e - \sigma_1 y^{e-1} + \sigma_2 y^{e-2} - \ldots + (-1)^e \sigma_e] - (y-f)^e \S.$$

But the practical application of any of these methods is very laborious when

* Disq. Arith., Art. 358, and Theor. Res. Biq., Arts. 14–22.

† See the memoir 'Sur la Théorie des Nombres,' in his 'Mémoires de Mathématique et de Physique,' pp. 121, 122. The notation of the memoir has been altered in the text. See also M. Lebesgue, in Liouville's Journal, vol. ii. p. 287, and vol. iii. p. 113.

‡ In this congruence $x_1, x_2, \ldots x_k$ are k terms (the same or different) of a *complete* system of residues for the modulus λ; and in counting the number of solutions, two solutions are to be considered as different in which the same places are not occupied by the same numbers. A simpler formula for S_{k+1} may be obtained by considering $x_1, x_2, \ldots x_k$ to represent terms of a system of residues prime to λ, and denoting by $e^k\gamma_k$ the number of solutions of M. Libri's congruence on this hypothesis. We thus find $S_{k+1} = \lambda\gamma - f^k$ (Liouville, vol. iii. p. 116).

§ Liouville, vol. iii. p. 119.

λ is a large number, chiefly on account of the determinations which they all require of the numbers of solutions of which certain congruences are susceptible. For $e=2$ the equation is

$$y^2+y+\tfrac{1}{4}\{1-(-1)^\mu\lambda\}=0,$$

or, putting $r=2y+1$, $r^2-(-1)^\mu\lambda=0$. The cubic and biquadratic equations corresponding to the cases $e=3$ and $e=4$ are also known from Gauss's investigations. The results assume the simplest forms if we put $r=ey+1$. We then have

(1) $e=3$, $4\lambda=M^2+27N^2$, $M\equiv 1$, mod 3; $r^3-3\lambda r-\lambda M=0$.

(2) $e=4$; $\lambda=A^2+B^2$; $A\equiv 1$, mod 4; $e=(-1)^f$.

$$[r^2+(1-2\epsilon)\lambda]^2-4\lambda(r-A)^2=0\ *.$$

Though these determinations are not required in M. Kummer's theory, we have nevertheless given them here, in order to facilitate arithmetical verifications of his results. The forms of the period-equations for the cases $e=8$ and $e=12$ can (it may be added) be elicited from the results given by Jacobi in his note on the division of the circle (Crelle, vol. xxx. pp. 167, 168).

44. *The Period-Equations considered as Congruences.*—An arithmetical property of the equation $F(y)=0$, which renders it of fundamental importance in the theory of complex numbers, is expressed in the following theorem:

'If q be a prime number satisfying the congruence $q^f\equiv 1$, mod λ, the congruence $F(y)=0$, mod q, is completely resoluble, *i.e.* it is possible to establish an indeterminate congruence of the form

$$F(y)\equiv(y-u_0)(y-u_1)\ldots(y-u_{e-1}),\ \text{mod } q,$$

$u_0, u_1, \ldots u_{e-1}$ denoting integral numbers, congruous or incongruous, mod q †.'

* M. Lebesgue, Comptes Rendus, vol. li. p. 9. Gauss has not exhibited this last equation in its explicit form. See Theor. Res. Biq. *l. c.*

† This theorem was first given by Schoenemann (Crelle, vol. xix. p. 306); his demonstration, however, supposes that $q\geqq e$,—a limitation to which the theorem itself is not subject. The following proof is, with a slight modification, that given by M. Kummer (Crelle, vol. xxx. p. 107, or Liouville, vol. xvi. p. 408). {See also Liouville (II), vol. v. p. 369.} From the indeterminate congruence of Lagrange (see Art. 10 of this Report),

$$x(x-1)(x-2)\ldots.(x-q+1)\equiv x^q-x,\ \text{mod } q,$$

it follows that

$$(y-\eta_k)(y-\eta_k-1)(y-\eta_k-2)\ldots(y-\eta_k-q+1)\equiv(y-\eta_k)^q-(y-\eta_k)$$
$$\equiv y^q-\eta_k{}^q-(y-\eta_k)\equiv y^q-y,\ \text{mod } q,$$

observing that $\eta_k{}^q\equiv\eta_{k+\text{Ind}\,q}$, and that, if Ind q be divisible by e (or, which is the same thing, if q

A particular case of this theorem, relating to the equation $\frac{x^\lambda - 1}{x-1} = 0$ (which may of course be regarded as the equation of the $\lambda - 1$ periods, consisting each of a single root), is due to Euler, and is included in his theory of the Residues of Powers; for it follows from that theory (see Art. 12 of this Report), that the binomial congruence $x^\lambda - 1 \equiv 0$ (and therefore also the congruence $\frac{x^\lambda - 1}{x-1} \equiv 0$, mod q) is completely resoluble for every prime of the form $m\lambda + 1$.

A remarkable relation subsists between the periods $\eta_0, \eta_1, \ldots \eta_{e-1}$ of the *equation* $F(y) = 0$, and the roots $u_0, u_1, u_2, \ldots, u_{e-1}$ of the *congruence* $F(y) \equiv 0$, mod q. This relation is expressed in the following theorem:—

'Every equation which subsists between any two functions of the periods, will subsist as a congruence for the modulus q when we substitute for the periods the roots of the congruence $F(y) \equiv 0$ taken in a certain order.'

It is immaterial which root of the congruence we take to correspond to any given root of the equation. But when this correspondence has once been established in a single case, we must attend to the sequence which exists among the roots of the congruence corresponding to the sequence of the periods. When $u_0, u_1, \ldots, u_{e-1}$ are all incongruous, their order of sequence is determined by the congruences

$$u_1 \equiv \phi(u_0), \quad u_2 \equiv \phi(u_1), \ \ldots\ldots, \ u_0 \equiv \phi(u_{e-1}), \text{ mod } q,$$

which correspond to the equations

$$\eta_1 = \phi(\eta_0), \quad \eta_2 = \phi(\eta_1), \ \ldots\ldots, \ \eta_0 = \phi(\eta_{e-1}),$$

and which are always significant, although the coefficients of ϕ are fractional, because it may be proved that their denominators are prime to the modulus q. When $u_0, u_1, \ldots, u_{e-1}$ are not all incongruous [an exceptional case which implies that q divides the *discriminant* of $F(y)$], a precisely similar relation subsists, though it cannot be fixed in the same manner, and though the number of incongruous solutions of the congruence is not equal to the number of the

---

satisfy the congruence $q^f \equiv 1$, mod λ), $\eta_{k+\text{Ind } q} = \eta_k$. Multiplying together the e congruences obtained by giving to k the e values of which it is susceptible in the formula

$$(y-\eta_k)(y-\eta_k-1)(y-\eta_k-2)\ldots(y-\eta_k-q+1) \equiv y^q - y, \text{ mod } q,$$

we find $$F(y)\,F(y-1)\,F(y-2)\ldots F(y-q+1) \equiv (y^q-y)^e, \text{ mod } q;$$

whence, by a principle to which we shall have occasion to refer subsequently (see Art. 69), it appears that $F(y)$ is congruous for the modulus q to a product of the form

$$(y-u_0)(y-u_1)\ldots(y-u_{e-1}).$$

periods. (See a paper by M. Kummer in Crelle's Journal, vol. liii. p. 142, in which he has established this fundamental proposition on a satisfactory basis*.)

45. *Conditions for the Divisibility of the Norm of a Complex Number by a Real Prime*†. Instead of the complex number

$$f(\alpha) = a_0 + a_1\alpha + a_2\alpha^2 + \dots + a_{\lambda-2}\alpha^{\lambda-2},$$

let us now, for a moment, consider the complex number

$$\psi(\eta_0) = c_0\eta_0 + c_1\eta_1 + c_2\eta_2 + \dots + c_{e-1}\eta_{e-1},$$

which, with its conjugates

$$\psi(\eta_1) = c_0\eta_1 + c_1\eta_2 + c_2\eta_3 + \dots + c_{e-1}\eta_0,$$
$$\psi(\eta_2) = c_0\eta_2 + c_1\eta_3 + c_2\eta_4 + \dots + c_{e-1}\eta_1,$$
$$\dots\dots\dots\dots\dots\dots$$
$$\psi(\eta_{e-1}) = c_0\eta_{e-1} + c_1\eta_0 + c_2\eta_1 + \dots + c_{e-1}\eta_{e-2},$$

is a function of the periods only, and is therefore a specialized form of the general complex number $f(\alpha)$; and let q still denote a real prime, satisfying the congruence $q^f \equiv 1$, mod λ. By means of the relation subsisting between the equation-roots $\eta_0, \eta_1, \dots, \eta_{e-1}$, and the congruence-roots $u_0, u_1, \dots, u_{e-1}$, M. Kummer has demonstrated the two following theorems:—

(i.) 'The necessary and sufficient condition that $\psi(\eta)$ should be divisible by q (*i.e.* that the coefficients $c_0, c_1, \dots, c_{e-1}$ should be all separately divisible by q) is that the e congruences

$$\psi(u_0) = c_0u_0 + c_1u_1 + c_2u_2 + \dots + c_{e-1}u_{e-1} \equiv 0, \text{ mod } q,$$
$$\psi(u_1) = c_0u_1 + c_1u_2 + c_2u_3 + \dots + c_{e-1}u_0 \equiv 0, \text{ mod } q,$$
$$\dots\dots\dots\dots\dots\dots$$
$$\psi(u_{e-1}) = c_0u_{e-1} + c_1u_0 + c_2u_1 + \dots + c_{e-1}u_{e-2} \equiv 0, \text{ mod } q,$$

should be simultaneously satisfied.'

(ii.) 'The necessary and sufficient condition that the norm of $\psi(\eta)$, taken with respect to the periods, *i.e.* the number $\psi(\eta_0).\psi(\eta_1)\dots\psi(\eta_{e-1})$, should be divisible by q, is that *one* of the e congruences

$$\psi(u_0) \equiv 0, \ \psi(u_1) \equiv 0, \dots \psi(u_{e-1}) \equiv 0, \text{ mod } q,$$

should be satisfied.'

---

* {If q does not divide the discriminant, it is true, conversely, that if $F(y) \equiv 0$ for any value of y, mod q, $q^f \equiv 1$, mod λ. For we readily find $(\eta_{\text{Ind}\,q} - \eta)^e \equiv 0$, mod q: that is q divides the discriminant, which is contrary to the hypothesis.

M. Kummer shows that if q divides $F(y)$, and y^f is not $\equiv 1$, mod p, then q is a residue of a power having with e some common divisor other than unity; therefore if e is a prime, q is a residue of an e-th power.}

† The outline of the theory of complex numbers contained in this and the subsequent Articles is chiefly derived from M. Kummer's mémoire in Liouville, vol. xvi. p. 411.

These results may be extended to *any* complex number $f(\alpha)$, by first reducing it to the form

$$f(\alpha) = \psi_0(\eta_0) + \alpha\psi_1(\eta_0) + \alpha^2\psi_2(\eta_0) + \ldots + \alpha^{f-1}\psi_{f-1}(\eta_0).$$

This is always possible; for, since the f roots which compose any one period, *e.g.* η_0, are the roots of an equation $\chi(\alpha) = 0$ of order f, the coefficients of which are complex integers involving the periods only*, we may simply divide $f(\alpha)$ by $\chi(\alpha)$, and the remainder will give us the expression of $f(\alpha)$ in the required form. Further, let q now denote a prime *appertaining to the exponent* f (not merely satisfying the congruence $q^f \equiv 1$, mod λ, but also satisfying no congruence of lower index and of the same form). The two preceding theorems are then replaced by the two following, which are analogous to them, and include them.

(i.) 'The necessary and sufficient condition that $f(\alpha)$ should be divisible by q, is that the congruences

$$\psi_0(u_k) \equiv 0,\ \psi_1(u_k) \equiv 0,\ \ldots,\ \psi_{f-1}(u_k) \equiv 0, \text{ mod } q,$$

should be simultaneously satisfied for every value of k.'

(ii.) 'And the condition that the norm of $f(\alpha)$ should be divisible by q, is that the same congruences should be satisfied for some one value of k.'

When the congruences $\psi_0(u_k) \equiv 0$, $\psi_1(u_k) \equiv 0$, ... $\psi_{f-1}(u_k) \equiv 0$, mod q, are simultaneously satisfied, $f(\alpha)$ is said to be *congruous to zero (mod q), for the substitution* $\eta_0 = u_k$. These f congruences may be replaced by a single congruence in either of two different ways. Thus, if we denote by $F(\eta_0)$ the complex number involving the periods only, which we obtain by multiplying together the f complex numbers

$$f(\alpha),\ f(\alpha^{\gamma^e}),\ f(\alpha^{\gamma^{2e}}), \ldots f(\alpha^{\gamma^{(f-1)e}}),$$

it may be proved that the single congruence $F(u_k) \equiv 0$, mod q, is precisely equivalent to the f congruences

$$\psi_0(u_k) \equiv 0,\ \psi_1(u_k) \equiv 0,\ \ldots,\ \psi_{f-1}(u_k) \equiv 0.$$

Or, again, if we denote by $\Psi(\eta_0)$ a complex number congruous to zero for every one of the substitutions $\eta_0 = u_1$, $\eta_0 = u_2$, ... $\eta_0 = u_{e-1}$, but not congruous to zero for the substitution $\eta_0 = u_0$ (such complex numbers, involving the periods only, can in every case be assigned) †, it is readily seen that the same f congruences are comprehended in the single formula

$$\Psi(\eta_{e-k}) f(\alpha) \equiv 0, \text{ mod } q.$$

* Disq. Arith., Art. 348.

† Crelle, vol. liii. p. 145. The number $\Psi(\eta)$ of this memoir possesses the property in question.

The utility of this latter mode of expressing the f congruences will appear in the sequel; the formula $F(u_k) \equiv 0$, mod q, is of importance, because it supplies an immediate demonstration of the important proposition, that if a product of two factors be congruous to zero for the substitution $\eta_0 = u_k$, one or other of the factors must be congruous to zero for that substitution.

46. *Definition of Ideal Prime Factors.*—To develope the consequences of the preceding theorems, let us consider a prime number q appertaining to the exponent f; and let us first suppose that it is capable of being expressed as the norm (taken with respect to the periods) of a complex number $\psi(\eta_0)$, which contains the periods of f terms only; so that

$$q = \psi(\eta_0) \,.\, \psi(\eta_1) \ldots \psi(\eta_{e-1}).$$

If the substitution of u_0 in ψ render $\psi(u_0) \equiv 0$, mod q, we may distinguish the e factors of q by means of the substitutions which respectively render them congruous to zero; so that, for example, $\psi(\eta_{e-k})$ is the factor *appertaining to the substitution* $\eta_0 = u_k$.

We thus obtain the theorem that if $f(\alpha)$ be congruous to zero, mod q, for any substitution $\eta_0 = u_0$, $f(\alpha)$ is divisible by the factor of q appertaining to that substitution. For if $\psi(\eta_0)$ be that factor of q,

$$\frac{f(\alpha)}{\psi(\eta_0)} = \frac{f(\alpha)\,\psi(\eta_1)\,\psi(\eta_2) \ldots \psi(\eta_{e-1})}{q};$$

but $f(\alpha)\,\psi(\eta_1)\,\psi(\eta_2) \ldots \psi(\eta_{e-1})$ is congruous to zero, mod q, for every one of the substitutions $\eta_0 = u_0$, $\eta_0 = u_1$, ... $\eta_0 = u_{e-1}$; it is consequently divisible by q; *i.e.* $f(\alpha)$ is divisible by $\psi(\eta_0)$. A useful particular case of this theorem is that $u_k - \eta_k \equiv 0$, mod $\psi(\eta_0)$, if $\psi(u_0) \equiv 0$, mod q.

Again, it may be shown that these complex factors of q are *primes* in the most proper sense of the word: *i.e.*, first, that they are incapable of resolution into any two complex factors, unless one of those factors be a complex unit; and secondly, that if any one of them divide the product of two factors, it necessarily divides one or other of the two factors separately. That $\psi(\eta_0)$ possesses the first property is evident, because its norm is a real prime, and that it possesses the second is a consequence of the last theorem of Art. 45. For if $\psi(\eta_0)$ divide $f_1(\alpha) \times f_2(\alpha)$, either $f_1(\alpha)$ or $f_2(\alpha)$, by virtue of that theorem, is congruous to zero (mod q) for the substitution $\eta_0 = u_0$; that is to say, either $f_1(\alpha)$ or $f_2(\alpha)$ is divisible by $\psi(\eta_0)$.

Now, if every prime q which appertains to the exponent f were actually capable of resolution into e complex factors composed of the e periods of f roots, these factors would represent to us all the true primes to be considered in the theory of the residues of λ-th powers. And for values of λ inferior to **11**, perhaps to **23**, this is, in fact, the case. But for higher values of λ, the real primes appertaining to the exponent f divide themselves into two different groups, according as they are or are not susceptible of resolution into e conjugate factors. Let, then, q represent any prime appertaining to the exponent f, whether susceptible or not of this resolution, and let $f(\alpha)$ still denote a complex number which is rendered congruous to zero by the substitution $\eta_0 = u_0$; $f(\alpha)$ is said by M. Kummer to contain *the ideal factor of q appertaining to the substitution* $\eta_0 = u_0$. This definition is admissible, because it is verified, as we have just seen, when q is actually resoluble into e conjugate factors; and its introduction is justified, as M. Kummer observes, by its utility. To obtain a definition of the multiplicity of an ideal factor, we may employ a complex number $\Psi(\eta)$ possessing the property indicated in the last article. If of the two congruences

$$[\Psi(\eta_0)]^n \ f(\alpha) \equiv 0, \bmod q^n,$$
$$[\Psi(\eta_0)]^{n+1} f(\alpha) \equiv 0, \bmod q^{n+1},$$

the former be satisfied, and the latter not, $f(\alpha)$ is said to contain n times precisely the ideal factor of q which appertains to the substitution $\eta_0 = u_0$.

47. *Elementary Theorems relating to Ideal Factors.*—The following propositions are partly restatements (in conformity with the definitions now introduced) of results to which we have already referred, and partly simple corollaries from them. They will serve to show that the elementary properties of ordinary integers may now be transferred to complex numbers.

(1.) A complex number is divisible by q when it contains all the ideal factors of q. If it contain all of those factors n times but not all of them $n+1$ times, it is divisible by q^n but not by q^{n+1}.

(2.) The norm of a complex number is divisible by q when the complex number contains one of the ideal factors of q. If (counting multiple factors) it contain, in all, k of the ideal factors of q, the norm is divisible by q^{kf}, but by no higher power of q (f denoting the exponent to which q appertains).

(3.) A product of two or more factors contains the same ideal divisors as its factors taken together.

(4.) The necessary and sufficient condition that one complex number should be divisible by another is, that the dividend should contain all the ideal factors of the divisor at least as often as the divisor.

(5.) Two complex numbers which contain the same ideal factors are identical, or else differ only by a unit factor.

(6.) Every complex number contains a finite number of ideal prime factors.

These ideal prime factors (as well as the multiplicity of each of them) are perfectly determinate.

The prime number λ is the only real prime excluded from the preceding considerations. Since $\lambda = (1-\alpha)(1-\alpha^2)\dots(1-\alpha^{\lambda-1})$, it appears that the norm of $1-\alpha$ is a real prime, and therefore $1-\alpha$ cannot be resolved into the product of two factors, unless one of them be a unit. Again, because the necessary and sufficient condition for the divisibility of a complex number by $1-\alpha$ is that the sum of the coefficients of the complex number should be congruous to zero for the modulus λ, and because the sum of the coefficients of a product of complex numbers is congruous, for the modulus λ, to the product of the sums of the coefficients of the factors, it appears that if the norm of a complex number is divisible by λ, the complex number is itself divisible by $1-\alpha$; and also that, if the product of two complex numbers be divisible by $1-\alpha$, one or other of the factors separately must be divisible by $1-\alpha$. Hence $1-\alpha$ is a true complex prime, and is the only prime factor of λ; in fact,

$$\lambda = (1-\alpha)(1-\alpha^2)\dots(1-\alpha^{\lambda-1}) = e(\alpha)(1-\alpha)^{\lambda-1},$$

if $e(\alpha)$ denote the complex unit

$$\frac{1-\alpha^2}{1-\alpha}\cdot\frac{1-\alpha^3}{1-\alpha}\cdots\cdots\frac{1-\alpha^{\lambda-1}}{1-\alpha}.$$

The theorems which have preceded enable us to give a definition of the norm of an ideal complex number. If the ideal number contain the factor $1-\alpha$ m times, and if it besides contain $k, k', k'', \dots$ prime factors of the primes $q, q', q'', \dots$ appertaining to the exponents $f, f', f'', \dots$ respectively, we are to understand by its norm the positive integral number

$$\lambda^m q^{kf} q'^{k'f'} q''^{k''f''} \dots;$$

a definition which, by virtue of the second proposition of this article, is exact in the case of an actually existing number.

It will be observed that the number of actual or ideal prime factors (compound of λ-th roots of unity) into which a given real prime can be decomposed, depends exclusively on the exponent to which the prime appertains for the modulus λ. If the exponent is f, the number of ideal factors is $\dfrac{\lambda-1}{f} = e$. Thus, if q be a primitive root of λ, q continues a prime in the complex theory; if it be a primitive root of the congruence $x^{\frac{1}{2}(\lambda-1)} \equiv 1$, mod λ, it is only resoluble into two conjugate prime factors. This dependence of the number of ideal prime factors of a given prime upon the exponent to which it appertains is a remarkable instance of an intimate and simple connexion between two properties of the same prime number, which appear at first sight to have no immediate connexion with one another.

It may be convenient to remark that the word Ideal is sometimes used so as to include, and sometimes so as to exclude, actually existent complex numbers; but it is not apprehended that any confusion can arise from this ambiguity, which it is not worth while to remove at the expense of introducing a new technical term.

48. *Classification of Ideal Numbers.*—An ideal number (using the term in its restricted sense) is incapable of being exhibited in an isolated form as a complex integer; as far as has yet appeared, it has no quantitative existence; and the assertion that a given complex number contains an ideal factor is only a convenient mode of expressing a certain set of congruential conditions which are satisfied by the coefficients of the complex number. Nevertheless we may, without fear of error, represent ideal numbers by the same symbols, $f(\alpha)$, $F(\alpha)$, $\phi(\alpha)$, ..., which we have employed to denote actually existing complex numbers, if we are only careful to remember that these symbols, when the numbers which they represent are ideal, admit of combination by multiplication or division, but not by addition or subtraction. Thus $f(\alpha) \times f_1(\alpha)$, $f(\alpha) \div f_1(\alpha)$, $[f(\alpha)]^m$, are significant symbols, and their interpretation is contained in what has preceded; but we have no general interpretation of a combination such as $f(\alpha) + f_1(\alpha)$, or $f(\alpha) - f_1(\alpha)$ *. This symbolic representation of ideal numbers is very convenient, and tends to abbreviate many demonstrations.

Every ideal number is a divisor of an actual number, and, indeed, of an infinite number of actual numbers. Also, if the ideal number $\phi(\alpha)$ be a divisor of the actual number $F(\alpha)$, the quotient $\phi_1(\alpha) = F(\alpha) \div \phi(\alpha)$ is always ideal; for if $\phi_1(\alpha)$ were an actual number, $\phi(\alpha)$, which is the quotient of $F(\alpha)$ divided by $\phi_1(\alpha)$, ought also to be an actual number. It appears, therefore, that there exists an infinite number of different ideal multipliers, which all render actual the same ideal number. It has, however, been shown by M. Kummer that a finite number of ideal multipliers are sufficient to render actual all ideal numbers whatever; so that it is possible (and that in an infinite number of different

* These symbols are, however, interpretable when $f(\alpha)$ and $f_1(\alpha)$ belong to the same *class*. Thus, if $\phi(\alpha) \times f(\alpha)$ and $\phi(\alpha) \times f_1(\alpha)$ be both actual, $f(\alpha) + f_1(\alpha)$ is the ideal quotient obtained by dividing $\phi(\alpha) \times f(\alpha) + \phi(\alpha) \times f_1(\alpha)$ by $\phi(\alpha)$.

ways) to assign a system of ideal multipliers, such that every ideal number is rendered actual by one of them, and one only. Ideal numbers are thus distributed into a certain finite number of classes,—a class comprehending those numbers which are rendered actual by the same multiplier; and this distribution into classes is independent of the particular system of multipliers by which it is effected, inasmuch as it is found that if two ideal numbers be rendered actual by the same multiplier, every other multiplier which renders one of them actual will also render the other actual. Ideal numbers which belong to the same class are said to be *equivalent*; so that two ideal numbers, which are each of them equivalent to a third, are equivalent to one another. We may regard actual numbers (which need no ideal multiplier) as forming the first or *principal* class in the distribution, and, consequently, as all equivalent to one another. If $f(a)$ be equivalent to $f_1(a)$, and $\phi(a)$ to $\phi_1(a)$, $f(a)\times\phi(a)$ is equivalent to $f_1(a)\times\phi_1(a)$,—a result which is expressed by saying that 'equivalent ideal numbers multiplied by equivalent numbers give equivalent products;' and the class of the product is said to be the class *compounded* of the classes of the factors.

49. *Representation of Ideal Numbers as the roots of Actual Numbers.*—An important conclusion is deducible from the theorem that the number of classes of ideal numbers is finite. Let $f(a)$ be any ideal number; and let us consider the series of ideal numbers $f(a), f(a)^2, f(a)^3, \ldots$ These numbers cannot all belong to different classes; we can therefore find two different powers of $f(a)$, for example $[f(a)]^m$ and $[f(a)]^{m+n}$, which are equivalent to one another. But the equivalence of these numbers implies that $[f(a)]^n$ is equivalent to the actual number $+1$; *i.e.* that $[f(a)]^n$ is itself an actual number. We may therefore enunciate the theorem, 'Every ideal number, raised to a certain power, becomes an actual number.'

The index of this power is the same for all ideal numbers of the same class, but may be different for different classes. By reasoning precisely similar to that employed by Euler in his 2nd proof of Fermat's Theorem *, it may be proved that the index of the first term in the series $f(a)$, $[f(a)]^2$, $[f(a)]^3 \ldots$, which is an actual number, is either equal to the whole number of classes, or to a submultiple of that number. This least index is said to be *the exponent to which the class of ideal numbers containing $f(a)$ appertains.* It would seem that for certain values of the prime λ, there exist classes of ideal numbers appertaining to the

* See Art. 10 of this Report.

exponent H, if H denote the number of classes of ideal numbers *. Such classes (when they exist) possess a property similar to that of the primitive roots of prime numbers; *i.e.*, by compounding such a class continually with itself we obtain all possible classes, just as by continually multiplying a primitive root by itself we obtain all residues prime to the prime of which it is a primitive root. It has, however, been ascertained by M. Kummer that these *primitive* classes do not in all cases, or even in general, exist.

The theorem of this article enables us to express ideal numbers as roots of actually existing complex numbers. Thus, if q be a prime appertaining to the exponent f for the modulus λ, and resoluble into the product of e conjugate ideal factors $\phi(\eta_0)$, $\phi(\eta_1)$, $\phi(\eta_2)$, ..., $\phi(\eta_{e-1})$, these ideal numbers, which will not in general belong to the same class, will nevertheless appertain to the same exponent h; so that $[\phi(\eta_0)]^h$, $[\phi(\eta_1)]^h$, ... will all be actual numbers. The power q^h is therefore resoluble into the product of e actually existing complex factors. If we effect this resolution, and represent the factors of q^h by $\Phi(\eta_0)$, $\Phi(\eta_1)$, ... the ideal numbers $\phi(\eta_0)$, $\phi(\eta_1)$, ... may be represented by the formulæ

$$\phi(\eta_0) = [\Phi(\eta_0)]^{\frac{1}{h}} \quad \phi(\eta_1) = [\Phi(\eta_1)]^{\frac{1}{h}}, \ldots.$$

50. *The Number of Classes of Ideal Numbers.*—The number of classes of ideal numbers was first determined by Dirichlet. He effected this determination by methods which he had previously introduced into the higher arithmetic, and which had already led him to a demonstration of the celebrated theorem, that every arithmetical progression, the terms of which are prime to their common difference, contains an infinite number of prime numbers; and to the determination of the number of non-equivalent classes of quadratic forms of a given determinant †. Dirichlet's investigation of the problem which we are here

* See on this subject M. Kummer's note 'on the Irregularity of Determinants' in the Monatsberichte of the Berlin Academy for 1853, p. 194. M. Kummer's investigation, however, is restricted to classes containing ideal numbers $f(\alpha)$ such that $f(\alpha) \times f(\alpha^{-1})$ is an actual number.

† See his memoirs on Arithmetical Progressions, in the Transactions of the Berlin Academy for the years 1837 (p. 45) and 1841 (p. 141), or in Liouville, vol. iv. p. 393, ix. p. 255. The first of these papers relates to progressions of real integers, the second to progressions of complex numbers of the form $a+bi$. In the memoir 'Recherches sur diverses applications de l'analyse infinitésimale à la Théorie des Nombres' (Crelle, vol. xix. p. 324, xxi. pp. 1 & 134), Dirichlet has applied his method to quadratic forms having real and integral coefficients; and in a subsequent memoir (Crelle, vol. xxiv. p. 291) he has extended this application to quadratic forms, of which the coefficients are complex numbers containing i. See also Crelle, vol. xviii. p. 259, xxi. p. 98 (or the Monatsberichte for 1840, p. 49), xxii. p. 375 (Monatsberichte for 1841, p. 190). We shall have occasion, in a later part of this Report, to give an abstract of the contents of this invaluable series of memoirs.

considering has never been published; but that since given by M. Kummer is probably in all essential respects the same, as it reposes on an extension of the principles developed in Dirichlet's earlier memoirs. Our limits compel us to omit the details of M. Kummer's analysis; the final result, however, is that, if H denote the number of non-equivalent classes of ideal numbers, $H=\frac{P}{(2\lambda)^{\mu-1}}\times\frac{D}{\Delta}$. In this formula P is a quantity defined by the equations

$$P=\phi(\beta).\phi(\beta^3).\phi(\beta^5)\ldots\phi(\beta^{\lambda-2}),$$

$$\phi(\beta)=1+\gamma_1\beta+\gamma_2\beta^2+\gamma_3\beta^3+\ldots+\gamma_{\lambda-2}\beta^{\lambda-2},$$

β representing a primitive root of the equation $\beta^{\lambda-1}=1$, γ a primitive root of the congruence $\gamma^{\lambda-1}\equiv 1$, mod λ, and $\gamma_1, \gamma_2, \gamma_3, \ldots$ the least positive residues of $\gamma, \gamma^2, \gamma^3, \ldots$ for the modulus λ; Δ is the logarithmic determinant (see Art. 42 of this Report) of any system of $\mu-1$ fundamental units, and D the logarithmic determinant of a particular system of independent but not fundamental units, $e(\alpha), e(\alpha^{\gamma}), e(\alpha^{\gamma^2}), \ldots, e(\alpha^{\gamma^{\mu-2}})$, defined by the equation

$$e(\alpha)=\sqrt{\frac{(1-\alpha^{\gamma})(1-\alpha^{-\gamma})}{(1-\alpha)(1-\alpha^{-1})}}=\pm\frac{\alpha^{\mu(\gamma-1)}(1-\alpha^{\gamma})}{1-\alpha}=\pm\frac{\sin\frac{k\gamma\pi}{\lambda}}{\sin\frac{\pi}{\lambda}},\text{ if }\alpha=e^{\frac{2ik\pi}{\lambda}};$$

so that

$$D=\begin{vmatrix} L.e(\alpha) & , L.e(\alpha^{\gamma}) & , L.e(\alpha^{\gamma^2}), \ldots, & L.e(\alpha^{\gamma^{\mu-2}}) \\ L.e(\alpha^{\gamma}) & , L.e(\alpha^{\gamma^2}) & , L.e(\alpha^{\gamma^3}), \ldots, & L.e(\alpha^{\gamma^{\mu-1}}) \\ L.e(\alpha^{\gamma^2}) & , L.e(\alpha^{\gamma^3}) & , L.e(\alpha^{\gamma^4}), \ldots, & L.e(\alpha^{\gamma^{\mu}}) \\ \cdot & \cdot & \cdot & \cdot \\ L.e(\alpha^{\gamma^{\mu-2}}), & L.e(\alpha^{\gamma^{\mu-1}}), & L.e(\alpha^{\gamma^{\mu}}), \ldots, & L.e(\alpha^{\gamma^{2\mu-4}}) \end{vmatrix}$$

Each of the two factors $\frac{P}{(2\lambda)^{\mu-1}}$ and $\frac{D}{\Delta}$, of which the value of H is composed, is separately an integral number. That $\frac{D}{\Delta}$ is integral is a consequence of the relation which exists between the logarithmic determinant of a system of fundamental units, and that of any system of independent units; that P is divisible by $(2\lambda)^{\mu-1}$ may be rendered evident from the nature of the expression P itself*. The factor $\frac{D}{\Delta}$, taken by itself, represents the number of classes that contain ideal numbers composed with the periods of two terms $\alpha+\alpha^{-1}, \alpha^2+\alpha^{-2}, \ldots$

* See the investigation in the next article.

only; or, which is the same thing, it represents the number of classes each of which contains the reciprocal $f(\alpha^{-1})$ of every ideal number $f(\alpha)$ comprehended in it; $\frac{P}{(2\lambda)^{\mu-1}}$, on the other hand, is the number of classes of those ideal numbers which become actual by multiplication with their own reciprocals *. The actual calculation of the factor $\frac{D}{\Delta}$ is extremely laborious, as it requires the preliminary investigation of a system of fundamental units. For the cases $\lambda=5$, $\lambda=7$, the *trigonometrical* units $e(\alpha)$, $e(\alpha^{\gamma})$, $e(\alpha^{\gamma^2})\ldots$ are themselves a fundamental system, so that in these two cases $D=\Delta$, and $\frac{D}{\Delta}=+1$. The computation of the first factor $\frac{P}{(2\lambda)^{\mu-1}}$ presents somewhat less difficulty; and M. Kummer (though not without great labour) has assigned its value for all primes inferior to 100. For the primes 3, 5, 7, 11, 13, 17, 19, that value is unity; for 23 it is 3, and then increases with extraordinary rapidity; so that for 97 it already amounts to $411322823001=3457\times 118982593$. The asymptotic law of this increase is expressed by the formula

$$\text{Lim}\left[\frac{P}{(2\lambda)^{\mu-1}}\div\frac{\lambda^{1+\frac{1}{2}\mu}}{2^{\mu-1}\pi^{\mu}}\right]=1,$$

when λ increases without limit †. It will be seen that the number of classes of ideal numbers for $\lambda=3$, $\lambda=5$, $\lambda=7$, is unity; *i.e.*, for those values of λ every complex prime is actual. In the absence of any determination of a system of fundamental units for $\lambda=11$, $\lambda=13$, $\lambda=17$, and $\lambda=19$, it is not possible to say whether this is or is not the case for these values also. But from and after the limit $\lambda=23$, the value of the factor $\frac{P}{(2\lambda)^{\mu-1}}$ indicates that a complex number is not necessarily a complex prime because it is irresoluble into factors.

51. *Criterion of the Divisibility of H by λ.*—The number of classes of ideal numbers, which we have symbolized by H, is not in general divisible by λ; but in certain cases it may happen that it is so. The quotient $\frac{D}{\Delta}$ is never divisible by λ, except when the other factor $\frac{P}{(2\lambda)^{\mu-1}}$ is also divisible by λ. And it has been found by M. Kummer that the necessary and sufficient condition for the

* See the note already cited, 'on the Irregularity of Determinants,' in the Monatsberichte for 1853, p. 195.

† Liouville, vol. xvi. p. 473. The formula is given without demonstration.

divisibility of $\frac{P}{(2\lambda)^{\mu-1}}$ by λ is that the numerator of one of the first $\mu-1$ functions of Bernoulli should be divisible by λ. The investigation of this singular criterion depends on a transformation of the function $\phi(\beta)$ which enters into the expression of P. If we represent the product

$$(\gamma\beta-1)\,\phi(\beta)=(\gamma\gamma_{\lambda-2}-1)+(\gamma-\gamma_1)\beta+(\gamma\gamma_1-\gamma_2)\beta^2+\ldots+(\gamma\gamma_{\lambda-3}-\gamma_{\lambda-2})\beta^{\lambda-2},$$

in which every coefficient is divisible by λ, by

$$\lambda\left[b_0+b_1\beta+b_2\beta^2+\ldots b_{\lambda-2}\beta^{\lambda-2}\right], \text{ or } \lambda\psi(\beta)$$

(b_m denoting the quotient $\frac{\gamma\gamma_{m-1}-\gamma_m}{\lambda}$, or $I\frac{\gamma\gamma_{m-1}}{\lambda}$, if I represent the greatest integer contained in the fraction before which it is placed), we obtain by multiplication the equality

$$(\gamma^\mu+1)\,P=\lambda^\mu\psi(\beta)\,.\,\psi(\beta^3)\ldots\psi(\beta^{\lambda-2})\,;$$

or, since $\gamma^\mu+1$ is divisible by λ, and may be supposed not divisible by λ^2 *,

$$\frac{C.P}{(2\lambda)^{\mu-1}}=\psi(\beta)\,.\,\psi(\beta^3)\ldots\psi(\beta^{\lambda-2}),$$

C denoting a coefficient prime to λ. The congruence $\frac{P}{(2\lambda)^{\mu-1}}\equiv 0$, mod λ, is therefore equivalent to the congruence

$$\psi(\beta)\,.\,\psi(\beta^3)\ldots\psi(\beta^{\lambda-2})\equiv 0, \text{ mod } \lambda,$$

which may, in its turn, be replaced by the following,

$$\psi(\gamma)\,.\,\psi(\gamma^3)\ldots\psi(\gamma^{\lambda-2})\equiv 0, \text{ mod } \lambda.$$

For, if there be an equation which, considered as a congruence for a given modulus λ, is completely resoluble for that modulus, any symmetrical function of the roots of the congruence is congruous, for the modulus λ, to the corresponding function of the roots of the equation. The function

$$\psi(\beta)\,.\,\psi(\beta^3)\ldots\psi(\beta^{\lambda-2}),$$

which is a symmetric function of $\beta, \beta^3, \ldots \beta^{\lambda-2}$, the roots of the equation $x^\mu+1=0$, is therefore congruous to $\psi(\gamma)\,.\,\psi(\gamma^3)\ldots\psi(\gamma^{\lambda-2})$, which is the same function of $\gamma, \gamma^3, \gamma^5, \ldots \gamma^{\lambda-2}$, the roots of the congruence $x^\mu+1\equiv 0$, mod λ.

---

* For $\gamma^\mu+1$ and $(\gamma+\lambda)^\mu+1$ are both of them divisible by λ; but only one of them can be divisible by λ^2, since their difference is not divisible by λ^2. We can therefore, without changing $\gamma_0, \gamma_1, \ldots, \gamma_{\lambda-2}$, determine γ in accordance with the supposition in the text.

Hence the necessary and sufficient condition for the divisibility of $\frac{P}{(2\lambda)^{\mu-1}}$ by λ is that one of the μ congruences included in the formula

$$\psi(\gamma^{2n-1}) \equiv 0, \text{ mod } \lambda, \; n=1, 2, 3, \ldots, \lambda, \quad \ldots \quad (a)$$

should be satisfied. Now

$$\gamma^{-(2n-1)}\psi(\gamma^{2n-1}) \equiv b_0\gamma_{\lambda-2}^{2n-1} + b_1\gamma_0^{2n-1} + b_2\gamma_1^{2n-1} + \ldots + b_{\lambda-2}\gamma_{\lambda-3}^{2n-1};$$

or, observing that $\gamma_0, \gamma_1, \gamma_2, \ldots \gamma_{\lambda-2}$ are the numbers $1, 2, 3, \ldots \lambda-1$, taken in a certain order, and introducing the values of $b_0, b_1, b_2, \ldots$

$$\gamma^{-(2n-1)}\psi(\gamma^{2n-1}) \equiv \sum_{x=1}^{x=\lambda-1} x^{2n-1} I\frac{\gamma x}{\lambda}, \text{ mod } \lambda.$$

This last expression may be further transformed as follows. If $f(x)$ denote any function of x, and $F(x) = \sum_{x=1}^{x=x} f(x)$, we have the identical equation

$$\sum_{x=1}^{x=\lambda-1} I\frac{\gamma x}{\lambda}\cdot f(x) + \sum_{x=1}^{x=\gamma-1} F\left(I\frac{\lambda x}{\gamma}\right) = (\gamma-1)\,F(\lambda-1),$$

γ and λ being any two numbers prime to one another. To verify this equation, we may construct a system of unit points in a plane; then the right-hand member is the sum of the values of $f(x)$ for all unit points in the *interior* of the parallelogram $(0, 0)$, $(\lambda, 0)$, (λ, γ), $(0, \gamma)$; while the two terms of the left-hand member represent similar sums for the two triangles into which the parallelogram is divided by its diagonal $\gamma x - \lambda y = 0$. Writing then in this identity x^{2n-1} for $f(x)$, and employing the symbol $F_{2n-1}(x)$ to represent the sum $\sum_{x=1}^{x=x} x^{2n-1}$, or rather the function

$$\frac{x^{2n}}{2n} + \tfrac{1}{2}x^{2n-1} + B_1\frac{\Pi(2n-1)}{\Pi(2n-2)\,\Pi(2)}x^{2n-2} - B_2\frac{\Pi(2n-1)}{\Pi(2n-4)\,\Pi(4)}x^{2n-4} + \ldots$$
$$\ldots + (-1)^n B_{n-1}\frac{\Pi(2n-1)}{\Pi(2)\,\Pi(2n-2)}x^2,$$

in which $B_1, B_2, \ldots, B_n$ are the functions of Bernoulli, and which, when x is an integral number, coincides with that sum, we find

$$\sum_{x=1}^{x=\lambda-1} x^{2n-1} I\frac{\gamma x}{\lambda} + \sum_{x=1}^{x=\gamma-1} F_{2n-1}\left[I\frac{\lambda x}{\gamma}\right] = (\gamma-1)\,F_{2n-1}(\lambda-1).$$

But $F_{2n-1}(\lambda-1) = F_{2n-1}(\lambda) - \lambda^{2n-1}$ is evidently divisible by λ; so that

$$\sum_{x=1}^{x=\lambda-1} x^{2n-1} I\frac{\gamma x}{\lambda} + \sum_{x=1}^{x=\gamma-1} F_{2n-1}\left[I\frac{\lambda x}{\gamma}\right] \equiv 0, \text{ mod } \lambda.$$

The congruences (a) may therefore be replaced by the congruences

$$\sum_{x=1}^{x=\gamma-1} F_{2n-1}\left[I\frac{\lambda x}{\gamma}\right] \equiv 0, \text{ mod } \lambda,$$

which may be written in the simpler form

$$\sum_{x=1}^{x=\gamma-1} F_{2n-1}\left(-\frac{x}{\gamma}\right) \equiv 0, \text{ mod } \lambda,$$

if we observe that (λ being prime to γ) the numbers $I\frac{\lambda}{\gamma}$, $I\frac{2\lambda}{\gamma}$, ..., $I\frac{(\gamma-1)\lambda}{\gamma}$ are congruous (mod λ) to the fractions $-\frac{1}{\gamma}$, $-\frac{2}{\gamma}$, ... $-\frac{\gamma-1}{\gamma}$, taken in a certain order. But, by a curious property of the function F_{2n-1}, demonstrated for the first time by M. Kummer,

$$\sum_{x=1}^{x=\gamma-1} F_{2n-1}\left(-\frac{x}{\gamma}\right) = \frac{(-1)^n B_n(\gamma^{2n}-1)}{2n\gamma^{2n-1}}.$$

The condition for the divisibility of H by λ is therefore that one of the μ congruences included in the formula $B_n(\gamma^{2n}-1) \equiv 0$, mod λ, should be satisfied. The last of these congruences, or $B_\mu(\gamma^{2\mu}-1) \equiv 0$, is never satisfied; for it is easily proved that the denominator of B_μ contains λ as a factor, while

$$\gamma^{2\mu}-1 = (\gamma^\mu+1)(\gamma^\mu-1),$$

though divisible by λ, is not divisible by λ^2. And since, if $n < \mu$, $\gamma^{2n}-1$ is prime to λ, that factor may be omitted in the remaining $\mu-1$ congruences; so that the condition at which we have arrived coincides with that enunciated at the commencement of this article.

We have exhibited M. Kummer's analysis of this problem with more fulness of detail than might seem warranted by the nature of this Report, not only on account of its elegance, but also because it exemplifies transformations and processes which are of frequent occurrence in arithmetical investigation *.

52. *'Exceptional' Primes.*—A prime number λ, which, like 37, 59, and 67 in the first hundred, divides the numerator of one of the first $\frac{1}{2}(\lambda-3)$ functions of

* In Liouville, vol. i. (New Series) p. 396, M. Kronecker has given a very simple demonstration of the congruence

$$2n\lambda\psi(\gamma^{2n-1}) \equiv (\gamma^{2n}-1)[1^{2n}+2^{2n}+\ldots+(\lambda-1)^{2n}], \text{ mod } \lambda^2,$$

which, combined with another easily demonstrated formula, viz.,

$$1^{2n}+2^{2n}+\ldots+(\lambda-1)^{2n} \equiv (-1)^{n-1}B_n\lambda, \text{ mod } \lambda^2 [n<\mu],$$

leads immediately to the theorem of M. Kummer.

Bernoulli, and which consequently divides the number of classes of ideal numbers composed with λ-th roots of unity, is termed by M. Kummer an *exceptional* prime. Such primes have to be excluded from the enunciation of several important propositions; and their theory presents difficulties which have not yet been overcome. Thus the following propositions are true for all primes other than the exceptional primes, but are not true for the exceptional primes.

(1.) The exponent to which any class of ideal numbers appertains (see Art. 49) is prime to λ.

(2.) The index of the lowest power of any unit which can be expressed as a product of *integral* powers of the trigonometric units is prime to λ. For that index is a divisor of $\frac{D}{\Delta}$ (see Art. 42).

(3.) Every complex unit which is congruous to a real integer for the modulus λ is a perfect λ-th power. (Whether λ be an exceptional prime or not, the λ-th power of any complex number is congruous, for the modulus λ, to a real integer, viz. to the sum of the coefficients of the complex number.)

(4.) If $f(\alpha)$ denote any (actual) complex number prime to λ (*i. e.* not divisible by $1-\alpha$), a complex unit $e(\alpha)$ can always be assigned, such that the product $F(\alpha)=e(\alpha)f(\alpha)$ shall satisfy the two congruences

$$F(\alpha)\,F(\alpha^{-1})\equiv[F(1)]^2, \text{ mod } \lambda,$$

$$F(\alpha)\equiv F(1), \text{ mod } (1-\alpha)^2.$$

A complex number satisfying these two congruential conditions is called a *primary* complex number; the product of two primary numbers is therefore itself primary. This definition, in the particular case $\lambda=3$, includes the primary numbers of Art. 37, taken either positively or negatively.

53. *Fermat's Theorem for Complex Primes.*—Let $\phi(\alpha)$ be an actual or ideal complex prime, and let $N=N.\phi(\alpha)$ represent its norm. A system of N actual numbers can always be assigned such that every complex number shall be congruous to one and only to one of them for the modulus $\phi(\alpha)$. These N numbers may be said therefore to form a complete system of residues for the modulus $\phi(\alpha)$; and by omitting the term divisible by $\phi(\alpha)$, we obtain a system of $N-1$ residues prime to $\phi(\alpha)$.

Let q be a prime appertaining to the exponent f, so that $N=q^f$, and let $\phi(\alpha)$ or $\phi_1(\eta_0)$ be the prime factor of q which appertains to the substitution $\eta_0=u_0$; the formula

$$a_0+a_1\alpha+a_2\alpha^2+\ldots+a_{f-1}\alpha^{f-1} \qquad \text{(A)}$$

will represent a complete system of residues for the modulus $\phi_1(\eta_0)$, if we assign to the coefficients $a_0, a_1, a_2, \ldots$ the values $0, 1, 2, \ldots, q-1$ in succession. For if

$$f(a) = \psi_0(\eta_0) + a\psi_1(\eta_0) + \ldots + a^{f-1}\psi_{f-1}(\eta_0)$$

be any complex number, $f(a)$ is congruous for the modulus $\phi_1(\eta_0)$ to

$$\psi_0(u_0) + a\psi_1(u_0) + \ldots + a^{f-1}\psi_{f-1}(u_0),$$

because $u_0 - \eta_0 \equiv 0$, mod $\phi_1(\eta_0)$: that is, $f(a)$ is congruous to one of the complex numbers included in (A); nor can any two numbers

$$a_0 + a_1 a + a_2 a^2 + \ldots + a_{f-1} a^{f-1} \quad \text{and} \quad b_0 + b_1 a + b_2 a^2 + \ldots + b_{f-1} a^{f-1}$$

included in that formula be congruous to one another; for the congruence

$$(a_0 - b_0) + a(a_1 - b_1) + a^2(a_2 - b_2) + \ldots + a^{f-1}(a_{f-1} - b_{f-1}) \equiv 0, \text{ mod } \phi_1(\eta_0),$$

involves, by M. Kummer's theory (see Art. 45), the coexistence of the f congruences

$$a_0 - b_0 \equiv 0, \text{ mod } q; \quad a_1 - b_1 \equiv 0, \text{ mod } q; \quad \ldots\ldots; \quad a_{f-1} - b_{f-1} \equiv 0, \text{ mod } q;$$

i.e. the identity of the complex numbers

$$a_0 + a\,a_1 + a^2 a_2 + \ldots + a^{f-1} a_{f-1}, \quad \text{and} \quad b_0 + a\,b_1 + a^2 b_2 + \ldots + a^{f-1} b_{f-1}.$$

It is worth while to notice that, if q be a prime appertaining to the exponent 1, for the modulus λ, that is if q be of the linear form $m\lambda + 1$, the real numbers $0, 1, 2, 3, \ldots, q-1$ will represent the terms of a complete system of residues for the modulus $\phi(a)$; but if $\phi(a)$ be a factor of a prime appertaining to any higher exponent than unity, a complete system will contain complex as well as real integral residues.

By applying the principle (see Art. 10) that a system of residues prime to the modulus, multiplied by a residue prime to the modulus, produces a system of residues prime to the modulus, we obtain the theorem, which here replaces Fermat's Theorem, that if $\psi(a)$ be any actual number prime to $\phi(a)$,

$$[\psi(a)]^{N-1} \equiv 1, \text{ mod } \phi(a).$$

If we combine with this theorem the principle of Lagrange (cited in Art. 11) which is valid for complex no less than for real prime modules, we may extend, *mutatis mutandis*, to the general complex theory the elementary propositions relating to the Residues of Powers, Primitive Roots, and Indices, which, as we have seen, exist in the case of complex primes formed with cubic or biquadratic roots of unity. In fact, these propositions are of a character of even greater generality, and may be extended, not only to complex numbers formed with roots of unity whose index is a composite number, but also to all complex numbers formed with the roots of equations having integral coefficients, as soon as the prime factors of those complex numbers are properly defined.

54. *M. Kummer's Law of Reciprocity.*—We can now enunciate M. Kummer's law of reciprocity. It appears, from the last article, or it may be proved immediately by dividing the $N-1$ residues of $\phi(\alpha)$ into λ groups of $\frac{N-1}{\lambda}$ terms, after the following scheme,

$$\begin{array}{llll}
(0) & r_1, & r_2, \ldots\ldots, & r_{\frac{N-1}{\lambda}}, \\
(1) & \alpha r_1, & \alpha r_2, \ldots\ldots, & \alpha r_{\frac{N-1}{\lambda}}, \\
(2) & \alpha^2 r_1, & \alpha^2 r_2, \ldots\ldots, & \alpha^2 r_{\frac{N-1}{\lambda}}, \\
 & \cdot\ \cdot\ \cdot & \cdot\ \cdot\ \cdot & \cdot\ \cdot\ \cdot \\
(\lambda-1) & \alpha^{\lambda-1} r_1, & \alpha^{\lambda-1} r_2, \ldots\ldots, & \alpha^{\lambda-1} r_{\frac{N-1}{\lambda}},
\end{array}$$

and proceeding as in Art. 33 of this Report, that if $\psi(\alpha)$ be any actual complex number prime to $\phi(\alpha)$, $\psi(\alpha)^{\frac{N-1}{\lambda}}$ is congruous for the modulus $\phi(\alpha)$ to a certain power α^k of α. This power of α may be denoted by the symbol $\left[\frac{\psi(\alpha)}{\phi(\alpha)}\right]_\lambda$; so that we have the congruence

$$[\psi(\alpha)]^{\frac{N-1}{\lambda}} \equiv \left[\frac{\psi(\alpha)}{\phi(\alpha)}\right]_\lambda \equiv \alpha^k, \bmod \phi(\alpha).$$

The symbol $\left[\frac{\psi(\alpha)}{\phi(\alpha)}\right]_\lambda$, which we may term the λ-tic character of $\psi(\alpha)$ with regard to $\phi(\alpha)$, is evidently of the same nature as the corresponding symbols with which we have already met in the quadratic, cubic, and biquadratic theories, and admits of an extension of meaning similar to that of which they are susceptible. Availing himself of this symbol, M. Kummer has expressed his law of reciprocity by the formula

$$\left[\frac{\psi(\alpha)}{\phi(\alpha)}\right]_\lambda = \left[\frac{\phi(\alpha)}{\psi(\alpha)}\right]_\lambda,$$

$\phi(\alpha)$ and $\psi(\alpha)$ denoting real or ideal primes. But, to interpret this equation rightly, it is important to attend to the following observations.

(1) When $\psi(\alpha)$ and $\phi(\alpha)$ are both actual numbers, the formula supposes that they are both *primary* prime numbers. The prime $1-\alpha$ is therefore excluded.

(2) The definition that we have given of the symbol $\left[\frac{\phi(\alpha)}{\psi(\alpha)}\right]_\lambda$ becomes un-

meaning when $\phi(\alpha)$ is ideal, because no signification can be assigned to an ideal number which presents itself, not as a modulus or divisor, but as a residue. Let, therefore, h denote the index of the lowest power of $\phi(\alpha)$ which is an actual number; *i.e.*, let h be the exponent to which the class of $\phi(\alpha)$ appertains; and let $[\phi(\alpha)]^h$ represent the actually existing *primary* complex number which contains the factor $\phi(\alpha)$ h times, but contains no other prime factor; then the symbol $\left[\frac{\phi(\alpha)^h}{\psi(\alpha)}\right]_\lambda$ has by the preceding definition a perfectly definite meaning. Let then $\left[\frac{\phi(\alpha)^h}{\psi(\alpha)}\right]_\lambda = \alpha^{k'}$; we may define the value of the symbol $\left[\frac{\phi(\alpha)}{\psi(\alpha)}\right]_\lambda$ by means of the equation

$$\left[\frac{\phi(\alpha)}{\psi(\alpha)}\right]_\lambda^h = \left[\frac{\phi(\alpha)^h}{\psi(\alpha)}\right]_\lambda = \alpha^{k'},$$

which, *if h be prime to* λ, always gives a determinate value α^k for $\left[\frac{\phi(\alpha)}{\psi(\alpha)}\right]_\lambda$, k being defined by the congruence $hk \equiv k'$, mod λ. For the symbol $\left[\frac{\phi(\alpha)}{\psi(\alpha)}\right]_\lambda$ so defined, the law of reciprocity still subsists, subject however to the condition that $[\phi(\alpha)]^h$ is primary.

It will be seen, therefore, that the exceptional primes of Art. 52 are excluded from M. Kummer's law of reciprocity, for a twofold reason:—first, because if λ be one of those numbers, the definition of a primary number is not in general applicable; and secondly, because, on the same supposition, the symbol $\left[\frac{\phi(\alpha)}{\psi(\alpha)}\right]_\lambda$ may become unmeaning.

55. *The Theorems complementary to M. Kummer's Law of Reciprocity.*— The prime $1-\alpha$, and its conjugate primes, as well as the complex units, are excluded from the law of reciprocity; but complementary theorems by which the λ-tic characters of these numbers may be determined have been given by M. Kummer. For a simple unit α^k, we have the formula

$$\left[\frac{\alpha^k}{\phi(\alpha)}\right]_\lambda = \alpha^{k\frac{N-1}{\lambda}}.$$

With regard to λ, which is the norm of $1-\alpha$, it may be observed that if $\phi(\alpha)$ be a prime factor of a real prime q appertaining, for the modulus λ, to any exponent f different from unity, *i.e.* if q be not of the linear form $m\lambda+1$, the character of every real integer, and therefore of λ, with respect to $\phi(\alpha)$ is $+1$, because, if $f>1$, $\frac{q^f-1}{\lambda}$ is divisible by $q-1$. But whatever be the linear form of q, the *characteristic* of λ or $\chi(\lambda)$ (for so we shall for brevity term the index of a in the equation $\left[\frac{\lambda}{\phi(a)}\right]_\lambda = a^k$) is determined by the congruence

$$\chi(\lambda) \equiv \frac{1}{\lambda} D_\lambda, \text{ mod } \lambda,$$

D_λ being the value (for $v=0$) of the differential coefficient $\frac{d^\lambda \log \phi(e^v)}{dv^\lambda}$, if $\phi(a)$ be an actually existent number, or of $\frac{1}{h}\frac{d^\lambda \log [\phi(e^v)]^h}{dv^\lambda}$ if it be ideal. To obtain the characteristics of the units, M. Kummer considers the system of independent units

$$E_1(a),\ E_2(a), \ldots\ldots, E_{\mu-1}(a),$$

defined by the formula

$$E_k(a) = e(a)\, e(a^\gamma)^{\gamma^{-2k}} e(a^{\gamma^2})^{\gamma^{-4k}} \ldots\ldots e(a^{\gamma^{\mu-1}})^{\gamma^{-2(\mu-1)k}},$$

in which $e(a)$ represents the trigonometrical unit of Art. 50, and γ is the same primitive root of λ which occurs in the expression of $e(a)$. We have then, for $\chi[E_k(a^n)]$ and $\chi(1-a^k)$, the formulae

$$\chi[E_k(a^n)] \equiv (-1)^k (\gamma^{2k}-1) \frac{B_k n^{2k}}{4k} D_{\lambda-2k}, \text{ mod } \lambda,$$

and

$$\chi(1-a^k) \equiv -\frac{D_\lambda}{\lambda} + \tfrac{1}{2}\frac{N-1}{\lambda} + B_1 D_{\lambda-2}\frac{k^2}{2} - B_2 D_{\lambda-4}\frac{k^4}{4} + \ldots + (-1)^\mu B_{\mu-1} D_3 \frac{k^{\lambda-3}}{\lambda-3},$$

N representing the norm of $\phi(a)$, $B_1, B_2, \ldots, B_{\mu-1}$ the functions of Bernoulli, and D_m the value of the differential coefficient

$$\frac{d^m \log \phi(e^v)}{dv^m} \left(\text{or } \frac{d^m \log [\phi(e^v)]^h}{h\, dv^m}\right) \text{ for } v=0.$$

These formulae do not in general hold for the *exceptional* prime numbers λ, which divide the numerator of one of the first $\mu-1$ functions of Bernoulli. This is evident from the occurrence in them of the coefficients D_m, which if $\phi(a)$ be ideal, and h be divisible by λ, may acquire denominators divisible by λ, thus rendering the congruences nugatory. It is sufficient to have determined the characteristics of the particular system of units $E_1(a), E_2(a), \ldots, E_{\mu-1}(a)$, because, as that system is independent, every other unit $\epsilon(a)$ is included in the formula

$$\epsilon(a) = E_1(a)^{m_1} E_2(a)^{m_2} \ldots\ldots E_{\mu-1}(a)^{m_{\mu-1}};$$

so that $\chi[\epsilon(a)]$ may be found from the congruence

$$\chi[\epsilon(a)] \equiv \sum_{k=1}^{k=\mu-1} m_k\, \chi[E_k(a)], \text{ mod } \lambda,$$

which cannot become unmeaning, except in the case of the exceptional primes;

because if D' be the logarithmic determinant of the system of units $E_1(\alpha)$, $E_2(\alpha), \ldots, E_{\mu-1}(\alpha)$, D and Δ retaining the meanings assigned to them in Art. 50, it may be shown that $\frac{D'}{D}$ is prime to λ, and therefore $\frac{D'}{\Delta} = \frac{D'}{D} \times \frac{D}{\Delta}$ is also prime to λ; *i.e.*, the denominators of the fractions $m_1, m_2, \ldots, m_{\mu-1}$ are prime to λ (see Art. 42). But M. Kummer has also given a formula which assigns directly the characteristic of any unit $\epsilon(\alpha)$ whatsoever. If Δ_k denote the value of the differential coefficient $\frac{d^k \log \epsilon(e^v)}{dv^k}$, for $v=0$, we have

$$\chi[\epsilon(\alpha)] \equiv \Delta_1 \frac{N-1}{\lambda} + \sum_{k=1}^{k=\mu-1} \Delta_{2k} D_{\lambda-2k}, \bmod \lambda\,*.$$

56. We have already observed (see Art. 39) that it is impossible to deduce a proof of the highest laws of reciprocity from the formulae which present themselves in the theory of the division of the circle. It is true (as we shall presently see) that the formulae IV. and V. of Art. 30 determine the decomposition of the real prime p (supposed to be of the form $k\lambda+1$) into its $\lambda-1$ complex prime factors; but it will be perceived that these complex factors occur, not isolated, but combined in a particular manner. From equation IV. of the article cited we infer that $p = \psi(\alpha)\,\psi(\alpha^{-1})$; let then

$$\psi(\alpha) = f(\alpha_1)\,f(\alpha_2)\ldots f(\alpha_\mu);$$

$\alpha_1, \alpha_2, \ldots, \alpha_\mu$ being μ different roots (of which no two are reciprocals) of the equation $\frac{\alpha^\lambda - 1}{\alpha - 1} = 1$; so that $f(\alpha_1), f(\alpha_2), \ldots, f(\alpha_\mu)$ are one-half of the complex primes of which p is composed; if $e(\alpha)$ be any *real* unit, satisfying the equation $e(\alpha) = e(\alpha^{-1})$, it is plain that

$$e(\alpha_1)^2\, e(\alpha_2)^2 \ldots e(\alpha_\mu)^2 = 1, \quad \text{or} \quad \psi(\alpha) = \pm e(\alpha_1) f(\alpha_1) \times e(\alpha_2) f(\alpha_2) \times \ldots \times e(\alpha_\mu) f(\alpha_\mu).$$

The consideration, therefore, of the number $\psi(\alpha)$ cannot supply us with any determination of the λ-tic character of $f(\alpha_1)$ which will not equally apply to $f(\alpha_1) \times e(\alpha_1)$. But for all values of λ greater than 3, the number of real complex units is, as we have seen, infinite; and the character of any complex prime $f(\alpha)$ with respect to any other complex prime evidently changes when $f(\alpha)$ is multiplied by a unit of which the λ-tic character is not unity. The inapplicability of the formulae of Art. 30 to any general demonstration of the law of reciprocity

* The formulae of this article are taken from M. Kummer's second memoir on the complementary theorems (Crelle, vol. lvi. p. 270).

is thus apparent. The only equation of reciprocity that has been elicited from them is the following:—

$$\left[\frac{\phi(a)}{q_1}\right]_\lambda \times \left[\frac{\phi(a)}{q_2}\right]_\lambda \times \ldots \times \left[\frac{\phi(a)}{q_e}\right]_\lambda = \left[\frac{q_1}{\phi(a)}\right]_\lambda \times \left[\frac{q_2}{\phi(a)}\right]_\lambda \times \ldots \times \left[\frac{q_e}{\phi(a)}\right]_\lambda,$$

in which $\phi(a)$ is a complex prime factor of a prime number p of the form $m\lambda+1$, and $q_1, q_2, \ldots, q_e$ are the e conjugate factors of a prime number q appertaining to the exponent f for the modulus λ. This equation, which, if we adopt the generalised meaning of the symbol of reciprocity, may be written more briefly thus, $\left[\frac{\phi(a)}{q}\right]_\lambda = \left[\frac{q}{\phi(a)}\right]_\lambda$, was first obtained by Eisenstein, who inferred it from M. Kummer's investigation of the ideal prime divisors of $\psi(a)$ (see a note addressed by Eisenstein to Jacobi, and communicated by Jacobi to the Berlin Academy, in the Monatsberichte for 1850, May 30, p. 189). In a later memoir (Crelle's Journal, vol. xxxix. p. 351), Eisenstein proposes an ingenious method—reposing, however, on an undemonstrated principle—for the discovery of the higher laws of reciprocity; but it would seem that the application of this method failed to lead him to any definite result; and it is unquestionably to M. Kummer alone that we are indebted for the enunciation as well as for the demonstration of the theorem.

57. M. Kummer appears to have waited until he had developed the theory of complex numbers with a certain approximation to completeness, before proceeding to apply the principles he had discovered to the purpose which he had in view throughout, the investigation of the law of reciprocity. He succeeded in discovering the law which we have enunciated, in the year 1847, and, after verifying it by calculated tables of some extent, he communicated it to Dirichlet and Jacobi in January 1848, and subsequently, in 1850, to the Berlin Academy, in a note which also contained the demonstration of the complementary theorems relating to the units, and the prime divisors of λ. From the analogy of the cubic theorem, it was natural to conjecture that the law of reciprocity would assume the simple form $\left[\frac{p_1}{p_2}\right]_\lambda = \left[\frac{p_2}{p_1}\right]_\lambda$ for primes p_1 and p_2 reduced, by multiplication with proper complex units, to a form satisfying certain congruential conditions. But to determine properly these conditions, *i.e.* to assign the true definition of a primary complex prime, was no doubt the principal difficulty that M. Kummer had to overcome in the discovery of his theorem. If $\lambda=3$, the single congruence $f(a)\equiv f(1)$, mod $(1-a)^2$, sufficiently characterises a primary number; and since, whatever prime be repre-

sented by λ, that congruence is satisfied by one, and one only, of the numbers included in the formula $\alpha^k f(\alpha)$, it was probable that it ought to form one of the congruential conditions included in the definition of a primary complex prime. In determining the second condition, M. Kummer appears to have been guided by a method which depends on the arithmetical properties of the logarithmic expansion of a complex number. If we develope $\log \frac{f(\alpha)}{f(1)}$ in ascending powers of $\frac{f(\alpha)-f(1)}{f(1)}$ and represent by $L\frac{f(\alpha)}{f(1)}$ the finite number of terms which remain in this expansion after rejecting those which are congruous to zero for the modulus λ, we are led, after some transformations, to the congruence

$$-L\frac{f(\alpha)}{f(\alpha)} \equiv D_1 X_1(\alpha) + D_2 X_2(\alpha) + \ldots + D_{\lambda-2} X_{\lambda-2}(\alpha), \text{ mod } \lambda,$$

where $X_k(\alpha)$ represents the function $\sum\limits_{s=0}^{s=\lambda-2} \gamma^{-sk} \alpha^{\gamma^s}$, and D_k denotes, as in Art. 55, the value (for $v=0$) of the differential coefficient $\frac{d^k \log f(e^v)}{dv^k}$. In this congruence the first coefficient alone is altered when $f(\alpha)$ is multiplied by a simple unit; and only the even coefficients are altered when $f(\alpha)$ is multiplied by a real unit. Now D_1 is rendered congruous to zero by the condition $f(\alpha) \equiv f(1)$, mod $(1-\alpha)^2$; and M. Kummer has shown that, by multiplying $f(\alpha)$ by a properly chosen real unit, $D_2, D_4, \ldots, D_{\lambda-3}$ may be similarly made to disappear, so that we obtain

$$-L\frac{f(\alpha)}{f(1)} \equiv D_3 X_3(\alpha) + D_5 X_5(\alpha) + \ldots + D_{\lambda-2} X_{\lambda-2}(\alpha), \text{ mod } \lambda,$$

a congruence which is proved to involve the second congruence of condition satisfied by a primary number, *i.e.* $f(\alpha) f(\alpha^{-1}) \equiv f(1)^2$, mod λ *.

58. The methods to which M. Kummer at first had recourse in order to obtain a demonstration of his theorem, consisted in extensions of the theory of the division of the circle. By such extensions he demonstrated the complementary theorems, and even a particular case of the law of reciprocity itself—that in which the two complex primes compared are conjugate. But, after repeated efforts, he found himself compelled to abandon these methods, and to seek elsewhere for more fertile principles. 'I turned my attention,' he says, 'to Gauss's second demonstration of the law of quadratic reciprocity, which depends on the theory of quadratic forms. Though the method of this demon-

* Crelle, vol. xliv. pp. 130–140.

stration had never been extended to any other than quadratic residues, yet its principles appeared to me to be characterised by such generality as led me to hope that they might be successfully applied to residues of higher powers; and in this expectation I was not disappointed *.'

M. Kummer's demonstration of the law of reciprocity was communicated to the Academy of Berlin in the year 1858, ten years after the date of his first discovery of it. An outline of the demonstration is contained in the Monatsberichte for that year; and it is exhibited with great clearness and fulness of detail in a memoir published in the Berlin Transactions for 1859, which contains what is for the present the latest result of science on a problem which, if we date from the first enunciation of the quadratic theorem by Euler, has been studied by so many eminent geometers for nearly a century. It would, however, be impossible, without exceeding the limits within which this Report is confined, to give an account of its contents, which should be intelligible to persons not already familiar with the subject to which it refers. Taken by itself the demonstration of the theorem is, indeed, sufficiently simple; but it is based on a long series of preliminary researches relating to the complex numbers that can be formed with the roots of the equation $w^\lambda = D(a)$, in which $D(a)$ itself denotes a complex number composed of λth roots of unity. To those researches, and to the demonstration of the law of reciprocity founded on them, we shall again very briefly refer, when we come to speak of the corresponding investigations in the theory of quadratic forms, an acquaintance with which is essential to a comprehension of the method adopted by M. Kummer in his memoir. We may add that M. Kummer has intimated that he has already obtained two other demonstrations of his law of reciprocity, which, though they also depend on the consideration of complex numbers containing w, yet do not require the same complicated preliminary considerations.

59. *Complex Numbers composed of Roots of Unity, of which the Index is not a Prime.*—In a special memoir (see the list in Art. 41, note, No. 16), M. Kummer has considered the theory of complex numbers composed with a root of the equation $\omega^n = 1$, in which n denotes a composite number. The *primitive* roots of this equation are the roots of an irreducible equation of the form

$$F(\omega) = \frac{(\omega^n - 1)\, \Pi\left(\omega^{\overline{\frac{n}{p_1 p_2}}} - 1\right) \ldots}{\Pi\left(\omega^{\frac{n}{p}} - 1\right) \Pi\left(\omega^{\overline{\frac{n}{p_1 p_2 p_3}}} - 1\right) \ldots} = 0,$$

* See the Berlin Transactions for 1859, p. 29.

$p_1, p_2, p_3, \ldots$ denoting the different prime divisors of n *. If $\psi(n)$ be the number of numbers less than n and prime to it, $F(\omega)$ is of the order $\psi(n)$, and every complex number containing ω can be reduced (and that in one way only) to the form

$$f(\omega) = a_0 + a_1\omega + a_2\omega^2 + \ldots + a_{\psi(n)-1}\omega^{\psi(n)-1}.$$

The numbers conjugate to $f(\omega)$ are the $\psi(n)$ numbers obtained by writing in succession for ω the $\psi(n)$ primitive roots of $\omega^n = 1$; and the norm of $f(\omega)$ is the real and positive integer produced by multiplying together the $\psi(n)$ conjugates. If q be a prime number not dividing n, the sum

$$\varpi_k = \omega^k + \omega^{kq} + \omega^{kq^2} + \ldots,$$

in which the series of terms is to be continued until it begins to repeat itself, is termed a period. The n periods $\varpi_1, \varpi_2, \ldots, \varpi_n$ remain unchanged if for ω we write ω^q, ω^{q^2}, etc. Hence, if q appertain to the exponent t for the modulus n (*i.e.* if q satisfy the congruence $q^t \equiv 1$, mod n, but no congruence of a lower order and similar form), the number of different numbers conjugate to a given complex number containing the periods only is at most $\frac{\psi(n)}{t}$. For brevity, a complex number containing the periods only—for example, the number

$$c_0 + c_1\varpi_1 + c_2\varpi_2 + \ldots + c_n\varpi_n,$$

may be symbolised by $f(\varpi_1)$, so that

$$f(\varpi_k) = c_0 + c_1\varpi_k + c_2\varpi_{2k} + \ldots + c_n\varpi_{nk}.$$

If $1, r_1, r_2, \ldots$ are a set of $\frac{\psi(n)}{t}$ numbers prime to n and such that the quotient of no two of them (considered as a congruential fraction †) is congruous for the modulus n to any power of q, the numbers conjugate to $f(\varpi)$ may be represented

---

* The irreducibility of the equation $\frac{x^n - 1}{x - 1} = 0$ when n is prime was first established by Gauss (Disq. Arith., Art. 341). For other and simpler demonstrations of the same theorem, see the memoirs of MM. Kronecker (Crelle, xxix. p. 280, and Liouville, 2nd series, vol. i. p. 399), Schönemann (Crelle, vol. xxxi. p. 323, vol. xxxii. p. 100, & vol. xl. p. 188), Eisenstein (Crelle, vol. xxxix. p. 166), and Serret (Liouville, vol. xv. p. 296). The principles on which these demonstrations depend suffice to establish the irreducibility of the equation $\frac{x^{p^m} - 1}{x^{p^{m-1}} - 1} = 0$, but they fail, as M. Kronecker has observed, to furnish the corresponding demonstration when n, as in the text, is a product of powers of different primes. This demonstration was first given by M. Kronecker (Liouville, vol. xix. p. 177), who has been followed by M. Dedekind (Crelle, vol. liv. p. 27), and by M. Arndt (*ib.* lvi. p. 178).

† For the definition of a congruential fraction see Art. 14.

by $f(\varpi_1)$, $f(\varpi_{r_1})$, $f(\varpi_{r_2})$, ... The periods are the roots of certain irreducible equations, each of which is completely resoluble when considered as a congruence for the modulus q; and the roots u_1, u_2, ... of the congruences are connected with the roots ϖ_1, ϖ_2, ... of the equations, by a relation precisely similar to that enunciated in Art. 44. This relation M. Kummer has established by introducing certain conjugate complex numbers* $\Psi(\varpi_1)$, $\Psi(\varpi_{r_1})$, $\Psi(\varpi_{r_2})$, ... involving the

---

* These complex numbers are defined as follows (see the memoir cited at the commencement of this article, sect. 3, and that in Crelle, vol. liii. p. 142):—Let ϖ_k be a period satisfying the irreducible equation $\phi(\varpi_k) = 0$, and let $a_1, a_2, \ldots$ be the incongruous roots of $\phi(y) \equiv 0$, mod q; $b_1, b_2, \ldots$ the remaining terms of a complete system of residues, mod q, so that $\phi(b_1)$, $\phi(b_2)$, ... are prime to q. Since $\varpi_k^q \equiv \varpi_{kq}$, mod q, and $\varpi_{kq} = \varpi_k$, we have, by Lagrange's indeterminate congruence (see Art. 10 of this Report),

$$(\varpi_k - a_1)(\varpi_k - a_2)\ldots(\varpi_k - b_1)(\varpi_k - b_2)\ldots \equiv 0, \text{ mod } q,$$

or, since $\varpi_k - b_1$ divides $\phi(b_1)$ etc.,

$$\phi(b_1)\,\phi(b_2)\ldots(\varpi_k - a_1)(\varpi_k - a_2)\ldots \equiv 0, \text{ mod } q;$$

$$i.e. \quad (\varpi_k - a_1)(\varpi_k - a_2)\ldots \equiv 0, \text{ mod } q.$$

We may now consider the n series of factors

$$\varpi_k - a_1, \quad \varpi_k - a_2, \quad \varpi_k - a_3, \quad \ldots,$$

corresponding to the n values of k [the numbers $a_1, a_2, \ldots$ are of course the same for two periods which satisfy the same irreducible equation, but not in general the same for any two periods], and, retaining among these factors only those which are different, we may take for $\Psi(\varpi_1)$ the complex number formed by combining as many of them as possible, in such a manner as to give a product which is not divisible by q, but which is rendered divisible by q by the accession of any one factor not already contained in it. It is evident that $\Psi(\varpi_1)$ cannot contain all the factors

$$\varpi_k - a_1, \quad \varpi_k - a_2, \quad \ldots;$$

let us then denote by $\varpi_k - u_k$ a factor which is not contained in $\Psi(\varpi_1)$; we thus obtain the relation

$$\Psi(\varpi_1)(\varpi_k - u_k) \equiv 0, \text{ mod } q,$$

or, changing the primitive root ω into ω^r,

$$\Psi(\varpi_r)(\varpi_{rk} - u_k) \equiv 0, \text{ mod } q.$$

The conjugates of $\Psi(\varpi_1)$ are all complex numbers formed according to the same law as $\Psi(\varpi_1)$ itself; and, besides $\Psi(\varpi_1)$ and its conjugates, no other complex number can be formed according to that law. Also the number u_k which corresponds to a given period ϖ_k is absolutely determined as soon as we have selected the multiplier $\Psi(\varpi_1)$; for if two of the factors $\varpi_k - a_1$, $\varpi_k - a_2$, ... were absent from $\Psi(\varpi_1)$ we should have

$$\Psi(\varpi_1)(\varpi_k - a_1) \equiv 0, \quad \Psi(\varpi_1)(\varpi_k - a_2) \equiv 0, \text{ mod } q;$$

and thence

$$(a_1 - a_2)\Psi(\varpi_1) \equiv 0, \text{ mod } q,$$

contrary to the hypothesis that a_1 and a_2 are incongruous, and that $\Psi(\varpi_1)$ is not divisible by q. The correspondence of the numbers $u_1, u_2, \ldots, u_n$, with the periods $\varpi_1, \varpi_2, \ldots, \varpi_n$, can thus be fixed in as many ways as there are numbers conjugate to $\Psi(\varpi_1)$, *i.e.* in $\dfrac{\psi(n)}{t}$ different ways.

periods only, not themselves divisible by q, but each satisfying the n congruences included in the formula

$$\Psi(\varpi_r)(\varpi_{kr} - u_k) \equiv 0, \text{ mod } q,$$
$$k = 1, 2, 3, \ldots, n.$$

From these congruences it is easy to infer that, if $f(\varpi_r, \varpi_{2r}, \ldots, \varpi_{nr}) = 0$ be any identical relation subsisting for the periods, a similar relation

$$f(u_1, u_2, \ldots, u_n) \equiv 0, \text{ mod } q,$$

will subsist for the numbers $u_1, u_2, \ldots, u_n$; for we find

$$\Psi(\varpi_r) f(\varpi_r, \varpi_{2r}, \ldots) \equiv \Psi(\varpi_r) f(u_1, u_2, \ldots), \text{ mod } q,$$

i.e. $f(u_1, u_2, \ldots) \equiv 0$, mod q. Another important property of the complex number $\Psi(\varpi_1)$ is that it is congruous to zero, mod q, for every one of the substitutions $\varpi_1 = u_1$, $\varpi_1 = u_{r_1}$, $\varpi_1 = u_{r_2}$, ... except the first: thus the congruences $\Psi(u_{r_1}) \equiv 0$, $\Psi(u_{r_2}) \equiv 0$ are satisfied, ... but not $\Psi(u_1) \equiv 0$, mod q. If, then, $f(\omega)$ be any complex number satisfying the congruence

$$\Psi(\varpi_r)^m f(\omega) \equiv 0, \text{ mod } q^m,$$

but not the congruence

$$\Psi(\varpi_r)^{m+1} f(\omega) \equiv 0, \text{ mod } q^{m+1},$$

$f(\omega)$ is said to contain m times precisely the ideal factor of q corresponding to the substitution $\varpi_{kr} = u_k$. Since it can be shown that the numbers conjugate to $\Psi(\varpi_1)$ are all different from one another, it follows from the definition, that the quotient $\frac{\psi(n)}{t}$ represents the number of conjugate ideal prime factors contained in the real prime q, appertaining to the exponent t. If q be a divisor of n, the definition of its ideal factors requires a certain modification, which we cannot here particularise. (See sect. 6 of M. Kummer's Memoir.) The two definitions, corresponding to the cases of q prime to n, and q a divisor of n, enable us, when they are taken together, to transfer to the general case when n is composite, the elementary theorems already shown to exist when n is prime (see Art. 47). We may add that it is easy to prove, in the general as in the special case (see Art. 48), that the number of *classes* of ideal numbers is finite.

60. *Application to the Theory of the Division of the Circle.*—We cannot quit the subject of complex numbers without mentioning certain important investigations in which they have been successfully employed. The first relates to the problem of the division of the circle. In this problem the

resolvent function of Lagrange $\sum_{s=0}^{s=p-2} \theta^s x^{\gamma^s}$ (see Art. 30) is, as is well known, of primary importance. Retaining, with a slight modification, the notation of Art. 30, and still representing by λ a prime divisor of $p-1$, and by α a root of the equation $\frac{\alpha^\lambda - 1}{\alpha - 1} = 0$, let us consider the function $F(\alpha, x)$, which is a particular case of the resolvent, and let us represent the quotient $\frac{F(\alpha, x)\, F(\alpha^k, x)}{F(\alpha^{k+1}, x)}$ by $\psi_k(\alpha)$. We thus find

$$[F(\alpha, x)]^s = \psi_1(\alpha).\psi_2(\alpha) \dots \psi_{s-1}(\alpha).F(\alpha^s, x), \quad \dots \dots \quad (1)$$

and in particular, observing that $F(\alpha, x)\, F(\alpha^{\lambda-1}, x) = p$,

$$[F(\alpha, x)]^\lambda = p\psi_1(\alpha).\psi_2(\alpha) \dots \psi_{\lambda-2}(\alpha), \quad \dots \dots \dots \quad (2)$$

a result which is in accordance with the known theorem that $[F(\alpha, x)]^\lambda$ is independent of x and is an integral function of α only. The resolution of the auxiliary equation of order λ, the roots of which are the λ periods of $\frac{p-1}{\lambda}$ roots of the equation $\frac{x^p - 1}{x - 1} = 0$, depends solely on the determination of the complex numbers $\psi_1(\alpha), \psi_2(\alpha), \dots, \psi_{\lambda-2}(\alpha)$. For when these complex numbers are known, we may equate $F(\alpha, x)$ to any λ-th roots of the expression $p\psi_1(\alpha).\psi_2(\alpha) \dots \psi_{\lambda-2}(\alpha)$; from the value of $F(\alpha, x)$, thus obtained, those of $F(\alpha^2, x)$, $F(\alpha^3, x)$, ... may be inferred by means of equation (1); and, lastly, from the values of $F(1, x)$, $F(\alpha, x)$, ..., $F(\alpha^{\lambda-1}, x)$, the values of the periods themselves are deducible by the solution of a system of linear equations. To determine the numbers $\psi_1(\alpha)$, $\psi_2(\alpha)$, ... M. Kummer assigns the ideal prime factors of which they are composed, employing for this purpose the results cited in Art. 30. The equation $\psi_k(\alpha)\, \psi_k(\alpha^{-1}) = p$ shows that $\psi_k(\alpha)$ contains precisely $\frac{1}{2}(p-1)$ ideal prime divisors of p, and no other complex prime. To distinguish the prime factors of p contained in $\psi_k(\alpha)$ from those contained in $\psi_k(\alpha^{-1})$ M. Kummer avails himself of the congruence V. of Art. 30, viz.,

$$\psi(\gamma) \equiv -\frac{\Pi(m+n)}{\Pi(m).\Pi(n)}, \bmod p.$$

Let $\lambda' = \frac{p-1}{\lambda}$, and $u \equiv \gamma^{\lambda'}$, mod p, so that $u, u^2, \dots, u^{\lambda-1}$ are the roots of $\frac{x^\lambda - 1}{x - 1} \equiv 0$, mod p; also, to adapt the formulae of Art. 30 to our present purpose, let $\theta^{-\lambda'} = \alpha$, $m = \lambda'$, $n = k\lambda'$; it will result from these substitutions, that $\psi_k(u^{-h}) \equiv 0$, mod p, if k and h satisfy the inequality $[h] + [kh] > \lambda$, where

$[h]$ and $[kh]$ are positive numbers less than λ, and congruous, mod λ, to h and kh respectively. If we represent by $f(\alpha)$ the ideal prime factor of p which appertains to the substitution $\alpha = u$, this may be expressed by saying that $\psi_k(\alpha)$ contains the factor $f(\alpha^{-h})$, if $\left[\frac{1}{h}\right] + \left[\frac{k}{h}\right] > \lambda$, the symbols $\left[\frac{1}{h}\right]$ and $\left[\frac{k}{h}\right]$ denoting the least positive numbers satisfying the congruences $hx \equiv 1$, mod λ, and $hx \equiv k$, mod λ. Assigning, therefore, to the number h every positive value less than λ compatible with this condition, we may write

$$\psi_k(\alpha) = \pm \alpha^s \Pi f(\alpha^{-h}),$$

$\pm \alpha^s$ being a simple unit which may be determined by the congruence

$$\psi_k(\alpha) \equiv -1, \text{ mod } (1-\alpha)^2 \, * :$$

it is not necessary to add a real complex unit, for a reason which has already appeared (see Art. 56, *supra*). From the expression for $\psi_k(\alpha)$ a still simpler formula for $F(\alpha, x)^\lambda$ may be obtained, viz.

$$[F(\alpha, x)]^\lambda = \pm \alpha^s \prod_{m=1}^{m=\lambda-1} [f(\alpha^{-m})]^{\left[\frac{1}{m}\right]} . \dagger$$

61. *Application to the Last Theorem of Fermat.*—The second investigation to which we shall advert relates to the celebrated proposition known as the 'Last Theorem of Fermat,' viz. that the equation $x^n + y^n = z^n$ is irresoluble, in integral numbers, for all values of n greater than 2 ‡. As Fermat himself

---

* The numbers $\psi_k(\alpha)$ are *primary* according to M. Kummer's definition (Art. 52); for

$$\psi_k(\alpha) = \frac{F(\alpha, x)\, F(\alpha^k, x)}{F(\alpha^{k+1}, x)} = \Sigma \alpha^{y_1 + k y_2},$$

the summation extending to every pair of values of y_1 and y_2 that satisfy the congruence

$$\gamma^{y_1} + \gamma^{y_2} \equiv 1, \text{ mod } p,$$

in which γ represents the same primitive root of p that occurs in the expression $F(\alpha, x)$. Hence

$$\psi_k(1) = p-2 \equiv -1, \text{ mod } \lambda, \quad \text{and} \quad \psi_k(\alpha)\,\psi_k(\alpha^{-1}) = p \equiv 1 \equiv [\psi_k(1)]^2, \text{ mod } \lambda.$$

Also $\psi_k(\alpha) - \psi_k(1)$ is divisible by $(1-\alpha)^2$; for

$$\psi'_k(1) = \Sigma(y_1 + k y_2) = \tfrac{1}{2}(1+k)(p-1)(p-2),$$

observing that y_1 and y_2 each receive all the values $1, 2, \ldots, p-2$ in succession. We have, therefore, the congruence $\psi'_k(1) \equiv 0$, mod λ, from which it follows (see a note on the next article) that

$$\psi_k(\alpha) \equiv \psi_k(1), \text{ mod } (1-\alpha)^2, \quad \text{or} \quad \psi_k(\alpha) \equiv -1, \text{ mod } (1-\alpha)^2, \quad \text{as in the text.}$$

† Liouville, vol. xvi. p. 448. M. Kummer has also extended his solution of this problem to the case in which n is any divisor of $p-1$. See the memoir quoted in the last article, sect. 11.

‡ Fermat's enunciation of this celebrated theorem is contained in the first of the MS. notes placed by him on the margin of his copy of Bachet's edition of Diophantus. It would seem that this copy is now lost; but in the year 1670 an edition of Bachet's Diophantus was published at Toulouse,

has left us a proof of the impossibility of this equation in the case of $n=4$, by a method which Euler has extended to the case of $n=3$, we may suppose, without

by Samuel de Fermat (the son of the great geometer), in which these notes are preserved (Diophanti Alexandrini Arithmeticorum libri sex, et de Numeris Multangulis liber unus, cum commentariis C. G. Bacheti V. C. et observationibus D. P. de Fermat senatoris Tolosani. Tolosæ 1670). The theorems contained in them are, with a few exceptions, enunciated without proof; and it may be inferred from the preface of S. Fermat that he found no demonstration of them among his father's papers. Nevertheless, in the case of several of these propositions, we have the assertion of Fermat himself that he was in possession of their demonstration; and although, when we consider the imperfect state of analysis in his time, it is surprising that he should have succeeded in creating methods which subsequent mathematicians have failed to rediscover, yet there is no ground for the suspicion that he was guilty of an untruth, or that he mistook an apparent for a real proof. In fact these suspicions are refuted, not only by the reputation for honour and veracity which he enjoyed among his contemporaries, and by the evidence of singular clearness of insight which his extant writings supply, but also by the facts of the case itself. {Gauss, vol. ii. p. 160, expresses himself unfavourably to Fermat: see especially p. 152.} It would be inexplicable, if his conclusions reposed on induction only, that he should never have adopted an erroneous generalization; and yet, with the exception of the 'Last Theorem' (the demonstration of which, after two centuries, is still incomplete), every proposition of Fermat's has been verified by the labours of his successors. There is, indeed, one other exception to this statement; but it is an exception which proves the rule. In the letter to Sir Kenelm Digby which concludes the 'Commercium Epistolicum etc.' edited by Wallis (Oxford, 1658), Fermat enunciates the proposition that the numbers contained in the formula $2^{2^n}+1$ are all primes, acknowledging, however, that, though convinced of its truth, he had not succeeded in obtaining its demonstration. This letter, which is undated, was written in 1658; but it appears, from a letter of Fermat's to M. de * * *, dated October 18, 1640, that even at that earlier date he was acquainted with the proposition, and had convinced himself of its truth (D. Petri de Fermat Varia Opera Mathematica, Tolosæ, 1679, p. 162). It was, however, subsequently observed by Euler that $2^{2^5}+1 = 4294967297 = 641 \times 6700417$, *i.e.* that the undemonstrated proposition is untrue (Op. Arith. collecta, vol. i. p. 356). The error, if it is an error, is a fortunate one for Fermat; it exemplifies his candour and veracity, and it shows that he did not mistake inductive probability for rigorous demonstration :—'Mais je vous advoue tout net,' are his words in the letter last referred to, '(car par advance je vous advertis que comme je ne suis pas capable de m'attribuer plus que je ne sçay, je dis avec même franchise ce que je ne sçay pas), que je n'ay peu encore démonstrer l'exclusion de tous diviseurs en cette belle proposition que je vous avois envoyée, et que vous m'avez confermée touchant les nombres 3, 5, 17, 257, 6553, &c. Car bien que je reduise l'exclusion à la pluspart des nombres, et que j'aye même des raisons probables pour le reste, je n'ay peu encore démonstrer nécessairement la vérité de cette proposition, de laquelle pourtant je ne doute non plus à cette heure que je faisois auparavant. Si vous en avez la preuve assurée, vous m'obligerez de me la communiquer: car après cela rien ne m'arrestera en ces matières.'

The 'Last Theorem' is enunciated by Fermat as follows :—

'Cubum autem in duos cubos, aut quadrato-quadratum in duos quadrato-quadratos, et generaliter nullam in infinitum ultrà quadratum potestatem in duos ejusdem nominis fas est dividere; cujus rei demonstrationem mirabilem sane detexi. Hanc marginis exiguitas non caperet.' (Fermat's Diophantus, p. 51.)

Fermat has also asserted that neither the sum (*ibid.* p. 258) nor the difference (*ibid.* p. 338) of loss of generality, that n is an uneven prime number λ greater than 3, and we may write the equation in the symmetrical form $x^\lambda+y^\lambda+z^\lambda=0$. The impos-

---

two biquadrates can be a square. Each of these propositions comprehends the theorem that the sum of two biquadrates cannot be a biquadrate; and of the second, we possess a very remarkable demonstration by Fermat himself (*ibid.* p. 338; and compare Euler, Elémens d'Algèbre, vol. ii. sect. 13; Legendre, Théorie des Nombres, vol. ii. p. 1). The essential part of this demonstration consists in showing that, from any supposed solution of the Diophantine equation $x^4-y^4=$ a square, another solution may be deduced in which the values of the indeterminates are not equal to zero, and yet are absolutely less than in the proposed solution, from which it immediately follows that the Diophantine equation is impossible. This method has been successively employed by Euler (*loc. cit.*) to demonstrate several negative Diophantine propositions, and in particular the theorem that the sum of two cubes cannot be a cube. The only arithmetical principles (not included in the first elements of the science) which are employed by Euler and Fermat in their applications of this method, relate to certain simple properties of the quadratic forms x^2+y^2, x^2+2y^2, x^2+3y^2; and as these principles seem inadequate to overcome the difficulties presented by the equation $x^n+y^n+z^n=0$, when n is >4, it is probable that Fermat's 'demonstratio mirabilis sane' of the general theorem was entirely different from that which he has incidentally given of the particular case.

The impossibility of the equation $x^n+y^n+z^n=0$ for $n=5$ was first demonstrated by Legendre (Mémoires de l'Académie des Sciences, 1823, vol. vi. p. 1, or Théorie des Nombres, vol. ii. p. 361. See also an earlier paper by Lejeune Dirichlet, Crelle, vol. iii. p. 354, with the addition at p. 368, and a later one by M. Lebesgue, Liouville, vol. viii. p. 49); for $n=14$, by Dirichlet (Crelle, vol. ix. p. 390); and for $n=7$, by M. Lamé (Mémoires des Savans Etrangers, vol. viii. p. 421, or Liouville, vol. v. p. 195. See also the Comptes Rendus, vol. ix. p. 359, and a paper by M. Lebesgue, Liouville, vol. v. pp. 276 and 348). But the methods employed in these researches are specially adapted to the particular exponents considered, and do not seem likely to supply a general demonstration. The proof in Barlow's Theory of Numbers, pp. 160–169, is erroneous, as it reposes (see p. 168) on an elementary proposition (cor. 2, p. 20) which is untrue. A memoir by M. Kummer on the equation $x^{2\lambda}+y^{2\lambda}=z^{2\lambda}$, in which complex numbers are not employed, and in which no single case of the theorem is demonstrated (Crelle, vol. xvii. p. 203), is nevertheless of great interest for the number of auxiliary propositions contained in it. Of the same character are the notes by MM. Lebesgue and Liouville, in Liouville's Journal, vol. v. pp. 184 and 360, and a few theorems given without demonstration by Abel, Œuvres, vol. ii. p. 264.

In the year 1847, M. Lamé presented to the Academy at Paris a memoir containing a general demonstration of Fermat's Theorem, based on the properties of complex numbers (Comptes Rendus, vol. xxiv. p. 310; Liouville, vol. xii. pp. 137 and 172). It was, however, observed by M. Liouville (Comptes Rendus, vol. xxiv. p. 315), that this demonstration is defective, as it assumes, without proof, the proposition that a complex number can be represented, and in one way only, as the product of powers of complex primes—a proposition which, as we have seen, is untrue, unless we admit ideal as well as actual complex primes. The discussion on M. Lamé's memoir attracted Cauchy's attention to Fermat's Theorem; and the 24th and 25th volumes of the Comptes Rendus contain several communications from him on the subject of complex numbers [or polynômes radicaux, as he has preferred to term them]. In the earlier papers of this series, Cauchy attempts to prove a proposition which, as we have already observed (see Art. 41), is untrue for complex numbers considered generally, viz. that the norm of the remainder in the division of one complex number by another can be rendered less than the norm of the divisor (see Comptes Rendus, vol. xxiv. pp. 517, 633, and 661). Elsewhere (*ibid.* p. 579) he sibility of solving this equation has been demonstrated by M. Kummer, first, for all values of λ not included among the exceptional primes *; and secondly, for all exceptional primes which satisfy the three following conditions :—

(1) That the first factor of H, though divisible by λ, is not divisible by λ^2 (see Art. 50).

(2) That a complex modulus can be assigned, for which a certain definite complex unit is not congruous to a perfect λ-th power.

(3) That $B_{\kappa\lambda}$ is not divisible by λ^3, B_κ representing that Bernoullian number $[\kappa \leqq \mu - 1]$ which is divisible by λ †.

Three numbers below 100, viz. 37, 59, 67, are, as we have seen, exceptional primes. But it has been ascertained by M. Kummer that the three conditions just given are satisfied in the case of each of those numbers; so that the impossibility of Fermat's equation has been demonstrated for all values of the exponent up to 100. Indeed, it would probably be difficult to find an exceptional prime not satisfying the three conditions, and consequently excluded from M. Kummer's demonstration.

We must confine ourselves here to an indication of the principles on which the demonstration rests in the case of the non-exceptional primes ‡.

---

assumes the proposition as a hypothesis, and deduces from it conclusions which are erroneous (pp. 581, 582). But at p. 1029 he recognises and demonstrates its inaccuracy. The results at which he arrives in his subsequent papers on the same subject are, for the most part, comprehended in M. Kummer's general theory (Comptes Rendus, vol. xxv. pp. 37, 46, 93, 132, 177). In one place, however (p. 181), he enunciates, though without demonstrating, the following important result :—

'If the equation $x^\lambda + y^\lambda + z^\lambda = 0$ be resoluble, x, y, z denoting integral numbers prime to λ, the sum

$$1^{\lambda-4} + 2^{\lambda-4} + 3^{\lambda-4} + \ldots + \{\tfrac{1}{2}(\lambda-1)\}^{\lambda-4}$$

is divisible by λ.'

(Compare M. Kummer's memoir in the Berlin Transactions for 1857, p. 64.)

The investigation of the Last Theorem of Fermat has been twice proposed as a prize-question by the Academy of Paris—first at some time previous to 1823 (see Legendre's memoir already cited, in vol. vi. of the Mémoires de l'Académie des Sciences, p. 2), and again in 1850 (Comptes Rendus, vol. xxx. p. 263): at neither time was the prize adjudged to any of the memoirs received. On the last occasion, after several postponements of the date originally fixed for the award, the prize was ultimately, in 1857 (*ib.* vol. xliv. p. 158), conferred on M. Kummer, who had not been a competitor, for his researches on complex numbers.

* Liouville, vol. xvi. p. 488, or Crelle, vol. xl. p. 131.

† See the memoir No. 15 in the list of Art. 41.

‡ When λ is not an exceptional prime, the equation $x^\lambda + y^\lambda + z^\lambda = 0$ is irresoluble not only in ordinary integral numbers, but also in any complex integers composed of λ-th roots of unity. The demonstration does not possess the same generality when λ is an exceptional prime satisfying the three conditions cited in the text. In this case M. Kummer has only shown that the equation

We may suppose that λ is greater than 3, and that no two of the numbers, x, y, z admit any common divisor. And first, let none of them be divisible by $1-a$, a still representing a root of the equation $\frac{a^\lambda - 1}{a-1} = 0$. Since for x we may write $a^s x$, we may assume that x, y, z are of the form

$$x = a + (1-a)^2 X,$$
$$y = b + (1-a)^2 Y,$$
$$z = c + (1-a)^2 Z,$$

a, b, c denoting integral numbers prime to λ, which evidently satisfy the congruence $a + b + c \equiv 0$, mod λ. The equation $x^\lambda + y^\lambda + z^\lambda = 0$ may then be written thus

$$(x + ay)(x + a^2 y)(x + a^3 y) \ldots\ldots (x + a^{\lambda-1} y) = -z^\lambda.$$

No two of the factors of which the left-hand member is composed can have any common divisor; each of them is therefore the product of a perfect λ-th power by a unit; so that we may write, $x + a^s y = a^\rho e(a) v^\lambda$, $e(a)$ denoting a real unit. Since v^λ is an actual number, it follows (remembering that λ is not an exceptional prime) that v is also actual; hence v^λ is congruous, mod λ, to a certain integral number m. Eliminating $m \times e(a)$ between the two congruences

$$x + a^s y \equiv m a^\rho e(a), \text{ and } x + a^{-s} y \equiv m a^{-\rho} e(a), \text{ mod } \lambda,$$

we find

$$a^{-\rho}(x + a^s y) - a^\rho (x + a^{-s} y) \equiv 0, \text{ mod } \lambda.$$

For the modulus $(1-a)$ this congruence is identically satisfied *. That it should be satisfied, mod $(1-a)^2$, we must have the relation $(a+b)\rho \equiv bs$, mod λ; whence, putting $\frac{b}{a+b} \equiv k$, mod λ, we have $\rho \equiv ks$, mod λ. Substituting this value for ρ, we find that the congruence

$$a^{-ks}(x + a^s y) - a^{ks}(x + a^{-s} y) = 0$$

is identically satisfied, mod $(1-a)^3$; but in order that it should be satisfied, mod $(1-a)^4$, we have the condition

$$s^3 b(2k-1)(k-1) - 3s\{(k-1)y'' + kx''\} \equiv 0, \text{ mod } \lambda,$$

$x^\lambda + y^\lambda + z^\lambda = 0$ is irresoluble when we suppose that x, y, z are ordinary integral numbers prime to λ, or else complex numbers containing the binary periods $a + a^{-1}$, one of which has a common divisor with λ.

* Since λ is divisible by $(1-a)^{\lambda-1}$, and since

$$\phi(a) = \phi(1) + (a-1)\phi'(1) + (a-1)^2 \frac{\phi''(1)}{1.2} + \ldots,$$

it is readily seen that, if $r \leqq \lambda - 1$, the conditions for the divisibility of $\phi(a)$ by $(1-a)^r$ are

$$\phi(1) \equiv 0, \ \phi'(1) \equiv 0, \ \ldots, \ \phi^{(r-1)}(1) \equiv 0, \text{ mod } \lambda.$$

where x'' and y'' are the values (for $\alpha=1$) of the second derived functions of x and y with respect to α. This conditional congruence must be satisfied for every value of s; either therefore $k\equiv 1$, mod λ, or $2k\equiv 1$, mod λ. The supposition $k\equiv 1$ is inadmissible; for it implies that $a\equiv 0$, mod λ, contrary to the hypothesis. Hence we must have $2k\equiv 1$, and $a\equiv b$, or, by parity of reasoning, $a\equiv b\equiv c$, mod λ. But also $a+b+c\equiv 0$, mod λ, whence we again infer the inadmissible conclusion $a\equiv b\equiv c\equiv 0$, mod λ.

Secondly, let one of the numbers x, y, z (for example, z) be divisible by $1-\alpha$; it will be convenient to consider the equation in the generalised form

$$x^\lambda+y^\lambda=E(\alpha)(1-\alpha)^{m\lambda}z^\lambda, \quad \dots\dots\dots \quad (1)$$

in which x, y, and z are all prime to $1-\alpha$, and $E(\alpha)$ is any unit. We may assume that the values of x and y are of the form

$$x=a+(1-\alpha)^2X,$$
$$y=b+(1-\alpha)^2Y,$$

a and b being prime to λ, but satisfying the relation $a+b\equiv 0$, mod λ. In the first place, m must be greater than 1. For since

$$x^\lambda\equiv a^\lambda, \text{ and } y^\lambda\equiv b^\lambda, \text{ mod } (1-\alpha)^{\lambda+1},$$

if $x^\lambda+y^\lambda$ be divisible by $(1-\alpha)^\lambda$, $a^\lambda+b^\lambda$ is divisible by λ^2, and therefore $x^\lambda+y^\lambda$ by $(1-\alpha)^{\lambda+1}$. Again, each of the factors $x+\alpha y$, $x+\alpha^2y$, ..., $x+\alpha^{\lambda-1}y$ is divisible once, and once only, by $1-\alpha$; whence it follows that $x+y$ is divisible by $(1-\alpha)^{m\lambda-\lambda+1}$, and that no two of the λ factors of $x^\lambda+y^\lambda$ have any other common divisor than $1-\alpha$. Hence the λ factors

$$\frac{x+y}{(1-\alpha)^{m\lambda-\lambda+1}}, \quad \frac{x+\alpha y}{1-\alpha}, \quad \dots\dots, \quad \frac{x+\alpha^{\lambda-1}y}{1-\alpha}$$

are relatively prime, and may be represented by expressions of the form

$$e_0(\alpha)\phi_0^\lambda, \quad e_1(\alpha)\phi_1^\lambda, \quad \dots\dots, \quad e_{\lambda-1}(\alpha)\phi_{\lambda-1}^\lambda,$$

$e_0(\alpha)$, $e_1(\alpha)$, ... representing units, and ϕ_0^λ, ϕ_1^λ, ... λ-th powers prime to $1-\alpha$. Eliminating x and y from the three equations

$$x+\ \ y=e_0(\alpha)\phi_0^\lambda(1-\alpha)^{m\lambda-\lambda+1},$$
$$x+\alpha^r y=e_r(\alpha)\phi_r^\lambda(1-\alpha),$$
$$x+\alpha^s y=e_s(\alpha)\phi_s^\lambda(1-\alpha),$$

we obtain a result of the form

$$\phi_r^\lambda+\epsilon(\alpha)\phi_s^\lambda=E_1(\alpha)(1-\alpha)^{(m-1)\lambda}\phi_0^\lambda, \quad \dots\dots\dots \quad (2)$$

$\epsilon(\alpha)$ and $E_1(\alpha)$ denoting two units. But, as in the former case, it may be shown that ϕ_r^λ and ϕ_s^λ are congruous, mod λ, to real integers, and

$(1-a)^{(m-1)\lambda} \equiv 0$, mod λ, because $m > 1$. Hence $\epsilon(a)$ is also congruous to a real integer for the modulus λ, and is therefore a perfect λ-th power by a property of every non-exceptional prime (see Art. 52). The equation (2) therefore assumes the form $x_1^\lambda + y_1^\lambda = E_1(a) z_1^\lambda (1-a)^{(m-1)\lambda}$.

If, therefore, the proposed equation (1) be possible, it will follow, by successive applications of this reduction, that the equation

$$x^\lambda + y^\lambda = E(a)(1-a)^\lambda z^\lambda$$

is also possible. But this equation has been shown to be impossible; the equation (1) is therefore also impossible.

62. *Application to the Theory of Numerical Equations.*—In the Monatsberichte for June 20, 1853 (see also the Monatsberichte for 1856, p. 203), M. Kronecker has enunciated the following theorem :—

'The roots of any Abelian equation, the coefficients of which are integral numbers, are rational functions of roots of unity.' The demonstration of this theorem (Monatsberichte for 1853, pp. 371–373) depends on a comparison of a certain form, of which the resolvent function of any Abelian equation is susceptible, with M. Kummer's expression for the resolvent function in the case of the equation of the division of the circle (see Art. 60). It thus involves considerations relating to ideal numbers.

Two propositions of a more special character, and closely connected with one another, have also been given by M. Kronecker (Crelle, vol. liii. p. 173). Their demonstration is immediately deducible from the principles of Dirichlet's theory of complex units :—

'If unity be the analytical modulus of every root of an equation, of which the first coefficient is unity and all the coefficients are integral numbers, the roots of the equation are roots of unity.'

'If all the roots of an equation (having its first coefficient unity and all its coefficients integral) be real and inferior in absolute magnitude to 2, so that they can be represented by expressions of the form $2\cos\alpha$, $2\cos\beta$, $2\cos\gamma, \dots$ the arcs α, β, γ are commensurable with the complete circumference.'

In the following proposition M. Kronecker has extended a theorem of M. Kummer's (Art. 42) relating to complex units composed with roots of unity of which the index is a prime, to complex units composed with any roots of unity (Crelle, vol. liii. p. 176) :—

'Every complex unit composed with the roots of the equation $\omega^n = 1$, can be rendered real by multiplication with a $4n$-th root of unity. If n be even, a $2n$-th root will always suffice; and if n be a power of a prime, an n-th root will suffice.'

The demonstration of this proposition is also deducible from Dirichlet's principles.

63. *Tables of Complex Primes.*—In M. Kummer's earliest memoir on complex numbers (Liouville, vol. xii. p. 206) he has given a table of the complex factors, composed of λ-th roots of unity, which are contained in real primes of the form $m\lambda+1$ inferior to 1000, λ representing one of the primes 5, 7, 11, 13, 17, 19, 23. This memoir was written before M. Kummer had considered the complex factors of primes of linear forms other than $m\lambda+1$, and before he had introduced the conception of ideal numbers. The complex prime factors of real primes of those other linear forms are, therefore, not exhibited in the Table; and the five numbers of the form $23m+1$, 47, 139, 277, 461, 967, each of which contains 22 ideal factors composed of 23rd roots of unity, are represented as products of 11 actual factors (each of which contains two reciprocal ideal factors). The tentative methods by which the complex factors were discovered are explained in sect. 9 of the memoir cited. Since the full development of M. Kummer's theory, Dr. Reuschle has undertaken to complete and extend the Table. He has already given tables containing the complex prime factors of *all* real primes less than 1000, composed of 5th, 7th, 11th, 13th, 17th, 23rd, and 29th roots of unity, together with the complete solution of the congruences corresponding to the equations of the periods (see the Monatsberichte for 1859, pp. 488 and 694, and for 1860, pp. 150 and 714). For 5, 7, 11, 13, 17, the complex primes are exhibited in a primary form; for 19, 23, and 29 they are exhibited in a form which satisfies the condition

$$f(\alpha)\equiv f(1), \text{ mod } (1-\alpha)^2,$$

but not the condition

$$f(\alpha)f(\alpha^{-1})\equiv [f(1)]^2, \text{ mod } \lambda.$$

The ideal factors Dr. Reuschle represents by their lowest actual powers; for 23 this power is the cube, for 29 it is the square; for 11, 13, 17, 19, as well as for 5 and 7, all complex prime factors of real primes less than 1000 are actual. It appears from the Table (and it has indeed been proved by M. Kummer), that 29 is an 'irregular determinant' (see Art. 49, note); for the number of classes is 8, while the square of every ideal number (occurring as a factor of a real prime inferior to 1000) is actual. The methods employed by Dr. Reuschle in the calculation of his tables have not yet been published by him. In some instances, as M. Kummer has observed, they have not led him to the simplest possible forms of the ideal primes.

A particular investigation relating to the ideal factors of 47, composed of 23rd roots of unity, has been given by Mr. Cayley (Crelle, vol. lv. p. 192, and lvi. p. 186).

64. The investigations relating to Laws of Reciprocity, which have so long occupied us in this report, have introduced us to considerations apparently so remote from the theory of the residues of powers of integral numbers, that it requires a certain effort to bear in mind their connexion with that theory. It will be remembered that the complex numbers to which our attention has been directed are not of that general kind to which we have referred in Art. 41, but are exclusively those which are composed of roots of unity. The theory of complex numbers, in the widest sense of that term, does indeed present to us an important generalisation of the theory of the residues of powers; for the theorem of Fermat (see Art. 53) subsists alike for every species of complex numbers. But the complex numbers of Gauss, of Jacobi, and of M. Kummer force themselves upon our consideration, not because their properties are generalisations of the properties of ordinary integers, but because certain of the properties of integral numbers can only be explained by a reference to them. The law of quadratic reciprocity does not, as we have seen, necessarily require for its demonstration any considerations other than those relating to ordinary integers; the real prime numbers of arithmetic are here the ultimate elements that enter into the problem. But when we come to binomial congruences of higher orders, we find that the true elements of the question are no longer real primes, but certain complex factors, composed of roots of unity, which are, or may be conceived to be, contained in real primes. For we find that the law which expresses the mutual relation (with respect to the particular kind of congruences considered) of two of these complex factors is a primary and simple one; while the corresponding relations between the real primes themselves are composite and derivative, and, in consequence, complicated. It thus becomes indispensable, for the investigation of the properties of real numbers, to construct an arithmetic of complex integers; and this is what has been accomplished by the researches, of which an account has been given in the preceding articles.

The higher laws of reciprocity (like that of quadratic residues) may be considered as furnishing a criterion for the resolubility or irresolubility of binomial congruences; and this, though not the only application of which they are susceptible, is that which most naturally suggests itself. When the binomial congruence is cubic or biquadratic, it is easy to resolve the real prime modulus into factors of the form $a+b\rho$, or $a+bi$ (Arts. 37 and 24), and equally easy to determine the value of the critical symbol of reciprocity by a uniform and elementary process (see Art. 36). For these, therefore, as well as for quadratic congruences, the criterion deducible from the laws of reciprocity is all that can be desired. But for binomial congruences of higher orders this criterion is not a satisfactory one, because of the difficulty of obtaining the resolution of a real prime into its complex factors, and also because of the impossibility of determining the value of the critical symbol by the conversion of an ordinary fraction into a continued fraction.

The only known criterion applicable to such congruences is the following, the demonstration of which is deducible from the elements of the theory of the residues of powers :—Let $x^n \equiv A$, mod p, represent the proposed congruence; it will be resoluble or irresoluble according as the index of A is or is not divisible by d, the greatest common divisor of n and $p-1$, *i.e.* according as the exponent to which A appertains is or is not a divisor of $\frac{p-1}{d}$ (see Arts. 14 and 15).

65. *Solution of Binomial Congruences.*—We now come to the problem of the actual solution of binomial congruences—a subject upon which our knowledge is confined within very narrow limits.

When a table of indices for the prime p has been constructed, the resolution of every binomial congruence, if it be resoluble, or, if not, the demonstration of its irresolubility, is implicitly contained in it. But to use a table of indices for the solution of a binomial congruence is, as we have already observed in a similar case (Art. 16), to solve a problem by means of a recorded solution of it. When the congruence $x^n \equiv A$, mod p, is resoluble, its solution may always be made to depend on that of a congruence of the form $x^d \equiv a$, mod p, where d is the greatest common divisor of n and $p-1$, and where $a \equiv A^s$, mod p, and $ns \equiv d$, mod $p-1$. We may therefore suppose that, in the congruence $x^n \equiv A$, mod p, n is a divisor of $p-1$. This congruence (if resoluble at all) will have as many roots as it has dimensions; if ξ be any one of them, and $1, \theta_1, \theta_2, \ldots, \theta_{n-1}$ be the roots of the congruence $x^n \equiv 1$, mod p, the roots of $x^n \equiv A$, mod p, will be $\xi, \xi\theta_1, \xi\theta_2, \ldots, \xi\theta_{n-1}$; so that the complete resolution of the congruence $x^n \equiv A$, mod p, requires, first, the determination of a single root of that congruence itself, and, secondly, the complete resolution of the congruence $x^n \equiv 1$, mod p. With regard to the first of these requisites, in the important case in which the exponent t to which A appertains is prime to n, a value of x satisfying the congruence $x^n \equiv A$, mod p, can be determined by a direct method (Disq. Arith., Arts. 66, 67). For, in this case, it will always happen that one value of x is a certain power A^k of A, where k is determined by the congruence $kn \equiv 1$, mod t. Nor is it necessary, in order to determine k, to know the exponent t to which A appertains; it is sufficient to have ascertained that it is prime to n; for, if we resolve $p-1$ into two factors prime to one another, and such that one of them is divisible by n and contains no prime not contained in n, the other will be divisible by t, and may be employed as modulus instead of t in the congruence $kn \equiv 1$, mod t. When this method is inapplicable, we can only investigate a root of the congruence $x^n \equiv A$, mod p (where A is different from 1), by tentative processes, which, however, admit of certain abbreviations (Disq. Arith., Arts. 67, 68). The work of Poinsot (Réflexions sur la Théorie des Nombres, cap. iv. p. 60) contains a very full and elegant exposition of the theory of binomial congruences; but neither he nor any other writer subsequent to Gauss has been able to add any other direct method to that which we have just mentioned.

66. *Solution of the Congruence* $x^n \equiv 1$, *mod* p.—When a single root of the congruence $x^n \equiv A$ is known, we may, as we have seen, complete its resolution by obtaining all the roots of the congruence $x^n \equiv 1$, mod p. The methods of Gauss, Lagrange, and Abel for the solution of the binomial equation $x^n - 1 = 0$ are in a certain sense applicable to binomial congruences of this special form. It is evident, from a comparison of several passages in the Disquisitiones Arithmeticæ *, that Gauss himself contemplated this arithmetical application of his theory of the division of the circle, and that he intended to include it in the 8th section of his work, which, however, has never been given to the world. In fact, the method of Abel † which comprehends that of Gauss, and which gives the solution of any Abelian equation, is equally applicable to any *Abelian* congruence; *i.e.* to any completely resoluble congruence of order m, the m roots of which (considered with regard to the prime modulus p) may be represented by the series of terms

$$r,\ \phi(r),\ \phi^2(r),\ \ldots,\ \phi^{m-1}(r),$$

the symbol ϕ denoting a given rational [fractional or integral] function. And as we can always express the roots of an Abelian equation by radicals (*i.e.* by

* See Disq. Arith., Arts. 61, 73, and especially Art. 335.

† See Abel's memoir, 'Sur une classe particulière d'équations résolubles algébriquement,' sect. 3. (Œuvres, vol. i. p. 114, or Crelle, vol. iv. p. 131), and M. Serret's Algèbre Supérieure, 26th and 27th lessons.

the roots of equations of two terms), so also the solution of an Abelian congruence depends ultimately on the solution of binomial congruences. When, for any prime modulus, an Abelian equation admits of being considered as an Abelian congruence, so precise is the correspondence of the equation and the congruence, that (as Poinsot has observed in a memoir in which he has occupied himself with the comparative analysis of the equation $x^n=1$, and the congruence $x^n \equiv 1$, mod p *) we may consider the analytical expression of the roots of the equation as also containing an expression of the roots of the congruence; and by giving a congruential interpretation † to the radical signs which occur in that expression, we may elicit from it the actual values of the roots of the congruence. An example taken from Poinsot's memoir will render this intelligible ‡. The six roots of the equation $\frac{x^7-1}{x-1}=0$ are comprised in the formula

$$x=\frac{-1+\sqrt{-7}}{6}+\frac{1}{3}\left[7-\frac{1}{2}\sqrt{-7}+\frac{3}{2}\sqrt{21}\right]^{\frac{1}{3}}+\frac{1}{3}\left[7-\frac{1}{2}\sqrt{-7}-\frac{3}{2}\sqrt{21}\right]^{\frac{1}{3}},$$

where the signs + and − are to be successively attributed to $\sqrt{-7}$, and where the product of the two cube roots is $+\sqrt{-7}$, or $-\sqrt{-7}$, according to the sign attributed to $\sqrt{-7}$. Considering the equation as a congruence with regard to the modulus 43, and observing that

$$\sqrt{-7}\equiv\pm 6, \text{ mod } 43, \quad \sqrt{21}\equiv\pm 8, \text{ mod } 43,$$

we obtain in the first place

$$x\equiv \frac{5}{6}+\frac{1}{3}\sqrt[3]{16}+\frac{1}{3}\sqrt[3]{-8}, \text{ mod } 43,$$

and

$$x\equiv -\frac{7}{6}+\frac{1}{3}\sqrt[3]{22}+\frac{1}{3}\sqrt[3]{-2}, \text{ mod } 43,$$

the product of the two cube roots being congruous to $+6$ in the first formula, and to -6 in the second; and finally, observing that

---

* 'Sur l'Application de l'Algèbre à la Théorie des Nombres,' Mémoires de l'Académie des Sciences, vol. iv. p. 99.

† Gauss employs the symbol $\sqrt[n]{A}$, mod p, to denote a root of the congruence $x^n\equiv A$, mod p, just as he employs the symbol $\frac{B}{A}$, mod p, to denote the root of the congruence $Ax\equiv B$, mod p The *congruential radical* $\sqrt[n]{A}$, mod p, has of course as many values as the congruence $x^n\equiv A$, mod p, has solutions; if that congruence be irresoluble, the symbol is impossible.

‡ See the memoir cited above, p. 125.

$$\sqrt[3]{16} \equiv 21, \quad -3, \quad -18, \text{ mod } 43,$$
$$\sqrt[3]{-8} \equiv 14, \quad -2, \quad -12, \text{ mod } 43,$$
$$\sqrt[3]{22} \equiv -15, \quad -4, \quad 19, \text{ mod } 43,$$
$$\sqrt[3]{-2} \equiv +9, \quad -20, \quad +11, \text{ mod } 43,$$

and attending to the limitation to which the cube roots are subject,

$$x \equiv -8, \ +11, \ +21, \text{ or, } -2, \ +4, \ +16; \text{ mod } 43.$$

Thus the complete solution of a congruence of the sixth order is obtained by means of binomial congruences of the second and third orders only.

An essential limitation to the usefulness of this method arises from the circumstance that it does not always (or even in general) happen that (as in the example just given) each surd entering into the expression of the root becomes separately rational. For that expression may itself acquire a rational value, while certain surds contained in it continue irrational, precisely as, in the irreducible case of cubic equations, a real quantity is represented by an imaginary formula. To illustrate this point by an example, let us consider the same congruence $\frac{x^7-1}{x-1} \equiv 0$ with respect to the modulus 29 *. Here in the expression

$$x = \frac{-1+\sqrt{-7}}{6} + \frac{1}{3}\rho\left[7 - \frac{1}{2}\sqrt{-7} + \frac{3}{2}\sqrt{21}\right]^{\frac{1}{3}} + \frac{1}{3}\rho^2\left[7 - \frac{1}{2}\sqrt{-7} - \frac{3}{2}\sqrt{21}\right]^{\frac{1}{3}},$$

where ρ denotes a cube root of unity, we have, putting $\sqrt{-7} \equiv +14$, and $\rho = 1$,

$$x \equiv \frac{13}{6} + \frac{1}{3}\left[\frac{3}{2}\sqrt{21}\right]^{\frac{1}{3}} + \frac{1}{3}\left[-\frac{2}{3}\sqrt{21}\right]^{\frac{1}{3}}, \ \equiv \frac{13}{6} \equiv 7, \text{ mod } 29,$$

the irrational cube roots disappearing of themselves. Again, putting

$$\rho = \frac{1}{2}(-1 \pm \sqrt{-3}),$$

we find

$$x \equiv 7 \pm \frac{1}{3}\sqrt{-3}\left(\frac{3}{2}\sqrt{21}\right)^{\frac{1}{3}} \equiv 7 \mp \left(\frac{1}{2}\sqrt{-7}\right)^{\frac{1}{3}}$$
$$\equiv 7 \mp (7)^{\frac{1}{3}} \equiv 7 \pm 16 \equiv -6 \text{ or } -9,$$

where every radical becomes rational of itself. Similarly taking the values

$$\sqrt{-7} \equiv -14, \ \rho = \frac{1}{2}(-1 \pm \sqrt{-3}),$$

we find $x \equiv -5$ or -13. But lastly, putting $\sqrt{-7} \equiv -14$, $\rho = 1$, we find

$$x \equiv 12 + \frac{1}{3}[14 + 7\sqrt{2}]^{\frac{1}{3}} + \frac{1}{3}[14 - 7\sqrt{2}]^{\frac{1}{3}}.$$

* *Ibid.* p. 132.

To rationalise this expression, we have to observe that $14+7\sqrt{2}$, relatively to the modulus 29, is the cube of a complex number of similar form; in fact, we have $(14\pm 7\sqrt{2})\equiv(5\pm 11\sqrt{2})^3$, mod 29, whence $x\equiv -4$. To elicit, therefore, the value of this root from the irrational formula, we are obliged to solve the cubic congruence $x^3\equiv 14+7\sqrt{2}$, which, although of lower dimensions than the proposed congruence, is probably less easy to solve tentatively, because 29 has $29^2-1=840$ residues of the form $a+b\sqrt{2}$, and only $29-1=28$ ordinary integral residues; so that practically the method fails. Theoretically, however, the relation between the analytical expression of the equation-roots and the values of the congruence-roots is of considerable importance, and the subject would certainly repay a closer examination than it has yet received. We may add that, if m be a divisor of $p-1$, the complete solution of an Abelian congruence of order m requires only two things,—1st, the complete solution of the congruence $x^m-1\equiv 0$, mod p, and, 2ndly, the determination of a single root of a certain congruence of the form $x^m-a\equiv 0$, mod p, in which a is an ordinary integer; so that in this case (which is that of the congruence $\frac{x^7-1}{x-1}\equiv 0$, mod 43) we obtain a real, and not only an apparent reduction of the proposed congruence *.

It should also be observed that the primitive roots of the equation $\frac{x^n-1}{x-1}=0$ furnish, when rationalised, the primitive roots of the congruence $\frac{x^n-1}{x-1}\equiv 0$, mod p. This, the only direct method that has ever been suggested for the determination of a primitive root, appears to be the same as that referred to by Gauss in the Disq. Arith. (Art. 73).

Poinsot expresses the conviction that this method of rationalisation is applicable to any congruence corresponding to an equation, the roots of which can be expressed by radicals †. With regard to equations of the second, third, and fourth orders this is certainly true. If, for example, the biquadratic equation $F_4(x)=0$ be completely resoluble when considered as a congruence

---

* This will be at once evident, if we observe that when the congruence $x^m\equiv 1$, mod p, is completely resoluble, its roots may be employed to replace, in Abel's method, the roots of the equation $x^m-1=0$.

† See the memoir cited above, p. 107, and M. Libri, Mémoires de Mathématique et Physique, p. 63.

for the modulus p, so that

$$F_4(x) \equiv (x-a_1)(x-a_2)(x-a_3)(x-a_4), \text{ mod } p,$$

it is plain that the four roots of $F(x)=0$, and the four numbers a_1, a_2, a_3, a_4 may be obtained by substituting, in the general formula which expresses the root of any biquadratic equation as an irrational function of its coefficients, the values of the coefficients of the functions $F(x)$ and

$$(x-a_1)(x-a_2)(x-a_3)(x-a_4)$$

respectively. But these two sets of coefficients differ only by multiples of p; *i.e.* the values of a_1, a_2, a_3, a_4 can be deduced from the expressions of the roots of $F(x)=0$ by adding multiples of p to the numbers which enter into those expressions. But this reasoning ceases to be applicable to equations of an order higher than the fourth, because no general formula exists representing the roots of an equation of the fifth or any higher order. If, therefore, $F(x)=0$ be an equation of the nth order, the roots of which can be expressed by a radical formula, and which is also completely resoluble when considered as a congruence for the modulus p, so that

$$F(x) \equiv (x-a_1)(x-a_2)\ldots(x-a_n), \text{ mod } p,$$

it will not necessarily follow that the formula which gives the roots of $F(x)=0$ is also capable (when we add multiples of p to the numbers contained in it) of giving the roots of

$$(x-a_1)(x-a_2)\ldots(x-a_n)=0,$$

i.e. the roots of the congruence $F(x)\equiv 0$, mod p; and thus the principle enunciated by M. Poinsot is, it would seem, not rigorously demonstrated.

67. *Cubic and Biquadratic Congruences.*—The reduction of cubic congruences to binomial ones has been treated of by Cauchy (Exercices de Mathématiques, vol. iv. p. 279), and more completely by M. Oltramare (Crelle, vol. xlv. p. 314). Some cases of biquadratic congruences are also considered by Cauchy in the memoir cited, p. 286. The following criteria for the resolubility or irresolubility of cubic congruences include the results obtained by M. Oltramare, *l. c.*, and appear sufficiently simple to deserve insertion here:—

Let the given cubic congruence be

$$a\theta^3+3b\theta^2+3c\theta+d\equiv 0, \text{ mod } p,$$

p denoting a prime greater than 3, which does not divide the discriminant of the congruence; *i.e.*, the number

$$D=-a^2d^2+6abcd-4ac^3-4db^3+3b^2c^2;$$

and, in connection with the congruence, consider the allied system of functions *

$$U=(a, b, c, d)(x, y)^3,$$
$$H=(ac-b^2, \tfrac{1}{2}(ad-bc), bd-c^2)(x, y)^2,$$
$$\Phi=(-a^2 d+3abc-2b^3, -abd+2ac^2+b^2 c, acd-2b^2 d+bc^2, ad^2-3bcd+2c^3)(x, y)^3,$$

which are connected by the equation

$$\Phi^2+DU^2=-4H^3;$$

let also u and ϕ denote the values of U and Φ corresponding to any given values of x and y, which do not render $H\equiv 0$, mod p. Then, if $\left(\frac{\frac{1}{3}D}{p}\right)=-1$, the congruence has always one and only one real root; if $\left(\frac{\frac{1}{3}D}{p}\right)=+1$, it has either three real roots, or none: viz., if $\left(\frac{\frac{1}{2}(\phi+u\sqrt{-D})}{p}\right)_3=+1$, it has three; if $\left(\frac{\frac{1}{2}(\phi+u\sqrt{-D})}{p}\right)_3=\rho$, or $=\rho^2$, it has none. The interpretation of the cubic symbol of reciprocity will present no difficulty if we observe that $\sqrt{-D}$, mod p, is a real integer if $p=3n+1$, *i.e.* if $\left(\frac{-3}{p}\right)=1$, and that, if $p=3n-1$, *i.e.* if $\left(\frac{-3}{p}\right)=-1$, we have

$$\sqrt{-D}\equiv\sqrt{-3}\times\sqrt{\tfrac{1}{3}D}\equiv(\rho-\rho^2)\sqrt{\tfrac{1}{3}D}, \text{ mod } p,$$

so that $\sqrt{-D}$, mod p, is a complex integer involving ρ. It will however be observed that the application of the criterion requires in either case the solution of a quadratic congruence, $r^2\equiv -D$, mod p, or $r^2\equiv\frac{1}{3}D$, mod p.

Similar, but of course less simple, criteria for the resolubility or irresolubility of biquadratic congruences may be deduced from the known formulae for the solution of biquadratic equations.

68. *Quadratic Congruences—Indirect Methods of Solution.*—The general form of a quadratic congruence is $ax^2+2bx+c\equiv 0$, mod p, where p denotes an uneven prime modulus, and a is a number prime to p. It may be immediately reduced to the binomial form $r^2\equiv D$, mod p, by putting

$$r\equiv ax+b,\quad D\equiv b^2-ac, \text{ mod } p.$$

The number of its solutions is 2, 0, or 1, according as D is a quadratic residue

* See a note by Mr. Cayley in Crelle's Journal, vol. l. p. 285.

or non-residue of p, or is divisible by p, and is therefore in every case expressed by the formula $1+\left(\frac{D}{p}\right)$.

If $p=4n+3$, and $\left(\frac{D}{p}\right)=1$, the congruence $r^2-D\equiv 0$, mod p, is satisfied by $r\equiv D^{n+1}$, and $r\equiv -D^{n+1}$, and is in fact resoluble by the direct method of Art. 65. But no direct method, applicable to the case when $p=4n+1$, is at present known. Two tentative methods are proposed in the sixth section of the Disquisitiones Arithmeticæ. They are both applicable to congruences with composite as well as with prime modules. This circumstance is important, because, when the modulus is a very great number, we may not be able to tell whether it is prime or composite, and, if composite, what the primes are of which it is composed, although, when the prime divisors of a composite modulus are known, it is simplest first to solve the congruence for each of them separately, and afterwards (by a method to which we shall hereafter refer) to deduce from these solutions the solution for the given composite modulus. To apply the first of Gauss's methods, the congruence is written in the form $r^2=D+Py$, P denoting the modulus. If in the formula $V=D+Py$ we substitute for y in succession all integral values which satisfy the inequality $-\frac{D}{P}<y<\frac{1}{4}P-\frac{D}{P}$, and select those values of V which are perfect squares, their roots (taken positively and negatively) will give us all the solutions of the congruence. We should thus have $I(\frac{1}{4}P)$ or $1+I(\frac{1}{4}P)$ trials to make, I denoting the greatest integer contained in the fraction before which it is placed. If, however, we take any number E, greater than 2, and prime to P (it is simplest to take for E a prime, or power of a prime), of which the quadratic non-residues are $a, b, c, \ldots$, and then determine the values of $\alpha, \beta, \gamma, \ldots$ in the congruences $a\equiv D+\alpha P$, mod E, $b\equiv D+\beta P$, mod E, &c., we shall find that every value of y contained in one of the linear forms $mE+\alpha$, $mE+\beta$, &c., gives rise to a value of V which is a quadratic non-residue of E, and which cannot, therefore, be a perfect square; so that we may at once exclude these values of y from the series of numbers to be tried. A second *excludent* E' may then be taken, and by its aid another set of linear forms may be determined, such that no value of y contained in them can satisfy the congruence. Thus the number of trials may be diminished as far as we please. The application of this method is still further facilitated by the circumstance that it is not necessary actually to solve the congruences $a\equiv D+\alpha P$, mod $E, \ldots$

but only the single congruence $D+Py\equiv 0$, mod E (Disq. Arith., Art. 322). Gauss's second method depends on the theory of quadratic forms; it supposes that the congruence is written in the form $r^2+D\equiv 0$, mod P. By a tentative process (abbreviated, as in the first method, by the use of excludents) Gauss obtains all possible primitive representations of P by the quadratic forms of determinant $-D$; whence the complete solution of the congruence $r^2+D\equiv 0$, mod P, is immediately deduced. This method involves the construction of a complete system of quadratic forms of determinant $-D$, or, if the prime factors of D be known, of one *genus* of forms of that system; it becomes therefore more difficult of application as D increases, whereas the first method is not affected by the increase of D. The second method, however, especially recommends itself when P is a very great number; in fact, if we do not employ any excludent, the number of trials required by the first method varies (approximately, and when P is a great number) as P, whereas, on the same supposition, the number of trials required by the second method varies as $\sqrt{D}\times\sqrt{P}$.

M. Desmarest (in his Théorie des Nombres) has proposed a method less scientific in its character than those of Gauss, but sometimes easily applicable in practice. He has shown that if the congruence $r^2+D\equiv 0$, mod P, be resoluble, we can always satisfy the equation $mP=x^2+Dy^2$ with a value of m inferior to $\frac{P}{16}+3$, and of y not superior to 3. The demonstration of this theorem is not very satisfactory, and the number of trials that it still leaves is very great, viz. $3\left(I\left(\frac{P}{16}\right)+3\right)$.

The application of Gauss's second method is rendered somewhat more uniform, and at the same time the necessity for constructing a system of quadratic forms of determinant $-D$ is avoided by the following modification of it:—By a known property of quadratic forms, whenever the congruence $r^2+D\equiv 0$, mod P, is resoluble, the equation $mP=x^2+Dy^2$ is resoluble for some value of $m<2\sqrt{\tfrac{1}{3}D}$. By assigning, therefore, to m all values in succession which are inferior to that limit, and which satisfy the condition $\left(\frac{m}{D}\right)=\left(\frac{P}{D}\right)$, and then obtaining (by Gauss's method) all prime representations of the resulting products by the form x^2+Dy^2, we shall have

$$r\equiv\pm\frac{x'}{y'},\quad r\equiv\pm\frac{x''}{y''},\ \ldots,\ \text{mod } P,$$

x', y', x'', y'', etc. denoting the different pairs of values of x and y in the equation $mP = x^2 + Dy^2$.

69. *General Theory of Congruences.*—We may infer from several passages in the Disquisitiones Arithmeticæ, that Gauss intended to give a general theory of congruences of every order in the 8th section of his work, and that, at the time of its publication, he was already in possession of the principal theorems relating to the subject *. These theorems were, however, first given by Evariste Galois †, in a note published in the Bulletin de Férussac for June, 1830 (vol. xiii. p. 438), and reprinted in Liouville's Journal, vol. xi. p. 398. An account of Galois's method (completed and extended in some respects) will be found in M. Serret's Cours d'Algèbre Supérieure, leçon 25. The theory has also been independently investigated by M. Schönemann, who seems to have been unacquainted with the earlier researches of Galois (see Crelle's Journal, vol. xxxi. p. 269, and vol. xxxii. p. 93). In several of Cauchy's arithmetical memoirs (see in particular Exercices de Mathématiques, vol. i. p. 160, vol. iv. p. 217; Comptes Rendus, vol. xxiv. p. 1117; Exercices d'Analyse et de Physique Mathématique, vol. iv. p. 87) we find observations and theorems relating to it. Lastly, in a memoir in Crelle's Journal (vol. liv. p. 1) M. Dedekind has given (with important accessions) an excellent and lucid résumé of the results obtained by his predecessors.

In the following account of the principles of this theory, the functional symbols F, ϕ, ψ, ... will represent (as in general throughout this Report) rational and integral functions having integral coefficients; we shall use p to denote a prime modulus, and x an absolutely indeterminate quantity. As we shall have to consider the functions $F(x)$, $f(x)$, $\psi(x)$, etc., only in relation to the modulus p, we shall consider two functions $F_1(x)$ and $F_2(x)$, which differ only by multiples of p, as identical, and we shall represent their identity by the congruence $F_1(x) \equiv F_2(x)$, mod p, which is equivalent to an identical equation of the form $F_1(x) = F_2(x) + p\phi(x)$. The designation 'modular function,'

* See Disq. Arith., Arts. 11 and 43.

† Galois was born October 26, 1811, and lost his life in a duel, May 30, 1832. He was consequently eighteen at the time of the publication of the note referred to in the text. His mathematical works are collected in Liouville's Journal, vol. xi. p. 381. Obscure and fragmentary as some of these papers are, they nevertheless evince an extraordinary genius, unparalleled, perhaps, for its early maturity, except by that of Pascal. It is impossible to read without emotion the letter in which, on the day before his death and in anticipation of it, Galois endeavours to rescue from oblivion the unfinished researches which have given him a place for ever in the history of mathematical science.

which has been introduced by Cauchy (Comptes Rendus, vol. xxiv. p. 1118), will serve (though, perhaps, not in itself very appropriate) to indicate that the function to which it is applied is thus considered in relation to a prime modulus. Since in any modular function we may omit those terms the coefficients of which are multiples of p, we shall always suppose that the coefficient of the highest power of x in the function is prime to p.

If $F(x) \equiv f_1(x) \times f_2(x)$, mod p, $f_1(x)$ and $f_2(x)$ are each of them said to be *divisors of* $F(x)$ *for the modulus* p, or, more briefly, modular divisors of $F(x)$, or even simply divisors of $F(x)$ when no ambiguity can arise from this elliptical mode of expression. If a be a function of order zero, *i.e.* an integral number prime to p, a is a divisor, for the modulus p, of every other modular function; so that we may consider the $p-1$ terms $a_1, a_2, a_3, \ldots, a_{p-1}$, of a system of residues prime to p, as the units of this theory, and, in any set of $p-1$ *associated* functions

$$a_1 F(x), \quad a_2 F(x), \quad \ldots, \quad a_{p-1} F(x),$$

we may distinguish that one as primary in which the highest coefficient is congruous to unity (mod p).

If $F(x)$ be a function which is divisible (mod p) by no other function (except the units and its own associates), $F(x)$ is said to be a prime or irreducible function for the modulus p. And it is a fundamental proposition in this theory, that every modular function can be expressed in one way, and one way only, as the product of a unit by the powers of primary irreducible modular functions. The demonstration of this theorem depends (precisely as in the case of ordinary integral numbers) on Euclid's process for finding the greatest common divisor, which, it is easy to show, is applicable to the modular functions we are considering here. For, if $\phi_1(x)$ and $\phi_2(x)$ be two such functions [the degree of $\phi_2(x)$ being not higher than that of $\phi_1(x)$], we can always form the series of congruences

$$\phi_1(x) \equiv q_1(x)\, \phi_2(x) + r_1\, \phi_3(x), \text{ mod } p,$$
$$\phi_2(x) \equiv q_2(x)\, \phi_3(x) + r_2\, \phi_4(x), \text{ mod } p,$$
$$\cdots\cdots\cdots\cdots$$

in which $r_1, r_2, \ldots$ denote integral numbers, $q_1(x), q_2(x), \ldots$ modular functions, and $\phi_3(x), \phi_4(x), \ldots$ primary modular functions, the orders of which are successively lower and lower, until we arrive at a congruence

$$\phi_k(x) \equiv q_k(x)\, \phi_{k+1}(x) + r_k\, \phi_{k+2}(x), \text{ mod } p,$$

in which $r_k \equiv 0$, mod p. The function $\phi_{k+1}(x)$ is then the greatest common divisor (mod p) of the given functions $\phi_1(x)$ and $\phi_2(x)$; and, in particular, if $\phi_{k+1}(x)$ be of order zero, those two functions are relatively prime. We may add that, if R be the *Resultant* of $\phi_1(x)$ and $\phi_2(x)$, the necessary and sufficient condition that these functions should have a common modular divisor of an order higher than zero is contained in the congruence $R \equiv 0$, mod p *—a theorem exactly corresponding to an important algebraical proposition. From the nature of the process by which the greatest common divisor is determined, we may infer the fundamental proposition enunciated above, by precisely the same reasoning which establishes the corresponding theorem in common arithmetic. Similarly, we may obtain the solution of the following useful problem:—'Given two relatively prime modular functions A_m and A_n, of the orders m and n, to find two other functions, of the orders $m-1$ and $n-1$ respectively, which satisfy the congruence

$$A_m X_{n-1} - A_n X_{m-1} \equiv 1, \text{ mod } p.\text{'}$$

The assertion, that $f(x)$ is a divisor of $F(x)$ for the modulus p, is for brevity expressed by the congruential formula

$$F(x) \equiv 0, \text{ mod } [p, f(x)],$$

which represents an equation of the form

$$F(x) = p\phi(x) + f(x)\,\psi(x).$$

Similarly the congruence

$$F_1(x) \equiv F_2(x), \text{ mod } [p, f(x)],$$

is equivalent to the equation

$$F_1(x) = F_2(x) + p\phi(x) + f(x)\,\psi(x).$$

If $f(x)$ be a function of order m, it is evident that any given function is congruous, for the *compound* modulus $[p, f(x)]$ to one, and one only, of the p^m functions contained in the formula $a_0 + a_1 x + \ldots + a_{m-1} x^{m-1}$, in which $a_0, a_1, \ldots, a_{m-1}$ may have any values from zero to $p-1$ inclusive. These p^m functions, therefore, represent a complete system of residues for the modulus $[p, f(x)]$.

A congruence $F(X) \equiv 0$, mod $[p, f(x)]$, is said to be solved when a functional value is assigned to X which renders the left-hand member divisible

* See Cauchy, Exercices de Mathématiques, vol. i. p. 160, or M. Libri, Mémoires de Mathématique et de Physique, pp. 73, 74. But a proof of this proposition is really contained in Lagrange's Additions to Euler's Algebra (sect. 4).

by $f(x)$ for the modulus p; and the number of solutions of the congruence is the number of functional values (incongruous mod $[p, f(x)]$) which may be attributed to X. The coefficients of the powers of X in the function $F(X)$ may be integral numbers or functions of x. The linear congruence

$$AX \equiv B, \text{ mod } [p, f(x)],$$

in which A and B denote two modular functions, is, in particular, always resoluble when A is prime to $f(x)$, mod p, and admits, in that case, of only one solution.

We shall now suppose that the function $f(x)$ in the compound modulus $[p, f(x)]$ is irreducible for the modulus p,—a supposition which involves the consequence that, if a product of two factors be congruous to zero for the modulus $[p, f(x)]$, one, at least, of those factors is separately congruous to zero for the same modulus. We thus obtain the principle (cf. Art. 11) that no congruence can have more solutions, for an irreducible compound modulus, than it has dimensions. For, if $X \equiv \xi$, mod $[p, f(x)]$, satisfy the congruence

$$F_m(X) \equiv 0, \text{ mod } [p, f(x)],$$

we find $$F_m(X) \equiv F_m(X) - F_m(\xi) \equiv (X - \xi)\, F_{m-1}(X), \text{ mod } [p, f(x)],$$

$F_{m-1}(X)$ denoting a new function of order $m-1$; whence it follows that if the principle be true for a congruence of $m-1$ dimensions, it is also true for one of m dimensions; *i.e.* it is true universally.

70. *Extension of Fermat's Theorem.*—Let θ denote any one of the $p^m - 1$ residues of the modulus $[p, f(x)]$ which are prime to $f(x)$]; it may be shown, by a proof exactly similar to Dirichlet's proof of Fermat's Theorem, that

$$\theta^{p^m-1} \equiv 1, \text{ mod } [p, f(x)]. \quad . \quad . \quad . \quad . \quad . \quad . \quad . \quad . \quad \text{(A)}$$

This result, which is evidently an extension of Fermat's theorem, involves several important consequences.

It implies, in the first place, the existence of a theory of residues of powers of modular functions, with respect to a compound modulus, precisely similar to the theory of the residues of the powers of integral numbers with regard to a common prime modulus. A single example (taken from M. Dedekind's memoir) will suffice to show the exact correspondence of the two theories. The modular function θ is or is not a quadratic residue of $f(x)$, for the modulus p, according as it is or is not possible to satisfy the quadratic congruence $X^2 \equiv \theta$, mod $[p, f(x)]$. In the former case θ satisfies the congruence $\theta^{\frac{1}{2}(p^m-1)} \equiv 1$, mod $[p, f(x)]$; in the latter, $\theta^{\frac{1}{2}(p^m-1)} \equiv -1$, mod $[p, f(x)]$. And, further, if θ_1 and θ_2 be two primary irreducible modular functions of the orders

m and n respectively, and if we use the symbols $\left[\frac{\theta_1}{\theta_2}\right]$ and $\left[\frac{\theta_2}{\theta_1}\right]$ to denote the positive or negative units which satisfy the congruences

$$\theta_1^{\frac{1}{2}(p^n-1)} \equiv \left(\frac{\theta_1}{\theta_2}\right), \text{ mod } (p, \theta_2), \quad \text{and} \quad \theta_2^{\frac{1}{2}(p^m-1)} \equiv \left(\frac{\theta_2}{\theta_1}\right), \text{ mod } (p, \theta_1),$$

respectively, these two symbols are connected by the law of reciprocity

$$\left[\frac{\theta_1}{\theta_2}\right] = (-1)^{mn} \left[\frac{\theta_2}{\theta_1}\right].$$

But the equation (A) admits also of an immediate application to the theory of ordinary congruences with a simple prime modulus.

In that equation let us assign to θ the particular value x; we conclude that the function $x^{p^m-1}-1$ is divisible for the modulus p by $f(x)$, *i.e.* by every irreducible modular function of order m. Further, if d be a divisor of m, $x^{p^m-1}-1$ is algebraically divisible by $x^{p^d-1}-1$; whence it appears that $x^{p^m-1}-1$ is divisible, for the modulus p, by every function of which the order is a divisor of m. But it is easily shown that $x^{p^m-1}-1$ is not divisible (mod p) by any other modular function, and that it cannot contain any multiple modular factors. Hence we have the indeterminate congruence

$$x^{p^m-1}-1 \equiv \Pi f(x), \text{ mod } p, \quad \ldots \ldots \quad \text{(B)}$$

in which $f(x)$ denotes any primary and irreducible function, the order of which is a divisor of m, and the sign of multiplication Π extends to every value of $f(x)$. This theorem, again, is a generalisation of Lagrange's indeterminate congruence (Art. 10). We may infer from it that, when m is >1, the number of primary functions of order m, which are irreducible for the modulus p, is

$$\frac{1}{m}\left[p^m - \Sigma p^{\frac{m}{q_1}} + \Sigma p^{\frac{m}{q_1 q_2}} - \Sigma p^{\frac{m}{q_1 q_2 q_3}} + \ldots\right],$$

$q_1, q_2, \ldots$ denoting the different prime divisors of m. As this expression is always different from zero, it follows that there exist functions of any given order, which are irreducible for the modulus p.

A congruence $F(x) \equiv 0$, mod p, may be considered resolved when we have expressed its left-hand member as a product of irreducible modular factors. The linear factors (if any) then give the real solutions; the factors of higher orders may be supposed to represent imaginary solutions. We have already observed that even when all the modular factors of $F(x)$ are linear, we possess no general and direct method by which they can be assigned; it is hardly necessary to add that the problem of the direct determination of modular factors of higher orders than the first, presents even greater difficulties. Nevertheless the congruence (B) enables us to advance one step toward the decomposition of $F(x)$ into its irreducible factors; for, by means of it, we can separate those divisors of $F(x)$ which are of the same order, not, indeed, from one another, but from all its other divisors. We may first of all suppose that $F(x)$ is cleared of its multiple factors, which may be done, as in algebra, by investigating the greatest common divisor of $F(x)$ and $F'(x)$ for the modulus p. The greatest common divisor (mod p) of $F(x)$ and $x^{p-1}-1$ will then give us the product of all the linear modular factors of $F(x)$; let $F(x)$ be divided (mod p) by that product, and let the quotient be $F_1(x)$; the greatest common divisor (mod p) of $F_1(x)$ and $x^{p^2-1}-1$ will give us the product of the irreducible quadratic factors of $F(x)$; and by continuing this process, we shall obtain the partial resolution of $F(x)$ to which we have referred.

71. *Imaginary Solutions of a Congruence.*—We have said that the non-linear modular factors of $F(x)\equiv 0$, mod p, may be considered to represent imaginary solutions. These imaginary solutions can be actually exhibited, if we allow ourselves to assign to x certain complex values. The following proposition, which shows in what manner this may be effected, is due to Galois:—

'If $f(x)$ represent an irreducible modular function of order m, the congruence

$$F(\theta)\equiv 0, \text{ mod } [p, f(x)],$$

is completely resoluble when $F(x)$ is an irreducible modular function of order m, or of any order the index of which is a divisor of m.'

To establish this theorem, write θ for x in equation (B); we find

$$\theta^{p^m-1}-1\equiv \Pi\, F(\theta), \text{ mod } p,$$

the sign of multiplication Π extending to every irreducible modular function having m or a divisor of m for the index of its order. But the congruence

$$\theta^{p^m-1}\equiv 1, \text{ mod } [p, f(x)],$$

admits of as many roots as it has dimensions; therefore also every divisor of $\theta^{p^m-1}-1$ (and, in particular, the function $F(\theta)$ considered as a congruence for the same compound modulus) admits of as many roots as it has dimensions. {Add that no two irreducible congruences whose indices divide m can have a root in common.}

Let the order of the congruence $F(\theta)\equiv 0$, mod $[p, f(x)]$, be δ, and let any one of its roots be represented by r; it may be shown that all its roots are represented by the terms of the series $r, r^p, r^{p^2}, \ldots, r^{p^{\delta-1}}$. For, if

$$F(r) \equiv 0, \bmod [p, f(x)],$$

we have also $F(r^p) \equiv [F(r)]^p \equiv 0, \bmod [p, f(x)]$,

and similarly $F(r^{p^2}) \equiv [F(r)]^{p^2} \equiv 0, \bmod [p, f(x)]$; and so on;

so that $r, r^p, r^{p^2}, \ldots, r^{p^{\delta-1}}$ are all roots of

$$F(\theta) \equiv 0, \bmod [p, f(x)].$$

It remains to show that these δ functions are all incongruous, mod $[p, f(x)]$. If possible let

$$r^{p^{k+k'}} \equiv r^{p^k}, \bmod [p, f(x)],$$

k and k' being less than δ; we have, raising each side of this congruence to the power $p^{\delta-k'}$,

$$r^{p^{\delta+k}} \equiv r^{p^\delta}, \bmod [p, f(x)],$$

i.e. $r^{p^k} \equiv r$, or $\quad r^{p^k-1} \equiv 1, \quad \bmod [p, f(x)]$,

observing that $\quad r^{p^\delta} \equiv r, \quad \bmod [p, f(x)]$,

because $r^{p^\delta-1} - 1$ is divisible by $F(r)$ for the modulus p. We conclude, therefore, that r is a root, mod $[p, f(x)]$, of some irreducible modular divisor of the function $\theta^{p^k-1} - 1$, *i.e.* of some irreducible function of an order lower than δ, because k is less than δ; r is therefore a root, mod $[p, f(x)]$, of two different irreducible modular functions, which is impossible *.

If, therefore, we suppose x to represent, not an indeterminate quantity, but a root of the equation $f(x) = 0$, we may enunciate Galois' theorem as follows:—

'Every irreducible congruence of order m is completely resoluble in complex numbers composed with roots of any congruence which is irreducible for the modulus p, and which has m or a multiple of m for the index of its order.

'And all its roots may be expressed as the powers of any one of them.'

72. *Congruences having Powers of Primes for their Modules.*—It remains for us to advert to the theory of congruences with composite modules—a subject to which (if we except the case of binomial congruences) it would seem that the attention of arithmeticians has not been much directed. We shall suppose, first, that the modulus is a power of a prime number.

The theorem of Lagrange (Art. 11), and the more general proposition of

---

* {Let ω = the common divisor of k and δ, since $r^{p^\delta\ 1} \equiv 1$, $r^{p^k-1} \equiv 1$, we can prove that $r^{p^\omega} \equiv 1$, mod $[p, f(x)]$, but $\theta^{p^\omega-1} - 1$ = a product of irreducible functions whose indices divide ω; therefore r is a root of two different irreducible functions whose indices divide m, and this is impossible.}

Art. 69, in which it is (as we have seen) included, cannot be extended to congruences having powers of primes for their modules.

Let the proposed congruence be $F(x) \equiv 0$, mod p^m; and let us suppose (what is here a restriction in the generality of the problem) that the coefficient of the highest power of x in $F(x)$ is prime to p, or, which comes to the same thing, that it is unity. Let $F(x) \equiv P \times Q \times R \ldots$ mod p, where $P, Q, R, \ldots$ are powers of different irreducible modular functions. It may then be shown that $F(x) \equiv P' \times Q' \times R' \ldots$, mod p^m, where $P', Q', R', \ldots$ are functions of the same order as $P, Q, R, \ldots$, respectively congruous to them for the modulus p, and deducible from them by the solution of linear congruences only. We have thus the theorem that $F(x)$, considered with respect to the modulus p^m, can always be resolved in one way and in one way only, into a product of modular functions, each of which is relatively prime (for the modulus p) to all the rest, and is congruous (for the same modulus p) to a power of an irreducible function. We may therefore replace the congruence $F(x) \equiv 0$, mod p^m, by the congruences $P' \equiv 0$, mod p^m, $Q' \equiv 0$, mod p^m, $R' \equiv 0$, mod p^m, ... But no general investigation appears to have been given of the peculiarities that may be presented by a congruence of the form $P' \equiv 0$, mod p^m, in the case in which P is a power of an irreducible function (mod p), and not itself such a function—a supposition which implies that the discriminant of $F(x)$ is divisible by p. If, however, P be itself an irreducible function, the congruence $P' \equiv 0$, mod p^m, gives us one and only one solution of the given congruence if P be linear; or, if P be not linear, it may be considered as representing as many imaginary solutions as it has dimensions. In particular, if we consider the case in which all the divisors $P, Q, R, \ldots$ are linear, we obtain the theorem:—

'Every congruence which, considered with respect to the modulus p, has as many *incongruous* solutions as it has dimensions, is also completely resoluble for the modulus p^m; and has as many roots as it has dimensions, and no more.'

If $x \equiv a_1$, mod p, be a solution of the congruence $F(x) \equiv 0$, mod p, and if that congruence have no other root congruous to a_1, the corresponding solution $x \equiv a_m$, mod p^m, of the congruence $F(x) \equiv 0$, mod p^m, may be obtained by the solution of linear congruences only—a proposition which is included in a preceding and more general observation. The process is as follows:— If, in the equation

$$F(a_1 + kp) = F(a_1) + kp\, F'(a_1) + \frac{k^2 p^2}{1.2} F''(a_1) + \ldots,$$

we determine k by the congruence

$$\frac{1}{p}F(a_1)+kF'(a_1)\equiv 0, \text{ mod } p,$$

(which is always possible, because the hypothesis that $(x-a_1)^2$ is not a divisor of $F(x)$, mod p, implies that $F'(a_1)$ is not divisible by p *), and then put $a_2\equiv a_1+kp$, mod p^2, we have $F(a_2)\equiv 0$, mod p^2. Similarly, from the expansion

$$F(a_2+kp^2)=F(a_2)+kp^2F'(a_2)+\dots,$$

a value of k may be deduced which satisfies the congruence $F(a_2+kp^2)\equiv 0$, or $F(a_3)\equiv 0$, mod p^3; and so on continually until we arrive at a congruence of the form $F(a_m)\equiv 0$, mod p^m. But when $F(x)$ is divisible (for the modulus p) by $(x-a)^2$ or a higher power of $x-a$, the congruence $F(x)\equiv 0$, mod p^m, is either irresoluble, or has a plurality of roots incongruous for the modulus p^m but all congruous to a for the modulus p. Thus the congruence

$$(x-a)^2+kp(x-b)\equiv 0, \text{ mod } p^2,$$

is irresoluble, unless $a\equiv b$, mod p; whereas if that condition be satisfied, it admits of p incongruous solutions, comprised in the formula

$$x\equiv a+\mu p, \text{ mod } p^2, \ \mu=0, 1, 2, 3, \dots, p-1.$$

73. *Binomial Congruences having a Power of a Prime for their Modulus.*— If M be any number, and $\psi(M)$ represent the number of terms in a system of residues prime to M, it will follow (from a principle to which we have already frequently referred: see Arts. 10, 26, 53, 70) that every residue of that system satisfies the congruence $x^{\psi(M)}\equiv 1$, mod M,—a proposition which is well known as Euler's generalisation of Fermat's theorem †. In particular, when $M=p^m$, we have $x^{(p-1)p^{m-1}}\equiv 1$, mod p^m. This congruence has, consequently, precisely as many roots as it has dimensions—a property which is also possessed by every congruence of the form $x^d\equiv 1$, mod p^m, d denoting a divisor of $(p-1)p^{m-1}$. This has been established by Gauss in the 3rd section of the Disquisitiones Arithmeticæ, by a particular and somewhat tedious method ‡. The simpler and more general demonstration which he intended to give in the 8th section §, was perhaps in principle identical with the following; we exclude the case $p=2$, to which indeed the theorem itself is inapplicable :—

* If $F(x)\equiv(x-a_1)\phi(x)$, mod p, where $\phi(a_1)$ is not divisible by p, we have

$$F'(x)\equiv\phi(x)+(x-a_1)\phi'(x), \text{ mod } p, \text{ or } F'(x)\equiv\phi(a_1), \text{ mod } p.$$

† Euler, Comment. Arith. vol. i. p. 284.

‡ Disquisitiones Arithmeticæ, Arts. 84–88. See also Poinsot, Reflexions sur la Théorie des Nombres, cap. iv. Art. 6.

§ Disquisitiones Arithmeticæ, Art. 84.

Let $d = \delta p^n$, δ representing a divisor of $p-1$, and n being $\leqq m-1$; and let us form the indeterminate congruence

$$x^\delta - 1 \equiv (x-a_1)(x-a_2)\dots(x-a_\delta), \text{ mod } p^{m-n},$$

which is always possible, because $x^\delta - 1 \equiv 0$, mod p, has δ incongruous roots. It is readily seen that, if A and B represent two numbers prime to p, and if $A \equiv B$, mod p^r, $A^{p^s} \equiv B^{p^s}$, mod p^{r+s}; and conversely, if $A^{p^s} \equiv B^{p^s}$, mod p^{r+s}, $A \equiv B$, mod p^r.* By applying this principle it may be shown that

$$x^{\delta p^n} - 1 \equiv (x^{p^n} - a_1^{p^n})(x^{p^n} - a_2^{p^n})\dots(x^{p^n} - a_\delta^{p^n}), \text{ mod } p^m.$$

For if we divide $x^{\delta p^n} - 1$ by $x^{p^n} - a_1^{p^n}$, the remainder is $a_1^{\delta p^n} - 1$. But, because $a_1^\delta \equiv 1$, mod p^{m-n}, $a_1^{\delta p^n} \equiv 1$, mod p^m; *i.e.* $x^{p^n} - a_1^{p^n}$ divides $x^{\delta p^n} - 1$ for the modulus p^m. Similarly $x^{\delta p^n} - 1$ is divisible (mod p^m) by $x^{p^n} - a_2^{p^n}$, etc.; and since all these divisors are relatively prime for the modulus p, $x^{\delta p^n} - 1$ is divisible (mod p^m) by their product; *i.e.*,

$$x^{\delta p^n} - 1 \equiv (x^{p^n} - a_1^{p^n})(x^{p^n} - a_2^{p^n})\dots(x^{p^n} - a_\delta^{p^n}), \text{ mod } p^m.$$

We have thus effected the resolution of $x^{\delta p^n} - 1$ into factors relatively prime, each of which is congruous (mod p) to a power of an irreducible function; since evidently $(x^{p^n} - a^{p^n}) \equiv (x-a)^{p^n}$, mod p. To investigate the solutions of

$$x^{\delta p^n} - 1 \equiv 0, \text{ mod } p^m,$$

we have therefore only to consider separately the δ congruences included in the formula $x^{p^n} \equiv a^{p^n}$, mod p^m. But each of these congruences (by virtue of the principle already referred to) admits precisely p^n solutions, viz. the p^n numbers (incongruous, mod p^m) which are congruous to a, mod p^{m-n}. The whole number of solutions of $x^{\delta p^n} - 1 \equiv 0$, mod p^m, is therefore equal to the index δp^n of the congruence. It further appears that the complete solution of the binomial congruence $x^{\delta p^n} - 1 \equiv 0$, may be obtained by a direct method, when the complete solution of the simpler congruence $x^\delta - 1 \equiv 0$, mod p, has been found. For we may first (by the method given in the last article) deduce the complete solution of $x^\delta - 1 \equiv 0$, mod p^{m-n}, from that of $x^\delta - 1 \equiv 0$, mod p; and then the roots of $x^{\delta p^n} - 1 \equiv 0$, mod p^m, can be written down at once.

74. *Primitive Roots of the Powers of a Prime.*—All the elementary pro-

---

* If $A \equiv B$, mod p^r, but not, mod p^{r+1}, we have $A = B + kp^r$, where k is prime to p. Hence

$$A^{p^s} = (B+kp^r)^{p^s} = B^{p^s} + kB^{p^s-1}p^{s+r} + Kp^{s+r},$$

K denoting a coefficient divisible by p; or $A^{p^s} \equiv B^{p^s}$, mod p^{s+r}, but not, mod p^{s+r+1}; because kB^{p^s-1} is prime to p. This result implies the principle enunciated in the text.

perties of the residues of powers, considered with regard to a modulus which is a power of a prime number, may be deduced from the theorem just proved. In particular, the demonstration of the existence and number of primitive roots (Art. 12) is applicable here also; so that we have the theorem :—

'There are $p^{m-2}(p-1)\psi(p-1)$ residues prime to p^m, the successive powers of any one of which represent all residues prime to p^m.' These residues are of course the primitive roots of p^m.

If γ be a primitive root of p, of the p numbers included in the formula $\gamma+kp$ (mod p^2), $p-1$ precisely will be primitive roots of p^2. For $\gamma+kp$ is a primitive root of p^2 unless $(\gamma+kp)^{p-1}\equiv 1$, mod p^2; and the congruence $x^{p-1}\equiv 1$, mod p^2, has always one, and only one, root congruous to γ for the modulus p. But every primitive root of p^2 is a primitive root of p^3, and of every higher power of p, as may be shown by an application of the principle proved in a note to the last article, or, again, by observing that every primitive root of p^{m+1} is necessarily congruous, for the modulus p^m, to some primitive root of p^m, and that there are p times as many primitive roots of p^{m+1} as of p^m. (See Jacobi's Canon Arithmeticus, Introduction, p. xxxiii; also a problem proposed by Abel in Crelle's Journal, vol. iii. p. 12, with Jacobi's answer, ibid. p. 211.)

75. *Case when the Modulus is a Power of* 2.—The powers of the even prime 2 are excepted from the demonstrations of the two last articles—in fact, if $m \gtreqless 3$, 2^m has no primitive roots. Gauss, however, has shown (Disq. Arith., Arts. 90, 91) that the successive powers of any number of the form $8n+3$ represent, for the modulus 2^m, all numbers of either of the forms $8n+3$ or $8n+1$; similarly all numbers of the forms $8n+5$ and $8n+1$ are represented by successive powers of any number of the form $8n+5$. If, therefore, we denote by γ any number of either of the two forms $8n+3$ or $8n+5$, we may represent all uneven numbers less than 2^m by the formula $(-1)^\alpha\gamma^\beta$, in which α is to receive the values 0 and 1, and β the values 1, 2, 3, ..., 2^{m-2}. A double system of indices may thus be used to replace the simple system supplied by a primitive root when such roots exist.

Tables of indices for the powers of 2, and of uneven primes inferior to 1000, have been appended by Jacobi to his Canon Arithmeticus.

76. *Composite Modules.*—No general theory has been given of the representation of rational and integral functions of an indeterminate quantity as products of modular functions with regard to a composite modulus divisible by more than one prime. And it is possible that no advantage would be

gained by considering the theory of congruences with composite modules from this general point of view. A few isolated theorems relating to particular cases have, however, been given by Cauchy (Comptes Rendus, vol. xxv. p. 36, 1847). Of these the following may serve as a specimen :—

'If the congruence $F(x) \equiv 0$, mod M, admit as many roots as it has dimensions, and if, besides, the differences of these roots be all relatively prime to M, we have the indeterminate congruence

$$F(x) \equiv k(x-r_1)(x-r_2)(x-r_3)\ldots(x-r_n), \text{ mod } M,$$

k denoting the coefficient of the highest power of x in $F(x)$.'

But if, instead of considering the modular decomposition of the function $F(x)$, we confine ourselves to the determination of the real solutions of the congruence $F(x) \equiv 0$, mod M, it is always sufficient to consider the congruences

$$F(x) \equiv 0, \text{ mod } A, \quad F(x) \equiv 0, \text{ mod } B, \quad F(x) \equiv 0, \text{ mod } C, \text{ etc.}, \quad . \; . \; . \quad \text{(G)}$$

where $A \times B \times C \ldots = M$, and $A, B, C, \ldots$ denote powers of different primes. For if $x \equiv a$, mod A, $x \equiv b$, mod B, $x \equiv c$, mod C, denote any solutions of the first, second, third, ... of those congruences respectively, it is evident that, if X be a number satisfying the congruences

$$X \equiv a, \text{ mod } A, \quad X \equiv b, \text{ mod } B, \quad X \equiv c, \text{ mod } C$$

(and such a number can always be assigned), we shall have $F(X) \equiv 0$ for each of the modules $A, B, C, \ldots$ separately, and therefore for the modulus M; and further, if the congruences (G) admit respectively $\alpha, \beta, \gamma, \ldots$ incongruous solutions, the congruence $F(x) \equiv 0$, mod M, will admit $\alpha \times \beta \times \gamma \times \ldots$ in all; for we can combine any solution of $F(x) \equiv 0$, mod A, with any solution of $F(x) \equiv 0$, mod B, and so on *.

77. *Binomial Congruences with Composite Modules.*—The investigation of the real solutions of binomial congruences depends (in the manner just stated) on the investigation of the real solutions of similar congruences the modules of which are the powers of primes. With regard to the relations by which these real solutions are connected with one another, little of importance has

* 'Infra [*i.e.* in the 8th section] congruentias quascumque secundum modulum e pluribus primis compositum, ad congruentias quarum modulus est primus aut primi potestas reducere, fusius docebimus' (Disq. Arith., Art. 92). It is difficult to see why Gauss should have employed the word 'fusius' if his investigations extended no further than the elementary observations referred to in the text. Nevertheless it is remarkable that Gauss in the 3rd section of the Disq. Arith. sometimes speaks of demonstrations as obscure, which are of extreme simplicity when compared with one in the 4th and several in the 5th section (see in particular Arts. 53, 55, 56).

been added to the few observations on this subject in the Disquisitiones Arithmeticæ (Art. 92). If the modulus M be $=p^a q^b r^c \ldots$, where $p, q, r, \ldots$ represent different primes, the congruence $x^{\psi(M)} \equiv 1$, mod M, possesses no primitive roots; for if n be the least common multiple of

$$p^{a-1}(p-1),\ q^{b-1}(q-1),\ r^{c-1}(r-1), \ldots,$$

n will be less than, and a divisor of, $\psi(M)$. But evidently, if x be any residue prime to M, the congruence $x^n - 1 \equiv 0$ will be satisfied separately for the modules p^a, q^b, $r^c, \ldots$, and therefore for the modulus M; *i.e.*, no residue exists, the first $\psi(M)$ powers of which are incongruous, mod M. If, however, $M = 2p^a$ this conclusion does not hold, since the least common multiple of $\psi(2)$ and $\psi(p^m)$ is $\psi(2p^m)$ itself; and we find accordingly that every uneven primitive root of p^m is a primitive root of $2p^m$. When, as is sometimes the case, it is convenient to employ indices to designate the residues prime to a given composite modulus, we must employ (as in the case of a power of 2) a system of multiple indices. To take the most general case, let $M = 2^\theta p^a q^b r^c \ldots$; let u be any number of either of the forms $8n+3$ or $8n+5$, and $P, Q, R, \ldots$ primitive roots of p^a, q^b, $r^c, \ldots$ respectively. Then, if n be any given number prime to M, it will always be possible to find a set of integral numbers $\epsilon_n, \omega_n, \alpha_n, \beta_n, \gamma_n, \ldots$ satisfying the conditions

$$\begin{aligned} (-1)^{\epsilon_n} u^{\omega_n} &\equiv n, \text{ mod } 2^\theta; \quad 0 \leqq \epsilon_n < 2, \quad 0 \leqq \omega_n < 2^{\theta-2}, \\ P^{\alpha_n} &\equiv n, \text{ mod } p^a; \quad 0 \leqq \alpha_n < p^{a-1}(p-1), \\ Q^{\beta_n} &\equiv n, \text{ mod } q^b; \quad 0 \leqq \beta_n < q^{b-1}(q-1), \\ R^{\gamma_n} &\equiv n, \text{ mod } r^c; \quad 0 \leqq \gamma_n < r^{c-1}(r-1); \end{aligned}$$

and these numbers form a system of indices by which the residue of n for each of the modules 2^θ, p^a, q^b, $r^c, \ldots$ (and consequently for the modulus M) is completely determined. (See Dirichlet's memoir on the Arithmetical Progression, sect. 7, in the Berlin Memoirs for 1837.)

78. *Primitive Roots of the Powers of Complex Primes.*—Dirichlet has shown * that, in the theory of complex numbers of the form $a+bi$, the powers of primes of the second species (see Art. 25) have primitive roots; in fact, if $a+bi$ be such a prime, and $N(a+bi) = a^2+b^2 = p$, every primitive root of p^m is a primitive root of $(a+bi)^m$. On the other hand, if q be a real prime of the form $4n+3$, q^m has no primitive roots in the complex theory. For in general, if M be any complex modulus, and $M = \mathfrak{a}^\alpha \mathfrak{b}^\beta \mathfrak{c}^\gamma \ldots$, $\mathfrak{a}, \mathfrak{b}, \mathfrak{c}, \ldots$ being

* See sect. 2 of the memoir, Untersuchungen über die Theorie der complexen Zahlen, in the Berlin Memoirs for 1841.

different complex primes, and if $A=N(a)$, $B=N(b)$, $C=N(c)$, etc., the number of terms in a system of residues prime to M, is

$$A^{\alpha-1}(A-1)\,B^{\beta-1}(B-1)\,C^{\gamma-1}(C-1)\ldots;$$

and if we denote this number by $\psi(M)$, every residue prime to M will satisfy the congruence $$x^{\psi(M)}\equiv 1, \text{ mod M},$$

which here corresponds to Euler's extension of Fermat's Theorem. If $M=q^m$, this congruence becomes $$x^{q^{2(m-1)}(q^2-1)}\equiv 1, \text{ mod } q^m;$$

but it is easily shown that every residue prime to q^m satisfies the congruence

$$x^{q^{m-1}(q^2-1)}\equiv 1, \text{ mod } q^m;$$

i.e., q^m has no primitive roots, because the exponent $q^{m-1}(q^2-1)$ is a divisor of, and less than, $q^{2(m-1)}(q^2-1)$. Nevertheless two numbers, γ and γ', can always be assigned, of which one appertains to the exponent $q^{m-1}(q^2-1)$ and the other to the exponent q^{m-1}, and which are such that no power of either of them can become congruous to a power of the other, mod q^m, without becoming congruous to unity; from which it will appear that every residue prime to q^m may be represented by the formula $\gamma^x\gamma'^y$, if we give to x all values from 0 to $(q^2-1)q^{m-1}-1$ inclusive, and to y all values from 0 to $q^{m-1}-1$ inclusive.

The corresponding investigations for other complex numbers besides those of the form $a+bi$ have not been given.

We here conclude our account of the Theory of Congruences. The further continuation of this Report will be occupied with the Theories of Quadratic and other Homogeneous Forms.

[The Additions to Arts. 16, 20, 22, 24, 25, 36, 37, and 38 to Part I of the Report (1859), inserted in their proper places, see footnote p. 57, were published at the end of this Part of the Report.]

VII.

REPORT ON THE THEORY OF NUMBERS.

PART III.

[Report of the British Association for 1861, pp. 292–340.]

(B) *Theory of Homogeneous Forms.*

79. *PROBLEM of the Representation of Numbers.*—A rational and integral homogeneous function (a *quantic* according to the nomenclature introduced by Mr. Cayley), of which the coefficients are integral numbers, is, in the Theory of Numbers, termed a *form* (Disq. Arith., Art. 266). The form is linear, quadratic, cubic, biquadratic or quartic, quintic, &c., according to its order in respect of the indeterminates it contains; and binary, ternary, quaternary, &c., according to the number of its indeterminates. Thus x^2+y^2 is a binary quadratic form, $x^3+y^3+z^3-3xyz$ a ternary cubic form. A form is considered to be given, when its coefficients are given numbers; and a number is said to be *represented* by a given form, when integral values are assigned to the indeterminates of the form, such that the form acquires the value of the number. If the values of the indeterminates are relatively prime, the representation is said to be *primitive*; if they admit any common divisor beside unity, it is a *derived* representation. Thus 13 and 8 can be represented by x^2+y^2; for $3^2+2^2=13$, $2^2+2^2=8$; and the first of these representations is primitive, the second is derived. The first general problem, then, that presents itself in this part of the Theory of Numbers, is the following, 'To find whether a given number is or is not capable of representation by a given form, and, if it is, to find all its representations by that form.' The number of different representations of a given number by a given form may be either finite or infinite; in the former case the complete solution of the problem of representation consists in

the actual exhibition of the different sets of values that can be given to the indeterminates of the form: in the latter case it consists in assigning general formulae, in which all those values are comprised. It is in either case sufficient to consider primitive representations only; for if the given form f be of order m, and the given number N be divisible by the mth powers $d_1^m, d_2^m, \ldots$, the derived representations of N by f coincide with the primitive representations of $\frac{N}{d_1^m}, \frac{N}{d_2^m}, \ldots$ by the same form.

80. *Problems of the Transformation and Equivalence of Forms.*—A form $f_2(x'_1, x'_2, \ldots, x'_n)$ is said to be *contained* in another form $f_1(x_1, x_2, \ldots, x_n)$, when f_2 arises from f_1 by a linear transformation of the type

$$\begin{aligned} x_1 &= a_{1,1}x'_1 + a_{1,2}x'_2 + \ldots\ldots + a_{1,n}x'_n, \\ x_2 &= a_{2,1}x'_1 + a_{2,2}x'_2 + \ldots\ldots + a_{2,n}x'_n, \\ &\cdots\cdots\cdots\cdots\cdots\cdots \\ x_n &= a_{n,1}x'_1 + a_{n,2}x'_2 + \ldots\ldots + a_{n,n}x'_n, \end{aligned}$$

in which the coefficients $a_{i,j}$ are integral numbers and the determinant is different from zero *. This transformation we may, for brevity, describe as the transformation $|a|$. When $|a|$ is a unit-transformation, *i.e.* when the determinant of $|a|$ is a positive or negative unit, the inverse transformation of $|a|$, which will be a transformation of the same type as $|a|$, will have all its coefficients integral numbers; so that in this case f_1, which contains f_2, is also contained in it. When each of two forms is thus contained in the other, they are said to be *equivalent*. If f_1 contain f_2, and f_2 contain f_3, f_1 will contain f_3; for if f_1 be changed into f_2 by the transformation $|a|$, and f_2 into f_3 by the transformation $|b|$, it is clear that f_1 will be changed into f_3 by a transformation $|T|$, of which the constituents are defined by the equation

$$T_{i,j} = a_{i,1}b_{1,j} + a_{i,2}b_{2,j} + \ldots\ldots + a_{i,n}b_{n,j}.$$

The transformation $|T|$ is said to be *compounded* of the transformations $|a|$ and $|b|$, and this composition is expressed by the symbolic equation

$$|T| = |a| \times |b|,$$

in which it is to be observed that the order of the symbols $|a|$ and $|b|$ is not, in general, convertible. When, in particular, f_1 is equivalent to f_2, and f_2 to

* Gauss says that f_2 is contained in f_1, even when the determinant of transformation is zero (Disq. Arith., Art. 215). But we shall find it more convenient to retain the restriction specified in the text.

f_3, f_1 is equivalent to f_3; *i.e.* forms which are equivalent to the same form are equivalent to one another. All the forms, therefore, which are equivalent to one and the same form, may be considered as forming a *class.* All the invariants of any two equivalent forms have the same values; but it is not true, conversely, that two forms which have the same invariants are necessarily equivalent. Nevertheless it may be conjectured that all forms of the same sort (*i.e.* of the same degree, and the same number of indeterminates), the invariants of which have the same values, distribute themselves into a finite number of classes; and this conjectural proposition is certainly true for binary forms of all orders, and for quadratic forms of any number of indeterminates. It is readily seen that if a number be capable of representation by one of two equivalent forms, it is also capable of representation by the other; and that the number of representations is either finite for both, or infinite for both, and, if finite, is the same for each. The general problem, therefore, of the representation of numbers (which we have already enunciated) suggests naturally the following, which we may term that of the equivalence of forms: 'Given two forms (of the same sort), of which the invariants have equal values, to find whether they are, or are not, equivalent, and if they are, to assign all the transformations of either of them into the other.' The number of transformations may be either finite or infinite; if finite, the transformations themselves, if infinite, general formulae containing them, are required for the complete solution of the problem.

When f_1 is not equivalent to, but contains f_2, the invariants of f_2 are derived from those of f_1 by multiplication with certain powers of the *modulus* (*i.e.* of the determinant) of the transformation by which f_1 is changed into f_2; viz. if I be an invariant of f_1, and if i and m be the orders of I, and of f_1 or f_2, the corresponding invariant of f_2 is $a^{\frac{mi}{n}}I$, a denoting the modulus of transformation, and the number $\frac{mi}{n}$ being always integral. This observation enables us to enunciate with precision a problem in which the preceding is included: 'Given two forms, of which the invariants have values consistent with the supposition that one of them contains the other, to find whether this supposition is true or not, and, if it is, to find all the transformations of the one form into the other.' But, in every case, the solution of the problem in this more general form may be made to depend on the solution of the problem of equivalence. For every transformation of order n, and modulus a, arises, in one way and in one only, from the composition of two transformations $|a|$ and $|v|$, of which the latter is a unit-transformation, and the former one of the finite number of transformations included in the formula

$$\begin{vmatrix} \mu_1, & k_{1,2}, & k_{1,3}, & \ldots\ldots, & k_{1,n} \\ 0, & \mu_2, & k_{2,3}, & \ldots\ldots, & k_{2,n} \\ 0, & 0, & \mu_3, & \ldots\ldots, & k_{3,n} \\ \cdot & \cdot & \cdot & \cdot\ \cdot\ \cdot & \cdot \\ 0, & 0, & 0, & \ldots\ldots, & \mu_n \end{vmatrix} \qquad (C)$$

in which $\mu_1 \times \mu_2 \times \ldots \times \mu_n = a$, and $0 \leqq k_{i,j} < \mu_i$ (Phil. Trans., vol. cli. p. 312). To determine, therefore, whether the form f_1 can be transformed into f_2 by a transformation of modulus a, we apply to f_1 all the transformations (C) in succession, obtaining a series of transformed forms $\phi_1, \phi_2, \ldots$. If none of the forms ϕ are equivalent to f_2, f_1 cannot contain f_2; but if one or more of the forms ϕ be equivalent to f_2, f_1 will contain f_2, and all its transformations into f_2 may be obtained as soon as the transformations of the forms ϕ into f_2 have been determined. This is the method proposed by Gauss for binary quadratic forms (Disq. Arith., Arts. 213, 214); it is evidently of universal application; but the following modification of it possesses a certain advantage. Instead of representing $|T|$ by the formula $|T| = |a| \times |v|$, we may employ the formula $|T| = |v| \times |a|$, in which $|v|$ is a unit-transformation as before, and $|a|$ is one of the transformations included in the formula (C), where, however, the inequality $0 \leqq k_{i,j} < \mu_i$ is to be replaced by $0 \leqq k_{i,j} < \mu_j$; the transformations thus defined we shall call the transformations (C′). If we now apply to f_2 the inverse of each transformation included in (C′), we shall obtain a series of forms $\phi_1, \phi_2, \phi_3, \ldots$ of which the coefficients will not necessarily be integral numbers, because the coefficients of the inverse transformations are not necessarily integral. If all the forms $\phi_1, \phi_2, \ldots$ be fractional, or if none of those which are integral be equivalent to f_1, f_1 cannot contain f_2; but if some of them be integral, and equivalent to f_2, it is plain that f_1 contains f_2, and that all the transformations of f_1 into f_2 may be obtained by means of the transformations of f_1 into those forms ϕ which are equivalent to it. The advantage above referred to consists in the circumstance that the rejection of the fractional forms ϕ diminishes the number of the problems of equivalence which must be solved to obtain the complete solution of the question proposed (compare Disq. Arith., Art. 284, and note.)

81. *Automorphic Transformations.*—The unit-transformations by which a form passes into itself are the automorphics of the form; thus $\begin{vmatrix} 2, & 3 \\ 1, & 2 \end{vmatrix}$ is an

automorphic of x^2-3y^2. When every invariant of a form is zero, the form may pass into itself by transformations of which the modulus is different from unity; for example, $x^2-4xy+4y^2$, a binary quadratic form of which the discriminant is zero, passes into itself by the transformation $\begin{vmatrix} 3, & 2 \\ 1, & 2 \end{vmatrix}$, of which the modulus is 4. In like manner it is to be observed that when two forms of the same sort have all their invariants equal to zero, it may happen that each of them passes into the other by transformations of which the modulus is not a unit. But in this Report we shall have no occasion to consider these exceptional cases, whether of equivalence or of automorphism, and we shall therefore employ these terms with reference to unit-transformations exclusively. If $|T_1|$ and $|T_2|$ be automorphics of a form f, $|T_1|\times|T_2|$ and $|T_2|\times|T_1|$ are also automorphics of f; so that, in particular, every power of an automorphic is also an automorphic. (The positive powers of a transformation are, of course, the transformations which arise from compounding it continually with itself; its negative powers are the positive powers of its inverse. See Mr. Cayley's Memoir on the Theory of Matrices, Phil. Trans., vol. cxlviii. p. 17.) Hence, if a form have a single automorphic, of which no two powers are identical, it will have an infinite number of automorphics. The importance of automorphic transformations in the solution of the problems of equivalence and transformation will be apparent from the following considerations. If f_1 and f_2 be two equivalent forms, $|h|$ a given transformation of f_1 into f_2, $|a_1|$ and $|a_2|$ the general formulae representing all the automorphics of f_1 and f_2 respectively, all the transformations of f_1 into f_2 will be represented by either of the formulae $|a_1|\times|h|$ or $|h|\times|a_2|$. And again, if f_1 contain f_2, and if we represent by $|h_1|$, $|h_2|$,... certain particular transformations of f_1 into f_2, obtained by compounding each transformation (C), which gives a form ϕ equivalent to f_2, with some one transformation of ϕ into f_2, then all the transformations of f_1 into f_2 will be comprised in a finite number of formulae of the type

$$|h_1|\times|a_2|,\quad |h_2|\times|a_2|,\quad |h_3|\times|a_2|,\ \ldots\ldots,$$

$|a_2|$ still denoting indefinitely any automorphic of f_2. Or, if we employ the second method of the preceding article, the same transformations will be represented by

$$|a_1|\times|h_1'|,\quad |a_1|\times|h_2'|,\quad |a_1'|\times|h_3'|,\ \ldots\ldots,$$

where $|a_1|$ is any automorphic of f_1, and $|h_1'|$, $|h_2'|$, $|h_3'|$, ... are certain particular transformations of f_1 into f_2, obtained in a manner sufficiently indicated by the method itself. It appears, therefore, that when we know all the automorphics either of f_1 or f_2, we can deduce all the transformations of f_1 into f_2, from one of those transformations when f_1 is equivalent to f_2, and from a certain finite number of them when f_1 contains, but is not equivalent to, f_2. We may add, that when one transformation of two equivalent forms, and the automorphics of either of them are known, those of the other are known also, for we evidently have the equation

$$|\alpha_2| = |h|^{-1} \times |\alpha_1| \times |h|.$$

82. *Problem of the Representation of Forms.*—We give the enunciation of one other general problem, which may be said to occupy a middle place between the problems of the representation of numbers, and of the equivalence of forms. By using a *defective* substitution of the type

$$\begin{aligned} x_1 &= a_{1,1}x'_1 + a_{1,2}x'_2 + \ldots\ldots + a_{1,n-2}x'_{n-r}, \\ x_2 &= a_{2,1}x'_1 + a_{2,2}x'_2 + \ldots\ldots + a_{2,n-2}x'_{n-r}, \\ &\ldots\ldots\ldots\ldots\ldots\ldots\ldots \\ x_n &= a_{n,1}x'_1 + a_{n,2}x'_2 + \ldots\ldots + a_{n,n-2}x'_{n-r}, \end{aligned}$$

a form $f_1(x_1, x_2, \ldots, x_n)$ may be changed into another $f_2(x'_1, x'_2, \ldots, x'_{n-r})$ of the same order but containing fewer indeterminates. The form f_2 is said to be *represented* by f_1; and the representation is *proper* or *improper* according as the determinants of the system do not, or do, admit of any common divisor besides unity. Our third general problem therefore is, 'Given two forms of the same order, of which the first contains more indeterminates than the second, to find whether the second can be represented (properly or improperly) by the first, and, if it can, to assign all the representations of which it is susceptible.' If the second form contain only one indeterminate (*i.e.* if it be an expression of the form Ax^n), the problem reduces itself to that of the representation of the number A by the form f_1. If, again, f_2 contains as many indeterminates as f_1, the problem becomes that of the transformation of f_1 into f_2. We may add that the problem of improper representation may be made to depend on that of proper representation, by methods analogous to those by which the problem of transformation depends on that of equivalence. (See Disq. Arith., Art. 284, where Gauss treats of the improper representation of binary by ternary quadratic forms.)

83. It is hardly necessary to state that what has been done towards obtaining a complete solution of these problems is but very little compared with what remains to be done. Our knowledge of the algebra of homogeneous forms (notwithstanding the accessions which it has received in recent times) is far

too incomplete to enable us even to attempt a solution of them co-extensive with their general expression. And even if our algebra were so far advanced as to supply us with that knowledge of the invariants and other concomitants of homogeneous forms which is an essential preliminary to an investigation of their arithmetical properties, it is probable that this arithmetical investigation itself would present equal difficulties. The science, therefore, has as yet had to confine itself to the study of particular sorts of forms; and of these (excepting linear forms, and forms containing only one indeterminate) the only sort of which our knowledge can be said to have any approach to completeness are the binary quadratic forms, the first in order of simplicity, as they doubtless are in importance. Of all other sorts of forms our knowledge, to say the least, is fragmentary.

We shall arrange the researches of which we have now to speak in the following order, according to the subjects to which they refer:—

1. Binary Quadratic Forms.
2. Binary Cubic Forms.
3. Other Binary Forms.
4. Ternary Quadratic Forms.
5. Other Quadratic Forms.
6. Forms of order n decomposable into n linear factors.

The theory of linear forms (*i.e.* of linear indeterminate equations) we shall refer to hereafter. That of forms containing only one indeterminate will not require any further notice.

(1) *Binary Quadratic Forms.*

84. Instead of confining our attention exclusively to the most recent researches in the Theory of Quadratic Forms, we propose, in the following articles, to give a brief but systematic *résumé* of the theory itself, as it appears in the Disq. Arith., introducing, in their proper places, notices, as full as our limits will admit, of the results obtained by later mathematicians. We adopt this method, partly to render the later researches themselves more easily intelligible, by showing their connexion with the whole theory; but partly also in the hope of facilitating to some persons the study of the Fifth Section of the Disq. Arith., which, probably owing to the obscurity of certain parts of it, is even now too much neglected by mathematicians. This section is composed, as Lejeune Dirichlet has observed (Crelle, vol. xix. p. 325), of two very distinct parts. The results contained in the former of the two (Arts. 153–222)

are for the most part those which had been already obtained by Euler, Lagrange, and Legendre; but they are completed in many respects; they are derived, in part at least, from different principles, and are expressed in a terminology which has been adopted by most subsequent writers. The second part (Arts. 223–307) is occupied, after some preliminary disquisitions (Arts. 223–233), with the ulterior researches of Gauss himself. We proceed then to give a summary of the definitions and theorems contained in the first of these two portions.

85. *Elementary Definitions.*—The quadratic form $ax^2+2bxy+cy^2$ is symbolised by the formula $(a, b, c)(x, y)^2$, or, when it is not necessary to specify the indeterminates, by the simpler formula (a, b, c). The second coefficient is always supposed to be even; and an expression of the form $px^2+qxy+ry^2$ (in which q is uneven) is not considered by Gauss as itself a quadratic form, but as the half of the quadratic form $(2p, q, 2r)$. The discriminant b^2-ac of the form (a, b, c) is called by Gauss the *determinant* of the form; an expression which at the present time it would be neither possible nor desirable to alter. When two forms are equivalent, they are said to be *properly* equivalent if the modulus of transformation is $+1$, but *improperly* equivalent if it is -1. Only those forms which are properly equivalent to one another are considered to belong to the same class; two forms which are only improperly equivalent are said to belong to *opposite* classes. This distinction between proper and improper equivalence is due to Gauss, and is of very great importance. In what follows, unless the contrary is expressly specified, we shall use the terms equivalence and automorphism to denote proper equivalence and proper automorphism. It is readily seen that the greatest common divisors of a, $2b$, c, and of a, b, c are the same for (a, b, c) and for every form equivalent to (a, b, c); if each of those greatest common divisors is unity, (a, b, c) is a *properly primitive* form, and the class of forms equivalent to (a, b, c) a properly primitive class; if the first greatest common divisor be 2, and the second 1, the form, and the class of forms equivalent to it, are termed *improperly primitive.* Every form which is not itself primitive, is a numerical multiple of some primitive form of a less determinant, and is therefore called a derived form. Thus x^2+3y^2 is a properly primitive form of det. -3, but $2x^2+2xy+2y^2$ is an improperly primitive form of the same determinant; while $2x^2+6y^2$, $4x^2+4xy+4y^2$ are derived forms of det. -12.

In all questions relating to the representation of numbers, or the equivalence of forms, it is sufficient to consider primitive forms, as the solution of these problems for derived forms is immediately deducible from their solution for primitive forms; but in certain investigations connected with the transformation of forms the consideration of derived forms is indispensable. (The problem of Art. 82 coincides with that of the representation of numbers, in the case of binary forms of any order.)

The nature of the quadratic form (a, b, c) depends very mainly on the value of its determinant, which we shall symbolise by D. (1) If $D=0$, the form (a, b, c) reduces itself to an expression of the type $m(px+qy)^2$, p and q denoting two numbers relatively prime, and m being the greatest common divisor of a, b, c. The arithmetical theory of such expressions, which are not binary forms at all, since they are adequately represented by a formula such as mX^2, is so simple, and at the same time diverges so much from that of true binary quadratic forms, that we shall not advert to it again in this Report, and in all that follows the determinant is supposed to be different from zero. (2) When D is a perfect positive square, the form (a, b, c) reduces itself to an expression of the type $m(p_1x+q_1y)(p_2x+q_2y)$, *i.e.* it becomes a product of two linear forms. Owing to this circumstance the theory of forms of a square determinant is so much simpler than that of other quadratic forms, that we shall not enter into any details with regard to them, though it is not necessary to exclude them (as is the case with forms of determinant zero) from those investigations which relate simultaneously to the two remaining kinds of quadratic forms; viz. (3) those of a negative determinant, and (4) those of a positive and not square determinant. An essential difference between these two kinds of forms is, that whereas both positive and negative numbers can be represented by any form of positive and not square determinant, forms of a negative determinant can represent either positive numbers only, or negative numbers only. For if the roots of $a+2b\theta+c\theta^2=0$ be real, it is clear that $ax^2+2bxy+cy^2$ will have values of different signs, when the ratio $y:x$ falls between the two roots and when it falls outside them; but if the roots be imaginary, the form will always obtain values having the same sign (viz. that of a or c), whatever the ratio $y:x$ may be. If (a, b, c) be a positive form (*i.e.* a form representing positive numbers only) of a negative determinant $D=-\Delta$, $(-a, -b, -c)$ is a negative form of the same determinant, and can represent negative numbers only. We see, therefore, that there are as many pos tive as negative classes for any negative determinant; and as everything that can be said about positive forms or classes may be transferred at once, *mutatis mutandis*, to negative forms and classes, we shall in what follows exclude the latter from consideration, and, when we are speaking of forms of a negative determinant, confine ourselves to the positive forms.

Since $x^2 - Dy^2$, or $(1, 0, -D)$, is a form of determinant D, we see that one class at least of properly primitive forms exists for every determinant; and the class containing the form $x^2 - Dy^2$ is called the principal class. Improperly primitive forms only exist for those determinants which satisfy the condition $D \equiv 1$, mod 4; since, if (a, b, c) be improperly primitive, we have $b \equiv 1$, mod 2, $a \equiv c \equiv 0$, mod 2. But for every determinant satisfying this condition, one class at least of improperly primitive forms exists; for $\left(2, 1, -\frac{D-1}{2}\right)$ is an improperly primitive form of determinant D, and the class containing it may be called the principal class of improperly primitive forms.

86. *Reduction of the Problem of Representation to that of Equivalence.*—The problem of the representation of numbers depends, first, on the solution of a quadratic congruence, and, secondly, on the solution of a problem of equivalence. This dependence is established by the two following theorems:—

(i.) 'When the number M admits of a primitive representation by (a, b, c), the quadratic congruence $x^2 - D \equiv 0$, mod M, is resoluble.'

For if $am^2 + 2bmn + cn^2 = M$ be a primitive representation of M, let μ, ν be two numbers satisfying the equation $m\nu - n\mu = 1$; we then find

$$(am^2 + 2bmn + cn^2)(a\mu^2 + 2b\mu\nu + c\nu^2) = (am\mu + b[m\nu + n\mu] + cn\nu)^2 - D;$$

or $$\Omega^2 \equiv D, \text{ mod } M; \text{ if } \Omega = am\mu + b[m\nu + n\mu] + cn\nu.$$

We have already referred to this result in Art. 68.

The representation $am^2 + 2bmn + cn^2$ of the number M by the form (a, b, c), is said by Gauss to appertain to the value Ω of the congruential radical $\sqrt{D}$, mod M. To understand this definition with precision, it is to be observed that if in the expression of Ω we replace μ and ν by any two other numbers satisfying the equation $m\nu - n\mu = 1$, the new value of Ω will be of the form $\Omega + kM$; and conversely, values for μ and ν can always be found which shall give to $am\mu + b[m\nu + n\mu] + cn\nu$ any assigned value of the form $Q + kM$. Two different representations of M appertaining to the same value of $\sqrt{D}$, mod M, are said to belong to the same *set*.

(ii.) 'If M admit of a primitive representation by the form (a, b, c) appertaining to the value Ω of $\sqrt{D}$, mod M, the two forms (a, b, c) and

$$\left(M, \Omega, \frac{\Omega^2 - D}{M}\right)$$

are equivalent; and conversely, if these two forms are equivalent, M admits of a primitive representation by (a, b, c) appertaining to the value Ω of $\sqrt{D}$, mod M.'

To establish the first part of this theorem, we observe that the assertion that M admits of a primitive representation by the form (a, b, c) appertaining to the value Ω of $\sqrt{D}$, mod M, implies the existence of four numbers m, n, μ, ν, satisfying the equations

$$\left.\begin{array}{l} m\nu - n\mu = 1, \\ am^2 + 2bmn + cn^2 = M, \\ am\mu + b\left[m\nu + n\mu\right] + cn\nu = \Omega. \end{array}\right\} \quad . \; . \; . \; . \; . \; . \; . \; . \; . \quad (k)$$

If, therefore, we apply to (a, b, c) the transformation $\begin{vmatrix} m, & \mu \\ n, & \nu \end{vmatrix}$, the resulting form will have M and Ω for its first and second coefficients respectively; its third coefficient will therefore be $\dfrac{\Omega^2 - D}{M}$, because its determinant must be D; *i.e.* the two forms (a, b, c) and $\left(M, \Omega, \dfrac{\Omega^2 - D}{M}\right)$ are equivalent. And, conversely, the equivalence of the two forms (a, b, c) and

$$\left(M, \Omega, \frac{\Omega^2 - D}{M}\right)$$

implies the existence of a transformation $\begin{vmatrix} m, & \mu \\ n, & \nu \end{vmatrix}$ of (a, b, c) into

$$\left(M, \Omega, \frac{\Omega^2 - D}{M}\right);$$

i.e. it implies the existence of four numbers m, n, μ, ν, satisfying the equations (k); or, finally, of a primitive representation of M by (a, b, c) appertaining to the value Ω of $\sqrt{D}$, mod M.

If (A, B, C) be a form equivalent to a form (a, b, c) by which

$$M = am^2 + 2bmn + cn^2$$

is represented, and if $\begin{vmatrix} \alpha, & \beta \\ \gamma, & \delta \end{vmatrix}$ be a transformation of (A, B, C) into (a, b, c), it is clear that

$$(A, B, C)\,(\alpha m + \beta n,\ \gamma m + \delta n)^2 = (a, b, c)\,(m, n)^2 = M.$$

Two such representations of M by equivalent forms are called *corresponding* representations; and we may enunciate the theorem, 'Corresponding representations of the same number M by equivalent forms appertain to the same value of the expression $\sqrt{D}$, mod M;' the truth of this is evident from the nature of the function $Am\mu + B[m\nu + n\mu] + Cn\nu$, which is a covariant (in respect of m, n and μ, ν) to $Ax^2 + 2Bxy + Cy^2$.

To obtain, therefore, all the primitive representations of a given number by a given form (a, b, c), we investigate all the values of the expression $\sqrt{D}$, mod M. If $\Omega_1, \Omega_2, \ldots$ be those values, we next compare each of the forms

$$\left(M, \Omega, \frac{\Omega^2 - D}{M}\right)$$

with (a, b, c). If none of them be equivalent to (a, b, c), M does not admit of primitive representation by (a, b, c); but if one or more of them, as

$$\left(M, \Omega_1, \frac{\Omega_1^2 - D}{M}\right),$$

be equivalent to (a, b, c), let $\begin{vmatrix} \alpha, & \beta \\ \gamma, & \delta \end{vmatrix}$ be the formula exhibiting all the transformations of (a, b, c) into

$$\left(M, \Omega_1, \frac{\Omega_1^2 - D}{M}\right);$$

then all the primitive representations of M by (a, b, c), which appertain to the value Ω_1 of $\sqrt{D}$, mod M, are contained in the formula

$$(a, b, c)\ (\alpha, \gamma)^2 = M.$$

87. *Determination of the number of Sets of Representations.*—It appears from what has preceded, that if S denote a system of representative forms of determinant D (*i.e.* a system of forms containing one form, and only one, for every class of forms of determinant D), the number of different sets of primitive representations of M by the forms of S is equal to the number of different solutions of the congruence $x^2 \equiv D$, mod M. If, in particular, M be uneven and prime to D, it is clear that M can only be represented by properly primitive forms; and in this case the number of solutions of the congruence $x^2 \equiv D$, mod M, *i.e.* the number of sets of primitive representations of M by the properly primitive forms contained in S, is expressed by either of the two formulae

$$\Pi\left(1 + \left(\frac{D}{p}\right)\right), \text{ or } \Sigma\left(\frac{D}{\delta}\right),$$

in which p and δ denote respectively the prime divisors of M, and those divisors of M which are divisible by no square; while $\left(\frac{D}{p}\right)$ and $\left(\frac{D}{\delta}\right)$ are the quadratic symbols of Lagrange and Jacobi (see Arts. 16, 17, 68, 76). If μ denote the number of different primes dividing M, the common value of the two expressions

$$\Pi\left(1 + \left(\frac{D}{p}\right)\right) \text{ and } \Sigma\left(\frac{D}{\delta}\right),$$

is 2^μ or zero, according as the condition $\left(\frac{D}{p}\right)=1$ is satisfied by every prime divisor of M, or is not satisfied by one or more of them. When D is $\equiv 1$, mod 4, S will certainly contain improperly primitive forms; and the unevenly even number $2M$ (where M is still supposed prime to D) will admit of primitive representation only by the improperly primitive forms contained in S (for if Ω denote any root of the congruence $x^2\equiv D$, mod $2M$, Ω will be uneven, $\frac{\Omega^2-D}{2M}$ even, and the form $\left(M, \Omega, \frac{\Omega^2-D}{2M}\right)$ will be improperly primitive). And the number of sets of primitive representations of $2M$ by these improperly primitive forms will be the same as the number of sets of primitive representations of M by the properly primitive forms in S.

The problem of obtaining the derived representations of M by (a, b, c) depends on that of finding the primitive representations of a given number by a given form (see Art. 79). Two derived representations of M are said to belong to the same set, when the greatest common divisor of the indeterminates, which we will symbolise by ω, is the same for each, and when the two primitive representations of $\frac{M}{\omega^2}$, from which they are derived, appertain to the same value of $\sqrt{D}$, mod $\frac{M}{\omega^2}$. Adopting this definition, we may enunçiate the theorem, 'If M be an uneven number prime to D, the whole number of sets of representations of M (and if $D\equiv 1$, mod 4, of $2M$) by a system of representative forms of determinant D is $\Sigma\left(\frac{D}{d}\right)$; d denoting any divisor of D.' We may add that, as before, M will be represented only by properly primitive forms; and, when $D\equiv 1$, mod 4, $2M$ only by improperly primitive forms *.

88. *Reduction of the Problem of Transformation to that of Equivalence.*— It has been shown in Art. 80, that the general problem, 'Given two forms of unequal determinants, to decide whether one of them contains the other, and if so, to find all the transformations of the containing into the contained form,'

---

* The theorems of this Article will not be found in the Disq. Arith. If, in their expression, we transform the symbols $\left(\frac{D}{\delta}\right)$, $\left(\frac{D}{d}\right)$ by the law of reciprocity, we obtain results which coincide with those given by Lejeune Dirichlet in his memoir, 'Recherches sur l'application etc.,' sect. 7 (Crelle, vol. xxi. pp. 1–6).

can be reduced to the simpler problem of the equivalence of forms. For the sake of clearness we shall here point out how the first of the two general methods of that article is to be applied to quadratic forms. If of two forms f and F the former contain the latter, the determinant of F is a multiple of that of f by a square number, viz. by the square of the modulus of transformation. Let the determinant of f be D, and that of F, De^2; also let m and μ be any two conjugate divisors of e, so that $m\mu = e$. Then every transformation of which the modulus is e may be expressed in one way, and one only, by the formula $\begin{vmatrix} m, & k \\ 0, & \mu \end{vmatrix} \times \begin{vmatrix} \alpha, & \beta \\ \gamma, & \delta \end{vmatrix}$, in which k denotes one of the numbers $0, 1, 2, 3, \ldots, m-1$, and $\begin{vmatrix} \alpha, & \beta \\ \gamma, & \delta \end{vmatrix}$ is any unit-transformation whatever. If, therefore, we apply to the form f all the transformations included in the formula $\begin{vmatrix} m, & k \\ 0, & \mu \end{vmatrix}$ (of which the number is equal to the sum of the divisors of e), we shall obtain a series of forms $\phi_1, \phi_2, \ldots$ of determinant De^2. If none of these forms be equivalent to F, F is certainly not contained in f; but if one or more of them, for example, ϕ, arising from the transformation $\begin{vmatrix} m, & k \\ 0, & \mu \end{vmatrix}$, is equivalent to F, let $\begin{vmatrix} \alpha, & \beta \\ \gamma, & \delta \end{vmatrix}$ represent indefinitely any transformation of ϕ into F; then f passes into F by any one of the transformations included in the formula $\begin{vmatrix} m, & k \\ 0, & \mu \end{vmatrix} \times \begin{vmatrix} \alpha, & \beta \\ \gamma, & \delta \end{vmatrix}$. If we take in succession for ϕ every form in the series $\phi_1, \phi_2, \ldots$ which is equivalent to F, it is readily seen that the transformations of f into F, which are thus obtained, are all different, and that they include all possible transformations of f into F.

We have supposed the number e to be positive, *i.e.* we have supposed that f contains F properly. To decide whether f contains F improperly, we have only to examine whether any of the forms $\phi_1, \phi_2, \ldots$ be improperly equivalent to F; and if any one of them be so, to combine the transformation of f into it, with its (improper) transformations into F.

89. *Problem of Equivalence.*—It remains to speak of the problem of equivalence. Of the three parts of which this problem consists, viz. (1) to decide whether two given forms are equivalent or not, (2) if they are, to obtain a single transformation of one form into the other, and (3) from a single transformation to deduce all the transformations, the last only admits of being treated by a method equally applicable to forms of a positive and negative determinant. We shall therefore consider it first. The solution which Gauss has given of it

(Disq. Arith., Art. 162) depends on principles which are concealed (as is frequently the case in the Disquisitiones Arithmeticæ) by the synthetical form in which he has expressed it. We shall not therefore repeat the details of his solution, but shall endeavour to point out the basis on which it rests.

Let $f=(a, b, c)(x, y)^2$ be transformed into $F=(A, B, C)(x, y)^2$ by two different, but *similar*, transformations, $\begin{vmatrix} \alpha_0, \beta_0 \\ \gamma_0, \delta_0 \end{vmatrix}$ and $\begin{vmatrix} \alpha_1, \beta_1 \\ \gamma_1, \delta_1 \end{vmatrix}$; *i.e.* by two transformations of which the determinants are equal in sign as well as in magnitude to the same positive or negative number e. Let also, for brevity,

$$X_0=\alpha_0 x+\beta_0 y, \quad Y_0=\gamma_0 x+\delta_0 y, \quad X_1=\alpha_1 x+\beta_1 y, \quad Y_1=\gamma_1 x+\delta_1 y,$$

so that $f(X_0, Y_0)=f(X_1, Y_1)=F(x, y)$; we have then the algebraical theorem—

'The homogeneous functions $F(x, y)$ and $X_0 Y_1 - X_1 Y_0$ differ only by a numerical factor, not containing x or y.'

The truth of this theorem is independent of the supposition that the coefficients of the given forms and given transformations are integral numbers. Its demonstration is implicitly contained in the formulae given by Gauss; or it may be verified more indirectly by the consideration, that if ω be a root of the equation $a+2b\omega+c\omega^2=0$, we have, simultaneously,

$$\omega=\frac{\gamma_0+\delta_0\Omega}{\alpha_0+\beta_0\Omega}, \qquad \omega=\frac{\gamma_1+\delta_1\Omega}{\alpha_1+\beta_1\Omega},$$

Ω denoting in each case the *same* root of the equation $A+2B\Omega+C\Omega^2=0$, an assertion which would not be true, if the equal determinants $\alpha_0\delta_0-\beta_0\gamma_0$ and $\alpha_1\delta_1-\beta_1\gamma_1$ were of opposite signs. Hence the equation

$$\frac{\gamma_0+\delta_0\Omega}{\alpha_0+\beta_0\Omega}=\frac{\gamma_1+\delta_1\Omega}{\alpha_1+\beta_1\Omega}$$

coincides with the equation

$$A+2B\Omega+C\Omega^2=0;$$

i.e. $X_0 Y_1 - X_1 Y_0$ is identical (if we neglect a factor not containing x or y) with $F(x, y)$.

Comparing this conclusion with the identity

$$\left.\begin{aligned}&[F(x, y)]^2=f(X_0, Y_0)\times f(X_1, Y_1)\\ &=[aX_0X_1+b(X_0Y_1+X_1Y_0)+cY_0Y_1]^2-D(X_0Y_1-X_1Y_0)^2,\end{aligned}\right\} \quad . \quad . \quad . \quad (h)$$

we obtain a second result of the same kind—

'The function $aX_0X_1+b(X_0Y_1+X_1Y_0)+cY_0Y_1$ differs from $F(x, y)$ only by a numerical factor not containing x or y.'

Let m be the greatest common divisor of A, $2B$, and C; U and T the greatest common divisors of the coefficients of x^2, xy, and y^2 in $X_0Y_1-X_1Y_0$ and $aX_0X_1+b(X_0Y_1+X_1Y_0)+cY_0Y_1$ respectively; m being a positive integer, but the signs of U and T being fixed by the equations

$$\frac{F(x,y)}{m}=\frac{X_0Y_1-X_1Y_0}{U}=\frac{aX_0X_1+b(X_0Y_1+X_1Y_0)+cY_0Y_1}{T}, \quad . \quad . \quad . \quad (k)$$

which are implied by the two algebraical theorems that have preceded; the numbers T, U, and m will satisfy the equation $T^2-DU^2=m^2$, which is obtained by combining the equations (h) and (k), and will serve to express the relation which subsists between the transformations $\begin{vmatrix}\alpha_0, \beta_0\\ \gamma_0, \delta_0\end{vmatrix}$ and $\begin{vmatrix}\alpha_1, \beta_1\\ \gamma_1, \delta_1\end{vmatrix}$. Solving the equations

$$X_0Y_1-X_1Y_0 \qquad =\frac{U}{m}F(x,y)=\frac{U}{m}f(X_0,Y_0),$$

$$aX_0X_1+b(X_0Y_1+X_1Y_0)+cY_0Y_1=\frac{T}{m}F(x,y)=\frac{T}{m}f(X_0,Y_0),$$

for X_1 and Y_1, we find

$$mX_1=(T-bU)X_0-\quad cU\ Y_0,$$
$$mY_1=\quad aU\ X_0+(T+bU)Y_0;$$

or, finally, equating the coefficients of x and y,

$$\begin{vmatrix}\alpha_1, \beta_1\\ \gamma_1, \delta_1\end{vmatrix}=\frac{1}{m}\times\begin{vmatrix}T\alpha_0-U(b\alpha_0+c\gamma_0), & T\beta_0-U(b\beta_0+c\delta_0)\\ T\gamma_0+U(a\alpha_0+b\gamma_0), & T\delta_0+U(a\beta_0+b\delta_0)\end{vmatrix}$$

$$=\frac{1}{m}\times\begin{vmatrix}T-bU, & -cU\\ aU, & T+bU\end{vmatrix}\times\begin{vmatrix}\alpha_0, \beta_0\\ \gamma_0, \delta_0\end{vmatrix}. \quad . \quad . \quad . \quad . \quad . \quad . \quad . \quad (\mathrm{C})$$

If we suppose the complete solution of the indeterminate equation $T^2-DU^2=m^2$ to be known, the formula (C) supplies us with a complete solution of the problem, 'Given one transformation of f into F, to deduce all the similar transformations of f into F.' For if we suppose in that formula that T and U denote indefinitely any two numbers satisfying the indeterminate equation, it will appear (1) that every transformation of f into F is contained in (C); (2) that every transformation contained in (C) is a transformation of f into F; (3) that no two transformations contained in (C), and corresponding to different values of T and U, are identical. Only it is to be observed that the transformations (C) are not, in general, all integral. They are so, however, when e, the modulus of transformation, is a unit, a supposition which we have not yet introduced; *i.e.* when the forms f and F are either properly or improperly equivalent; because $\frac{a}{m}$, $\frac{2b}{m}$, and $\frac{c}{m}$ are then evidently integral; whence it may be inferred that $\frac{T+bU}{m}$ and $\frac{T-bU}{m}$ are so too.

90. *Expression for the Automorphics of a Quadratic Form.*—To find the automorphics of any quadratic form it is sufficient to consider the case of a primitive form. Putting then $f=F$, and taking for $\begin{vmatrix} \alpha_0, \beta_0 \\ \gamma_0, \delta_0 \end{vmatrix}$ the identical transformation $\begin{vmatrix} 1, 0 \\ 0, 1 \end{vmatrix}$, we obtain from the formula (C) the following general expression for the automorphics of f,

$$\begin{vmatrix} \alpha, \beta \\ \gamma, \delta \end{vmatrix} = \frac{1}{m} \times \begin{vmatrix} T-bU, & -cU \\ aU, & T+bU \end{vmatrix}, \quad \ldots \ldots \quad (D)$$

where $m=1$, or 2, according as f is properly or improperly primitive. The nature of this expression for the automorphics depends on the value of D. If D be positive and not square, let us represent the least positive numbers satisfying the equation $T^2-DU^2=m^2$ by T_1 and U_1; we then have, by a known theorem, the following formula for all the solutions in which T is positive,

$$\frac{T_k+U_k\sqrt{D}}{m} = \left(\frac{T_1+U_1\sqrt{D}}{m}\right)^k,$$

k denoting any positive or negative integral number.

From this we can infer that if $\begin{vmatrix} \alpha_1, \beta_1 \\ \gamma_1, \delta_1 \end{vmatrix}$ be the automorphic in the formula (D), arising from the values T_1, U_1 of T and U, all the other proper automorphics are powers of $\begin{vmatrix} \alpha_1, \beta_1 \\ \gamma_1, \delta_1 \end{vmatrix}$, and are included in the formula

$$|\epsilon| \times \begin{vmatrix} \alpha_1, \beta_1 \\ \gamma_1, \delta_1 \end{vmatrix}^k,$$

$|\epsilon|$ representing one or other of the identical transformations

$$\begin{vmatrix} 1, 0 \\ 0, 1 \end{vmatrix} \text{ and } \begin{vmatrix} -1, & 0 \\ 0, & -1 \end{vmatrix}.$$

If D be a negative number, the only solutions of the equation

$$T^2-DU^2=m^2$$

(except in two cases presently to be noticed) are $T=\pm m$, $U=0$. Hence the only proper automorphics of a form of negative determinant are the two identical transformations $\begin{vmatrix} 1, 0 \\ 0, 1 \end{vmatrix}$ and $\begin{vmatrix} -1, & 0 \\ 0, & -1 \end{vmatrix}$. The two excepted cases are

$$(1)\ D=-1,\ m=1; \qquad (2)\ D=-3,\ m=2.$$

In the former case we have for T and U the four values ± 1, 0, and 0, ± 1; whence the proper automorphics of a form of det. -1 are the four transformations supplied by the formula $\begin{vmatrix} -b, & -c \\ a, & b \end{vmatrix}^k$. If $D = -3$, $m = 2$, the solutions of $T^2 + 3U^2 = 4$ are six in all, viz. $\pm 2, 0$; $\pm 1, 1$; and $\pm 1, -1$; whence six automorphics, comprised in the formula

$$\begin{vmatrix} \frac{1}{2}(1-b), & -\frac{1}{2}c \\ \frac{1}{2}a \quad , & \frac{1}{2}(1+b) \end{vmatrix}^k,$$

exist for an improperly primitive form of det. -3. We may add that in each of these two cases, in addition to the proper automorphics we have found, there exist an equal number of improper automorphics.

From the formula (C), compared with the theory of representation contained in Art. 86, it follows that if $(a, b, c)\ (\alpha, \gamma)^2 = M$ be any representation of M by (a, b, c), all the representations of the same set are included in the formula

$$\left[\frac{T\alpha - U(b\alpha + c\gamma)}{m}, \quad \frac{T\gamma + U(a\alpha + b\gamma)}{m}\right].$$

For forms of a positive and not square determinant the number of representations in each set is therefore infinite. For forms of a negative determinant the number of representations in each set is, in general, two; and if $[\alpha, \gamma]$ be one of them, the other is $[-\alpha, -\gamma]$. But if the determinant be -1, or if the form be derived from a form of det. -1, the number of representations in each set is four; and if the form be an improperly primitive form of det. -3, or be derived from such a form, the number of representations in each set is six.

91. *Expression for the Automorphics—Method of Lejeune Dirichlet.*—We have inferred the expression (D) of the automorphics of f, from the formula (C) of which it is a particular case. But it is plain, from the general theory of Art. 81, that, when f and F are equivalent, we can conversely infer the formula (C) from (D). This method has been preferred by Lejeune Dirichlet, who obtains the automorphics of a primitive form $f = (a, b, c)$, of which the determinant is not a positive square, by the following process (Crelle, vol. xxiv. p. 324). If $\begin{vmatrix} \alpha, & \beta \\ \gamma, & \delta \end{vmatrix}$ be any rational automorphic of f, we have evidently,

$$\begin{aligned} a(ax^2 + 2bxy + cy^2) &= [ax + (b + \sqrt{D})y] \quad [ax + (b - \sqrt{D})y], \\ &= [(a\alpha + [b + \sqrt{D}]\gamma)x + (a\beta + [b + \sqrt{D}]\delta)y] \times \\ &\quad\ [(a\alpha + [b - \sqrt{D}]\gamma)x + (a\beta + [b - \sqrt{D}]\delta)y], \end{aligned}$$

an equation which, for brevity, we may write

$$(p_1x + q_1y)\ (p_2x + q_2y) = (P_1x + Q_1y)\ (P_2x + Q_2y),$$

and which implies one or other of the two following systems:—

$$(1)\quad p_1 p_2 = P_1 P_2; \qquad \frac{p_1}{P_1} = \frac{q_1}{Q_1}; \qquad \frac{p_2}{P_2} = \frac{q_2}{Q_2},$$

$$(2)\quad p_1 p_2 = P_1 P_2; \qquad \frac{p_1}{P_2} = \frac{q_1}{Q_2}; \qquad \frac{p_2}{P_1} = \frac{q_2}{Q_1}.$$

If (1) be the system which is satisfied by $\begin{vmatrix} \alpha, \beta \\ \gamma, \delta \end{vmatrix}$, let

$$\frac{p_1}{P_1} = \frac{1}{m}[T + U\sqrt{D}], \qquad \frac{p_2}{P_2} = \frac{1}{m}[T - U\sqrt{D}],$$

T and U denoting rational numbers, and m still representing the greatest common divisor of a, $2b$, c. These assumptions are legitimate, because $\frac{p_1}{P_1}$ and $\frac{p_2}{P_2}$ contain no irrationality but $\sqrt{D}$, and are conjugate with regard to $\sqrt{D}$. Substituting in the equations

$$\frac{P_1}{p_1} = \frac{Q_1}{q_1} = \frac{1}{m}(T + U\sqrt{D}),$$

$$\frac{P_2}{p_2} = \frac{Q_2}{q_2} = \frac{1}{m}(T - U\sqrt{D}),$$

for p_1, p_2; q_1, q_2; P_1, P_2; Q_1, Q_2, the expressions which these letters represent, and equating the rational and irrational parts, we find

$$\begin{vmatrix} \alpha, \beta \\ \gamma, \delta \end{vmatrix} = \frac{1}{m} \times \begin{vmatrix} T - bU, & -cU \\ aU, & T + bU \end{vmatrix}.$$

In this expression T and U satisfy the equation $T^2 - DU^2 = m^2$, because $p_1 p_2 = P_1 P_2$. From this we infer that $\alpha\delta - \beta\gamma = 1$; further, if we now introduce the condition that $\alpha, \beta, \gamma, \delta$ are to be integral and not merely rational numbers, it will follow, because γ, $\delta - \alpha$, $-\beta$ are integral, that $\frac{a}{m}U$, $\frac{2b}{m}U$, $\frac{c}{m}U$ are also integral; *i.e.* that U itself, and consequently T, is integral; so that the formula at which we have arrived coincides exactly with the formula (D). The system (2), treated in a similar manner, leads to the conclusion $\alpha\delta - \beta\gamma = -1$; whence it follows that that system can be satisfied by no proper automorphic of f.

This method, as Dirichlet observes, has the advantage of putting in a clear light the difference between proper and improper automorphism. A proper automorphic changes each of the two factors, into which the form may be decomposed, into a multiple of itself by a complex unit of the form $\frac{1}{m}[T + U\sqrt{D}]$;

whereas improper automorphics, which only exist for particular kinds of forms, change each factor into a multiple of the other. A similar distinction subsists between proper and improper equivalence; the radical $\sqrt{D}$ entering with the same sign, or with opposite signs, into the factors which are transformed into one another, according as the transformation is proper or improper.

92. *Problem of Equivalence—Forms of a Negative Determinant.*—To complete the solution of the problem of equivalence, we consider, first, forms of a negative, and then those of a positive and not square determinant.

A form (a, b, c) of a negative determinant $D = -\Delta$, which satisfies the conditions enunciated in the following Table, is called a *reduced* form. The symbols $[2b]$ etc. are used to denote the absolute values of the quantities enclosed within the brackets.

General Conditions.	Special Conditions.
1. $[2b] \leqq [a]$.	1. If $a = c$, $b \geqq 0$.
2. $[2b] \leqq [c]$.	2. If $[2b] = [a]$, $b \geqq 0$.
3. $[a] \leqq [c]$.	

The essential character of a reduced form is sufficiently expressed by the two symmetrical conditions $[2b] \leqq [a]$, and $[2b] \leqq [c]$. The third general condition (which combined with the first implies the second), and the special conditions, are, it may be said, artificial restrictions, intended to enable us to enunciate with precision the theorem that 'every class contains one, and only one, reduced form.'

To show that one reduced form always exists in any given class, we select from the given class all those forms in which the coefficient of x^2 is the least; and again, from those forms we select that one form, (a, b, c), or those two forms, (a, b, c) and $(a, -b, c)$, in which the coefficient of y^2 is the least. The single form (a, b, c), or the two forms (a, b, c), $(a, -b, c)$, thus obtained, will, it is easy to see, satisfy the general conditions; and since, if $a = c$, or again if $[2b] = [a]$, the *opposite* forms (a, b, c) and $(a, -b, c)$, each of which satisfies the general conditions, are equivalent, and therefore both belong to the given class, it is clear that a form always exists satisfying the special conditions proper to these cases. That only one reduced form exists in each class may be proved by employing a principle due to Legendre (Théorie des Nombres, vol. i. p. 77):—

'If $f = (a, b, c)$ be a form satisfying the general conditions for a reduced form, $f(1, 0)$, or a is the least number (other than zero) which can be represented by f; and $f(0, 1)$ or c is the least number which can be represented by f, with any value of the second indeterminate different from zero.'

For, if we wish to find the least numbers that can be represented by f, it will be sufficient to attribute positive values to x and y in the formula

$$f = ax^2 - 2bxy + cy^2,$$

in which we suppose b positive as well as a and c. But

$$f(x-1, y) = f(x, y) - 2b(x-y) - (a-2b)x - a(x-1),$$
$$f(x, y-1) = f(x, y) - 2b(y-x) - (c-2b)y - c(y-1);$$

from which equations it appears that if in the formula $f(x, y)$ we diminish by a unit that indeterminate which is not less than the other, we diminish, or at least we do not increase, the value of $f(x, y)$; a conclusion which leads immediately to the principle enunciated by Legendre.

From this principle it follows that a form satisfying the general conditions of reduction is the form, or one of the two opposite forms, to which we are led by the process of selection above described. If, therefore, there be two reduced forms in the same class, they must be two opposite forms (a, b, c) and $(a, -b, c)$. But it is easily proved that two such opposite forms, each satisfying the general conditions of reduction, cannot be equivalent, unless either $[2b] = a$, or $a = c$; in which cases only one of the two forms satisfies the special conditions. In every case, therefore, there exists one, and only one, reduced form in each class.

To obtain the reduced form equivalent to a given form, we form a series of *contiguous* forms, beginning with the given form and ending with the reduced form (Disq. Arith., Art. 171). Two forms of the same determinant, (a, b, c) and (a', b', c'), are said to be *contiguous* when

$$c = a', \text{ and } b + b' \equiv 0, \text{ mod } a'.$$

Two contiguous forms are always equivalent; for if $b + b' = \mu a'$, the former passes into the latter by the transformation $\begin{vmatrix} 0, & -1 \\ 1, & \mu \end{vmatrix}$.

Let, then, (a_0, b_0, a_1) be the given form of det. $-\Delta$, which is supposed not to satisfy the general conditions for a reduced form. Let $b_0 + b_1 = \mu_1 a_1$, $-b_1$ denoting the *minimum* residue of b_0, mod a_1, so that $[2b_1] \leqq a_1$; and let a_2 represent the integral number $\dfrac{b_1^2 + \Delta}{a_1}$. The form (a_1, b_1, a_2) will be contiguous, and therefore equivalent, to (a_0, b_0, a_1). Let a third form, (a_2, b_2, a_3), be similarly derived from (a_1, b_1, a_2), and let the series of contiguous forms (a_0, b_0, a_1), (a_1, b_1, a_2), (a_2, b_2, a_3), ... be continued until we arrive at a form

(a_n, b_n, a_{n+1}), in which $a_{n+1} \geqq a_n$. We shall certainly arrive at such a form, or we should have a series of numbers $a_1, a_2, a_3, \ldots$ all represented by the form (a_0, b_0, a_1), and yet continually decreasing for ever; whereas a form of negative determinant can acquire only a finite number of values inferior to any given limit. The form (a_n, b_n, a_{n+1}), in which $a_n \leqq a_{n+1}$, satisfies the general conditions for a reduced form. For by the law of the series of forms $[2b_n] \leqq a_n$; and since $a_n \leqq a_{n+1}$, we have also

$$[2b_n] \leqq a_{n+1}.$$

Again, the process can always be terminated in such a manner as to give a form satisfying the special conditions for a reduced form. If $a_n = a_{n+1}$, and b_n is negative, instead of stopping at the form (a_n, b_n, a_n), we continue the process one step further and obtain the reduced form $(a_n, -b_n, a_n)$. If $-2b_n = a_n$, instead of the form (a_n, b_n, a_{n+1}), we take the form $(a_n, -b_n, a_{n+1})$, which is contiguous to (a_{n-1}, b_{n-1}, a_n), for the concluding form of the series.

The transformation $|T_n|$ by which (a_0, b_0, a_1) passes into the equivalent reduced form (a_n, b_n, a_{n+1}), is

$$\begin{vmatrix} 0, & -1 \\ 1, & \mu_1 \end{vmatrix} \times \begin{vmatrix} 0, & -1 \\ 1, & \mu_2 \end{vmatrix} \times \ldots\ldots\ldots \times \begin{vmatrix} 0, & -1 \\ 1, & \mu_n \end{vmatrix},$$

where $$\mu_i = \frac{b_{i-1} + b_i}{a_i};$$

or if we represent the successive convergents to the continued fraction

$$-\cfrac{1}{\mu_1 - \cfrac{1}{\mu_2 - \cfrac{1}{\mu_3 - \cfrac{1}{\mu_4 - \ldots}}}}$$

by $\dfrac{P_0}{Q_0}, \dfrac{P_1}{Q_1}, \dfrac{P_2}{Q_2}, \ldots,$ so that

$$P_0 = 0, \quad P_1 = -1, \quad P_2 = -\mu_2, \ldots, \quad Q_0 = 1, \quad Q_1 = \mu_1, \quad Q_2 = \mu_1\mu_2 - 1, \ldots,$$

we may express $|T_n|$ by the formula $|T_n| = \begin{vmatrix} P_{n-1}, & P_n \\ Q_{n-1}, & Q_n \end{vmatrix}$.

The theory of the reduction of quadratic forms was first given by Lagrange. (See his 'Recherches d'Arithmétique' in the Nouveaux Mémoires de l'Académie de Berlin for **1773**; see also his additions to Euler's Algebra, Art. **32**; a memoir of Euler's, 'De insigni promotione scientiæ numerorum,' Opusc. Anal., vol. ii. p. **273**, or Comment. Arith., vol. ii. p. **140**; Legendre, Théorie des Nombres, première partie, sect. viii; Disq. Arith., Arts. **171–173**; M. Hermite in Crelle's Journal, vol. xli. p. **193**.) The method is applicable to forms of a positive, as well as to those of a negative determinant; but when the determinant is positive, the reduced forms are not, in general, all non-equivalent. When the determinant is negative, it is as applicable to forms, of which the coefficients are any real quantities whatever, as to those of which the coefficients are integral numbers. We shall revert hereafter to the consequences which M. Hermite has deduced from this important observation.

We have now a complete solution of the problem of equivalence for forms of a negative determinant. To decide whether two forms f_1 and f_2 of the same negative determinant are equivalent or not, we have only to investigate the reduced forms ϕ_1 and ϕ_2 equivalent to f_1 and f_2: according as ϕ_1 and ϕ_2 are or are not identical, f_1 and f_2 are or are not equivalent; and if they are equivalent, all the transformations of f_1 into f_2 are obtained, by compounding the reducing transformation of f_1, first, with the automorphics of ϕ_2, and then with the inverse of the reducing transformation of f_2.

93. *Problem of Equivalence for Forms of a Positive and not Square Determinant.*—The solution of the problem of equivalence for forms of a positive and not square determinant occupies a considerable space in the Disq. Arith. (Arts. 183–196). But, as Lejeune Dirichlet has observed, in a memoir which he has devoted to this problem ('Vereinfachung der Theorie der binären quadratischen Formen,' in the Memoirs of the Academy of Berlin for 1854, or in Liouville, New Series, vol. ii. p. 353), the demonstrations relating to it may be greatly abbreviated by employing certain known results of the theory of continued fractions. The following method does not differ materially from that proposed by Lejeune Dirichlet; nor indeed is it, in principle, very distinct from that of Gauss, the connexion of which with the theory of continued fractions he has suppressed.

We shall suppose that the forms which we consider are primitive—a supposition which involves no loss of generality; and we shall understand, in what follows, by a 'quadratic equation,' an equation of the form $a_0+2b_0\theta+a_1\theta^2=0$, in which $b_0^2-a_0a_1$ is positive, and a_0, b_0, a_1 are integral numbers without any common divisor. Such a quadratic equation we shall symbolise by the formula $[a_0, b_0, a_1]$, and we shall regard the two quadratic equations $[a_0, b_0, a_1]$, $[-a_0, -b_0, -a_1]$ as different. If $\sqrt{D}$ denote the positive square root of $b_0^2-a_0a_1$, it is convenient to call

$$\frac{-b_0-\sqrt{D}}{a_1}, \text{ and } \frac{-b_0+\sqrt{D}}{a_1}$$

the first and second roots of $[a_0, b_0, a_1]$ respectively; so that if we change the sign of the equation throughout, we change at the same time the denomination of the roots. Whenever, therefore, a root of a quadratic equation, and the denomination of the root, are given, the quadratic equation itself is given. It is readily seen that if two forms (a_0, b_0, a_1), (A_0, B_0, A_1) be properly or improperly equivalent, so that $\begin{vmatrix} \alpha, \beta \\ \gamma, \delta \end{vmatrix}$ transforms (a_0, b_0, a_1) into (A_0, B_0, A_1), the *corresponding* roots of the quadratics

$$a_0 + 2b_0\omega + a_1\omega^2 = 0, \qquad A_0 + 2B_0\Omega + A_1\Omega^2 = 0,$$

i.e. those which are connected by the relation $\omega = \dfrac{\gamma + \delta\Omega}{\alpha + \beta\Omega}$, are of the same, or of opposite, denominations, according as the equivalence is proper or improper. Let the first root of the equation $[a_0, b_0, a_1]$ be developed in a continued fraction, of which all the integral quotients are positive except the first, which has the same sign as the root. In this process we obtain a perfectly determinate series of transformed equations, each having a complete quotient of the development for its first or second root, according as it occupies an uneven or an even place in the series, counting from the proposed equation inclusive. The complete quotients eventually form a period of an even number of terms; there exists therefore a corresponding period of transformed quadratic equations, which will be of the type

$$[a_0, \beta_0, a_1], \ [a_1, \beta_1, a_2], \ [a_2, \beta_2, a_3], \ \ldots\ldots, [a_{2k-1}, \beta_{2k-1}, a_0].$$

Every equation of the period has one of its roots positive and greater than unity, the other negative and less in absolute magnitude than unity; and if we suppose (as we shall do) that we begin the period with an equation occupying an uneven place in the series of transformed equations, the positive root of any equation of the period will be its first or second root, according as it occupies an uneven or an even place in the period.

To apply what has preceded to our present problem, we require the following lemma (see sect. 2 of Dirichlet's memoir, or M. Serret in Liouville, vol. xv. p. 153).

'If ω and Ω be two irrational quantities connected by the relation $\omega = \dfrac{\gamma + \delta\Omega}{\alpha + \beta\Omega}$, where α, β, γ, δ are integral and $\alpha\delta - \beta\gamma = \pm 1$, the developments of ω and Ω in a continued fraction will ultimately coincide, and the same quotient will occupy an even or an uneven place in both developments alike, if $\alpha\delta - \beta\gamma = +1$, but an even place in the one, and an uneven place in the other, if $\alpha\delta - \beta\gamma = -1$.

A quadratic form $(\alpha_0, \beta_0, \alpha_1)$ of positive determinant is said to be *reduced* * when the roots of $[\alpha_0, \beta_0, \alpha_1]$ are of opposite signs; the absolute value of the first root being greater, that of the second less than unity. A series of reduced forms equivalent to any proposed form (a_0, b_0, a_1) can always be found. For, if the first root of $[a_0, b_0, a_1]$ be developed in a continued fraction, and if its period of equations (beginning with an equation occupying an uneven place in the series of transformed equations) be represented as before by

$$[\alpha_0, \beta_0, \alpha_1], [\alpha_1, \beta_1, \alpha_2], \ldots\ldots, [\alpha_{2k-1}, \beta_{2k-1}, \alpha_0],$$

the forms $$(\alpha_0, \beta_0, \alpha_1), (\alpha_1, -\beta_1, \alpha_2), \ldots, (\alpha_{2k-1}, -\beta_{2k-1}, \alpha_0)$$

will be all reduced and all equivalent to (a_0, b_0, a_1). These forms, so deduced from the development of the first root of the equation $[a_0, b_0, a_1]$, we shall term the period of forms equivalent to (a_0, b_0, a_1), or, more briefly, the period of (a_0, b_0, a_1). It will be seen that each form of the period is contiguous to that which precedes it, and that the first is contiguous to the last.

We can now obtain a complete solution of our problem. If (a_0, b_0, a_1) and (A_0, B_0, A_1) are equivalent, the first roots of $[a_0, b_0, a_1]$ and $[A_0, B_0, A_1]$ will be corresponding roots, and the developments of these two roots will ultimately coincide, giving one and the same period of complete quotients. And, since the same complete quotient will occur in an even or in an uneven place alike in each development, it will be a root of the same denomination in the quadratic equation determining it in each development. The period of equations will therefore be precisely the same for each development; and the same equation may be taken as the first equation of each period. Consequently the periods of (a_0, b_0, a_1), (A_0, B_0, A_1) are identical. Two forms therefore are or are not equivalent, according as their periods are or are not identical. To obtain the transformations of (a_0, b_0, a_1) into (A_0, B_0, A_1), when these two forms are equivalent, let the complete quotients in the development of the first root of $[a_0, b_0, a_1]$ be $\omega_1, \omega_2, \ldots$, and let the convergent immediately preceding ω_{n+1} be $\frac{q_n}{p_n}$. Similarly, let Ω_{n+1} and $\frac{Q_n}{P_n}$ be a complete quotient and a convergent in the development of the first root of $[A_0, B_0, A_1]$. Then, if $\omega_\mu = \Omega_M$ (where $\mu \equiv M$, mod 2), all the transformations of (a_0, b_0, a_1) into (A_0, B_0, A_1) are comprised in the formula

$$\begin{vmatrix} p_{\mu-1}, & p_\mu \\ q_{\mu-1}, & q_\mu \end{vmatrix} \times |T| \times \begin{vmatrix} P_{M-1}, & P_M \\ Q_{M-1}, & Q_M \end{vmatrix}^{-1},$$

* These reduced forms are not to be confounded with the reduced forms of the last article.

$|T|$ denoting any automorphic of the form corresponding to the equation of which $\omega_{\mu+1}$ or Ω_{M+1} is a root.

It should be observed that a reduced form is always a form of its own period. To prove this, we remark that reduced forms are of two kinds; they are either such as (a_0, β_0, a_1), where the first root of $[a_0, \beta_0, a_1]$ is positive, or such as $(a_1, -\beta_1, a_2)$, where the first root of $[a_1, -\beta_1, a_2]$ is negative. Now a reduced form such as (a_0, β_0, a_1) is evidently a form of its own period, for the equation $[a_0, \beta_0, a_1]$ is itself an equation of the period in the development of its first root. And a reduced form such as $(a_1, -\beta_1, a_2)$ is also a form of its own period.* For if we develope the second root of $[a_1, \beta_1, a_2]$, we obtain a period of equations of which $[a_1, \beta_1, a_2]$ is itself one. Let $[a_2, \beta_2, a_3]$ be the equation immediately following $[a_1, \beta_1, a_2]$ in this period; then $[a_1, \beta_1, a_2]$ is an equation occupying an even place in the period of equations arising from the development of the first root of $[a_2, \beta_2, a_3]$, and consequently $(a_1, -\beta_1, a_2)$ is a form in the period of (a_2, β_2, a_3); *i.e.* it is a form in its own period, because it is equivalent to (a_2, β_2, a_3).

It follows from this that no reduced form can be equivalent to a given form, unless it occur in the period of that form.

The inequalities satisfied by the roots of any equation of a period give rise to certain inequalities which are satisfied by its coefficients. These inequalities (which are not all independent) are,

$$\text{(i)}\quad [a_0] < 2\sqrt{D};\quad [\beta_0] < \sqrt{D};\quad [a_1] < 2\sqrt{D};$$

$$\text{(ii)}\quad \sqrt{D}-[\beta_0] < [a_0] < \sqrt{D}+[\beta_0];$$

$$\text{(iii)}\quad \sqrt{D}-[\beta_0] < [a_1] < \sqrt{D}+[\beta_0].$$

The same inequalities are, of course, satisfied by the coefficients of a reduced form; its middle coefficient is, moreover, positive. And, conversely, every form whose middle coefficient is positive and whose coefficients satisfy these inequalities is a reduced form.

---

* {Or thus [May, 1876]:—Since $(a_1, -\beta_1, a_2)$ is properly equivalent to (a_0, β_0, a_1), the first root of $[a_1, -\beta_1, a_2]$ gives the same period as the first root of $[a_0, \beta_0, a_1]$, the same equations occupying even or uneven places in both periods alike. Hence the period of $(a_1, -\beta_1, a_2)$ is the same as the period of (a_0, β_0, a_1).

[July, 1876.] By developing the first root of $[a_1, -\beta_1, a_2]$ we do indeed obtain the period of $(a_1, -\beta_1, a_2)$: but not immediately. We ought, therefore, here to prove the theorem—that a reduced form has always one antecedent and one consequent reduced form contiguous to it.}

94. *Improper Equivalence—Ambiguous Forms and Classes.*—If it be required to find whether two forms (a, b, c) and (a', b', c') of the same positive or negative determinant are or are not improperly equivalent, it will suffice to change one of them, as (a, b, c), into its opposite $(a, -b, c)$, and then to solve the problem of proper equivalence for $(a, -b, c)$ and (a', b', c'). If it be found that these two forms are properly equivalent, let $|T|$ represent any transformation of the first into the second; then the improper transformations of (a, b, c) into (a', b', c') will be represented by the formula $\begin{vmatrix} 1, & 0 \\ 0, & -1 \end{vmatrix} \times |T|$.

It may happen that two forms are both properly and improperly equivalent to one another; when this is the case, each of the two forms, and every form of the class to which they belong, is improperly equivalent to itself, *i.e.* admits of improper automorphics. A class consisting of such forms is said to be ambiguous (*classis anceps—classe ambigue*). An *ambiguous form* is a form (a, b, c) in which $2b$ is divisible by a; if $2b = \mu a$, the ambiguous form is transformed into itself by the improper automorphic $\begin{vmatrix} 1, & \mu \\ 0, & -1 \end{vmatrix}$; and if $|T|$ be the general expression of its proper automorphics, all its improper automorphics are included by the formula $\begin{vmatrix} 1, & \mu \\ 0, & -1 \end{vmatrix} \times |T|$. Every ambiguous form belongs to an ambiguous class, and, as we shall presently see, every ambiguous class contains ambiguous forms.

To complete the theory of equivalence, we shall here briefly indicate the solution of the problem, 'To decide whether a given form is improperly equivalent to itself or not, and if it is, to find its improper automorphics.'

When the determinant is negative, it follows from the principle that two reduced forms cannot be equivalent, that no reduced form, the opposite of which is different from it and is also a reduced form, can be improperly equivalent to itself. Hence the only reduced forms which have improper automorphics are those in which $b = 0$, or $2b = a$, or $a = c$. In the two former cases the reduced form is ambiguous, in the latter it has the improper automorphic $\begin{vmatrix} 0, & 1 \\ 1, & 0 \end{vmatrix}$, and is moreover contiguous and therefore equivalent to the ambiguous form $(2a-2b,\ a-b,\ a)$. These considerations supply a sufficient criterion for deciding whether a form of negative determinant is {improperly} equivalent to itself or not. If it is, its improper automorphics are given by the formula $|T| \times |\tau| \times |T|^{-1}$; $|T|$ denoting the reducing transformation of the given form, and $|\tau|$ any improper automorphic of the reduced form. For forms of a positive determinant,* we observe that if

$$(a_0, \beta_0, a_1), (a_1, -\beta_1, a_2), \ldots\ldots, (a_{2k-1}, -\beta_{2k-1}, a_0)$$

be the period of (a, b, c), the period of $(a, -b, c)$ is

$$(a_0, -\beta_{2k-1}, a_{2k-1}), (a_{2k-1}, \beta_{2k-2}, a_{2k-2}), \ldots\ldots, (a_1, \beta_0, a_0).$$

For $(a, -b, c)$ is equivalent to $(a_0, -\beta_{2k-1}, a_{2k-1})$, because (a, b, c) is equivalent to $(a_{2k-1}, -\beta_{2k-1}, a_0)$; and, by a known theorem, the period of equations in the development of the second root of (a, b, c) is

$$[a_0, -\beta_{2k-1}, a_{2k-1}], [a_{2k-1}, -\beta_{2k-2}, a_{2k-2}], \ldots\ldots, [a_1, -\beta_0, a_0],$$

the equation $[a_0, -\beta_{2k-1}, a_{2k-1}]$ occupying an even place in the development; this period is therefore the period of equations in the development of the first root of $[a_0, -\beta_{2k-1}, a_{2k-1}]$; *i.e.* the period

$$(a_0, -\beta_{2k-1}, a_{2k-1}), (a_{2k-1}, \beta_{2k-2}, a_{2k-2}), \ldots\ldots, (a_1, \beta_0, a_0)$$

is the period of $(a_0, -\beta_{2k-1}, a_{2k-1})$, or, which is the same thing, of $(a, -b, c)$. If we now suppose that (a, b, c) is improperly equivalent to itself, it will be properly equivalent to $(a, -b, c)$; and these two forms will have the same period, which we shall represent by (p_0, q_0, p_1), (p_1, q_1, p_2), &c. If $(p_\lambda, q_\lambda, p_{\lambda+1})$ be any form of this period, the *associate* of $(p_\lambda, q_\lambda, p_{\lambda+1})$, *i.e.* the form $(p_{\lambda+1}, q_\lambda, p_\lambda)$, will also be a form of the period, and the indices of these two forms in the period will differ by an uneven number, because the signs of the numbers $p_\lambda, p_{\lambda+1}, \ldots$ are alternate. From this we can infer that there will be two different forms in the period, each of which will be immediately preceded by its own associate; so that the type of the period will be

$$(p_0, q_0, p_1), (p_1, q_1, p_2), \ldots\ldots, (p_{k-1}, q_{k-1}, p_k),$$
$$(p_k, q_{k-1}, p_{k-1}), (p_{k-1}, q_{k-2}, p_{k-2}), \ldots\ldots, (p_1, q_0, p_0),$$

where for simplicity we have supposed that (p_0, q_0, p_1) is one of the two forms which is preceded by its associate; the other is (p_k, q_{k-1}, p_{k-1}). These two forms are ambiguous, for it follows from the contiguity of each form to that which precedes it, that $2q_0 \equiv 0$, mod p_0; $2q_{k-1} \equiv 0$, mod p_k. We arrive therefore at the conclusion that the period of every ambiguous class contains two ambiguous forms; either of which enables us, as in the case of forms of a negative determinant, to obtain all the improper automorphics of any form of the class.

* {Here, again, it is not necessary to recur to the definition of the period of reduced forms equivalent to a given form; the associated period is a period of reduced forms equivalent to $(a, -b, c)$}.

Gauss has shown (Disq. Arith., Art. 164), by an analysis which it is not necessary to explain here, that if f contain F both properly and improperly, an ambiguous form contained in f, and containing F, can always be assigned. This theorem comprehends the result which we have incidentally obtained in this article, that every ambiguous class contains ambiguous forms. (See also a note by Dirichlet, in Liouville, New Series, vol. ii. p. 273.)

95. The important theorem, that for every positive or negative determinant the number of classes is finite, is a consequence of the theory of reduction. To establish its truth, it is sufficient to employ the reduction of Lagrange (Art. 92), which is applicable to forms of a positive determinant having integral coefficients no less than to forms of a negative determinant, and which shows that in every class of forms of determinant D there exists one form at least the coefficients of which satisfy the inequalities $[2b] \leqq [a]$, $[2b] \leqq [c]$. These inequalities give, if D be negative, $ac \leqq -\frac{4}{3}D$, $[b] \leqq \sqrt{-\frac{1}{3}D}$; and if D be positive, $[ac] \leqq D$, $[b] \leqq \sqrt{\frac{1}{5}D}$. The number of forms whose coefficients satisfy these inequalities is evidently limited; therefore, *à fortiori*, the number of non-equivalent classes is finite.

To construct a system of representative forms of det. D, we have only to write down all the forms of det. D whose coefficients satisfy the preceding inequalities, to which we may add $[a] \leqq [c]$. If the determinant be negative, it only remains to reject the forms which do not satisfy the special conditions; if it be positive, we must examine whether any of the forms which we have written down are equivalent; and, if so, retaining only one form out of each group of equivalent forms, we shall have the representative system required.

A few particular cases of the theory merit attention from their simplicity.

If $D = -1$, there is but one class of forms, represented by $x^2 + y^2$; and by the theorems of Arts. 87 and 90, the number of representations of any uneven (or unevenly even) number by the form $x^2 + y^2$ is the quadruple of the excess of the number of its divisors of the form $4n+1$, above the number of its divisors of the form $4n+3$. (See Jacobi in Crelle's Journal, vol. xii. p. 169; Dirichlet, *ibid.* vol. xxi. p. 3. In counting the solutions of the equation $x^2 + y^2 = 2p$, Jacobi considers two solutions, such as $x_1^2 + y_1^2 = 2p$ and $x_2^2 + y_2^2 = 2p$, to be identical, when $x_1^2 = x_2^2$, $y_1^2 = y_2^2$; the number of solutions is thus a fourth part of the number of representations.) In particular every prime of the form $4n+1$ (and the double of every such prime) is capable of decomposition in one way, and one only, into two squares relatively prime; and, conversely, every uneven number capable of such decomposition in one way only is a prime of the form $4n+1$.

If $D = -2$, x^2+2y^2 represents the only class of forms; and every uneven number can be represented by x^2+2y^2, in twice as many ways as it has divisors of either of the forms $8n+1$, or $8n+3$, in excess of divisors of the forms $8n+5$, or $8n+7$. (Dirichlet, *loc. cit.*) In particular every prime of either of the forms $8n+1$ or $8n+3$ is decomposable in one way, and in one only, into a square and the double of a square.

Again, for each of the determinants -3 and -7, there is but one properly and one improperly primitive class, which may be represented by the forms (1, 0, 3) and (2, 1, 2); (1, 0, 7) and (2, 1, 4). Uneven numbers are therefore represented by x^2+3y^2, in twice as many ways as they have divisors of the form $3n+1$, in excess of divisors of the form $3n-1$; and by x^2+7y^2 in twice as many ways as they have divisors of the forms $7n+1$, 2, 4, in excess of divisors of the forms $7n+3$, 5, 6. Similarly, x^2+4y^2 represents the only primitive class of det. -4.

For each of the eleven positive determinants of the first century 2, 5, 13, 17, 29, 41, 53, 61, 73, 89, 97, there is but one properly primitive class; there is also for each of the ten uneven determinants one improperly primitive class. Representing any one of these eleven numbers by D, by $[T, U]$ the least solution of $T^2-DU^2=1$, and by M, an uneven positive number prime to D, we may enunciate the theorem,

'The equation $x^2-Dy^2=M$ is capable of as many solutions in positive numbers x and y, satisfying the conditions $x \leqq T\sqrt{M}$, $y \leqq U\sqrt{M}$, as M has divisors of which D is a quadratic residue in excess of divisors of which D is a quadratic non-residue.'

Thus the number of solutions of the equation $x^2-2y^2=M$, where M is an uneven number, and $0<x\leqq 3\sqrt{M}$, $0<y\leqq 2\sqrt{M}$, is the excess of the divisors of M of the forms $8n\pm 1$ above its divisors of the forms $8n\pm 3$.

The conditions $0<x\leqq T\sqrt{M}$, $0<y\leqq U\sqrt{M}$, which are satisfied by one representation, and only one, in each set, are obtained by considerations to which we shall hereafter refer (Art. 100).

96. *The Pellian Equation.*—The two indeterminate equations,

$$T^2-DU^2=1 \quad \text{and} \quad T^2-DU^2=4,$$

are, as we have seen, of primary importance in the theory of quadratic forms of a positive and not square determinant. When the complete solution of these equations is known, we can deduce, from a single representation of a number by a form, every representation of the same set; and, from a single transformation of either of two equivalent forms into the other, every similar transformation. The same equations also present themselves in the solution in integral numbers of the general equation of the second degree containing two indeterminates, and enable us in the principal case in which it admits an infinite number of solutions to deduce them all from a certain finite number. This fundamental importance of the equation $T^2 - DU^2 = 1$ was first recognised by Euler, who has left several memoirs relating to it (see Comment. Arith., vol. i. pp. 4, 316; vol. ii. p. 35; also Euler's Algebra, vol. ii. cap. vii.); but the equation itself had already given rise to a discussion which forms a well-known passage in the scientific history of the seventeenth century. Its solution was proposed by Fermat (see the Commercium Epistolicum of Wallis, Ep. 8) as a challenge to the English mathematicians, and especially to Wallis. The problem was at first misunderstood by Lord Brouncker and Wallis, who each gave a method for its solution in fractional numbers; not attending to the restriction to integral numbers implied, though not expressed, in Fermat's enunciation, without which the problem is of a very elementary character. Ultimately, however, they obtained a complete solution by a method, which Wallis describes in the Comm. Epist., Epp. 17 (postscript) and 19, and in his Algebra, capp. xcviii. and xcix., attributing it to Lord Brouncker, though he seems himself to have had some share in its invention. This method is the same as that which is given by Euler in his Algebra, and in the first of the memoirs above cited, and which is attributed by him to Pell *. It differs, in form at least, from that now employed, and was evidently suggested by the artifices of substitution employed in Diophantine problems. It is most easily explained by an example. If $T^2 - 13U^2 = 1$ be the equation proposed, the process would stand thus:—

$$
\begin{array}{llll}
(1) & 3U < T < 4U; & \text{let } T = 3U + v_1; & -4U^2 + 6Uv_1 + v_1^2 = 1, \\
(2) & v_1 < U < 2v_1; & \text{let } U = v_1 + v_2; & 3v_1^2 - 2v_1v_2 - 4v_2^2 = 1, \\
(3) & v_2 < v_1 < 2v_2; & \text{let } v_1 = v_2 + v_3; & -3v_2^2 + 4v_2v_3 + 3v_3^2 = 1, \\
(4) & v_3 < v_2 < 2v_3; & \text{let } v_2 = v_3 + v_4; & 4v_3^2 - 2v_3v_4 - 3v_4^2 = 1, \\
(5) & v_4 < v_3 < 2v_4; & \text{let } v_3 = v_4 + v_5; & -v_4^2 + 6v_4v_5 + 4v_5^2 = 1, \\
(6) & 6v_5 < v_4 < 7v_5; & \text{let } v_4 = 6v_5 + v_6; & 4v_5^2 - 6v_5v_6 - v_6^2 = 1,
\end{array}
$$

* There does not seem to be any ground for attributing either the problem or its solution to Pell; and it is possible that Euler may have been misled by a confused recollection of the contents of Wallis's Algebra, in which an account is given of the method employed by Pell in solving Diophantine problems. Nevertheless the equation $T^2 - DU^2 = 1$ is often called the Pellian equation after him, probably upon Euler's authority.

$$\begin{array}{llll}
(7) & v_6 < v_5 < 2v_6\,; & \text{let } v_5 = v_6 + v_7\,; & -3v_6^2 + 2v_6v_7 + 4v_7^2 = 1, \\
(8) & v_7 < v_6 < 2v_7\,; & \text{let } v_6 = v_7 + v_8\,; & 3v_7^2 - 4v_7v_8 - 3v_8^2 = 1, \\
(9) & v_8 < v_7 < 2v_8\,; & \text{let } v_7 = v_8 + v_9\,; & -4v_8^2 + 2v_8v_9 + 3v_9^2 = 1, \\
(10) & v_9 < v_8 < 2v_9\,; & \text{let } v_8 = v_9 + v_{10}\,; & v_9^2 - 6v_9v_{10} - 4v_{10}^2 = 1.
\end{array}$$

In the last equation we may put $v_9 = 1$, $v_{10} = 0$; whence $T = 649$, $U = 180$. It will be seen that the success of the method depends on its leading at last to an equation in which the coefficient of one of the indeterminates is $+1$. Wallis does not prove that such an equation will always occur; and the demonstration which he has given of the resolubility of the equation $T^2 - DU^2 = 1$ is inconclusive. (See his Algebra, cap. xcix; the reader will find the paralogism which vitiates his reasoning in the proof of the lemma, upon which it depends; see also Lagrange's criticism in the 8th paragraph of the Additions to Euler's Algebra; and Gauss, Disq. Arith., Art. 202, note.) It is evident that the method of solution employed by Wallis really consists in the successive determination of the integral quotients in the development of $\frac{T}{U}$ in a continued fraction; in addition to this, Euler observed that $\frac{T}{U}$ is itself necessarily a convergent to the value of $\sqrt{D}$; so that to obtain the numbers T and U it suffices to develope $\sqrt{D}$ in a continued fraction. It is singular, however, that it never seems to have occurred to him that, to complete the theory of the problem, it was necessary to demonstrate that the equation is always resoluble, and that all its solutions are given by the development of $\sqrt{D}$. His memoir (Comment. Arith., vol. i. p. 316) contains all the elements necessary to the demonstration, but here, as in some other instances, Euler is satisfied with an induction which does not amount to a rigorous proof. The first admissible proof of the resolubility of the equation was given by Lagrange in the Mélanges de la Société de Turin, vol. iv. p. 41. He there shows that in the development of $\sqrt{D}$, we shall obtain an infinite number of solutions of some equation of the form $T^2 - DU^2 = A$, and that, by multiplying together a sufficient number of these equations, we can deduce solutions of the equation $T^2 - DU^2 = 1$. But the simpler demonstration of its solubility, which is now to be found in most books on algebra, and which depends on the completion of the theory (left unfinished by Euler) of the development of a quadratic surd in a continued fraction, was first given by Lagrange in the Hist. de l'Académie de Berlin for 1767 and 1768, vol. xxiii. p. 272, vol. xxiv. p. 236; and, in a simpler form, in the Additions to Euler's Algebra, Art. 37. Lastly, Gauss, who in the Disq. Arith. avoids the use of continued fractions, has shown that if we form, by the method which he indicates, the period of any quadratic form of det. D, we may infer the complete solution of the equation

$$T^2 - DU^2 = 1, \text{ or } = 4,$$

from the automorphics of any reduced form, according as the form is properly or improperly primitive. (Disq. Arith., Arts. 198 202.)

To express conveniently the principal theorems relating to these equations, we employ the following notation *. The numerator of the continued fraction

$$q_1 + \frac{1}{q_2 +}\frac{1}{q_3 + \ldots}\frac{1}{q_n}$$

is called the *cumulant* of the numbers $q_1, q_2, \ldots, q_n$, and is represented by the symbol $(q_1, q_2, q_3, \ldots, q_n)$; the denominator is evidently the cumulant $(q_2, q_3, \ldots, q_n)$. Accents are sometimes employed to indicate that the first or last quotient of a cumulant is to be omitted; thus

$$'(q_1, q_2, q_3, \ldots, q_n) = (q_2, q_3, \ldots, q_n), \quad (q_1, q_2, q_3, \ldots, q_n)' = (q_1, q_2, q_3, \ldots, q_{n-1}),$$
$$'(q_1, q_2, \ldots, q_n)' = (q_2, q_3, \ldots, q_{n-1}).$$

A *periodic* cumulant is represented by the notation $(\dot{q}_1, q_2, \ldots, \dot{q}_n)_x$, the suffix indicating the number of times which the period is repeated, and a point being placed over the first and last quotients of the period. In what follows m represents 1 or 2, according as we are considering the equation

$$T^2 - DU^2 = 1, \text{ or } = 4.$$

(i.) If $\mu_1, \mu_2, \ldots, \mu_{2k}$ be the period of integral quotients in the development of either root of a quadratic equation of determinant D, which we suppose properly or improperly primitive according as $m = 1$, or $m = 2$, the positive numbers T_x and U_x which satisfy the equation $T^2 - DU^2 = m^2$ are all contained in the formulae

$$\frac{T_x}{m} = \tfrac{1}{2}(A_x + \Delta_x),$$

$$\frac{U_x}{m} = \frac{B_x}{-a_0} = \frac{A_x - \Delta_x}{-2\beta_0} = \frac{\Gamma_x}{a_1};$$

where $$A_x = (\dot{\mu}_1, \mu_2, \ldots, \dot{\mu}_{2k})_x, \qquad B_x = (\dot{\mu}_1, \mu_2, \ldots, \dot{\mu}_{2k})'_x,$$
$$\Gamma_x = '(\dot{\mu}_1, \mu_2, \ldots, \dot{\mu}_{2k})_x, \qquad \Delta_x = '(\dot{\mu}_1, \mu_2, \ldots, \dot{\mu}_{2k})'_x,$$

* This notation is due to Euler (see Nov. Comm. Pet. vol. ix. p. 53, and the memoir already cited, 'De usu novi algorithmi in Problemate Pelliano solvendo.' Comment Arith., vol. i. p. 316). The convenient term 'cumulant' has been introduced by Professor Sylvester (Phil. Trans., vol. cxliii. p. 474), who has also suggested the use of accents to indicate the omission of initial or final quotients.

and $a_0+2\beta_0\theta+a_1\theta^2=0$ is the quadratic equation determining the quotient μ_1, in which we suppose for simplicity that a_1 is positive.

If, in particular, we consider the quadratic equation $\theta^2-D=0$, or rather $a^2-D-2a\theta+\theta^2=0$, where $a^2<D<(a+1)^2$, we have $m=1$, $\mu_1=2a$, and we find, by the symmetry of the period in this case,

$$T_x=\tfrac{1}{2}(A_x+\Delta_x)=(a, \mu_2, \mu_3, \ldots, \mu_{2k}, 2\dot{a}, \mu_2, \ldots, \dot{\mu}_{2k})_{x-1},$$

$$U_x=(\mu_2, \mu_3, \ldots, \mu_{2k}, 2\dot{a}, \mu_2, \ldots, \dot{\mu}_{2k})_{x-1},$$

which are Euler's formulae for the solution of the equation $T^2-DU^2=1$.

(ii.) We have already observed (Art. 90) that when T_1 and U_1 are known, T_x and U_x are defined by the equation

$$\frac{T_x+U_x\sqrt{D}}{m}=\left[\frac{T_1+U_1\sqrt{D}}{m}\right]^x.$$

Either from this equation, or from the cumulantive formulae for T_x, U_x, we infer that T_x and U_x satisfy the equation of finite differences,

$$v_{x+2}-\frac{2T_1}{m}v_{x+1}+v_x=0;$$

so that the two series, of which T_x and U_x are the general terms, are each a recurring series, the scale of relation being $1, -\dfrac{2T_1}{m}, 1$.

It is convenient to observe that $T_{-x}=T_x$; but $U_{-x}=-U_x$.

(iii.) If we denote by ψ the imaginary arc

$$\frac{1}{i}\log\left(\frac{T_1+U_1\sqrt{D}}{m}\right),$$

we have evidently

$$\frac{T_1}{m}=\cos\psi,\quad \frac{U_1\sqrt{D}}{mi}=\sin\psi,\quad \frac{T_x}{m}=\cos x\psi,\quad \frac{U_x\sqrt{D}}{mi}=\sin x\psi.$$

The analogy implied by these formulae enables us to transform many trigonometrical identities into formulae containing T_x and U_x. For example, from the formulae

$$\cos(\phi\pm\theta)=\cos\phi\cos\theta\mp\sin\phi\sin\theta,$$

$$\sin(\phi\pm\theta)=\sin\phi\cos\theta\pm\sin\theta\cos\phi,$$

we have, putting $\phi=x\psi$, $\theta=y\psi$, where x and y are any positive or negative integers,

$$T_{x\pm y}=\frac{1}{m}[T_xT_y\pm DU_xU_y],$$

$$U_{x\pm y}=\frac{1}{m}[T_xU_y\pm T_yU_x].$$

(iv.) It is also found that

$$\frac{T_x}{m} = (-1)^{\frac{1}{2}x(x-1)} \quad \left(\frac{T_1}{m}, -\frac{2T_1}{m}, \frac{2T_1}{m}, \ldots, (-1)^{x-1}\frac{2T_1}{m}\right),$$

$$\frac{U_x}{U_1} = (-1)^{\frac{1}{2}(x-1)(x-2)}\left(\frac{2T_1}{m}, -\frac{2T_1}{m}, \ \ldots\ldots\ldots, (-1)^{x-1}\frac{2T_1}{m}\right).$$

(v.) If q be any integral number whatever, we can always find a solution $[T_\lambda, U_\lambda]$ satisfying the congruences $T_\lambda \equiv T_0 = m$, mod q, and $U_\lambda \equiv U_0 = 0$, mod q. If $[T_\lambda, U_\lambda]$ be the least solution satisfying these congruences, λ will be less than $2q$, and the residues (mod q) of the terms of the two series T_x and U_x will each form a period of λ terms, so that we shall always have

$$T_{x+n\lambda} \equiv T_x, \quad U_{x+n\lambda} \equiv U_x, \text{ mod } q.$$

If $U_{\lambda'}$ be the first number of its series which is divisible by q, we shall have either $\lambda' = \lambda$, or $2\lambda' = \lambda$. In either case, the only numbers U which are divisible by q, are those whose indices are divisible by λ'; and the formula $\left[T_{m\lambda'}, \frac{U_{m\lambda'}}{q}\right]$ comprises all the solutions of the equation $T^2 - Dq^2U^2 = m^2$. Thus, in solving the equation $T^2 - DU^2 = m^2$, we can always substitute for D its quotient when divided by its greatest square divisor. (See Lagrange, Additions to Euler's Algebra, Art. 78. Gauss, Disq. Arith., Art. 201, Obs. 3 and 4.)

We may add, that if q be a prime (an uneven prime when $m=2$), and if q^κ and q^μ be the highest powers of q dividing U_λ and n respectively, $q^{\kappa+\mu}$ will be the highest power of q dividing $U_{n\lambda}$. (Dirichlet, in Liouville's Journal, New Series, vol. i. p. 76.)

(vi.) The methods of Lagrange and Gauss are applicable to the equation $T^2 - DU^2 = 4$, only when $D \equiv 1$, mod 4; because they suppose the existence of an improperly primitive form of det. D. In all other cases the equation $T^2 - DU^2 = 4$ may be divided by 4, and reduced to the form $T^2 - DU^2 = 1$: viz. if $D \equiv 0$, mod 4, T is even; and if $D \equiv 2$, or $\equiv 3$, mod 4, T and U are both even. A similar reduction takes place if $D \equiv 1$, mod 8; the equation $T^2 - DU^2 = 4$ admitting in that case only even solutions. But if $D \equiv 5$, mod 8, $T^2 - DU^2 = 4$ may or may not have uneven solutions; and no criterion is known for distinguishing *à priori* these two cases. If $T^2 - DU^2 = 4$ admit of uneven solutions, its least solution $[T_1, U_1]$ will be uneven; its even solutions will be comprised in the formula $[T_{3n}, U_{3n}]$, and consequently $[\frac{1}{2}T_{3n}, \frac{1}{2}U_{3n}]$ will represent the solutions of $T^2 - DU^2 = 1$.

(vii.) The equations $T^2 - DU^2 = -4$, $T^2 - DU^2 = -1$ are not resoluble for all values of D, but only for those values for which -1 is capable of representation by the principal form of det. D. Whenever the period of integral quotients in the development of $\sqrt{D}$ consists of an uneven number of terms, these equations will be resoluble, and conversely. This will always happen when D is a prime number of the form $4n+1$, and may happen in many other cases, but never can happen when D is divisible by any prime of the form $4n+3$. If $T^2-DU^2=-1$ be resoluble and $[T_1, U_1]$ be its least solution, the formula $[T_{2n+1}, U_{2n+1}]$ contains all its solutions, and $[T_{2n}, U_{2n}]$ all the solutions of $T^2-DU^2=1$. If, in addition to the supposition that $T^2-DU^2=-1$ is resoluble, we suppose that $T^2-DU^2=4$ admits of uneven solutions, $T^2-DU^2=-4$ will also admit of uneven solutions; and if $[T_1, U_1]$ be its least solution,

$$[T_{2n+1}, U_{2n+1}],\quad [T_{2n}, U_{2n}],\quad [\tfrac{1}{2}T_{6n+3}, \tfrac{1}{2}U_{6n+3}],\quad [\tfrac{1}{2}T_{6n}, \tfrac{1}{2}U_{6n}]$$

will represent all the solutions of $T^2-DU^2=-4$, $=4$, $=-1$, and $=1$, respectively. It is evident that these considerations will frequently serve to abbreviate the process of finding the least solution of $T^2-DU^2=1$. (See a memoir of Euler's in the Comment. Arith., vol. ii. p. 35.)

(viii.) The 'Canon Pellianus' of Degen (Havniæ, 1817) contains a Table, giving for every not square value of D less than 1000, the least solution of the equation $T^2-DU^2=1$, together with the development of $\sqrt{D}$ in a continued fraction. Its arrangement will be seen in the following specimens:—

357	18, 1, 8, (2) 1, 33, 4, 17 180 3401
97	9, 1, 5, 1, 1, (1, 1) 1, 16, 3, 11, 8, (9, 9) 6377352, 62809633.

The numbers in the third and fourth rows are the least values of U and T in the equation $T^2-DU^2=1$. The first row of numbers is the period of integral quotients in the development of $\sqrt{D}$: it is continued only as far as the middle quotient, or the two middle quotients, after which the same quotients recur in an inverse order. Thus,

$$180 = (1, 8, 2, 8, 1);$$
$$3401 = (18, 1, 8, 2, 8, 1);$$
$$6377352 = (1, 5, 1, 1, 1, 1, 1, 1, 5, 1, 18, \ 1, 5, 1, 1, 1, 1, 1, 1, 5, 1);$$
$$62809633 = (9, 1, 5, 1, 1, 1, 1, 1, 1, 5, \ 1, 18, 1, 5, 1, 1, 1, 1, 1, 1, 5, 1).$$

The numbers in the second row are the denominators of the complete quotients; *i.e.* taken alternately positively and negatively, they are the extreme coefficients in the equations of the period. Thus the period of equations for $\surd 357$ is $[-33,-18,1]$, $[1,18,-33]$, $[-33,-15,4]$, $[4,17,-17]$, $[-17,-17,4]$, $[4,15,-33]$. The first half of the period of equations for $\surd 97$ is $[-16,-9,1]$, $[1,+9,-16]$, $[-16,-7,3]$, $[3,8,-11]$, $[-11,-3,8]$, $[8,5,-9]$, $[-9,-4,9]$, $[9,+5,-8]$, $[-8,-3,11]$, $[11,8,-3]$, $[-3,-7,16]$; the second half being composed of the same equations in the same order but with their signs changed. The middle coefficients of the equations are not given in the Table; but if

$$[a_\lambda,\ \beta_\lambda,\ a_{\lambda+1}],\quad [a_{\lambda+1},\ \beta_{\lambda+1},\ a_{\lambda+2}]$$

be two consecutive equations, of which the former determines the integral quotient μ_λ, they may be successively calculated by the formula

$$\beta_{\lambda+1}=\mu_\lambda\, a_{\lambda+1}+\beta_\lambda.$$

Lagrange has proved that if $x^2-Dy^2=H$, and H be $<\surd D$, $\frac{x}{y}$ is always a convergent to $\surd D$; so that a number less than $\surd D$ is or is not capable of representation by the principal form of det. D, according as it is or is not included among the numbers of the second row.

The second Table of the 'Canon' contains the least solution of the equation $T^2-DU^2=-1$ for those values of D less than 1000 for which that equation is resoluble.

Mr. Cayley (Crelle, vol. liii. p. 369) has calculated the least solution of the equation $T^2-DU^2=4$, or $T^2-DU^2=-4$, for every number D of the form $8n+5$ less than 1000, for which those equations are resoluble in uneven numbers. This Table, as well as Degen's second Table, is implicitly contained in the first Table of the 'Canon,' as appears from the theorem of Lagrange just cited.

(ix.) The theory of the equations $T^2-DU^2=1$ and $=4$ is connected in a remarkable manner with that of the division of the circle*. Let $\lambda=2\mu+1$ represent an uneven number divisible by k unequal primes, but having no square divisor; let also the numbers less than λ and prime to it be represented by a or b, according as they satisfy the equation $\left(\frac{a}{\lambda}\right)=1$, or $\left(\frac{b}{\lambda}\right)=-1$;

* See Dirichlet, 'Sur la manière de résoudre l'équation $t^2-pu^2=1$ au moyen des fonctions circulaires,' Crelle, vol. xvii. p. 286. Also Jacobi's note on the division of the circle, Crelle, vol. xxx. p. 173.

and let $X=0$ be the equation of the primitive λ-th roots of unity. The form of this equation (see Art. 59) implies that

$$\Sigma e^{\frac{2ai\pi}{\lambda}} + \Sigma e^{\frac{2bi\pi}{\lambda}} = (-1)^k;$$

we have also the relation

$$\Sigma e^{\frac{2ai\pi}{\lambda}} - \Sigma e^{\frac{2bi\pi}{\lambda}} = i^{\mu^2}\sqrt{\lambda},$$

which is easily deducible from the formulae of Gauss (see Arts. 20 and 104 of this Report, or Dirichlet, Crelle, vol. xxi. pp. 141, 142). From these values of $\Sigma e^{\frac{2ai\pi}{\lambda}}$ and $\Sigma e^{\frac{2bi\pi}{\lambda}}$ we infer that $2\Pi(x-e^{\frac{2ai\pi}{\lambda}})$ and $2\Pi(x-e^{\frac{2bi\pi}{\lambda}})$ are two quantities of the form $Y+i^{\mu^2}Z\sqrt{\lambda}$, and $Y-i^{\mu^2}Z\sqrt{\lambda}$, Y and Z denoting integral functions of x with integral coefficients; *i.e.* that $4X=Y^2-(-1)^{\mu}\lambda Z^2$. From this equation, which is a generalisation of that obtained by Gauss for the case when λ is a prime (Disq. Arith., Art. 357), we can deduce a solution of the equation $T^2-\lambda Y^2=4$. In the formula $2\Pi(x-e^{\frac{2ai\pi}{\lambda}})=Y+i^{\mu^2}Z\sqrt{\lambda}$, let us first write i for x, and then $-i$ for i, and let us denote by X_i, Y_i, Z_i, X_{-i}, Y_{-i}, Z_{-i} the values which X, Y, and Z acquire when i and $-i$ are written for x. We thus find, denoting the number of numbers less than λ and prime to it by λ′,

$$4\Pi.(i-e^{\frac{2ai\pi}{\lambda}})(-i-e^{\frac{2ai\pi}{\lambda}}) = 2^{\lambda'+2}\Pi\cos^2\left(\frac{\pi}{4}+\frac{a\pi}{\lambda}\right)$$
$$=[Y_i Y_{-i}+\lambda Z_i Z_{-i}]+\sqrt{\lambda}[i^{\mu^2}Z_i Y_{-i}+i^{-\mu^2}Z_{-i}Y_i];$$

or, writing

$$T \text{ for } \tfrac{1}{2}[Y_i Y_{-i}+\lambda Z_i Z_{-i}], \quad U \text{ for } \tfrac{1}{2}[i^{\mu^2}Z_i Y_{-i}+i^{-\mu^2}Z_{-i}Y_i],$$

and observing that $X_i X_{-i}=1$,

$$\tfrac{1}{2}(T+U\sqrt{\lambda})=2^{\lambda'}\Pi\cos^2\left(\frac{\pi}{4}+\frac{a\pi}{\lambda}\right), \quad T^2-\lambda U^2=4,$$

where it is easily seen that T and U are integral numbers. When μ is even, we may obtain a solution of the equation more simply by writing $+1$ or -1 for x. (See the notes of Jacobi and Dirichlet already referred to.)

It is to be observed, however, that the solution obtained by these methods is not in general the least solution. Its ordinal place in the series of solutions depends (as we shall hereafter see) on the number of classes of forms of det. D.

97. *Solution of the General Indeterminate Equation of the second degree.*—The solution of the indeterminate equation

$$ax^2+2bxy+cy^2+2dx+2ey+f=0$$

depends on the problem of the representation of a given number by a quadratic form. We confine ourselves to the case which presents the greatest complexity, that in which $b^2-ac=D$ is a positive and not square number. The methods of solution contained in Euler's Memoirs relating to it (see Comment. Arith., vol. i. pp. 4, 297, 549, 570, vol. ii. p. 263; and the Algebra, vol. ii. cap. vi.) are incomplete in several respects: first, because Euler always assumes that a single solution is known, and only proposes to deduce all the solutions from it; secondly, because it is not possible, from a given solution, to deduce any other solutions than those which belong to the same set with the given solution, whereas the equation may admit of solutions belonging to different sets; and lastly, because he gives no method for distinguishing between the integral and fractional values contained in the formulae by which x and y are expressed. The first complete solution of the problem was given by Lagrange in his Memoir 'Sur la solution des Problèmes Indéterminés du second degré' (Hist. de l'Académie de Berlin for 1767, vol. xxiii. pp. 165–311). But the following method of solution, which is different in some respects and much simpler, will be found in a subsequent memoir, 'Nouvelle méthode pour résoudre les problèmes indéterminés en nombres entiers' (Hist. de l'Académie de Berlin for 1767, vol. xxiv. p. 181); and in the Additions to Euler's Algebra (paragraph 7). If we multiply by aD and write p for $ax+by+d$, q for $(b^2-ac)(y+bd-ae)$, M for $(bd-ae)^2-(b^2-ac)(d^2-af)$, the given equation becomes $q^2-Dp^2=M$. Confining ourselves to the primitive representations of M by q^2-Dp^2 (the derived representations, corresponding to the different square divisors of M, are to be treated separately by the same method), we see that, since p and M are prime, q is of the form $Mr+\Omega p$, where r and Ω are two new indeterminates of which the latter may be supposed $<[\frac{1}{2}M]$. On substituting this value for q, it will appear that $N=\dfrac{\Omega^2-D}{M}$ is necessarily integral, *i.e.* that Ω is one of the roots of the congruence $\Omega^2-D=0$, mod M; and the equation will assume the form

$$Np^2+2\Omega pr+Mr^2=1,$$

in which every admissible value of Ω is to be employed in succession. The development of either root of the equation $N+2\Omega\theta+M\theta^2=0$ will give all the values of p and r which satisfy the equation

$$Np^2+2\Omega pr+Mr^2=1,$$

because 1 is the *minimum* value which the form (N, Ω, M) can assume. (See the Additions, paragraph 2, and especially Arts. 33–35.) Or again, if we apply the transformation of Art. 92 to the form (N, Ω, M), we obtain an equation of the type $Px'^2+2Qx'y'+Ry'^2=1$, in which $Q^2-PR=D$, and $P<\sqrt{D}$; whence, if $x''=Px'+Qy'$, we finally deduce $x''^2-Dy'^2=P$, all the solutions of which (see Art. 96, viii.) are necessarily given by the development of $\sqrt{D}$ in a continued fraction. Applying either of these methods (the latter is not given in the Memoir, but only in the Additions to Euler's Algebra) to every equation of the form $Np^2+2\Omega p^2+Mr^2=1$ which can be deduced from the equation $q^2-Dp^2=M$, or from the equations of similar form obtained by replacing M by the quotient which it leaves when divided by any one of its square divisors, we obtain a finite number of formulae of the type

$$x=\frac{\alpha T+\beta U+\gamma}{\delta},\qquad y=\frac{\alpha' T+\beta' U+\gamma'}{\delta'},$$

$[T, U]$ denoting any solution of the equation $T^2-DU^2=1$. These formulae are fractional; but by attending to the principle of Art. 96, v., we can ascertain for each pair of formulae whether they contain any integral values or not, and if they do contain any, we can substitute for the single pair of fractional formulae a finite number of pairs not containing any fraction.

The form in which the solution of this problem has been exhibited by Gauss is remarkable for its elegance. Let

$$\begin{vmatrix} a, & b, & d \\ b, & c, & e \\ d, & e, & f \end{vmatrix}=\Delta,$$

and, representing by δ the greatest common divisor of b^2-ac, $cd-be$, $ae-bd$, let

$$\frac{D}{\delta}=D',\qquad \frac{\Delta}{\delta}=\Delta',\qquad \frac{cd-be}{\delta}=p,\qquad \frac{ae-bd}{\delta}=q,$$

then, putting $D'x=X+p$, $D'y=Y+q$, we find $aX^2+2bXY+cY^2=D'\Delta'$. If $[X_n, Y_n]$ denote indefinitely any representation of $D'\Delta'$ by (a, b, c), we have only to separate (by Lagrange's method) those values of X_n, Y_n which satisfy the congruences $X_n+p\equiv 0$, $Y_n+q\equiv 0$, mod D', from those which do not, and we shall obtain a finite number of formulae, exhibiting the complete solution required.

98. *Distribution of Classes into Orders and Genera.*—The classes of forms of any given positive or negative determinant D are divided by Gauss into Orders, and the classes belonging to each order into Genera. Two classes, represented by the forms (a, b, c), (a', b', c'), belong to the same order, when the greatest common divisors of a, b, c and $a, 2b, c$ are respectively equal to those of a', b', c', and of $a', 2b', c'$. Thus the properly primitive

classes form an order by themselves; and the improperly primitive classes form another order. To obtain the subdivision of orders into genera, it is only necessary to consider the primitive classes; because we can deduce the subdivision of a derived order of classes from the subdivision of the primitive order from which it is derived. The subdivision into genera of the order of properly primitive classes depends on the principles contained in the following equations, in which q is an uneven prime dividing D, m and m' uneven numbers prime to q, and capable of representation by the same properly primitive form of determinant D.

(i.) $\left(\frac{m}{q}\right) = \left(\frac{m'}{q}\right)$.

(ii.) If $D \equiv 3$, mod 4, $(-1)^{\frac{1}{2}(m-1)} = (-1)^{\frac{1}{2}(m'-1)}$.

(iii.) If $D \equiv 2$, mod 8, $(-1)^{\frac{1}{8}(m^2-1)} = (-1)^{\frac{1}{8}(m'^2-1)}$.

(iv.) If $D \equiv 6$, mod 8, $(-1)^{\frac{1}{2}(m-1)+\frac{1}{8}(m^2-1)} = (-1)^{\frac{1}{2}(m'-1)+\frac{1}{8}(m'^2-1)}$.

(v.) If $D \equiv 4$, mod 8, $(-1)^{\frac{1}{2}(m-1)} = (-1)^{\frac{1}{2}(m-1)}$.

(vi.) If $D \equiv 0$, mod 8,

$$(-1)^{\frac{1}{2}(m-1)} = (-1)^{\frac{1}{2}(m'-1)}; \text{ and } (-1)^{\frac{1}{8}(m^2-1)} = (-1)^{\frac{1}{8}(m'^2-1)}.$$

The interpretation of these symbolic formulae is very simple. Thus, the formula (i.) expresses that—

'The numbers prime to any prime divisor q of D which can be represented by f, the same properly primitive form of det. D, are either all quadratic residues of q, or else all quadratic non-residues.'

Again, the formula (iv.) expresses that—

'If D be of the form $8n+6$, the uneven numbers that can be represented by f are either all included in one of the two forms $8n+1$, $8n+3$, or else in one of the two forms $8n-1$, $8n-3$.'

All the formulae are deducible by the most elementary considerations from the three equations

$$m = ax^2 + 2bxy + cy^2, \qquad m' = ax'^2 + 2bx'y' + cy'^2,$$

$$(ax^2+2bxy+cy^2)(ax'^2+2bx'y'+cy'^2) = \{(axx'+b[xy'+x'y]+cyy')^2 - D(xy'-x'y)\}^2.$$

Thus we find immediately

$$\left(\frac{mm'}{q}\right) = 1, \quad \text{or} \quad \left(\frac{m}{q}\right) = \left(\frac{m'}{q}\right).$$

And again, if $D \equiv 6$, mod 8, the last equation shows that $axx' + b\,[xy' + x'y] + cyy'$

is uneven ; and consequently

$$mm' \equiv 1 - 6(xy' - x'y)^2, \text{ mod } 8, \quad i.e.\ mm' \equiv +3, \text{ or } \equiv +1, \text{ mod } 8,$$

according as $xy' - x'y$ is uneven or even; whence m and m' are either both of the forms $8n+1$, $8n+3$, or else both of the forms $8n-1$, $8n-3$.

The form f is said to have the *particular character*

$$\left(\frac{f}{q}\right) = +1, \quad \text{or} \quad \left(\frac{f}{q}\right) = -1,$$

according as the numbers (prime to q) which are represented by it satisfy the equation

$$\left(\frac{m}{q}\right) = 1, \quad \text{or} \quad \left(\frac{m}{q}\right) = -1;$$

and we are to understand in the same way the expressions that f has the particular character $(-1)^{\frac{1}{8}(f^2-1)} = +1$, or $= -1$, &c.
Every particular character of a form belongs equally to all forms of the same class, and is therefore termed a particular character of the class. The complex of the particular characters of a form or class constitutes its *complete* or *generic character*; and those classes which have the same complete character are considered to belong to the same genus: so that the complete character of a form is possessed not only by every form of the same class, but by every form of any class belonging to the same genus.

To enable the reader to form with facility the complete character of any given properly primitive class, we add the following Table, taken from Dirichlet (Crelle, vol. xix. p. 338), in which S^2 denotes the greatest square dividing D; P or $2P$ is the quotient $\frac{D}{S^2}$, according as that quotient is uneven or even; $p, p', \ldots$ are the prime divisors of P; and r, r' the uneven primes dividing S, but not P.

I. $D = PS^2$, $P \equiv 1$, mod 4.

(α) $S \equiv 1$, mod 2.

$$\left(\frac{f}{p}\right), \left(\frac{f}{p'}\right), \ldots, \quad \Big| \quad \left(\frac{f}{r}\right), \left(\frac{f}{r'}\right), \ldots.$$

(β) $S \equiv 2$, mod 4.

$$\left(\frac{f}{p}\right), \left(\frac{f}{p'}\right), \ldots, \quad \Big| \quad (-1)^{\frac{1}{2}(f-1)}, \left(\frac{f}{r}\right), \left(\frac{f}{r'}\right), \ldots.$$

(γ) $S \equiv 0$, mod 4.

$$\left(\frac{f}{p}\right), \left(\frac{f}{p'}\right), \ldots, \quad \Big| \quad (-1)^{\frac{1}{2}(f-1)}, (-1)^{\frac{1}{8}(f^2-1)}, \left(\frac{f}{r}\right), \left(\frac{f}{r'}\right), \ldots.$$

II. $D=PS^2$, $P\equiv 3$, mod 4.

(α) $S\equiv 1$, mod 2.

$$(-1)^{\frac{1}{2}(f-1)},\ \left(\frac{f}{p}\right),\ \left(\frac{f}{p'}\right),\ldots,\ \Big|\ \left(\frac{f}{r}\right),\ \left(\frac{f}{r'}\right),\ldots.$$

(β) $S\equiv 2$, mod 4.

$$(-1)^{\frac{1}{2}(f-1)},\ \left(\frac{f}{p}\right),\ \left(\frac{f}{p'}\right),\ldots,\ \Big|\ \left(\frac{f}{r}\right),\ \left(\frac{f}{r'}\right),\ldots.$$

(γ) $S\equiv 0$, mod 4.

$$(-1)^{\frac{1}{2}(f-1)},\ \left(\frac{f}{p}\right),\ \left(\frac{f}{p'}\right),\ldots,\ \Big|\ (-1)^{\frac{1}{8}(f^2-1)},\ \left(\frac{f}{r}\right),\ \left(\frac{f}{r'}\right),\ldots.$$

III. $D=2PS^2$, $P\equiv 1$, mod 4.

(α) $S\equiv 1$, mod 2.

$$(-1)^{\frac{1}{8}(f^2-1)},\ \left(\frac{f}{p}\right),\ \left(\frac{f}{p'}\right),\ldots,\ \Big|\ \left(\frac{f}{r}\right),\ \left(\frac{f}{r'}\right),\ldots.$$

(β) $S\equiv 0$, mod 2.

$$(-1)^{\frac{1}{8}(f^2-1)},\ \left(\frac{f}{p}\right),\ \left(\frac{f}{p'}\right),\ldots,\ \Big|\ (-1)^{\frac{1}{2}(f-1)},\ \left(\frac{f}{r}\right),\ \left(\frac{f}{r'}\right),\ldots.$$

IV. $D=2PS^2$, $P\equiv 3$, mod 4.

(α) $S\equiv 1$, mod 2.

$$(-1)^{\frac{1}{2}(f-1)+\frac{1}{8}(f^2-1)},\ \left(\frac{f}{p}\right),\ \left(\frac{f}{p'}\right),\ldots,\ \Big|\ \left(\frac{f}{r}\right),\ \left(\frac{f}{r'}\right),\ldots.$$

(β) $S\equiv 0$, mod 2.

$$(-1)^{\frac{1}{2}(f-1)},\ (-1)^{\frac{1}{8}(f^2-1)},\ \left(\frac{f}{p}\right),\ \left(\frac{f}{p'}\right),\ldots,\ \Big|\ \left(\frac{f}{r}\right),\ \left(\frac{f}{r'}\right),\ldots.$$

It appears from this Table, that if μ be the number of uneven primes which divide D, the total number of generic characters that can be formed by combining the particular characters in every possible way is 2^μ when $D\equiv 1$ or 5, mod 8; $2^{\mu+2}$ when $D\equiv 0$, mod 8; and $2^{\mu+1}$ in every other case. But it follows from the law of quadratic reciprocity, that one-half of these complete characters are impossible; *i.e.* that no quadratic form characterised by them can exist. To see this, we observe that if m be a positive and uneven number prime to D, and capable of primitive representation by f, the congruence $\Omega^2-D\equiv 0$, mod m, is resoluble; and consequently $\left(\frac{D}{m}\right)=+1$. Therefore also $\left(\frac{P}{m}\right)=1$, or $\left(\frac{2P}{m}\right)=1$, according as D is of the form PS^2 or $2PS^2$. In the first case we have

$$\left(\frac{m}{P}\right)=(-1)^{\frac{1}{4}(m-1)(P-1)}, \text{ or } \left(\frac{m}{p}\right)\left(\frac{m}{p'}\right)\ldots=(-1)^{\frac{1}{4}(m-1)(P-1)};$$

in the other case $\left(\frac{2}{m}\right)\left(\frac{P}{m}\right)=1$; *i.e.*

$$\left(\frac{m}{P}\right)=(-1)^{\frac{1}{4}(m-1)(P-1)+\frac{1}{8}(m^2-1)}, \text{ or } \left(\frac{m}{p}\right)\left(\frac{m}{p'}\right)\ldots=(-1)^{\frac{1}{4}(m-1)(P-1)+\frac{1}{8}(m^2-1)}.$$

A comparison of these equations with the preceding Table will show that the product of the particular characters which stand before the line of division in the Table is equal to $+1$ in the case of any really existing genus; *i.e.* that precisely one-half of the whole number of complete generic characters are impossible. We shall hereafter see that the remaining half of the generic characters correspond to actually existing genera, and that each genus contains an equal number of classes. That genus, every particular character of which is a positive unit, is called the *principal* genus; it evidently contains the principal class, and is therefore, in every case, an actually existing genus.

Since the extreme coefficients of a form are numbers represented by it, and since, further, if the form be properly primitive, one or other of them is prime to 2 and to any prime divisor of the determinant, we see that the generic character of a form can always be ascertained by considering the values of its first and last coefficients. Thus the complete character of the form (11, 2, 15), of which the det. is $-161=-7\times 23$ (case II. (*a*) in the Table), is

$$\left(\frac{f}{7}\right)=1, \quad \left(\frac{f}{23}\right)=-1, \quad (-1)^{\frac{1}{2}(f-1)}=-1;$$

that of (5, 2, 33), of the same determinant, is

$$\left(\frac{f}{7}\right)=-1, \quad \left(\frac{f}{23}\right)=-1, \quad (-1)^{\frac{1}{2}(f-1)}=+1.$$

Two forms, which have different generic characters, cannot be equivalent; nor can a number be represented by a form if its character is incompatible with the generic character of the form. It is therefore convenient, in any problem of equivalence or representation, to begin by comparing the generic characters of the given forms with one another, or with the characters of the given numbers.

The uneven numbers prime to the determinant, which are represented by forms of the same genus, are contained in one or other of a certain number of linear forms. If R denote the product of the primes $r, r', \ldots$ already defined, and if θ be any term of a system of residues prime to 2^kPR, where k is $=1$, when $D \equiv 1$ or 5, mod 8, is $= 3$ when $D \equiv 2$, 6, or 0, and is $= 2$ in every other case, the numbers contained in the formula $2^k PR + \theta$ can be represented only by forms belonging to that genus the character of which coincides with the character of the number θ. It is clear that one half of the linear forms, included in the formula $2^k PR + \theta$, do not satisfy the condition of possibility indicated in the Table, and are therefore incompatible with any quadratic form of determinant D; while the remaining half of those linear forms will be equally distributed among the actually existing genera; so that there will be either

$$\Pi \{\tfrac{1}{2}(p-1) . \tfrac{1}{2}(r-1)\} \quad \text{or} \quad 2\Pi \{\tfrac{1}{2}(p-1) . \tfrac{1}{2}(r-1)\}$$

linear forms proper to each genus. But although no number contained in any one of the first-named linear forms can be represented by a form of determinant D, yet it is not to be inferred that every number m contained in the other half of the linear forms is capable of such representation; for from the linear form of m, we can indeed infer the equation $\left(\frac{D}{m}\right) = 1$; but, if m be not a prime, or at least the product of a prime by a square, we cannot from this equation infer the resolubility of the congruence $\Omega^2 \equiv D$, mod m, or of any congruence of the form $\Omega^2 \equiv D$, mod $\frac{m}{d^2}$. We may add that if we assume the theorem that every arithmetic progression, the terms of which are prime to their common difference, contains prime numbers, the consideration of the case in which m is a prime establishes the actual existence of every genus the character of which satisfies the condition of possibility. (Crelle, vol. xviii. p. 269.)

If m be an uneven number not divisible by q, a prime divisor of D, and if the double of m can be represented by an improperly primitive form f of det. D, we attribute to f the particular character $\left(\frac{f}{q}\right) = +1$, or $= -1$, according as $\left(\frac{m}{q}\right) = +1$, or $= -1$; and to form the complete character of f, we may use the Table

$$D = PS^2, \quad P \equiv 1, \text{ mod } 4, \quad S \equiv 1, \text{ mod } 2.$$

$$\left(\frac{f}{p}\right), \left(\frac{f}{p'}\right), \ldots, \quad \Big| \quad \left(\frac{f}{r}\right), \left(\frac{f}{r'}\right), \ldots *.$$

---

* All the results of this article are given in the Disq. Arith., Arts. 223–232; but as Gauss does not employ the symbol of reciprocity, we have preferred to follow the notation of Dirichlet. It is also to be noticed that Gauss does not use the law of quadratic reciprocity to demonstrate the impossibility

99. In the preceding Articles we have briefly recapitulated the definitions and principles which constitute the elements of the theory of quadratic forms. We have hitherto followed closely the 5th section of the Disq. Arith. (Arts. 153–222 and 223–233); but before we proceed to an examination of the remainder of that section, it will be convenient to place before the reader an account of the method employed by Lejeune Dirichlet in his great memoir, 'Recherches sur diverses applications de l'analyse infinitésimale à la théorie des nombres,' for the determination of the number of quadratic forms of a given positive or negative determinant.

It appears from the Additamenta to Art. 306, X. of the Disq. Arith., that Gauss, at the time of the publication of that work, had already succeeded in effecting this determination; and the method by which he effected it will at length appear in the second volume of the complete edition of his works, the publication of which is now promised by the Society of Göttingen. Nevertheless the originality of Dirichlet in this celebrated investigation is unquestionable, as there is nothing whatever in the Disq. Arith. to suggest either the form of the result, or the method by which it is obtained *.

---

of one-half of the generic characters; for, as we shall hereafter see, this impossibility is proved in the Disq. Arith. (Art. 261) independently of the law of reciprocity, and is then employed to establish that law. (Gauss's second demonstration, see Disq. Arith., Art. 262.) There is also an unimportant difference between Dirichlet and Gauss with respect to the definition of the generic character of an improperly primitive form; for Gauss obtains the generic character (see Art. 232) by considering the numbers represented by the form, and not the halves of those numbers. But he also observes (Arts. 227, and 256, VI.) that each improperly primitive class is connected in a particular manner (to which we shall again refer) with one or with three properly primitive classes; and that this consideration may be employed to divide the improperly primitive classes into genera. And it will be found that the complete character which Dirichlet's definition attributes to an improperly primitive form is, in fact, the complete character of the properly primitive class or classes with which it is connected.

* The following is a list of the papers of Lejeune Dirichlet which relate to the theory of quadratic forms:—

1. Sur l'usage des séries infinies dans la théorie des nombres.—Crelle, vol. xviii. p. 259.

2. Recherches sur diverses applications de l'analyse infinitésimale à la théorie des nombres.—Crelle, vol. xix. p. 324, and xxi. pp. 1, 134.

3. Auszug aus einer der Akademie der Wissenschaften zu Berlin am 5 März 1840 vorgelesenen Abhandlung. (Crelle, vol. xxi. p. 98, or the Monatsberichte for 1840, p. 49.)

This paper is an abstract of an unpublished memoir containing the demonstration of the theorem that every properly primitive form represents an infinite number of primes.

4. Untersuchungen über die Theorie der complexen Zahlen. (Crelle, vol. xxii. p. 375, or in the Monatsberichte for 1841, p. 190) An abstract of the following memoir.

We propose, in what follows, to give as full an analysis as our limits will permit of the contents of the memoir. Its first section contains certain principles relative to the theory of series.

(i.) 'If $k_1 \leqq k_2 \leqq k_3 \leqq k_4 \ldots$ be a series of continually increasing positive quantities; and if the ratio $\frac{n}{k_n}$ continually tend to a finite limit a (that is to say, if, δ denoting a given positive quantity, however small, we can always assign a finite value of $n = N$, such that for all values of n surpassing N the inequalities

$$a - \delta < \frac{n}{k_n} < a + \delta$$

are satisfied), the limit of the expression $\rho \sum_{n=1}^{n=\infty} \frac{1}{k_n^{1+\rho}}$, when the positive quantity ρ is diminished without limit, is a.' *

For
$$\rho \sum_{n=1}^{n=\infty} \frac{1}{k_n^{1+\rho}} = \rho \sum_{n=1}^{n=N} \frac{1}{k_n^{1+\rho}} + \rho \sum_{n=N+1}^{n=\infty} \frac{1}{k_n^{1+\rho}},$$

N denoting a finite number; and by virtue of the inequalities written above

$$\sum_{n=N+1}^{n=\infty} \frac{(a-\delta)^{1+\rho}}{n^{1+\rho}} < \sum_{n=N+1}^{n=\infty} \frac{1}{k_n^{1+\rho}} < \sum_{n=N+1}^{n=\infty} \frac{(a+\delta)^{1+\rho}}{n^{1+\rho}}.$$

Observing that $\lim \rho \sum_{n=N+1}^{n=\infty} \frac{1}{n^{1+\rho}}$ is intermediate between

$$\lim \rho \int_{N+1}^{\infty} \frac{dx}{x^{1+\rho}} \text{ and } \lim \rho \int_{N}^{\infty} \frac{dx}{x^{1+\rho}},$$

---

5. Recherches sur les formes quadratiques à coëfficients et à indéterminés complexes.—Crelle, vol. xxiv. p. 291.

6. Sur un théorème relatif aux séries. (Liouville, New Series, vol. i. p. 80, or Crelle, vol. liii. p. 130.)

7. Sur une propriété des formes quadratiques à déterminant positif. (Monatsberichte for July 16, 1855, or Liouville, New Series, vol. i. p. 76, or Crelle, vol. liii. p. 127.)

8. Vereinfachung der Theorie der binären quadratischen Formen von positiver Determinante. (Memoirs of the Berlin Academy for 1854, p. 99, or, with additions by the author, in Liouville, New Series, vol. ii. p. 353.)

9. Démonstration nouvelle d'une proposition relative à la théorie des formes quadratiques.—Liouville, New Series, vol. ii. p. 273.

10. De formarum binarium secundi gradus compositione.—Crelle, vol. xlvii. p. 155.

The three last papers contain important simplifications of theories which appear in a very complicated form in the Disq. Arith. To two of them we have already referred (Arts. 93, 94).

* This theorem is a generalisation of that in the memoir (Crelle, vol. xix. p. 326). It is given by Dirichlet in No. 6. of the preceding list.

and is consequently unity, we infer from the last inequalities that

$$\lim \rho \sum_{n=N+1}^{n=\infty} \frac{1}{k_n^{1+\rho}},$$

and therefore also

$$\lim \rho \sum_{n=1}^{n=\infty} \frac{1}{k_n^{1+\rho}},$$

which is identical with it, because

$$\lim \rho \sum_{n=1}^{n=N} \frac{1}{n^{1+\rho}} = 0$$

differs from a by a quantity comminuent with δ; *i.e.*

$$\lim \rho \sum_{n=1}^{n=\infty} \frac{1}{k_n^{1+\rho}} = a,$$

since by hypothesis δ is a quantity as small as we please.

(ii.) A convergent infinite series may be convergent in two very different ways. It may be convergent, and always have the same sum irrespective of the arrangement of its terms; or it may be convergent for certain arrangements of its terms, giving the same or different sums for these different arrangements, and divergent for other arrangements. We suppose, however, that we consider only such different arrangements of the terms of a series as are compatible with the condition that any term which occupies a *finitesimal* place in any one arrangement should occupy a *finitesimal* place in every other arrangement *. Thus the series

$$\frac{1}{1^{1+\rho}} + \frac{1}{2^{1+\rho}} + \frac{1}{3^{1+\rho}} + \dots, \ \rho > 0,$$

is convergent, and has the same sum in whatever order we sum its terms; but of the two series

* This condition is necessary, because without it the sum of no series whatever would be independent of the arrangement of its terms, if by the sum of a series we understand the limit to which we approximate by the continual addition of its terms in the order in which they are given. For example, the series cited in the text,

$$\frac{1}{1^{1+\rho}} + \frac{1}{2^{1+\rho}} + \frac{1}{3^{1+\rho}} + \dots, \ \rho>0,$$

is convergent, and its sum is irrespective of the arrangement of its terms, provided that arrangement satisfy the condition enunciated in the text. But if we were to arrange the terms of the series in an order regulated (say) by the number of primes dividing their denominators, the limit to which we should continually approach by adding together the terms taken in their new order would be $\Sigma \frac{1}{p^{1+\rho}}$, in which p denotes any prime, and not $\Sigma \frac{1}{n^{1+\rho}}$, in which n denotes any integer.

$$1-\frac{1}{2^{\frac{1}{2}}}+\frac{1}{3^{\frac{1}{2}}}-\frac{1}{4^{\frac{1}{2}}}+\frac{1}{5^{\frac{1}{2}}}-\frac{1}{6^{\frac{1}{2}}}+\dots,$$

$$1+\frac{1}{3^{\frac{1}{2}}}-\frac{1}{2^{\frac{1}{2}}}+\frac{1}{5^{\frac{1}{2}}}+\frac{1}{7^{\frac{1}{2}}}-\frac{1}{4^{\frac{1}{2}}}+\dots,$$

only the first is convergent; while the two series

$$1-\frac{1}{2}+\frac{1}{3}-\frac{1}{4}+\frac{1}{5}-\frac{1}{6}+\dots,$$

$$1+\frac{1}{3}-\frac{1}{2}+\frac{1}{5}+\frac{1}{7}-\frac{1}{6}+\dots$$

are both convergent, but have two very different sums *.

These observations will show the importance of the following proposition †:—

'If c_n be a periodic function of n, satisfying the equations

$$c_{n+k}=c_n,$$

$$c_1+c_2+c_3+\dots+c_k=0,$$

the series $\sum\limits_{n=1}^{n=\infty}\frac{c_n}{n^s}$ in which the terms are taken in their natural order, is convergent for all values of s superior to zero, and its sum is a continuous function of s.'

For if we add together the k consecutive terms

$$\frac{c_1}{(km+1)^s}+\frac{c_2}{(km+2)^s}+\dots+\frac{c_k}{(km+k)^s},$$

we obtain a fraction of which the denominator is of the order ks in respect of m, while the numerator is only of the order $(k-1)s-1$, because the coefficient of $m^{(k-1)s}$ is zero. We may therefore replace the given series by a series of the form $\sum\limits_{m=1}^{m=\infty}\frac{1}{\phi(m)}$, in which $\phi(m)$ is a function of the order $1+s$ in respect of m. This series is always convergent for positive values of s; its convergence is irrespective of the arrangement of its terms, and its sum is a continuous function of s, because $\phi(m)$ is a continuous function of s. The given series is therefore also convergent, and its sum is a continuous function of s.

100. The second section of the memoir refers to the symbols of reciprocity of Jacobi and Legendre (Arts. 15, 16, and 17 of this Report).

---

* These illustrations are taken from the Memoir on the Arithmetical Progression in the Berlin Memoirs for 1837, pp. 48 and 49.

† The demonstration in the text is a little simpler than that given by Dirichlet, who uses the function Γ to express the sum of the series.

The third and fourth sections contain the principal theorems relating to the generic characters of quadratic forms, and to the representation of numbers. There is only one of these theorems to which we need direct our attention here, as the others have already come before us in the preceding articles.

Let (a, b, c) be a primitive form of the positive determinant D; let also $(a, b, c)(x_0, y_0) = M$ a positive number represented by (a, b, c); m the greatest common divisor of $a, 2b, c$; $[T, U]$ the least positive solution of $T^2 - DU^2 = m^2$; so that if

$$x_n = \frac{1}{m}[T_n x_0 - U_n(bx_0 + cy_0)], \qquad y_n = \frac{1}{m}[T_n y_0 + U_n(ax_0 + by_0)],$$

the two formulae $[x_n, y_n]$ and $[-x_n, -y_n]$ will together express every representation of M, which belongs to the same set as $[x_0, y_0]$. Similarly, let $[x'_n, y'_n]$, $[-x'_n, -y'_n]$ denote a complete set of representations of the positive number M' by (a, b, c).

If we trace the hyperbola represented by the equation $ax^2 + 2bxy + cy^2 = 1$ referred to rectangular axes, the diameters included in the formula $y = \frac{y_k}{x_k}x$, in which k is to receive all values from $-\infty$ to $+\infty$, will form a pencil of lines, which all meet the curve, and which, commencing with the asymptote $y = -\frac{a}{\sqrt{D}+b}x$, continually recede from it, and approximate to the asymptote $y = \frac{a}{\sqrt{D}-b}x$. The sectorial area contained between any two consecutive lines of this pencil and either branch of the hyperbola is constant and equal to $\frac{1}{2} \cdot \frac{1}{\sqrt{D}} \log \frac{(T+U\sqrt{D})}{m}$; as may be ascertained by employing polar coordinates. Since the same observations apply to the pencil $y = \frac{y'_n}{x'_n}x$, we infer that the lines of these two pencils lie alternately, unless the two pencils coincide. Let us now suppose that in the form (a, b, c), a is positive and c negative; so that the axis of x does, and the axis of y does not cut the curve. On this supposition the values of $\frac{y_n}{x_n}$ and of $\frac{y'_n}{x'_n}$ continually increase from $-\frac{a}{\sqrt{D}+b}$ to $\frac{a}{\sqrt{D}-b}$ as n increases from $-\infty$ to $+\infty$. The alternate position of the lines of the two pencils gives, in this case, the theorem,—

'The inequalities

$$\frac{y'_k}{x'_k} < \frac{y_n}{x_n} \leq \frac{y'_{k+1}}{x'_{k+1}},$$

in which k represents any given number, are satisfied for one value of n, and one only.' If, taking a for M' and [1, 0] for $[x'_0, y'_0]$, we put $k=0$, we obtain the conclusion,—

'Each set of representations of the positive number M by the form (a, b, c), in which a is positive and c negative, contains one and only one representation which satisfies the inequalities

$$x_n > 0,\ y_n > 0,\ y_n \leqq \frac{aU}{T-bU} x_n.\text{'}$$

It is in this form that the theorem appears in Dirichlet's memoir. We may add that any values of x and y which satisfy these inequalities will give a positive value to (a, b, c); for such a pair of values will correspond to a point situated in the internal angle between the asymptotes of the hyperbola.

The fifth section contains the demonstration of the theorem, that if Δ denote the absolute value of D, and $\psi(2\Delta)$ be the number of numbers less than 2Δ and prime to it, a properly primitive form of determinant D will acquire a value prime to $2D$, if its indeterminates x and y satisfy any one of a certain set of $2\Delta\psi(2\Delta)$ congruential conditions included among the $4\Delta^2$ conditions represented by the formulae

$$x=\alpha, \text{ mod } 2\Delta; \quad y=\beta, \text{ mod } 2\Delta,$$

in which both α and β represent any term of a complete system of residues, mod 2Δ; but will acquire a value not prime to $2D$, if x and y satisfy any of the other congruential conditions.

If the form be improperly primitive, the number of congruential conditions that will render its value unevenly even and prime to Δ will be $\Delta\psi(\Delta)$, or $3\Delta\psi(\Delta)$, according as $D\equiv 1$, or $\equiv 5$, mod 8.

These theorems are easily demonstrated by considering separately the prime divisors of Δ. For example, if the form (a, b, c) be improperly primitive, and p be a prime divisor of D, since either a or c is prime to p, let a be prime to p; then $(ax+by)^2-Dy^2$ will be prime to p, when $ax+by$ is so; *i.e.* it will be prime to p, for $p(p-1)$ combinations of the residues (mod p) of x and y; or, if p^n be the highest power of p dividing D, for $p^{2n-1}(p-1)$ combinations of the residues of x and y, mod p^n. Again, the 4 combinations of residues for the modulus 2 will give $\frac{1}{2}(a, b, c)$ the values

$$0,\ \tfrac{1}{2}a_2,\ \tfrac{1}{2}c_2,\ \tfrac{1}{2}a+b+\tfrac{1}{2}c,$$

of which it is easily seen that one or three will be uneven, according as $ac\equiv 0$, or 4, mod 8; *i.e.* according as $D \equiv 1$, or 5, mod 8. The combination of these results will give Dirichlet's theorem.

101. *Series expressing the number of Primitive Classes.*—The sixth section of the memoir contains the demonstration of the formulae which express in the form of an infinite series the number of classes of properly and improperly primitive quadratic forms of a given determinant. We shall abbreviate the demonstration of these formulae by using the theorem of Art. 87.

Let h be the number of properly primitive classes of determinant D; we shall first suppose D to be negative, and $= -\Delta$; let also

$$(a_1, b_1, c_1), \quad (a_2, b_2, c_2), \ldots, (a_h, b_h, c_h)$$

be a system of forms representing the properly primitive classes of that determinant; and let us consider the sum $S=$

$$\Sigma_1 \frac{1}{(a_1x^2+2b_1xy+c_1y^2)^s} + \Sigma_2 \frac{1}{(a_2x^2+2b_2xy+c_2y^2)^s} + \ldots + \Sigma_h \frac{1}{(a_hx^2+2b_hxy+c_hy^2)^s},$$

the sign of summation Σ_k extending to all values of x and y from $-\infty$ to $+\infty$, which give the form (a_k, b_k, c_k), a value prime to Δ. By the theorem of Art. 87, any uneven number n prime to Δ is capable of $2\Sigma\left(\frac{D}{d}\right)$ representations by the properly primitive forms of determinant D (for there are $\Sigma\left(\frac{D}{d}\right)$ sets of representations, and each set contains two*.) We have therefore the equation

$$S = 2\Sigma\left[\Sigma\left(\frac{D}{d}\right)\frac{1}{n^s}\right] \quad . \quad . \quad . \quad . \quad . \quad . \quad . \quad . \quad . \quad (a)$$

(the inner sign of summation referring to every divisor d of n; and the outer sign extending to every positive value of n prime to 2Δ). If we write n for d, and nn' for n, so that n and n' each represent any positive number prime to 2Δ, this equation assumes the simpler form

$$S = 2\Sigma\left(\frac{D}{n}\right)\frac{1}{(nn')^s}, \quad . \quad . \quad . \quad . \quad . \quad . \quad . \quad . \quad . \quad . \quad (b)$$

the sign Σ indicating two independent summations with respect to n and n'; or, if we perform the two summations separately, and omit the accent,

$$S = 2\Sigma\frac{1}{n^s}\,\Sigma\left(\frac{D}{n}\right)\frac{1}{n^s}. \quad . \quad . \quad . \quad . \quad . \quad . \quad . \quad . \quad . \quad . \quad (c)$$

* If $\Delta = 1$, each set contains four representations. To obtain a correct result in this case, we must therefore double the right-hand members of the equations (a), (b), (c), and (A).

To deduce an expression for h from this equation, we write $1+\rho$ for s, and multiplying each side by ρ, we suppose ρ to be positive and to diminish without limit. In order to find the limit of ρS on this supposition, we consider separately the partial sums, such as $\rho\Sigma\dfrac{1}{(ax^2+2bxy+cy^2)^{1+\rho}}$, of which it is composed.

If $\dfrac{1}{k_n^{1+\rho}}$ be the n-th term of the series $\Sigma\dfrac{1}{(ax^2+2bxy+cy^2)^{1+\rho}}$, in which we suppose that the terms are so arranged that no term surpasses any that precedes it, it can be shown that $\lim\dfrac{n}{k_n}=\dfrac{\pi\psi(2\Delta)}{2\Delta\sqrt{\Delta}}$. For if $2\Delta\xi+\xi_0$, $2\Delta\eta+\eta_0$ represent generally any one of the $2\Delta\psi(2\Delta)$ systems of values that can be attributed to x and y consistently with the condition that (a, b, c) assumes a value prime to 2Δ, the number of terms up to k_n inclusive (*i.e.* the number n) is evidently equal to the number of points having coordinates of any one of the forms $[2\Delta\xi+\xi_0,\ 2\Delta\eta+\eta_0]$ that lie within the ellipse $ax^2+2bxy+cy^2=k_n$, together with one, or all, or some of the similar points lying on the contour of the ellipse, according as $\dfrac{1}{k_n^{1+\rho}}$ is the first or the last, or neither the first nor the last of the terms equal to it in the series. The area of the ellipse is $\dfrac{\pi k_n}{\sqrt{\Delta}}$; whence, if n be very great, the number of the points we have defined is approximately $\dfrac{2\Delta\psi(2\Delta)\pi k_n}{4\Delta^2\sqrt{\Delta}}$, the error being of the same order as $\sqrt{k_n}$; *i.e.*

$$\lim\frac{n}{k_n}=\frac{\pi\psi(2\Delta)}{2\Delta\sqrt{\Delta}}.$$

Hence by Dirichlet's first Lemma (Art. 99),

$$\lim\rho S=\frac{\pi\psi(2\Delta)}{2\Delta\sqrt{\Delta}}h.$$

Again, by the same Lemma, the expression $\rho\Sigma\dfrac{1}{n^{1+\rho}}$ has $\dfrac{\psi(2\Delta)}{2\Delta}$ for its limit, when ρ diminishes without limit. And, lastly, the limit of the series $\Sigma\left(\dfrac{D}{n}\right)\dfrac{1}{n^{1+\rho}}$ is the series $\Sigma\left(\dfrac{D}{n}\right)\dfrac{1}{n}$, in which the terms are taken in their natural order. To establish this, we observe that the symbol $\left(\dfrac{D}{n}\right)$ is a periodic function of n, and that the sum of the terms of which one of its periods is composed is zero. Using the notation of Art. 98, and attributing the value $+1$ or -1 to the symbol δ according as $P\equiv 1$ or $\equiv 3$, mod 4, and to the symbol ϵ according as

$D = PS^2$ or $= 2PS^2$, we have, by Jacobi's law of reciprocity,

$$\left(\frac{D}{n}\right) = \delta^{\frac{1}{2}(n-1)}\,\epsilon^{\frac{1}{8}(n^2-1)}\left(\frac{n}{P}\right).$$

Hence $\left(\frac{D}{n}\right) = \left(\frac{D}{n'}\right)$, if $n \equiv n'$, mod $2^k PR$*; or $\left(\frac{D}{n}\right)$ is a periodic function of n. Again, if a and b denote the general terms of a system of residues prime to 2^k and p respectively, we find

$$\Sigma\,\delta^{\frac{1}{2}(n-1)}\,\epsilon^{\frac{1}{8}(n^2-1)}\left(\frac{n}{P}\right) = \Sigma\,\delta^{\frac{1}{2}(a-1)}\,\epsilon^{\frac{1}{8}(a^2-1)} \times \Pi\,.\,\Sigma\left(\frac{b}{p}\right) \times \Pi\,(r-1),$$

where in the left-hand member the summation extends to every value of n prime to $2^k PQ$ and less than it, while in the right-hand member the signs of summation refer to a and b, and the signs of multiplication to p and r respectively. This equation is easily verified; for if

$$n \equiv a, \text{ mod } 2^k, \quad = b, \text{ mod } p, \quad \equiv b', \text{ mod } p', \ldots,$$

we have

$$\delta^{\frac{1}{2}(n-1)}\,\epsilon^{\frac{1}{8}(n^2-1)}\left(\frac{n}{P}\right) = \delta^{\frac{1}{2}(a-1)}\,\epsilon^{\frac{1}{8}(a^2-1)}\left(\frac{b}{p}\right)\left(\frac{b'}{p'}\right)\ldots;$$

so that each member of the equation consists of the same units. But one at least of the factors of which the right-hand member is composed is zero; unless we have simultaneously $\delta = 1$, $\epsilon = 1$, $P = 1$, a supposition which is inadmissible, because it implies that D is a perfect square. We infer therefore that $\Sigma\left(\frac{D}{n}\right) = 0$, *i.e.* that the sum of the terms of a period of the symbol $\left(\frac{D}{n}\right)$ is equal to zero. If, then, we suppose the terms of the series $\Sigma\left(\frac{D}{n}\right)\frac{1}{n^{1+\rho}}$ to be taken in their natural order, it will follow from Dirichlet's second Lemma (Art. 99) that its sum represents a finite and continuous function of ρ for all values of ρ superior to -1; *i.e.* the limit of the series $\Sigma\left(\frac{D}{n}\right)\frac{1}{n^{1+\rho}}$, for $\rho = 0$ is the series $\Sigma\left(\frac{D}{n}\right)\frac{1}{n}$, in which the terms are taken in their natural order. We thus obtain the equation

$$h = \frac{2\sqrt{\Delta}}{\pi}\,\Sigma\left(\frac{D}{n}\right)\frac{1}{n}. \qquad \text{(A)}$$

---

* The index k is not the same as in Art. 98; it is 1 when $\delta = 1$, $\epsilon = 1$; 2 when $\delta = -1$, $\epsilon = 1$, and 3 when $\epsilon = -1$.

Secondly, let the determinant D be positive; and let us retain the same notation as in the former case. If in the series $S=$

$$\Sigma_1 \frac{1}{(a_1 x^2+2b_1 xy+c_1 y^2)^s} + \Sigma_2 \frac{1}{(a_2 x^2+2b_2 xy+c_2 y^2)^s} + \ldots + \Sigma_h \frac{1}{(a_h x^2+2b_h xy+c_h y^2)^s}$$

(in which it is convenient to suppose that the forms (a_k, b_k, c_k), representing the properly primitive classes of determinant D, have their first coefficients positive, and their last coefficients negative) we suppose the sign of double summation Σ_k to extend only to those integral values of x and y which render the value of the form (a_k, b_k, c_k) prime to $2D$, and which further satisfy the inequalities

$$x > 0, \quad y > 0, \quad y \leqq \frac{a_k U}{T-b_k U} x,$$

we obtain, by a comparison of Arts. 87 and 100, the equation

$$S=\Sigma \frac{1}{n^s} \Sigma \left(\frac{D}{n}\right) \frac{1}{n^s}; \quad \ldots \ldots \ldots \quad (c')$$

in which n denotes any positive number prime to $2D$, and which corresponds to equation (c).

If $\dfrac{1}{k_n^{1+\rho}}$ be the n-th term of the series

$$\Sigma \frac{1}{(ax^2+2bxy+cy^2)^{1+\rho}},$$

n is equal to the number of points having coordinates of any one of the forms

$$[2\Delta\xi+\xi_0, \; 2\Delta\eta+\eta_0],$$

which lie in the interior of the sectorial area, bounded by the positive axis of x, the arc of the hyperbola $ax^2+2bxy+cy^2=k_n$, and the straight line

$$y=\frac{aU}{T-bU} x;$$

together with one, all, or some of the similar points on the contour of the sector. The area of the sector is

$$\frac{k_n}{2\sqrt{D}} \log (T+U\sqrt{D});$$

whence, reasoning as before, we find

$$h=\frac{2\sqrt{D}}{\log [T+U\sqrt{D}]} \Sigma \left(\frac{D}{n}\right) \frac{1}{n} \quad \ldots \ldots \ldots \quad (B)$$

for the number of properly primitive forms of a positive determinant D. The corresponding formulae for improperly primitive forms are obtained by a precisely equivalent process. The results are, if $D=-\Delta$,

$$[2-(-1)^{\frac{1}{8}(D^2-1)}]\,h' = \frac{2}{\pi}\sqrt{\Delta}\,\Sigma\left(\frac{D}{n}\right)\frac{1}{n}*, \quad . \quad . \quad . \quad . \quad . \quad . \quad (C)$$

and if $D=+\Delta$,

$$[2-(-1)^{\frac{1}{8}(D^2-1)}]\,h' = \frac{2\sqrt{D}}{\log\frac{1}{2}(T'+U'\sqrt{D})}\,\Sigma\left(\frac{D}{n}\right)\frac{1}{n}, \quad . \quad . \quad . \quad . \quad (D)$$

$[T', U']$ denoting the least solution of the equation $T^2-DU^2=4$.

102. *Proof that each Genus contains the same number of Classes.*—The sixth section of the memoir also contains a demonstration of the proposition to which we have already referred (Art. 98), that all the possible genera actually exist, and contain an equal number of classes. This demonstration is not deduced from the expression for the number of properly primitive forms, but depends on an equation between two infinite series similar to the equation (a) of the last article. Let χ denote any one of the particular characters proper to the determinant, and let ϕ be any term in the product $\Pi(1+\chi)$, with the exception of the first term, which is unity, and also of that particular combination of the values of χ, the value of which, by the condition of possibility, is also a positive unit. If λ be the number of particular characters, $2^\lambda-2$ will be the number of expressions symbolised by ϕ. Let H and H' be the numbers of classes satisfying the conditions $\phi=1$ and $\phi=-1$ respectively. It can be shown, as follows, that $H=H'$. Confining ourselves, for perspicuity, to the case of forms of a negative determinant, we have, by the principle of Art. 87,

$$\Sigma_1\frac{\phi_1}{(a_1x^2+2b_1xy+c_1y^2)^s}+\Sigma_2\frac{\phi_2}{(a_2x^2+2b_2xy+c_2y^2)^s}+\ldots+\Sigma_h\frac{\phi_h}{(a_hx^2+2b_hxy+c_hy^2)^s}$$
$$=2\Sigma\left[\Sigma\left(\frac{D}{d}\right)\right]\left(\frac{\phi}{n}\right)\frac{1}{n^s}, \quad . \quad . \quad (d)$$

where in the right-hand member $\left(\frac{\phi}{n}\right)$ is $+1$ or -1, according as the number n satisfies the condition $\phi=1$ or $\phi=-1$; and similarly, in the left-hand member $\phi_k=-1$ or $=+1$, according as the generic character of the form (a_k, b_k, c_k) satisfies the condition $\phi=1$ or $\phi=-1$. In this equation the signs of summation have the same signification as in the similar equation (a) of the last article; and, as in that equation, the right-hand member may be expressed in the simpler form

$$\Sigma\left(\frac{\phi}{n}\right)\frac{1}{n^s}\;\Sigma\left(\frac{D}{n}\right)\left(\frac{\phi}{n}\right)\frac{1}{n^s}.$$

---

* If $\Delta=3$, we must triple the right-hand member of this equation; as each set of representations of a number by a form of determinant -3 contains six representations, instead of two.

If we now write $1+\rho$ for s, and, multiplying by ρ, allow ρ to converge to zero, the limit of the left-hand number is $(H-H')\dfrac{\pi\psi(2\Delta)}{2\Delta\sqrt{\Delta}}$. The series $\Sigma\left(\dfrac{D}{n}\right)\left(\dfrac{\phi}{n}\right)\dfrac{1}{n^{1+\rho}}$ converges to a finite limit; for $\left(\dfrac{D}{n}\right)$ and $\left(\dfrac{\phi}{n}\right)$ are each of them expressions of the form $\delta^{\frac{1}{2}(n-1)}\epsilon^{\frac{1}{8}(n^2-1)}\left(\dfrac{n}{Q}\right)$, δ and ϵ denoting positive or negative units, and Q an uneven number composed of unequal primes dividing D; their product is therefore another expression of the same form, in which δ, ϵ, and Q are not simultaneously equal to 1, because we have expressly excluded that combination of the particular characters which causes $\left(\dfrac{\phi}{n}\right)$ to coincide with $\left(\dfrac{D}{n}\right)$. It can therefore be shown, by reasoning as in the last article, that the second Lemma of Art. 99 is applicable to the series, and that it converges to the finite limit $\Sigma\left(\dfrac{D}{n}\right)\left(\dfrac{\phi}{n}\right)\dfrac{1}{n}$. Similarly, it may be shown that $\Sigma\left(\dfrac{\phi}{n}\right)\dfrac{1}{n^{1+\rho}}$ converges to a finite limit. The limit of the right-hand member of the equation (d) is consequently zero on account of the evanescent factor ρ; from which it follows that $H=H'$. Let $G_1, G_2, \ldots$ be the different possible genera; $h_1, h_2, \ldots$ the number of classes they severally contain; $\left(\dfrac{\phi}{G}\right)$ the value of ϕ for the genus G.

The equation $H-H'=0$ comprises $2^\lambda-2$ equations of the type

$$\left(\frac{\phi}{G_1}\right)h_1+\left(\frac{\phi}{G_2}\right)h_2+\ldots=0,$$

corresponding to the $2^\lambda-2$ different expressions symbolised by ϕ. If we multiply each of these equations by the coefficient of h_k in it, and add the products to the equation

$$2h_1+2h_2+2h_3+\ldots=2h,$$

we arrive at the conclusion $2^\lambda h_k=2h$. For the coefficient of h_r in the resulting equation is the product $\Pi\left[1+\left(\dfrac{\chi}{G_r}\right)\left(\dfrac{\chi}{G_k}\right)\right]$; and this product is 2^λ, if G_r and G_k are identical, but is zero in every other case, as one at least of the factors will be zero.

103. The seventh section (Crelle, vol. xxi. p. 1) commences with the proof of the theorem that the number of sets of representations of any number M

prime to $2D$ by quadratic forms of determinant D, is equal to the excess of the number of those divisors d of M which satisfy the equation

$$\delta^{\frac{1}{2}(d-1)}\,\epsilon^{\frac{1}{8}(d^2-1)}\left(\frac{d}{P}\right)=1,$$

above the number of those divisors which satisfy the equation

$$\delta^{\frac{1}{2}(d-1)}\,\epsilon^{\frac{1}{8}(d^2-1)}\left(\frac{d}{P}\right)=-1,$$

the symbols δ and ϵ having the same signification as in Art. 101. Of this theorem, which coincides with that of Art. 87, since

$$\left(\frac{D}{d}\right)=\delta^{\frac{1}{2}(d-1)}\,\epsilon^{\frac{1}{2}(d^2-1)}\left(\frac{d}{P}\right),$$

two demonstrations are given, one purely arithmetical, the other derived from the equation (*b*) of Art. 101, the proof of which in Dirichlet's memoir does not involve the theorem of Art. 87, but is deduced from the arithmetical principles on which that theorem itself depends. We have already referred (Art. 95) to some of the particular results which can be deduced from the general theorem.

It is evident from the mode of formation of the equation (*b*), or of the corresponding equation for a positive determinant, that it may be generalised by taking instead of the power $(ax^2+2bxy+cy)^{-s}$, any function of

$$ax^2+2bxy+cy^2$$

which renders the two members of the equation convergent; *i.e.* we may write, in the case of a negative determinant,

$$\Sigma_1.\,\phi\,(a_1x^2+2b_1xy+c_1y^2)+\Sigma_2.\,\phi\,(a_2x^2+2b_2xy+c_2y^2)+\ldots=2\Sigma\left(\frac{D}{n}\right)\phi\,(nn').$$

Dirichlet illustrates this observation by giving to ϕ the exponential form q^z, which satisfies the condition of convergence, if the analytical modulus of q be inferior to unity. Each double sum, such as $\Sigma q^{ax^2+2bxy+cy^2}$ in the left-hand member of the equation

$$\Sigma_1 q^{a_1x^2+2b_1xy+c_1y^2}+\Sigma_2 q^{a_2x^2+2b_2xy+c_2y^2}+\ldots=2\Sigma\left(\frac{D}{n}\right)q^{nn'},$$

can then be replaced by $2a\Delta\psi(2\Delta)$ (or sometimes by fewer) products of the form

$$\sum_{v=-\infty}^{v=\infty} q^{\frac{1}{a}(2a\Delta v+v_0)^2}\times\sum_{v=-\infty}^{v=\infty} q^{\frac{\Delta}{a}(2a\Delta v+v_1)^2},$$

in which each simple series such as

$$\sum_{v=-\infty}^{v=\infty} q^{\frac{1}{a}(2a\Delta v+v_0)^2}$$

can be expressed by means of the elliptic function Θ; the right-hand member can also be expressed by means of elliptic series. If, for example, $D=-3$, we have the equation

$$\sum_{v=-\infty}^{v=\infty} q^{(6v+1)^2} \times \sum_{v=-\infty}^{v=+\infty} q^{3\cdot(2v)^2} + \sum_{v=-\infty}^{v=+\infty} q^{(6v+2)^2} \times \sum_{v=-\infty}^{v=\infty} q^{3(2v+1)^2}$$

$$= \sum_{k=0}^{k=\infty} \frac{q^{6k+1}+q^{5(6k+1)}}{1-q^{6(6k+1)}} - \sum_{k=0}^{k=\infty} \frac{q^{6k+5}+q^{5(6k+5)}}{1-q^{6(6k+5)}}.$$

It does not appear that this remarkable transformation, which is only very briefly noticed by Dirichlet, has been further examined. (See a note by Mr. Cayley in the Cambridge and Dublin Mathematical Journal, vol. ix. p. 163.)

In the eighth section Dirichlet assigns the relation between the numbers of properly and improperly primitive classes. When the determinant is negative we find, by a comparison of the formulae (A) and (C), $h=h'$, or $h=3h'$, according as $D\equiv 1$, or $\equiv 5$, mod 8; observing only that if $D=-3$ we have, exceptionally, $h=h'$. When the determinant is positive, we infer from the formulae (B) and (D),

$$h=\frac{\log \frac{1}{2}(T'+U'\sqrt{D})}{\log (T+U\sqrt{D})}h',$$

or $$h=\frac{3\log \frac{1}{2}(T'+U'\sqrt{D})}{\log (T+U\sqrt{D})}h',$$

according as $D\equiv 1$ or $\equiv 5$, mod 8. Comparing these expressions with the observations in Art. 96 (vi.), we find, if $D\equiv 1$, mod 8, $h=h'$; and if $D\equiv 5$, mod 8, $h=h'$, or $h=3h'$, according as the least solution of the equation $T^2-DU^2=4$ is uneven or even.

Dirichlet also deduces from the formulae (A) and (B) the relation which subsists between the numbers of properly primitive classes for any two determinants which are to one another as two square numbers. It is sufficient to consider two determinants such as D and DS^2, of which the former is not divisible by any square. If h and H be the numbers of classes for these two determinants, we have evidently, when the determinants are negative,

$$\frac{H}{h}=S\frac{\Sigma\left(\frac{D}{n}\right)\frac{1}{n}}{\Sigma\left(\frac{D}{n}\right)\frac{1}{n}};$$

the two series in the numerator and denominator not being identical, because in the one n is any number prime to $2DS^2$, in the other any number prime to $2D$. But, by a principle due to Euler,

$$\Sigma\left(\frac{D}{n}\right)\frac{1}{n}=\Pi\frac{1}{1-\left(\frac{D}{n}\right)\frac{1}{p}},$$

p representing any prime, except those dividing $2DS^2$ or $2D$. Hence

$$H=hS\Pi\left(1-\left(\frac{D}{s}\right)\frac{1}{s}\right),$$

if s denote any prime dividing S but not dividing D. For a positive determinant we find

$$H=hS\Pi\left(1-\left(\frac{D}{s}\right)\right)\frac{\log(T+U\sqrt{D})}{\log(T'+U'\sqrt{D})},$$

$[T', U']$ denoting the least solution of the equation $T^2-DS^2U^2=1$; *i.e.* the least solution $[T_k, U_k]$ of $T^2-DU^2=1$, which satisfies the condition $U_k\equiv 0$, mod S; so that we may write

$$H=h\frac{S}{k}\Pi\left(1-\left(\frac{D}{s}\right)\frac{1}{s}\right).$$

In a subsequent note (No. 7 in the list) Dirichlet infers from this expression that, given any positive determinant D, we can always deduce from it an infinite number of determinants of the form DS^2 having each the same number of classes. For if we attribute to S a series of values of the form $\Pi . s^a$, all composed of the same prime numbers s, and having continually increasing numbers for the indices of those primes, it appears from a remark to which we have already referred (see Art. 96, (v.)), that the quotient $\frac{S}{k}$ will eventually be constant; *i.e.* there will exist an infinite series of determinants, all composed of the same primes, and all having the same number of properly primitive classes. As it is possible to find determinants contained in a series of this kind, and having only one class in each genus, it appears that the number of the positive determinants, which have only one class in each genus, is infinite. This result, which was anticipated by Gauss (Disq. Arith., Art. 304), is remarkable, because it is probable, from the result of a very extensive induction, that there are but 65 negative determinants, of which the greatest is -1848, having the same property.

104. *Summation of the series expressing the number of Properly Primitive Classes.*—It appears from the last article that, to obtain expressions in a finite

form for the number of classes, we may confine our attention to the order of properly primitive forms, and may suppose that the determinant is not divisible by any square. To sum the series $\Sigma\left(\frac{D}{n}\right)\frac{1}{n}$ upon this supposition, Dirichlet employs the formulae given by Gauss in his memoir, 'Summatio Serierum quarundam singularium,' to which we have already referred in this Report (Art. 20). The ninth section is occupied with the demonstration of these formulae; in the tenth they are applied to the summation of the series $\Sigma\left(\frac{D}{n}\right)\frac{1}{n}$. Two different methods are given by Dirichlet, by either of which this summation can be effected.

(i.) If k be the index of periodicity of $\left(\frac{D}{n}\right)$, so that

$$\left(\frac{D}{n+k}\right)=\left(\frac{D}{n}\right), \quad \text{and} \quad \sum_1^k\left(\frac{D}{n}\right)=0,$$

the summation indicated by the symbol $\sum\limits_1^k$ extending to all values of n prime to $2D$ from 1 to k, we have, writing V for $\Sigma\left(\frac{D}{n}\right)\frac{1}{n}$,

$$V=-\int_0^1\frac{\frac{f(x)}{x}}{x^k-1}\,dx,$$

where $f(x)=\sum\limits_1^k\left(\frac{D}{n}\right)x^n$, so that $f(1)=0$. Integrating by the ordinary method of decomposition into partial fractions, we find

$$-kV=\sum_{m=1}^{m=k-1}f\left(e^{\frac{2m\pi i}{k}}\right)\int_0^1\frac{dx}{x-e^{\frac{2m\pi i}{k}}}$$

$$=\sum_{m=1}^{m=k-1}f\left(e^{\frac{2m\pi i}{k}}\right)\left[\log\left(2\sin\frac{m\pi}{k}\right)+\tfrac{1}{2}i\pi\left(1-\frac{2m}{k}\right)\right].$$

To simplify this complicated expression, it is requisite to transform the symbol $\left(\frac{D}{n}\right)$ by the law of reciprocity, and to consider separately the eight cases which arise from every possible combination of the hypotheses, (α) D positive or negative, (β) D even or uneven, (γ) D, or $\frac{1}{2}D$, $\equiv 1$, mod 4, or $\equiv 3$, mod 4. As an example of the process, we shall take the two cases in which $D\equiv 3$, mod 4, so that

$$\left(\frac{D}{n}\right)=(-1)^{\frac{1}{2}(n-1)}\left(\frac{n}{\Delta}\right), \quad k=4\Delta,$$

Δ still denoting the absolute value of D. The value of $f(e^{\frac{2m\pi i}{4\Delta}})$ is assigned by the formulae of Gauss; it is

$$2i^{\frac{1}{4}(1+\Delta)^2}(-1)^{\frac{1}{2}(m-1)}\left(\frac{m}{\Delta}\right)\sqrt{D},$$

or zero, according as m is, or is not, prime to 4Δ *. We thus find

---

* If p be any prime divisor of Δ, an uneven number admitting of no square divisor, and if, for brevity, $P=\frac{\Delta}{p}$; we have, by Gauss's formula,

$$\sum_{k=1}^{k=p-1}\left(\frac{k}{p}\right)e^{\frac{2kmP\pi i}{p}}=\left(\frac{m}{p}\right)\left(\frac{P}{p}\right)i^{\frac{1}{4}(p-1)^2}\sqrt{p}, \text{ or } =0, \quad . \quad . \quad . \quad . \quad . \quad . \quad . \quad . \quad (1)$$

according as m is or is not prime to p. If we multiply together the equations of this type, corresponding to every prime divisor of Δ, and observe

(1) that $\theta=\Sigma\,.\,kP^2$ represents a system of residues prime to Δ,

(2) that $\left(\frac{\theta}{\Delta}\right)=\left(\frac{\theta}{p_1}\right)\left(\frac{\theta}{p_2}\right)\cdots=\left(\frac{k_1}{p_1}\right)\left(\frac{k_2}{p_2}\right)\cdots$,

(3) that $\Pi\left(\frac{P}{p}\right)i^{\frac{1}{4}(p-1)^2}=(-1)^{\Sigma\frac{1}{4}(p_1-1)(p_2-1)}\times i^{\Sigma\frac{1}{4}(p-1)^2}=i^{\frac{1}{4}[\Sigma(p-1)]^2}=i^{\frac{1}{4}(\Delta-1)^2}$,

we find
$$\Sigma\left(\frac{\theta}{\Delta}\right)e^{\frac{2\theta m\pi i}{\Delta}}=\left(\frac{m}{\Delta}\right)i^{\frac{1}{4}(\Delta-1)^2}\sqrt{\Delta}, \text{ or } =0, \quad . \quad . \quad . \quad . \quad . \quad . \quad . \quad . \quad . \quad (2)$$

according as m is or is not prime to Δ. We have already met with this equation in Art. 96, (ix.). If in the equations (1) we write $4P$ for P, and join to them the equation

$$\Sigma(-1)^{\frac{1}{2}(k-1)}e^{\frac{1}{2}km\Delta\pi i}=2i(-1)^{\frac{1}{2}(m-1)+\frac{1}{2}(\Delta-1)}, \ (m \text{ uneven}), \text{ or } =0 \ (m \text{ even}),$$

in which k is either term of a system of residues prime to 4, we obtain after multiplication the equation which is employed in the text. And similarly may the function $f\left(e^{\frac{2mni}{k}}\right)$ be evaluated, whatever be the form of D.

The formulae (A) and (A′) of Art. 20 are only particular cases of the general result obtained by Gauss in the 'Summatio Serierum &c.' The general formula, including (A), is

$$\sum_{k=0}^{k=n-1}r^{hk^2}=\left(\frac{h}{n}\right)i^{\frac{1}{4}(n-1)^2}\sqrt{n},$$

h denoting any number prime to n. When n is even, the formula (A′) of Art. 20 is similarly include in the following,

$$\Sigma r^{hk^2}=0, \text{ or } =\left(\frac{n}{h}\right)i^{-\frac{1}{4}(h-d)^2}(1+i)\sqrt{n}, \quad \{=\left(\frac{n}{h}\right)(1+i^h)\sqrt{n}\},$$

according as n is unevenly or evenly even.

When n is uneven and not divisible by any square, the two sums

$$\sum_{k=0}^{k=n-1}r^{hk^2} \text{ and } \Sigma\left(\frac{\theta}{n}\right)r^{h\theta}$$

are identical, as appears from a comparison of (2) with the generalisation of (A), and has been already observed in the case when n is a prime (Art. 21).

{[Aug. 8, 1877.] The generalised formulae (A) and (A′) here given coincide with the formulae

$$-4\Delta V = 2i^{\frac{1}{4}(1+\Delta)^2}\sqrt{\Delta}\,\Sigma(-1)^{\frac{1}{2}(m-1)}\left(\frac{m}{\Delta}\right)\left[\log\left(2\sin\frac{m\pi}{4\Delta}\right)+\frac{\pi}{2}i\left(1-\frac{m}{2\Delta}\right)\right],$$

the summation extending to all values of m prime to 4Δ and less than it. In this expression the sum $\Sigma(-1)^{\frac{1}{2}(m-1)}\left(\frac{m}{\Delta}\right)$ is zero, because the terms correspond to m and $2\Delta+m$ destroy one another; so that

$$-4\Delta V = 2i^{\frac{1}{4}(1+\Delta)^2}\sqrt{\Delta}\,\Sigma(-1)^{\frac{1}{2}(m-1)}\left(\frac{m}{\Delta}\right)\left[\log\left(\sin\frac{m\pi}{4\Delta}\right)-i\,\frac{m\pi}{4\Delta}\right].$$

Distinguishing the two cases $D=\Delta$, and $D=-\Delta$, and observing that the imaginary parts vanish identically, as they ought to do, because V is real, we have, finally, if $D=\Delta$,

$$-4DV = 2\sqrt{D}\,\Sigma(-1)^{\frac{1}{2}(m-1)}\left(\frac{m}{D}\right)\log\sin\left(\frac{m\pi}{4\Delta}\right)$$
$$= 2\sqrt{D}\,\Sigma\left(\frac{D}{m}\right)\log\sin\frac{m\pi}{4\Delta};$$

and if $D=-\Delta$,

$$-4\Delta V = \frac{\pi}{2\sqrt{\Delta}}\,\Sigma(-1)^{\frac{1}{2}(m-1)}\left(\frac{m}{\Delta}\right)m$$
$$= \frac{\pi}{2\sqrt{\Delta}}\,\Sigma\left(\frac{D}{m}\right)m.$$

(ii.) The series $\Sigma\left(\frac{D}{n}\right)\frac{1}{n}$ can also be summed by substituting for $\left(\frac{D}{n}\right)$ its trigonometrical value deducible from the formulae of Gauss. We will take as an example the case in which $D=-\Delta\equiv 3$, mod 4. Writing n for m, and m for n, in the equation

$$f\left(e^{\frac{2m\pi i}{4\Delta}}\right) = 2i^{\frac{1}{4}(1+\Delta)^2}(-1)^{\frac{1}{2}(m-1)}\left(\frac{m}{\Delta}\right)\sqrt{\Delta},$$

we find, observing that $\frac{1}{2}(1+\Delta)$ is uneven,

$$\left(\frac{D}{n}\right) = (-1)^{\frac{1}{2}(n-1)}\left(\frac{n}{\Delta}\right) = \frac{1}{2i\sqrt{\Delta}}\,\Sigma\left(\frac{D}{m}\right)e^{\frac{2mn\pi i}{4\Delta}}$$
$$= \frac{1}{2\sqrt{\Delta}}\,\Sigma\left(\frac{D}{m}\right)\sin\left(\frac{2mn\pi}{4\Delta}\right),$$

---

of M. Lebesgue (Liouville (I), vol. xii. p. 509). If $n=P=ap$, $h=Q=bq$, a and b being powers of 2, p and q uneven, we have

$$\Sigma = \tfrac{1}{2}\left(\frac{a}{q}\right)\left(\frac{b}{p}\right)\left(\frac{q}{p}\right)(1+i^{-P})(1+i^{-P+bp+aq}):$$

this is wrong when P is even, and $p\equiv -1$, mod 4. We have, however, in every case,

$$\Sigma = \tfrac{1}{2}\left(\frac{P}{q}\right)\left(\frac{b}{p}\right)(1+i^{-qP})(1+i^{q})\sqrt{P}\}.$$

the summation extending to every value of m prime to 4Δ and less than it. Substituting this expression for $\left(\frac{D}{n}\right)$ in V, we have

$$V=\frac{1}{2\sqrt{\Delta}}\Sigma\left(\frac{D}{m}\right)\Sigma\frac{1}{n}\sin\left(\frac{2mn\pi}{4\Delta}\right).$$

Since the expression which we have substituted for $\left(\frac{D}{n}\right)$ is zero, when n is not prime to 4Δ, we may attribute to n, in the series

$$\Sigma\frac{1}{n}\sin\left(\frac{2mn\pi}{4\Delta}\right),$$

either all uneven values, or all integral values. The sum of the series

$$\frac{\sin x}{1}+\frac{\sin 3x}{3}+\frac{\sin 5x}{5}+\dots$$

is, by a known theorem, $\frac{1}{4}\pi$ or $-\frac{1}{4}\pi$, according as $0<x<\pi$, or $\pi<x<2\pi$. Hence attributing to n only uneven values, and denoting by m' and m'' the values of m inferior and superior to 2Δ,

$$V=\frac{\pi}{8\sqrt{\Delta}}\Sigma\left[\left(\frac{D}{m'}\right)-\left(\frac{D}{m''}\right)\right]$$

$$=\frac{\pi}{4\sqrt{\Delta}}\Sigma\left(\frac{D}{m'}\right),$$

because $$\left(\frac{D}{m'}\right)=-\left(\frac{D}{2\Delta+m'}\right).$$

If we attribute to n all integral values, the equation

$$\tfrac{1}{2}(\pi-x)=\frac{\sin x}{1}+\frac{\sin 2x}{2}+\frac{\sin 3x}{3}+\dots,$$

which subsists for all positive values of x less than 2π, will give the value already obtained for V by the former method, viz.,

$$-4\Delta V=\frac{\pi}{2\sqrt{\Delta}}\Sigma\left(\frac{D}{m}\right)m.$$

The mode of application of this method may be still further varied; for, instead of substituting for $(-1)^{\frac{1}{2}(n-1)}\left(\frac{n}{\Delta}\right)$, we may leave the factor $(-1)^{\frac{1}{2}(n-1)}$ unchanged, and substitute for $\left(\frac{n}{\Delta}\right)$, by means of the equation

$$\left(\frac{n}{\Delta}\right)=\frac{1}{\sqrt{\Delta}}\Sigma\left(\frac{m}{\Delta}\right)\cos\left(\frac{2mn\pi}{\Delta}\right),$$

which, as well as the substitution which we have employed, is deducible from the formulae of Gauss *. We should thus obtain a third expression for V, different in form from both of those which we have already found.

The forms which the expression of h can assume are very numerous; we select the following as examples, D still denoting a determinant not divisible by any square.

I. If $D \equiv 1$, mod 4.

For a positive determinant, $D = \Delta$,

$$h = \left(\frac{2}{D}\right) \frac{2}{\log (T + U\sqrt{D})} \Sigma \left(\frac{D}{m}\right) \log \tan \left(\frac{m\pi}{2D}\right).$$

For a negative determinant, $D = -\Delta$,

$$h = -\left(\frac{2}{\Delta}\right) \Sigma \left(\frac{D}{m}\right),$$

the summations extending to every uneven value of m prime to Δ and less than Δ.

II. If D be not $\equiv 1$, mod 4.

For a positive determinant,

$$h = -\frac{1}{\log (T + U\sqrt{D})} \Sigma \left(\frac{D}{m}\right) \log \sin \left(\frac{m\pi}{4D}\right).$$

For a negative determinant,

$$h = -\frac{1}{4\Delta} \Sigma \left(\frac{D}{m}\right) m = +\tfrac{1}{2} \Sigma \left(\frac{D}{m'}\right),$$

the summations with respect to m and m' extending to all values prime to 2Δ, and inferior to 4Δ and 2Δ respectively.

Dirichlet observes that when the determinant is positive, the coefficient of $\dfrac{1}{\log (T + U\sqrt{D})}$ is a logarithm of the form $\log (T_h + U_h \sqrt{D})$; (T_h, U_h) being one of those solutions of the equation $T^2 - DU^2 = 1$ which are deducible from the theory of the division of the circle. Thus h is in fact determined as the index of the place occupied in the series of solutions of $T^2 - DU^2 = 1$, by an assigned trigonometrical solution. (See a note by M. Arndt in Crelle, vol. lvi. p. 100.)

* See equation (2) of the preceding note.

In the particular case in which the determinant is a prime of the form $4n+3$ taken negatively, an expression for the number of classes had already been given by Jacobi (Crelle, vol. ix. p. 189). It would seem, from his note on the division of the circle (Crelle, vol. xxx. p. 166), that the unpublished method, by which his result was obtained, formed a part of that theory.

VIII.

REPORT ON THE THEORY OF NUMBERS.

PART IV.

[Report of the British Association for 1862, pp. 503–526.]

---

105. *GENERAL Theorems relating to Composition.*—The theory of the composition of quadratic forms occupies an important place in the second part of the 5th section of the 'Disquisitiones Arithmeticæ,' and is the foundation of nearly all the investigations which follow it in that section. In accordance with the plan which we have followed in this portion of our Report, we shall now briefly resume the theory as it appears in the 'Disquisitiones Arithmeticæ,' directing our special attention to the additions which it has received from subsequent mathematicians. We premise a few general remarks on the Problem of composition.

If $F_1(x_1, x_2, \ldots, x_n)$ be a form of order m, containing n indeterminates, which, by a bipartite linear transformation of the type

$$\left.\begin{array}{r} x_\alpha = \Sigma\, a_{\alpha,\beta,\gamma}\, y_\beta\, z_\gamma, \\ \alpha = 1, 2, 3, \ldots, n, \\ \beta = 1, 2, 3, \ldots, n, \\ \gamma = 1, 2, 3, \ldots, n, \end{array}\right\}$$

is changed into the product of two forms $F_2(y_1, y_2, \ldots, y_n)$ and $F_3(z_1, z_2, \ldots, z_n)$ of the same order, and containing the same number of indeterminates, F_1 is said to be *transformable into the product of F_2 and F_3*; and, in particular, if the determinants of the matrix

$$| a_{\alpha,\beta,\gamma} |,$$

which is of the type $n \times n^2$, be relatively prime, F_1 is said to be *compounded of F_2 and F_3*. Adopting this definition, we may enunciate the theorem—'If F_1

be transformable into $F_2 \times F_3$, and if F_1, G_2, G_3 be contained in G_1, F_2, F_3 respectively, G_1 is transformable into $G_2 \times G_3$; and, in particular, if F_1 be compounded of F_2 and F_3, and the forms F_1, G_2, G_3 be equivalent to the forms G_1, F_2, F_3 respectively, G_1 is compounded of G_2 and G_3.'

It is only in certain cases that the multiplication of two forms gives rise to a third form, transformable into their product. Supposing that F_2 and F_3 are irreducible forms, *i.e.* that neither of them is resoluble into rational factors, let I_1, I_2, I_3, be any corresponding invariants of F_1, F_2, F_3, and let us represent by B and C the determinants

$$\left|\frac{dx_\alpha}{dy_\beta}\right| \qquad \left.\begin{array}{l}\alpha=1,\ 2,\ 3,\ \ldots,\ n,\\ \beta=1,\ 2,\ 3,\ \ldots,\ n,\end{array}\right\}$$

and

$$\left|\frac{dx_\alpha}{dz_\gamma}\right| \qquad \left.\begin{array}{l}\alpha=1,\ 2,\ 3,\ \ldots,\ n,\\ \gamma=1,\ 2,\ 3,\ \ldots,\ n.\end{array}\right\}$$

The transformation of F_1 into $F_2 \times F_3$ then gives rise to the relations

$$I_1 \times B^{\frac{mi}{n}} = I_2 \times F_3^i,$$

$$I_1 \times C^{\frac{mi}{n}} = I_3 \times F_2^i,$$

i denoting the order of the invariants I_1, I_2, I_3. If one of the two numbers I_2 and I_3 be different from zero, we infer that m is a divisor of n. For if $\frac{\mu}{\nu}$ be the fraction $\frac{m}{n}$ reduced to its lowest terms, the equations

$$I_1^\nu \times B^{\mu i} = I_2^\nu \times F_3^{\nu i},$$

$$I_1^\nu \times C^{\mu i} = I_3^\nu \times F_2^{\nu i}$$

imply that F_2 and F_3 (cleared of the greatest numerical divisors of all their terms) are perfect powers of the order μ; *i.e.*, $\mu = 1$, or m divides n, since F_2 and F_3 are by hypothesis irreducible. We thus obtain the theorem (which however applies only to irreducible forms having at least one invariant different from zero)—'No form can be transformed into the product of two forms of the same sort, unless the number of its indeterminates is a multiple of its order.' For example, there is no theory of composition for any binary forms, except quadratic forms, nor for any quadratic forms of an uneven number of indeterminates.

Again, when m is a divisor of n, let $n = km$, and let b, c, d_2, d_3 represent the greatest numerical divisors of B, C, F_2, F_3 respectively; we find

$$\frac{I_1}{I_2} = \left(\frac{d_3^k}{b}\right)^{\frac{mi}{n}}, \quad \left(\frac{I_1}{I_3}\right) = \left(\frac{d_2^k}{c}\right)^{\frac{mi}{n}}, \quad \frac{B}{b} = \left(\frac{F_3}{d_3}\right)^k, \quad \frac{C}{c} = \left(\frac{F_2}{d_2}\right)^k.$$

The first two of these equations show that the invariants of the three forms F_1, F_2, F_3 are so related to one another, that we may imagine them to have been all derived by transformation from one and the same form (see Art. 80); the last two (which, it is to be observed, present an ambiguity of sign when $\frac{mi}{n}$ is even) show that the forms B and F_3^k, C and F_2^k, are respectively identical, if we omit a numerical factor.

Lastly, let Φ_1, Φ_2, Φ_3 be any corresponding covariants of F_1, F_2, F_3. The relation of covariance gives rise to the equations

$$\Phi_1(x_1, x_2, \ldots, x_n) \times B^{\frac{mp-q}{n}} = \Phi_2(y_1, y_2, \ldots, y_n) \times F_3^p(z_1, \ldots, z_n),$$

$$\Phi_1(x_1, x_2, \ldots, x_n) \times C^{\frac{mp-q}{n}} = \Phi_3(z_1, z_2, \ldots, z_n) \times F_2^p(y_1, \ldots, y_n),$$

where p and q are the orders of the covariants in the coefficients and in the indeterminates respectively. Combining with these equations the values of B and C already given, we see that $\Phi_2 \times F_3^{\frac{q}{m}}$ and $\Phi_3 \times F_2^{\frac{q}{m}}$ are identical, excepting a numerical factor; *i.e.* that Φ_2 and Φ_3 are either identically zero, or else numerical multiples of powers of F_2 and F_3. If therefore two forms can be combined by multiplication so as to produce a third form transformable into their product, their covariants are all either identically zero or else are powers of the forms themselves. There is, consequently, no *general* theory of composition for any forms other than quadratic forms, because all other sorts of forms have covariants which cannot be supposed equal to zero, or to a multiple of a power of the form itself, without particularizing the nature of the form. And even as regards quadratic forms, we may infer that composition is possible only in cases of continually increasing particularity, as the number of indeterminates increases.

106. *Composition of Quadratic Forms.—Preliminary Lemmas.*—The following lemma is given by Gauss as a preliminary to the theory of the composition of binary quadratic forms (Disq. Arith., Art. 234):—

(i.) ‘If the two matrices

$$\begin{vmatrix} A \\ B \end{vmatrix} = \begin{vmatrix} A_1\, A_2 \ldots A_\mu \\ B_1\, B_2 \ldots B_\mu \end{vmatrix}$$

and

$$\begin{vmatrix} a \\ b \end{vmatrix} = \begin{vmatrix} a_1\; a_2 \ldots a_\mu \\ b_1\; b_2 \ldots b_\mu \end{vmatrix}$$

be connected by the equation $\begin{vmatrix} A \\ B \end{vmatrix} = k \begin{vmatrix} a \\ b \end{vmatrix}$, in which the sign of equality refers to corresponding determinants in the two matrices; and if the determinants of $\begin{vmatrix} a \\ b \end{vmatrix}$ admit of no common divisor beside unity; the equation

$$\begin{vmatrix} A \\ B \end{vmatrix} = \begin{vmatrix} k \end{vmatrix} \times \begin{vmatrix} a \\ b \end{vmatrix},$$

in which the sign of equality refers to corresponding constituents in the two matrices, is always satisfied by a matrix $|k|$ of the type 2×2, of which the determinant is k, and the constituents integral numbers.' *

The subsequent analysis of Gauss can be much abbreviated if to this lemma we add three others.

In their enunciations we represent by X, Y, x, y, four functions, homogeneous and linear in respect of each of the n binary sets, $\xi_1 \eta_1, \xi_2 \eta_2, \ldots, \xi_n \eta_n$; by $\begin{vmatrix} A \\ B \end{vmatrix}$ and $\begin{vmatrix} a \\ b \end{vmatrix}$ the matrices composed of the coefficients of X, Y and x, y, respectively; by (P, Q, R), (P', Q', R') quadratic forms of which the coefficients are any quantities whatever; and by k an integral number.

(ii.) 'If X, Y, x, y, satisfy the n equations included in the formula

$$\frac{dX}{d\xi_i}\frac{dY}{d\eta_i} - \frac{dX}{d\eta_i}\frac{dY}{d\xi_i} = k\left(\frac{dx}{d\xi_i}\frac{dy}{d\eta_i} - \frac{dx}{d\eta_i}\frac{dy}{d\xi_i}\right),$$

the matrices $\begin{vmatrix} A \\ B \end{vmatrix}$ and $\begin{vmatrix} a \\ b \end{vmatrix}$ satisfy the equation

$$\begin{vmatrix} A \\ B \end{vmatrix} = k \begin{vmatrix} a \\ b \end{vmatrix}.'$$

(iii.) 'The greatest numerical common divisor of the n resultants

$$\frac{dX}{d\xi_i}\frac{dY}{d\eta_i} - \frac{dX}{d\eta_i}\frac{dY}{d\xi_i}$$

is equal to the greatest common divisor of the determinants of $\begin{vmatrix} A \\ B \end{vmatrix}$.'

(iv.) 'If the n resultants of X and Y be not all identically equal to zero, the equation $\quad PX^2 + 2QXY + RY^2 = P'X^2 + 2Q'XY + R'Y^2$ implies the equations $P = P'$, $Q = Q'$, $R = R'$.'

107. *Gauss's Six Conclusions.*—Taking F, f, f' to represent the forms $(A, B, C)(X, Y)^2$, $(a, b, c)(x, y)^2$, $(a', b', c')(x', y')^2$, of which the determinants are

* For a generalisation of this theorem, see a paper by M. Bazin, in Liouville, vol. xix. p. 209; or Phil. Trans., vol. cli. p. 295.

D, d, d'; let also M, m, m' be the greatest common divisors of $A, 2B, C$, of $a, 2b, c$, and of $a', 2b', c'$; $\mathfrak{M}, \mathfrak{m}, \mathfrak{m}'$ the greatest common divisors of A, B, C, of a, b, c, and of a', b', c', respectively. Supposing that F is transformed into $f \times f'$ by the substitution

$$X = p_0\, xx' + p_1\, xy' + p_2\, x'y + p_3\, yy', \quad Y = q_0\, xx' + q_1\, xy' + q_2\, x'y + q_3\, yy',$$

let us represent the two resultants

$$\frac{dX}{dx}\frac{dY}{dy} - \frac{dX}{dy}\frac{dY}{dx} \quad \text{and} \quad \frac{dX}{dx'}\frac{dY}{dy'} - \frac{dX}{dy'}\frac{dY}{dx'}$$

by Δ and Δ'; the six determinants of the matrix $\begin{vmatrix} p_0, & p_1, & p_2, & p_3 \\ q_0, & q_1, & q_2, & q_3 \end{vmatrix}$ (taken in their natural order) by P, Q, R, S, T, U; the greatest common divisor of these six numbers by k, and the greatest numerical divisors of Δ and Δ' by δ and δ' respectively, so that (Lemma 3) k is the greatest common divisor of δ and δ'.

From the invariant property of the determinants of F, f and f' we infer

$$\Delta'^2 = \frac{d'}{D} f^2, \quad \Delta^2 = \frac{d}{D} f'^2, \quad D\delta'^2 = d'm^2, \quad D\delta^2 = dm'^2.$$

Hence *the quotients* $\frac{d'}{D}$ *and* $\frac{d'}{D}$ *are squares.* (Gauss's 1st conclusion.) Also *D divides $d'm^2$ and dm'^2.* (Gauss's 2nd conclusion.) But k is the greatest common divisor of δ and δ'; therefore *Dk^2 is the greatest common divisor of $d'm^2$ and dm^2.* (Gauss's 4th conclusion.) Let $\frac{d}{D} = n^2$, $\frac{d'}{D} = n'^2$, and let the signs of n and n' be so taken that $\Delta' = n'f$, $\Delta = nf''$; these two equations are equivalent to the six following:—

$$\frac{P}{a} = \frac{R-S}{2b} = \frac{U}{c} = n'; \quad \frac{Q}{a'} = \frac{R+S}{2b'} = \frac{T}{c'} = n. \qquad (\Omega)$$

(Gauss's 3rd conclusion.)

Multiplying together the two resultants Δ and Δ', we obtain an identity, which we shall write at full:

$$\begin{aligned} &[(p_0 q_1 - p_1 q_0)\, x^2 + (p_0 q_3 - p_3 q_0 + p_2 q_1 - p_1 q_2)\, xy + (p_2 q_3 - p_3 q_2)\, y^2] \\ &\quad \times [(p_0 q_2 - p_2 q_0)\, x'^2 + (p_0 q_3 - p_3 q_0 + p_1 q_2 - p_2 q_1)\, x'y' + (p_1 q_3 - p_3 q_1)\, y'^2] \\ &\quad = (q_1\, q_2 - q_0\, q_3)\, (p_0\, xx' + p_1\, xy' + p_2\, x'y + p_3\, yy')^2 \\ &\quad + (q_0 p_3 + p_0 q_3 - q_1 p_2 - q_2 p_1)\, (p_0\, xx' + p_1\, xy' + p_2\, x'y + p_3\, yy') \qquad (\mathrm{I}) \\ &\quad \times (q_0\, xx' + q_1\, xy' + q_2\, x'y + q_3\, yy') \\ &\quad + (p_1 p_2 - p_0 p_3)\, (q_0\, xx' + q_1\, xy' + q_2\, x'y + q_3\, yy')^2. \end{aligned}$$

Comparing this identity with the equation $\Delta\Delta' = nn'ff' = nn'F$, we find by Lemma 4

$$\frac{q_1 q_2 - q_0 q_3}{A} = \frac{q_0 p_3 + p_0 q_3 - q_1 p_2 - q_2 p_1}{2B} = \frac{p_1 p_2 - p_0 p_3}{C} = nn'. \quad . \quad . \quad (\Omega')$$

The 5th and 6th conclusions relate to the order of the form compounded of two given forms. The equation

$$AX^2 + 2BXY + CY^2 = (ax^2 + 2bxy + cy^2) \times (a'x'^2 + 2b'x'y' + c'y'^2)$$

shows that M divides mm'. But also mm' divides Mk^2. For operating on the equation just written with $\frac{d^2}{dx^2}, \frac{d^2}{dxdy}, \frac{d^2}{dy^2}$ successively, we find

$$A\frac{dX^2}{dx^2} + 2B\frac{dX}{dx}\frac{dY}{dx} + C\frac{dY^2}{dx^2} \equiv 0, \text{ mod } mm',$$

$$2\left[A\frac{dX}{dx}\frac{dX}{dy} + B\left(\frac{dX}{dx}\frac{dY}{dy} + \frac{dX}{dy}\frac{dY}{dx}\right) + C\frac{dY}{dx}\frac{dY}{dy}\right] \equiv 0, \text{ mod } mm', \quad . \quad (j)$$

$$A\frac{dX^2}{dy^2} + 2B\frac{dX}{dy}\frac{dY}{dy} + C\frac{dY^2}{dy^2} \equiv 0, \text{ mod } mm'.$$

Whence $A\Delta^2$, $2B\Delta^2$, $C\Delta^2$, and consequently $M\delta^2$, are congruous to zero, mod mm'. Similarly $M\delta'^2 \equiv 0$, mod mm'; i.e. mm' divides Mk^2. If then $k=1$, *i.e. if F be compounded of f and f', $M = mm'$.* (Gauss's 5th conclusion.)

Again, if in the congruences (j) we take $m'\mathfrak{m}$ as modulus instead of mm', we may omit the factor 2 in the second congruence, and may infer that $A\Delta^2$, $B\Delta^2$, $C\Delta^2$ are all divisible by $m'\mathfrak{m}$, *i.e.* that $\mathfrak{m}m'$ divides $\mathfrak{M}k^2$, or $\mathfrak{M}$, when F is compounded of f and f'. It is also readily seen that $\mathfrak{M}$ divides $m\mathfrak{m}'$ and $\mathfrak{m}m'$; whence observing that $\mathfrak{m} = m$ or $\frac{1}{2}m$, $\mathfrak{m}' = m'$ or $\frac{1}{2}m'$, $\mathfrak{M} = M$, or $\frac{1}{2}M$, according as f, f', and F are derived from properly or improperly primitive forms, we conclude that *if f and f' be both derived from properly primitive forms, the form compounded of them is also derived from a properly primitive form; but if either f or f' be derived from an improperly primitive form, the form compounded of them is derived from a similar form.* (Gauss's 6th conclusion.)

In the transformation of F into $f \times f'$, the form f is said to be taken directly or inversely, according as the fraction n is positive or negative. And similarly for f' and n'.

108. *Solution of the Problem of Composition.*—It appears from the identity (I) that if $A, B, C, p_0, p_1, p_2, p_3, q_0, q_1, q_2, q_3$, be integral numbers satisfying the nine equations (Ω), the form $(A, B, C)(X, Y)^2$ will be transformed into the product of the two forms $(a, b, c)(x, y)^2$ and $(a', b', c')(x', y')^2$ by the substitution

$$X = p_0 xx' + p_1 xy' + p_2 yx' + p_3 yy', \qquad Y = q_0 xx' + q_1 xy' + q_2 yx' + q_3 yy'.$$

In order, therefore, to find a form, F, compounded directly or inversely of two given forms of which the determinants are to one another as two squares, we have to find eleven integral and two fractional numbers, satisfying the equations (Ω) and (Ω'), in which a, b, c, a', b', c', and the signs of n and n', are alone given; the numbers $p_0, p_1, p_2, p_3, q_0, q_1, q_2, q_3$, being further subject to the condition that the determinants of the matrix $\begin{vmatrix} p_0, p_1, p_2, p_3 \\ q_0, q_1, q_2, q_3 \end{vmatrix}$ are to admit of no common divisor. To determine n and n', we observe that the six determinants satisfy the identical relation $PU - QT + RS = 0$; from which we infer, first, that $P, Q, R-S, R+S, T, U$ must be relatively prime, if P, Q, R, S, T, U are to be so; and secondly, substituting for the determinants their values given by the first six of the equations (Ω), that $dn'^2 = d'n^2$. Denoting by δ' and δ the greatest common divisors of $P, R-S, U$ and of $Q, R+S, T$, so that δ and δ' are relatively prime, we have evidently $n' = \pm \frac{\delta'}{m}$, $n = \pm \frac{\delta}{m'}$; the positive or negative signs being taken according as f' and f enter the composition directly or inversely; and the absolute values of δ and δ' being determined by the equation $\delta^2 d'm^2 = \delta'^2 dm'^2$. The fractions n and n' being thus ascertained, the values of P, Q, R, S, T, U are known from the equations (Ω): these values are all integral: for $P, Q, R-S, R+S, T, U$, this is evident from the equations (Ω), and may be proved for R and S by means of the identity $PU - QT + RS = 0$. We have next to assign such values to the constituents of the matrix $\begin{vmatrix} p_0, p_1, p_2, p_3 \\ q_0, q_1, q_2, q_3 \end{vmatrix}$, that its determinants may acquire the known values of P, Q, R, S, T, U. To do so, it is sufficient * to obtain a *fundamental* set of solutions of the indeterminate system,

$$\left.\begin{array}{r} x_1 U - x_2 T + x_3 S = 0, \\ -x_0 U \qquad + x_2 R - x_3 Q = 0, \\ x_0 T - x_1 R \qquad + x_3 P = 0, \\ -x_0 S + x_1 Q - x_2 P \qquad = 0, \end{array}\right\} \quad \ldots\ldots\ldots \quad (S)$$

which is equivalent to only two independent equations. From the skew sym-

* For a solution of the general problem, 'To find all the matrices of a given type, of which the determinants have given values,' see a paper by M. Bazin, in Liouville, vol. xvi. p. 145; or Phil. Trans., vol. cli. p. 302. For the definition of a *fundamental* set of solutions of an indeterminate system, see *ibid.* p. 297. It may be observed that the analysis of Gauss, which is exhibited in the text, is applicable to any matrix of the type $n \times (n+2)$.

metrical form of the matrix of this system, it appears that if $\theta_0, \theta_1, \theta_2, \theta_3$ be any multipliers whatever, any four numbers (x_0, x_1, x_2, x_3) proportional to

$$\left.\begin{array}{l} \theta_1 P+\theta_2 Q+\theta_3 R, \\ -\theta_0 P \quad\quad +\theta_2 S+\theta_3 T, \\ -\theta_0 Q-\theta_1 S \quad\quad +\theta_3 U, \\ -\theta_0 R-\theta_1 T-\theta_2 U \quad\quad , \end{array}\right\} \quad . \quad . \quad . \quad (\Sigma)$$

will satisfy the system (S), and in addition the equation

$$\theta_0 x_0+\theta_1 x_1+\theta_2 x_2+\theta_3 x_3=0.$$

Assigning, then, to $\theta_0, \theta_1, \theta_2, \theta_3$ any arbitrary values whatever, let q_0, q_1, q_2, q_3 be four numbers relatively prime, and proportional to the four numbers (Σ); let also $\pi_0 q_0+\pi_1 q_1+\pi_2 q_2+\pi_3 q_3=1$; and employing $\pi_0, \pi_1, \pi_2, \pi_3$ in the place of $\theta_0, \theta_1, \theta_2, \theta_3$, let us represent by p_0, p_1, p_2, p_3 the solution of (S) thus obtained. We have thus two solutions of (S), satisfying respectively the relations

$$\pi_0 p_0+\pi_1 p_1+\pi_2 p_2+\pi_3 p_3=0, \text{ and } \pi_0 q_0+\pi_1 q_1+\pi_2 q_2+\pi_3 q_3=1,$$

which prove that the two solutions form a fundamental set, *i.e.* that the determinants

$$\left|\begin{array}{l} p_0, p_1, p_2, p_3 \\ q_0, q_1, q_2, q_3 \end{array}\right|=[P, Q, R, S, T, U].$$

It only remains to show that the values of A, B, C, which are now supplied by the equations (Ω'), are integral. Operating on the identity (I) with $\frac{d^2}{dx^2}, \frac{d^2}{dx\,dy}, \frac{d^2}{dy^2}$, and also with $\frac{d^2}{dx'^2}, \frac{d^2}{dx'dy'}, \frac{d^2}{dy'^2}$, we find, by reasoning similar to that which we have employed to establish the 5th conclusion, that $2Ann', 2Bnn', 2Cnn'$, which are certainly integral numbers, are divisible by $2\delta\delta'$ if $\frac{R+S}{\delta}$ and $\frac{R-S}{\delta}$ are both even, and by $\delta\delta'$ if either of these numbers is uneven. In the former case A, B, C are evidently integral; in the latter, either $\frac{2b}{m}$ or $\frac{2b'}{m'}$ is uneven, *i.e.* either m or m' is even, and the quotients of $2Ann', 2Bnn', 2Cnn'$ divided by $\delta\delta'$ are $\frac{2A}{mm'}, \frac{2B}{mm'}, \frac{2C}{mm'}$; whence, again, A, B, C are integral.*

---

* Gauss shows that A, B, C are integral by substituting the values of $p_0, \ldots, q_0, \ldots$ in

$$q_1 q_2-q_0 q_3, \quad \tfrac{1}{2}(q_0 p_3+p_0 q_3-q_1 p_2-p_1 q_2), \quad p_1 p_2-p_0 p_3,$$

and observing that the results, after division by nn', are integral. The values of $p_0, \ldots$ are always obtained free from any common divisor by the process in the text; but Gauss has to determine four new multipliers, $\theta_0, \theta_1, \theta_2, \theta_3$, to obtain from the formulae (Σ) the exact values of $q_0, \ldots$, and not equimultiples of those values. M. Schläfli (Crelle, vol. lvii. p. 170) has shown that Gauss's demonstration is connected with a remarkable symbolical formula.

109. *Composition of several Forms.*—It will now be convenient to extend the definition of composition to the case in which more than two forms are compounded. If a quadratic form, F, be changed by a substitution, linear in respect of n binary sets, into the product of n quadratic forms, $f_1, f_2, \ldots, f_n$, so that

$$F(X, Y) = \prod_{i=1}^{i=n} (a_i x_i^2 + 2bx_i y_i + c_i y_i^2),$$

we shall say that F is transformable into $f_1 \times f_2 \times \ldots \times f_n$; and if the determinants of the matrix of the transformation are relatively prime, we shall say that F is compounded of $f_1, f_2, \ldots, f_n$. We shall retain, with an obvious extension, the notation of Art. 107. The invariant property of the determinant of F supplies the n equations

$$\Delta_i^2 f_i^2 = \frac{d_i}{D}[\Pi f]^2;$$

from which we infer (1) that $D, d_1, d_2, \ldots$ are to one another as square numbers, (2) that Dk^2 is the greatest common divisor of the n numbers $\dfrac{d_i}{m_i^2}\Pi m^2$. According as the equation $\Delta_i f_i = \sqrt{\dfrac{d_i}{D}} \cdot \Pi f$ is satisfied by a positive or negative value of the radical, we shall say that f_i is taken directly or inversely. Adopting this definition, we can enunciate the theorem—

'If F be compounded of $f_1, f_2, \ldots, f_n$, and F' be transformable into

$$f_1 \times f_2 \times \ldots \times f_n,$$

the forms being similarly taken in each case, F' contains F.' For we infer from (2) that $D'k'^2 = D$, whence $\Delta'_i = k'\Delta_i$, or by the Lemmas 2 and 1 of Art. 107,

$$X' = \alpha X + \beta Y, \quad Y' = \gamma X + \delta Y,$$

$\alpha, \beta, \gamma, \delta$ denoting integral numbers which satisfy the equation $\alpha\delta - \beta\gamma = k'$. We thus obtain the equation

$$F'(\alpha X + \beta Y, \ \gamma X + \delta Y) = F(X, Y),$$

whence, by Lemma 4, F' is transformed into F by $\begin{vmatrix} \alpha, \beta \\ \gamma, \delta \end{vmatrix}$.

If F be compounded of $f_1, f_2, \ldots, f_n$, and a single transformation of F into $f_1 \times f_2 \times \ldots \times f_n$ be given, we may, by the same principles, find all the transformations of F into the product of $f_1, f_2, \ldots, f_n$, taken as in the given transformation. For if $F(X_0, Y_0) = \Pi f$ represent the given transformation, and $F(X, Y) = \Pi f$ be any other transformation, we find

$$X = \alpha X_0 + \beta Y_0, \quad Y = \gamma X_0 + \delta Y_0, \quad \alpha\delta - \beta\gamma = +1,$$

and consequently $F(\alpha X_0+\beta Y_0,\ \gamma X_0+\delta Y_0)=F(X_0, Y_0)$;

or $\begin{vmatrix} \alpha, \beta \\ \gamma, \delta \end{vmatrix}$ is, by Lemma 4, a proper automorphic of F. The formula

$$X=\alpha X_0+\beta Y_0,\quad Y=\gamma X_0+\delta Y_0,$$

in which $\begin{vmatrix} \alpha, \beta \\ \gamma, \delta \end{vmatrix}$ is an automorphic of F, will therefore represent all the transformations required.

If F be transformable into $f_1 \times f_2 \times \dots \times f_n$, and Φ contain F, while $f_1, f_2, \dots, f_n$ contain $\phi_1, \phi_2, \dots, \phi_n$, Φ will be transformable into $\phi_1 \times \phi_2 \times \dots \times \phi_n$. This follows from a preceding general observation (Art. 105); but we must add here that if T, τ_i denote any positive or negative units, according as the transformations of Φ into F, and f_i into ϕ_i are proper or improper, while v_i denotes a positive or negative unit according as f_i is taken directly or inversely, ϕ_i will be taken directly or inversely according as $T \times \tau_i \times v_i$ is positive or negative. This is apparent if we observe that the sign of the quantity $\dfrac{f_i \Delta_i}{\Pi f}$ is altered by an improper transformation of X, Y, or x_i, y_i, but is not altered by a transformation of any of the other sets.

The theorem that 'forms compounded of equivalent forms, similarly taken, are themselves equivalent' is included in the preceding. We may, therefore, speak of the class compounded of any number of given classes.

It is an important and not a self-evident proposition, that if F be compounded of $\phi, f_3, f_4, \dots, f_n$, and ϕ be compounded of f_1, f_2, F is compounded of $f_1, f_2, \dots, f_n$. Let $\phi=\alpha\xi^2+2\beta\xi\eta+\gamma\eta^2$, let μ be the greatest common divisor of α, 2β, γ, and ∇ the determinant of ϕ; let also X, Y transform F into

$$\phi \times f_3 \times f_4 \times \dots \times f_n.$$

Writing in X and Y for ξ and η the bipartite expressions linear in $x_1 y_1$, $x_2 y_2$, by which ϕ is transformed into $f_1 \times f_2$, we obtain a transformation of F into $f_1 \times f_2 \times \dots \times f_n$. If k be the greatest common divisor of the determinants of the matrix of this transformation, Dk^2 is the greatest common divisor of the n numbers $\dfrac{d_i}{m_i^2}\Pi m^2$. But this common divisor is the same as the greatest common divisor of $\nabla \times \prod_{i=3}^{i=n} m_i^2$, and the $n-2$ numbers

$$\frac{d_i \mu^2}{m_i^2}\prod_{s=3}^{s=n} m_s^2, \qquad i=3, \dots, n;$$

because ∇ is the greatest common divisor of $d_1 m_2^2$ and $d_2 m_1^2$ (4th conclusion), and because $\mu = m_1 m_2$ (5th conclusion); *i.e.*, $Dk^2 = D$, or $k^2 = 1$, and F is compounded of $f_1, f_2, \ldots, f_n$. Also, if $i > 2$, f_i is similarly taken in both compositions, for

$$\frac{\Delta_i f_i}{f_1 \times f_2 \times \ldots \times f_n} \quad \text{and} \quad \frac{\Delta_i f_i}{\phi \times f_3 \times \ldots \times f_n}$$

are identical; and if $i = 1$, or 2,

$$\Delta_i = \frac{dX}{dx_i}\frac{dY}{dy_i} - \frac{dX}{dy_i}\frac{dY}{dx_i} = \left(\frac{dX}{d\xi}\frac{dY}{d\eta} - \frac{dX}{d\eta}\frac{dY}{d\xi}\right) \times \left(\frac{d\xi}{dx_i}\frac{d\eta}{dy_i} - \frac{d\xi}{dy_i}\frac{d\eta}{dx_i}\right),$$

whence, if Ω and ω_i be positive or negative units, according as ϕ and f_i are taken directly or inversely in the composition of F and ϕ respectively, f_i will be taken directly or inversely in the composition of F, according as $\Omega \times \omega_i$ is positive or negative.

By this theorem, the problem of finding a form compounded of any number of given forms is reduced to the problem of finding a form compounded of two given forms. For if $f_1, f_2, \ldots, f_n$ be the given forms, we may compound the first with the second, the resulting form with the third, and so on until we have gone through all the forms, when the form finally obtained will be compounded of the given forms, as will immediately appear from successive applications of the preceding theorem. We also see that we may compound the forms in any order that we please, or we may divide them into sets in any way we please, and compounding first the forms of each set, afterwards compound the resulting forms. If any of the given forms are to be taken inversely, we may substitute for them their *opposites* (Art. 92) taken directly. We may thus, without any loss of generality, and with some gain in point of simplicity, avoid the consideration of inverse composition altogether; and, for the future, when we speak of the form compounded of given forms, or the class compounded of given classes, we shall understand the form or class compounded directly of the given forms or classes.

110. The solution of the problem of composition given in Art. 108 may be put into a form better suited to actual computation.

The system (S) is evidently satisfied by $[0, P, Q, R]$, and also by $[P, 0, -S, -T]$; and these solutions are independent, because the determinants of their matrix cannot all be zero unless $P = 0$, a supposition which may be rejected as it implies that $a = 0$, *i.e.* that d is a square. From this set of independent solutions a set of fundamental solutions is deduced, as follows. Let μ be the greatest common divisor of P, Q, R; and let k be determined by the congruences

$$k\frac{Q}{\mu} - S \equiv 0, \quad k\frac{R}{\mu} - T \equiv 0, \text{ mod } \frac{P}{\mu},$$

which are simultaneously possible, because $\frac{Q}{\mu}$ and $\frac{R}{\mu}$ have no common divisor with the modulus, while the determinant

$$\frac{1}{\mu}(RS-QT)=-U\frac{P}{\mu}$$

is divisible by it. The solutions

$$\left[\mu,\ k,\ \frac{kQ-\mu S}{P},\ \frac{kR-\mu T}{P}\right],\quad \left[0,\ \frac{P}{\mu},\ \frac{Q}{\mu},\ \frac{R}{\mu}\right]$$

are then a fundamental set, and may be taken for $[p_0, p_1, p_2, p_3]$, $[q_0, q_1, q_2, q_3]$, respectively. We thus find

$$Ann'=\frac{PQ}{\mu^2},\ \text{or}\ A=\frac{aa'}{\mu^2};\qquad 2Bnn'=R+S-2k\frac{Q}{\mu}.$$

Multiplying this equation by $\frac{P}{\mu}$, $\frac{Q}{\mu}$, $\frac{R}{\mu}$ in succession, and attending to the congruences satisfied by k, we obtain the congruences

$$\frac{P}{\mu}B\equiv\frac{ab'}{\mu},\quad \frac{Q}{\mu}B\equiv\frac{a'b}{\mu},\quad \frac{R}{\mu}B\equiv\frac{bb'+Dnn'}{\mu},\ \text{mod}\ A;$$

which determine B for the modulus A, because $\frac{P}{\mu}$, $\frac{Q}{\mu}$, $\frac{R}{\mu}$ are relatively prime. These determinations [viz. of A, and of B, mod A] are sufficient for our purpose; because, if $B'=B+\lambda A$, the forms

$$\left(A,\ B,\ \frac{B^2-D}{A}\right)\quad\text{and}\quad\left(A,\ B',\ \frac{B'^2-D}{A}\right)$$

are equivalent. To obtain, therefore, the form compounded of two given forms (a, b, c), (a', b', c'), we first take the greatest common divisor of $d'm^2$ and dm'^2 for D (giving to D the sign of d or d'); we then determine n and n' by the equations $n=\sqrt{\frac{d}{D}}$, $n'=\sqrt{\frac{d'}{D}}$, and, representing by μ the greatest common divisor of an', $a'n$, $bn'+b'n$, we obtain A, B, C, from the system

$$A=\frac{aa'}{\mu^2},$$

$$\left.\begin{aligned}\frac{an'}{\mu}B&\equiv\frac{ab'}{\mu},\\ \frac{a'n}{\mu}B&\equiv\frac{a'b}{\mu},\\ \frac{bn'+b'n}{\mu}B&\equiv\frac{bb'+Dnn'}{\mu},\end{aligned}\right\}\ \text{mod}\ A.$$

$$C=\frac{B^2-D}{A}.$$

These formulae, which are applicable to every case of composition, and are therefore more general than the analogous formulae given by Gauss (Disq. Arith., Art. 243), are due to M. Arndt *, who has also given an independent investigation of them, though our limits have compelled us here to deduce them from Gauss's general solution of the problem of composition. That (A, B, C) is transformed into $(a, b, c)\times(a', b', c')$ by the substitution

$$\frac{1}{\mu}X = xx' + \frac{b'-Bn'}{a'}xy' + \frac{b-Bn}{a}x'y + \frac{bb'+Dnn'-B(bn'+b'n)}{aa'}yy',$$

$$\mu Y = \qquad an'xy' + \qquad a'nx'y + \qquad (b'n+bn')yy',$$

may be inferred from the values of $p_0, \ldots, q_0, \ldots$; or may be verified directly by observing that

$$\mu[AX+(B+\sqrt{D})Y] = [ax+(b+n\sqrt{D})y]\times[a'x'+(b'+n'\sqrt{D})y'].$$

111. *Composition of Forms—Method of Dirichlet.*—Lejeune Dirichlet, in an academic dissertation ('De formarum binariarum secundi gradus compositione,' Crelle, vol. xlvii. p. 155), has deduced the theory of the composition of forms from that of the representation of numbers. The principles of this method are applicable to any case of composition; but Dirichlet has restricted his investigation to properly primitive forms of the same determinant D. Let (a, b, c), (a', b', c') be two such forms; let M and M' be two numbers prime to $2D$, and capable of the primitive representations

$$M = am^2 + 2bmn + cn^2, \qquad M' = a'm'^2 + 2b'm'n' + c'n'^2,$$

by the forms (a, b, c) and (a', b', c') respectively; also let these representations appertain to the values ω and ω' of $\sqrt{D}$, so that

$$\omega^2 \equiv D, \text{ mod } M, \quad \omega'^2 \equiv D, \text{ mod } M',$$

---

* Crelle's Journal, vol. lvi. p. 64. In the new edition of the Disq. Arith. (Göttingen, 1863), a MS. note of Gauss is printed at p. 263, containing the congruences by which B is determined in the case of the direct composition of two forms of the same determinant.

The account of the theory of composition in the preceding Articles (106–109) differs from that in the Disq. Arith. (Arts. 234–243) chiefly in the use which is here made of the invariant property of the determinant. A different mode of treatment of Gauss's analysis is adopted by M. Bazin, in Liouville, vol. xvi. p. 161.

In Arts. 108 and 110 we have endeavoured to supply the analysis of a problem which Gauss, as is not unusual with him, has treated in a purely synthetical manner (Disq. Arith., Arts. 236 and 242, 243); and it is for this reason that we have introduced the consideration of *fundamental* sets of solutions of indeterminate systems, which are not explicitly mentioned in the Disq. Arith. It is perhaps singular that Gauss does not employ the identity $PU-QT+RS=0$; it was first given by M. Poullet Delisle, in a note on Art. 235 in his Translation of the Disq. Arith.

and so that the forms (a, b, c), (a', b', c') are respectively equivalent to the forms

$$\left(M,\ \omega,\ \frac{\omega^2-D}{M}\right),\quad \left(M',\ \omega',\ \frac{\omega'^2-D}{M'}\right).$$

If the values ω and ω' are *concordant, i.e.* if it is possible to find a number Ω satisfying the three congruences

$$\Omega^2 \equiv D, \text{ mod } MM',\quad \Omega\equiv\omega, \text{ mod } M,\quad \Omega\equiv\omega', \text{ mod } M',$$

(in which case the solution Ω of the congruence $\Omega^2\equiv D$, mod MM', may be said to *comprehend* the solutions ω and ω' of the congruences $\omega^2\equiv D$, mod M, and $\omega'^2\equiv D$, mod M',) the form

$$\left(MM',\ \Omega,\ \frac{\Omega^2-D}{MM'}\right)$$

will be a properly primitive form of determinant D, and will belong to one and the same class (which may be termed the class compounded of the classes containing (a, b, c) and (a', b', c')) whatever two numbers (subject to the conditions prescribed) are taken for M and M'. To prove this, a few preliminary remarks are necessary. (1) If the solutions ω and ω' are concordant, there is but one solution Ω (incongruous mod MM') comprehending them. (2) The necessary and sufficient condition for the concordance of ω and ω' is $\omega\equiv\omega'$, for every prime modulus dividing both M and M'. (3) If Ω, ω, ω' satisfy the congruence $x^2\equiv D$ for the modules MM', M, and M' respectively; and if, besides, $\Omega\equiv\omega$, $\Omega\equiv\omega'$, for every prime divisor of M and M' respectively, ω and ω' are concordant, and Ω is the solution comprehending them. (4) The value of $\sqrt{D}$, to which any given primitive representation (such as $M=am^2+2bmn+cn^2$) appertains, may be defined by congruences, without employing the numbers μ and ν which satisfy the equation $m\nu-n\mu=1$ (see Art. 86); in fact, we find

$$am+(b+\omega)n\equiv 0, \text{ mod } M,\quad (b-\omega)m+cn\equiv 0, \text{ mod } M;$$

whence also $\omega\equiv -b$, mod d, $\omega\equiv +b$, mod d', if d and d' are common divisors of M and m and of M and n.

We may suppose that the given forms (a, b, c) and (a', b', c') are so prepared*

---

* It is readily proved that a properly primitive form can represent numbers prime to any given number; thus a form can always be found equivalent to a given properly primitive form, and having its first coefficient prime to a given number. This transformation will be frequently employed in the sequel: in the present instance, we have only to substitute for the given forms any two forms respectively equivalent to them and having their first coefficients relatively prime.

that the representations of a and a' by them appertain to concordant values of $\sqrt{D}$; *i.e.* that we can find a number B satisfying the congruences

$$B^2 \equiv D, \text{ mod } aa', \quad B \equiv b, \text{ mod } a, \quad B \equiv b', \text{ mod } a'.$$

Let $\frac{B^2 - D}{aa'} = C$; the forms (a, b, c), (a', b', c') are then equivalent to $(a, B, a'C)$, (a', B, aC) respectively; and if

$$X = xx' - Cyy', \quad Y = axy' + a'x'y + 2Byy',$$

we find by actual multiplication

$$aa'X^2 + 2BXY + CY^2 = (ax^2 + 2Bxy + a'Cy^2) \times (a'x'^2 + 2Bx'y' + aCx'^2).$$

From this equation (which is included as a particular case in the formulae of M. Arndt) it appears that MM' is capable of representation by (aa', B, C); it can also be shown (1) that this representation is primitive; (2) that it appertains to a value of $\sqrt{D}$, mod MM', comprehending the values ω and ω', to which the representations of M and M' by (a, b, c) and (a', b', c') respectively appertain. (1) If x, y, x', y', and X, Y are the values of the indeterminates in the representations of M, M', and MM' by $(a, B, a'C)$, (a', B, aC), and (aa', B, C), the hypothesis that X and Y admit of a common prime divisor p is expressed by the simultaneous congruences

$$xx' - Cyy' \equiv 0, \quad axy' + a'x'y + 2Byy' \equiv 0, \text{ mod } p.$$

These congruences are linear in respect of the relatively prime numbers x' and y'; their coexistence implies, therefore, that p divides their determinant M; similarly it may be shown that p divides M'; so that $\omega \equiv \omega'$, mod p, because ω and ω' are concordant. The congruences satisfied by ω and ω' now give the relations

$$ax + (B + \omega) y \equiv 0, \quad a'x' + (B + \omega) y' \equiv 0, \text{ mod } p;$$

whence, eliminating x and x' from the congruence $Y \equiv 0$, and observing that 2ω is prime to M, and therefore to p, we find $yy' \equiv 0$, mod p. If y is divisible by p, we infer, from the congruence $X \equiv 0$, that x' is also divisible by p; but the congruences satisfied by ω and ω' give in this case the contradictory results $\omega \equiv +B$, $\omega \equiv -B$; *i.e.* y is not divisible by p, and similarly it may be shown that y' is not divisible by p. The congruence $yy' \equiv 0$, mod p, is therefore impossible; or the representation of MM' by (aa', B, C) is primitive. (2) Let Ω' be the value of $\sqrt{D}$, to which this representation appertains; and let p be any divisor of M; then Ω' satisfies the congruences

$$aa'X + (B + \Omega') Y \equiv 0, \quad (B - \Omega') X + CY \equiv 0, \text{ mod } p;$$

and it will be found, on substituting the values of X and Y, that these congruences are also satisfied by ω; whence it follows, since either X or Y is prime to p, that $\Omega' \equiv \omega$, mod p. Similarly, if p be a prime divisor of M', $\Omega' \equiv \omega'$, mod p; or Ω' is a solution of the congruence $\Omega'^2 \equiv D$, mod MM', comprehending the solutions ω and ω'. Hence $\Omega' \equiv \Omega$, mod MM', and the form

$$\left(MM', \Omega, \frac{\Omega^2 - D}{MM'}\right)$$

is equivalent to (aa', B, C), because either of them is equivalent to

$$\left(MM', \Omega', \frac{\Omega'^2 - D}{MM'}\right).$$

The equivalence of all the forms included in the expression

$$\left(MM', \Omega, \frac{\Omega^2 - D}{MM'}\right)$$

is therefore demonstrated.

It will be seen that Dirichlet's method may be applied to the composition of any number of forms, and that the theorems of Art. 109 present themselves as immediate consequences of his definition of composition.

112. *Composition of Classes of the same Determinant.*—We shall now consider more particularly the composition of classes of the same determinant D. We represent these classes by the letters $f, \phi, \ldots$, and we use the signs of equality and of multiplication to denote equivalence and composition respectively*. The following theorems are then immediately deducible from the six conclusions of Art. 107, and from the formulae of Art. 110.

(i.) 'If f be a properly primitive class, $f \times \Phi$ is of the same order as Φ.'

(ii.) 'A class is unchanged by composition with the principal class.' In consequence of this property, it is sometimes convenient to represent the principal class by 1.

(iii.) 'The composition of two opposite† properly primitive classes produces the principal class.'

If, then, f denote any properly primitive class, we may denote its opposite by f^{-1}, and we may write $f \times f^{-1} = 1$.

---

* Gauss uses the sign of addition instead of that of multiplication; thus $f \times \phi$ is $f + \phi$ in the Disq. Arith., and f^n is nf. The change appears to have been introduced by his French translator, and to have been acquiesced in by subsequent writers.

† Two classes which are improperly equivalent are called opposite, because they contain opposite forms (see Art. 92).

(iv.) 'If f be a given properly primitive class, and Φ any given class, the equation $F \times f = \Phi$ is always satisfied by one class, F, and by one only; viz. by the class $F = \Phi \times f^{-1}$.'

(v.) 'If $\Phi_1, \Phi_2, \ldots$ be all different classes, and f be a properly primitive class, $f \times \Phi_1, f \times \Phi_2, \ldots$ are all different classes.'

(vi.) 'A properly primitive ambiguous class produces by its duplication the principal class;' for an ambiguous class is its own opposite. Conversely, if $\phi^2 = 1$, *i.e.* if ϕ be a class which, by its duplication, produces the principal class, ϕ is a properly primitive ambiguous class; for we find $\phi^2 \times \phi^{-1} = \phi^{-1}$, whence $\phi = \phi^{-1}$, or ϕ and its opposite are properly equivalent.

(vii.) 'The class compounded of the opposites of two or more forms is the opposite of the class compounded of those forms.' It follows from this, or from (vi.), that a class compounded of ambiguous classes is itself ambiguous.

(viii.) Let $\Phi_0, \Phi_1, \ldots, \Phi_{\omega-1}$ represent all the classes of det. D, and of a given order Ω; and let $1, f_1, f_2, \ldots, f_{n-1}$ represent the properly primitive classes of the same determinant; it may then be shown that ω is a divisor of n, and that, given two classes of the order Ω, there always exist $\frac{n}{\omega}$ properly primitive classes, which, compounded with one of them, produce the other. Assuming, for a moment, that a form Φ_0 exists, such that the ω equations included in the formula $\Phi_0 \times f = \Phi_\mu$ can all be satisfied, we see that each of these equations is satisfied by the same number of properly primitive classes f; for if the equation $\Phi_0 \times f = \Phi_0$ be satisfied by k primitive classes, $1, \phi_1, \phi_2, \ldots, \phi_{k-1}$, the equation $\Phi_0 \times f = \Phi_\mu$, which is, by hypothesis, satisfied by a single class, f_μ, is also satisfied by the $k-1$ classes $f_\mu \times \phi_1, \ldots, f_\mu \times \phi_{k-1}$, but by no other class. Since, then, the classes $\phi_0 \times f$, of which the number is n, represent every class of the order Ωk times, we have evidently $n = k\omega$. It is also readily seen that every equation of the type $\Phi_\nu \times f = \Phi_\mu$ admits of k solutions; and thus it only remains to justify the assumption on which the preceding proof depends. If the order Ω be derived by the multiplier m from a properly primitive class of determinant $\Delta = \frac{D}{m^2}$, we may take for Φ_0 the class represented by the form $(m, 0, -\Delta m)$; if Ω be derived from an improperly primitive class, we take for Φ_0 the class represented by the form $(2m, m, -\frac{1}{2}m(\Delta - 1))$. Representing Φ_μ in the first case by the form (ma, mb, mc), and in the second by the form $(2ma, mb, 2mc)$, and supposing (as we may do) that a in each case is prime to $2D$, we see that the forms (a, mb, m^2c) and $(a, bm, 4cm^2)$ are properly primitive; and we find by the formulae of composition (Art. 110),

$$(m, 0, -\Delta m) \times (a, bm, cm^2) = (ma, mb, mc),$$
$$(2m, m, -\tfrac{1}{2}m(\Delta - 1)) \times (a, bm, 4cm^2) = (2ma, mb, 2mc);$$

i.e. the equation $\Phi_0 \times f = \Phi_\mu$ can be satisfied for every value of μ.

113. *Comparison of the numbers of Classes of different Orders.*—To determine the quotient $\frac{n}{\omega}$ of the last Article, Gauss investigates the properly primitive classes of det. D, which, compounded with the classes

$$(m, 0, -\Delta m) \quad \text{and} \quad (2m, m, -\tfrac{1}{2}m(\Delta - 1)),$$

reproduce those classes themselves. He thus employs the theory of composition to compare the number of properly primitive classes of a given determinant with the number of classes contained in any other order of the same determinant; or, which comes to the same thing, to compare the numbers of classes, of any given orders, of two determinants which are to one another as square numbers (Disq. Arith., Arts. 253–256). We have already seen (Art. 103) that the infinitesimal analysis of Dirichlet supplies a complete solution of this problem; whereas, in the case of a positive determinant, the result in its simplest form was not obtained by Gauss. It has, however, been recently shown by M. Lipschitz (Crelle, vol. liii. p. 238) that the formulae of Dirichlet may be deduced, in a very elementary manner, from the theory of transformation. We propose in this place to give an account of this investigation, and to point out its relation to the method pursued by Gauss. We begin with the theorem—

'Every properly primitive class of determinant De^2 is contained in one, and only one, properly primitive class of determinant D.'

Let (A, B, C) be a properly primitive form of det. De^2, in which A is prime to e; let B' be determined by the congruence $eB' \equiv B$, mod A, and C' by the equation $C' = \frac{B'^2 - D}{A}$; then the forms (A, B, C) and $(A, B'e, C'e^2)$ are equivalent; but $(A, B'e, C'e^2)$ is contained in (A, B', C'), therefore also (A, B, C) is contained in (A, B', C'), that is, in a properly primitive form of determinant D. Again, if (a, b, c), (a', b', c') are two forms of det. D, each containing (A, B, C) these two forms are equivalent. For applying to (A, B, C) the system of transformations of modulus e, included in the formula $\begin{vmatrix} m, & k \\ 0, & \mu \end{vmatrix}$ (Art. 88), we readily find that, of the resulting forms, one, and only one, will have its coefficients divisible by e^2 *; therefore the class represented by (A, B, C) contains one, and only one, class of det. De^4, and of the type (e^2p, e^2q, e^2r). But, applying to (A, B, C) the transformations inverse to those by which (a, b, c) and (a', b', c') are changed into (A, B, C), then (A, B, C) is changed thereby into (e^2a, e^2b, e^2c) and (e^2a', e^2b', e^2c'); these two forms are therefore equivalent; *i.e.* (a, b, c) and (a', b', c') are equivalent.

We have next to ascertain how many different properly primitive classes of determinant De^2 are contained in the class represented by (a, b, c), a properly primitive form of det. D, in which a may be supposed prime to e. Applying to (a, b, c) a complete system of transformations of modulus e, we inquire, in the first place, how many of the resulting forms are properly primitive. For this purpose we observe that if $e = e_1 \times e_2 \times e_3 \times \ldots$, ($e_1, e_2, \ldots$ representing factors of which no two have any common divisor), a complete system of transformations for the modulus e is obtained by compounding, in any definite order, the systems of transformations for the modules $e_1, e_2, \ldots$; *i.e.* if $|e_1|, |e_2|, \ldots$ be symbols representing complete systems of transformations for the modules $e_1, e_2, \ldots$, every transformation of modulus e is equivalent by post-multiplication † to one, and only one, of the transformations $|e_1| \times |e_2| \times |e_3| \times \ldots$. It will, therefore, be sufficient to determine the number of properly primitive forms obtained by applying to a properly primitive form a complete system of transformations for a modulus which is the power of a prime. Let p be an uneven prime, and let (a, b, c) be changed into (A, B, C) by $\begin{vmatrix} p^{\alpha-\gamma}, & k \\ 0, & p^{\gamma} \end{vmatrix}$, a formula which will represent a complete system of transformations for the modulus p^{α}, if γ receive every value from 0 to α inclusive, and if k be the general term of a complete system

---

* If $\begin{vmatrix} m, & k \\ 0, & \mu \end{vmatrix}$ transform (A, B, C) into (P, Q, R), we have

$$P = Am^2, \quad Q = m(Ak + B\mu), \quad R = Ak^2 + 2Bk\mu + C\mu^2.$$

Observing that A is prime to e, we infer from the congruence $P \equiv 0$, mod e^2, that $m = e$, $\mu = 1$; the congruence $Q \equiv 0$, mod e^2, then becomes $Ak + B \equiv 0$, mod e, giving one, and only one, value of k, mod e; and this value satisfies the remaining congruence $R \equiv 0$, mod e^2, since $AR = (Ak + B)^2 - De^2$.

† If $|A|$ and $|B|$ are two transformations connected by the symbolic equation

$$|B| = |A| \times |V|,$$

in which $|V|$ is a unit transformation, $|A|$ and $|B|$ are said to be equivalent by post-multiplication, or to belong to the same set. A complete system of transformations for any modulus contains one transformation belonging to each set.

of residues, mod $p^{\alpha-\gamma}$; we find

$$A = ap^{2(\alpha-\gamma)}, \quad B = (ak + bp^\gamma)\, p^{\alpha-\gamma}, \quad C = ak^2 + 2bkp^\gamma + cp^{2\gamma};$$

whence, if $\gamma = \alpha$, (A, B, C) is properly primitive; and if so, or not, for every other value of γ, according as C is not, or is, divisible by p. If $\gamma = 0$, we have $C \equiv 0$, for $p^{\alpha-1}\left[1 + \left(\frac{D}{p}\right)\right]$ values of k, incongrouous mod p^α; if γ have any value intermediate between 0 and α, we have $C \equiv 0$, for $p^{\alpha-\gamma-1}$ values of k, incongruous mod $p^{\alpha-\gamma}$. Hence the number of properly primitive forms is

$$[1 + p + p^2 + \ldots + p^\alpha] - p^{\alpha-1}\left[1 + \left(\frac{D}{p}\right)\right] - [p^{\alpha-2} + p^{\alpha-3} + \ldots + 1]$$
$$= p^\alpha\left[1 - \frac{1}{p}\left(\frac{D}{p}\right)\right];$$

and similarly if $p = 2$ it will be found that the number of properly primitive forms is 2^α. Hence the number N of properly primitive forms, arising from the application of a complete system of transformations of modulus e to the form (a, b, c), is $e\Pi\left[1 - \frac{1}{p}\left(\frac{D}{p}\right)\right]$, p denoting any uneven prime dividing e. It remains to determine the number of non-equivalent classes in which these N forms are contained. For brevity, we consider the case of a positive determinant. Let $[T_x, U_x]$ represent any solution of the equation $T^2 - DU^2 = 1$, and let σ be the index of the least solution of that equation which is also a solution of $T^2 - e^2DU^2 = 1$, *i.e.* let σ be the index of the first number in the series $U_1, U_2, \ldots$ which is divisible by e; also let (A, B, C) represent any one of the N properly primitive forms into which (a, b, c) is transformed. The transformations of modulus e by which (a, b, c) is changed into (A, B, C) belong to σ different sets, the transformations of the same set being equivalent by post-multiplication, but those of different sets not being so equivalent. For if $\begin{vmatrix} \alpha, & \beta \\ \gamma, & \delta \end{vmatrix}$ be a transformation of (a, b, c) into (A, B, C), any other transformation is represented (Art. 89) by the formula

$$\begin{vmatrix} T_x - bU_x, & -cU_x \\ aU_x, & T_x + bU_x \end{vmatrix} \times \begin{vmatrix} \alpha, & \beta \\ \gamma, & \delta \end{vmatrix},$$

and these two transformations will or will not belong to the same set, according as a unit transformation $\begin{vmatrix} \lambda, & \mu \\ \nu, & \rho \end{vmatrix}$, satisfying the equation

$$\begin{vmatrix} \alpha, & \beta \\ \gamma, & \delta \end{vmatrix} \times \begin{vmatrix} \lambda, & \mu \\ \nu, & \rho \end{vmatrix} \times \begin{vmatrix} T_x - bU_x, & -cU_x \\ aU_x, & T_x + bU_x \end{vmatrix} \times \begin{vmatrix} \alpha, & \beta \\ \gamma, & \delta \end{vmatrix},$$

does or does not exist. Premultiplying each side of this equation by $\begin{vmatrix} \delta, & -\beta \\ -\gamma, & \alpha \end{vmatrix}$, we find

$$e \times \begin{vmatrix} \lambda, & \mu \\ \nu, & \rho \end{vmatrix} = \begin{vmatrix} eT_x - BU_x, & -CU_x \\ AU_x, & eT_x + BU_x \end{vmatrix},$$

whence, observing that A, B, C are relatively prime, we see that λ, μ, ν, ρ are or are not integral according as U_x is, or is not, divisible by e; a conclusion which implies that the transformations of (a, b, c) into (A, B, C) are contained in σ different sets. It thus appears that, of the N transformations which applied to (a, b, c) give properly primitive forms, there are σ which give forms equivalent to (A, B, C); *i.e.* the number of properly primitive classes of det. De^2, contained in (a, b, c), a properly primitive class of det. D, is

$$\frac{N}{\sigma} = \frac{e}{\sigma} \Pi \left[1 - \frac{1}{p}\left(\frac{D}{p}\right)\right];$$

a result which is in accordance with the formula of Dirichlet (Art. 103). If D be negative, we have only to put $\sigma = 1$, as is sufficiently apparent from the preceding proof; if, however, $D = -1$, $\sigma = 2$.

The properly primitive classes of det. De^2, into which a given properly primitive class (a, b, c) of det. D is transformable, are always such that, compounded with the class $(e, 0, -De)$, they produce the class (ea, eb, ec). For let (a, b, c) be transformable into (A, B, C) of det. De^2, and let us take a form of the type $(A, B'e, C'e^2)$, equivalent to (A, B, C); then (a, b, c) and (A, B', C') are equivalent. But

$$(e, 0, -De) \times (A, B'e, C'e^2) = (eA, eB', eC'),$$

therefore also $$(e, 0, -De) \times (A, B, C) = (ea, eb, ec).$$

And, conversely, the classes which, compounded with $(e, 0, -De)$, produce (ea, eb, ec) are precisely the classes into which (a, b, c) is transformable. Thus the properly primitive classes of det. De^2, which, compounded with $(e, 0, -De)$, reproduce that class itself, are no other than the properly primitive classes of det. De^2 into which $(1, 0, -D)$ is transformable. And it is by this substitution of a problem of transformation for a problem of composition that M. Lipschitz has simplified and completed the analysis of Gauss.

A method similar in principle is applicable to the comparison of the numbers of properly and improperly primitive classes. We can first show that if $D \equiv 1$, mod 4, the double of every properly primitive class of det. D arises by a transformation of modulus 2 from one, and only one, improperly primitive class of the same determinant; viz. if (a, b, c) is a given properly primitive form *, in which a and b are uneven, $(2a, b, \frac{1}{2}c)$ is improperly primitive, and is changed into $(2a, 2b, 2c)$ by $\begin{vmatrix} 1, 0 \\ 0, 2 \end{vmatrix}$; and, again, if $(2p, q, 2r)$, $(2p', q', 2r')$ are two improperly primitive forms, each of which is transformable into $(2a, 2b, 2c)$, these two forms are equivalent, because (a, b, c) is transformable into $(4p, 2q, 4r)$ and also into $(4p', 2q', 4r')$, while it can be shown that (a, b, c) is transformable into the double of only one improperly primitive class. Also, applying the system of transformations, $\begin{vmatrix} 1, 0 \\ 0, 2 \end{vmatrix}$, $\begin{vmatrix} 2, 0 \\ 0, 1 \end{vmatrix}$, $\begin{vmatrix} 2, 1 \\ 0, 1 \end{vmatrix}$, to the improperly primitive form $(2p, q, 2r)$, we obtain, if $D \equiv 1$, mod 8, the double of only one properly primitive form: in this case therefore the numbers of properly and improperly primitive classes are equal. If $D \equiv 5$, mod 8, we obtain the doubles of three properly primitive forms; and we have to decide to how many different classes these three forms belong. It appears from Art. 89, that if $\begin{vmatrix} \alpha, \beta \\ \gamma, \delta \end{vmatrix}$ be a transformation of $(2p, q, 2r)$ into the double of a properly primitive form (a, b, c), all the transformations are included in the formula

$$\begin{vmatrix} \frac{1}{2}(T_x - qU_x), & -rU_x \\ pU_x, & \frac{1}{2}(T_x + qU_x) \end{vmatrix} \times \begin{vmatrix} \alpha, \beta \\ \gamma, \delta \end{vmatrix},$$

$[T_x, U_x]$ denoting any solution of the equation $T^2 - DU^2 = 4$. Taking the case of a positive determinant, and employing the same reasoning as before, we infer that if U_σ be the first of the numbers U_1, U_2, ... which is even, these transformations are contained in σ different sets. But σ is either 1 or 3 according as U_1 is even or uneven (see Art. 96, (vi.)); the three forms will therefore represent three classes or one, according as U_1 is even or uneven; and the number of properly primitive classes, in these two cases respectively, will be three times the number of improperly primitive classes, or equal to it. If D be negative, the three forms will belong to different classes; and there will be three times as many properly as improperly primitive classes. From this statement, however, we must except the determinant -3, which has one properly and one improperly primitive class.

It will be found that the properly primitive class or classes, into the double

---

* {If (a, b, c) is properly primitive, a and c uneven, b even, $\begin{vmatrix} 1, & \pm 1 \\ \mp, & 1 \end{vmatrix}$ transforms (a, b, c) into $2 \times$ an improperly primitive form; *i.e.* when there are improperly primitive forms at all, which implies $a + c \equiv 0$, mod 4. Either sign may be taken.}

of which a given improperly primitive class can be transformed, and which in turn can be transformed into the double of the given class, are also the class or classes which, compounded with the class $(2, 0, -\frac{1}{2}(D-1))$, produce the given class. Thus every improperly primitive class is connected either with one or three properly primitive classes (see Art. 98, note, and Art. 118).

114. *Composition of Genera.*—Let f and f' be two properly primitive classes of det. D, m and m' two numbers prime to one another and to $2D$, and represented by f and f' respectively; then mm' is represented by $f \times f'$. Hence the generic character of $f \times f'$ is obtained by multiplying together the values of the particular characters of f and f'. For those generic characters which are expressed by quadratic symbols this is evident, since

$$\left(\frac{mm'}{p}\right) = \left(\frac{m}{p}\right) \times \left(\frac{m'}{p}\right);$$

and it is equally true for the supplementary characters, since it will be found that

$$(-1)^{\frac{1}{2}(mm'-1)} = (-1)^{\frac{1}{2}(m-1)} \times (-1)^{\frac{1}{2}(m'-1)},\quad (-1)^{\frac{1}{8}(m^2m'^2-1)} = (-1)^{\frac{1}{8}(m^2-1)} \times (-1)^{\frac{1}{8}(m'^2-1)}.$$

The genus Γ, in which $f \times f'$ is contained, is said to be compounded of the genera γ and γ', in which f and f' are contained; and this composition is expressed by the symbolic equation $\Gamma = \gamma \times \gamma'$. It will be seen that the composition of any genus with itself gives the principal genus.

The same considerations may be extended to improperly primitive classes. Thus, if f and f' be respectively properly and improperly primitive, m and m' uneven numbers prime to one another and to D, represented by f and $\frac{1}{2}f'$, the genus of the improperly primitive class, $f \times f'$, may be inferred from the number mm', *i.e.* it is obtained by the composition of the generic characters of f and f'. Or, again, if f and f' be both improperly primitive, so that the class compounded of them is the double of an improperly primitive class, the generic character of this improperly primitive class is obtained by compounding those of the two given classes.

It follows, from these principles, that the number of classes in any two genera of the same order is the same. For if $\Phi_1, \Phi_2, \ldots, \Phi_n$ be all the classes of any genus of properly or improperly primitive forms, F_1 a class belonging to any other genus of the same order, and ϕ a properly primitive class satisfying the equation $\Phi_2 \times \phi = F_1$, the classes $\Phi_1 \times \phi, \ldots, \Phi_n \times \phi$ are all different, and all belong to the genus (F); consequently (F) has at least as many classes as (Φ), and *vice versâ* (Φ) has at least as many as (F), *i.e.* they both contain the same number of classes.

115. *Determination of the Number of Ambiguous Classes, and Demonstration of the Law of Quadratic Reciprocity.*—The number of actually existing genera of properly primitive forms cannot exceed the number of properly primitive ambiguous classes. For let n be the number of classes in each genus, k the number of actually existing genera, so that kn is the number of properly primitive classes; let also $1, A_1, A_2, \ldots, A_{h-1}$ be the properly primitive ambiguous classes. Every class produces, by its duplication, a class of the principal genus; and if K be a class of the principal genus produced by the duplication of X, K is also produced by the duplication of

$$X \times A_1, X \times A_2, \ldots, X \times A_{h-1},$$

but by the duplication of no other class. If, therefore, there be n' classes in the principal genus which can be produced by duplication, the whole number of properly primitive classes is $h \times n'$, *i.e.* $hn' = kn$. But $n' \leqq n$, therefore $k \leqq h$.

It may be inferred from Art. 112, (vii.), that all genera which contain any ambiguous classes contain an equal number of them. We shall immediately see that the number of ambiguous classes is equal to the number of genera, and is consequently a power of 2. The number of ambiguous classes in any genus is, therefore, either zero or a power of 2; and if any genus contain 2^κ ambiguous classes, such classes will exist only in $\frac{k}{2^\kappa}$ genera.

Gauss determines the number h of properly primitive ambiguous classes by very elementary reasoning. He first finds the number of properly primitive ambiguous forms of one or other of the two types $(A, 0, C)$ and $(2B, B, C)$, and then assigns the number of non-equivalent classes in which these forms are contained. Let D be divisible by μ different primes; and let us except the case $D = -1$. Resolving D in every possible manner into two positive or negative factors, having no common divisor but unity, we find $2^{\mu+1}$ properly primitive forms of the type $(A, 0, C)$; but we shall diminish this number by one-half by rejecting one of the two equivalent forms $(A, 0, C)$ and $(C, 0, A)$, viz. that in which $[A] > [C]$. There are no properly primitive forms of the type $(2B, B, C)$ unless $D \equiv 3$, mod 4, or $D \equiv 0$, mod 8; for one or other of these congruences is implied by the equation $D = B(B - 2C)$, because C is uneven. Resolving D into any two factors relatively prime, if $D \equiv 3$, mod 4, and having 2 for their greatest common divisor, if $D \equiv 0$, mod 8, we take one of them for B, the other for $B - 2C$; and we obtain, in either case, $2^{\mu+1}$ properly primitive forms of the type $(2B, B, C)$. If $BB' = -D$, it is easily seen that the forms $(2B, B, C)$ and $(2B', B', C')$* are equivalent. We may thus diminish by one-half the number of forms of the type $(2B, B, C)$, rejecting those in which $[B] > \sqrt{[D]}$. We conclude, therefore, that if we now denote by μ the number of *uneven* primes dividing D, we have in all $2^{\mu+2}$ ambiguous forms when $D \equiv 0$, mod 8, 2^{μ} when $D \equiv 1$, or $\equiv 5$, mod 8, and $2^{\mu+1}$ in every other case. These ambiguous forms we shall call Ω, and we observe that their number is equal to the whole number of assignable generic characters (Art. 98).

To find the number of non-equivalent classes in which these forms are contained, we consider separately the case of a positive and of a negative determinant. For a negative determinant, we diminish by one-half the number of the forms by rejecting the negative forms. The remaining forms, if of the type $(A, 0, C)$, are evidently reduced, because $A < C$; if of the type $(2B, B, C)$, they are also reduced, unless $2B > C$, an inequality which implies that $(C, C-B, C)$, to which $(2B, B, C)$ is equivalent, is reduced (Art. 92). The number of [positive] ambiguous classes is, therefore, one-half the number of the ambiguous forms Ω.

For a positive determinant, we deduce from the forms Ω an equal number of reduced ambiguous forms. Thus $(A, 0, C)$ is equivalent to (A, kA, C'); and because $[A] < \sqrt{D}$, this form is reduced, if kA be positive and be the greatest multiple of $[A]$ not surpassing $\sqrt{D}$. Similarly $(2B, (2k+1)B, C')$ is equivalent to $(2B, B, C)$, and is reduced if $(2k+1)B$ be positive, and be the greatest uneven multiple of $[B]$ not surpassing $\sqrt{D}$. There are, therefore, as many reduced ambiguous forms as there are forms in Ω; and there are no more, because it is readily seen that every reduced ambiguous form is included in one or other of the two series of forms (A, kA, C') and $(2B, (2k+1)B, C')$ which we have obtained. But every ambiguous class contains two reduced ambiguous forms (Art. 94); we infer, therefore, that for positive as well as for negative determinants the number of ambiguous classes is one-half the number of the forms Ω, *i.e.* one-half of the number of assignable generic characters.

Combining this result with the theorem at the commencement of this

---

* When the first two coefficients of a form are given, the third is given also; thus C' is here used for $\frac{B'^2-D}{2B'}$. Similar abbreviations will be employed occasionally in the sequel. The symbols $[A]$, &c. are used, as in Art. 92, to denote the absolute values of the quantities enclosed within the brackets.

article, we obtain a proof of the impossibility of at least one-half of the assignable generic characters. As this proof is independent of the law of quadratic reciprocity, we may employ the result to demonstrate that law. [Gauss's second demonstration, Disq. Arith., Art. 262.] Let p and q be two primes, and first let one of them, as p, be of the form $4n+1$. If $\left(\frac{q}{p}\right)=-1$, we infer that $\left(\frac{p}{q}\right)=-1$; for if $\left(\frac{p}{q}\right)=+1$, we should have $\omega^2 \equiv p$, mod q, and consequently there would exist a form $\left(q, \omega, \frac{\omega^2-p}{q}\right)$ of det. p, of which the character would be $\left(\frac{f}{p}\right)=-1$, *i.e.* there would be 2 genera of forms of determinant p. Similarly, if $\left(\frac{q}{p}\right)=+1$, we have $\omega^2 \equiv \pm q$, mod p; and $\left(p, \omega, \left(\frac{\omega^2 \mp q}{p}\right)\right.$ is a form of det. $\pm q$. If $\pm q$ be of the form $4n+1$, there will be but one genus of forms, *i.e.* the principal genus; whence $\left(\frac{p}{q}\right)=+1$. These two conclusions are sufficient to establish the theorem of reciprocity when one of the two primes is of the form $4n+1$. If both p and q be of the form $4n+3$, there are four assignable characters for the determinant pq. Of these

$$\left(\frac{f}{p}\right)=1, \quad \left(\frac{f}{q}\right)=1; \qquad \left(\frac{f}{p}\right)=-1, \quad \left(\frac{f}{q}\right)=-1;$$

are possible, as is shown by the existence of the forms

$$(1, 0, -pq), \quad (-1, 0, pq);$$

the other two are therefore impossible. Hence in the form $(p, 0, -q)$ we must have either

$$\left(\frac{p}{q}\right)=1=\left(\frac{-}{p}\right), \quad \text{or} \quad \left(\frac{p}{q}\right)=-1=\left(\frac{-q}{p}\right),$$

which expresses the theorem of reciprocity for this case. The supplementary theorems relating to 2 and -1 can be similarly proved.

116. *Equality of the Number of Genera and of Ambiguous Classes.*—In the preceding article it has only been shown that k cannot exceed h. But, as we have already seen (Art 102) that the number of actually existing genera is one-half the whole number of assignable generic characters, we know that $k=h$. To prove this, by the principles of the composition of forms, it is sufficient to show that $n=n'$, *i.e.* that the problem 'to find a class which by its duplication shall produce a given class of the principal genus' is always resoluble. This problem Gauss actually solves (Disq. Arith., Arts. 286, 287); he shows, first, that any proposed binary form, belonging to the principal genus of its own determinant, can be represented by the ternary

quadratic form X^2-2YZ; and, secondly, that from this representation we can always deduce a binary form, which shall produce by its duplication the proposed form. This solution implies a previous investigation of the theory of ternary quadratic forms, and cannot be properly introduced here.

A more elementary method, however, has been given by M. Arndt (Crelle, lvi. p. 72). Let $D=\Delta S^2$, S^2 representing any square dividing D; M. Arndt observes that the ratio of the number of actually existing genera to the whole number of assignable generic characters is the same for each of the two determinants D and Δ. To prove this we make use of the following subsidiary proposition :—

'If $f=(a, b, c)$ be a properly primitive form of any det. D, and if $8M$ and θ be two numbers relatively prime, the necessary and sufficient condition for the resolubility of the congruence

$$ax^2+2bxy+cy^2\equiv\theta, \text{ mod } 8M \quad . \quad . \quad . \quad . \quad . \quad . \quad . \quad . \quad \text{(A)}$$

is that the supplementary characters of f (if any), and the particular characters of f (if any) which relate to uneven primes dividing both M and D, should coincide with the corresponding characters of θ.'

We may add (though this is not necessary for our present purpose), that if θ_1 and θ_2 be two values of θ for each of which the congruence (A) is resoluble, it is resoluble for each an equal number of times.

On reference to the Table in Art. 98, it will be seen that the particular characters proper to the determinant Δ are included among the particular characters proper to D. Let then (Γ) and (Γ, Γ') represent any two complete generic characters for the determinants Δ and D, the particular characters common to the two complete characters having the same values attributed to them in each. It may then be shown that the genus (Γ, Γ') is or is not an existent genus, according as (Γ) is or is not existent. For (1) if (Γ, Γ') be actually existent, let θ be a number prime to $2D$ and capable of primitive representation by some class of that genus; the congruence $\omega^2\equiv D$, mod θ is therefore resoluble; *i.e.* the congruence $\omega^2\equiv\Delta$, mod θ, is resoluble, so that θ can be represented by a class of properly primitive forms of det. Δ, or the genus (Γ) is actually existent. And (2) if (Γ) be an existing genus, let f be a form included in (Γ), and θ a number prime to $2D$ and satisfying the generic character (Γ, Γ'). It appears from the subsidiary proposition that some number Θ of the linear form $8mD+\theta$ is capable of representation by f; if δ be the greatest common divisor of the indeterminates in the representation of Θ by f, the congruence $\omega^2\equiv\Delta$, and consequently the congruence

$\omega^2 \equiv D$, is resoluble for the modulus $\frac{\Theta}{\delta^2}$; *i.e.* $\frac{\Theta}{\delta^2}$, the character of which coincides with the character of θ, and therefore with that of the genus (Γ, Γ'), is capable of representation by a form of det. D, or (Γ, Γ') is an actually existing genus.

If, then, κ be the number of particular characters contained in (Γ, Γ') and not in (Γ), the numbers of actually existing genera and assignable generic characters for the det. D are each 2^κ times the corresponding numbers for the det. Δ.

It appears from this result that it will be sufficient for our present purpose to consider determinants not divisible by any square. If (a, b, c) be a form of the principal genus of such a determinant (we suppose that a is prime to D), the equation $ax^2 + 2bxy + cy^2 = \omega^2$ is resoluble with values of ω prime to D; for if $a = a'\delta^2$, δ^2 representing the greatest square divisor of a, the equation

$$\xi^2 - D\eta^2 = a'\zeta^2$$

is certainly resoluble in relatively prime integers, by virtue of a celebrated theorem of Legendre *; and the values of ζ which satisfy it are prime to D; whence, if

$$x = \mu\frac{\xi - b\eta}{a}, \quad y = \mu\eta, \quad \omega = \mu\frac{\zeta}{\delta},$$

μ denoting a multiplier, which renders the values of x, y and ω integral and relatively prime, the equation $ax^2 + 2bxy + cy^2 = \omega^2$ will be satisfied, and the values of ω will be prime to D. The form (a, b, c) is therefore equivalent to a form of the type (ω^2, λ, ν); and this form is produced by the duplication of $(\omega, \lambda, \nu\omega)$ if ω be uneven, and of $(2\omega, \lambda + \omega, \nu')$ if ω be even.

117. *Arrangement of the Classes of the Principal Genus.*—If C be a class of the principal genus, the classes $C, C^2, C^3, \ldots$ will all belong to that genus. And it will be found, by reasoning similar to that employed in Euler's second proof of Fermat's theorem (see Art. 10 of this Report), (1) that the classes $1, C, C^2, \ldots$ are all different until we arrive at a class C^μ, equivalent to the principal class; (2) that μ is either equal to, or a divisor of, the number n of classes in the principal genus; (3) that if $C^r = 1$, r is a multiple of μ. The μ classes $C, C^2, C^3, \ldots, C^{\mu-1}, 1$, are called the period † of the class C; C is said to appertain to the exponent μ; and the determinant is

---

* Théorie des Nombres, ed. 3, vol. i. p. 41; Disq. Arith., Art. 294.

† These periods of non-equivalent classes are not to be confounded with the periods of equivalent reduced forms of Art. 93.

regular or *irregular* according as classes do or do not exist which appertain to the exponent n. With the former case we may compare the theory of the residues of powers for a prime modulus; with the latter the same theory for a modulus composed of different primes (see Art. 77).

(i.) When the determinant is regular, we may take any class appertaining to the exponent n as a basis, and may represent all the classes of the principal genus (to which we at present confine ourselves) as its powers. It will then appear (1) that if d be a divisor of n, the number of classes appertaining to the exponent d is $\psi(d)$; so that, for example, the number of classes that may be taken for a base is $\psi(n)$: (2) that if $ef=n$, the equation $X^e=1$ will be satisfied by e classes of the principal genus; and if these classes be represented by $A_1, A_2, \ldots, A_e$, each of the equations $X^f=A$ will be satisfied by f different classes of the same genus: (3) that the only classes of the principal genus which satisfy the equation $X^k=1$ are those which satisfy the equation $X^d=1$, where d is the greatest common divisor of k and n.

It will be seen in particular that the equation $X^2=1$ admits of only one, or only two solutions, according as n is uneven or even; *i.e.* the principal genus of a regular determinant cannot contain more than two ambiguous classes.

To obtain a class appertaining to the exponent n, Gauss employs the same method which serves to find a primitive root of a prime number (Art. 13; Disq. Arith., Arts. 73, 74), and which reposes on the observation, that if A and B be two classes appertaining to the exponents α and β, neither of which divides the other, and if M, the least common multiple of α and β, be resolved into two factors p and q, relatively prime and such that p divides α and q divides β, the class $A^{\frac{\alpha}{p}} \times B^{\frac{\beta}{q}}$ will appertain to the exponent M.

(ii.) When the determinant is irregular, the classes of the principal genus cannot be represented by the simple formula C^i, and we must employ an expression of the form $C_1^{i_1} \times C_2^{i_2} \times C_3^{i_3} \ldots$. To obtain an expression thus representing all the classes of the principal genus, we take for C_1 a class appertaining to the greatest exponent θ_1 to which any class can appertain; and in general for C_μ we take a class appertaining to the greatest exponent θ_μ to which any class can appertain when its period contains no class, except the principal class, capable of representation by the formula $C_1^{i_1} \times C_2^{i_2} \times \ldots \times C_{\mu-1}^{i_{\mu-1}}$. The number $\frac{n}{\theta_1} = \theta_2 \times \theta_3 \times \ldots$ is called by Gauss the exponent of irregularity; and similarly we might term $\frac{n}{\theta_1\theta_2}$, $\frac{n}{\theta_1\theta_2\theta_3}$, &c., the second, third, &c., exponents of irregularity. From the mode in which the formula $C_1^{i_1} \times C_2^{i_2} \times \ldots$ is obtained, it can be inferred that θ_1 is divisible by θ_2, θ_2 by θ_3, and so on; whence it appears that a determinant cannot be irregular unless n be divisible by a square; nor can it have r indices of irregularity unless n be divisible by a power of order $r+1$. Moreover, whenever the principal genus contains but one ambiguous class, the determinant is either regular or has an uneven exponent of irregularity; if, on the contrary, the principal genus contain more than two ambiguous classes, the determinant is certainly irregular, and the index of irregularity even; if it contain 2^κ ambiguous classes, the irregularity is at least of order κ, and the κ exponents of irregularity are all even.

A few further observations are added by Gauss. Irregularity is of much less frequent occurrence for positive than for negative determinants; nor had Gauss found any instance of a positive determinant having an uneven index of irregularity (though it can hardly be doubted that such determinants exist). The negative determinants included in the formulae,

$$-D = 216k + 27, \quad = 1000k + 75, \quad = 1000k + 675,$$

except -27 and -75, are irregular, and have an index of irregularity divisible by 3. In the first thousand there are five negative determinants (576, 580, 820, 884, 900) which have 2 for their exponent of irregularity, and eight (243, 307, 339, 459, 675, 755, 891, 974) which have 3 for that exponent; the numbers of determinants having these exponents of irregularity are 13 and 15 for the second thousand, 31 and 32 for the tenth. Up to 10,000 there are, possibly, no determinants having any other exponents of irregularity; but it would seem that beyond that limit the exponent of irregularity may have any value.

118. *Arrangement of the other Genera.*—In the preceding article we have attended to the classes of the principal genus only; to obtain a natural arrangement of all the properly primitive classes, we observe that, if the number of genera be 2^μ, the terms of the product

$$(1+\Gamma_1)(1+\Gamma_2)(1+\Gamma_3)\ldots(1+\Gamma_\mu),$$

in which Γ_i represents any genus not already included in the product of the $i-1$ factors preceding $1+\Gamma_i$, will represent all the genera. If, then, $A_1, A_2, \ldots, A_\mu$ represent any classes of the genera $\Gamma_1, \Gamma_2, \ldots, \Gamma_\mu$ respectively, and $|C|$ be the formula representing all the classes of the principal genus, the expression

$$|K| = |C| \times (1+A_1)(1+A_2)\dots(1+A_\mu)$$

supplies a type for a simple arrangement of all the classes of the given determinant. When every genus contains an ambiguous class, it is natural to take for $A_1, A_2, \dots, A_\mu$, the ambiguous classes contained in the genera $\Gamma_1, \Gamma_2, \dots, \Gamma_\mu$ respectively. When the principal genus contains two ambiguous classes (and when, consequently, one-half of the genera contain no such classes), let C_1 be the class taken as base (or, if the determinant be irregular, as first of the bases) in the arrangement of the classes of the principal genus, and let $\Omega_1^2 = C_1$; it may then be shown that Ω_1 will belong to a genus containing no ambiguous class, and that the formula

$$|K| = |C| \times (1+\Omega_1)(1+A_2)\dots(1+A_\mu),$$

in which $A_2, \dots, A_\mu$, are ambiguous classes, represents all the classes *. In general, if the principal genus contain 2^κ ambiguous classes (a supposition which implies that the determinant is irregular, having κ even exponents of irregularity, and that there are only $2^{\mu-\kappa}$ genera containing ambiguous classes) let

$$\Omega_1^2 = C_1, \quad \Omega_2^2 = C_2, \dots, \quad \Omega_\kappa^2 = C_\kappa;$$

it will be found that all the classes are represented by the formula

$$|K| = |C| \times (1+\Omega_1)(1+\Omega_2)\dots(1+\Omega_\kappa)(1+A_{\kappa+1})\dots(1+A_\mu),$$

in which $A_{\kappa+1}, \dots, A_\mu$ are ambiguous classes, and $\Omega_1, \Omega_2, \dots, \Omega_\kappa$ classes belonging to genera containing no ambiguous class †.

A similar arrangement of the improperly primitive classes (when such classes exist) is easily obtained. Let Σ denote the principal class of improperly primitive forms, *i.e.* the class containing the form

$$(2, 1, -\tfrac{1}{2}(D-1));$$

we have seen (Art. 113) that the number of properly primitive classes which,

* Gauss employs a class Ω_1 producing C_1 by its duplication, both when one and when two ambiguous classes are contained in the principal genus. The number of classes requisite for the construction of the complete system of classes is therefore μ in either case, since C_1 may be replaced by Ω_1^2.

† The principles employed by Gauss for the arrangement of the classes of a regular determinant are extended in the text to irregular determinants. If the determinant have κ' *uneven* exponents of irregularity, the number of classes requisite for the construction of the complete system of classes is $\mu+\kappa'$.

compounded with Σ, produce Σ, is either one or three. When there is only one such class, the number of improperly primitive classes is equal to that of properly primitive classes; and if $|K|$ be the general formula representing the properly primitive classes, the improperly primitive classes will be represented by $\Sigma \times |K|$. When there are three properly primitive classes, which, compounded with Σ, produce Σ, the principal class will be one of them, and if ϕ be another of them, ϕ^2 will be the third; also ϕ and ϕ^2 will belong to the principal genus, and will appertain to the exponent 3. When the determinant is regular, instead of the complete period of classes of the principal genus, $1, C, C^2, \ldots, C^{n-1}$, we take the same series as far as the class $C^{\frac{1}{3}n}$ exclusively; when the determinant is irregular, we can always choose the bases $C_1, C_2, \ldots$ in such a manner that the period of one of them shall contain ϕ and ϕ^2, and this period we similarly reduce to its third part by stopping just before we come to ϕ or ϕ^2. Employing these truncated periods, instead of the complete ones, in the general expression for the properly primitive classes, we obtain an expression, which we shall call $|K'|$, representing a third part of the properly primitive classes, and such that $\Sigma \times |K'|$ represents all the improperly primitive classes.

119. *Tabulation of Quadratic Forms.*—In Crelle's Journal, vol. lx. p. 357, Mr. Cayley has tabulated the classes of properly and improperly primitive forms for every positive and negative determinant (except positive squares) up to 100. The classes are represented by the simplest forms contained in them *; the generic character of each class, and, for positive determinants, the period of reduced forms (Art. 93) contained in it, are also given. The arrangement of the genera and classes is in accordance with the construction of Gauss, explained in the preceding articles; and the position of each class in the arrangement is indicated by placing opposite to it, in a separate column, the term to which it corresponds in the symbolic formula (such as $|K|$ or $\Sigma \times |K|$) which forms the type of the arrangement. To the two Tables of positive and negative determinants Mr. Cayley has added a third, containing the thirteen irregular negative determinants of the first thousand.

---

* The simplest form contained in a class is that form which has the least first coefficient of all forms contained in the class, and the least second coefficient of all forms contained in the class and having the least first coefficient. If a choice presents itself between two numbers differing only in sign, the positive number is preferred. In the case of an ambiguous class of a positive determinant, the simplest ambiguous form contained in the class is taken as its representative.

In a letter addressed to Schumacher, and dated May 17, 1841, Gauss expresses a decided opinion of the uselessness of an extended tabulation of quadratic forms. 'If, without having seen M. Clausen's Table, I have formed a right conjecture as to its object, I shall not be able to express an opinion in favour of its being printed. If it is a canon of the classification of binary forms for some thousand determinants, that is to say, if it is a Table of the reduced forms contained in every class, I should not attach any importance to its publication. You will see, on reference to the Disq. Arith., p. 521 (note), that in the year 1800 I had made this computation for more than four thousand determinants' [viz., for the first three and tenth thousands, for many hundreds here and there, and for many single determinants besides, chosen for special reasons]; 'I have since extended it to many others; but I have never thought it was of any use to preserve these developments, and I have only kept the final result for each determinant. For example, for the determinant $-11{,}921$, I have not preserved the whole system, which would certainly fill several pages*, but only the statement that there are 8 genera, each containing 21 classes. Thus, all that I have kept is the simple statement viii. 21, which in my own papers is expressed even more briefly. I think it quite superfluous to preserve the system itself, and much more so to print it, because (1) any one, after a little practice, can easily, without much expenditure of time, compute for himself a Table of any particular determinant, if he should happen to want it, especially when he has a means of verification in such a statement as viii. 21; (2) because the work has a certain charm of its own, so that it is a real pleasure to spend a quarter of an hour in doing it for one's self; and the more so, because (3) it is very seldom that there is any occasion to do it. . . . . . . My own abbreviated Table of the number of genera and classes I have never published, principally because it does not proceed uninterruptedly.'† Probably the third of Gauss's three reasons will commend itself most to mathematicians who do not possess his extraordinary powers of computation. An abbreviated Table of the kind he describes, extending from $-10{,}000$ to $+10{,}000$, would occupy only a very limited space, and might be computed from Dirichlet's formulae for the number of classes (see Art. 104), without constructing systems of representative forms. But it would, perhaps, be desirable (nor would it

* Mr. Cayley's Table of the first hundred negative determinants occupies about four pages of Crelle's Journal; the determinant $-11{,}921$ would occupy about one page.

† Briefwechsel zwischen C. F. Gauss und H. C. Schumacher, vol. iv. p. 30.

increase the bulk of the Table to any enormous extent) to give for each determinant not only the number of genera, and of classes in each genus, but also the elements necessary for the construction, by composition only, of a complete system of all the classes. For this purpose it would not be necessary to specify (by means of representative forms) more than 5 or 6 classes, in the case of any determinant within the limits mentioned.

IX.

REPORT ON THE THEORY OF NUMBERS.

PART V.

[Report of the British Association for 1863, pp. 768–786.]

---

120. *GEOMETRICAL Representation of Forms of a Negative Determinant.*—Before quitting the subject of binary quadratic forms, we have still to mention several investigations of great interest, relating chiefly to forms of a negative determinant. We shall first refer to the geometrical considerations which Gauss has employed to illustrate the nature of these forms *.

Let an infinite plane area be divided by two systems of parallel lines into similar and equal parallelograms. The vertices of these parallelograms we shall call nodes; and we observe that every system of nodes possesses the characteristic property, that if it be displaced without rotation in its own plane, so as to bring any one node into a position originally occupied by any other node, then every node will also occupy a position originally occupied by another node; and the system in its second position will entirely coincide with the system in its original position. From this property we infer that the system of nodes admits of an infinite number of parallelisms besides the given parallelism; *i.e.* that it may be regarded, in an infinite number of different ways, as dividing the plane area into similar and equal parallelograms. For let O and O' be any two nodes such that no node lies on OO' between O and O'; let P be one of those nodes which lie the nearest to the line OO' produced indefinitely both ways, and let PP' be drawn parallel and equal to OO'; then

* See Gauss's review of Seeber's 'Untersuchungen über die Eigenschaften der ternären quadratischen Formen,' in the Göttingen 'Gelehrte Anzeigen' for 1831, No. 108, or in Crelle's Journal, vol. xx. p. 312; also Lejeune Dirichlet, Crelle, vol. xl. p. 209.

P' is a node, and $OO'P'P$ is a parallelogram of which the vertices are nodes, and which has no other node either on its contour or in its interior; such a parallelogram we shall call an elementary parallelogram. It is then evident from the characteristic property of the system, that every elementary parallelogram supplies us with a parallelism of the system; also we can obtain an infinite number of dissimilar elementary parallelograms; for if Ox and Oy are the two lines of the given parallelism which intersect in O, and if m and n are any two integers relatively prime, the intersection of the mth parallel from Ox with the nth parallel from Oy will give a node O' such that no node can lie on OO' between O and O'; and, again, instead of P in the preceding construction, we may take any node lying on either of the two lines of the system which are the nearest to OO'. The areas, however, of all elementary parallelograms are equal. To prove this, we observe that if AOB is an elementary triangle (*i.e.* a triangle of which the vertices are nodes, but which has no other node either on its contour or inside it), the parallelogram $OAO'B$, obtained by drawing parallels to any two of its sides OA and OB through the opposite vertices B and A, is an elementary parallelogram. For if AO and BO are produced to A' and B', so that O bisects AA' and BB', A' and B' are nodes, and the triangle $A'OB'$ is elementary; because if there were a node x' (other than its vertices) in $A'OB'$, we could immediately construct a node x (other than its vertices) in AOB. But $A'OB'$ can be made to coincide with $BO'A$ by a displacement without rotation; therefore $BO'A$ is elementary as well as AOB; or the parallelogram $AOBO'$ is elementary. Hence, if two elementary triangles have a common base, they are certainly equal. For if through the vertex of either triangle we draw a parallel to the base, an elementary parallelogram will be contained between that parallel and the base; that is, the altitude of either triangle will be the distance of the base from the parallel nearest to the base; or the triangles will be equal. Again, let AOB, aOb, be any two elementary triangles, which we may suppose to have a common vertex; if BOa is an elementary triangle, they are each of them equal to it and to one another; if not, let x be that node contained in BOa which lies the nearest to OB, then BOx is elementary, and has the side BO in common with AOB; by proceeding in this manner we shall form a series of elementary triangles, of which the first is AOB, and the last aOB, each triangle having a side in common with that preceding it, whence $AOB = aOb$; *i.e.* any two elementary parallelograms are equal.

We shall next show that it is always possible to find a *reduced* parallelogram, *i.e.* an elementary parallelogram, the sides of which are not greater than its diagonals. Let O be any node; A a node as near to O as any other; B a node on one of the parallels nearest to OA, and as near to O as any node on either of those parallels; complete the elementary parallelogram $OAO'B$; it will have the property required. Produce $O'B$ to O'', making $BO''=O'B$; then $AB=OO'$; but by hypothesis $OA \leqq OB$, and $OB \leqq OO'$, $OB \leqq OO''$; *i.e.* the sides of $OAO'B$ are not greater than its diagonals.

Again, if $OAO'B$ is a reduced parallelogram in which $OA \leqq OB$, it can be proved that no node lies nearer to O than A, and that no node, out of the line OA, lies nearer to O than B; for, first, no node on the line $O''BO'$ lies nearer to O than B, because by hypothesis $OB \leqq OO'$, $OB \leqq OO''$, and because the extremity of the perpendicular drawn from O to $O''O'$ falls between the points of bisection of the segments $O''B$ and BO', or on one of those points: secondly, no node on any parallel beyond $O''BO'$ can lie as near to O as B, for the limits of the angle AOB are evidently 60° and 120°; whence the perpendicular distance of OA from the parallel nearest to it but one is $\geqq OB\sqrt{3}$; *i.e.* the distance of any node on that parallel from O is $> OB$.

If then we join any node O, first to a node A, which lies as near to O as any other node, and, secondly, to a node B, which lies as near to O as any node out of the line OA, the joining lines are adjacent sides of a reduced parallelogram; for, by what precedes, B must lie on one or other of the parallels nearest to OA.

In general, a system of nodes has but one reduced parallelism, because in general there is a pair of opposite nodes AA', each of which is nearer to O than any other node whatever, and a second pair of opposite nodes BB', not lying in the line AOA', each of which is nearer to O than any node not lying in that line. Even if A and B are equidistant from O, provided only that their common distance from O is less than the distance of any other node from O, the system has but one reduced parallelism. But there are two special cases in which a nodal system admits of more than one reduced parallelism.

1. If there is one pair of opposite nodes AA' nearer to O than any other node, and two pairs BB', bb', equidistant from O, not lying in the line AOA', and nearer to O than any other node not in that line, the system admits of two reduced parallelisms, having one set of parallels in common, and having their common set of parallels equally inclined to the other two sets.

2. If there are three pairs of points at the minimum distance from O, the system of nodes forms a system of equilateral triangles; and, suppressing

in turn each one of the three systems of parallel lines by which these triangles are formed, we obtain the three reduced parallelisms of which the system admits.

That, in these two cases, the reduced parallelisms are such as we have described, and that, except in these two cases, there is but one reduced parallelism, may be inferred from the existence of a reduced parallelogram in every system, and from the properties which have been shown to belong to it.

To apply these results to the theory of quadratic forms, let $ax^2+2bxy+cy^2$ be a form of the negative determinant $-\Delta$; let $\cos\omega=\frac{b}{\sqrt{ac}}$, and with a pair of axes inclined to one another at an angle ω, let us construct all the points whose coordinates are integral multiples of $\sqrt{a}$ and $\sqrt{c}$ respectively; thus forming a nodal system. The expression $ax^2+2bxy+cy^2$ will then represent the square of the distance between any two nodes, the differences of whose coordinates are $x\sqrt{a}$ and $y\sqrt{b}$: and the area of an elementary parallelogram will be $\sqrt{\Delta}$. If the transformation $x=\alpha X+\beta Y$, $y=\gamma X+\delta Y$, where $\alpha\delta-\beta\gamma=\pm 1$, change $ax^2+2bxy+cy^2$ into $AX^2+2BXY+CY^2$; and if, in the same plane as before, we construct a nodal system corresponding to the latter form—the directions of rotation from the axis of X to the axis of Y, and from the axis of x to that of y, being the same—it will be found that the two systems may be made to coincide. For if we consider the point in the first system whose coordinates are $x\sqrt{a}$, $y\sqrt{c}$ as corresponding to the point in the second system whose coordinates are $X\sqrt{A}$, $Y\sqrt{C}$, the distance between any two points of the first system is equal to that between the corresponding points of the second system; therefore the two systems are identical, and are either similarly situated, *i.e.* are capable of being made to coincide by moving either of them about in their common plane, or else are symmetrically situated, *i.e.* are capable of being made to coincide after the plane of one of them has been turned over and applied again to the plane of the other. On comparing any two corresponding triangles in the two systems, for example the triangle obtained by giving to X and Y the values (0, 0), (1, 0), (0, 1), with the triangle obtained by giving to x and y the values (0, 0), (α, γ), (β, δ), it will be seen that the two systems are similarly or symmetrically situated, according as $\alpha\delta-\beta\gamma=+1$, or $=-1$.

It thus appears that a class of quadratic forms of a negative determinant may be considered to represent a nodal system, and that each form of the class corresponds to a parallelism of the system. Conversely, to each parallelism of the system a form of the class corresponds. For let Ox, Oy be lines of any parallelism of the system, and OX, OY lines of any other parallelism, the directions of rotation from Ox to Oy and from OX to OY being the same; let also $\sqrt{a}$, $\sqrt{c}$ be the lengths of the sides of an elementary parallelogram in the first system, and $\frac{b}{\sqrt{ac}}$ the cosine of the angle between them; and let $\sqrt{A}$, $\sqrt{C}$, $\frac{B}{\sqrt{AC}}$ have the same signification with regard to the second system; then, if

$$(x\sqrt{a},\ y\sqrt{c}),\quad (X\sqrt{A},\ Y\sqrt{C})$$

are the coordinates of the same node P, we must have two equations of the form

$$x=\alpha X+\beta Y,\qquad y=\gamma X+\delta Y,$$

in which x and y are integral if X and Y are so, and *vice versâ*; hence α, β, γ, δ are integral, and $\alpha\delta-\beta\gamma=+1$; the sign of the unit being determined by the supposition we have made as to the situation of the axes with respect to one another. Also

$$OP^2=ax^2+2bxy+cy^2=AX^2+2BXY+CY^2;$$

or the two given parallelisms are represented by two properly equivalent forms.

The theorem that in every nodal system a reduced parallelism exists, has for its arithmetical expression, 'In every class a form exists in which $[2b]\leqq[a]$, $[2b]\leqq c$.' We thus obtain an independent proof of the theory of reduction of Art. 92; the geometrical signification of the special conditions in the definition of a reduced form is as follows:—If $a=c>[2b]$, the corresponding nodal system has only one reduced parallelism; but either of the two directions in this reduced parallelism may be taken for the axis of x, consistently with the condition that the rotation from Ox to Oy should have a given direction; the condition $2b\geqq 0$ implies that if the angle between the axes is not right, that direction is to be assumed for the axis of x which renders the angle between Ox and Oy acute. Similarly, if $a<c$, but $a=[2b]$, the system has two reduced parallelisms, and the condition $2b\geqq 0$ distinguishes one of them from the other. If $a=[2b]=c$, the system has three reduced parallelisms, which are identical and similarly placed; the condition $2b>0$ does not distinguish between these, but only between the two modes in which any one of them can be taken.

The number of automorphics of a class may be ascertained by causing the nodal system which represents it to revolve in its own plane round one of its nodes and examining the number of positions in which it coincides with its original position. After a revolution of 180° it will always do so; but in order that it should do so in any other position, the first and second sides of its reduced parallelogram must be equal, and must include an angle of 90° or 60°, *i.e.* the system must be one of squares or of equilateral triangles. Hence we infer (Art. 90) that there are in general but two automorphics for a form of a negative determinant, but that for the classes containing the forms x^2+y^2 and $2x^2+2xy+2y^2$ (or multiples of those forms) there are four and six respectively.

Similarly we may investigate the conditions for the ambiguity of a class. In order that a class should be ambiguous, the nodal system representing it must be symmetrically equivalent to itself. If therefore there is but one reduced parallelogram, that parallelogram must be symmetrically equivalent to itself, *i.e.* it must be either a rectangle or a rhombus. When there are two reduced parallelograms, we have seen that they are symmetrically equivalent to one another; and when there are three, they are each of them rhombs. We thus obtain the conclusion that if (a, b, c) is the reduced form of an ambiguous class, either $b=0$, or $a=c$, or $a=2b$ (Art. 94).

121. *Application of Formulae relating to the Division of the Circle to the Theory of Quadratic Forms.*—We have already referred to the trigonometrical solutions of the equation $T^2-DU^2=1$ (Art. 96, (ix).) and to the connexion existing between them, and the number of classes of quadratic forms of determinant D (Art. 104).

If p is a prime of the form $3n+1$ or $4n+1$, the coefficients of the cubic, or biquadratic, equation of the periods depend on the values of the indeterminates in the equation $4p=x^2+3y^2$, or $p=x^2+y^2$ (Art. 43). Thus in these two cases, if, for any given value of p, we calculate the equation of the periods, we obtain, by a direct though tedious process, the values of the indeterminates in certain simple quadratic decompositions of $4p$ or p. But the theory of the division of the circle supplies a method equally direct and of more general application for the investigation of such decompositions in certain cases. The principles of this method were discovered by Gauss, who deduced from them the first of the three following theorems:—

'If $p=4n+1=x^2+y^2$,

$$x \equiv \tfrac{1}{2}\,\frac{\Pi 2n}{\Pi n\,.\,\Pi n},\ \text{mod } p;\quad x\equiv 1,\ \text{mod } 4;$$

$$y \equiv \pm\,\frac{\Pi 2n\,.\,\Pi 2n}{\Pi n\,.\,\Pi n},\ \text{mod } p.\text{'}$$

(Gauss, Theor. Res. Biq. Comm. prima, Art. 23.)

'If $p=3n+1$, $4p=x^2+3y^2$,

$$x \equiv -\frac{\Pi 2n}{\Pi n \,.\, \Pi n}, \text{ mod } p; \quad x \equiv 1, \text{ mod } 3;$$

$$y \equiv 0, \text{ mod } 3.\text{'}$$

(Jacobi, Crelle, vol. ii. p. 69; Stern, *ib.* vol. vii. p. 104, vol. ix. p. 97, vol. xviii. p. 375*; Clausen, *ib.* vol. viii. p. 140.)

'If $p=8n+1=x^2+2y^2$,

$$x \equiv \tfrac{1}{2}\frac{\Pi 5n}{\Pi n \,.\, \Pi 4n}, \text{ mod } p; \quad x \equiv 1, \text{ mod } 4.\text{'}$$

(Jacobi, Crelle, vol. xxx. p. 168; Stern, *ib.* vol. xxxii. p. 89.)

In all these formulae the absolute value of x is evidently $< \tfrac{1}{2}p$; so that x is determined without ambiguity as the minimum residue for the modulus p of the binomial coefficient. And the combination of the two congruences satisfied by x gives rise in each case to a remarkable property of the coefficient: thus, from the two congruences satisfied by x in the first theorem, we infer that 'if p is a prime of the form $4n+1$, the minimum residue of $\tfrac{1}{2}\frac{\Pi 2n}{\Pi n \,.\, \Pi n}$ for the modulus p is of the form $4m+1$.'

To show, by an example, how these formulae are obtained, we shall consider the last of them in particular. Resuming the notation of Art. 30, let θ be a primitive root of the equation $x^{p-1}-1=0$; and let

$$\theta^n = \theta^{\frac{1}{8}(p-1)} = \omega, \quad \theta^{2n} = \omega^2 = i, \quad F(\omega) = \sum_{s=0}^{s=p-2} \omega^s x^{\gamma^s},$$

x representing a root of the equation $\dfrac{x^p-1}{x-1}=0$, and γ a primitive root of the congruence $x^{p-1} \equiv 1$, mod p. Then

$$\psi(\omega) = \frac{F(\theta^{-n})\,F(\theta^{-4n})}{F(\theta^{-5n})}$$

is an integral function of ω only (Art. 30, III.); let

$$\psi(\omega) = a + b\omega + c\omega^2 + d\omega^3.$$

The function $\dfrac{F(\theta^{-n})\,F(\theta^{-4n})}{F(\theta^{-5n})}$ is not changed, if for θ^{-n} we write θ^{-3n}; therefore $\psi(\omega)=\psi(\omega^3)$; *i.e.* $(b-d)(1-i)\,\omega + 2ci = 0$,

or $c=0$, $b=d$, and $\psi(\omega)=a+b(1+i)\,\omega$, $\psi(\omega^{-1})=a+b(1-i)\,\omega^{-1}$;

so that $p=\psi(\omega)\times\psi(\omega^{-1})=a^2+2b^2$ (Art. 30, IV.).

---

* {This reference relates to $p = 8n+1 = x^2+y^2$.}

Again, $$\psi(\gamma^n) = a + b(\gamma^n + \gamma^{3n}) \equiv -\frac{\Pi 5n}{\Pi n \,.\, \Pi 4n}, \text{ mod } p \text{ (Art. 30, v.)},$$

and $$\psi(\gamma^{5n}) = a - b(\gamma^n + \gamma^{3n}) \equiv -\frac{\Pi 9n}{\Pi 4n \,.\, \Pi 5n}, \text{ mod } p;$$

whence $$a \equiv -\tfrac{1}{2}\frac{\Pi 5n}{\Pi n \,.\, \Pi 4n}.$$

To show that $a \equiv -1$, mod 4, we observe that by the definition of the function ψ, $\psi(\omega) = \Sigma \omega^{-y_1 - 4y_2}$, y_1 and y_2 representing any two numbers of the series $1, 2, \ldots, p-2$, which satisfy the congruence $\gamma^{y_1} + \gamma^{y_2} \equiv 1$, mod p. Hence $a = \Sigma(-1)^{y_2 + \eta_1}$, where η_1 is one of the numbers $1, 2, \ldots, 2n-1$, and η_1, y_2 satisfy the congruence $\gamma^{4\eta_1} + \gamma^{y_2} \equiv 1$, mod p. Let σ be any one of the numbers $1, 2, \ldots, n-1$, and let A, B be the values of γ^{y_2} corresponding to the values $n-\sigma$, $n+\sigma$ of η_1; then

$$A \times B \equiv (1 - \gamma^{4(n-\sigma)}) \times (1 - \gamma^{4(n+\sigma)}) \equiv -\gamma^{-4(n+\sigma)} \times (1 - \gamma^{4(n+\sigma)})^2, \text{ mod } p;$$

therefore $A \times B$ is a quadratic residue of p, and the values of y_2 corresponding to the values $n-\sigma$, $n+\sigma$ of η_1 are either both even or else both uneven; also, if $\eta_1 = n$, $\gamma^{y_2} \equiv 2$, mod p, and y_2 is even, because $\left(\frac{2}{p}\right) = 1$. Let k be the number of values of η_1, included in the series $1, 2, \ldots, n-1$, for which $y_2 + \eta_1$ is uneven; then

$$a = \Sigma(-1)^{y_2 + \eta_1} = 2(n-1) - 4k + (-1)^n; \quad i.e. \ a \equiv -1, \text{ mod } 4.$$

We might also determine x in the equation $p = x^2 + 2y^2$ by the congruence

$$x \equiv (-1)^n \tfrac{1}{2}\frac{\Pi 4n}{\Pi n \,.\, \Pi 3n}, \text{ mod } p,$$

or by the congruence $$x \equiv 2^{2n} \times \tfrac{1}{2}\frac{\Pi 2n}{\Pi n \,.\, \Pi n}, \text{ mod } p.$$

These determinations, which have been given by M. Stern, may either be obtained directly by considering the functions

$$\frac{F(\theta^{-n})\, F(\theta^{-3n})}{F(\theta^{-4n})}, \quad \frac{[F(\theta^{-n})]^2}{F(\theta^{-2n})},$$

or may be deduced from the formula of Jacobi. The formulae for the determination of x in the first two theorems also admit of various modifications. It will be observed that, in the first, y is determined by a congruence as well as x. This determination is obtained by a comparison of the two congruence

$$1 + \frac{y^2}{x^2} \equiv 0, \text{ mod } p, \quad 1 + (\Pi 2n)^2 \equiv 0, \text{ mod } p,$$

(the latter arising from Sir J. Wilson's theorem); with regard to it Gauss observes, 'quum insuper noverimus quo signo affecta prodeat radix quadrati imparis, eo scilicet ut semper fiat formae $4m+1$, attentione perdignum est, quod simile criterium generale respectu radicis quadrati paris hactenus inveniri non potuerit. Quale si quis inveniat et nobiscum communicet magnam de nobis gratiam feret.'

These congruential determinations possess great interest, not only because direct methods of solution present themselves very rarely in the theory of numbers, but also on account of the singular connexion which they establish between certain binomial coefficients and certain quadratic decompositions of primes. Nor is it less remarkable that the properties of the resolvent function of Lagrange form the intermediate links in this connexion; although it is proper to observe that Gauss has exhibited his demonstration of the theorem relating to the equation $p=x^2+y^2$ in a form in which its connexion with the theory of the division of the circle is disguised.

Results of a more general character have been obtained by Jacobi and Cauchy. Cauchy has treated of the subject with great fulness of detail in his Memoir on the Theory of Numbers, in the 17th volume of the Memoirs of the Academy of Sciences (pp. 249–768); while Jacobi has barely indicated his method in his note on the division of the circle (Crelle, vol. xxx. p. 166); nevertheless, as in some respects it seems preferable to that employed by Cauchy, we shall endeavour to adhere to it in what follows.

Retaining the other notations which we have employed in this article, let

$$\frac{F(\theta^{-m})\,F(\theta^{-n})}{F(\theta^{-(m+n)})}=\psi(m,n,\theta) \quad\text{or}\quad \psi(m,n),$$

when there is no occasion to consider θ explicitly; we observe that

$$\psi(m,n)=\psi(n,m); \quad \psi(0,n)=\psi(m,0)=\psi(0,0)=-1;$$

also $$\psi(m',n')=\psi(m,n), \text{ if } m'\equiv m, \bmod p-1, \ n'\equiv n, \bmod p-1;$$

$$\psi(m,n)=(-1)^{m+1}p=(-1)^{n+1}p, \quad \text{if } m+n\equiv 0, \bmod p-1,$$

but m and n are not $\equiv 0$, mod $p-1$. Let $m_1, m_1', \ldots, m_1^{(\sigma)}$ be any set of $\sigma+1$ numbers, each of which satisfies the conditions $0\leqq m_1<p-1$; let

$$m_1+m_1'+\ldots+m_1^{(\sigma)}=n_1(p-1)+s_1,$$

where $0\leqq s_1<p-1$; and put

$$F(\theta^{-m_1})\,F(\theta^{-m_1'})\ldots F(\theta^{-m_1^{(\sigma)}})=\chi(\theta)\,F(\theta^{-s_1}).$$

Writing, for brevity,

$$\mu_1\equiv m_1, \ \mu_1'\equiv m_1+m_1', \ \mu_1''\equiv m_1+m_1'+m_1'', \ldots, \bmod p-1,$$

and determining $\mu_1, \mu_1', \mu_1'', \ldots$ so as to satisfy the conditions $0 \leqq \mu_1 < p-1$, we find
$$\chi(\theta) = \psi(\mu_1, m_1')\,\psi(\mu_1', m_1'') \ldots \psi(\mu_1^{(\sigma-1)}, m_1^{(\sigma)}).$$
In this expression if $\mu_1^{(i)} + m_1^{(i+1)} > p-1$, we write for $\psi(\mu_1^{(i)}, m_1^{(i+1)})$ its equivalent $p \div \psi(p-1-\mu_1^{(i)}, p-1-m_1^{(i+1)})$; and if $\mu_1^{(i)} + m_1^{(i+1)} = p-1$, we write for $\psi(\mu_1^{(i)}, m_1^{(i+1)})$ its equivalent $(-1)^{1+m_1^{(i+1)}} p$. It is evident that the condition $\mu^{(i)} + m^{(i+1)} \gtreqq p-1$ will be satisfied n_1 times precisely; so that $\chi(\theta)$ assumes the form $p^{n_1}\dfrac{\Phi_1(\theta)}{\Psi_1(\theta)}$, $\Phi_1(\theta)$ and $\Psi_1(\theta)$ denoting products of factors of the form $\psi(h, h')$, in each of which $h+h' < p-1$. It will now be found that
$$\frac{\Phi_1(\gamma)}{\Psi_1(\gamma)} \equiv (-1)^{\sigma-n_1} \frac{\Pi s_1}{\Pi m_1\, \Pi m_1'\, \Pi m_1'' \ldots}, \text{ mod } p.$$
For (1), if $\mu_1^{(i)} + m_1^{(i+1)} < p-1$, we have
$$\psi(\mu_1^{(i)}, m_1^{(i+1)}, \gamma) \equiv -\frac{\Pi\, \mu_1^{(i+1)}}{\Pi \mu_1^{(i)} . \Pi m_1^{(i+1)}}, \text{ mod } p;$$
(2), if $\mu_1^{(i)} + m_1^{(i+1)} > p-1$, we have $\dfrac{1}{\psi(p-1-\mu_1^{(i)}, p-1-m_1^{(i+1)}, \gamma)}$
$$\equiv -\frac{\Pi(p-1-\mu_1^{(i)}) . \Pi(p-1-m_1^{(i+1)})}{\Pi(2p-2-\mu_1^{(i)}-m_1^{(i+1)})} \equiv \frac{\Pi \mu_1^{(i+1)}}{\Pi \mu_1^{(i)} . \Pi m_1^{(i+1)}}, \text{ mod } p,$$
since, by Sir J. Wilson's theorem,
$$\Pi(p-1-j) \equiv -\frac{(-1)^j}{\Pi j}, \text{ mod } p, \text{ if } j < p-1;$$
(3), if $\mu_1^{(i)} + m_1^{(i+1)} = p-1$, we have
$$(-1)^{1+m_1^{(i+1)}} \equiv \frac{\Pi \mu_1^{(i+1)}}{\Pi \mu_1^{(i)} . \Pi m_1^{(i+1)}}, \text{ mod } p,$$
because
$$\Pi \mu_1^{(i)}\, \Pi m_1^{(i+1)} \equiv (-1)^{1+m_1^{(i+1)}}, \text{ mod } p,$$
by Sir J. Wilson's theorem, while $\Pi \mu_1^{(i+1)} = 1$, since $\mu_1^{(i+1)} = 0$; whence, multiplying and writing s_1 for $\mu_1^{(\sigma)}$, we obtain the congruence written above. Let r represent any term of a system of residues prime to $p-1$; let the numbers $m_r, m_r', \ldots, m_r^{(\sigma)}$ be determined by the congruences $m_r^{(i)} \equiv m_1^{(i)} r$, mod $(p-1)$, combined with the condition $0 \leqq m_r^{(i)} < p-1$; and let
$$m_r + m_r' + \ldots + m_r^{(\sigma)} = n_r(p-1) + s_r,$$
where again $0 \leqq s_r < p-1$: we have for every value of r an equation of the form
$$\chi(\theta^r) = p^{n_r} \times \frac{\Phi_r(\theta)}{\Psi_r(\theta)},$$
and a congruence of the form
$$\frac{\Phi_r(\gamma)}{\Psi_r(\gamma)} \equiv (-1)^{\sigma-n_r} \frac{\Pi s_r}{\Pi m_r\, \Pi m_r'\, \Pi m_r'' \ldots}, \text{ mod } p.$$

Let $$\chi(\theta) = A_0 + A_1\theta + \ldots + A_k\theta^k,$$
$k+1$ denoting the number of terms in a system of residues prime to $p-1$; let n_ν be the least of the numbers $n_1, \ldots, n_{k+1}$, and j the exponent of the highest power of p dividing $A_0, A_1, \ldots, A_k$: then shall $j = n_\nu$. For, first, if $j > n_\nu$, from the equation $\Psi_\nu(\theta)\dfrac{\chi(\theta^\nu)}{p^{n_\nu}} = \Phi_\nu(\theta)$, in which the coefficients of the powers of θ are integral numbers, we infer the congruence

$$\Psi_\nu(\gamma)\frac{\chi(\gamma^\nu)}{p^{n_\nu}} \equiv \Phi_\nu(\gamma), \text{ mod } p;$$

but $$\frac{\chi(\gamma^\nu)}{p^{n_\nu}} \equiv 0, \text{ mod } p; \quad \text{therefore, } \Phi_\nu(\gamma) \equiv 0, \text{ mod } p,$$

which is impossible. Secondly, if $j < n_\nu$, writing A_i' for $A_i \div p^j$, and observing that $\Psi_r(\gamma)$ is prime to p, for every value of r, we find

$$A_0' + A_1'\gamma^r + A_2'\gamma^{2r} + \ldots + A_k'\gamma^{kr} \equiv 0, \text{ mod } p^{n_\nu - j},$$

for every value of r: but the determinant of this system is prime to p, therefore $A_0' \equiv 0$, $A_1' \equiv 0$, ..., $A_k' \equiv 0$, mod $p^{n_\nu - j}$, which is contrary to the hypothesis that $j < n_\nu$, and that p^j is the highest power of p dividing $A_0, A_1, \ldots, A_k$.

The application of these results leads to the following general theorems; in the enunciations of which p is an uneven prime, and Δ a number not divisible by any square.

'If $\Delta = 4m+3$, $p = \Delta n + 1$, and if we represent by a and b numbers less than Δ and prime to Δ, respectively satisfying the equations

$$\left(\frac{a}{\Delta}\right) = 1, \quad \left(\frac{b}{\Delta}\right) = -1, \text{ we have}$$

$$\text{(A)}, \quad \begin{cases} 4p^{\frac{\Sigma b - \Sigma a}{\Delta}} = x^2 + \Delta y^2, \\ x \equiv -(-1)^{\frac{\Sigma b}{\Delta}} \dfrac{1}{\Pi_a[\Pi an]}. \end{cases}$$

'If $p = 4\Delta n + 1$, Δ being of any other linear form than $4m+3$, and if we represent by a and b numbers less than 4Δ and prime to 4Δ, respectively satisfying the equations $\left(\dfrac{-\Delta}{a}\right) = +1$, $\left(\dfrac{-\Delta}{b}\right) = -1$, we have

$$\text{(B)}, \quad \begin{cases} p^{\frac{\Sigma b - \Sigma a}{\Delta}} = x^2 + \Delta y^2, \\ x \equiv -(-1)^{\frac{\Sigma b}{4\Delta}} \dfrac{1}{2} \dfrac{1}{\Pi_a[\Pi an]}. \end{cases}$$

In these formulae the signs of summation extend to every value of a and b

respectively; and in the expression $\Pi_a[\Pi an]$ the exterior sign of multiplication Π_a extends to every value of a, while the interior sign is the factorial symbol, so that $\Pi an = 1.2.3 \ldots an$. The number 3 is excluded from the first formula; the numbers 1 and 2 from the second.

It will suffice to show how the first of these two theorems is to be demonstrated. For this purpose we consider the product $\Pi F(\theta^{-an})$; taking an, $a'n$, $a''n$, ... for m_1, m_1', ... we find $\chi(\theta) = -\Pi F(\theta^{-an})$; because (as may easily be proved) $\Sigma a \equiv 0$, mod Δ, whence $\Sigma an \equiv 0$, mod $p-1$. We shall now show that $\chi(\theta)$ is of the form $A\Sigma\theta^{an} + B\Sigma\theta^{bn}$. Actually multiplying the expressions $F(\theta^{-an})$, $F(\theta^{-a'n})$, ..., the coefficient, in the product, of any term such as $x^k\theta^{mn}$ is equal to the number N of the solutions of the simultaneous congruences

$$\gamma^y + \gamma^{y'} + \gamma^{y''} + \ldots \equiv k, \text{ mod } p, \qquad ay + a'y' + a''y'' + \ldots \equiv -m, \text{ mod } \Delta.$$

If r is a number prime to Δ, and satisfying the equation $\left(\frac{r}{\Delta}\right) = +1$, N will not be changed, if we write rm, ra, ra', ... (or rather the least positive residues of those numbers, mod Δ) for m, a, a', ... Hence, in $\chi(\theta)$ all powers of θ whose exponents are of the form an have the same coefficient A', and all powers of θ whose exponents are of the form bn have the same coefficient B'. Again, consider a power of θ of which the exponent is of the form $\alpha\delta n$; δ representing a given divisor of Δ (other than 1 or Δ), and α representing any number less than $\frac{\Delta}{\delta}$, and prime to $\frac{\Delta}{\delta}$; all such powers of θ will have the same coefficient. For we can always find a number r prime to Δ, satisfying the equation $\left(\frac{r}{\Delta}\right) = 1$, and yet congruous, for the modulus $\frac{\Delta}{\delta}$, to any given number prime to $\frac{\Delta}{\delta}$; whence it follows that the number N will remain the same for all values of m included in the formula $\alpha\delta$. But a sum of the form $\Sigma_\alpha \theta^{\alpha\delta n}$ is equal to $+1$ or -1, according as the number of primes dividing $\frac{\Delta}{\delta}$ is even or uneven, because it is the sum of the primitive roots of the equation $x^{\frac{\Delta}{\delta}} = 1$. Thus, the function $\chi(\theta)$ assumes the form $A'\Sigma\theta^{an} + B'\Sigma\theta^{bn} + C'$, whence, attending to the equation $\Sigma\theta^{an} + \Sigma\theta^{bn} = (-1)^\lambda$, in which λ is the number of primes dividing Δ, we find, as has been said,

$$-\Pi F(\theta^{-an}) = \chi(\theta) = A\Sigma\theta^{an} + B\Sigma\theta^{bn}.$$

If we write θ^{-1} for θ in this equation, it becomes

$$-\Pi F(\theta^{an}) = \chi(\theta^{-1}) = A\Sigma\theta^{bn} + B\Sigma\theta^{an},$$

since

$$\left(\frac{-a}{\Delta}\right) = -1, \left(\frac{-b}{\Delta}\right) = +1.$$

Multiplying the two equations together, and observing that

$F(\theta^{-an})\,F(\theta^{an})=(-1)^{an}p=p$, because n is even, we obtain

$$4p^{\frac{1}{2}\psi_0(\Delta)}=[(-1)^\lambda(A+B)+(A-B)(\Sigma\theta^{an}-\Sigma\theta^{bn})]$$
$$\times[(-1)^\lambda(A+B)+(A-B)(\Sigma\theta^{bn}-\Sigma\theta^{an})],$$

or, since $(\Sigma\theta^{an}-\Sigma\theta^{bn})^2=-\Delta$*,

$$4p^{\frac{1}{2}\psi_0(\Delta)}=(A+B)^2+\Delta(A-B)^2,$$

$\psi_0(\Delta)$ representing the number of numbers less than Δ and prime to Δ. We have next to determine the highest power of p dividing $A+B$ and $A-B$, or, which is the same thing, A and B. By the principles indicated above, we have

$$A\,\Sigma\theta^{an}+B\,\Sigma\theta^{bn}=p^{\frac{\Sigma a}{\Delta}}\frac{\Phi_1(\theta)}{\Psi_1(\theta)},$$

$$A\,\Sigma\theta^{bn}+B\,\Sigma\theta^{an}=p^{\frac{\Sigma b}{\Delta}}\frac{\Phi_{-1}(\theta)}{\Psi_{-1}(\theta)}.$$

Writing in these equations γ for θ, and observing that the determinant

$$(\Sigma\gamma^{an})^2-\Sigma(\gamma^{bn})^2\ †,$$

as well as the four numbers

$$\Phi_1(\gamma),\quad \Phi_{-1}(\gamma),\quad \Psi_1(\gamma),\quad \Psi_{-1}(\gamma),$$

is prime to p, we infer that the exponent of the highest power of p dividing A and B is the less of the two numbers $\frac{\Sigma a}{\Delta}$, $\frac{\Sigma b}{\Delta}$. Of these the former is the less ‡; if therefore we write x and y for

$$(-1)^\lambda(A+B)\,p^{-\frac{\Sigma a}{\Delta}},\ \text{and}\ (A-B)\,p^{-\frac{\Sigma a}{\Delta}}$$

---

* See Art. 96, (ix.) of this Report, or the note on Art. 104.

† Since $\Sigma\theta^{an}+\Sigma\theta^{bn}=(-1)^\lambda$, we have $\Sigma\gamma^{an}+\Sigma\gamma^{bn}\equiv(-1)^\lambda$, mod p;
and since $(\Sigma\theta^{an}-\Sigma\theta^{bn})^2=-\Delta$, we have $(\Sigma\gamma^{an}-\Sigma\gamma^{bn})^2\equiv-\Delta$, mod p.
Thus the two factors of the determinant are each of them prime to p.

The principle that any rational equation containing only powers of θ and integral numbers may be changed into a congruence for the modulus p, if γ be written in it for θ, has already been employed in this Article. Its truth is evident, if we observe that the irreducible equation satisfied by θ, if considered as a congruence for the modulus p, is satisfied by γ. This principle is of more general application than a similar one which has been already employed in Art. 51 of this Report; but its proof supposes the irreducibility of the equation of the primitive roots, which is not necessary to the proof of the principle of Art. 51.

‡ $\Sigma b-\Sigma a$ is certainly positive, because $\frac{\Sigma b-\Sigma a}{\Delta}$ is equal to the number of improperly primitive classes of the negative determinant $-\Delta$. Or (as it is desirable to avoid making use of this result

respectively, our equation becomes

$$4p^{\frac{\Sigma b-\Sigma a}{\Delta}}=x^2+\Delta y^2.$$

Also, since
$$\frac{\Phi_1(\gamma)}{\Psi_1(\gamma)}\equiv-(-1)^{\frac{\Sigma b}{\Delta}}\frac{1}{\Pi_a[\Pi an]},$$

we have
$$Ap^{-\frac{\Sigma a}{\Delta}}\Sigma\gamma^{an}+Bp^{-\frac{\Sigma a}{\Delta}}\Sigma\gamma^{bn}\equiv-(-1)^{\frac{\Sigma b}{\Delta}}\frac{1}{\Pi_a[\Pi an]},\text{ mod } p,$$
$$Ap^{-\frac{\Sigma a}{\Delta}}\Sigma\gamma^{bn}+Bp^{-\frac{\Sigma a}{\Delta}}\Sigma\gamma^{an}\equiv 0,\text{ mod } p,$$

---

here) $\Sigma b-\Sigma a$ is positive because $\frac{\pi}{\Delta\sqrt{\Delta}}(\Sigma b-\Sigma a)$ is the sum of the series $\sum_1^\infty\left(\frac{n}{\Delta}\right)\frac{1}{n}$, the summation extending to every value of n prime to Δ, and the terms being taken in their natural order. This series is positive, because the series $\sum_1^\infty\left(\frac{n}{\Delta}\right)\frac{1}{n^{1+\rho}}$, of which it is the limit, when ρ is diminished without limit, is certainly positive, being the reciprocal of the product

$$\Pi\left[1-\left(\frac{q}{\Delta}\right)\frac{1}{q^{1+\rho}}\right],$$

in which the sign of multiplication extends to every prime q not dividing Δ, and in which every factor is positive. The series $\sum_1^\infty\left(\frac{n}{\Delta}\right)\frac{1}{n}$ is one of those summed by Dirichlet in the memoir 'Recherches sur diverses applications &c.' (Crelle, vol. xxi. p. 141 *et seq.*): for the case in which Δ is a prime, he had already summed it in the memoir on the Arithmetical Progression (Memoirs of the Academy of Berlin for 1837, p. 55). Cauchy (Mémoires de l'Académie des Sciences, vol. xvii. p. 673 *et seq.*) inverts Dirichlet's process, and transforms sums of the form $\Sigma f(a)-\Sigma f(b)$ into infinite series. The transformation is effected by substituting for $f(x)$, in the expression

$$\sum_1^{\Delta-1}\left(\frac{x}{\Delta}\right)f(x)=\Sigma f(a)-\Sigma f(b),$$

the equivalent infinite series

$$\frac{1}{\Delta}\int_0^\Delta f(s)\,ds+\frac{2}{\Delta}\sum_{m=1}^{m=\infty}\int_0^\Delta\cos\frac{2m\pi(x-s)}{\Delta}f(s)\,ds;$$

whence, observing that

$$\sum_{x=1}^{x=\Delta-1}\left(\frac{x}{\Delta}\right)\sin\frac{2mx\pi}{\Delta}=\left(\frac{m}{\Delta}\right)\sqrt{\Delta},\quad\text{and}\quad\sum_{x=1}^{x=\Delta-1}\left(\frac{x}{\Delta}\right)\cos\frac{2mx\pi}{\Delta}=0,$$

we obtain
$$\tfrac{1}{2}\sqrt{\Delta}\,(f(a)-f(b))=\sum_{m=1}^{m=\infty}\left(\frac{m}{\Delta}\right)\int_0^\Delta\sin\frac{2ms\pi}{\Delta}f(s)\,ds;$$

a formula from which Dirichlet's result is immediately deducible, by putting $f(x)=x$, and performing the integrations. It is a remarkable fact that the inequality $\Sigma b>\Sigma a$ has never been proved by elementary considerations, or without the use of infinite series (see the Memoir on the Arithmetical Progression, p. 57). If Δ is a prime, $\Sigma b-\Sigma a$ is certainly not zero, for $\Sigma b+\Sigma a$ is uneven (because Δ is of the form $4n+3$); but even this remark cannot be extended to the case in which Δ is composite.

whence by addition

$$x \equiv -(-1)^{\frac{\Sigma b}{\Delta}} \frac{1}{\Pi_a[\Pi an]}, \text{ mod } p.$$

If Δ is a prime, x also satisfies the congruence $\frac{1}{2}x \equiv 1$, mod Δ; for the sum of the coefficients in any function $\psi(m, n)$ is $\equiv -1$, mod $p-1$, and therefore mod Δ; whence the sum of coefficients in $\chi(\theta)$, which is a product of an even number of such functions, is $\equiv 1$, mod Δ; because the reduction of $\chi(\theta)$ to the form $A\Sigma\theta^{an} + B\Sigma\theta^{bn}$ is effected only by means of the equations

$$\theta^{n\Delta} - 1 = 0, \quad \frac{\theta^{n\Delta} - 1}{\theta^n - 1} = 0;$$

whereof the former does not alter the sum of the coefficients at all, and the latter alters them only by a multiple of Δ. Consequently

$$\tfrac{1}{2}(\Delta - 1)(A + B) \equiv 1, \text{ mod } \Delta,$$

or, since $p^{\frac{\Sigma a}{\Delta}} \equiv 1$, mod Δ, and $(-1)^\lambda = -1$, $\frac{1}{2}x \equiv 1$, mod Δ.

It will be observed that if Δ is of the form $8m + 7$, whether Δ is a prime or not, x and y are necessarily even in the equation (A); whence, dividing by 4, we may put the equation in the form

$$p^{\frac{\Sigma b - \Sigma a}{\Delta}} = x^2 + \Delta y^2$$

$$x \equiv -(-1)^{\frac{\Sigma b}{\Delta}} \tfrac{1}{2} \frac{1}{\Pi_a[\Pi an]}, \text{ mod } p.$$

Ex. Let $\Delta = 7$, $p = 7n + 1$; the values of a are 1, 2, 4; of b, 3, 5, 6: hence

$$p = x^2 + 7y^2, \quad x \equiv -\tfrac{1}{2} \frac{1}{\Pi n . \Pi 2n . \Pi 4n} \equiv \tfrac{1}{2} \frac{\Pi 3n}{\Pi n . \Pi 2n}, \text{ mod } p; \quad \text{also } x \equiv 1, \text{ mod } 7.$$

(Jacobi, Crelle, vol. ii. p. 69.)

Whenever the exponent of p is 1, the formulae (A) and (B) completely determine the value of x; when the exponent of p is 2, we can only be sure that the absolute value of x is less than p, so that x is not completely determined, but is either the least positive or the least negative residue of the binomial coefficient; though in this case if Δ be a prime of the form $4n + 3$, the ambiguity may be removed by the congruence $\frac{1}{2}x \equiv 1$, mod Δ. But when the exponent of p is > 2, x is never completely determined by the congruence for the modulus p.

It is very remarkable that the exponent of p in the formula (A) is precisely the number of improperly primitive classes of determinant $-\Delta$, and in the formula (B) is precisely the number of properly primitive classes of determinant $-\Delta$*.

---

* See Art. 104 of this Report. When Δ is of the form $4n+3$, the two expressions given by Dirichlet for the number of properly primitive classes of determinant $-\Delta$ are

$$\left(2-\left(\frac{2}{\Delta}\right)\right)\frac{(\Sigma b-\Sigma a)}{\Delta} \quad \text{and} \quad A-B,$$

where A and B represent the numbers of residues inferior to $\frac{1}{2}\Delta$, and satisfying the conditions $\left(\frac{a}{\Delta}\right)=+1$ and $\left(\frac{b}{\Delta}\right)=-1$ respectively. Hence $\frac{\Sigma b-\Sigma a}{\Delta}$ is the number of improperly primitive classes; because that number is equal to or is one-third of the number of properly primitive classes, according as $\Delta\equiv 7$, or $\equiv 3$, mod 8 (see Art. 103 or 113).

There is no difficulty in showing that Dirichlet's two expressions are identical. If $\left(\frac{2}{\Delta}\right)=1$, the congruence $2b'\equiv b$, mod Δ, is always resoluble; and if b receive in succession all positive values less than Δ which satisfy the condition $\left(\frac{b}{\Delta}\right)=-1$, b' will obtain the same values in a different order. Hence

$$\Sigma\frac{b}{\Delta}=2\frac{\Sigma b'}{\Delta}-\Sigma\frac{b}{\Delta}=\Sigma\frac{2b'-b}{\Delta}.$$

But if $\quad b'<\frac{1}{2}\Delta,\ 2b'-b=0;\ \text{ if } b'>\frac{1}{2}\Delta,\ 2b'-b=\Delta,\ i.e.\ \Sigma\frac{b}{\Delta}=A,$

for there are A values of b greater than $\frac{1}{2}\Delta$. Similarly $\Sigma\frac{a}{\Delta}=B$, so that $\frac{\Sigma b-\Sigma a}{\Delta}=A-B$. In precisely the same manner it may be shown, if $\left(\frac{2}{\Delta}\right)=-1$, by considering the congruences

$$2b'+b\equiv 0,\ \text{mod } \Delta, \qquad 2a'+a\equiv 0,\ \text{mod } \Delta,$$

that

$$3\frac{\Sigma b}{\Delta}=\Sigma\frac{2b'+b}{\Delta}=2A+B \quad \text{and} \quad \frac{3\,\Sigma a}{\Delta}=\Sigma\frac{2a'+a}{\Delta}=2B+A;$$

whence

$$3\,\frac{\Sigma b-\Sigma a}{\Delta}=A-B.$$

Also the expression given in Art. 104 coincides with $A-B$. For that expression may be written in the form

$$-\left(\frac{2}{\Delta}\right)\Sigma\left(\frac{\Delta-2a'}{\Delta}\right)-\frac{2}{\Delta}\Sigma\left(\frac{\Delta-2b'}{\Delta}\right)=\Sigma\left(\frac{a'}{\Delta}\right)-\Sigma\left(\frac{b'}{\Delta}\right)=A-B,$$

a' and b' representing numbers less than $\frac{1}{2}\Delta$.

If we consider, as Cauchy has done, the product $\frac{\Pi\,[F(\theta^{-an})]^2}{\Pi\,F(\theta^{-2an})}$, the exponent of p in the formula (A), will be $A-B$. That product is evidently equal to $\Pi\,F(\theta^{-an})$, or to $\frac{\Pi\,[F(\theta^{-an})]^2}{\Pi\,F[\theta^{-bn})}$, according as $\left(\frac{2}{\Delta}\right)=+1$ or $=-1$; a result which is in accordance with the equation

$$A-B=\left(2-\left(\frac{2}{\Delta}\right)\right)\frac{(\Sigma b-\Sigma a)}{\Delta}.$$

In the formula (B), the exponent of p, obtained by the consideration of the same product, is

$$A'-B'=2\times\frac{(\Sigma b-\Sigma a)}{4\Delta};$$

A' and B' denoting respectively the numbers of residues of the classes a and b respectively, which are inferior to 2Δ.

Before Dirichlet's discovery of the formulae expressing the number of classes of quadratic forms of a given determinant, Jacobi, having succeeded in determining the exponent of p in the formula (A), for the case in which Δ is a prime number, was led with singular sagacity to conjecture that $\frac{\Sigma b - \Sigma a}{\Delta}$ must represent the number of improperly primitive classes of determinant $-\Delta$*. If h is the number of classes in the principal genus of improperly primitive forms of determinant $-\Delta$, it follows from the theory of composition of quadratic forms that $2p^h$ can always be represented primitively by the principal form in that genus, *i.e.* by the form $(2, 1, \frac{1}{2}(\Delta+1))$, and that the exponent of the lowest power of p which is capable of such representation is either h or a submultiple of h. Again, the equation

$$4p^{\frac{\Sigma b - \Sigma a}{\Delta}} = x^2 + \Delta y^2,$$

if we write in it $2X + Y$ for x, and Y for y, becomes

$$2p^{\frac{\Sigma b - \Sigma a}{\Delta}} = (2, 1, \tfrac{1}{2}(\Delta+1))\,(X, Y)^2,$$

the values of X and Y being integral. Assuming, then, that there exist primes of the linear form $n\Delta+1$, the doubles of which are capable of representation by a class appertaining to the exponent h (an assumption which implies that $-\Delta$ is not an irregular determinant, at least in respect of its improperly primitive classes), we see that in the case in which Δ is a prime of the form $4n+3$, and in which therefore there is but one genus of improperly primitive forms, $\frac{\Sigma b - \Sigma a}{\Delta}$ must be equal either to the number of improperly primitive classes, or to a multiple of that number; and as Jacobi found, upon a sufficient induction, that h was always equal to $\frac{\Sigma b - \Sigma a}{\Delta}$, he did not scruple to enunciate the theorem as true. We know, however, from an account which Dirichlet has given of a communication made to him by Jacobi, that Jacobi never obtained a demonstration of the theorem; and, indeed, it

---

* Crelle, vol. ix. p. 189. Jacobi counts the classes of the prime determinant $-\Delta$ on the principle of Legendre, not distinguishing opposite classes from one another. If n is the number of improperly primitive classes so counted, we have $h = 2n-1$, because there is but one improperly primitive ambiguous class. When Δ is of the form $8n+7$, Jacobi enunciates the theorem with reference to the number of properly primitive classes, which in this case is equal to the number of improperly primitive classes.

would seem probable, as has been observed by Dirichlet, that its demonstration requires other principles (Crelle, vol. lii. p. 206).

It is hardly necessary to add that when there is more than one genus of forms of determinant $-\Delta$, *i.e.* in every case except when Δ is a prime of the form $4n+3$, the exponent of p in the formulae (A) and (B) is always a multiple of the least exponent for which those formulae can be satisfied.

122. *Extension of the preceding Theorem by Eisenstein.*—In the theory of which an account has been given in the last article, the prime number p is throughout supposed of the linear form $n\Delta+1$ or $4n\Delta+1$; thus in the equations $p=x^2+7y^2$, $p=x^2+8y^2$, we have supposed p to be of the forms $7n+1$ and $8n+1$ respectively. But we know that some power of every prime of which $-\Delta$ is a quadratic residue is capable of representation by the form $x^2+\Delta y^2$; and, in particular, that primes of the form $8n+3$ are capable of representation by x^2+2y^2, and primes of either of the forms $7n+2$ or $7n+4$ by x^2+7y^2. M. Stern found by induction that the value of x in the equation

$$p=8n+3=x^2+2y^2$$

satisfies the congruences

$$x\equiv -\tfrac{1}{2}\frac{\Pi\, 4n+1}{\Pi n\,.\,\Pi\, 3n+1}, \text{ mod } p, \quad x\equiv(-1)^n, \text{ mod } 4\,*;$$

and Eisenstein succeeded in demonstrating this theorem, as well as the two following †:—

'If $p=7n+2=x^2+7y^2$,

$$x\equiv\tfrac{1}{2}\frac{\Pi\, 3n}{\Pi n\,.\,\Pi\, 2n}, \text{ mod } p; \quad x\equiv 3, \text{ mod } 7;\text{'}$$

'If $p=7n+4=x^2+7y^2$,

$$x\equiv\tfrac{1}{2}\frac{\Pi\, 3n+1}{\Pi n\,.\,\Pi\, 2n+1}, \text{ mod } p; \quad x\equiv 2, \text{ mod } 7.\text{'}$$

These demonstrations are obtained by expressing the prime number p as the product of two complex factors, composed of 8th or 7th roots of unity. But the decomposition of p is no longer supplied by the formula of Art. 30; nor are the complex factors included in the definition of the functions ψ, which have been considered in Art. 30 and in the last Article.

If $p=8n+3$ is a real prime, p is also a prime in the theory of complex

---

* Crelle, vol. xxxii. p. 89. We enunciate the latter part of the theorem in the form in which it has been given by Eisenstein.

† Crelle, vol. xxxvii. p. 97.

numbers of the form $a+bi$; let γ be a primitive root of p in that theory, and let $\gamma^y \equiv 1+iz$, mod p, z representing one of the real integers, $0, 1, \ldots, p-1$. Also let $\psi(\omega)=\Sigma\omega^y$, ω denoting a primitive 8th root of unity, and the summation extending to every value of y. Eisenstein establishes the equations

$$\psi(\omega)\,\psi(\omega^{-1})=p, \quad \psi(\omega)=\psi(\omega^3);$$

whence $\psi(\omega)$ is of the form $A+B(1+i)\,\omega$, and

$$p=\psi(\omega)\,\psi(\omega^{-1})=A^2+2B^2.$$

To find the residue of A, mod p, let

$$e=\tfrac{1}{8}(p^2-1)=3n+1+np;$$

and write successively γ^e and γ^{5e} for ω in the function $\psi(\omega)$. We find

$$\psi(\gamma^e)=\Sigma\gamma^{ey}\equiv\Sigma(1+iz)^e\equiv\Sigma(1+iz)^{3n+1}(1-iz)^n, \text{ mod } p,$$

because in general $(a+bi)^p\equiv(a-bi)$, mod p. In this expression no power of z has an exponent divisible by $p-1$; but

$$\sum_{z=0}^{z=p-1} z^\theta \equiv 0, \text{ mod } p,$$

unless θ is different from zero, and is a multiple of $p-1$; therefore

$$\psi(\gamma^e)\equiv 0, \text{ mod } p.$$

Again, because $5e=7n+2+(5n+1)\,p$,

$$\psi(\gamma^{5e})\equiv\Sigma(1+zi)^{7n+2}(1-zi)^{5n+1}, \text{ mod } p;$$

in this expression the coefficient C of z^{p-1} is

$$\Sigma_\mu(-1)^{1+\mu'}\frac{\Pi 7n+2\,.\,\Pi 5n+1}{\Pi\mu\,.\,\Pi 7n+2-\mu\,.\,\Pi\mu'\,.\,\Pi 5n+1-\mu'},$$

where $\mu+\mu'=p-1$, and the summation extends from $\mu=3n+1$ to $\mu=7n+2$. Writing $3n+1+\nu$ for μ, $5n+1-\nu$ for μ', and observing that

$$\Pi\mu\,.\,\Pi\mu'\equiv(-1)^{1+\mu'}, \text{ mod } p,$$

we find

$$C\equiv\sum_{\nu=0}^{\nu=4n+1}\frac{\Pi 7n+2\,.\,\Pi 5n+1}{\Pi 4n+1-\nu\,.\,\Pi\nu}\equiv\frac{\Pi 7n+2\,.\,\Pi 5n+1}{\Pi 4n+1}\times\sum_{\nu=0}^{\nu=4n+1}\frac{\Pi 4n+1}{\Pi\nu\,.\,\Pi 4n+1-\nu}$$

$$\equiv\frac{\Pi 7n+2\,.\,\Pi 5n+1}{\Pi 4n+1}\times 2^{4n+1}\equiv\frac{\Pi 4n+1}{\Pi n\,.\,\Pi 3n+1}, \text{ mod } p,$$

observing that $2^{4n+1}\equiv -1$, mod p, and transforming each of the three factorials by Sir J. Wilson's theorem. Hence, finally,

$$A\equiv\tfrac{1}{2}\psi(\gamma^e)+\tfrac{1}{2}\psi(\gamma^{5e})\equiv-\tfrac{1}{2}C\equiv-\tfrac{1}{2}\frac{\Pi 4n+1}{\Pi n\,.\,\Pi 3n}, \text{ mod } p,$$

in accordance with the enunciation of M. Stern. The congruence

$$A \equiv (-1)^n, \text{ mod } 4,$$

is inferred by Eisenstein from the values of

$$\psi(1), \quad \psi(-1), \quad \psi(i), \quad \psi(-i);$$

but we may omit these determinations here.

If $p = 7n + 2$, or $7n + 4$, Eisenstein considers the complex numbers formed with the roots of the equation $\eta^3 - 21\eta - 7 = 0$. If ω is an imaginary seventh root of unity, and $\eta_k = 3(\omega^k + \omega^{-k}) + 1$, the roots of this equation are η_1, η_2, η_3; and every complex number formed with them is of the type $a + b\eta_1 + c\eta_2$, a, b, c denoting real integral numbers. Let γ be a primitive root of p in this complex theory (p is a prime of the theory, because the congruence

$$\eta^2 - 21\eta - 7 \equiv 0, \text{ mod } p,$$

is irresoluble: see Art. 44 of this Report), and let

$$\gamma^y \equiv 1 + z_1\eta_1 + z_2\eta_2, \text{ mod } p,$$

z_1 and z_2 each representing any term of the series $0, 1, 2, \ldots, p-1$. The function $\psi(\omega) = \Sigma\omega^y$ (the summation extending to all the p^2 values of y) is shown by Eisenstein to satisfy the equations

$$\psi(\omega) = \psi(\omega^2) = \psi(\omega^4), \quad \psi(\omega^3) = \psi(\omega^5) = \psi(\omega^6), \quad \psi(\omega) \times \psi(\omega^{-1}) = p;$$

whence $\psi(\omega)$ is of the form $a + b(\omega + \omega^2 + \omega^4) + c(\omega^3 + \omega^5 + \omega^6)$, and $p = A^2 + 7B^2$; if $A = a - \frac{1}{2}(b + c)$, $B = \frac{1}{2}(b - c)$. The equation $p^2 = a + 3b + 3c$, considered as a congruence, mod 7, becomes $A \equiv p^2$, mod 7; *i.e.* $A \equiv 4$, or $\equiv 2$, mod 7, according as p is of the form $7n + 2$ or $7n + 4$. To obtain the congruence, mod p, which is satisfied by A, we consider the congruence

$$2A \equiv \psi(\gamma^e) + \psi(\gamma^{3e}), \text{ mod } p;$$

in which

$$e = \tfrac{1}{7}(p^3 - 1) = \alpha + \beta p + \gamma p^2,$$

α, β, γ representing positive integers less than p, of which the sum will be found to be $p - 1$. Now $\psi(\gamma^e) = \Sigma\gamma^{ey}$

$$\equiv \sum_0^{p-1}\sum_0^{p-1} (1 + z_1\eta_1 + z_2\eta_2)^\alpha \times (1 + z_1\eta_p + z_2\eta_{2p})^\beta \times (1 + z_1\eta_{p^2} + z_2\eta_{2p^2})^\gamma, \text{ mod } p_0;$$

because in general

$$[f(\eta_1)]^p \equiv f(\eta_p), \text{ mod } p.$$

Hence $\psi(\gamma^e) \equiv 0$, mod p, because $\alpha + \beta + \gamma = p - 1$, and because

$$\sum_0^{p-1}\sum_0^{p-1} z_1^{\theta_1} z_2^{\theta_2} \equiv 0, \text{ mod } p,$$

unless θ_1 and θ_2 are both different from zero, and both divisible by $p - 1$.

Again, if $3e = \alpha' + \beta' p + \gamma' p^2$, we find $\alpha' + \beta' + \gamma' = 2(p-1)$; and omitting terms in which the sum of the indices of z_1 and z_2 is inferior to $2(p-1)$,

$$\psi(\gamma^{3e}) \equiv \sum_0^{p-1} \sum_0^{p-1} (z_1 \eta_1 + z_2 \eta_2)^{\alpha'} (z_1 \eta_p + z_2 \eta_{2p})^{\beta'} (z_1 \eta_{p^2} + z_2 \eta_{2p^2})^{\gamma'}, \text{ mod } p.$$

Substituting for $z_1 \eta_{p^2} + z_2 \eta_{2p^2}$ its value $-z_1 \eta_1 - z_2 \eta_2 - z_1 \eta_p - z_2 \eta_{2p}$, we obtain

$$\psi(\gamma^{3e}) \equiv \Pi\alpha'.\Pi\beta'.\Pi\gamma'.\sum_0^{p-1} \sum_0^{p-1} (z_1 \eta_1 + z_2 \eta_2)^{p-1} (z_1 \eta_p + z_2 \eta_{2p})^{p-1}, \text{ mod } p;$$

because every such sum as

$$\sum_0^{p-1} \sum_0^{p-1} (z_1 \eta_1 + z_2 \eta_2)^{p-1+\theta} (z_1 \eta_p + z_2 \eta_{2p})^{p-1-\theta},$$

in which θ is one of the numbers $1, 2, 3, \ldots, p-1$, taken positively or negatively, is certainly $\equiv 0$, mod p, as may be seen by substituting $(z_1 \eta_p + z_2 \eta_{2p})$ for $(z_1 \eta_1 + z_2 \eta_2)^p$. Lastly, the coefficient C of $(z_1 z_2)^{p-1}$ in the expression

$$\sum_0^{p-1} \sum_0^{p-1} (z_1 \eta_1 + z_2 \eta_2)^{p-1} (z_1 \eta_p + z_2 \eta_{2p})^{p-1}$$

is evidently
$$\sum_{\mu=0}^{\mu=p-1} K_\mu^2 (\eta_1 \eta_{2p})^\mu (\eta_2 \eta_p)^{p-1-\mu},$$

K_μ representing the coefficient of x^μ in the expansion of $(1+x)^{p-1}$. Hence

$$C \equiv (\eta_1 \eta_{2p} - \eta_2 \eta_p)^{p-1} \equiv 1, \text{ mod } p,$$

because $\eta_1 \eta_{2p} - \eta_2 \eta_p = \pm 21$; so that finally

$$A \equiv \tfrac{1}{2} \Pi\alpha'.\Pi\beta'.\Pi\gamma'.\sum_0^{p-1} \sum_0^{p-1} (z_1 z_2)^{p-1} \equiv \tfrac{1}{2} \Pi\alpha'.\Pi\beta'.\Pi\gamma', \text{ mod } p;$$

an expression which, on substituting for α', β', γ' their values in the two cases $p = 7n+2$, $p = 7n+4$, will be found to coincide with the formulae given by Eisenstein.

There can be no doubt that the principles of this method are capable of many other applications; but nothing has as yet been added to these researches of Eisenstein.

123. *Applications of Continued Fractions to the Theory of Quadratic Forms.* Representations of a number by quadratic forms are in certain cases deducible from the development of its square root in a continued fraction. If A is any number not a square, $\dfrac{J_n + \sqrt{A}}{D_n}$ the $(n+1)$th complete quotient in the development of $\sqrt{A}$, and $\dfrac{p_n}{q_n}$ the convergent fraction immediately preceding that complete

quotient, so that $p_n^2 - Aq_n^2 = (-1)^n D_n$, then the form $(q_n^2, -p_n, A)$, of which the determinant is $(-1)^n D_n$, is either properly or improperly primitive, and belongs in either case to the principal genus of its order. If we investigate the transformation by which this form is reduced to the simplest form in its class, we shall obtain, by an operation exempt from all tentative processes, a representation of A by that simplest form. The following proposition, however, supplies a method by which, when q_n is uneven, and $(q_n^2, -p_n, A)$ belongs to the principal class of properly primitive forms, or when q_n is even, and $(\frac{1}{2}q_n^2, -p_n, 2A)$ belongs to the principal class of improperly primitive forms, we can frequently infer from the development of $\sqrt{A}$ itself the solution of the equations

$$X^2 - (-1)^n D_n Y^2 = A, \quad 2X^2 + 2XY + \tfrac{1}{2}\{1 + (-1)^{n+1} D_n\} Y^2 = A.$$

'If (a, b, c), (a', b', c') are two primitive forms of the determinants D and D', whose joint invariant $ac' - 2bb' + ca'$ is zero, and if m and m' are the greatest common divisors of a, $2b$, c; a', $2b'$, c'; then $m^2 D'$ and $m'^2 D$ are capable of primitive representation by the duplicates of (a, b, c) and (a', b', c') respectively.'

Thus if (a', b', c') is properly primitive and ambiguous, D can be represented primitively by $(1, 0, -D')$; if (a', b', c') is improperly primitive and ambiguous, $2D$ can be represented by $(2, 1, \frac{1}{2}(1 - D'))$. For (a, b, c) and (a', b', c') let us take $(1, 0, -A)$ and $(q_n, -p_n, q_n A)$, whose joint invariant is zero, and of which the first is properly primitive; while the second is properly or improperly primitive according as q_n is uneven or even, and has for its duplicate in the former case $(q_n^2, -p_n, A)$, in the latter $2 \times (\frac{1}{2}q_n^2, -p_n, 2A)$: so that it is ambiguous in both cases alike. Further, let us represent by $(\epsilon_s, -\delta_s, \epsilon_{s-1})$ the form into which $(q_n, -p_n, q_n A)$ is transformed by $\begin{vmatrix} p_s, & p_{s-1} \\ q_s, & q_{s-1} \end{vmatrix}$; we infer, from the property of the invariants, the equations

$$(-1)^{n+1} D_n = \epsilon_s \epsilon_{s-1} - \delta_s^2, \quad \epsilon_{s-1} D_s - 2\delta_s J_s - \epsilon_s D_{s-1} = 0.$$

Let us first suppose that n is uneven, so that $(-1)^n D_n$ is a negative determinant which we shall call $-\Delta$; since

$$q_n (q_n x^2 - 2p_n xy + q_n A y^2) = (q_n x - p_n y)^2 + \Delta y^2,$$

it is evident that when $q_n x^2 - 2p_n xy + q_n A y^2$ attains its minimum value, $\frac{x}{y}$ is a convergent to $\frac{p_n}{q_n}$; not, we may add, the last convergent, if the last integral quotient in the development of $\frac{p_n}{q_n}$ is unity. If therefore $(q_n, -p_n, q_n A)$

is properly primitive and of the principal class, we shall have, for some value of s, $\epsilon_s=1$; whence

$$D_{s-1}=-2\delta_s J_s+\epsilon_{s-1}D_s, \quad \text{and} \quad A=J_s^2+D_s D_{s-1}=(J_s-\delta_s D_s)^2+\Delta D_s^2.$$

If $(q_n, -p_n, q_n A)$ is improperly primitive, and of the principal class of its order, we shall have for some value of s, $\epsilon_s=2$, $D_{s-1}=-\delta_s J_s+\frac{1}{2}\epsilon_{s-1}D_s$,

$$2A=2(J_s-\tfrac{1}{2}(\delta_s+1)D_s)^2+2(J_s-\tfrac{1}{2}(\delta_s+1)D_s)D_s+\tfrac{1}{2}(\Delta+1)D_s^2.$$

We may therefore enunciate the theorem: 'If $\frac{p_n}{q_n}$ is an inferior convergent to $\sqrt{A}$, and $\Delta=q_n^2A-p_n^2$; when $(q_n, -p_n, q_n A)$ is of the principal class of forms of determinant $-\Delta$, A is of the form $X^2+\Delta Y^2$, and Y is the denominator of a complete quotient in the development of $\sqrt{A}$; when $(q_n, -p_n, q_n A)$ is of the principal class of improperly primitive forms of determinant $-\Delta$, A is of the form $2X^2+2XY+\frac{1}{2}(\Delta+1)Y^2$, and Y is the denominator of a complete quotient in the development of $\sqrt{A}$.'

When $(q_n, -p_n, q_n A)$ is ambiguous and properly primitive, but of some other class than the principal class, we must distinguish between two cases, that in which the reduced form equivalent to $(q_n, -p_n, q_n A)$ is itself an ambiguous form, and that in which it is of the type (a, b, a), In the former case we shall arrive at a form $(\epsilon_s, -\delta_s, \epsilon_{s-1})$, in which ϵ_s, being the least number which can be represented by $(q_n, -p_n, q_n A)$, is a divisor of $2\delta_s$, and consequently of D_s and 2Δ; and we shall find

$$A=\left(J_s-\delta_s\frac{D_s}{\epsilon_s}\right)^2+\Delta\frac{D_s^2}{\epsilon_s^2}.$$

In the latter case we shall, in the series of forms $(\epsilon_s, -\delta_s, \epsilon_{s-1})$, arrive at a sequence of one or other of the three types:

(1) $(2[a-b], -[a-b], a), \; (a, (a-b), 2[a-b])$;
(2) $(a, -[a-b], 2[a-b]), \; (a, b, a)$;
(3) $(a, -b, a), \; (a -b, a)$;

i.e. we shall arrive at a form in which ϵ_s is the least number but one, which can be represented by $(q_n, -p_n, q_n A)$, and is a divisor of D_s and 2Δ; we shall then find

$$(1) \quad A=(J_s-\tfrac{1}{2}D_s)^2+\Delta\frac{D_s^2}{\epsilon_s^2};$$

$$(2) \quad A=(J_{s+1}+\tfrac{1}{2}D_s)^2+\Delta\frac{D_s^2}{\epsilon_s^2};$$

$$(3) \quad A=(J_s+\tfrac{1}{2}D_s)^2+\Delta\frac{D_s^2}{\epsilon_s^2}.$$

Similar results may be enunciated for the case in which $(q_n, -p_n, q_n A)$ is improperly primitive and ambiguous, but not of the principal class.

In applying the preceding formulae to particular cases, the following theorem of Goepel's is very useful. Since

$$\epsilon_s = q_n p_s^2 - 2p_n p_s q_s + A q_n q_s^2, \quad -\delta_s = q_n p_s p_{s-1} - p_n(p_s q_{s-1} + p_{s-1} q_s) + A q_n q_s q_{s-1},$$

we find, if μ_s is the integral quotient immediately succeeding $\frac{p_s}{q_s}$, that $\delta_{s+1} = \delta_s - \mu_s \epsilon_s$. Hence $\delta_1, \delta_2, \ldots$ form a continually decreasing series. But $\delta_1 = p_n$ is positive, and $\delta_n = -\Delta q_{n-1}$ is negative; there exists, therefore, a pair of consecutive terms δ_s and δ_{s+1}, of which the former is positive, or zero, and the second negative; Goepel shows that $-\delta_s \delta_{s+1} < \Delta$. For we find

$$q_s \epsilon_{s-1} + q_{s-1}\delta_s = (-1)^s (q_n p_{s-1} - p_n q_{s-1}), \quad q_s \delta_s + q_{s-1}\epsilon_s = (-1)^{s+1}(q_n p_s - p_n q_s);$$

i.e. $$\frac{q_s \epsilon_{s-1} + q_{s-1}\delta_s}{q_s \delta_s + q_{s-1}\epsilon_s} = \mu_s + \frac{1}{\mu_{s+1}} + \ldots;$$

whence $$q_s \epsilon_{s-1} + q_{s-1}\delta_s > \mu_s(q_s \delta_s + q_{s-1}\epsilon_s);$$

or multiplying by ϵ_s, $$\Delta q_s > -\delta_s \delta_{s+1} q_s - \epsilon_s \delta_{s+1} q_{s-1};$$

that is, $\Delta > -\delta_s \delta_{s+1}$, because δ_{s+1} is negative.

Thus if $\Delta = 1$, we have necessarily

$$\delta_s = 0, \quad \epsilon_s = \epsilon_{s-1} = 1, \quad D_{s-1} = D_s, \quad A = J_s^2 + D_s^2.$$

If $\Delta = 2$, we have either

(1) $\delta_s = 0, \quad \epsilon_{s-1} = 2, \quad \epsilon_s = 1, \quad D_{s-1} = 2D_s, \quad A = J_s^2 + 2D_s^2;$

or (2) $\delta_s = 0, \quad \epsilon_{s-1} = 1, \quad \epsilon_s = 2, \quad D_s = 2D_{s-1}, \quad A = J_s^2 + 2D_{s-1}^2;$

or (3) $\delta_s = 1, \quad \delta_{s+1} = -1, \quad \mu_s = 2, \quad \epsilon_s = 1, \quad \epsilon_{s-1} = \epsilon_{s+1} = 3;$

$$D_{s-1} = 3D_s - 2J_s, \quad A = (J_s - D_s)^2 + 2D_s^2 = (J_{s+1} - D_s)^2 + 2D_s^2.$$

If $\Delta = 3$, we have either

(1) $\delta_s = 0, \quad \epsilon_{s-1} = 3, \quad \epsilon_s = 1, \quad D_{s-1} = 3D_s, \quad A = J_s^2 + 3D_s^2;$

or (2) $\delta_s = 0, \quad \epsilon_{s-1} = 1, \quad \epsilon_s = 3, \quad D_s = 3D_{s-1}, \quad A = J_s^2 + 3D_{s-1}^2;$

or (3) $\delta_s = 1, \quad \delta_{s+1} = -2, \quad \mu_s = 3, \quad \epsilon_s = 1, \quad \epsilon_{s-1} = 4, \quad D_{s-1} = 4D_s - 2J_s,$
$A = (J_s - D_s)^2 + 3D_s^2;$

or (4) $\delta_s = 2, \quad \delta_{s+1} = -1, \quad \mu_s = 3, \quad \epsilon_s = 1, \quad \epsilon_{s+1} = 4, \quad D_{s+1} = 4D_s - 2J_{s+1},$
$A = (J_{s+1} - D_s)^2 + 3D_s^2;$

or (5) $\delta_s = 1, \quad \delta_{s+1} = -1, \quad \mu_s = 2, \quad \epsilon_s = 1, \quad \epsilon_{s-1} = \epsilon_{s+1} = 4, \quad D_{s-1} = 4D_s - 2J_s,$
$A = (J_s - D_s)^2 + 3D_s^2;$

or (6) $\delta_s = 1, \quad \delta_{s+1} = -1, \quad \mu_s = 1, \quad \epsilon_s = \epsilon_{s-1} = \epsilon_{s+1} = 2; \quad D_{s-1} = D_s - J_s = J_{s+1},$
$D_{s+1} = D_s - J_{s+1} = J_s, \quad A = J_s^2 - D_s J_s + D_s^2 = J_{s+1}^2 - D_s J_{s+1} + D_s^2;$

the last case occurring always and only when q_n is even. If $\Delta=7$, and if we suppose q_n even, so that $(q_n, -p_n, q_n A)$ is improperly primitive, we shall certainly arrive at a form $(\epsilon_s, -\delta_s, \epsilon_{s-1})$, in which $\delta_s=\pm 1$, and either $\epsilon_s=2$, $\epsilon_{s-1}=4$, or *vice versâ* $\epsilon_{s-1}=2$, $\epsilon_s=4$; so that there are four cases

$$(1, 2)\quad \pm J_s=2D_{s-1}-D_s,\quad A=J_s^2\mp J_s D_{s-1}+2D_{s-1}^2,$$
$$(3, 4)\quad \pm J_s=2D_s-D_{s-1},\quad A=J_s^2\mp J_s D_s+2D_{s-1}^2.$$

Let us next suppose that n is even, so that $(-1)^n D_n=\Delta$ is a positive determinant. Then it is evident $\delta_1, \delta_2, \ldots$ are all positive, for

$$q_n \delta_s=-(q_n p_s-p_n q_s)(q_n p_{s-1}-p_n q_{s-1})+\Delta q_s q_{s-1},$$

of which both parts are positive. Again, the numbers $\epsilon_1, \epsilon_2, \ldots$ form a continually decreasing series; for $q_n \epsilon_s=(q_n p_s-p_n q_s)^2-\Delta q_s^2$; of which the positive part continually decreases, and the negative increases in absolute magnitude. But $\epsilon_1=q_n$, and $\epsilon_n=-\Delta q_n$; there exists, therefore, a term ϵ_{s-1} which is positive, while the following term ϵ_s is negative; whence $\delta_s^2=\Delta+\epsilon_s\epsilon_{s-1}<\Delta$. Thus if $\Delta=2$, we shall have

$$\delta_s=1,\quad \epsilon_{s-1}=1,\quad \epsilon_s=-1,\quad 2J_s=D_s+D_{s-1},$$
$$A=(J_s+D_s)^2-2D_s^2=(J_s+D_{s-1})^2-2D_{s-1}^2.$$

If $\Delta=3$, we shall have either

(1) $\delta_s=1,\ \epsilon_{s-1}=1,\ \epsilon_s=-2,\ 2J_s=D_s+2D_{s-1},\ A=(J_s+D_{s-1})^2-3D_{s-1}^2$;

or (2) $\delta_s=1,\ \epsilon_{s-1}=2,\ \epsilon_s=-1,\ 2J_s=2D_s+D_{s-1},\ A=(J_s+D_s)^2-3D_s^2$.

If a is the integral number immediately inferior to $\sqrt{A}$, the period of integral quotients in the development of $\sqrt{A}$ is of the type

$$\mu_1, \mu_2, \ldots, \mu_{k-1}, b, \mu_{k-1}, \mu_{k-2}, \ldots, \mu_1, 2a;$$

and it is sometimes possible to assign *à priori* the value of D_k, the denominator of the complete quotient corresponding to b; for that denominator is always a divisor of $2A$, and is besides $<2\sqrt{A}$. Thus if A is a prime, $D_k=1$ or 2; if $\frac{1}{2}A$ is a prime, $D_k=1$, 2, or 4. Hence if A or $\frac{1}{2}A$ is a prime of the form $4n+1$, $(-1)^k D_k=-1$; for the equations $x^2-Ay^2=\pm 2$, $=\pm 4$ are impossible on the supposition that x and y are relatively prime, and the equation $x^2-Ay^2=1$ is inadmissible, because b is not the last quotient of a period. Similarly if A or $\frac{1}{2}A$ is a prime of the form $4m+3$, $(-1)^k D_k=2$ or -2, according as the prime is of the form $8m+7$ or $8m+3$; if $\frac{1}{3}A$ is a prime of the form $4m+3$, $(-1)^k D_k=+3$ or -3, according as the prime is of the form $12m+11$ or $12m+7$; and, in general, if λ and $\dfrac{A}{\lambda}$ are each of them a prime of the form $4n+3$, and if $2\lambda < \sqrt{A}$, $(-1)^k D_k = \lambda$, or $-\lambda$, according as λ is or is not a quadratic residue of $\frac{A}{\lambda}$. We thus obtain a direct method for the representation of primes of the forms $4m+1$, $8m+3$, $8m+7$, or the doubles of such primes, by the forms x^2+y^2, x^2+2y^2, x^2-2y^2: when λ is a prime of the form $12m+7$, the developments of $\sqrt{\frac{1}{3}\lambda}$ and $2\sqrt{\frac{1}{3}\lambda}$ will give representations of 3λ by the forms x^2-xy+y^2, x^2+3y^2: when λ is a prime of one of the forms $28m+11$, $28m+15$, $28m+23$, the development of $\sqrt{\frac{1}{7}\lambda}$ will give a representation of 7λ by the form $x^2-xy+2y^2$, &c.

The theorem relating to primes of the form $4n+1$ is very celebrated; it was established independently by Gauss and Legendre, and it no doubt suggested the researches of Goepel in his doctoral dissertation 'De quibusdam aequationibus indeterminatis secundi gradûs' (Crelle, vol. xlv. pp. 1–13). Goepel confined his investigation to the case $D_n=2$, though his method, which in the main is that here described, is of a much more general character. The theorems relating to the case $\Delta=-3$ were first given by M. Stern, who employs Goepel's method with very little modification (Crelle, vol. liii. pp. 87–98). A paper by M. Hermite, which appeared in Crelle's Journal (vol. xlv. p. 191) prior to the republication there of Goepel's dissertation, contains a method (see pp. 211–213) which is very similar to that of Goepel, but which does not connect itself so readily with the common theory of continued fractions. In these researches of M. Hermite the invariant $ac'-2bb'+a'c$ appears explicitly; which is not the case in Goepel's paper.

X.

REPORT ON THE THEORY OF NUMBERS.

PART VI.

[Report of the British Association for 1865, pp. 322–375.]

---

124. *APPLICATION of the Theory of Elliptic Functions to Quadratic Forms.—The Theta Functions of Jacobi.*—It will be for the convenience of the reader to give in this place a brief statement of a few principles and results which belong to the theory of elliptic functions, and to which we shall have occasion to refer in the following articles.

The Theta functions of Jacobi are defined by the equation

$$\theta_{\mu,\nu}(x,\omega)=\sum_{m=-\infty}^{m=+\infty}(-1)^{m\nu}\,e^{i\pi\left[(2m+\mu)\frac{x}{a}+\frac{1}{4}(2m+\mu)^2\omega\right]},$$

if $e^{i\pi\omega}=q$, by the equation

$$\theta_{\mu,\nu}(x,\omega)=\sum_{m=-\infty}^{m=\infty}(-1)^{m\nu}\,q^{\frac{1}{4}(2m+\mu)^2}e^{(2m+\mu)\frac{i\pi x}{a}}.$$

In these equations, μ and ν are given integral numbers; ω is an imaginary constant, having for the coefficient of i in its imaginary part a quantity different from zero and positive; so that the analytical modulus of q is inferior to unity, and the series defining the Theta functions is convergent for all values of x real or imaginary; lastly, a is a constant at present undetermined, but to which we shall hereafter assign a particular value depending on that of ω. When it is not necessary to specify the value of ω, we shall write $\theta_{\mu,\nu}(x)$, instead of $\theta_{\mu,\nu}(x,\omega)$. The following equations are immediate consequences of the definition of the Theta functions:.

$$\theta_{\mu+2,\nu}(x) = (-1)^{\nu}\ \theta_{\mu,\nu}(x). \quad (1)$$

$$\theta_{\mu,\nu+2}(x) = \theta_{\mu,\nu}(x). \quad (2)$$

$$\theta_{\mu,\nu}(-x) = (-1)^{\mu\nu}\theta_{\mu,\nu}(x). \quad (3)$$

$$\theta_{\mu,\nu}(x+a) = (-1)^{\mu}\ \theta_{\mu,\nu}(x). \quad (4)$$

$$\theta_{\mu,\nu}(x+a\omega) = (-1)^{\nu}\ \theta_{\mu,\nu}(x)\, e^{-i\pi\left(2\frac{x}{a}+\omega\right)}. \quad (5)$$

$$\theta_{\mu+\mu',\nu+\nu'}(x) = \theta_{\mu,\nu}\left(x+\tfrac{1}{2}(\mu'\omega+\nu')a\right)\times e^{\pi i\left[\mu'\frac{x}{a}+\frac{1}{4}\mu'^2\omega-\frac{1}{2}\mu\nu'\right]}. \quad (6)$$

Thus there are only four different Theta functions, $\theta_{0,0}(x)$, $\theta_{0,1}(x)$, $\theta_{1,0}(x)$, $\theta_{1,1}(x)$ (equations 1 and 2); of these, the first three are even functions, the last an uneven function (equation 3); they are all periodic, having a or $2a$ for their period, according as μ is even or uneven (equation 4); the quotient $\theta_{\mu,\nu}(x)\div\theta_{\mu',\nu'}(x)$ is doubly periodic, having $a\omega$ or $2a\omega$ for its second period, according as $\nu-\nu'$ is even or uneven (equation 5); finally, any one of the four can be expressed as the product of any other by an exponential factor (equation 6).

The identical equations

$$\left.\begin{aligned} 1+q\ (v+v^{-1})+q^4(v^2+v^{-2})+q^9(v^3+v^{-3})+q^{16}(v^4+v^{-4})+\ldots \\ =(1-q^2)(1-q^4)(1-q^6)\ldots \\ \times(1+qv)(1+q^3v)(1+q^5v)\ldots \\ \times(1+qv^{-1})(1+q^3v^{-1})(1+q^5v^{-1})\ldots, \end{aligned}\right\} \quad (7)$$

$$\left.\begin{aligned} q^{\frac{1}{4}}(v+v^{-1})+q^{\frac{9}{4}}(v^3+v^{-3})+q^{\frac{25}{4}}(v^5+v^{-5})+\ldots \\ =(1-q^2)(1-q^4)(1-q^6)\ldots\times q^{\frac{1}{4}}(v+v^{-1}) \\ \times(1+q^2v^2)(1+q^4v^2)(1+q^6v^2)\ldots \\ \times(1+q^2v^{-2})(1+q^4v^{-2})(1+q^6v^{-2})\ldots, \end{aligned}\right\} \quad (8)$$

in which v is any quantity whatever, and q any quantity of which the analytical modulus is inferior to unity, express an important property of the Theta functions. Elementary demonstrations of the first have been given by Jacobi and Cauchy*; the second is immediately deducible from it, by writing

* Jacobi, Fundamenta Nova, pp. 176–183; Crelle's Journal, vol. xxxvi. p. 75; Cauchy, Comptes Rendus, vol. xvii. pp. 523, 567. See also the note (by M. Hermite), 'Sur la Théorie des Fonctions Elliptiques' in the 6th edition (Paris, 1862) of Lacroix, Traité Elémentaire du Calcul Différentiel, vol. ii. p. 397.

{In (7), $v=e^{\frac{2i\pi x}{a}}$; in (8), $v=e^{\frac{i\pi x}{a}}$. Hence it would be better in (7) to write v^2 for v: then to obtain (8) by putting $q^{\frac{1}{2}}v$ for v, and multiplying by $q^{\frac{1}{4}}v$. This transformation of (7) into (8) is immediately suggested by a comparison of

$$\theta_{0,0}(x)=\Sigma q^{m^2}e^{\frac{2mi\pi x}{a}}, \quad \text{and} \quad \theta_{1,0}(x)=\Sigma q^{m^2}\left(qe^{\frac{2i\pi x}{a}}\right)^m.q^{\frac{1}{4}}e^{\frac{i\pi x}{a}}.\}$$

qv^2 for v, and multiplying by $q^{\frac{1}{4}}v$. We infer from these identities the four formulae*:

$$\theta_{0,0}(x)\left\{=\vartheta_3\left(\frac{\pi x}{a}\right)\right\}=\sum_{-\infty}^{+\infty}{}_m q^{m^2}\cos\frac{2m\pi x}{a}$$

$$=\prod_1^\infty{}_m(1-q^{2m})\prod_1^\infty{}_m\left(1+2q^{2m-1}\cos\frac{2\pi x}{a}+q^{4m-2}\right); \quad . \quad . \quad . \quad . \quad . \quad . \quad (9)$$

$$\theta_{0,1}(x)\left\{=\vartheta\left(\frac{\pi x}{a}\right)\right\}=\sum_{-\infty}^{+\infty}(-1)^m q^{m^2}\cos\frac{2m\pi x}{a}$$

$$=\prod_1^\infty{}_m(1-q^{2m})\prod_1^\infty{}_m\left(1-2q^{2m-1}\cos\frac{2\pi x}{a}+q^{4m-2}\right); \quad . \quad . \quad . \quad . \quad . \quad . \quad (10)$$

$$\theta_{1,0}(x)\left\{=\vartheta_2\left(\frac{\pi x}{a}\right)\right\}=\sum_{-\infty}^{+\infty}{}_m q^{\frac{1}{4}(2m+1)^2}\cos(2m+1)\frac{\pi x}{a}$$

$$=2q^{\frac{1}{4}}\cos\frac{\pi x}{a}\prod_1^\infty{}_m(1-q^{2m})\frac{\pi x}{a}\prod_1^\infty{}_m\left(1+2q^{2m}\cos\frac{2\pi x}{a}+q^{4m}\right); \quad . \quad . \quad (11)$$

$$\frac{1}{i}\theta_{1,1}(x)\left\{=\vartheta_1\left(\frac{\pi x}{a}\right)\right\}=\sum_{-\infty}^{+\infty}(-1)^m q^{\frac{1}{4}(2m+1)^2}\sin(2m+1)\frac{\pi x}{a}$$

$$=2q^{\frac{1}{4}}\sin\frac{\pi x}{a}\prod_1^\infty{}_m(1-q^{2m})\prod_1^\infty{}_m\left(1-2q^{2m}\cos\frac{2\pi x}{a}+q^{4m}\right); . \quad . \quad . \quad . \quad (12)$$

by which the Theta functions are expressed as convergent products of an infinite number of factors.

Other important consequences are deducible from the equation

$$\left.\begin{aligned}
&2\theta_{\mu_1,\nu_1}(x_1)\,\theta_{\mu_2,\nu_2}(x_2)\,\theta_{\mu_3,\nu_3}(x_3)\,\theta_{\mu_4,\nu_4}(x_4)\\
&=\theta_{\sigma-\mu_1,\sigma'-\nu_1}(s-x_1)\times\theta_{\sigma-\mu_2,\sigma'-\nu_2}(s-x_2)\times\theta_{\sigma-\mu_3,\sigma'-\nu_3}(s-x_3)\times\theta_{\sigma-\mu_4,\sigma'-\nu_4}(s-x_4)\\
&+\theta_{\sigma-\mu_1,\sigma'-\nu_1+1}(s-x_1)\times\theta_{\sigma-\mu_2,\sigma'-\nu_2+1}(s-x_2)\times\theta_{\sigma-\mu_3,\sigma'-\nu_3+1}(s-x_3)\\
&\qquad\qquad\times\theta_{\sigma-\mu_4,\sigma'-\nu_4+1}(s-x_4)\\
&+(-1)^{\sigma'}\theta_{\sigma-\mu_1+1,\sigma'-\nu_1}(s-x_1)\times\theta_{\sigma-\mu_2+1,\sigma'-\nu_2}(s-x_2)\times\theta_{\sigma-\mu_3+1,\sigma'-\nu_3}(s-x_3)\\
&\qquad\qquad\times\theta_{\sigma-\mu_4+1,\sigma'-\nu_4}(s-x_4)\\
&+(-1)^{1+\sigma'}\theta_{\sigma-\mu_1+1,\sigma'-\nu_1+1}(s-x_1)\times\theta_{\sigma-\mu_2+1,\sigma'-\nu_2+1}(s-x_2)\\
&\qquad\qquad\times\theta_{\sigma-\mu_3+1,\sigma'-\nu_3+1}(s-x_3)\times\theta_{\sigma-\mu_4+1,\sigma'-\nu_4+1}(s-x_4),
\end{aligned}\right\}\quad(13)$$

---

* {This ϑ-notation is that employed by Jacobi in his Lectures; see Enneper, Sect. 15, p. 78. It gives $\vartheta_3(x)=\sum_{-\infty}^{\infty}q^{m^2}\cos 2mx$. (See also p. 95.) The notation further gives

$$\sin\operatorname{am}\frac{2K(x)}{\pi}=\frac{1}{\sqrt{\kappa}}\frac{\vartheta_1(x)}{\vartheta(x)}.$$

Perhaps it might be best to use it with double suffixes. To these notations we must now add that of Glaisher's tables; $\Theta=\Theta$, $\Theta_1=\frac{1}{\sqrt{\kappa}}\mathrm{H}$, $\Theta_2=\frac{\sqrt{\kappa'}}{\sqrt{\kappa}}\mathrm{H}_1$, $\Theta_3=\sqrt{\kappa'}\,\Theta_1$.}

which contains four independent arguments, x_1, x_2, x_3, x_4, and in which

$$2s = x_1+x_2+x_3+x_4, \quad 2\sigma = \mu_1+\mu_2+\mu_3+\mu_4, \quad 2\sigma' = \nu_1+\nu_2+\nu_3+\nu_4;$$

the numbers $\mu_1, \mu_2, \mu_3, \mu_4$ and $\nu_1, \nu_2, \nu_3, \nu_4$ being subject to the restriction that their sums are respectively even, so that σ and σ' are integral *. Let $\sqrt{\kappa}$, $\sqrt{\kappa'}$ be two quantities defined by the equations

$$\sqrt{\kappa} = \frac{\theta_{1,0}(0)}{\theta_{0,0}(0)}; \qquad \sqrt{\kappa'} = \frac{\theta_{0,1}(0)}{\theta_{0,0}(0)}; \quad . \quad . \quad . \quad . \quad . \quad . \quad . \quad . \quad (14)$$

attributing in (13) to the elements

$$\left| \begin{matrix} x_1, & x_2, & x_3, & x_4 \\ \mu_1, & \mu_2, & \mu_3, & \mu_4 \\ \nu_1, & \nu_2, & \nu_3, & \nu_4 \end{matrix} \right|$$

the systems of values

$$\text{(i)} \left| \begin{matrix} 0, & 0, & 0, & 0 \\ 0, & 0, & 0, & 0 \\ 0, & 0, & 0, & 0 \end{matrix} \right|$$

$$\text{(ii)} \left| \begin{matrix} x, & x, & 0, & 0 \\ 1, & 1, & 0, & 0 \\ 1, & 1, & 0, & 0 \end{matrix} \right|$$

$$\text{(iii)} \left| \begin{matrix} x, & x, & 0, & 0 \\ 0, & 0, & 0, & 0 \\ 1, & 1, & 0, & 0 \end{matrix} \right|$$

---

* This very symmetrical formula is, it would seem, nearly the same as that employed by Jacobi in his Lectures on Elliptic Functions at the University of Königsberg (see his letter to M. Hermite in Crelle's Journal, vol. xxxii. p. 177). It may be proved by actually multiplying the four Theta series, and transforming the indices of -1, $e^{i\pi\omega}$, and $e^{\frac{i\pi}{a}}$ in the general term of the product by means of the elementary formulae

$$a^2+b^2+c^2+d^2 = (s-a)^2+(s-b)^2+(s-c)^2+(s-d)^2,$$

$$a\alpha+b\beta+c\gamma+d\delta = (s-a)(\Sigma-\alpha)+(s-b)(\Sigma-\beta)+(s-c)(\Sigma-\gamma)+(s-d)(\Sigma-\delta),$$

where $2s = a+b+c+d$, $2\Sigma = \alpha+\beta+\gamma+\delta$.

{Put (i) $\begin{cases}\mu_1=\mu_2=\mu_3=\mu_4=0, \\ \nu_1=\nu_2=\nu_3=\nu_4=0;\end{cases}$ (ii) $\begin{cases}\mu_1=\mu_2=\mu_3=\mu_4=1, \\ \nu_1=\nu_2=\nu_3=\nu_4=0;\end{cases}$

and add; we find

$$\prod_{\sigma=1}^{\sigma=4} \theta_{0,0}(x_\sigma) + \prod_{\sigma=1}^{\sigma=4} \theta_{1,0}(x_\sigma) = \prod_{\sigma=1}^{\sigma=4} \theta_{0,0}(s-x_\sigma) + \prod_{\sigma=1}^{\sigma=4} \theta_{1,0}(s-x_\sigma).$$

This is given by Rosenhain as Jacobi's Fundamental Formula (Mémoires des Savants Etrangers, vol. xi. p. 61.)}

[The formula (13) forms the subject of Professor Smith's paper 'On a formula for the multiplication of four Theta Functions, No. XVI.]

we obtain successively

$$\kappa^2+\kappa'^2=1, \qquad (15)$$

$$\kappa'\theta^2_{1,0}(x)=\theta^2_{1,1}(x)+\kappa\theta^2_{0,1}(x), \qquad (16)$$

$$\kappa'\theta^2_{0,0}(x)=\theta^2_{0,1}(x)+\kappa\theta^2_{1,1}(x). \qquad (17)$$

Again, attributing to the same elements the values

$$\left|\begin{matrix} x+y, & x-y, & 0, & 0 \\ 0\ , & 1\ , & 1, & 0 \\ 1\ , & 1\ , & 0, & 0 \end{matrix}\right|$$

we find

$$\theta_{1,1}(x-y)\,\theta_{0,1}(x+y)\,\theta_{1,0}(0)\,\theta_{0,0}(0)$$
$$=\theta_{1,1}(x)\,\theta_{0,1}(x)\,\theta_{0,0}(y)\,\theta_{1,0}(y)$$
$$-\theta_{0,0}(x)\,\theta_{1,0}(x)\,\theta_{1,1}(y)\,\theta_{0,1}(y).$$

Dividing by y, and diminishing y without limit, we obtain

$$\frac{d}{dx}\left(\frac{\theta_{1,1}(x)}{\theta_{0,1}(x)}\right)=\frac{\theta_{0,1}(0)\,\theta'_{1,1}(0)}{\theta_{1,0}(0)\,\theta_{0,0}(0)}\,\frac{\theta_{1,0}(x)\,\theta_{0,0}(x)}{\theta^2_{0,1}(x)}. \qquad (18)$$

Similarly, we might form the differential coefficient of any other quotient of two Theta functions; of these we require only the two following:

$$\frac{d}{dx}\left(\frac{\theta_{1,0}(x)}{\theta_{0,1}(x)}\right)=\frac{\theta_{0,0}(0)\,\theta'_{1,1}(0)}{\theta_{1,0}(0)\,\theta_{0,1}(0)}\,\frac{\theta_{1,1}(x)\,\theta_{0,0}(x)}{\theta^2_{0,1}(x)}, \qquad (18\,a)$$

$$\frac{d}{dx}\left(\frac{\theta_{0,0}(x)}{\theta_{0,1}(x)}\right)=\frac{\theta_{1,0}(0)\,\theta'_{1,1}(0)}{\theta_{0,0}(0)\,\theta_{0,1}(0)}\,\frac{\theta_{1,0}(x)\,\theta_{1,1}(x)}{\theta^2_{0,1}(x)}. \qquad (18\,b)$$

We shall now attribute to a, which has hitherto been left indeterminate, the value $2\mathrm{K}$, K being a constant, the square root of which is determined by the equation

$$\sqrt{\frac{2\mathrm{K}}{\pi}}=\theta_{0,0}(0)=\sum_{+\infty}^{-\infty}{}_m\, q^{m^2}=\prod_{\infty}^{1}{}_m\,(1-q^{2m})\,(1+q^{2m-1})^2\,; \qquad (19)$$

we shall also write K' for $\dfrac{\omega \mathrm{K}}{i}$. Attending to the values of $\sqrt{\kappa'}$ and $\sqrt{\kappa}$, we find from (10) and (11),

$$\sqrt{\frac{2\kappa'\mathrm{K}}{\pi}}=\theta_{0,1}(0)=\sum_{-\infty}^{+\infty}{}_m\,(-1)^m q^{m^2}=\prod_{1}^{\infty}{}_m\,(1-q^{2m})\,(1-q^{2m-1})^2, \qquad (20)$$

$$\sqrt{\frac{2\kappa\mathrm{K}}{\pi}}=\theta_{1,0}(0)=\sum_{-\infty}^{+\infty}{}_m\, q^{\frac{1}{4}(2m+1)^2}=2q^{\frac{1}{4}}\prod_{1}^{\infty}{}_m\,(1-q^{2m})\,(1+q^{2m})^2. \qquad (21)$$

Multiplying together the infinite products (19), (20), (21), and reducing by an identity of Euler's,

$$\prod_{1}^{\infty}(1-q^{2m-1})=\prod_{1}^{\infty}{}_m\,\frac{1}{1+q^m}, \qquad (\mathrm{E})$$

we obtain also

$$\sqrt{\frac{2\kappa\kappa' K}{\pi}} = \frac{\pi}{K} q^{\frac{1}{4}} \Pi_m^{1}{}_{\infty} (1-q^{2m})^3 = \frac{\pi}{2K} \Sigma_{m}{}_{-\infty}^{+\infty} (-1)^m (2m+1) q^{\frac{1}{4}(2m+1)^2} = \frac{1}{i}\theta'_{1,1}(0). \quad (22)$$

These equations (19–22) are of great importance in the arithmetical applications of the theory.

The constant a having the particular value $2K$, the functions

$$\theta_{0,0}(x), \quad \theta_{0,1}(x), \quad \theta_{1,0}(x), \quad \frac{1}{i}\theta_{1,1}(x)$$

are denoted by Jacobi by the symbols $\Theta_1(x)$, $\Theta(x)$, $H_1(x)$, $H(x)$; we shall find it convenient occasionally to employ this notation.

The elliptic functions (properly so called), $\sin \mathrm{am}\, x$, $\cos \mathrm{am}\, x$, $\Delta \mathrm{am}\, x$, are defined by the equations

$$\sin \mathrm{am}\, x = \frac{1}{\sqrt{\kappa}} \frac{H(x)}{\Theta(x)}; \quad \cos \mathrm{am}\, x = \frac{\sqrt{\kappa'}}{\sqrt{\kappa}} \frac{H_1(x)}{\Theta(x)}; \quad \Delta \mathrm{am}\, x = \sqrt{\kappa'} \frac{\Theta_1(x)}{\Theta(x)}. \quad (23)$$

These functions are all doubly periodic, having for their periods $4K$, $2iK'$; $\{2(K \pm iK')\}$; $2K$, $4iK$ respectively; introducing them into the equations (16–18), we obtain

$$\left.\begin{aligned} \cos^2 \mathrm{am}\, x + \sin^2 \mathrm{am}\, x &= 1, \\ \Delta^2 \mathrm{am}\, x + \kappa^2 \sin^2 \mathrm{am}\, x &= 1, \end{aligned}\right\} \quad (24)$$

$$\left.\begin{aligned} \frac{d \,.\, \sin \mathrm{am}\, x}{dx} &= \cos \mathrm{am}\, x \,\Delta \mathrm{am}\, x, \\ \frac{d \,.\, \cos \mathrm{am}\, x}{dx} &= -\sin \mathrm{am}\, x \,\Delta \mathrm{am}\, x, \\ \frac{d \,.\, \Delta \mathrm{am}\, x}{dx} &= -\kappa^2 \sin \mathrm{am} x \cos \mathrm{am} x. \end{aligned}\right\} \quad (25)$$

From these formulae it appears that if $y = \sin \mathrm{am}\, x$, x is *one* of the values of the integral $\int_0^y \frac{dy}{\sqrt{(1-y^2)(1-\kappa^2 y^2)}}$. All the values of that integral are represented by the formula $x + 4mK + 2m'iK'$, in which m and m' represent any integral numbers whatever. Since $\sin \mathrm{am}\, K = 1$, K is one of the values of the integral $\int_0^1 \frac{dy}{\sqrt{(1-y^2)(1-\kappa^2 y^2)}}$; and it can be proved that K' is one of the values of the integral $\int_0^1 \frac{dy}{\sqrt{(1-y^2)(1-\kappa'^2 y^2)}}$. When the real part of ω vanishes (in which case q, K, K', κ, κ' are real and positive, and κ, κ' less than unity), K and K' are the *ordinary* values of those definite integrals; *i.e.* the values obtained by causing y to pass from the inferior to the superior limit, through a series of real values.

The well-known formulae of Addition and Subtraction which express the elliptic functions of the sum or difference of two arguments in terms of the elliptic functions of the arguments themselves, are easily deduced from (13). But as we shall not require these formulae in the following articles, we may omit them here.

125. *The Modulus and its Complement.—The Theory of Transformation.*— In the arithmetical application of the theory, the functions κ and κ', which are respectively termed the *modulus* of the elliptic functions, and the *complement* of the modulus are of primary importance. They are respectively fourth powers of the quantities

$$u = \sqrt{2}\, q^{\frac{1}{8}} \prod_1^\infty{}_m \frac{1+q^{2m}}{1+q^{2m-1}}\,; \qquad u' = \prod_1^\infty{}_m \frac{1-q^{2m-1}}{1+q^{2m-1}}\,; \quad . \quad . \quad . \quad . \quad . \quad (26)$$

which are themselves perfectly determinate functions of ω, if we understand the positive square root of 2 by $\sqrt{2}$, and $e^{\frac{1}{8}i\pi\omega}$ by $q^{\frac{1}{8}}$. Of these functions, which we shall designate by $\phi(\omega)$ and $\psi(\omega)$, the following equivalent expressions have been given by Jacobi (Crelle's Journal, vol. xxxvii. pp. 75–77):

$$\left.\begin{aligned}
u &= \sqrt{2}\, q^{\frac{1}{8}} \prod \frac{1-q^{4m}}{(1+q^{2m-1})(1-q^{2m})} = \sqrt{2}\, q^{\frac{1}{8}} \frac{\Sigma(-1)^m q^{6m^2+2m}}{\Sigma(-1)^{\frac{1}{2}m(m+1)} q^{\frac{1}{2}(3m^2+m)}}\,;\\
u &= \sqrt{2}\, q^{\frac{1}{8}} \prod \frac{(1+q^{4m-2})(1-q^{8m})}{(1+q^{2m-1})(1-q^{4m})} = \sqrt{2}\, q^{\frac{1}{8}} \frac{\Sigma q^{4m^2+2m}}{\Sigma q^{2m^2+m}} = \sqrt{2}\, \frac{\theta_{1,0}(0,\omega)}{\theta_{1,0}(0,\frac{1}{2}\omega)}\,;\\
u &= \sqrt{2}\, q^{\frac{1}{8}} \prod \frac{(1-q^{2m-1})(1-q^{4m})}{(1-q^{4m-2})^2(1-q^{4m})} = \sqrt{2}\, q^{\frac{1}{8}} \frac{\Sigma(-1)^m q^{2m^2+m}}{\Sigma(-1)^m q^{2m^2}}\\
&\qquad = \frac{e^{-\frac{1}{8}i\pi}}{\sqrt{2}} \frac{\theta_{1,0}(0,\frac{1}{2}(\omega+1))}{\theta_{0,1}(0,2\omega)}\,;\\
u &= \sqrt{2}\, q^{\frac{1}{8}} \prod \frac{(1+q^{2m-1})(1-q^{4m})}{(1+q^{2m-1})^2(1-q^{2m})} = \sqrt{2}\, q^{\frac{1}{8}} \frac{\Sigma q^{2m^2+m}}{\Sigma q^{m^2}} = \frac{1}{\sqrt{2}} \frac{\theta_{1,0}(0,\frac{1}{2}\omega)}{\theta_{0,0}(0,\omega)}\,;
\end{aligned}\right\} \quad (27)$$

$$\left.\begin{aligned}
u' &= \prod \frac{1-q^m}{(1+q^{2m-1})(1-q^{2m})} = \frac{\Sigma(-1)^m q^{\frac{1}{2}(3m^2+m)}}{\Sigma(-1)^{\frac{1}{2}m(m+1)} q^{\frac{1}{2}(3m^2+m)}}\,;\\
u' &= \prod \frac{(1-q^{2m-1})(1-q^{4m})}{(1+q^{2m-1})(1-q^{4m})} = \frac{\Sigma(-1)^m q^{2m^2+m}}{\Sigma q^{2m^2+m}} = e^{-\frac{1}{8}i\pi} \frac{\theta_{1,0}(0,\frac{1}{2}(\omega+1))}{\theta_{1,0}(0,\frac{1}{2}\omega)}\,;\\
u' &= \prod \frac{(1-q^{2m-1})^2(1-q^{2m})}{(1-q^{4m-2})^2(1-q^{4m})} = \frac{\Sigma(-1)^m q^{m^2}}{\Sigma(-1)^m q^{2m^2}} = \frac{\theta_{0,1}(0,\omega)}{\theta_{0,1}(0,2\omega)}\,;\\
u' &= \prod \frac{(1-q^{4m-2})^2(1-q^{4m})}{(1+q^{2m-1})^2(1-q^{2m})} = \frac{\Sigma(-1)^m q^{2m^2}}{\Sigma q^{m^2}} = \frac{\theta_{0,1}(0,2\omega)}{\theta_{0,0}(0,\omega)}\,.
\end{aligned}\right\} \quad (28)$$

These expressions of u and u' may be verified by a comparison of their general factors with the general factors in the formulae (26): for some of them, this comparison requires the Eulerian identity already cited (E). Limits of Π and Σ are $1, +\infty$, and $-\infty, +\infty$; the transformation of the products into sums is effected by means of (7).

If $\omega = a + bi$, and if the positive quantity b increases without limit, a remaining finite, we infer, from (26), that

$$\lim \psi(a+bi) = +1, \quad \lim \frac{\phi(a+bi)}{\sqrt{2}\, e^{-\frac{1}{8}b\pi}} = \cos \tfrac{1}{8} a\pi + i \sin \tfrac{1}{8} a\pi.$$

We shall presently see that $\phi(\omega) = \psi\left(-\frac{1}{\omega}\right)$; hence if $\omega = \frac{i}{b}$ and b increase without limit,

$$\lim \phi\left(\frac{i}{b}\right) = \lim \psi(bi) = +1, \quad \lim \psi\left(\frac{i}{b}\right) = \lim \phi(ib) = 0, \quad \lim \frac{\psi\left(\frac{i}{b}\right)}{\sqrt{2}\, e^{-\frac{1}{8}b\pi}} = 1.$$

The principal properties of $\phi(\omega)$ and $\psi(\omega)$ are deducible from the Theory of the Transformation of Elliptic Functions. The general problem considered in that theory is 'Given $\omega = \frac{c + d\Omega}{a + b\Omega}$, where a, b, c, d are positive or negative integral numbers, to express the Theta functions containing Ω by means of the Theta functions containing ω.' The determinant $ad - bc$ must be different from zero and positive, because the coefficients of i in the imaginary parts of ω and Ω must both be different from zero and positive; if $ad - bc = n$, the transformation is said to be of order n. Let $\Lambda, \Lambda', \lambda, \lambda', v, v'$ be the same functions of Ω that $K, K', \kappa, \kappa', u, u'$ are of ω; since $\Omega = i\frac{\Lambda'}{\Lambda}$, $\omega = i\frac{K'}{K}$, the equation $\omega = \frac{c + d\Omega}{a + b\Omega}$ implies the existence of two others of the form

$$\left.\begin{aligned} \frac{1}{M} K &= a\Lambda + bi\Lambda', \\ \frac{1}{M} iK' &= c\Lambda + di\Lambda'; \end{aligned}\right\} \qquad (29)$$

in which M is a coefficient termed the *multiplier*; when Λ has been found, M is determined by the equation

$$\frac{1}{M} = \frac{\Lambda}{K}(a + b\Omega) = \frac{\Lambda}{iK'}(c + d\Omega); \qquad (30)$$

it also satisfies the relation

$$M^2 = \frac{1}{n} \frac{\lambda(1-\lambda^2)}{\kappa(1-\kappa^2)} \cdot \frac{d\kappa}{d\lambda} {}^{*}. \qquad (31)$$

* Fundamenta Nova, p. 75.

If $n=1$, the theory of the transformations of the first order has been comprised by M. Hermite in the single formula *,

$$\theta_{\mu,\nu}\left(\frac{x}{\mathrm{M}},\Omega\right)=\frac{J\,\delta_{\mu,\nu}}{\sqrt{-i(a+b\Omega)}}\,e^{-\frac{\frac{1}{4}i\pi b x^2}{\mathrm{K\Lambda M}}}\,\theta_{m,n}(x,\omega), \quad . \quad . \quad . \quad . \quad (32)$$

in which

$$m=a\mu+b\nu+ab,$$
$$n=c\mu+d\nu+cd,$$
$$\delta_{\mu,\nu}=e^{-\frac{1}{4}i\pi(ac\mu^2+2bc\mu\nu+bd\nu^2+2abc\mu+2abd\nu+ab^2c)},$$
$$J=\frac{1}{\sqrt{b}}\sum_{0}^{b-1}{}_s\, e^{-\frac{i\pi}{b}a(s-\frac{1}{2}b)^2}$$
$$=\left(\frac{b}{a}\right)i^{-\frac{1}{2}a}, \text{ if } a \text{ is uneven,}$$

or

$$=\left(\frac{a}{b}\right)i^{-\frac{1}{2}a}\times i^{-\frac{1}{2}(a-1)(b-1)}, \text{ if } b \text{ is uneven †;}$$

---

* Liouville, New Series, vol. iii. p. 26; and, with less detail, in the Comptes Rendus, vol. xlvi. p. 171.

† These determinations of the value of J coincide with those given by M. Hermite in Liouville's Journal, vol. iii. p. 29; where, however, it would seem that the formulae relating to the two cases of 'a pair' and 'a impair' ought to be transposed.

{Observe that, if we denote $i^{-\mu\nu}\,\theta_{\mu,\nu}$ by $\theta_{\mu,\nu}$, (32) acquires on the right-hand side the factor $i^{mn-\mu\nu}$; and

$$i^{mn-\mu\nu}\times\delta_{\mu,\nu}=e^{\frac{1}{4}i\pi\left\{ac\mu^2+2bc\mu\nu+bd\nu^2+2acd\mu+2bcd\nu+abc(2d-b)\right\}}.$$

Observe also that $\sqrt{-i(a+b\Omega)}=i^{-\frac{1}{2}}\sqrt{a+b\Omega}$, the real parts in both radicals being positive. It is convenient to divide by $i^{-\frac{1}{2}}$; so that

$$J_1=i^{\frac{1}{2}}J=\left(\frac{b}{a}\right)i^{-\frac{1}{2}(a-1)},\quad a \text{ uneven,}$$
$$=\left(\frac{a}{b}\right)i^{-\frac{1}{2}b(a-1)},\ b \text{ uneven,}$$

and

$$\vartheta_{\mu,\nu}\left([a+b\Omega]\frac{\pi x}{h},\ \Omega\right)=\frac{J_1\,\delta_{\mu,\nu}}{\sqrt{(a+b\Omega)}}\times e^{-i\pi b(a+b\Omega)\frac{x^2}{h^2}}\times\vartheta_{m,n}\left(\frac{\pi x}{h},\ \omega\right),$$

or, since

$$\vartheta_{\mu,\nu}\left(\frac{\pi x}{2\Lambda},\ \Omega\right)=\theta_{\mu,\nu}(x,\Omega),$$
$$\vartheta_{m,n}\left(\frac{\pi x}{2\mathrm{K}},\ \omega\right)=\theta_{\mu,\nu}(x,\omega),$$
$$a+b\Omega=\frac{\mathrm{K}}{\mathrm{M}\Lambda};$$

putting $h=2\mathrm{K}$,

$$\theta_{\mu,\nu}\left(\frac{x}{\mathrm{M}},\ \Omega\right)=C\times e^{-\frac{\frac{1}{4}i\pi bx^2}{\mathrm{K\Lambda M}}}\times\theta_{m,n}(x,\omega),$$

as in the text.}

the radical $\sqrt{-i(a+b\Omega)}$ represents that square root of $-i(a+b\Omega)$, of which the real part is positive; lastly, Λ is determined by the equation

$$\sqrt{\frac{2\Lambda}{\pi}} = \theta_{0,0}(0, \Omega) = \frac{J\delta_{0,0}}{\sqrt{-i(a+b\Omega)}}\theta_{ab,cd}(0, \omega), \quad . \quad . \quad . \quad . \quad (33)$$

which is a particular case of the formula (32); and M by the equation

$$\frac{1}{\mathrm{M}} = iJ^2\delta^2_{0,0}\frac{\theta^2{}_{ab,cd}(0, \omega)}{\theta^2{}_{0,0}(0, \omega)}. \quad . \quad . \quad . \quad . \quad . \quad . \quad . \quad . \quad . \quad (34)$$

The formula supposes that b is different from zero and positive; if $b=0$, we may suppose $a=d=1$, so that $\omega=c+\Omega$, and the formula of transformation is

$$\theta_{\mu,\nu}\left(\frac{x}{\mathrm{M}}, \Omega\right) = e^{-\frac{1}{4}i\pi c\mu^2}\theta_{\mu, c\mu+c+\nu}(x, \omega), \quad . \quad . \quad . \quad . \quad . \quad . \quad (35)$$

where
$$\frac{1}{\mathrm{M}} = \frac{\theta^2{}_{0,c}(0, \omega)}{\theta^2{}_{0,0}(0, \omega)}.$$

The equations of the annexed Table, which, for any transformation of the first order, express the relation subsisting between the given and the transformed modulus, are also due to M. Hermite, and are of great importance in the theory of the functions $\phi(\omega)$ and $\psi(\omega)$ *. They may be obtained by applying the formula of transformation (32) to the expressions of $\phi(\omega)$ given by Jacobi (27). There are six cases, answering to the six solutions, of which the congruence $ad-bc\equiv 1$, mod 2 is susceptible. We add, in each case, the value of the multiplier †.

---

* 'Sur la resolution de l'équation du cinquième degré,' Comptes Rendus, vol. xlvi. p. 508; or in a separate reprint (including other memoirs from vols. xlvi. and xlviii.) with the title 'Sur la théorie des équations modulaires, et la résolution de l'équation du cinquième degré,' p. 4.

† [The column giving the transformations of $\psi(\omega)$ was added in manuscript by Professor Smith. He mentions that the values in this column were taken from Koenigsberger, Clebsch's Annalen, vol. iii. p. 10, and verified by

$$\psi(\omega) = \phi\left(-\frac{1}{\omega}\right).$$

The subject-matter of §§ 124 and 125 is considered in much greater detail by Professor Smith in his 'Memoir on the Theta and Omega Functions,' on which he was engaged at the time of his death.]

TABLE A.

$$\omega = \frac{c+d\,\Omega}{a+b\,\Omega}, \qquad ad-bc=1, \qquad \sigma = e^{\frac{1}{8}i\pi}.$$

	$a \equiv$	$b \equiv$	$c \equiv$	$d \equiv$	$\phi(\omega) =$	$\psi(\omega) =$	$\frac{1}{M} =$
I.	1	0	0	1	$\left(\frac{2}{d}\right) \sigma^{cd}\,\phi(\Omega)$	$\left(\frac{2}{a}\right) \sigma^{-ab}\,\psi(\Omega)$	$(-1)^{\frac{1}{8}(a-1)}$
II.	0	1	1	0	$\left(\frac{2}{c}\right) \sigma^{-cd}\,\psi(\Omega)$	$\left(\frac{2}{b}\right) \sigma^{ab}\,\phi(\Omega)$	$(-1)^{\frac{1}{8}(b-1)}\,i$
III.	1	1	0	1	$\left(\frac{2}{d}\right) \sigma^{-cd}\,\frac{1}{\phi(\Omega)}$	$\sigma^{-ab}\,\frac{\psi(\Omega)}{\phi(\Omega)}$	$(-1)^{\frac{1}{8}(a+c-1)}\,\kappa$
IV.	1	1	1	0	$\left(\frac{2}{c}\right) \sigma^{cd}\,\frac{1}{\psi(\Omega)}$	$\sigma^{ab}\,\frac{\phi(\Omega)}{\psi(\Omega)}$	$(-1)^{\frac{1}{8}(c+1)}\,i\kappa$
V.	1	0	1	1	$\sigma^{cd}\,\frac{\phi(\Omega)}{\psi(\Omega)}$	$\left(\frac{2}{a}\right) \sigma^{ab}\,\frac{1}{\psi(\Omega)}$	$(-1)^{\frac{1}{8}(a-1)}\,\kappa'$
VI.	0	1	1	1	$\sigma^{-cd}\,\frac{\psi(\Omega)}{\phi(\Omega)}$	$\left(\frac{2}{b}\right) \sigma^{-ab}\,\frac{1}{\phi(\Omega)}$	$(-1)^{\frac{1}{8}(b-1)}\,i\kappa'$

It would be easy to write these equations so as to express $\phi(\Omega)$ in terms of $\phi(\omega)$, thus completing the solution of the Problem of Transformation of the first order; but it is more convenient to retain them in their actual form. Similar formulae exist expressing $\psi(\omega)$, $\dfrac{\phi(\omega)}{\psi(\omega)}$, in terms of $\phi(\Omega)$ and $\psi(\Omega)$ *.

The propositions implied in the equations of the Table may also be enunciated conversely. Thus to case I. corresponds the theorem 'If ω and Ω are imaginaries in which the coefficient of i is positive, and if $\phi^{2\nu}(\omega)=\phi^{2\nu}(\Omega)$, four integral numbers a, b, c, d can be found satisfying the relations

$$\omega = \frac{c+d\Omega}{a+b\Omega}; \quad ad-bc=1; \quad a\equiv d\equiv 1, \text{ mod } 2; \quad b\equiv 0, \text{ mod } 2; \quad c\equiv 0, \text{ mod } 2^{4-\nu}.\text{'}$$

* M. Hermite has also shown that the function

$$\chi(\omega) = \sqrt[6]{2}\,.\,q^{\frac{1}{24}}(1-q)(1+q^2)(1-q^3)(1+q^4)\ldots,$$

which is a cube root of $\phi(\omega)\times\psi(\omega)$, possesses a similar property; viz. if $\omega = \dfrac{c+d\Omega}{a+b\Omega}$, $ad-bc=1$, $\chi(\omega)$ can be expressed in terms of $\chi(\Omega)$, $\phi(\Omega)$, and $\psi(\Omega)$. (Sur la théorie des équations Modulaires, p. 15.)

If $\phi(\omega) = \phi(\Omega)$, four integral numbers a, b, c, d can be found satisfying the relations $\omega = \dfrac{c+d\Omega}{a+b\Omega}$; $ad-bc=1$; $b \equiv 0$, mod 2; and *either* $a \equiv d \equiv \pm 1$, mod 8; $c \equiv 0$, mod 16, *or* $a \equiv d \equiv \pm 3$, mod 8, $c \equiv 8$, mod 16.'

These converse propositions may be demonstrated by means of the differential equations satisfied by the elliptic functions; by a similar process we obtain the following equally important theorem :—

'If A is any quantity, real or imaginary, other than zero or positive unity, there exist values of ω, having the coefficient of i in their imaginary parts different from zero and positive, which satisfy the equation $\phi^8(\omega) = A$.'

When n is an uneven integer other than 1, the formula of tranformation is

$$\theta_{\mu,\nu}\left(\frac{x}{\mathrm{M}}, \Omega\right) = Te^{-\frac{\frac{1}{4}ib\pi x^2}{\mathrm{K\Lambda M}}}\theta_{m,n}(x, \omega), \quad . \quad . \quad . \quad . \quad . \quad . \quad (36)$$

in which m and n are determined as before, and T is a homogeneous function of order $\frac{1}{2}(n-1)$ of the squares of two of the functions $\theta_{\mu,\nu}(x, \omega)$. We need not occupy ourselves here with the determination of Λ and T, but shall confine ourselves to the consideration of the modulus and multiplier alone. Representing by $\Phi(n)$ the sum of the divisors of n, every binary matrix of order n is included in the formula $\begin{vmatrix} a, & b \\ c, & d \end{vmatrix} = |\Delta| \times |\epsilon|$, in which $|\epsilon|$ is an unit matrix, and $|\Delta|$ one of the $\Phi(n)$ matrices $\begin{vmatrix} \gamma, & 0 \\ k, & \gamma' \end{vmatrix}$, γ and γ' being conjugate divisors of n, and k representing any term of a complete system of residues, mod γ'. It is thus sufficient to consider a system of $\Phi(n)$ transformations of order n, since all others arise from compounding transformations of the first order with the transformations of that system. If we take, in particular, the system of transformations, $\omega = \dfrac{-16k+\gamma'\Omega}{\gamma}$, corresponding to the matrices $\begin{vmatrix} \gamma, & 0 \\ -16k, & \gamma' \end{vmatrix}$ (since n, and therefore γ' is uneven, we may take a system of residues, mod γ', of which every term is divisible by 16), we have for the determination of the transformed modulus, the fundamental theorem *,

* M. Hermite, Sur la théorie des équations Modulaires, p. 36; M. Joubert, Comptes Rendus, vol. l. p. 774; or, in a separate reprint with the title 'Sur la Théorie des Fonctions Elliptiques, et son application a la Théorie des Nombres,' p. 21. The demonstration of this theorem for the case in which n is a prime, is contained in Sohnke's important memoir 'Æquationes modulares pro transformatione functionum ellipticarum,' Crelle, vol. xvi. p. 97. From this particular case, the truth of the theorem for any value of n is inferred without difficulty.

'The quantities $\left(\frac{2}{\gamma}\right)\phi(\Omega)=\left(\frac{2}{\gamma}\right)\phi\left(\frac{\gamma\omega+16k}{\gamma'}\right)$ are the roots of an equation of order $\Phi(n)$, in which the first coefficient is unity, and the other coefficients are rational and integral functions of $\phi(\omega)$ having integral coefficients.'

This equation is termed the modular equation of the transformation of the nth order; designating $\phi(\omega)$ by u, and $\left(\frac{2}{\gamma}\right)\phi\left(\frac{\gamma\omega+16k}{\gamma'}\right)$ by v, we shall represent it by $f(n, u, v)=0$, or more simply by $f(u, v)=0$. The function $f(u, v)$ is characterized by the following, among many other properties,

$$\left.\begin{aligned} f(u, v) &= (-1)^{\Phi(n)}\,\Pi\left(\frac{2}{\gamma}\right)\times f\left(\left(\frac{2}{n}\right)v, u\right), \\ f(u, v) &= (-1)^{\Phi(n)}\,\Pi\left(\frac{2}{\gamma}\right)\times (uv)^{\Phi(n)} f\left(\frac{1}{u}, \frac{1}{v}\right). \end{aligned}\right\} \quad . \quad . \quad . \quad . \quad (37)$$

If in the equation $f(u, v)=0$ we put $u=\psi(\omega)$, the roots are represented by

$$\left(\frac{2}{\gamma'}\right)\psi\left(\frac{\gamma\omega+16k}{\gamma'}\right).$$

If we put $u=e^{\frac{1}{4}i\pi s}\phi(\omega)$, where s is any integral number, the roots are represented by

$$\left(\frac{2}{\gamma}\right)e^{\frac{1}{4}i\pi ns}\phi\left(\frac{\gamma\omega+16k}{\gamma'}\right).$$

If $u=e^{\frac{1}{4}i\pi s}$, there are γ' roots represented by $\left(\frac{2}{\gamma}\right)e^{\frac{1}{4}i\pi ns}$, γ denoting any divisor of n. If we put $u=e^{\frac{1}{8}i\pi s}\frac{\phi(\omega)}{\psi(\omega)}$, where s is any uneven number, the roots are represented by

$$\left(\frac{2}{n}\right)e^{\frac{1}{8}i\pi ns}\frac{\phi\left(\frac{\gamma\omega+16k}{\gamma'}\right)}{\psi\left(\frac{\gamma\omega+16k}{\gamma'}\right)}\,*.$$

The equations whose roots are respectively the squares, fourth powers, and eighth powers of the roots of $f(u, v)=0$, contain only the squares, fourth powers, and eighth powers of u; we shall represent these modular equations by

$$f_2(u^2, v^2)=0,\ f_4(u^4, v^4)=0,\ \text{or}\ f_4(\kappa, \lambda)=0,\ \text{and}\ f_8(u^8, v^8)=0,\ \text{or}\ f_8(\kappa^2, \lambda^2)=0.$$

---

* It is easily seen that $v=\phi(\omega)$ is one of the roots of $f\left[\left(\frac{2}{\gamma'}\right)\phi\left(\frac{\gamma\omega+16k}{\gamma'}\right), v\right]=0$: this establishes the first of the equations (37). The other properties given in the text are deducible from the equations $u=\phi(\omega)$, $v=\left(\frac{2}{\gamma}\right)\phi\left(\frac{\gamma\omega+16k}{\gamma'}\right)$, by applying to ω different transformations of the first order, and employing the formulae of the Table A.

The last equation (by what has preceded) remains unchanged if we write (1) κ for λ, and *vice versâ*, (2) $1-\kappa^2$ for κ^2, $1-\lambda^2$ for λ^2, (3) $\frac{1}{\kappa}$ for κ, $\frac{1}{\lambda}$ for λ.

If n admits of a square divisor δ^2, $f(n, u, v)$ is divisible by $f\left(\frac{n}{\delta^2}, u, \left(\frac{2}{\delta}\right) v\right)$; for if $\gamma\gamma' = \frac{n}{\delta^2}$, $v = \left(\frac{2}{\gamma}\right) \phi\left(\frac{\gamma\omega + 16k}{\gamma'}\right)$ is a root of $f\left(\frac{n}{\delta^2}, u, v\right) = 0$, and $\left(\frac{2}{\gamma\delta}\right) \phi\left(\frac{\gamma\delta\omega + 16k\delta}{\delta\gamma'}\right) = \left(\frac{2}{\delta}\right) v$ is a root of $f(n, u, v) = 0$.

It is sometimes convenient to suppose that the modular equation has been freed by divison from the factors corresponding to the quotients of n divided by its square divisors; its degree will then be

$$\Phi(n) - \Sigma\Phi\left(\frac{n}{p_1^2}\right) + \Sigma\Phi\left(\frac{n}{p_1^2 p_2^2}\right) - \dots$$

if $p_1, p_2, \dots$ represent the primes whose squares divide n, or $n\Pi\left(1+\frac{1}{p}\right)$, if p represent any prime dividing n. The roots of this reduced modular equation are expressed by the same formula as before; only that γ, γ', and k are now subject to the condition that they must not have any common divisor.

With regard to transformations of an even order, we shall only have occasion to consider the case in which n is a power of 2. If $n=2$, we have the modular equations,

$$v^4 = \frac{2u^2}{1+u^4}, \quad v'^4 = \frac{1-u^4}{1+u^4}, \quad \dots \dots \dots (38)$$

of which, if $u = \phi(\omega)$, the roots are given by the equations

$$v^4 = \phi^4\left(\tfrac{1}{2}\omega\right), \quad v'^4 = \psi^4\left(\tfrac{1}{2}\omega\right).$$

If we represent the modular equation of order 2^μ, when cleared of fractions, by $f(2^\mu, u, v_\mu) = 0$, the modular equation of order $2^{\mu+1}$, or $f(2^{\mu+1}, u, v_{\mu+1}) = 0$, is obtained by eliminating v_μ from the two equations

$$f(2^\mu, u, v_\mu) = 0, \quad \text{and} \quad v^4{}_{\mu+1} = \frac{2u^2{}_\mu}{1+v^4{}_\mu}.$$

We may thus successively calculate the modular equations of the orders 4, 8, 16, ...; and, attending to the expression, by means of the transcendent ϕ, of the roots of the equation $v^4 = \frac{2u^2}{1+u^4}$, we may establish the following properties:—the function $f(2^\mu, u, v_\mu)$ is of the order $2^{\mu-2}$ in v^8, or of the order 2^μ in u^2; the coefficient of $u^{2^{\mu+1}}$ is $v^{2^{\mu+1}}$, and the equation is not altered by writing

$\frac{1}{u^2}$ for u^2 and multiplying by $u^{2^\mu+1}$; if $u=\phi(\omega)$, the values of v are given by the equation

$$v^8=\phi^8\left(\frac{\omega+8k}{2^\mu}\right), \quad \ldots \ldots \quad (39)$$

in which k represents any term of a complete system of residues, mod $2^{\mu-2}$, and correspond to the transformations defined by the formula

$$\omega=\frac{c+d\Omega}{a+b\Omega}, \quad \begin{vmatrix} a, b \\ c, d \end{vmatrix} = \begin{vmatrix} 1, 0 \\ -8k, 2^\mu \end{vmatrix} \times \begin{vmatrix} 1, 0 \\ 2h, 1 \end{vmatrix},$$

where h is any term of a system of residues, mod 8; if $v=\phi(\Omega)$, the values of u are given by the equation

$$u^2=\phi^2\left(\frac{2^\mu\Omega}{1+2h\Omega}\right), \quad \ldots \ldots \quad (40)$$

where h is any term of a system of residues, mod 2^μ.

For the determination of the multiplier in a transformation of an uneven order n, we have the theorem,

'If M is the multiplier corresponding to the transformation $\omega=\frac{-16k+\gamma'\Omega}{\gamma}$, the $\Phi(n)$ quantities $z=(-1)^{\frac{1}{2}(\gamma-1)}\frac{1}{\mathrm{M}}$ satisfy an equation of order $\Phi(n)$, in which the coefficient of the highest power of z is unity, and the coefficients of the other powers of z are rational and integral functions with integral coefficients of κ^2; the absolute term, in particular, being $\pm n$' *.

126. *The Complex Multiplication of the Argument.*—The problem of the *multiplication of the argument* is 'Given an integral number n, to express the Theta functions of nx and ω by means of the Theta functions of x and ω.' The solution of this problem may be made to depend on that of the addition of arguments; for to add n equal arguments is to multiply the argument by n. The problem is also included in that of transformation; for if we consider the transformation of order n^2, of which the matrix is $\begin{vmatrix} n, 0 \\ 0, n \end{vmatrix}$, we have

$$\Omega=\omega, \quad \Lambda=\mathrm{K}, \quad \Lambda'=\mathrm{K}', \quad \frac{1}{\mathrm{M}}=n.$$

When ω is not the root of a quadratic equation having integral coefficients,

---

* Jacobi in Crelle's Journal, vol. iii. p. 308. M. Joubert (Comptes Rendus, vol. xlvii. p. 341) has calculated the equations of the multiplier for the orders 3, 5, 7, 11. See also M. Brioschi in Tortolini's Annals, vol. i. (New Series) p. 175, M. Hermite, Equations Modulaires, pp. 12 and 31. No complete demonstration of the theorem appears to have been given.

the transformations, of any square order n^2, and of the type $\begin{vmatrix} n, & 0 \\ 0, & n \end{vmatrix}$ are the only transformations which do not alter the value of ω. For if $\omega = \dfrac{c+d\Omega}{a+b\Omega}$, and $\omega=\Omega$, we have $b\omega^2+(a-d)\omega-c=0$. But, by hypothesis, ω is not the root of any quadratic equation having integral coefficients; neither is ω rational; therefore the three numbers b, $a-d$, and c are all zero, and the matrix $\begin{vmatrix} a, & b \\ c, & d \end{vmatrix}$ is of the type $\begin{vmatrix} n, & 0 \\ 0, & n \end{vmatrix}$. But if ω is the root of a quadratic equation having integral coefficients, an infinite number of transformations, other than those included in the formula $\begin{vmatrix} n, & 0 \\ 0, & n \end{vmatrix}$, can be assigned, which do not alter the value of ω. Let $A+2B\omega+C\omega^2=0$ be the equation satisfied by ω; and let $AC-B^2=\Delta$; then Δ is different from zero and positive; also A and C are of the same sign, and may be supposed to be positive, so that $\omega = \dfrac{-B+i\sqrt{\Delta}}{C}$; lastly, let $\theta=1$, or $=2$, according as (A, B, C) is properly or improperly primitive. Let n be any number such that $\theta^2 n$ admits of representation by $(1, 0, \Delta)$; and let σ, τ be the values of the indeterminates in any such representation; then the transformation

$$\begin{vmatrix} \dfrac{1}{\theta}(\sigma+\tau B), & \dfrac{\tau C}{\theta} \\ -\dfrac{\tau A}{\theta}, & \dfrac{1}{\theta}(\sigma-\tau B) \end{vmatrix}$$

of order n will not alter the value of ω, because $\omega = \dfrac{-\tau A+(\sigma-\tau B)\omega}{\sigma+\tau B+\tau C\omega}$, and will have for the reciprocal of its multiplier

$$\frac{1}{\theta}[\sigma+\tau B+\tau C\omega] = \frac{1}{\theta}(\sigma+i\tau\sqrt{\Delta}).$$

The transformations derived from different values of n, or from different representations of the same value, are all different; and every transformation of order n which does not alter the value of ω, is derived from some representation of $\theta^2 n$ by $(1, 0, \Delta)$; so that the transformations and representations correspond to one another one by one, and are equal in number. It will be observed that the multiplier corresponding to any of these transformations is a complex factor (composed with $\sqrt{-\Delta}$) of the number expressing the order of the transformation; so that the transformation is equivalent to a complex multiplication of the argument. And the Theta functions containing ω do, or do not, admit of complex multiplication, according as ω is or is not a quadratic surd.

If we consider the values of ω contained in Theta functions admitting of multiplication with $i\sqrt{\Delta}$, we see that these values are infinite in number; each form of determinant $-\Delta$ supplying one. But the values of $\phi^8(\omega)$, corresponding to these values, are finite in number, being six times as many as the classes of forms of det. $-\Delta$; provided that in the enumeration of the classes a class of det. -1, or a class derived from a class of det. -1, is counted as $\frac{1}{2}$ instead of 1; and an improperly primitive class of det. -3, or a class derived from such a class, is counted as $\frac{1}{3}$ instead of 1. For it appears from the Table (A) that the values of $\phi^8(\omega)$ corresponding to two equivalent forms, are equal or not, according as the transformation, by which one form passes into the other, is or is not of the type $\begin{vmatrix} 1, & 0 \\ 0, & 1 \end{vmatrix}$, mod 2. We have therefore only to ascertain how many subclasses each class contains, a subclass consisting of forms equivalent by transformations of the type $\begin{vmatrix} 1, & 0 \\ 0, & 1 \end{vmatrix}$. A simple discussion shows that the number of subclasses is six (corresponding to the six types of binary matrices for the modulus 2); except in the two cases just referred to, when the number of subclasses is reduced to 3 and 2 respectively, owing to the existence in those two cases of automorphics which are not of the type $\begin{vmatrix} 1, & 0 \\ 0, & 1 \end{vmatrix}$, mod 2. Thus the whole number of values of $\phi^8(\omega)$ is $6G(\Delta)$, $G(\Delta)$ representing the number of classes of det. $-\Delta$, counted in the manner stated above*. It will be seen that the six values of $\phi^8(\omega)$ corresponding to the forms of the same class are of the type $\kappa^2, \dfrac{1}{\kappa^2}, \dfrac{1}{1-\kappa^2}, 1-\kappa^2, \dfrac{\kappa^2-1}{\kappa^2}, \dfrac{\kappa^2}{\kappa^2-1}$ (being in fact related to one another as the six anharmonic ratios of four points). The three values corresponding to the forms of det. -1 are -1, 2, $\frac{1}{2}$; and the two values corresponding to the improperly primitive forms of det. -3 are the imaginary cube roots of -1.

It is an important theorem (to which we shall again refer) that the $6G(\Delta)$ values of $\phi^8(\omega)$ satisfy an equation of that order, of which the coefficients are integral numbers (but the first coefficient not, in general, unity).

The whole number of values of $\phi(\omega)$, corresponding to the forms of determinant $-\Delta$, is $48G(\Delta)$. For if a be the value of $\phi(\omega)$ corresponding to any form of a given subclass, and η be any eighth root of unity, ηa will be a value of $\phi(\omega)$ corresponding to another form of the same subclass.

* $G(\Delta)$ is the sum of the *densities* of the classes of det. $-\Delta$; the density of a class, according to the definition of Eisenstein, being the reciprocal of the number of its automorphics.

127. *Jacobi's Formulae for the number of decompositions of a number into squares.*—The first applications of elliptic formulae to the theory of numbers were made by Jacobi. The developments, in series proceeding by powers of q, of the squares, fourth, sixth, and eighth powers of the functions

$$\sqrt{\frac{2\mathrm{K}}{\pi}} = 1+2q+2q^4+2q^9+2q^{16}+\ldots,$$

$$\sqrt{\frac{2\kappa\mathrm{K}}{\pi}} = \qquad 2q^{\frac{1}{4}}+2q^{\frac{9}{4}}+2q^{\frac{25}{4}}+\ldots,$$

which are found in the 'Fundamenta Nova' (sections 40–42, and 65, 66), are the analytical expression of arithmetical propositions relating to the composition of numbers by the addition of two, four, six, and eight squares. In these developments n represents any number from 1 to ∞, ν any uneven number from 1 to ∞; d is any divisor of n, δ any uneven divisor of n or any divisor of ν; d' and δ' are the divisors conjugate to d and δ; and the summations indicated by Σ_n, Σ_ν, Σ_d, and Σ extend to every value of n, ν, d, and δ respectively.

(1) $$\frac{2\mathrm{K}}{\pi} = 1+4\Sigma_\nu(-1)^{\frac{1}{2}(\nu-1)}\frac{q^\nu}{1-q^\nu} = 1+4\Sigma_n\frac{q^n}{1+q^{2n}}$$
$$= 1+4\Sigma\,\Sigma_\delta(-1)^{\frac{1}{2}(\delta-1)}q^n.$$

(2) $$\frac{2\kappa\mathrm{K}}{\pi} = 4\Sigma_\nu(-1)^{\frac{1}{2}(\nu-1)}\frac{q^{\frac{1}{2}\nu}}{1-q^\nu} = 4\Sigma_\nu\frac{q^{\frac{1}{2}\nu}}{1+q^\nu}$$
$$= 4\Sigma_\nu\,\Sigma_\delta(-1)^{\frac{1}{2}(\delta-1)}q^{\frac{1}{2}\nu}.$$

(3) $$\frac{4\mathrm{K}^2}{\pi^2} = 1+8\Sigma_n\frac{nq^n}{1+(-1)^n q^n} = 1+8\Sigma_n\frac{q^n}{(1+(-1)^n q^n)^2}$$
$$= 1+24\Sigma_n\Sigma_\delta\,\delta q^n - 16\Sigma_\nu\Sigma\delta q^\nu = 1+8[2+(-1)^n]\,\Sigma_n\Sigma_\delta\,\delta q^n.$$

(4) $$\frac{4\kappa^2\mathrm{K}^2}{\pi^2} = 16\Sigma_\nu\frac{\nu q^\nu}{1-q^{2\nu}} = 16\Sigma_\nu\frac{q^\nu(1+q^{2\nu})}{(1-q^{2\nu})^2}$$
$$= 16\Sigma_\nu\Sigma_\delta\,\delta q^\nu.$$

(5) $$\frac{8\mathrm{K}^3}{\pi^3} = 1+16\Sigma_n\frac{n^2q^n}{1+q^{2n}} - 4\Sigma_\nu(-1)^{\frac{1}{2}(\nu-1)}\frac{\nu^2q^\nu}{1-q^\nu}$$
$$= 1+4\Sigma_n\Sigma_\delta(-1)^{\frac{1}{2}(\delta-1)}(4\delta'^2-\delta^2)\,q^n.$$

(6) $$\frac{8\kappa^3\mathrm{K}^3}{\pi^3} = 4\Sigma_\nu\frac{\nu^2q^{\frac{1}{2}\nu}}{1+q^\nu} - 4\Sigma_\nu(-1)^{\frac{1}{2}(\nu-1)}\frac{\nu^2q^{\frac{1}{2}\nu}}{1-q^\nu}$$
$$= 4\Sigma_\nu\Sigma_\delta\left[(-1)^{\frac{1}{2}(\delta'-1)}-(-1)^{\frac{1}{2}(\delta-1)}\right]\delta^2q^{\frac{1}{2}\nu}.$$

$$(7)\qquad \frac{16\,K^4}{\pi^4} = 1 + 16\,\Sigma_n \frac{n^3 q^n}{1-(-1)^n q^n}$$
$$= 1 + 16\,\Sigma_n\,\Sigma_d\,(-1)^{n+d}\,d^3 q^n.$$

$$(8)\qquad \frac{16\,\kappa^4 K^4}{\pi^4} = 256\,\Sigma_n \frac{n^3 q^{2n}}{1-q^{4n}}$$
$$= 256\,\Sigma_n\,\Sigma\delta'^3\,q^{2n}.$$

Of these formulae, the first two are the analytical expression of the principal theorems relating to the composition of numbers by the addition of two squares (see Art. 95 of this Report); the others may be paraphrased as follows *.

(3) 'The number of representations of any number N as a sum of four squares is eight times the sum of its divisors if N is uneven, twenty-four times the sum of its uneven divisors if N is even.'

(4) 'The number of compositions of the quadruple of any uneven number N by the addition of four uneven squares is equal to the sum of the divisors of N.'

(5) 'The number of representations of any number N as a sum of six squares is $4\,\Sigma\,(-1)^{\frac{1}{2}(\delta-1)}\,(4\,\delta'^2-\delta^2)$, δ denoting any uneven divisor of N, δ' its conjugate divisor. In particular if $N\equiv 1$, mod 4, the number of representations is $12\,\Sigma\,(-1)^{\frac{1}{2}(\delta-1)}$; if $N\equiv -1$, mod 4, it is $-20\,\Sigma\,(-1)^{\frac{1}{2}(\delta-1)}\,\delta^2$.'

(6) 'The number of compositions of the double of any uneven number N by the addition of six uneven squares is

$$\tfrac{1}{16}\,\Sigma\left[(-1)^{\frac{1}{2}(\delta'-1)}-(-1)^{\frac{1}{2}(\delta-1)}\right]\delta^2\,;$$

if $N\equiv 1$, mod 4, this number is zero; if $N\equiv -1$, mod 4, it is

$$-\tfrac{1}{8}\,\Sigma\,(-1)^{\frac{1}{2}(\delta-1)}\,\delta^2.\text{'}$$

(7) 'The number of representations of any uneven number as a sum of eight squares is sixteen times the sum of the cubes of its divisors; for an even number it is sixteen times the excess of the cubes of the even divisors above the cubes of the uneven divisors.'

(8) 'If N is any number whatever, the number of compositions of $8\,N$ by the addition of eight uneven squares is equal to the sum of the cubes of those divisors of N whose conjugates are uneven.'

---

* The expansions of $(1)\times(2)$, $(1)\times(4)$, $(3)\times(2)$, $(3)\times(4)$, are also given in sections 40 and 41 of the 'Fundamenta'; and may be similarly interpreted.

In counting the number of compositions by addition of squares, two compositions are to be considered as different if, and only if, the same places in each are not occupied by the same squares; but in counting the number of representations we have to attend also to the signs of the roots of the squares. Thus each composition by the addition of four squares, none of which is zero, is equivalent to sixteen representations. Only one or two of the preceding theorems are enunciated in the published writings of Jacobi: see Crelle's Journal, vol. iii. p. 191; vol. xii. p. 167. Some of the others have been given by Eisenstein (Crelle, vol. xxxv. p. 135), who had also obtained purely arithmetical demonstrations of them from the theory of quadratic forms containing several indeterminates. 'In my investigations,' he says 'these theorems are proved by purely arithmetical considerations, and appear as special cases of more general theorems; at the same time we see why these developments close with the eighth power; since, in fact, eight is the greatest number of indeterminates for which only one class of forms, represented by a sum of squares, appertains to the determinant -1.'

In the second of the notes to which we have just referred (Crelle, vol. xii. p. 167), Jacobi has given an arithmetical demonstration of the theorem (4). It consists in a kind of translation of the analytical proof into an arithmetical one; and is of great interest and importance, as the first example of a new method, and as having suggested important researches to MM. Liouville and Kronecker (see Liouville's Journal, New Series, vol. vii. p. 48; M. Kronecker, 'Monatsberichte,' May 26, 1862, p. 307).

The doubly periodic functions of argument $\frac{2\mathrm{K}x}{\pi}$ obtained by dividing any Theta function by any other, or the product of any two of them, by the product of the other two, admit of development in series proceeding by sines or cosines of multiples of the argument x. These developments, which, unlike the developments of the Theta functions themselves, are not convergent for all values of x, real or imaginary, will be found for the most part in section 39 of the 'Fundamenta Nova'; and the complete system has been given by M. Hermite (Comptes Rendus, July 7, 1862). One, which we require in this place, will serve as an example of the rest,

$$\left.\begin{aligned}\frac{\kappa\mathrm{K}}{2\pi}\sin\operatorname{am}\frac{2\mathrm{K}x}{\pi} &= \Sigma_\nu \frac{q^{\frac{1}{2}\nu}\sin\nu x}{1-q^\nu} \\ &= \Sigma_\nu\Sigma_\delta \sin\delta x\,.\,q^{\frac{1}{2}\nu}.\end{aligned}\right\} \quad . \quad . \quad . \quad . \quad . \quad . \quad (\mathrm{A})$$

It is from these developments that the expansions (1) ... (8) of the powers

of $\sqrt{\frac{2K}{\pi}}$ and $\sqrt{\frac{2\kappa K}{\pi}}$ are deduced. Thus, writing $\frac{1}{2}\pi$ for x in (A), we find, since sin am K = 1,
$$\frac{\kappa K}{2\pi} = \Sigma_\nu (-1)^{\frac{1}{2}(\nu-1)} \frac{q^{\frac{1}{2}\nu}}{1-q^\nu},$$
which is the formula (2). We shall now show how the equation (4) can be obtained by squaring this formula. For this purpose we represent by α and β any two unequal positive uneven numbers congruous to one another for the modulus 4, and by α' and β' any two positive uneven numbers not congruous to one another for the modulus 4. We then have

$$\tfrac{1}{4}\frac{\kappa^2 K^2}{\pi^2} = \Sigma_\nu \frac{q^\nu}{(1-q^\nu)^2} + \Sigma_\alpha \Sigma_\beta \frac{q^{\frac{1}{2}(\alpha+\beta)}}{(1-q^\alpha)(1-q^\beta)} - \Sigma_{\alpha'}\Sigma_{\beta'} \frac{q^{\frac{1}{2}(\alpha'+\beta')}}{(1-q^{\alpha'})(1-q^{\beta'})},$$
$$= P + Q - R, \text{ for brevity.}$$

Here
$$P = \Sigma_\nu \frac{q^\nu}{(1-q^\nu)^2} = \Sigma_n \,.\, \Sigma\delta' \,.\, q^n = \Sigma_n \frac{nq^n}{1-q^{2n}};$$
again in Q, if we double each term we may suppose $\beta > \alpha$; let $\beta = \alpha + 4n$; observing that α may be any positive uneven number, and n any positive number whatever, we find
$$Q = 2\Sigma_n \Sigma_\nu \frac{q^{2n+\nu}}{(1-q^\nu)(1-q^{4n+\nu})}$$
$$= 2\Sigma_n \Sigma_\nu \frac{q^{2n}}{1-q^{4\nu}} \left[\frac{q^\nu}{1-q^\nu} - \frac{q^{4n+\nu}}{1-q^{4n+\nu}}\right]$$
$$= 2\Sigma_n \sum_{1}^{4n-1}{}_\nu \frac{q^{2n+\nu}}{(1-q^{4n})(1-q^\nu)}.$$

Lastly, in R let $\alpha' + \beta' = 4n$; so that
$$R = \Sigma_n \sum_{1}^{4n-1}{}_\nu \frac{q^{2n}}{(1-q^\nu)(1-q^{4n-\nu})}$$
$$= \Sigma_n \sum_{1}^{4n-1}{}_\nu \frac{q^{2n}}{1-q^{4n}} \left[\frac{q^\nu}{1-q^\nu} + \frac{q^{4n-\nu}}{1-q^{4n-\nu}} + 1\right]$$
$$= 2\Sigma_n \sum_{1}^{4n-1}{}_\nu \frac{q^{2n+\nu}}{(1-q^{4n})(1-q^\nu)} + \Sigma_n \frac{2nq^{2n}}{1-q^{4n}}.$$

Consequently
$$\tfrac{1}{4}\frac{\kappa^2 K^2}{\pi^2} = P + Q - R$$
$$= \Sigma_n \frac{nq^n}{1-q^{2n}} - \Sigma_n \frac{2nq^{2n}}{1-q^{4n}} = \Sigma \frac{\nu q^\nu}{1-q^{2n}},$$
which is the formula (4).

Thus by a purely analytical process we deduce from an equation which exhibits the number of compositions of the double of an uneven number by the addition of two uneven squares, an equation exhibiting the number of compositions of the quadruple of an uneven number by the addition of four uneven squares. This analysis Jacobi has expressed arithmetically as follows. Representing by N an uneven number, by $[4N]$ the number of compositions of $4N$ by the addition of four uneven squares, we resolve $4N$ in every possible way into two unevenly even numbers $2N_1$ and $2N_2$, and each of these in every possible way into two uneven squares; we thus obtain the equation

$$[4N]=\Sigma[2N_1=(2x+1)^2+(2y+1)^2]\times[2N_2=(2x+1)^2+(2y+1)^2],$$

in which the summation extends to every pair of uneven numbers N_1 and N_2 which satisfy the equation $2N=N_1+N_2$, and the square brackets represent the number of solutions in positive integers of the equations included in them. Observing that $[2N_1=(2x+1)^2+(2y+1)^2]$ is the excess of the number of divisors of N_1 which are of the form $4k+1$, above the number of its divisors which are of the form $4k-1$, retaining the signification of α, β, α', β', and denoting by a and b any positive uneven numbers, we may transform the expression of $[4N]$ into the following,

$$[4N]=[2N=(a+b)\,a]+[2N=a\alpha+b\beta]-[2N=a\alpha'+b\beta'],$$

in which the square brackets still retain the same signification. Supposing, as before, $\beta>\alpha$, and $\beta=\alpha+4n$, we have

$$[2N=a\alpha+b\beta]=2[2N=\alpha(a+b)+4nb];$$

or, putting $\alpha=\nu+4kn$, ν being less than $4n$,

$$[2N=a\alpha+b\beta]=2[2N=\nu(a+b)+4n(\overline{a+b}k+b)]=2[N=\nu x+2ny],$$

y being uneven and $\nu<2n$. Again, if in $[2N=a\alpha'+b\beta']$ we write $4n$ for $\alpha'+\beta'$, and suppose $a>b$ (the supposition $a=b$ is inadmissible as it would render N even), we have

$$[2N=a\alpha'+b\beta']=2[N=\alpha'.\tfrac{1}{2}(a-b)+2nb]=2[N=\nu x+2ny],$$

as before. Hence

$$[2N=a\alpha+b\beta]-[2N=a\alpha'+b\beta']=0, \quad \text{and} \quad [4N]=[2N=(a+b)\,a],$$

i.e. $[4N]$ is the sum of the divisors of N. In this arithmetical process we determine the coefficient of q^N in P, Q, R, instead of determining those functions themselves; and as the difference $Q-R=-\Sigma_n\dfrac{2nq^{2n}}{1-q^{4n}}$ is an even function in the analytical process, so the difference $[2N=a\alpha+b\beta]-[2N=a\alpha'+b\beta']$ vanishes in the arithmetical one.

Lejeune Dirichlet, in a letter addressed to M. Liouville (Liouville's Journal, New Series, vol. i. p. 210), has put Jacobi's demonstration into a form in which it is more easily followed, but is a little further removed from the analysis. He shows that to every solution of the equation $a\alpha+b\beta=2N$, in which $\alpha>\beta$, there corresponds a solution of the equation $a'\alpha'+b'\beta'=2N$, in which $\alpha'>\beta'$, and *vice versâ*, the two solutions being connected by the relation

$$\begin{vmatrix} \alpha', & b' \\ \beta', & -a' \end{vmatrix} = \begin{vmatrix} x+1, & x+2 \\ x\ , & x+1 \end{vmatrix} \times \begin{vmatrix} b, & \alpha \\ a, & -\beta \end{vmatrix},$$

where x is the integral number immediately inferior to $\dfrac{\beta}{a-\beta}$, or, which is the same thing, to $\dfrac{\beta'}{a'-\beta'}$. Hence, as before, $[2N=a\alpha+b\beta]=[2N=a'\alpha'+b'\beta']$, and $[4N]$ is equal to the sum of the divisors of N.

128. *Theorems of Jacobi on Simultaneous Quadratic Forms.*—In an elaborate memoir 'On Series whose Exponents are of two Quadratic forms'*, Jacobi has established a great number of elliptic formulae, which are the analytical expression of theorems relating to the representation of numbers by certain quadratic forms. A comparison of the two criteria of Gauss for the biquadratic character of 2 with respect to a prime p of the linear form $8k+1$, leads to a result which will serve as an example of these theorems. By the first criterion, 2 is or is not a biquadratic residue of a prime p of the form $8k+1$ according as a is even or uneven in the equation $p=(4a+1)^2+8b^2$; by the second, 2 is or is not a biquadratic residue of p according as β is even or uneven in the equation $p=(4\alpha+1)^2+16\beta^2$ †. We infer therefore that $a+\beta$

* Crelle's Journal, vol. xxxvii. pp. 61 and 221; or Mathematische Werke, vol. ii. p. 67.

† Theoria Residuorum Biquadraticorum, Arts. 13–21. To the second criterion we have already referred in this Report (Art. 24, and in the additions to Part I., printed at the end of Part II.); the first is more elementary, and is inferred from the equation $p=(4a+1)^2+8b^2$. in which p is a prime of the form $8k+1$. Raising each side of the congruence

$$-8b^2\equiv(4a+1)^2, \text{ mod } p,$$

to the power $\frac{1}{4}(p-1)$, and observing that

$$2^{\frac{1}{2}(p-1)}\equiv\left(\frac{2}{p}\right)=1,\quad (-1)^{\frac{1}{4}(p-1)}=1,\quad \text{we find}\quad 2^{\frac{1}{4}(p-1)}\left(\frac{b}{p}\right)\equiv\left(\frac{4a+1}{p}\right).$$

But if $b=2^\gamma\beta$, where β is uneven,

$$\left(\frac{b}{p}\right)=\left(\frac{\beta}{p}\right)=\left(\frac{p}{\beta}\right)=1,$$

because $p\equiv(4a+1)^2$, mod β; and

$$\left(\frac{4a+1}{p}\right)=\left(\frac{p}{4a+1}\right)=\left(\frac{8b^2}{4a+1}\right)=\left(\frac{2}{4a+1}\right)=(-1)^a.$$

Hence $2^{\frac{1}{4}(p-1)}\equiv(-1)^a$, mod p, which is Gauss's first criterion.

is even, or (since $a+b+\alpha$ is even by virtue of the congruence

$$(4a+1)^2+8b^2 \equiv (4\alpha+1)^2, \text{ mod } 16)$$

that $\alpha+\beta+b$ is even. The result is thus generalized by Jacobi:

'For any number P the sum $\Sigma(-1)^b$, *i.e.*, the excess of the number of solutions of the equation $P=(4a+1)^2+8b^2$ in which b is even above the number of solutions in which b is uneven, is equal to the sum $\Sigma(-1)^{\alpha+\beta}$, *i.e.* to the excess of the number of solutions of the equation $P=(4\alpha+1)^2+16\beta^2$, in which $\alpha+\beta$ is even above the number of solutions in which $\alpha+\beta$ is uneven.'

The generalized theorem is expressed analytically by the equation

$$\Sigma(-1)^n q^{(4m+1)^2+8n^2} = \Sigma(-1)^{m+n} q^{(4m+1)^2+16n^2}, \quad . \quad . \quad . \quad . \quad . \quad (\Sigma)$$

in which the summations extend to all values of m and n from $-\infty$ to $+\infty$. But this equation is an elliptic formula; for, on dividing by q, and writing q for q^8, it becomes

$$\Sigma(-1)^n q^{n^2} \times \Sigma q^{2m^2+m} = \Sigma(-1)^n q^{2n^2} \Sigma(-1)^m q^{2m^2+m},$$

which is included in the equations (28) of Art. 125, and is therefore a corollary from the fundamental property of the Theta functions expressed in equation (7) of Art. 124. We infer at the same time, from the equations (28), that either of the sums

$$\Sigma(-1)^{m+n} q^{(4m+1)^2+16n^2} \quad \text{or} \quad \Sigma(-1)^m q^{(4m+1)^2+8n^2}$$

is equal to the infinite product

$$q \prod_{m=1}^{m=\infty} (1-q^{8m})(1-q^{16m}).$$

We thus arrive at an analytical proof of Jacobi's theorem, including, as a particular case, a proof of the identity of Gauss's two criteria. But the continuation of Jacobi's memoir was intended to contain direct arithmetical demonstrations (which, however, have never been published) of the theorems of which the equation $\Sigma(-1)^b = \Sigma(-1)^{\alpha+\beta}$ is an example. He says, 'Though these arithmetical demonstrations of results obtained analytically present no essential difficulty, yet they are sometimes of a complicated character, and require peculiar classifications of numbers which perhaps may be of use in other researches. We have here a certain amount of freedom in the choice of methods, so that the proofs can easily be varied'*. Probably one of these methods was that employed by Dirichlet in his earliest arithmetical memoir,

---

* Mathematische Werke, vol. ii. p. 73.

to which Jacobi expressly refers. In this memoir* (written when only the enunciations of Gauss's criteria for the biquadratic character of 2 had been published) Dirichlet gives a demonstration of the first criterion, which does not differ from that subsequently given by Gauss (Theor. Res. Biq. Comm. prima, Art. 13), and then deduces the second criterion, as follows, from the first. Since

$$p=(4a+1)^2+8b^2=(4a+1)^2+16\beta^2,$$

we have

$$[4(a+\beta)+1]\times[4(a-\beta)+1]=(4a+1)^2-8b^2.$$

No common divisor of $4a+1$ and b can also be a common divisor of

$$4(a+\beta)+1 \quad \text{and} \quad 4(a-\beta)+1,$$

i.e. of $4a+1$ and b; for p is not divisible by any square. The greatest common divisor of $(4a+1)^2$ and b^2 must therefore be a product of two relatively prime uneven squares δ^2 and δ'^2, dividing $4(a+\beta)+1$ and $4(a-\beta)+1$ respectively; $\dfrac{4(a+\beta)+1}{\delta^2}$ is thus a divisor of the quadratic form x^2-8y^2, in which x and y are relatively prime; it is, consequently, itself of that quadratic form, and

$$4(a+\beta)+1\equiv 1, \text{ mod } 8;$$

this congruence implies that $a+\beta\equiv 0$, mod 2, or, which comes to the same thing, that $b\equiv a+\beta$, mod 2. It will be seen that this demonstration of the congruence $b\equiv a+\beta$, mod 2, applies to any two representations of any number P by the forms $f=(4a+1)^2+8b^2$ and $\phi=(4a+1)^2+16\beta^2$,

provided that in the two representations the four numbers $4a+1$, $4a+1$, b, β have no common divisor. To prove, for every uneven value of P, the truth of Jacobi's equation $\Sigma(-1)^a=\Sigma(-1)^{a+\beta}$, we observe, first of all, that the equation is evidently true if P is not $\equiv 1$, mod 8, or if P contain an uneven power of a prime of the linear form $8k+7$; for in these cases there are no representations of P by either form. We may therefore suppose that P is of the linear form $8k+1$; then the equation is true if P contains an uneven power of any prime p of either of the linear forms $8k\pm 3$; thus if $P=p^{2\nu+1}P'$, where P' is prime to p, and $p\equiv P'\equiv 3$, mod 8, there are no representations of P by ϕ, so that $\Sigma(-1)^{a+\beta}=0$; let the equations

$$p^{2\nu+1}=x^2+2y^2, \qquad P'=X^2+2Y^2$$

denote generally those representations of $p^{2\nu+1}$ and P' by the form (1, 0, 2)

* Crelle's Journal, vol. iii. p. 35.

in which the first indeterminate is $\equiv 1$, mod 4; then the representations of $P = p^{2\nu+1} \times P'$ by f will be comprised in the formula

$$P = (2yY - xX)^2 + 8\{\tfrac{1}{2}(yX + xY)\}^2 \text{*};$$

but of the two numbers $\frac{1}{2}(yX + xY)$, $\frac{1}{2}(yX - xY)$ (both being values of the second indeterminate), one is uneven and the other even; whence

$$\Sigma(-1)^b = 0 = \Sigma(-1)^{\alpha+\beta}.$$

Similarly if $P = p^{2\nu+1}P'$, where P' is prime to p, and $p \equiv P' \equiv 5$, mod 8, there are no representations of P by f, and it may be shown that

$$\Sigma(-1)^{\alpha+\beta} = 0 = \Sigma(-1)^b.$$

We may therefore confine ourselves to the case in which P is composed of any powers of primes of the linear form $8k+1$, and of even powers of primes of the forms $8k+3$, 5, 7. If, on this supposition, $P = P' \times P''$, where P' and P'' are relatively prime, and each is $\equiv 1$, mod 8, the sums $\Sigma(-1)^a$ and $\Sigma(-1)^{\alpha+\beta}$, relative to P, are the products of the corresponding sums relative to P' and P''. This may be proved by observing that the representations of P by f [or ϕ] may be obtained by compounding the representations of P' and P'' by that form, and that each representation of P has the character of an even or uneven b [or $\alpha + \beta$] according as the representations of P' and P'' of which it is compounded agree or differ in respect of that character. Thus it is sufficient to consider the four cases in which

(1) $P = p^\nu$, $p \equiv 1$, mod 8; (2) $P = p^{2\nu}$, $p \equiv 3$, mod 8;
(3) $P = p^{2\nu}$, $p \equiv 5$, mod 8; (4) $P = p^{2\nu}$, $p \equiv 7$, mod 8.

In the last of these cases it is evident that

$$\Sigma(-1)^b = +1 = \Sigma(-1)^{\alpha+\beta};$$

in the others, the proof is supplied by Dirichlet's method. (i) If $P = p^\nu$, $p \equiv 1$, mod 8, there are two primitive and $\nu - 1$ derived representations of P by each form; and the application of Dirichlet's method shows that, for *every* representation of P by ϕ, $(-1)^{\alpha+\beta}$ has the same value as $(-1)^b$ in either *primitive* representation of P by f, and, conversely, that for *every* representation of P by f, $(-1)^b$ has the same value as $(-1)^{\alpha+\beta}$ in either *primitive* representation of P by ϕ; whence the units $(-1)^b$ and $(-1)^{\alpha+\beta}$ have all the same value, and

$$\Sigma(-1)^b = \Sigma(-1)^{\alpha+\beta} = \pm(\nu + 1).$$

---

* For $2yY - xX \equiv 1$, mod 4; and the representations comprised in the formula are all different, their number being equal to the number of sets of representations of P by (1, 0, 8).

The ambiguous sign is that of $(-1)^b$ in the primitive representation of P by f, and will be found (by reasoning similar to that which establishes Gauss's first criterion) to coincide with $(-1)^{\frac{1}{8}(P-1)}\epsilon^\nu$, where ϵ is the unit satisfying the congruence $2^{\frac{1}{4}(p-1)}\equiv\epsilon$, mod p. (ii) If $P=p^{2\nu}$, $p\equiv 3$, mod 8, there is but one representation of P by ϕ, and $\Sigma(-1)^{\alpha+\beta}=(-1)^\nu$; there are $2\nu+1$ representations of P by f, of which two are primitive, $2(\nu-1)$ are derived from the primitive representations of $p^2, p^4, \ldots, p^{2(\nu-1)}$, and in the remaining one $b=0$. Applying Dirichlet's method to the equation

$$q^{2\sigma}=(4a+1)^2+8b^2=(4\alpha+1)^2+16\beta^2$$

in which $\sigma=1, 2, \ldots, \nu$, $\beta=0$, $4\alpha+1=(-1)^\sigma q^\sigma$, and the representation by f is primitive), we find $(-1)^b=(-1)^\sigma$; whence inasmuch as the character $(-1)^b$ is the same in a derived representation, and in the representation from which it is derived

$$\Sigma(-1)^b=1+2\sum_1^\nu(-1)^\sigma=(-1)^\nu=\Sigma(-1)^{\alpha+\beta}.$$

(iii) Lastly, if $P=p^{2\nu}$, $p\equiv 5$, mod 8, there is but one representation by f, and $\Sigma(-1)^b=+1$; there are $2\nu+1$ representations by ϕ. Applying Dirichlet's method as in the preceding case, we find that for any primitive representation of an even power of p by ϕ, $(-1)^{\alpha+\beta}=+1$; whence, for a derived representation in which the greatest common divisor of the indeterminates is q^σ, $(-1)^{\alpha+\beta}=(-1)^\sigma$. Consequently

$$\Sigma(-1)^{\alpha+\beta}=2\sum_0^{\nu-1}(-1)^\sigma+(-1)^\nu=+1=\Sigma(-1)^b.$$

This completes the demonstration of Jacobi's theorem.

Let P be any uneven number and $\chi(P)$ the positive numerical transcendent defined by the equation

$$\chi^2(P)\times\psi(P)=\Sigma(-1)^{\frac{1}{2}(d-1)}\times\Sigma(-1)^{\frac{1}{8}(d^2-1)}\times\Sigma(-1)^{\frac{1}{2}(d-1)+\frac{1}{8}(d^2-1)},$$

where $\psi(P)$ is the number of divisors of P, and d is any divisor of P. It will be seen that $\chi(P)=0$, except when P is capable of representation both by ϕ and f: when P is capable of such simultaneous representation, let $P=(4\alpha+1)^2+16\beta^2$ be a representation of P by ϕ in which the greatest common divisor of $4\alpha+1$ and β is the least possible; let $\varpi=4\alpha+1+4i\beta$, and let $\left[\frac{1+i}{\varpi}\right]$ represent the quadratic character (Art. 27) of $1+i$ with respect to ϖ; the equation

$$\Sigma(-1)^b=\Sigma(-1)^{\alpha+\beta}=\left[\frac{1+i}{\varpi}\right]\chi(P)$$

will hold in each of the cases considered separately above; but the numerical functions occurring in this equation satisfy the condition

$$\phi(P_1) \times \phi(P_2) = \phi(P_1 P_2),$$

where P_1, P_2 are relatively prime; the equation is therefore universally true for every uneven number P, and implies the identity

$$\Sigma(-1)^n q^{(4m+1)^2+8n^2} = \Sigma(-1)^{m+n} q^{(4m+1)^2+16n^2} = \Sigma\left[\frac{1+i}{\varpi}\right]\chi(P)\, q^P.$$

From the nature of the identity (Σ) it is evident that we may substitute any function whatever (which renders the two series convergent) for the exponential of q. Thus, for example, we find

$$\Sigma\frac{(-1)^n}{[(4m+1)^2+8n^2]^{1+\rho}} = \Sigma\frac{(-1)^{m+n}}{[(4m+1)^2+16n^2]^{1+\rho}} = \Sigma\left[\frac{1+i}{\varpi}\right]\frac{\chi(P)}{P^{1+\rho}}$$

$$= \Pi\frac{1}{\left(1-\dfrac{\eta}{p_1^{1+\rho}}\right)^2}\quad \Pi\frac{1}{1+\dfrac{1}{p_2^{2(1+\rho)}}}\quad \Pi\frac{1}{1-\dfrac{1}{p_3^{2(1+\rho)}}},$$

where p_1, p_2, p_3 are primes of the forms $8n+1$; $8n+3$; $8n+5$ or 7, respectively; and η is a positive or negative unit determined by the congruence

$$(-1)^{\frac{1}{8}(p_1-1)}\, 2^{\frac{1}{4}(p_1-1)} \equiv \eta, \bmod p_1.$$

It would seem that the method of Dirichlet which we have here described may be employed to prove all the theorems of Jacobi's memoir in which the two forms compared have different determinants. Those in which the two forms compared have the same determinant, or determinants differing only by a square factor, are of a more elementary character, and are capable of immediate verification. But Dirichlet's method may also be extended to cases in which one or both of the forms compared has a positive determinant. One example will suffice. If $P=(2a+1)^2+8b^2=(2a+1)^2-8\beta^2$, we have

$$\Sigma(-1)^n \phi[(2m+1)^2+8n^2] = \Sigma(-1)^{m+n}\phi[(2m+1)^2-8n^2],$$

ϕ representing any function whatever which renders the series convergent, and the limits of m and n in the first sum being 0, ∞, and $-\infty$, $+\infty$; in the second sum 0, ∞, and 1, $\frac{1}{3}(2m+1)$.

129. We proceed to indicate very briefly the origin of the principal formulae in Jacobi's memoir. Three of them are distinguished from the rest as general, being deduced from the equation (7) of Art. 124, without any specialization. If in that formula we write successively $+z$ and $-z$, for v, and multiply the results together, the left-hand member becomes $\Sigma(-1)^m q^{m^2+n^2} z^{m+n}$;

the right-hand member may be written in the form

$$\prod_1^\infty (1-q^{4m-2})^2 (1-q^{4m}) \times \prod_1^\infty (1-q^{4m})(1-q^{4m-2}z^2)(1-q^{4m-2}z^{-2}),$$

where the second infinite product, by the equation (7), is equal to

$$\Sigma(-1)^m q^{2m^2} z^{2m}, \text{ and the first to } \Sigma(-1)^m q^{2m^2}.$$

Hence $$\Sigma(-1)^m q^{m^2+n^2} z^{m+n} = \Sigma(-1)^{m+n} q^{2(m^2+n^2)} z^{2m}, \quad \ldots \ldots \quad \text{(A)}$$

which is one of Jacobi's general formulae. The other two general formulae, and most of the special ones, are obtained in like manner by considering infinite products which are capable of being expressed in more ways than one as the product of two Theta functions. To arrive at his special formulae, Jacobi transforms the equation (7) by writing q^a, where a is positive, for q, and $\pm q^b$ for v. He thus obtains the equations

$$\prod_1^\infty (1-q^{2ma-a-b})(1-q^{2ma-a+b})(1-q^{2ma}) = \Sigma(-1)^m q^{am^2+mb},$$

$$\prod_1^\infty (1+q^{2ma-a-b})(1+q^{2ma-a+b})(1-q^{2ma}) = \Sigma\, q^{am^2+mb}.$$

Any infinite product of either of the types occurring in these equations he calls an elliptic product; and every infinite product which can be formed in more ways than one by the multiplication of two elliptic products, leads directly to one of his special formulae. The five following elliptic products are of great importance in the theory; they correspond to the suppositions

$$a=1,\ b=0; \quad a=2,\ b=1; \quad a=\tfrac{3}{2},\ b=\tfrac{1}{2}.$$

$$\left.\begin{aligned}
&\prod_1^\infty (1-q^{2m-1})^2\, \Pi(1-q^{2m}) = \Sigma(-1)^m q^{m^2},\\
&\prod_1^\infty (1+q^{2m-1})^2\, \Pi(1-q^{2m}) = \Sigma\, q^{m^2},\\
&\prod_1^\infty (1-q^{2m-1})\ \Pi(1-q^{4m}) = \Sigma(-1)^m q^{2m^2+m},\\
&\prod_1^\infty (1+q^{2m-1})\ \Pi(1-q^{4m}) = \Sigma\, q^{2m^2+m},\\
&\qquad\prod_1^\infty (1-q^m) = \Sigma(-1)^m q^{\frac{1}{2}(3m^2+m)};
\end{aligned}\right\} \quad \ldots \ldots \quad \text{(B)}$$

the first two are the equations (19 and (20) of Art. 124; the last is a celebrated formula due to Euler.

The infinite products in the numerators and denominators of the fractions equal to u and u' (equations **27** and 28, Art. **125**) are all elliptic products of one or other of these five types, in some cases with q^2, or q^3, or $-q$ substituted for q. Hence a comparison of any two of the fractions equal to u or to u' gives immediately one of Jacobi's special formulae. The demonstration of the formula (Σ) in Art. **128** will serve as an example of this process.

Again, Jacobi has shown that the Eulerian product (Art. **124**, E.)

$$\prod_1^\infty (1+q^m) = \prod_1^\infty \frac{1}{1-q^{2m-1}},$$

which Euler had himself represented by the fraction

$$\prod_1^\infty \frac{1-q^{2m}}{1-q^m} = \frac{\Sigma(-1)^m q^{3m^2+m}}{\Sigma(-1)^m q^{\frac{1}{2}(3m^2+m)}}, \quad . \quad . \quad . \quad . \quad . \quad . \quad . \quad (\mathrm{C})$$

can be represented by six other fractions of which both the numerators and denominators are elliptic products; either the numerator or denominator, or both, being of one of the types (B). Thus, for example,

$$\prod_1^\infty (1+q^m) = \frac{\prod_1^\infty (1+q^{3m-2})(1+q^{3m-1})(1-q^{3m})}{\prod_1^\infty (1-q^{6m-3})^2(1-q^{6m})} = \frac{\Sigma q^{\frac{1}{2}(3m^2+m)}}{\Sigma(-1)^m q^{3m^2}}. \qquad (\mathrm{D})$$

Here again a comparison of any two of the seven equal fractions gives one of the special formulae: thus writing q^{24} for q in the two fractions (C) and (D), we find

$$\Sigma(-1)^n q^{(6m-1)^2+(6n+1)^2} = \Sigma(-1)^{m+n} q^{2(6m+1)^2+2(6n)^2},$$

which, however, is only a particular case of the general formula (A).

The Eulerian product is also of importance in the theory of the partition of numbers. If it be developed in a series proceeding by powers of q, the coefficient $C(m)$ of the mth power of q in the development, expresses the number of ways in which m can be composed by the addition of unequal numbers, or by the addition of equal or unequal uneven numbers. Euler observed that his fractional expression of the product furnishes a recurring formula for the calculation of $C(m)$, and the same thing is true of each of Jacobi's fractions; the simplest of the seven recurring formulae being that arising from the fraction (D), viz.,

$$\Sigma(-1)^s C(m-3s^2) = \epsilon,$$

the summation extending to all positive or negative values of s for which $m-3s^2$ is not negative, and ϵ representing 1, or 0, according as m is or is not of the form $\frac{1}{2}(3n^2 \pm n)$.

The equation $\prod_{1}^{\infty}(1-q^m)=\sum_{-\infty}^{+\infty}(-1)^m q^{\frac{1}{2}(3m^2+m)}$ is memorable historically as the earliest example of the introduction of a Theta function into analysis *. It expresses the theorem

'The excess of the number of ways in which a given number can be composed by the addition of an even number of unequal numbers above the number of ways in which it can be composed by the addition of an uneven number of unequal numbers is $(-1)^m$ or 0, according as the given number is, or is not, of the form $\frac{1}{2}(3m^2 \pm m)$.'

Of this theorem Jacobi has given an arithmetical demonstration, reproducing Euler's proof of the analytical formula.

The logarithmic differential of $\prod_{1}^{\infty}(1-q^m)$ is $-\frac{1}{q}\sum_{1}^{\infty}\Phi(m)q^m$, where $\Phi(m)$, as in Art. 125, is the sum of the divisors of m: Euler thus obtained the equation

$$\sum_{1}^{\infty}\Phi(m)q^m \times \sum_{-\infty}^{+\infty}(-1)^m q^{\frac{1}{2}(3m^2+m)}=\frac{1}{2}\sum_{-\infty}^{+\infty}(-1)^{m+1}(3m^2+m)q^{\frac{1}{2}(3m^2+m)},$$

which supplies a recurring formula for the calculation of $\Phi(m)$, viz.,

$$\Sigma(-1)^s\,\Phi(m-\tfrac{1}{2}(3s^2+s))=E(m),$$

the summation extending to all positive or negative values of s for which $m-\frac{1}{2}(3s^2+s)$ is positive, and $E(m)$ representing $(-1)^{s+1}m$, or 0, according as m is, or is not, of the form $\frac{1}{2}(3s^2 \pm s)$ †.

The cube of the Eulerian product is equal to the series

$$\tfrac{1}{2}\Sigma(-1)^m(2m+1)q^{\frac{1}{2}(m^2+m)}$$

(Art. 124, equation 22); so that

$$[\Sigma(-1)^m q^{\frac{1}{2}(3m^2+m)}]^3=\tfrac{1}{2}\Sigma(-1)^m(2m+1)q^{\frac{1}{2}(m^2+m)}, \quad . \quad . \quad . \quad . \quad (F)$$

---

* In the year 1750 or 1751. Nov. Comm. Petropol., vol. iii. p. 155.

† On the equations

$$\Pi(1+q^m)=\Pi\frac{1}{1-q^{2m-1}}, \qquad \Pi(1-q^m)=\Sigma(-1)^m q^{\frac{1}{2}(3m^2+m)},$$

and their connexion with the partitions and divisors of numbers, see Euler, Nov. Comm. Petropol. vol. iii. p. 125, vol. v. p. 59 and p. 75; Acta Petropol. vol. iv. Part I. p. 47 and p. 56 (or Commentationes Arithmeticæ Collectæ, Nos. IX., XI., XVI., L.; the first memoir in vol. iv. of the Acta is omitted in the collection); Introductio in Analysin Infinitorum, part 4. cap. 16; Waring, Philosophical Transactions for 1788, p. 388; Legendre, Théorie des Nombres, ed. 3, vol. ii. p. 128; Jacobi, Fundamenta Nova, p. 185, Crelle, vol. xxxii. p. 164, vol. xvxvii. pp. 67, 73 (or Mathematische Werke, vol. i. p. 345, vol. ii. pp. 73, 79).

a result which in an earlier memoir (Crelle, vol. xxi. p. 13, or translated in Liouville, First Series, vol. vii. p. 85) Jacobi describes as 'hitherto unparalleled in analysis.' Writing q^{24} for q, and multiplying by q^3, it becomes

$$\left[\Sigma_2\left(\frac{3}{a}\right)q^{a^2}\right]^3=\Sigma_1(-1)^{\frac{1}{2}(b-1)}\,bq^{3b^2},$$

the summations Σ_1 and Σ_2 extending respectively to all positive uneven numbers, and to all positive uneven numbers prime to 3. In this form it expresses the theorem

'The sum $\Sigma\left(\frac{3}{a_1a_2a_3}\right)$ extended to all compositions of any number N by the addition of three uneven squares a_1^2, a_2^2, a_2^3, all of which are prime to 3, is $(-1)^{\frac{1}{2}(m-1)}m$ or 0 according as N is or is not the triple of an uneven square.'

Differentiating logarithmically, we find

$$\Sigma_1\Sigma_2(-1)^{\frac{1}{2}(b-1)}\left(\frac{3}{a}\right)b\,(a^2-b^2))\,q^{a^2+3b^2}=0.$$

This equation (in which all the exponents have the same quadratic form) admits of immediate verification, elementary considerations sufficing to show that the sum $\Sigma(-1)^{\frac{1}{2}(b-1)}\left(\frac{3}{a}\right)b\,(a^2-b^2)$ extended to every solution of the equation $N=a^2+3b^2$, is zero. Jacobi thus obtains a direct arithmetical proof of the formula (F). (Crelle, vol. xxi. pp. 15–18.)

The square and the cube of the Eulerian product can also each of them be represented in two different ways as the quotient of two elliptic products.

Other formulae of Jacobi's are inferred from the fundamental equation (7) in a somewhat more complicated way. Replacing v in that equation by certain roots of unity, and multiplying two or more of the results together, Jacobi obtains products which can be expressed in more than one way by means of elliptic products; the formulae thus deduced are remarkable chiefly because they lead to equations, not between two, but between three or more series, the exponents of which have certain quadratic forms.

Lastly, a few additional equalities are derived not from the fundamental equation, but from the modular equations of the third and seventh orders. The modular equation of the third order was brought by Legendre into the form $\sqrt{\kappa'\lambda'}+\sqrt{\kappa\lambda}=1$; whence evidently

$$\phi^2(\omega)\,\phi^2(3\omega)+\psi^2(\omega)\,\psi^2(3\omega)=1\,;$$

writing for the functions ϕ^2 and ψ^2 their values given by equation (14), Art. 124, and changing q into q^4, we find

$$\Sigma\,(1-(-1)^{m+n})\,q^{4(m^2+3n^2)} = 4\,\Sigma\,q^{(4m+1)^2+3(4n+1)^2}.$$

The equation of the seventh order, in the form in which it has been put by M. Gutzlaff*,

$$\sqrt[4]{k'\lambda'}+\sqrt[4]{k\lambda}=1,$$

admits of similar treatment, and furnishes as many as seven formulae on account of the variety of expressions which the equations (27) and (28) allow us to substitute for ϕ and ψ in the equation

$$\phi\,(\omega)\;\phi\,(7\,\omega)+\psi\,(\omega)\;\psi\,(7\,\omega)=1.$$

It is only necessary to observe that we must choose for $\phi\,(\omega)$ and $\psi\,(\omega)$, and similarly for $\phi\,(7\,\omega)$ and $\psi\,(7\,\omega)$, expressions having the same denominator.

At the beginning of his memoir Jacobi says that the formulae to which it relates are probably finite in number. It would seem that when he expressed himself thus, he had not yet found his three general formulae, each of which contains an infinite number of equations between series having their exponents contained in the same quadratic form. But it is certainly very unlikely that equations between series whose exponents are contained in different quadratic forms, exist for any but a few of the simplest forms, or for them in infinite number.

130. *The Formulae of M. Kronecker.*—We now come to an important series of results, discovered within the last few years by M. Kronecker, which form a memorable accession to our knowledge of quadratic forms, and which have opened an entirely new field of arithmetical inquiry. Their demonstration requires considerations of a very complicated kind; and as they are certainly among the most interesting, so also they must be reckoned among the most abstruse of arithmetical truths. Unfortunately, in the brief notices† which

---

* Crelle's Journal, vol. xii. p. 173.

† The following are the memoirs of M. Kronecker on the application of the theory of elliptic functions to quadratic forms.

(1) 'Ueber elliptische Functionen und Zahlen-Theorie,' Monatsberichte, Oct. 29, 1857; and translated in Liouville, New Series, vol. iii. p. 265.

(2) 'Ueber die Anzahl der verschiedenen Klassen von quadratischen Formen von negativer Determinante,' Crelle, vol. lvii. p. 248; and translated in Liouville, vol. v. p. 289.

(3) 'Ueber eine neue Eigenschaft der quadratischen Formen von negativer Determinante,' Monatsberichte, May 26, 1862.

(4) 'Ueber die complexe Multiplication der elliptischer Functionen,' Ibid, June 26, 1862.

(5) 'Auflösung der Pellschen Gleichung mittelst elliptischer Functionem,' Ibid, Jan. 22, 1863.

M. Kronecker has given of his investigations, his methods are indicated only in a very general manner; and, notwithstanding the light which has been thrown on them in the subsequent memoirs of MM. Hermite and Joubert *, it is occasionally difficult to rediscover them. Nevertheless, as a mere enumeration of formulae, unaccompanied by any explanation of the methods by which they have been obtained, would be of little use to the reader, we shall attempt in the next article a complete demonstration of one or two of them, which may serve as specimens of the rest.

The following (with an unimportant change in the notation) are the eight equations given by M. Kronecker (Crelle, vol. lvii. p. 248; Liouville, New Series, vol. v. p. 289).

$$\text{I.}\quad F(2^\mu m)+2F(2^\mu m-1^2)+2F(2^\mu m-2^2)+\dots$$
$$=2\Phi(m)+\Phi(2^{\mu-2}m)+\Psi(2^{\mu-2}m).$$

$$\text{II.}\quad F(2m)+2F(2m-1^2)+2F(2m-2^2)+2F(2m-3^2)+\dots$$
$$=2\Phi(m).$$

$$\text{III.}\quad F(2m)-2F(2m-1^2)+2F(2m-2^2)-2F(2m-3^2)+\dots$$
$$=0.$$

$$\text{IV.}\quad 3G(m)+6G(m-1^2)+6G(m-2^2)+6G(m-3^2)+\dots$$
$$=\Phi(m)+3\Psi(m).$$

$$\text{V.}\quad 2F(m)+4F(m-1^2)+4F(m-2^2)+4F(m-3^2)+\dots$$
$$=\Phi(m)+\Psi(m).$$

$$\text{VI.}\quad 2F(m)-4F(m-1^2)+4F(m-2^2)-4F(m-3^2)+\dots$$
$$=(-1)^{\frac{1}{2}(m-1)}[\Phi(m)-\Psi(m)].$$

$$\text{VII.}\quad 2F(m)-4F(m-4^2)+4F(m-8^2)-4F(m-12^2)+\dots$$
$$=(-1)^{\frac{1}{8}(m-7)}\,\Phi'(m)-\Psi'(m)].$$

$$\text{VIII.}\quad 4\Sigma_s(-1)^{\frac{1}{16}(m-s^2)}\left[2F\left(\frac{m-s^2}{16}\right)-3G\left(\frac{m-s^2}{16}\right)\right]$$
$$=(-1)^{\frac{1}{8}(m-1)}[\Phi'(m)-\Psi'(m)].$$

In these formulae m is any positive uneven number; in the 1st, μ is $\geqq 2$;

* M. Hermite, 'Sur la théorie des équations Modulaires'; M. Joubert, 'Sur la Théorie des Fonctions Elliptiques et son application à la Théorie des Nombres,' already cited in the note on Art. 125.

in the 7th, m is $\equiv -1$, mod 8; in the 8th, m is $\equiv +1$, mod 8, and the summation extends to all values of s for which $\frac{1}{16}(m-s^2)$ is integral and not negative; similarly, the series in the first seven formulae are to be continued until the numbers affected with the signs F and G become negative. If n is any positive number, even or uneven, $\Phi(n)$ is the sum of the divisors of n, $\Psi(n)$ the excess of those divisors of n which surpass $\sqrt{n}$ above those divisors which are surpassed by $\sqrt{n}$; $\Phi'(n)$ is the sum $\Sigma\left(\frac{2}{d}\right)d$ extended to all the divisors of n; $\Psi'(n)$ the excess of the sum $\Sigma\left(\frac{2}{d}\right)d$ extended to all the divisors of n which surpass $\sqrt{n}$, above the sum $\Sigma\left(\frac{2}{d}\right)d$ extended to all divisors of n which are surpassed by $\sqrt{n}$. Lastly, $F(n)$ is the number of uneven classes, $G(n)$ the whole number of classes, of forms of determinant $-n$; the classes (1, 0, 1), (1, 1, 1), and their derived classes, being counted as $\frac{1}{2}$ and $\frac{1}{3}$ respectively; to $F(0)$ we attribute the value 0, to $G(0)$ the value $-\frac{1}{12}$ *.

The arithmetical functions $F(n)$ and $G(n)$ satisfy the equations

$$F(4n)=2F(n);\ G(4n)=F(4n)+G(n);\ G(n)=F(n), \text{ if } n\equiv 1, \text{ or } 2, \text{ mod } 4;$$

$$G(n)=2F(n), \text{ if } n\equiv 7, \text{ mod } 8;\quad G(n)=\tfrac{4}{3}F(n), \text{ if } n\equiv 3, \text{ mod } 8.$$

With the help of these relations (which may be demonstrated by elementary considerations [see Art. 113 of this Report], but which may also be inferred from the theory of elliptic functions) the formulae I.—VIII. may be transformed and combined in various ways, so as to afford new and interesting results. Of these derived formulae M. Kronecker has given two,

$$\text{IX.}\quad F(n)+F(n-1.2)+F(n-2.3)+F(n-3.4)+\ldots$$
$$=\tfrac{1}{8}\Psi(4n+1).$$

$$\text{X.}\quad E(n)+2E(n-1^2)+2E(n-2^2)+2E(n-3^2)+\ldots$$
$$=\tfrac{2}{3}[2+(-1)^n]X(n),$$

---

* The right-hand members of the formulae I.—VIII. are rendered simpler by this conventional estimation of a class of det. -1 as $\frac{1}{2}$, and of an improperly primitive class of det. -3 as $\frac{1}{3}$. We have already seen that this convention is a natural one (Art. 126, note); it is, however, less easy to interpret the assumption $G(0)=-\frac{1}{12}$. M. Kronecker has given his formulae in their complete expression when these conventional estimations are disregarded; in his subsequent notes, however, he seems to prefer the simpler form, which we have adopted in the text.

where n represents any positive integer, $X(n)$ the sum of its uneven divisors, and $E(n) = 2F(n) - G(n)$, so that $E(n)$ is a function satisfying the equations

$$E(4n) = E(n)\,; \quad E(n) = 0, \text{ if } n \equiv 7, \text{ mod } 8\,; \quad E(n) = \tfrac{2}{3}F(n), \text{ if } n \equiv 3, \text{ mod } 8.$$

The first of these formulae is obtained by subtracting VI. from V.; for

$$F(n - k\,.\,\overline{k+1}) = \tfrac{1}{2}F(4n + 1 - [2k+1]^2).$$

The other, if n is uneven, coincides with $[\text{V.}] - \frac{1}{3}[\text{IV.}]$; and with II. if n is unevenly even. If n is the quadruple of an uneven number, its left-hand member may be written in the form

$$\begin{aligned}&[E(\tfrac{1}{4}n) + 2E(\tfrac{1}{4}n - 1^2) + 2E(\tfrac{1}{4}n - 2^2) + \ldots]\\ &+ [\tfrac{4}{3}F(n - 1^2) + \tfrac{4}{3}F(n - 3^2) + \tfrac{4}{3}F(n - 5^2) + \ldots];\end{aligned}$$

the sum of the first of these series is $\frac{2}{3}X(\frac{1}{4}n) = \frac{2}{3}X(n)$; the second series, coinciding with $\frac{2}{3}[\text{I.}] - \frac{2}{3}[\text{V.}]$, has for its sum $\frac{4}{3}X(n)$; the two series together are therefore equal to $2X(n)$. Lastly, the formula, if true for any even number, is also true for its quadruple; for if $n \equiv 0$, mod 8,

$$\begin{aligned}&E(n) + 2E(n - 1^2) + 2E(n - 2^2) + \ldots\\ &= [E(\tfrac{1}{4}n) + 2E(\tfrac{1}{4}n - 1^2) + 2E(\tfrac{1}{4}n - 2^2) + \ldots]\\ &\quad + [2E(n - 1^2) + 2E(n - 3^2) + 2E(n - 5^2) + \ldots],\end{aligned}$$

and every term of the second series is zero, because the numbers $n - 1^2$, $n - 3^2$, ... are all $\equiv 7$, mod 8.

Of the preceding formulae those which contain the functions Ψ and Ψ' are to be regarded as of a more abstruse character than those which only contain X, Φ, and Φ'. The latter, in fact, are deducible from known theorems of arithmetic. Thus, if we multiply the formula X. by 12, the right-hand member becomes $8[2 + (-1)^n]X(n)$, or the number of representations of n as a sum of four squares (see Art. 127). Consequently $12E(n)$ is the number of representations of n as a sum of three squares; for, assuming that this is so for 1, 2, 3, ..., $n-1$, we may infer from the formula X. that it is so for n. Thus the celebrated theorem of Gauss *, which connects the number of representations of a number n as a sum of three squares, with the number of classes of quadratic forms of det. $-n$, is contained in the formula X.; and conversely, that formula is itself deducible from the theorem of Gauss combined with the other and more

* Disq. Arith. Art. 291. Legendre had discovered particular cases of the theorem by induction. Hist. de l'Ac. de Paris, 1785, p. 530 *sqq.* Théorie des Nombres, ed. 3, vol. i. troisième partie.

elementary theorem, which connects the number of representations of n as a sum of four squares with the sum of the uneven divisors of n.

131. *Demonstration of the Formulae of M. Kronecker.*—We shall first demonstrate the formula V. For this purpose, we consider the equation $f_8(x, 1-x)=0$, obtained by writing x for κ^2, and $1-x$ for λ^2 in the modular equation $f_8(\kappa^2, \lambda^2)=0$ of an uneven order n (Art. 125). We shall determine the order of this equation by two different methods; first, by ascertaining the dimensions of $f_8(x, 1-x)$, when x is increased without limit; secondly, by assigning its roots, and the multiplicity of each of them; a comparison of these two determinations will give the formula V.

(i.) Let $x=\phi^8(\theta)$; then

$$f_8(x, 1-x)=\Pi\left[\psi^8(\theta)-\phi^8\left(\frac{\gamma\theta+2k}{\gamma'}\right)\right],$$

because (Art. 125)

$$f_8(\phi^8(\theta), \lambda^2)=\Pi\left[\quad\lambda^2-\phi^8\left(\frac{\gamma\theta+2k}{\gamma'}\right)\right]\dagger,$$

the sign of multiplication Π extending to every system of values of γ, γ', and k.

Let $\theta=1+\dfrac{i}{\sigma}$, σ representing a real positive quantity; we obtain

$$\phi^8\left(1+\frac{i}{\sigma}\right)=-\frac{\psi^8(\sigma i)}{\phi^8(\sigma i)}; \qquad \psi^8\left(1+\frac{i}{\sigma}\right)=\frac{1}{\phi^8(\sigma i)}$$

(Table A, vi. Art. 125). Again, if δ is the greatest common divisor of $\gamma+2k$ and γ', and if a, b, δ' are determined by the equations

$$a=-\frac{\gamma+2k}{\delta}, \qquad b=\frac{\gamma'}{\delta}, \qquad \delta'=\frac{n}{\delta},$$

while c and d are two numbers, of which x is uneven, satisfying the equation $ad-bc=1$, we find

$$\frac{c+d\dfrac{\gamma\left(1+\frac{i}{\sigma}\right)+2k}{\gamma'}}{a+b\dfrac{\gamma\left(1+\frac{i}{\sigma}\right)+2k}{\gamma'}}=\frac{\delta\sigma i+d\gamma}{\delta'};$$

---

† Since $\phi^8(\omega+2)=\phi^8(\omega)$ (equation i. Table A, Art. 125), the systems of values represented by

$$\phi^8\left(\frac{\gamma\omega+2k}{\gamma'}\right) \text{ and } \phi^8\left(\frac{\gamma\omega+16k}{\gamma'}\right)$$

are identical, k denoting any term of a complete system of residues, mod γ'.

whence (Table A, iii.)

$$\phi^8\left(\frac{\gamma\left(1+\frac{i}{\sigma}\right)+2k}{\gamma'}\right)=\phi^{-8}\left(\frac{\delta\sigma i+d\gamma}{\delta'}\right)=\phi^{-8}\left(\frac{\delta\sigma i+2l+1}{\delta'}\right),$$

if $2l+1\equiv d\gamma$, mod δ'. If we give to γ, γ', and k in succession all the $\Phi(n)$ systems of values of which they are susceptible, $\phi^8\left(\frac{\gamma\left(1+\frac{i}{\sigma}\right)+2k}{\gamma'}\right)$ will acquire in succession $\Phi(n)$ values, which (except for particular values of σ) are all different; $\phi^{-8}\left(\frac{\delta\sigma i+2l+1}{\delta'}\right)$ will therefore also acquire the same number of different values; *i.e.* δ will represent in succession every divisor of n, and $2l+1$ every residue of its conjugate divisor δ'. We thus obtain the equation

$$f_8(x,\,1-x)=\Pi\left[\phi^{-8}(\sigma i)-\phi^{-8}\left(\frac{\delta\sigma i+2l+1}{\delta'}\right)\right],$$

the sign of multiplication extending to every combination of the values of δ, δ', and $2l+1$. Now let σ increase without limit, so that $x=-\frac{\psi^8(\sigma i)}{\phi^8(\sigma i)}$ increases without limit, and is of the same dimensions as $e^{\sigma\pi}$ (Art. 125). Observing that the factor

$$\phi^{-8}(\sigma i)-\phi^{-8}\left(\frac{\delta\sigma i+2l+1}{\delta'}\right)$$

has a finite ratio to $e^{\sigma\pi}$, if $1\geqq\frac{\delta}{\delta'}$, and to $e^{\frac{\delta}{\delta'}\sigma\pi}$, if $1\leqq\frac{\delta}{\delta'}$, we see that the product

$$\Pi\left[\phi^{-8}(\sigma i)-\phi^{-8}\left(\frac{\delta\sigma i+2l+1}{\delta'}\right)\right]$$

has a finite ratio to

$$e^{[\Phi(n)+\Psi(n)]\sigma\pi},\ \text{that is to}\ \left[\frac{\psi(\sigma i)}{\phi(\sigma i)}\right]^{8\Phi(n)+8\Psi(n)}$$

Hence
$$f_8(x,\,1-x)\div x^{\Phi(n)+\Psi(n)}$$
is finite, when x increases without limit, or $f_8(x,\,1-x)$ is of the order

$$\Phi(n)+\Psi(n).$$

(ii.) Neither 0 nor 1 is a root of the equation $f_8(x,\,1-x)=0$; for $f_8(0,\,1)=1$, $f_8(1,\,0)=1$, therefore (Art. 125) any one of its roots can be represented by $\phi^8(\omega)$, ω denoting an imaginary quantity, in which the coefficient of i is different from zero and positive. But if $x=\phi^8(\omega)$,

$$f_8(x,\,1-x)=\Pi\left[\psi^8(\omega)-\phi^8\left(\frac{\gamma\omega+2k}{\gamma'}\right)\right];$$

hence the supposition that $\phi^8(\omega)$ is a root of the equation $f_8(x, 1-x)=0$ implies that

$$\psi^8(\omega)=\phi^8\left(\frac{\gamma\omega+2k}{\gamma'}\right),$$

for one system (at least) of values of γ, γ', and k; *i.e.* (Art. 125) that there exists a unit matrix $\begin{vmatrix} a, & b \\ c, & d \end{vmatrix}$ satisfying the equation

$$\frac{\gamma\omega+2k}{\gamma'}=\frac{c+d\omega}{a+b\omega}, \quad \ldots \ldots \ldots \quad \text{(A)}$$

and the congruence

$$\begin{vmatrix} a, & b \\ c, & d \end{vmatrix} \equiv \begin{vmatrix} 0, & 1 \\ 1, & 0 \end{vmatrix}, \text{ mod } 2. \quad \ldots \ldots \ldots \quad \text{(A')}$$

Thus, if $\phi^8(\omega)$ is a root of $f_8(x, 1-x)=0$, ω is the root of a quadratic equation

$$2ak-c\gamma'+2\left(bk-\tfrac{1}{2}d\gamma'+\tfrac{1}{2}a\gamma\right)\omega+b\gamma\omega^2=0,$$

whose extreme coefficients are both uneven, and whose determinant, if

$$\sigma=-bk+\tfrac{1}{2}(a\gamma+d\gamma'),$$

is σ^2-n, a number necessarily negative, because ω is imaginary. Conversely, if ω is the root of a quadratic equation, of which the extreme coefficients are both uneven, and of which the determinant is negative and included in the formula σ^2-n, $\phi^8(\omega)$ is a root of $f_8(x, 1-x)=0$. Or, more precisely, if ω is the root of a properly primitive quadratic equation, of which the determinant $-\Delta$ is negative and the extreme coefficients are both uneven, and if n can be represented by the form $(1, 0, \Delta)$ with a positive and uneven value of the second indeterminate, $\phi^8(\omega)$ will be a root of $f_8(x, 1-x)=0$, and the multiplicity of this root will be equal to the number of such representations of n *. To establish this, we shall show (α) that ω annuls as many of the factors

$$\psi^8(\omega)-\phi^8\left(\frac{\gamma\omega+2k}{\gamma'}\right)$$

as there are representations of n; (β) that $f_8(x, 1-x)$ is divisible by $x-\phi^3(\omega)$ as often as there are factors annulled by ω. (α) Let $A+2B\omega+C\omega^2=0$ be the equation satisfied by ω, and let $n=\sigma^2+\Delta\tau^2$; A, C, and τ being positive and uneven; the four equations

$$\left.\begin{array}{ll} 2ak-c\gamma'=\tau A, & bk-\tfrac{1}{2}d\gamma'+\tfrac{1}{2}a\gamma=\tau B, \\ b\gamma'=\tau C, & -bk+\tfrac{1}{2}d\gamma'-\tfrac{1}{2}a\gamma=\sigma \end{array}\right\} \quad \ldots \quad \text{(B)}$$

---

* The method by which the multiplicity of the roots of the equation $f_8(x, 1-x)=0$ is here determined is chiefly taken from M. Joubert's Memoir, 'Sur la Théorie des Fonctions Elliptiques &c.', pp. 22–24.

will supply one, and only one, system of values for γ, γ', and k, and one, and only one, unit-matrix $\begin{vmatrix} a, & b \\ c, & d \end{vmatrix}$ satisfying the equation (A) and congruence (A′). For the equations $a\gamma = \tau B + \sigma$, $b\gamma = \tau C$, show that γ is the greatest common divisor of $\tau B + \sigma$ and τC; this common divisor is a divisor of n, because

$$n = \sigma^2 + \Delta\tau^2 = (\sigma - B\tau)(\sigma + B\tau) + AC\tau^2;$$

thus a, b, γ, and $\gamma' = \frac{n}{\gamma}$ are determined. Again, the congruences

$$\left.\begin{aligned} 2ak &\equiv \tau A \\ 2bk &\equiv \tau B - \sigma \end{aligned}\right] \text{mod } \gamma,$$

determine the value of k, because $2a$ and $2b$ have no common divisor with the modulus, while the determinant

$$2(-\tau aB + \tau bA + \sigma a) = \frac{2}{\gamma}[\sigma^2 - \tau^2 B^2 + \tau^2 AC] = 2\frac{n}{\gamma} = 2\gamma'$$

is divisible by it; when k is determined, the equations

$$c\gamma' = 2ak - \tau A, \qquad d\gamma' = 2bk - (\tau B - \sigma),$$

will supply integral values of c and d; the matrix $\begin{vmatrix} a, & b \\ c, & d \end{vmatrix}$ thus obtained is a unit matrix, because

$$ad - bc = \frac{1}{n}(a\gamma \times d\gamma' - b\gamma \times c\gamma') = \frac{1}{n}[(\sigma + \tau B)(\sigma - \tau B) + \tau^2 AC] = 1;$$

it also satisfies the congruence $\begin{vmatrix} a, & b \\ c, & d \end{vmatrix} \equiv \begin{vmatrix} 0, & 1 \\ 1, & 0 \end{vmatrix}$, mod 2, because, from the equations (B), taken as congruences for the modulus 2, we find $b \equiv c \equiv 1$, mod 2, $a \equiv d$, mod 2; but also $ad \equiv 0$, mod 2, so that $a \equiv d \equiv 0$, mod 2; lastly, the equation $\frac{\gamma\omega + 2k}{\gamma'} = \frac{c + d\omega}{a + b\omega}$ is satisfied by virtue of the first three of the equations (B). Thus to each representation of n there corresponds one, and only one, evanescent factor; conversely to each evanescent factor there corresponds one, and only one, representation of n. For, if ω annuls the factor

$$\psi^8(\omega) - \phi^8\left(\frac{\gamma\omega + 2k}{\gamma'}\right),$$

the equation (A) and congruence (A′) are satisfied by a unit matrix $\begin{vmatrix} a, & b \\ c, & d \end{vmatrix}$, in which $b > 0$, but, even if $\Delta = -1$, by only one such matrix: so that the equations (B) determine the values of σ and τ without ambiguity. The number of factors annulled by ω is therefore equal to the number of representations of n.

(β) Writing $\phi^8(\theta)$ for x, we have

$$f_8(x, 1-x) = \Pi\left[\psi^8(\theta) - \phi^8\left(\frac{\gamma\theta+2k}{\gamma'}\right)\right] = A\,\Pi_1\left[\phi^8(\theta) - \phi^8(\omega)\right],$$

where A is the coefficient of the highest power of x in $f_8(x, 1-x)$, and Π_1 extends to every root $\phi^8(\omega)$ of $f_8(x, 1-x) = 0$, each root having its proper multiplicity. M. Joubert has proved that, if $\psi^8(\theta) - \phi^8\left(\frac{\gamma\theta+2k}{\gamma'}\right)$ vanish when $\theta = \omega$, $\lim \dfrac{\psi^8(\theta) - \phi^8\left(\frac{\gamma\theta+2k}{\gamma'}\right)}{\phi^8(\theta) - \phi^8(\omega)}$ is neither infinite nor zero. For this limit is, by the usual rule,

$$-1 - \left[\frac{\frac{d}{d\theta}\cdot\phi^8\left(\frac{\gamma\theta+2k}{\gamma'}\right)}{\frac{d}{d\theta}\cdot\phi^8(\theta)}\right]_{\theta=\omega} = -1 - \frac{1}{n\mathrm{M}^2},$$

where M is the multiplier appertaining to the transformation $\omega = \dfrac{\gamma'\Omega - 2k}{\gamma}$, since (equation 31, Art. 125),

$$\mathrm{M}^2 = \frac{1}{n}\,\frac{\phi^8\left(\frac{\gamma\omega+2k}{\gamma'}\right)\,\psi^8\left(\frac{\gamma\omega+2k}{\gamma'}\right)}{\phi^8(\omega)\,\psi^8(\omega)}$$

$$\times\left[\frac{\frac{d}{d\theta}\cdot\phi^8(\theta)}{\frac{d}{d\theta}\cdot\phi^8\left(\frac{\gamma\theta+2k}{\gamma'}\right)}\right]_{\theta=\omega} = \frac{1}{n}\left[\frac{\frac{d}{d\theta}\cdot\phi^8(\theta)}{\frac{d}{d\theta}\cdot\phi^8\left(\frac{\gamma\theta+2k}{\gamma'}\right)}\right]_{\theta=\omega}.$$

The determination of M is effected as follows: from the equation

$$\frac{\gamma\omega+2k}{\gamma'} = \frac{c+d\omega}{a+b\omega},$$

or

$$\omega = \frac{\gamma'c - 2ka + (\gamma'd - 2kb)\,\omega}{\gamma a + \gamma b\omega},$$

it appears (Art. 126) that the multiplier corresponding to the compounded transformations $\begin{vmatrix} \gamma, & 0 \\ -2k, & \gamma' \end{vmatrix}$ and $\begin{vmatrix} a, & b \\ c, & d \end{vmatrix}$, applied to ω, is $[\sigma + i\tau\sqrt{\Delta}]^{-1}$; while that corresponding to the second of these transformations is simply $(-1)^{\frac{1}{2}(b-1)}\,i$ (Table A, II.); therefore $\dfrac{1}{\mathrm{M}^2} = -(\sigma + i\tau\sqrt{\Delta})^2$, and the limit above written becomes $-1+\dfrac{\sigma+i\tau\sqrt{\Delta}}{\sigma-i\tau\sqrt{\Delta}}$, which is certainly neither infinite nor zero. Hence $f_8(x, 1-x)$ is divisible by $x-\phi^8(\omega)$ precisely as often as there are factors $\psi^8(\theta)-\phi^8\left(\dfrac{\gamma\theta+2k}{\gamma'}\right)$ which vanish when $\theta=\omega$; *i.e.* the multiplicity of the root $x=\phi^8(\omega)$ in $f_8(x, 1-x)=0$ is precisely equal to the number of solutions of the equation $n=\sigma^2+\Delta\tau^2$, τ being positive and uneven. It remains to assign all the values of ω, which, annulling one or more of the factors $\psi^8(\theta)-\phi^8\left(\dfrac{\gamma\theta+2k}{\gamma'}\right)$, give different values to $\phi^8(\omega)$. It is evident that values of ω, arising from equations associated to properly primitive forms of different determinants, or of the same determinant and different classes, give different values to $\phi^8(\omega)$; again, of the six subclasses, contained in any one class, the extreme coefficients are uneven in only two; so that from each class we obtain two and only two values of $\phi^8(\omega)$. In the particular case in which the determinant is -1, there are but three subclasses, and but one subclass in which the extreme coefficients are uneven; so that to such a class there corresponds but one value of $\phi^8(\omega)$. Denoting then by $h(\Delta)$ the number of properly primitive classes of determinant $-\Delta$ (we count $\frac{1}{2}$ instead of 1 for a class of determinant -1), and by (n, Δ) the number of solutions of the equation $n=\sigma^2+\Delta\tau^2$, in which τ is positive and uneven, we have, for the number of unequal roots of $f_8(x, 1-x)=0$, the expression

$$2\Sigma h(\Delta),$$

and for the whole number of its roots, when each root is reckoned with its proper multiplicity,

$$2\Sigma(n, \Delta)\, h(\Delta),$$

the summations in each case extending to every value of Δ, for which the equation $n=\sigma^2+\Delta\tau^2$ is resoluble with an uneven value of τ.

We have now obtained the equation *

$$2\Sigma(n, \Delta)\, h(\Delta)=\Phi(n)+\Psi(n),$$

of which the left-hand member may be written in another form. Instead of counting the roots which appertain to the same value of Δ, and then summing with respect to Δ, we may count the roots which appertain to the same value of σ, and then sum with respect to σ. If $F(N)$ is the number of uneven

* M. Kronecker (Crelle, vol. lvii. p. 250) has exhibited each of the equations I.—VIII. in a similar form.

classes (primitive or derived) of determinant $-N$ (we again count $\frac{1}{2}$ for a class of det. -1, or for a class derived from such a class), $2F(n-\sigma^2)$ will be the number of roots appertaining to either of the values $+\sigma$ or $-\sigma$. We thus obtain, finally,

$$2F(n)+4F(n-1^2)+4F(n-2^2)+4F(n-3^2)+\ldots=\Phi(n)+\Psi(n),$$

which is the formula V.

132. We shall also demonstrate (but with less detail) the formula VII. Writing x for u, and $\frac{1}{x}$ for v, in the function $f(n, u, v)$, where $n\equiv -1$, mod 8, and multiplying by $x^{\Phi(n)}$, we obtain an integral function of order $2\Phi(n)$, which we shall designate by $f(x)$. This function is not divisible by x, for $f(0)=1$; but we shall now show that it is divisible by

$$(x^8-1)^{\frac{1}{2}[\Phi(n)-\Psi(n)+\Phi'(n)-\Psi'(n)]},$$

and that the quotient is prime to x^8-1. For this purpose we shall first determine the index λ, for which $\lim\left[\frac{f(x)}{(x-1)^\lambda},\ x=1\right]$ is finite and different from zero. Let $x=\phi\left(\frac{i}{\sigma}\right)$, so that

$$f(x)=\Pi\left[1-\left(\frac{2}{\gamma}\right)\phi\left(\frac{i}{\sigma}\right)\phi\left(\frac{\gamma\frac{i}{\sigma}+16k}{\gamma'}\right)\right],$$

and let the positive quantity σ increase without limit; then, ultimately,

$$\phi\left(\frac{i}{\sigma}\right)=\psi(i\sigma)=1-\tfrac{1}{8}\phi^8(i\sigma),\quad\text{and}\quad 1-x=\tfrac{1}{8}\phi^8(i\sigma).$$

Also, if δ is the greatest common divisor of $16k$ and γ', and if a, b, δ' are determined by the equations

$$a=-\frac{16k}{\delta},\quad b=\frac{\gamma'}{\delta},\quad \delta'=\frac{n}{\delta},$$

while c and d are two numbers (of which d is divisible by 2) satisfying the equation $ad-bc=1$, we find

$$\frac{c+d\,\dfrac{\gamma\frac{i}{\sigma}+16k}{\gamma'}}{a+b\,\dfrac{\gamma\frac{i}{\sigma}+16k}{\gamma'}}=\frac{d\gamma+\delta i\sigma}{\delta'};$$

whence solving for $\dfrac{\gamma\dfrac{i}{\sigma}+16k}{\gamma'}$, and applying the formula II. of Table A,

$$\phi\left(\frac{\gamma\dfrac{i}{\sigma}+16k}{\gamma'}\right)=\left(\frac{2}{c}\right)\psi\left(\frac{d\gamma+\delta i\sigma}{\delta'}\right)=\left(\frac{2}{c}\right)\psi\left(\frac{\delta i\sigma+2l}{\delta'}\right),$$

if $2l\equiv d\gamma$, mod δ'. But $\left(\frac{2}{c}\right)=\left(\frac{2}{b}\right)$, because $bc\equiv-1$, mod 8, and

$$\left(\frac{2}{b}\right)=\left(\frac{2}{\delta^2 b}\right)=\left(\frac{2}{\gamma'\delta}\right)=\left(\frac{2}{\gamma\delta}\right),$$

because $\gamma\gamma'\equiv-1$, mod 8; so that

$$\left(\frac{2}{\gamma}\right)\phi\left(\frac{\gamma\dfrac{i}{\sigma}+16k}{\gamma'}\right)=\left(\frac{2}{\delta}\right)\psi\left(\frac{\delta i\sigma+2l}{\delta'}\right)=\left(\frac{2}{\delta}\right)\left[1-\tfrac{1}{8}\phi^8\left(\frac{\delta i\sigma+2l}{\delta'}\right)\right],$$

ultimately; and

$$\lim\frac{f(x)}{(1-x)^\lambda}=8^\lambda\lim\frac{\Pi\left[1-\left(\frac{2}{\delta}\right)+\frac{1}{8}\left(\frac{2}{\delta}\right)\left(\phi^8(i\sigma)+\phi^8\left(\frac{\delta i\sigma+2l}{\delta'}\right)\right)\right]}{\phi^{8\lambda}(i\sigma)};$$

the sign of multiplication extending to every divisor δ of n, and to every term l of a complete system of residues of its conjugate divisor δ'. Observing that every factor of the numerator, in which $\left(\frac{2}{\delta}\right)=-1$, is finite, and that every factor, in which $\left(\frac{2}{\delta}\right)=+1$, is evanescent, and is of the same dimensions as $e^{-\sigma}$ or $e^{-\frac{\delta}{\delta'}\sigma}$, according as $\delta>\delta'$, or $\delta<\delta'$, we see that, in order to obtain a finite value for $\lim\dfrac{f(x)}{(x-1)^\lambda}$, we must take for λ twice the sum of those divisors of n which satisfy simultaneously the equation $\left(\frac{2}{\delta}\right)=+1$, and the inequality $\delta<\sqrt{n}$, so that we shall have

$$\lambda=\tfrac{1}{2}[\Phi(n)-\Psi(n)+\Phi'(n)-\Psi'(n)].$$

Further, if η is any eighth root of unity, it will be found that, when

$$n\equiv-1,\ \text{mod } 8,\quad f\left(\frac{u}{\eta},\ v\eta\right)=f(u,v),$$

whence $f\left(\frac{x}{\eta}\right)=f(x)$, or $f(x)$ contains only powers of x having exponents divisible by 8. Consequently $f(x)$ is divisible by

$$(x^8-1)^{\frac{1}{2}[\Phi(n)-\Psi(n)+\Phi'(n)-\Psi'(n)]},$$

and the quotient is prime to x^8-1.

Representing, as we may now do, any root of the equation $\dfrac{f(x)}{(x^8-1)^\lambda}=0$ by $\phi(\omega)$, we find that ω must satisfy the equation

$$\frac{1}{\phi(\omega)}=\left(\frac{2}{\gamma}\right)\phi\left(\frac{\gamma\omega+16k}{\gamma'}\right),$$

for one system at least of values of γ, γ', and k; that is (Table A, III. Art. 125), ω satisfies a quadratic equation of the form

$$\frac{c+d\omega}{a+b\omega}=\frac{\gamma\omega+16k}{\gamma'},$$

or

$$16ak-\gamma'c+(a\gamma'-d\gamma'+16kb)\,\omega+b\gamma\omega^2=0,$$

where $\begin{vmatrix} a, & b \\ c, & d \end{vmatrix} \equiv \begin{vmatrix} 1, & 1 \\ 0, & 1 \end{vmatrix}$, mod 2; $c\equiv 0$, mod 8; $(-1)^{\frac{1}{8}c}\left(\frac{2}{d}\right)=\left(\frac{2}{\gamma}\right)$.

The determinant of this equation, if $\sigma=\frac{1}{2}(a\gamma+d\gamma')-8kb$, is σ^2-n. For brevity, let us suppose that $n\equiv -1$, mod 16; we shall now prove that in this case $\sigma\equiv 0$, mod 8. Since $ad\equiv 1$, mod 8, it follows that $a\equiv d$, mod 8; let $a=8\alpha+\mu$, $d=8\delta+\mu$; substituting these values in the equation $ad-bc=1$, considered as a congruence for the modulus 16, we infer that

$$8(\alpha+\delta)\equiv c+\mu^2-1, \text{ mod } 16.$$

Again, since $\gamma\gamma'\equiv -1$, mod 8, let $\gamma=8\theta+\nu$, $\gamma'=8\theta'-\nu$; substituting these values in the congruence $\gamma\gamma'\equiv -1$, mod 16, we find

$$8(\theta+\theta')\equiv \nu^2-1, \text{ mod } 16.$$

But $$2\sigma\equiv a\gamma+d\gamma'\equiv 8[\alpha+\delta+\theta+\theta']\equiv c+\mu^2-1+\nu^2-1\equiv 0, \text{ mod } 16,$$

because $$(-1)^{\frac{1}{8}(c+\mu^2-1+\nu^2-1)}=(-1)^{\frac{1}{8}c}\left(\frac{2}{d}\right)\left(\frac{2}{\gamma}\right)=1.$$

Therefore σ is divisible by 8, and the quantity ω is the root of an equation

$$A+2B\omega+C\omega^2=0,$$

in which A is even, C uneven, and of which the determinant is included in the series of negative numbers,

$$-n,\quad -n+8^2,\quad -n+16^2,\ \ldots\ldots$$

An application (which we need not here repeat) of the method already employed to prove the formula V. will show that every quadratic equation satisfying these conditions supplies a value of $\phi(\omega)$. Sixteen different values of $\phi(\omega)$ will be obtained from the quadratic equations associated to the forms of any uneven class of a determinant included in the above series; because the conditions with respect to the extreme coefficients are satisfied in only two of the six subclasses contained in each class, and because each of these two subclasses supplies (Art. 126) eight values of $\phi(\omega)$. Lastly, the multiplicity of the root $\phi(\omega)$ of the equation $\frac{f(x)}{(x^8-1)^\lambda}=0$ is ascertained, by an application of the method of M. Joubert (which also we need not here repeat), to be equal to the number of factors

$$1-\left(\frac{2}{\gamma}\right)\phi(\omega)\,\phi\left(\frac{\gamma\omega+16k}{\gamma'}\right)$$

which are annulled by ω; or, which comes to the same thing, to the number of representations of n by the form $\sigma^2+\Delta\tau^2$, the first indeterminate being divisible by 8, the second being uneven and positive, and $-\Delta$ representing the determinant of the primitive equation by which ω is determined. Denoting by (n, Δ) the number of such representations, we have for the whole number of roots of the equation $\frac{f(x)}{(x^8-1)^\lambda}=0$, each root being taken with its proper multiplicity, the expression $16\,\Sigma(n,\Delta)\,h(\Delta)$; whence, by a transformation already employed,

$$\left.\begin{aligned}&16\,F(n)+32\,F(n-8^2)+32\,F(n-16^2)+\ldots\\&\quad=2\,\Phi(n)-4\left[\Phi(n)-\Psi(n)+\Phi'(n)-\Psi'(n)\right].\end{aligned}\right\}\quad\ldots\quad(\text{A})$$

Considering, instead of the function $f(x)$, the function $x_{\Phi(n)}f\left(x,-\frac{1}{x}\right)$, we obtain, by reasoning precisely similar, the formula

$$\left.\begin{aligned}&32\,F(n-4^2)+32\,F(n-12^2)+32\,F(n-20^2)+\ldots\\&\quad=2\,\Phi(n)-4\left[\Phi(n)-\Psi(n)+\Psi'(n)-\Phi'(n)\right],\end{aligned}\right\}\quad\ldots\quad(\text{B})$$

whence, by subtraction,

$$\begin{aligned}&2\,F(n)-4\,F(n-4^2+4\,F(n-8^2)-4\,F(n-12^2)+\ldots\\&\quad=\Psi'(n)-\Phi'(n),\end{aligned}$$

in accordance with M. Kronecker's formula VII.

If we had supposed $n\equiv 7$, mod 16, the left-hand members of the formulae (A) and (B) would have been interchanged, and the right-hand member of the formula resulting from them by subtraction would consequently become $\Phi'(n)-\Psi'(n)$.

133. We shall only indicate the origin of the remaining formulae. Of these, the formula I. requires the simultaneous consideration of the modular equations

$f(2^\mu, u^2, v^8)=0$, and $f_8(n, 1-u^8, v^8)=0$ (Art. 125). Writing x for v^8, and eliminating u^2 dialytically from these two equations, we obtain a resultant $R(x)$ of the order $2^{\mu+1}\Phi(n)$ in x, as appears from the theory of elimination. Writing $\phi^8(\omega)$ for x, and observing that the coefficient of the highest power of u in $f(u^2, v^8)$ is $v^{2\mu+1}$, and in $f_8(1-u^8, v^8)$ is unity, we find

$$R(x)=[\phi^8(\omega)]^{2^\mu\Phi(n)}\,\Pi\left[\phi^8\left(\frac{2^\mu\omega}{1+2h\omega}\right)-\psi^8\left(\frac{\gamma\omega+2k}{\gamma'}\right)\right],$$

an equation which expresses the resultant in terms of the roots of the two equations, and in which the sign of multiplication Π extends to every combination of two roots. Since all the roots of $f_8(1-u^8, 1)=0$ are zero, while none of the roots of $f(u^2, 1)$ are zero, no root of the equation $R(x)=0$ is a positive unit. But the equations $f_8(1-u^8, 0)=0$, $f(u^2, 0)=0$ have common roots; so that $x=0$ is a root of $R(x)=0$. To determine its multiplicity, write σi for ω in the expression of $R(x)$, and increase σ without limit. The quantity $\phi^8\left(\frac{2^\mu\omega}{1+2\omega}\right)$ which occurs in $\Phi(n)$ of the factors of $R(x)$ is equal to $\phi^{-8}(2^\mu\sigma i)$, and therefore increases without limit; but since $\lim[\phi^8(\sigma i)]^{2^\mu}\times[\phi^{-8}(2^\mu\sigma i)]=1$, these $\Phi(n)$ factors are cancelled by the initial factor $[\phi^8(\sigma i)]^{2^\mu\Phi(n)}$. Evaluating the remaining factors by the method of Art. 131, we find that

$$\lim\frac{R(x)}{x^{2\Phi(2^{\mu-2}n)-2\Psi(2^{\mu-2}n)}}$$

is finite when x diminishes without limit; so that the order of $R(x)$, after division by the highest power of x contained in it, is

$$2^{\mu+1}\Phi(n)-2\Phi(2^{\mu-2}n)+2\Psi(2^{\mu-2}n),$$

or, since $(2^{\mu-1}-1)\Phi(n)=\Phi(2^{\mu-2}n)$,

$$4\Phi(n)+2\Phi(2^{\mu-2}n)+2\Psi(2^{\mu-2}n).$$

The formulae II. and III. are obtained by successively combining with the equation $f_4(n, u^4, v^4)=0$ (Art. 125), the equations

$$v^4u^4+v^4+u^4-1=0,\quad\text{and}\quad v^4u^4-v^4+u^4+1=0,$$

the first of which is equivalent to the system $v^4=\phi^4(\omega)$, $u^4=\psi^4(\tfrac{1}{2}\omega)$; the second to the system $v^4=\phi^4(\omega)$, $v^4=-\psi^4(\tfrac{1}{2}\omega)$. The resultant of the elimination of u^4 is, in each case, an equation of the order $2\Phi(n)$ in $x=v^4$, and is not divisible by x, $x-1$, or $x+1$.

The following Table indicates the highest powers of x and of the divisors of x^8-1 by which the functions specified in it are divisible.

TABLE B.

	Function.	Order.	Extraneous Factor.	Order after Division.
	$f_8(x, x)$	$\Phi(n)+\Psi(n)$	$(x^2-x)^{\Phi(n)-\Psi(n)}$	$3\Psi(n)-\Phi(n)$
	$f_8(x, 1-x)$	$\Phi(n)+\Psi(n)$	. . . . .	$\Phi(n)+\Psi(n)$
	$x^{\Phi(n)} f_8\left(x, \frac{1}{x}\right)$	$2\Phi(n)$	$(x-1)^{\Phi(n)-\Psi(n)}$	$\Phi(n)+\Psi(n)$
	$(1-x)^{\Phi(n)} f_8\left(x, \frac{1}{1-x}\right)$	$2\Phi(n)$	. . . . .	$2\Phi(n)$
	$(x-1)^{\Phi(n)} f_8\left(x, \frac{x}{x-1}\right)$	$2\Phi(n)$	$x^{\Phi(n)-\Psi(n)}$	$\Phi(n)+\Psi(n)$
	$x^{\Phi(n)} f_8\left(x, \frac{x-1}{x}\right)$	$2\Phi(n)$	. . . . .	$2\Phi(n)$
	$x^{\Phi(n)} f_4\left(x, \frac{1}{x}\right)$	$2\Phi(n)$	$(x^2-1)^{\Phi(n)-\Psi(n)}$	$2\Psi(n)$
	$x^{\Phi(n)} f_4\left(x, -\frac{1}{x}\right)$	$2\Phi(n)$	. . . . .	$2\Phi(n)$
$n\equiv3$, mod 4	$x^{\Phi(n)} f_2\left(x, \frac{1}{x}\right)$	$2\Phi(n)$	$x^4-1)^{\Phi(n)-\Psi(n)}$	$4\Psi(n)-2\Phi(n)$
$n\equiv3$, mod 4	$x^{\Phi(n)} f_2\left(x, -\frac{1}{x}\right)$	$2\Phi(n)$	. . . . .	$2\Phi(n)$
$n\equiv1$, mod 4	$x^{\Phi(n)} f_2\left(x, \frac{1}{x}\right)$	$2\Phi(n)$	$(x^2-1)^{\Phi(n)-\Psi(n)}$	$2\Psi(n)$
$n\equiv1$, mod 4	$x^{\Phi(n)} f_2\left(x, -\frac{1}{x}\right)$	$2\Phi(n)$	$(x^2+1)^{\Phi(n)-\Psi(n)}$	$2\Psi(n)$
$n\equiv-1$, mod 8	$x^{\Phi(n)} f\left(x, \frac{1}{x}\right)$	$2\Phi(n)$	$(x^8-1)^{\frac{1}{2}[\Phi(n)-\Psi(n)+\Phi'(n)-\Psi'(n)]}$	
$n\equiv-1$, mod 8	$x^{\Phi(n)} f\left(x, -\frac{1}{x}\right)$	$2\Phi(n)$	$(x^8-1)^{\frac{1}{2}[\Phi(n)-\Psi(n)-\Phi'(n)+\Psi'(n)]}$	
$n\equiv1$, mod 8	$f(x, x)$	$\Phi(n)+\Psi(n)$	$x^{\Phi(n)-\Psi(n)}(x^8-1)^{\frac{1}{2}[\Phi(n)-\Psi(n)+\Phi'(n)-\Psi'(n)]}$	
$n\equiv1$, mod 8	$f(x, -x)$	$\Phi(n)+\Psi(n)$	$x^{\Phi(n)-\Psi(n)}(x^8-1)^{\frac{1}{2}[\Phi(n)-\Psi(n)+\Psi'(n)-\Phi'(n)]}$	

The formulae in this Table relating to $f_8(x, x)$, $f(x, x)$, and $f(x, -x)$ require a certain modification, when n is a perfect square *. If, on this hypothesis, we represent by $\mathrm{f}_8(x, x)$ the function obtained by writing $x = v^8 = u^8$ in $\frac{f_8(u^8, v^8)}{v^8 - u^8}$, $\mathrm{f}_8(x, x)$ is of the order $\Phi(n) + \Psi(n) - 1$, and is divisible by

$$(x^2 - x)^{\Phi(n) - \Psi(n) - 1}.$$

Again, if $n = \nu^2$ and $\left(\frac{2}{\nu}\right) = 1$, we represent

$$\lim \frac{f(u, v)}{v - u}\,[u = v = x] \text{ by } \mathrm{f}(x, x);$$

this function is of the order $\Phi(n) + \Psi(n) - 1$, and is divisible by

$$x^{\Phi(n) - \Psi(n) - 1} \times (x^8 - 1)^{\frac{1}{2}[\Phi(n) - \Psi(n) + \Phi'(n) - \Psi'(n)] - 1};$$

if $\left(\frac{2}{\nu}\right) = -1$, we represent

$$\lim \frac{f(u, -v)}{v - u}\,[u = v = x] \text{ by } \mathrm{f}(x, -x),$$

and this function is of the order $\Phi(n) - \Psi(n) - 1$, and is divisible by

$$x^{\Phi(n) + \Psi(n) - 1} \times (x^8 - 1)^{\frac{1}{2}[\Phi(n) - \Psi(n) + \Psi'(n) - \Phi'(n)] - 1}.$$

The formula IV. may be deduced from the equation

$$f_8(x, x) \times f_8(x, 1 - x) \times x^{\Phi(n)} f_8\left(x, \frac{1}{x}\right) \times (1 - x)^{\Phi(n)} f_8\left(x, -\frac{x}{1 - x}\right) = 0,$$

of which the order (after division by powers of x and $x - 1$) is shown by the Table to be $2\Phi(n) + 6\Psi(n)$.

In proving the formula V., we might have employed the equation

$$x^{\Phi(n)} f_8\left(x, \frac{1}{x}\right) = 0 \text{ instead of } f_8(x, 1 - x) = 0.$$

If, instead of the former equation, we employ the two

$$x^{\Phi(n)} f_4\left(x, \frac{1}{x}\right) = 0, \quad x^{\Phi(n)} f_4\left(x, -\frac{1}{x}\right) = 0,$$

we obtain the formulae V. and VI. simultaneously,

Lastly, the formulae VIII. depends on the equations

$$f(x, x) = 0, \text{ and } f(x, -x) = 0.$$

---

* The necessity for a corresponding modification of the formulae IV. and VII. is obviated by the assumption $G(0) = -\frac{1}{12}$.

134. *Connexion of the formulae of M. Kronecker with Elliptic series.—Researches of M. Hermite.*—M. Kronecker has given a remarkable analytical expression of the formulae IX. and X. (Art. 130). He employs the identical equations

$$\sum_1^\infty [2+(-1)^n] X(n)\, q^n = \sum_1^\infty \frac{q^n}{[1+(-1)^n q^n]^2},$$

$$\tfrac{1}{4}\sum_1^\infty \Psi(4n+1)\, q^{\frac{1}{4}(4n+1)} = q^{\frac{1}{4}} \sum_1^\infty \frac{q^{n^2+n-1}}{(1-q^{2n-1})^2}$$

$$(1+2q\ +2q^4+2q^9\ +\ldots)\times \sum_0^\infty . E(n)\, q^n$$

$$= \sum_0^\infty [E(n)+2E(n-1^2)-2E(n-2^2)+\ldots]\, q^n,$$

$$(2q^{\frac{1}{4}}+2q^{\frac{9}{4}}+2q^{\frac{25}{4}}\times\ldots)\times \sum_1^\infty F(n)\, q^n$$

$$= 2q^{\frac{1}{4}} \sum_1^\infty [F(n)+F(n-1.2)+F(n-2.3)+\ldots]\, q^n,$$

of which the first two are immediately verified by expanding their right-hand members, the last two by multiplying together the series in their left-hand members. Combining these identities with the formulae IX. and X., and attending to the equation $E(0)=\frac{1}{12}$, we obtain

$$\sum_1^\infty [F(n)\, q^n = \frac{q^{\frac{1}{4}}}{\theta_{1,0}(0)} \sum_1^\infty \frac{q^{n^2+n-1}}{(1-q^{2n-1})^2}, \quad \ldots\ldots\ldots \quad \text{(A)}$$

$$12\sum_0^\infty [E(n)\, q^n = \frac{1}{\theta_{0,0}(0)}\left[1+8\sum_1^\infty \frac{q^n}{[1+(-1)^n q^n]}\right]; \quad \ldots\ldots \quad \text{(B)}$$

of which the second (Art. 127, equation 3) may be written in the form

$$12\sum_0^\infty . E(n)\, q^n = (1+2q+2q^4+2q^9+\ldots)^3, \quad \ldots\ldots\ldots \quad \text{(B')}$$

in which it expresses the arithmetical theorem of Gauss, to which we have already referred (Art. 130).

It appears from the equations (A) and (B) that the generating functions of $F(n)$ and $E(n)$ are elliptic series; and M. Hermite, in two important memoirs (Comptes Rendus, Aug. 5, 1861, or Liouville, New Series, vol. vii. p. 25, and Comptes Rendus, July 7, 1862) suggested, as it would seem, by these equations, has succeeded in deducing the second of them, and others of the same character, from the general expansions of elliptic functions, without having occasion to consider the special modules which admit of complex multiplication. He has thus discovered a new and comparatively elementary method of arriving at the formulae of M. Kronecker; to whom indeed this method was already known, as his expressions of the generating functions of $E(n)$ and $F(n)$ indicate, and as he has himself expressly stated in a note published after the appearance of M. Hermite's first memoir (Monatsberichte, May 26, 1862, pp. 307, 308). M. Hermite's method is an extension of that employed by Jacobi (see Art. 127), and depends on the developments of doubly periodic functions in series proceeding by sines or cosines of the multiples of the argument. To this set of developments, however, M. Hermite adds a second obtained by dividing the product of two Theta functions by a third. A series of the first set, and one of the second (both alike containing only sines, or only cosines, and only even, or only uneven multiples of the argument), are then multiplied together. The non-periodic part of the product (or its integral, taken from the limit 0 to π) is a function of q only, and if the product can be formed in more than one way, we obtain different expressions of this function, a comparison of which supplies in each case an arithmetical formula. We take the following example from M. Hermite's first memoir; and, with him, we write for brevity, Θ, Θ_1, H, H_1 for $\Theta\left(\frac{2Kx}{\pi}\right)$, $\Theta_1\left(\frac{2Kx}{\pi}\right)$, $\mathrm{H}\left(\frac{2Kx}{\pi}\right)$, $\mathrm{H}_1\left(\frac{2Kx}{\pi}\right)$, and θ, θ_1, η_1 for $\Theta(0)$, $\Theta_1(0)$, $\mathrm{H}_1(0)$*.

Multiplying together the three pairs of series

$$\left.\begin{aligned} \eta_1^2\,\theta_1^2\,\frac{\mathrm{H}^2}{\Theta^2} &= 8\sum_1^\infty \frac{nq^n}{1-q^{2n}} - 8\sum_1^\infty \frac{nq^n\cos 2nx}{1-q^{2n}} \\ \Theta_1 &= \sum_{-\infty}^{+\infty} q^{n^2}\cos 2nx, \end{aligned}\right\}\quad\ldots\quad\text{(i)}$$

$$\left.\begin{aligned} \eta_1\,\theta\,\theta_1^2\,\frac{\mathrm{H}\Theta_1}{\Theta^2} &= 4\sum_0^\infty \frac{(2n+1)\,q^{\frac{1}{2}(2n+1)}}{1+q^{2n+1}}\sin(2n+1)x, \\ \mathrm{H} &= 2\sum_0^\infty (-1)^n\,q^{\frac{1}{4}(2n+1)^2}\sin(2n+1)x, \end{aligned}\right\}\quad\ldots\quad\text{(ii)}$$

$$\left.\begin{aligned} \eta_1\,\theta_1\,\frac{\mathrm{H}}{\Theta} &= 4\sum_0^\infty \frac{q^{\frac{1}{2}(2n+1)}}{1-q^{2n+1}}\sin(2n+1)x, \\ \theta_1\,\frac{\mathrm{H}\Theta_1}{\Theta} &= 2\sum_0^\infty Q_n\,q^{\frac{1}{4}(2n+1)^2}\sin(2n+1)x, \\ \text{where}\quad Q_n &= 1+2q^{-1}+2q^{-4}+\ldots+2q^{-n^2}; \end{aligned}\right\}\quad\ldots\quad\text{(iii)}$$

* The developments of elliptic functions, in series proceeding by sines or cosines of the multiples of the argument, which are employed in this article, will be found in the Fundamenta Nova (sections 40–42), or in M. Hermite's second memoir (Comptes Rendus, July 7, 1862).

and designating the definite integral $\eta_1 \theta_1^2 \int_0^\pi \frac{\mathrm{H}^2 \Theta_1}{\Theta^2} dx$ by πJ, we obtain

$$\eta_1 J = 8 \sum_1^\infty \frac{nq^n}{1-q^{2n}} - 8 \sum_1^\infty \frac{nq^{n^2+n}}{1-q^{2n}},$$

$$\theta J = 4 \sum_0^\infty (-1)^n \frac{(2n+1)\, q^{\frac{1}{4}(2n+1)(2n+3)}}{1+q^{2n+1}},$$

$$J = 4 \sum_0^\infty Q_n \frac{q^{\frac{1}{4}(2n+1)(2n+3)}}{1-q^{2n+1}}.$$

Let $\Delta(n)$ represent the sum of those divisors of n whose conjugates are uneven, and $\Gamma_1(n)$ the sum of those divisors of n which do not surpass $\sqrt{n}$, and which are even or uneven, according as their conjugate divisors are uneven or even; we find immediately

$$\eta_1 J = 8 \sum_1^\infty [\Delta(n) - \Gamma_1(n)]\, q^n;$$

also

$$\theta J = 2 \sum_0^\infty (-1)^n [\Phi(4n+3) - \Psi(4n+3)]\, q^{\frac{1}{4}(4n+3)},$$

$$J = 4 \sum_0^\infty{}_n \sum_{-n}^{+n}{}_\mu \sum_0^\infty{}_s\, q^{\frac{1}{4}[(2n+1)(2n+4s+3)-4\mu^2]}$$

$$= 4 \sum_0^\infty F'(4n+3)\, q^{\frac{1}{4}(4n+3)},$$

if $F'(4n+3)$ represent the number of solutions of the equation $4n+3 = ac - b^2$, in which a and c are positive and uneven, a is less than c, b is even, and less in absolute magnitude than a. But M. Hermite has shown that

$$F'(4n+3) = F(4n+3).$$

For $F'(4n+3)$ is evidently the number of quadratic forms (a, b, c) of determinant $-(4n+3)$, in which the second coefficient is even, and less than either extreme coefficient, and in which also the first coefficient is less than the third. But each uneven reduced form is equivalent to one, and only to one, of the forms (a, b, c). For the reducing transformation of a form (a, b, c) is necessarily one of the five following:—

$$\begin{vmatrix} 1, & 0 \\ 0, & 1 \end{vmatrix}, \quad \begin{vmatrix} 1, & \pm 1 \\ 0, & 1 \end{vmatrix}, \quad \begin{vmatrix} \pm 1, & 1 \\ -1, & 0 \end{vmatrix};$$

therefore, conversely, a reduced form can be transformed into a form (a, b, c) only by one or more of the transformations,

$$\begin{vmatrix} 1, & 0 \\ 0, & 1 \end{vmatrix}, \quad \begin{vmatrix} 1, & \mp 1 \\ 0, & 1 \end{vmatrix}, \quad \begin{vmatrix} 0, & -1 \\ 1, & \pm 1 \end{vmatrix};$$

and upon trial it will be found that there is always one, and only one, among them which applied to a reduced form, produces a form (a, b, c). The number of forms (a, b, c) is therefore equal to the number of reduced forms of determinant $-(4n+3)$; *i.e.* $F'(4n+3)=F(4n+3)$. Eliminating J, we obtain the first and third formulae of M. Hermite's memoir,

$$\eta_1 \sum_0^\infty F(4n+3)\, q^{\frac{1}{4}(4n+3)} = 2\sum_1^\infty [\Delta(n) - \Gamma_1(n)]\, q^n,$$

$$\theta \sum_0^\infty F(4n+3)\, q^{\frac{1}{4}(4n+3)} = \tfrac{1}{2}\sum_0^\infty (-1)^n [\Phi(4n+3) - \Psi(4n+3)]\, q^{\frac{1}{4}(4n+3)},$$

or, equating coefficients,

$$F(4n-1^2)+F(4n-3^2)+F(4n-5^2)+\dots$$
$$= \Delta(n) - \Gamma_1(n),$$
$$F(4n+3)-2F(4n+3-2^2)+2F(4n+3-4^2)-\dots$$
$$=(-1)^n \tfrac{1}{2}[\Phi(4n+3)-\Psi(4n+3)].$$

In his second memoir M. Hermite occupies himself with the demonstration of the equation (B′). Multiplying together the two series,

$$\theta_1^2 \frac{\Theta \mathrm{H}_1}{\Theta_1 \mathrm{H}} = \cot x - \sum_1^\infty (-1)^n \frac{4q^n \sin 2nx}{1+(-1)^n q^n},$$

$$\theta_1 \frac{\mathrm{H}\Theta_1}{\mathrm{H}_1} = \tan x + 2\sum_1^\infty (-1)^{n-1} Q_{n-1}\, q^{n^2} \sin 2nx,$$

and employing the formulae

$$\int_0^\pi \sin 2nx \cot x\, dx = \pi,$$

$$\int_0^\pi \sin 2nx \tan x\, dx = (-1)^{(n-1)}\pi,$$

he finds

$$\theta_1^3 = \frac{1}{\pi}\,\theta_1^3 \int_0^\pi \Theta\, dx = \frac{1}{\pi}\int_0^\pi \theta_1^2 \frac{\Theta \mathrm{H}_1}{\Theta_1 \mathrm{H}} \times \theta_1 \frac{\Theta_1 \mathrm{H}}{\mathrm{H}_1}\, dx$$
$$= 1 + 4\sum_1^\infty \frac{q^n}{1+(-1)^n q^n} - 2\sum_1^\infty (-1)^n Q_{n-1}\, q^{n^2}$$
$$+ 4\sum_1^\infty \frac{Q_{n-1}\, q^{n^2+n}}{1+(-1)^n q^n},$$

an expression which, by a detailed discussion, is shown to be equivalent to

$$12\sum_0^\infty E(n)\, q^n.$$

In the note of May 26, 1862, to which we have already referred, M. Kronecker has given other examples of the use of this method. Multiplying together the three pairs of series,

$$\left.\begin{aligned} \eta_1^2\,\theta_1^2\,\frac{\mathrm{H}^2}{\Theta^2} &= 8\sum_1^\infty \frac{nq^n}{1-q^{2n}} - 8\sum_1^\infty \frac{nq^n\cos 2nx}{1-q^{2n}}, \\ \mathrm{H}_1\cos x &= q^{\frac14} + \sum_1^\infty \left[q^{\frac14(2n-1)^2} + q^{\frac14(2n+1)^2}\right]\cos 2nx, \end{aligned}\right\} \quad \ldots\ldots \text{(i)}$$

$$\left.\begin{aligned} \eta_1^2\,\theta\,\theta_1\,\frac{\mathrm{H}\,\mathrm{H}_1}{\Theta^2} &= 8\sum_1^\infty \frac{nq^n\sin 2nx}{1+q^{2n}}, \\ \mathrm{H}\cos x &= \sum_1^\infty (-1)^n\left[q^{\frac14(2n+1)^2} - q^{\frac14(2n-1)^2}\right]\sin 2nx, \end{aligned}\right\} \quad \ldots\ldots \text{(ii)}$$

$$\left.\begin{aligned} \eta_1\theta_1\,\frac{\mathrm{H}}{\Theta}\cos x &= 2\sum_1^\infty \left[\frac{q^{\frac12(2n-1)}}{1-q^{2n-1}} + \frac{q^{\frac12(2n+1)}}{1-q^{2n+1}}\right]\sin 2nx, \\ \eta_1\,\frac{\mathrm{H}\,\mathrm{H}_1}{\Theta} &= 4\sum_1^\infty R_n\,q^{n^2}\sin 2nx, \end{aligned}\right\} \quad \ldots\ldots \text{(iii)}$$

where $$R_n = q^{-\frac14} + q^{-\frac94} + q^{-\frac{25}{4}} + \ldots + q^{-\frac14(2n-1)^2},$$

and designating the definite integral $\theta_1\,\eta_1^2\int_0^\pi \frac{\mathrm{H}^2\,\mathrm{H}_1}{\Theta^2}\cos x\,dx$ by $\pi\,\mathrm{I}$, we find

$$\theta_1\,\mathrm{I} = 4\,q^{\frac14}\sum_1^\infty n\left[\frac{q^{n^2-n} - 2 + q^{n^2+n}}{q^n - q^{-n}}\right],$$

$$\theta\,\mathrm{I} = 4\,q^{\frac14}\sum_1^\infty (-1)^n\,nq^{n^2}\,\frac{q^n - q^{-n}}{q^n + q^{-n}},$$

$$\mathrm{I} = 4\sum_1^\infty R_n\,q^{n^2}\left[\frac{q^{\frac12(2n-1)}}{1-q^{2n-1}} + \frac{q^{\frac12(2n+1)}}{1-q^{2n+1}}\right].$$

Let $\Gamma(n)$ represent the sum of those divisors of n which do not surpass $\sqrt{n}$, and which are even, or uneven, according as their conjugate divisors are even or uneven; and let $\Gamma'(n)$ represent the sum of the same divisors, each divisor being taken positively or negatively according as the sum of itself and its conjugate is unevenly or evenly even; if n is a perfect square, we are to replace $\sqrt{n}$ by $\frac12\sqrt{n}$ in the sums $\Gamma(n)$ and $\Gamma'(n)$; we then obtain the expansions

$$\theta_1\,\mathrm{I} = 8\,q^{\frac14}\sum_1^\infty\left[\Delta(n) - \Gamma(n)\right]q^n, \quad \theta\,\mathrm{I} = 8\,q^{\frac14}\sum_1^\infty \Gamma'(n)\,q^n,$$

$$\begin{aligned} \mathrm{I} &= 4\,q^{\frac14}\sum_{1}^{\infty}{}_n \sum_{1}^{n}{}_\mu \sum_{0}^{\infty}{}_s \left[q^{\frac14[(2n-1)(2n-1+4s+4)-(2\mu-1)^2]} + q^{\frac14[(2n+1)(2n+1+4s)-(2\mu-1)^2]}\right] \\ &= 4\,q^{\frac14}\sum_1^\infty F(4n)\,q^n = 8\,q^{\frac14}\sum_1^\infty F(n)\,q^n, \end{aligned}$$

because the coefficient of q^n in the expansion of $\frac{1}{4}q^{-\frac{1}{4}}\text{I}$ is a sum containing two units for every solution of the equation $4n = ac - b^2$, in which, a, b, c being positive and uneven, $b < a < c$, and one unit for every such solution in which either $b = a < c$, or $b < a = c$; and because (by reasoning similar to that already employed in this article) it is ascertained that this sum is equal to the number of reduced forms of determinant $-4n$, *i.e.* to $F(4n)$. Eliminating I, we obtain, finally,

$$\theta_1 \Sigma F(n)\, q^n = \sum_1^\infty [\Delta(n) - \Gamma(n)]\, q^n, \quad \theta \Sigma F(n)\, q^n = \sum_1^\infty \Gamma'(n)\, q^n,$$

or

$$\text{XI.} \quad F(n) + 2F(n-1^2) + 2F(n-2^2) + \ldots = \Delta(n) - \Gamma(n),$$

$$\text{XII.} \quad F(n) - 2F(n-1^2) + 2F(n-2^2) - \ldots = \Gamma'(n).$$

These equations are equivalent to the formulae I., II., III., V., VI. of M. Kronecker; these five, therefore (and with them, according to M. Kronecker, the remaining three, IV., VII., VIII.), are deducible by analytical transformations from the single equation

$$\frac{\theta_1 \eta_1^2}{8\pi q^{\frac{1}{4}}} \int_0^\pi \frac{\text{H}^2 \text{H}_1}{\Theta^2} \cos x dx = \sum_1^\infty F(n)\, q^n.$$

135. M. Kronecker asserts that the formulae I.—VIII. are independent, *i.e.*, that none of them can be deduced from the others by means of the elementary equations satisfied by the functions F and G; and that all the similar relations, which are supplied by the theory of complex multiplication, may be obtained, with the help of those elementary equations, by combining the eight formulae. And it is certain that the system of the eight formulae does, in this sense, explicitly contain all the relations of similar form, which have been subsequently given by MM. Hermite and Joubert. Thus, many of these relations are particular cases of the formulae XI. and XII., or of the combinations (XI.) $\pm$ (XII.) (in M. Joubert's memoir, the formulae 1, 2, 3, those of page 28, and the first of page 29; also the first two formulae in M. Hermite's memoir (Liouville, New Series, vol. vii. p. 25) are of this kind); others, again, are immediately deducible from the two formulae

$$4F(n) + 8F(n-4^2) + 8F(n-8^2) + \ldots = \Phi(n), \quad n \equiv 3, \text{ mod } 8,$$

$$8F(n-2^2) + 8F(n-6^2) + 8F(n-10^2) + \ldots = \Phi(n), \quad n \equiv 7, \text{ mod } 8,$$

combined by addition or subtraction with V., VI., and VII. But each of these two formulae results from the combination $\frac{9}{2}$(V.) $+ \frac{3}{2}$(VI.) $- 2$(IV.), simplified by means of the elementary equations satisfied by F and G. In this way the formulae 4–9 of M. Joubert's memoir, and the third formula in M. Hermite's memoir (Liouville, *ibid.* p. 36), may be obtained. Lastly, the equation

$$6\,G\left(\frac{n-1^2}{4}\right)+6\,G\left(\frac{n-3^2}{4}\right)+\ldots=\tfrac{1}{2}\left(3\,\Psi(n)-\Phi(n)\right),\quad n\equiv 1,\ \text{mod } 4,$$

(M. Joubert, p. 30) arises from the combination (IV.) $-\frac{3}{2}$ (V.).

M. Joubert's formulae, however, as they are given in his memoir, are not immediately comparable to those of M. Kronecker. He rejects from the modular equation of the uneven order n, the factors due to the square divisors of n (see Art. 125 of this report), and, in consequence, those derived classes whose coefficients have any common divisor with n are excluded from his enumerations. At the same time, the numerical functions, depending on the divisors of n, which occur in the right-hand members of his formulae, are rendered somewhat more complicated than those of M. Kronecker. It is always possible to pass from one of M. Joubert's formulae to the corresponding formulae of M. Kronecker, by an elementary process, of which M. Joubert has himself given an example (at p. 25 of his memoir).

One formula, however, has been obtained by M. Hermite from his investigation of the discriminant of the modular equation, which is entirely distinct in form, and as it would seem in substance, from those of M. Kronecker. Taking a modular equation of a prime order n, M. Hermite shows that its discriminant is of the form

$$u^{n+1}\,(1-u^8)^{n+\left(\frac{2}{n}\right)}\,\theta^2(u^8),$$

where $\theta(u)$ is a reciprocal polynomial, prime to u and to $1-u^8$, containing no equal factors, and of order $\frac{1}{8}(n^2-1)-\frac{1}{2}\left[n+\left(\frac{2}{n}\right)\right]$. From the nature of a discriminant, if ω renders two of the quantities $\left(\frac{2}{\gamma}\right)\phi\left(\frac{\gamma\omega+16k}{\gamma'}\right)$ equal to one another, $\phi(\omega)$ is a root of the equation $\theta(u)=0$, and conversely. It is thus possible, by a method of which we have already given examples, to assign a system of quadratic equations (or quadratic forms) having integral coefficients, which shall correspond, one by one, to the roots of the equation $\theta(u)=0$. Equating the number of these quadratic forms to the index of the polynomial $\theta(u)$, M. Hermite obtains a formula which is essentially limited to the case when n is a prime, and which, translated into the notation of M. Kronecker, is as follows,

$$2\,\Sigma_1 F(\Delta)+2\,\Sigma_2 F(\Delta)+6\,\Sigma_3 G(\Delta)=\tfrac{1}{8}(n^2-1)-\tfrac{1}{2}\left[n+\left(\frac{2}{n}\right)\right],$$

the summations Σ_1, Σ_2, Σ_3 extending respectively to all values of δ which give positive values to the numbers

$$\Delta = (8\delta - 3n)(n - 2\delta), \quad \Delta = 8\delta(n - 8\delta), \quad \Delta = \delta(n - 16\delta).$$

The difference between these series of determinants, and those which occur in M. Kronecker's formulae, is very remarkable.

136. *Arithmetical Demonstrations of the Formulae of M. Kronecker.*—M. Kronecker informs us that, when he had connected his formulae, in the manner already described, with the expansions of certain elliptic functions, he directed his attention to the process (Art. 127) by which Jacobi transformed the analytical proof of the 'theorem of four squares' into an arithmetical one *. Applying a similar transformation to the analytical proof of the equation of

$$\frac{\theta_1 \eta_1^2}{8\pi q^{\frac{1}{4}}} \int_0^\pi \frac{\mathrm{H}^2\,\mathrm{H}_1}{\Theta^2} \cos x dx = \sum_1^\infty F(n)\, q^n,$$

he succeeded, after many reductions, in obtaining a purely arithmetical proof of the formulae I. II. and V. which are included in XI. This important investigation has not yet been published: instead, M. Kronecker has given a remarkable theorem which appears (as he observes) to contain the germ of another, and very different, arithmetical demonstration of his formulae. He has enunciated the theorem for prime numbers only, remarking, however, that it admits of extension to composite numbers also. The result is simplest in the case of a prime number p of the form $4m+3$.

'Let (a, b, c) represent in succession every uneven reduced form of the determinants $-p$, $-p+1^2$, $-p+2^2$, ...; only, if $a = c$, let the reduced form satisfy the special condition (Art. 92) $b < 0$, instead of $b > 0$; the roots of the congruences

$$a\omega^2 + 2b\omega + c \equiv 0, \text{ mod } p,$$

of which roots the number is

$$F(p) + 2F(p - 1^2) + 2F(p - 2^2) + \ldots,$$

are a complete system of residues for the modulus p.'

As it appears from the formula V. that the number of these congruence-roots is equal to p, it is only necessary to prove that they are all different; the demonstration of this very difficult point M. Kronecker has effected by showing that the contrary supposition is inconsistent with the inequalities satisfied by the coefficients of the reduced forms. A proof, independent of the formula V.,

* Monatsberichte for 1862, p. 307.

that every residue of p is a root of one of the congruences, would of course supply a direct arithmetical proof of that formula, for the case in which n is a prime of the form $4m+3$.

Arithmetical demonstrations of the formulae of M. Kronecker have also been obtained by M. Liouville. These demonstrations depend on the principles introduced by him into arithmetic in the series of memoirs 'Sur quelques formules générales qui peuvent être utiles dans la théorie des nombres'*, and originally suggested (as he himself informs us) by Jacobi's arithmetical proof of the theorem of four squares. M. Liouville has given, as an example of his method, a proof of the equation (XI.)—(XII.), or

$$4F(n-1^2)+4F(n-3^2)+4F(n-5^2)+\ldots=\Delta(n)-\Gamma(n)-\Gamma'(n),$$

for the two cases in which n is unevenly, and evenly, even †. We shall confine our attention to the latter and somewhat simpler case. It requires two preliminary Lemmas, both included as very particular cases in M. Liouville's general formulae.

I. Let m represent a given uneven number, a a given positive exponent other than zero, $f(x)$ an even function, so that $f(x)=f(-x)$; we have the equation $$\Sigma\left[f(d'-d'')-f(d'+d'')\right]=2^{a-1}\,\Sigma d\left[f(0)-f(2^a d)\right],$$
the summations in the left and right-hand members extending respectively to all solutions of the equations

$$2^a m = d'\delta' + d''\delta'', \quad m = d\delta,$$

the indeterminates $d', d'', \delta', \delta''$ in the first equation, and d, δ in the second, being positive and uneven, and two solutions of either equation being regarded as different, unless the indeterminates of the two solutions are the same and in the same order.

To establish this equation, we consider the system

$$\left.\begin{aligned} d'\delta'+d''\delta'' &= 2^a m,\\ d'+d'' &= 2\mu,\\ \delta'-\delta'' &= 2\nu, \end{aligned}\right\} \quad \ldots\ldots\ldots\ldots \quad \text{(a)}$$

in which μ and ν are given positive integers. The solutions of this system are equal in number to the solutions of the system

$$\left.\begin{aligned} d'\delta'+d''\delta'' &= 2^a m,\\ d'-d'' &= -2\mu,\\ \delta'+\delta'' &= 2\nu. \end{aligned}\right\} \quad \ldots\ldots\ldots\ldots \quad \text{(a}'\text{)}$$

* Liouville's Journal, vols. iii.–viii. (New Series). † Liouville, New Series, vol. vii. p. 44.

For, eliminating δ' and d'', we find that (a) has as many solutions as $2^{\alpha-1} m = \nu d' + \mu \delta''$ has solutions in which $d' < 2\mu$; and (a') has as many solutions as the same equation has solutions in which $\delta'' < 2\nu$; *i.e.* (a) has as many solutions as (a'); inasmuch as to every solution of $2^{\alpha-1} m = \nu d' + \mu \delta''$ in which $d' < 2\mu$, but $\delta'' > 2\nu$, there corresponds a solution, in which $d' > 2\mu$, but $\delta'' < 2\nu$, and *vice versâ*; for example, if $d' < 2\mu$, but $\delta'' > 2\nu$, let $2k\nu$ be the multiple of 2ν next inferior to δ'', then

$$2^{\alpha} m = \nu (d' + 2k\mu) + \mu (\delta' - 2k\nu)$$

is a solution of the equation, in which $d' + 2k\mu > 2\mu$, but $\delta' - 2k\nu < 2\nu$.

Similarly it will be seen that the solutions of the systems

$$\left.\begin{aligned} d'\delta' + d''\delta'' &= 2^{\alpha} m, \\ d' + d'' &= 2\mu, \\ \delta' - \delta'' &= -2\nu, \end{aligned}\right\} \quad . \quad . \quad . \quad . \quad . \quad . \quad . \quad . \quad . \quad . \quad (b)$$

$$\left.\begin{aligned} d'\delta' + d''\delta'' &= 2^{\alpha} m, \\ d' - d'' &= 2\mu, \\ \delta' + \delta'' &= 2\nu, \end{aligned}\right\} \quad . \quad . \quad . \quad . \quad . \quad . \quad . \quad . \quad . \quad . \quad (b')$$

are equal in number.

Also the number of solutions of either of the systems

$$\left.\begin{aligned} d'\delta' + d''\delta'' &= 2^{\alpha} m, \\ d' + d'' &= 2^{\alpha} d, \\ \delta' - \delta'' &= \delta, \end{aligned}\right\} \quad . \quad . \quad . \quad . \quad . \quad . \quad . \quad . \quad . \quad . \quad (c)$$

$$\left.\begin{aligned} d'\delta' + d''\delta'' &= 2^{\alpha} m, \\ d' = d'' &= \delta, \\ \delta' + \delta'' &= 2^{\alpha} d, \end{aligned}\right\} \quad . \quad . \quad . \quad . \quad . \quad . \quad . \quad . \quad . \quad . \quad (c')$$

in each of which d, δ are two given conjugate divisors of m, is $2^{\alpha-1} d$.

Let us now attribute to μ, ν, d, δ in the systems (a), (b), (c), (a'), (b'), (c'), all values, in succession, for which those systems are resoluble. We shall evidently obtain the sum $\Sigma f(d' + d'')$, which occurs in the equation to be proved, by extending the summation, first, to all solutions of the various systems (a), secondly, to all the solutions of the various systems (b), and lastly, to all solutions of the various systems (c). Similarly, we shall obtain the sum $\Sigma f(d' - d'')$ by extending the summation to all solutions of the systems (a'), (b'), (c'). But the terms $f(d' + d'')$ arising from any one of the systems (a) or (b), are cancelled in the difference

$$\Sigma f(d' - d'') - \Sigma f(d' + d'')$$

by the terms $f(d'-d'')$ arising from the corresponding system (a′) or (b′). That difference is, therefore, equal to the excess of $\Sigma f(d'-d'')$, extended to all solutions of the systems (c′), above $\Sigma f(d'+d'')$ extended to all solutions of the systems (c); so that, finally,

$$\Sigma[f(d'-d'')-f(d'+d'')]=2^{a-1}\Sigma d[f(0)-f(2^a d)].$$

II. Let m be an uneven number, and $f(x)$ an uneven function; we have the equation

$$\Sigma f(d'+2m')=\Sigma f\{\tfrac{1}{2}(d_1+\delta_1)\},$$

the summations in the left- and right-hand members extending respectively to all solutions of the equations

$$m=2m'^2+d'\delta',$$
$$2m=\ m_1^2+d_1\delta_1,$$

the indeterminates d', δ', d_1, δ_1, m_1 being positive and uneven, but m' being even or uneven, positive, negative, or zero.

If we write $2m'+d'=x$, $\delta'-2m'=y$, $2m'+d'-\delta'=z$, so that conversely $2m'=x-y-z$, $d'=y+z$, $\delta'=x-z$, the equation $m=2m'^2+d'\delta'$ becomes $2m=x^2+y^2-z^2$, the indeterminates being subject to the conditions

$$y+z>0, \quad z<x.$$

If in addition $z+x<0$, the conditions are satisfied by the two solutions $[x, y, z]$, $[-x, y, z]$; which give rise, in the sum $\Sigma f(d''+2m')$, to two terms which cancel one another. We need only therefore consider those solutions, which satisfy the inequalities, $y+z>0$, $x+z>0$, $z<x$, or, which is the same thing, if $[z]$ represent the absolute value of z,

$$y+z>0, \quad x>0, \quad [z]<x. \quad . \quad . \quad . \quad . \quad . \quad . \quad . \quad . \quad (\text{d})$$

Again, if we write $\frac{1}{2}(d_1+\delta_1)=x$, $m_1=y$, $\frac{1}{2}(d_1-\delta_1)=z$, the equation $2m=m_1^2+d_1\delta_1$ becomes $2m=x^2+y^2-z^2$, the indeterminates being subject to the conditions

$$y>0, \quad x>0, \quad [z]<x. \quad . \quad . \quad . \quad . \quad . \quad . \quad . \quad . \quad . \quad (\text{d}')$$

To establish the proposed equation it is now only necessary to show that the equation $2m=x^2+y^2-z^2$ admits of equal numbers of solutions satisfying the inequalities (d) and (d′). But this is evident; for if $[x, y, z]$ satisfy one of the two sets of inequalities, but not both, $[x, -y, -z]$ satisfies the other, but not both.

By combining these two lemmas it may be proved that four times the number of solutions of the equation

$$2^{a+1}m=m_1^2+d_2\delta_2+(d_2+\delta_2)\delta_3 \quad . \quad . \quad . \quad . \quad . \quad . \quad . \quad . \quad (\text{A})$$

(in which m is a given uneven number, and α a given exponent > 0) is

$$\Delta(2^{\alpha+1}m) - \Gamma(2^{\alpha+1}m) - \Gamma'(2^{\alpha+1}m),$$

the indeterminates $m_1, d_2, \delta_2, \delta_3$ being all positive and uneven and $d_2 - \delta_2$ being evenly even. Representing by m' any number whatever, and by d', δ', δ_3 positive uneven numbers, let us consider the two equations

$$2^{\alpha+1}m - 2d_3\delta_3 = m_1^2 + d_2\delta_2, \quad \text{(e)}$$
$$2^{\alpha}\ m - d_3\delta_3 = 2m'^2 + d'\delta', \quad \text{(e')}$$

and let $f(x)$ be an even function, so that $f(x-d_3) - f(x+d_3)$ is an uneven one, and may be used instead of $f(x)$ in the second lemma. Applying that lemma to the two equations (e) and (e′), and afterwards summing for every value of d_3, we find

$$\Sigma[f\{\tfrac{1}{2}(d_2+\delta_2) - d_3\} - f\{\tfrac{1}{2}(d_2+\delta_2)+\delta_3\}]$$
$$= \Sigma[f(2m'+d'-d_3) - f(2m'+d'+d_3)],$$

the summations extending to all solutions of (e) and (e′) respectively. Observing that if $m'=0$, $f(2m'+x)$ is an even function of x, and that if m' is not $=0$, $f(-2m'+x)+f(2m'+x)$ is an even function of x, we transform the second member by the first lemma, and we obtain

$$\Sigma[f\{\tfrac{1}{2}(d_2+\delta_2) - \delta_3\} - f\{\tfrac{1}{2}(d_2+\delta_2)+\delta_3\}]$$
$$= 2^{\gamma-1}\Sigma d\,[f(2m') - f(2^{\gamma}d + 2m')],$$

the second summation extending to every solution of the equation

$$2^{\alpha}m - 2m'^2 = 2^{\gamma}d\,\delta,$$

d and δ being positive and uneven, and 2^{γ} representing the highest power of 2 contained in $2^{\alpha}m - 2m'^2$. Let $f(x)=1$, if $x=0$, but let $f(x)=0$ for every other value of x; the sum

$$\Sigma[f\{\tfrac{1}{2}(d_2+\delta_2) - \delta_3\} - f\{\tfrac{1}{2}(d_2+\delta_2)+\delta_3\}]$$

will then represent the number of solutions of the equation (A); the sum $2^{\gamma-1}\Sigma d f(2m')$ will become $2^{\alpha-1}\Sigma d$, the summation extending to all solutions of the equation $m = d\delta$; and the sum $2^{\gamma-1}\Sigma d f(2m'+2^{\gamma}d)$ will become $\Sigma 2^{\gamma-1}d$, the summation extending to all solutions of the equation

$$2^{\alpha-1}m = 2^{\gamma-1}d\,(2^{\gamma-1}d+\delta).$$

Of these sums $2^{\alpha-1}\Sigma d$ is evidently $\Delta(2^{\alpha-1}m) = \tfrac{1}{4}\Delta(2^{\alpha+1}m)$; and $\Sigma 2^{\gamma-1}d$ is the sum of those divisors of $2^{\alpha-1}m$, which are less than $\sqrt{2^{\alpha-1}m}$, and which are not of the same parity as their conjugates, a sum which is identical with

$\frac{1}{4}\Gamma(2^{\alpha+1}m)+\frac{1}{4}\Gamma'(2^{\alpha+1}m)$; as may be seen by considering separately the cases in which $\alpha=1$, and $\alpha>1$.

A second determination of the number of solutions of the equation (A) is obtained as follows. Write $2\theta+1$ for $\frac{1}{2}(d_2+\delta_2)$ and $4a$ for $d_2-\delta_2$; it becomes

$$2^{\alpha+1}m-m_1^2=(2\theta+1)(2\theta+1+2\delta_3)-4a^2,$$

which is of the same form as that considered by M. Hermite (see Art. 134). If then we attribute to m_1 any particular value, the number of solutions of the equation (A) is $F(2^{\alpha+1}m-m_1^2)$; its whole number of solutions is therefore

$$F(2^{\alpha+1}m-1^2)+F(2^{\alpha+1}m-3^2)+F(2^{\alpha+1}m-5^2)+\dots;$$

equating this expression to that which we have already found, we obtain the formula (XI.)—(XII.).

M. Liouville tells us that, until M. Hermite's discussion of the equation

$$4n+3=(2\theta+1)(2\theta+1+2\delta_3)-4a^2,$$

he had not observed that the number of solutions of the equation

$$2^{\alpha+1}m-m_1^2=d_2\delta_2+(d_2+\delta_2)\delta_3, \quad d_2\equiv\delta_2, \text{ mod } 4,$$

is equal to the number of classes of quadratic forms of det $.m_1^2-2^{\alpha+1}m$; but that with this exception all the principles of the preceding demonstration were in his possession; so that he had already arrived at formulae identical with those of M. Kronecker, but referring to the numbers of solutions of certain indeterminate equations instead of to the numbers of quadratic forms of certain determinants. We also learn from him that formulae exist, analogous to those of M. Kronecker, in which the series of determinants are of the type $2s^2-n$, $3s^2-n$, ... instead of s^2-n.

137. *Equations satisfied by the Modules which admit of Complex Multiplication.*—We have already observed (Art. 126) that the $6G(\Delta)$ values of $\phi^8(\omega)$ corresponding to the quadratic forms of det. $-\Delta$, are the roots of an equation of that order, having rational coefficients. Several important properties of this equation have been indicated by M. Kronecker; but, notwithstanding their intimate connexion with the theory of quadratic forms, we can only offer an imperfect account of them.

We resume the notation of Art. 131; and we shall begin by showing that if n is an uneven number, greater than 3, the values of $\phi^8(\omega)$, corresponding to the properly primitive classes of det. $-n$, satisfy one or other of three equations, each of the order $2h(n)$, and each having rational coefficients. We have already seen in Art. 131, that every value of $\phi^8(\omega)$, corresponding to a form of which the extreme coefficients are uneven, and of which the determinant is $\sigma^2 - n$, is a root of the equation $f_8(x, 1-x) = 0$, and that this equation has no other roots. Again, if $\chi_2\left(\frac{1}{M^2}, \kappa^2\right) = 0$ is the equation satisfied by the squares of the multipliers appertaining to the $\Phi(n)$ transformations of order n, the equation $\chi_2(n, x) = 0$ will be satisfied by those roots of the equation $f_8(x, 1-x) = 0$ which correspond to quadratic forms of det. $-n$, but not by the other roots of that equation. For, if $\phi^8(\omega)$ is a root of $f_8(x, 1-x) = 0$, $\phi^8(\omega)$ is transformed into $\psi^8(\omega)$ by one of the $\Phi(n)$ transformations of order n; and if $\phi^8(\omega)$ corresponds to a quadratic form of det. $-n$, the multiplier appertaining to this transformation is $\pm \frac{1}{\sqrt{n}}$; whereas if $\phi^8(\omega)$ corresponds to a quadratic form of det. $-\Delta = \frac{\sigma^2 - n}{\tau^2}$, the multiplier is $\pm[\tau\sqrt{\Delta} \pm i\sigma]^{-1}$ (see Art. 131). Forming then the greatest common divisor of the two functions

$$f_8(x, 1-x) \quad \text{and} \quad \chi_2(n, x),$$

we obtain an equation of which the roots are, exclusively, those values of $\phi^8(\omega)$ which correspond to quadratic forms of det. $-n$ *. Let $\psi_1(n, x)$ represent this greatest common divisor, and denoting by $p_1, p_2, \ldots$ the different primes, of which the squares are divisors of n, let us form the expression

$$\Psi_1(n, x) = \frac{\psi_1(n, x) \times \Pi\, \psi_1\left(\frac{n}{p_1^2 p_2^2}, x\right) \times \ldots}{\Pi\, \psi_1\left(\frac{n}{p_1^2}, x\right) \times \ldots}.$$

If (A, B, C) symbolize a system of quadratic forms, having their extreme coefficients uneven, and representing the properly primitive classes of det. $-n$, the roots of the equation $\Psi_1(n, x) = 0$ are those values of $\phi^8(\omega)$ which correspond to the systems of quadratic equations

$$A + 2B\omega + C\omega^2 = 0, \quad C - 2B\omega + A\omega^2 = 0.$$

Thus the order of the equation is $2h(n)$: if $x = \phi^8(\omega)$ is a root, $1 - x = \phi^8\left(-\frac{1}{\omega}\right)$ is also a root: the first coefficient is a power of 2, and the last coefficient is

---

* It is here assumed that if $\pm \frac{1}{\sqrt{n}}$ is not the multiplier appertaining to any of the transformations of order n by which x is changed into $1-x$, it is also not the multiplier appertaining to any of the $\Phi(n)$ transformations of the order n.

unity, because $\Psi_1(n, x)$ divides $f_8(x, 1-x)$, of which the first and last coefficients are respectively a power of 2 and unity*. From the equation $\Psi_1(n, x_2) = 0$, we may deduce two others, $\Psi_2(n, x_2) = 0$, $\Psi_3(n, x_3) = 0$, by the substitutions

$$x_1 = \frac{1}{x_2}, \qquad x_2 = \frac{x_3}{x_3 - 1}:$$

these equations will have for their roots the $4h(n)$ values of $\phi^8(\omega)$ corresponding to properly primitive forms of det. $-n$, not included in the subclasses (A, B, C), $(C, -B, A)$. The roots of the equation $\Psi_2(n, x) = 0$ are the reciprocals of the roots of $\Psi_1(n, x) = 0$: its first coefficient is therefore unity and its last a power of 2; the equation $\Psi_3(n, x) = 0$ is a reciprocal equation, and its first and last coefficients are units.

Each of the three functions $\Psi_1(n, x)$, $\Psi_2(n, x)$, $\Psi_3(n, x)$, can be decomposed into two factors, of the order $h(n)$, and containing no irrationality but $\sqrt{n}$. If $n \equiv 3$, mod 4, the value of $\frac{(-1)^{\frac{1}{2}(\gamma-1)}}{\mathrm{M}}$ (Art. 125) corresponding to one of the two forms (A, B, C), $(C, -B, A)$ is $+\sqrt{n}$, and that corresponding to the other $-\sqrt{n}$; if $n \equiv 1$, mod 4, the values of $\frac{(-1)^{\frac{1}{2}(\gamma-1)}}{\mathrm{M}}$ corresponding to those two forms are both $\sqrt{n}$ or both $-\sqrt{n}$, according as the generic character of the two forms is $(-1)^{\frac{1}{2}(f-1)} = +1$, or $(-1)^{\frac{1}{2}(f-1)} = -1$; in either case, therefore, the decomposition of $\Psi_1(n, x)$ into two factors, can be effected by comparing it with the equations

$$\chi_1(\sqrt{n}, x) = 0, \quad \chi_1(-\sqrt{n}, x) = 0, \quad \text{if } \chi_1\left(\frac{(-1)^{\frac{1}{2}(\gamma-1)}}{\mathrm{M}}, \kappa^2\right) = 0$$

is the equation satisfied by the multipliers appertaining to the transformations of order n.

But M. Kronecker has shown that $\Psi_1(n, x)$ admits of a more profound decomposition, when n is a composite number. In fact, if ν is the number of the primes $p_1, p_2, p_3, \ldots$ dividing n, $\Psi_1(n, x)$ can be resolved into 2^ν factors, each of the order $\frac{1}{2^{\nu-1}} h(n)$, and containing no irrationalities but $\sqrt{p_1}$, $\sqrt{p_2}$, $\sqrt{p_3}, \ldots$. If $n \equiv 3$, mod 4, the order of each factor is precisely equal to the number of classes in a properly primitive genus of det. $-n$; and the roots of each factor are the values of $\phi^8(\omega)$ corresponding to forms of a determinate genus. If

* The limit of $f_8(x, 1-x) \div x^{\Phi(n)+\Psi(n)}$, when x increases without limit, is $2^{4\Psi(n)}$, or $-2^{4\Psi(n)+1}$, according as n is not, or is, a perfect square.

$n \equiv 1$, mod 4, instead of the equation $\Psi_1(n, x) = 0$, we consider the equation, of which the roots are the $h(n)$ quantities $x(1-x)$; this equation can, as before, be resolved into 2^{ν} generic factors, each corresponding to a determinate genus. In either case the factors only differ from one another in respect of the signs of the radicals $\sqrt{p_1}$, $\sqrt{p_2}$, ..., and these signs are determined for each factor by the characters of the corresponding genus with respect to the primes $p_1, p_2, \ldots$. For an account of the method by which this resolution into generic factors is obtained, we must refer to the original note of M. Kronecker (Monatsberichte, June 26, 1862, p. 370).

In one of Abel's memoirs (Œuvres Complètes, vol. i. p. 272) we find the theorem that the modules which admit of complex multiplication are capable of expression by radicals. At an earlier period Abel had doubted of the truth of this result, and he has given no indication of the method by which he obtained it. M. Kronecker's researches on complex multiplication were originally suggested by Abel's theorem; and they have led him to a complete demonstration of it. Let $n \equiv 3$, mod 4, and let $\phi^8(\omega_1)$, $\phi^8(\omega_2)$, ... be the roots of one of the generic equations into which $\Psi_1(x, n) = 0$ is resolved, the imaginaries $\omega_1, \omega_2, \ldots$ being determined by the equations

$$A_1 + 2B_1\omega_1 + C_1\omega_1^2 = 0, \quad A_2 + 2B_2\omega_2 + C_2\omega_2^2 = 0, \ \ldots\ldots$$

so that the forms (A_1, B_1, C_1), (A_2, B_2, C_2) belong to the same genus. If the determinant $-n$ is regular (Art. 117), it can be proved by the theory of composition, that it is possible so to select the representative forms (A_1, B_1, C_1), (A_2, B_2, C_2) as to satisfy the equations $\omega_2 = \theta\omega_1$, $\omega_3 = \theta\omega_2$, ..., θ denoting an uneven number prime to n, or indeed a prime not dividing n. We may thus represent the roots of the generic equation by the expressions

$$\phi^8(\omega_1), \quad \phi^8(\theta\omega_1), \quad \phi^8(\theta^2\omega_1), \ \ldots\ldots.$$

But it can also be proved (by a comparison of the equations relating to the transformations of the orders n and θ) that $\phi^8(\theta\omega_1) = \chi[\phi^8(\omega_1)]$, χ representing a rational and integral function of which the coefficients contain no irrationalities but i, $\sqrt{p_1}$, $\sqrt{p_2}$, ... The generic equation is therefore Abelian, and of course resoluble by radicals. If the determinant is irregular, having only one index of irregularity, the general expression of the roots of the generic equation is $\phi^8(\theta_1^{a_1}\theta_2^{a_2}\omega_1)$, and, as before, it will be found that

$$\phi^8(\theta_1\omega_1) = \chi_1[\phi^8(\omega_1)], \quad \phi^8(\theta_2\omega_1) = \chi_2[\phi^8(\omega_1)],$$

the functions χ_1 and χ_2 still containing no irrationalities but i, $\sqrt{p_1}$, $\sqrt{p_2}$, ... and further satisfying the equation $\chi_1[\chi_2(x)] = \chi_2[\chi_1(x)]$. The generic equation is therefore in this case also resoluble by radicals; and the same conclusion is true whatever be the order of irregularity of the determinant $-n$. If $n \equiv 1$, mod 4, we should have to consider the generic equation of which the roots are of the form $\phi^8(\omega) \times \psi^8(\omega)$, and the demonstration of its resolubility is to be obtained in a similar manner.

We have for brevity confined ourselves to the case of properly primitive forms of an even determinant; but the values of $\phi^8(\omega)$ corresponding to the improperly primitive classes of a determinant of the form $-(4m+3)$, or to the properly primitive classes of even determinants, are also the roots of arithmetical equations possessing analogous properties. A method for the formation of these equations, different from that of M. Kronecker, and not requiring the use of the equation of the multiplier, has been given by M. Hermite (Théorie des Equations Modulaires, p. 44); but this method does not supply the ulterior decomposition of the equations into their generic factors.

138. *Application of the Theta Functions to the Pellian Equation.*—One more application of the Theta functions to arithmetic (inferior in importance to none that have been mentioned) has also been discovered by M. Kronecker (Monatsberichte, Jan. 22, 1863, p. 44). Let a, b, c, σ, τ represent real quantities, of which a and c are positive and a, b, c satisfy the inequality

$$ac - b^2 = \Delta > 0,$$

and let the positive quantity ρ be diminished without limit, the limit of the sum

$$\sum_{x=-\infty}^{x=+\infty} \sum_{y=-\infty}^{y=+\infty} \frac{e^{2\pi i(\sigma x + \tau y)}}{[ax^2 + 2bxy + cy^2]^{1+\rho}}$$

(in which, however, the value 0 is not to be attributed to x and y simultaneously) has been found by M. Kronecker to be

$$\left.\begin{aligned} &\frac{2\sigma^2\pi^2}{a} + \frac{\pi}{3\sqrt{\Delta}} \log\left[\frac{1}{4\pi^2}\theta'(0, \omega) \times \theta'(0, \omega')\right] \\ &\qquad - \frac{\pi}{\sqrt{\Delta}} \log[-\theta(\tau+\sigma\omega, \omega) \times \theta(-\tau+\sigma\omega', \omega')], \end{aligned}\right\} \quad . \quad . \quad . \quad . \quad \text{(i)}$$

where $$\omega = \frac{-b+i\sqrt{\Delta}}{a}, \qquad \omega' = \frac{b+i\sqrt{\Delta}}{a},$$

$$\theta(x, \omega) = \frac{1}{i}\theta_{1,1}(2Kx, \omega), \quad \theta'(x, \omega) = \frac{d.\theta(x, \omega)}{dx},$$

$\sqrt{\Delta}$ being positive, and the logarithms real.

Again, let $a_1 c_1 - b_1^2 = \Delta = ac - b^2$, a_1, b_1, c_1 being real and a_1, c_1 positive; and let δ represent an evanescent quantity, we find

$$\Sigma\Sigma \frac{1 - e^{2\pi i\delta\left(x\sqrt{a} + y\frac{b}{\sqrt{a}}\right)}}{[ax^2 + 2bxy + cy^2]^{1+\rho}} = \Sigma\Sigma \frac{1 - e^{2\pi i\delta\left(x\sqrt{a_1} + y\frac{b_1}{\sqrt{a_1}}\right)}}{[a_1x^2 + 2b_1xy + c_1y^2]^{1+\rho}};$$

and hence, by the theorem of M. Kronecker,

$$\lim\left[\Sigma\Sigma \frac{1}{[ax^2 + 2bxy + cy^2]^{1+\rho}} - \Sigma\Sigma \frac{1}{[a_1x^2 + 2b_1xy + c_1y^2]^{1+\rho}}\right]$$

$$= \lim\left[\Sigma\Sigma \frac{e^{2\pi i\delta\left[x\sqrt{a} + y\frac{b}{\sqrt{a}}\right]}}{[ax^2 + 2bxy + cy^2]^{1+\rho}} - \Sigma\Sigma \frac{e^{2\pi i\delta\left[x\sqrt{a_1} + y\frac{b_1}{\sqrt{a_1}}\right]}}{[a_1x^2 + 2b_1xy + c_1y^2]^{1+\rho}}\right]$$

$$= \frac{2\pi}{3\sqrt{\Delta}} \log \frac{a\sqrt{a}\,\theta'(0, \omega_1)\,\theta'(0, \omega_1')}{a_1\sqrt{a_1}\,\theta'(0, \omega)\,\theta'(0, \omega')}. \quad \text{(ii)}$$

This result M. Kronecker has applied, in the following manner, to the solution of the Pellian equation.

Let P and Q be positive integers not divisible by any square, of which P is >1, and let m and n represent positive integers prime to $2P$ and $2Q$ respectively, the limits of the sums

$$\sum_{m=1}^{m=\infty} \left(\frac{P}{m}\right) \frac{1}{m^{1+\rho}} \quad \text{and} \quad \sum_{n=1}^{n=\infty} \left(\frac{-Q}{n}\right) \frac{1}{n^{1+\rho}}$$

are known from the researches of Dirichlet (Crelle, vol. xix. pp. 360 and 364; or Art. 101 of this Report), and are respectively

$$\frac{h(P)\log[T + U\sqrt{P}]}{2\sqrt{P}} \quad \text{and} \quad \frac{\pi h(-Q)}{2\sqrt{Q}},$$

$h(P)$ and $h(-Q)$ denoting the number of properly primitive classes of the determinants P and $-Q$, and T, U being the least positive numbers which satisfy the equation $T^2 - PU^2 = 1$. Multiplying the two results together, and designating PQ by D, we find

$$\frac{\pi}{4\sqrt{D}} h(P) \,.\, h(-Q) \,.\, \log(T + U\sqrt{P}) = \lim \Sigma\Sigma \left(\frac{P}{m}\right)\left(\frac{-Q}{n}\right) \frac{1}{(mn)^{1+\rho}}. \quad \text{(iii)}$$

If P and Q are relatively prime, and congruous to one another, mod 4, so that D is not divisible by any square, and is $\equiv 1$, mod 4, the series

$$\tfrac{1}{2}\left[1 - \left(\frac{2}{R}\right)\frac{1}{2^{1+\rho}}\right] \Sigma \left(\frac{f}{R}\right) \Sigma\Sigma \frac{1}{(ax^2 + 2bxy + cy^2)^{1+\rho}} \quad \text{(A)}$$

(in which $R=P$, or $R=Q$, according as

$$P \equiv Q \equiv 1, \text{ mod } 4, \quad \text{or} \quad P \equiv Q \equiv 3, \text{ mod } 4\,;$$

f or (a, b, c) denotes any one of a set of representative forms of det. $-D$; $\left(\frac{f}{R}\right)$ is the particular character of f with respect to R, and the first sign of summation extends to every form of the representative system) is identical with the series

$$\Sigma\Sigma \left(\frac{P}{m}\right)\left(\frac{-Q}{n}\right)\frac{1}{(mn)^{1+\rho}} \quad . \quad . \quad . \quad . \quad . \quad . \quad . \quad . \quad (B)$$

To verify this we observe (1) that, because $D \equiv 1$, mod 4, the numbers of sets of representations (Art. 87) of N and $2N$ by forms of det. $-D$ are equal, N denoting any number whatever; (2) that, because D is not divisible by any square, the number of sets of representations of N by the forms of det. $-D$ is $\Sigma\left(\frac{-D}{d}\right)$, N denoting any uneven number, and d any divisor of N which is prime to D; (3) that the generic character of a form f of det. $-D$ may be ascertained from any number whatever N which is represented by f; in fact if p is a prime divisor of D, and if $N=N'p^{2\nu}$, N' being prime to p, we have

$$\left(\frac{f}{p}\right) = \left(\frac{N'}{p}\right);$$

if $N=N'p^{2\nu+1}$, $D=D'p$, N' and D' being prime to p,

$$\left(\frac{f}{p}\right) = \left(\frac{N'}{p}\right)\left(\frac{-D'}{p}\right).$$

From (1) and (3) we infer that the series (A) is equal to the series

$$\Sigma\left(\frac{f}{R}\right)\Sigma'\Sigma'\frac{1}{(ax^2+2bxy+cy^2)^{1+\rho}} \quad . \quad . \quad . \quad . \quad . \quad . \quad . \quad (C)$$

(in which the summations $\Sigma'\Sigma'$ extend only to those values of x and y for which f acquires uneven values); from (2) and (3), considering separately the two cases in which $R=P$, and $R=Q$, we infer that the coefficient of $\frac{1}{N^{1+\rho}}$ is the same in (B) and (C), N representing any uneven number; *i.e.* that the two series (A) and (B) are identical.

Diminishing ρ without limit in the equation (A) = (B), and employing the equations (ii) and (iii), we find immediately

$$h(P)\,.\,h(-Q)\,.\log(T+U\sqrt{P}) = \tfrac{2}{3}\left[2-\left(\frac{2}{R}\right)\right]\Sigma\left(\frac{f}{R}\right)\log\frac{a\sqrt{a}}{\theta'(0,\omega)\,\theta'(0,\omega')}, \quad \text{(iv)}$$

a remarkable equation which connects the least solution of the Pellian equation with the theory of the Theta functions.

If we suppose (as we may do) that the form (a, b, c) is reduced, so that $a < 2\sqrt{\tfrac{1}{3}D}$, we may approximate to the values of $\theta'(0, \omega)$ and $\theta'(0, \omega')$ by omitting in their developments all terms after the first, and writing

$$\theta'(0, \omega) \times \theta'(0, \omega') = 4\pi^2 e^{-\frac{1}{2}\pi\frac{\sqrt{D}}{a}} + \ldots.$$

Substituting in (iv), we obtain the approximative equation

$$h(P) \,.\, h(-Q) \,.\, \log(T + U\sqrt{P}) = \left[2 - \left(\frac{2}{R}\right)\right] \Sigma \left(\frac{a}{R}\right) \left[\frac{\pi\sqrt{D}}{3a} + \log a\right]. \quad \text{(v)}$$

The following examples of this formula are given by M. Kronecker. If $Q = 1$, the exponential $\frac{1}{8}e^{\frac{1}{2}\pi\sqrt{D}}$ is approximately equal to

$$2 + \sqrt{5}, \quad 18 + 5\sqrt{13}, \quad 882 + 145\sqrt{37}$$

when we attribute to D the values 5, 13, 37; again,

$$\tfrac{2}{9}e^{\frac{5}{18}\pi\sqrt{17}} = 4 + \sqrt{17}, \quad \tfrac{2}{49}e^{\frac{17}{42}\sqrt{97}} = 5604 + 569\sqrt{97}, \quad \text{if } D = 17, 97;$$

lastly, if $D = 85$, and we give to Q in succession the values 1, 5, 17, we find

$$\tfrac{1}{8}e^{\frac{3}{10}\pi\sqrt{85}} = 378 + 41\sqrt{85}, \quad \frac{1}{\sqrt{5}}e^{\frac{1}{10}\pi\sqrt{85}} = 4 + \sqrt{17}, \quad e^{\frac{1}{20}\pi\sqrt{85}} = 2 + \sqrt{5}.$$

These approximate representations of quadratic surds by exponentials are very remarkable; a similar observation had, however, already been made by M. Hermite*. If $D \equiv 3$, mod 8, the equation of which the roots are the values of $\phi^8(\omega)$ corresponding to the improperly primitive classes of det. $-D$, resolves itself into factors of the form $(x^2 - x + 1)^3 + \alpha(x^2 - x)^2 = 0$; and, in particular, if $(2, 1, \frac{1}{2}(D+1))$ is the only such class, α is an integral number. Attributing to ω the value $\frac{1}{2}(-1 + i\sqrt{D})$, and substituting for x, or $\phi^8(\omega)$, its approximate value $x = 2^4(e^{i\pi\omega} - 8e^{2i\pi\omega} + 44e^{3i\pi\omega})$ (equation (14) or (26), Art. 124), we find

$$\alpha = \frac{e^{\pi\sqrt{D}} - 744}{256}, \text{ nearly.}$$

Thus if $D = 43$, M. Hermite has found that

$$e^{\pi\sqrt{43}} = 884736743.9997775\ldots;$$

and that if $D = 163$, the decimal part of $e^{\pi\sqrt{163}}$ commences with twelve nines.

---

* Théorie des Equations Modulaires, p. 48.

M. Kronecker states that the formula (iv) implies the resolubility of $\Psi_1(n, x)$ into its generic factors, and that conversely the resolution of that equation into its factors implies the possibility of expressing certain solutions (though not necessarily the least solution) of the Pellian equation by means of the modules which admit of complex multiplication. For the case when $D \equiv 5$, mod 8, he has given the singularly elegant formula

$$\Pi \frac{\sin \frac{a\pi}{D}}{\sin \frac{b\pi}{D}} = \Pi \sqrt[3]{4\kappa^2\kappa'^2},$$

the first sign of multiplication extending to all numbers less than D and prime to it which satisfy the equations $\left(\frac{a}{D}\right) = 1$, $\left(\frac{b}{D}\right) = -1$; the second to a certain sixth part of the modules which admit of complex multiplication with $\sqrt{-D}$. The formula may also be written in the form

$$[T - U\sqrt{D}]^{h(D)} = \Pi \,.\, 4\kappa^2\kappa'^2$$

(see Dirichlet, Crelle, vol. xxi. p. 151).

INDEX OF NAMES.

The figures refer to the Articles: the more important references are indicated by heavy figures. Foot-notes are not separately referred to.

XI.

ON SYSTEMS OF INDETERMINATE LINEAR EQUATIONS.

[Report of the British Association for 1860. Sectional Proceedings, p. 6.]

THE object of this communication was to point out the connexion which exists between particular solutions of indeterminate linear equations, and their most general solution. The principle upon which this connexion depends may be explained in a very particular case. Let the system of indeterminate equations reduce itself to the single equation

$$Ax+By+Cz=0, \qquad \qquad (1)$$

in which we may suppose A, B, C to have no common divisor; let also a, b, c and a', b', c' be two different solutions of that equation in integral numbers; then *if the three numbers*

$$bc'-b'c, \quad ca'-ac', \quad ab'-a'b \qquad \qquad (2)$$

admit of no common divisor, the complete solution of the indeterminate equation is contained in the formulae

$$\left.\begin{aligned} x &= at+a'u, \\ y &= bt+b'u, \\ z &= ct+c'u, \end{aligned}\right\} \qquad \qquad (3)$$

in which t and u are absolutely indeterminate integral numbers; but if the condition (2) be not satisfied, the formulae (3) will not represent *all*, but only *some* of the solutions of the equation (1). If, therefore, by any method, as for example that of Euler, we have arrived at formulae of the type of the formulae (3), which demonstrably contain the complete solution of the indeterminate equation, we may be certain that the three numbers analogous to the numbers (2) admit of no common divisor. Thus, by applying Euler's method of solution, which is explained in most books of algebra, to the indeterminate equation

$$Ax + By + Cz = 0,$$

we obtain the solution of a celebrated problem, first considered by Gauss in the 'Disquisitiones Arithmeticae,' of which the following is the enunciation:—

'Given three numbers A, B, C, to find six others,

$$a, \ b, \ c,$$
$$a', \ b', \ c',$$

such that $A = bc' - b'c, \quad B = ca' - ac', \quad C = ab' - a'b$.'

Other methods, more symmetrical and perhaps not more tedious than that of Euler, were also suggested in this paper for the treatment of indeterminate equations, and for the resolution of an important class of arithmetical problems which depend on those equations in the manner just explained.

XII.

ON SYSTEMS OF LINEAR INDETERMINATE EQUATIONS AND CONGRUENCES.

[Philosophical Transactions, vol. cli. pp. 293–326. Received January 17; Read January 31, 1861.]

---

1. THE theory of the solution, in positive or negative integral numbers, of systems of linear indeterminate equations requires the consideration of rectangular matrices the constituents of which are integral numbers. It will therefore be convenient to explain the meaning which we shall attach to certain phrases and symbols relating to such matrices.

A matrix containing p constituents in every horizontal row, and q in every vertical column, is a matrix of the type $q \times p$. We shall employ the symbol $\left\| \begin{matrix} q\times p \\ A \end{matrix} \right\|$, or (when it is not necessary that the type of the matrix should be indicated in its symbol) the simpler symbol $\|A\|$ to represent the matrix

$$\left\| \begin{matrix} A_{1,1}, & A_{1,2}, & \ldots, & A_{1,p} \\ A_{2,1}, & A_{2,2}, & \ldots, & A_{2,p} \\ \cdot & \cdot & \cdot & \cdot \\ A_{q,1}, & A_{q,2}, & \ldots, & A_{q,p} \end{matrix} \right\|$$

If $\|A\|$ and $\|B\|$ be two matrices of the same type, the equation $\|A\| = \|B\|$ indicates that the constituents of $\|A\|$ are respectively equal to the constituents of $\|B\|$; whereas the equation $|A| = |B|$ will merely express that the determinants of $\|A\|$ are equal to the corresponding determinants of $\|B\|$. The determinants of a matrix are, of course, the determinants of the greatest square matrices contained in it; similarly, its minor determinants of order i are the determinants of the square matrices of the type $i \times i$ that are contained in it. Matrices of the types $n \times (m+n)$ and $m \times (m+n)$ are said to be of complementary types; if $\|A\|$ and $\|B\|$ be two such matrices, we shall employ the equation

$$|A| = |B|$$

to express that each determinant of $\|A\|$ is equal to that determinant of $\|B\|$ by which it is multiplied in the development of the determinant of the square matrix $\left\|\begin{matrix} A \\ B \end{matrix}\right\|$. When m and n are both uneven numbers, the signs of the determinants $\left|\begin{matrix} A \\ B \end{matrix}\right|$ and $\left|\begin{matrix} B \\ A \end{matrix}\right|$ are different: this occasions a certain ambiguity of sign in the interpretation of the equation $|A| = |B|$, which, however, will occasion no inconvenience. If $m = n$, the matrices $\|A\|$ and $\|B\|$ are at once of the same, and of complementary types; so that, in this case, the equation $|A| = |B|$ may stand for either of two very different sets of equations; but this also is an imperfection of the notation here employed which it is sufficient to have pointed out. If k denote any quantity whatever, it is hardly necessary to state that the equality

$$|A| = k \times |B|$$

implies that the determinants of $\|A\|$ are respectively k times the corresponding determinants of $\|B\|$.

Let $\|P\|$ be a square matrix of the type $n \times n$, and $\|Q\|$ a matrix of the type $n \times (n+m)$ (where $m \geqq 0$), we shall understand by the matrix compounded of $\|P\|$ and $\|Q\|$, the matrix $\|X\|$ of the same type as $\|Q\|$, the constituents of which are defined by the equation

$$X_{i,j} = P_{i,1} Q_{1,j} + P_{i,2} Q_{2,j} + \ldots + P_{i,n} Q_{n,j},$$

and we shall write

$$\|X\| = \|P\| \times \|Q\|;$$

in this equation $\|Q\|$ is said to be *premultiplied* by $\|P\|$, and $\|P\|$ to be *postmultiplied* by $\|Q\|$. This definition will suffice for our present purpose; as the only case of composition which we shall have to consider is that in which the vertical dimensions of the matrices to be compounded are all equal, and in which every premultiplying matrix is square, so that, if an oblong matrix present itself at all in a series of matrices to be compounded, it will occupy the last place in the series.

By the greatest divisor of a matrix we are to understand the greatest common divisor of the determinants of the matrix. If the matrix be square, its greatest divisor is, consequently, the determinant of the matrix. A *prime matrix* is one of which the greatest divisor is unity; *i.e.* the determinants of which are relatively prime. A prime square matrix (*i.e.* a matrix of which the determinant is unity) we shall call a *unit-matrix*.

In any system of linear equations, whether defective or redundant or neither, we shall understand by the matrix of the system the matrix formed by the coefficients of the unknown quantities. If to this matrix we add a vertical column, composed of the absolute terms of the equations, the resulting matrix we shall term (for brevity) the *augmented* matrix of the system.

Lastly, when we have occasion to consider square matrices the constituents of which, excepting those on the principal diameter, are zero, we shall represent them by symbols of the form

$$\| q_1, q_2, q_3, \ldots, q_n \|,$$

where $q_1, q_2, \ldots, q_n$ are the constituents of the principal diameter.

2. If every determinant of the augmented matrix of a redundant system of linear equations is equal to zero, while the determinants of the unaugmented matrix are not all equal to zero, the system admits of one solution, and one only. And, in particular, if the matrix of the system be a prime matrix, the values of the unknown quantities which satisfy the system are integral numbers. For these values may be expressed as fractions having for their denominators any one of the determinants of the matrix, and these determinants are relatively prime.

Let $\|A\|$ be a given prime matrix of the type $n \times (n+m)$, $\|K\|$ a given matrix of the same type connected with $\|A\|$ by the equation

$$|K| = k \times |A|, \quad \ldots \ldots \quad (1)$$

which implies that k is the greatest divisor of $\|K\|$; then the symbolic equation

$$\|K\| = \|k\| \times \|A\|, \quad \ldots \ldots \quad (2)$$

in which $\|k\|$ denotes a square matrix of the type $n \times n$, will admit of one solution, and one only.

For, to determine $k_{r,1}, k_{r,2}, \ldots, k_{r,n}$, the constituents of the rth horizontal row of $\|k\|$, we have the redundant system

$$\left. \begin{array}{c} K_{r,i} = A_{1,i} k_{r,1} + A_{2,i} k_{r,2} + \ldots + A_{n,i} k_{r,n} \\ i = 1, 2, 3, \ldots, n+m \end{array} \right\}, \quad \ldots \ldots \quad (3)$$

which is involved in the symbolic equation (2). The matrix of this system is the prime matrix $\|A\|$; and the determinants of its augmented matrix are all equal to zero; for, by virtue of equation (1), they are equal to the determinants

$$-\frac{1}{k}\times\begin{vmatrix} K_{r,1}, & K_{r,2}, & \dots, & K_{r,n+m} \\ K_{1,1}, & K_{1,2}, & \dots, & K_{2,n+m} \\ K_{2,1}, & K_{2,2}, & \dots, & K_{2,n+m} \\ \cdot & \cdot & \cdot & \cdot \\ K_{n,1}, & K_{n,2}, & \dots, & K_{n,n+m} \end{vmatrix},$$

in which two horizontal rows are identical. Thus the system (3), and consequently the equation (2), admits of one solution, and one only. It is evident that the determinant of $\|k\|$ is k. The case in which $m=0$ is not included in this demonstration; its proof, however, presents no difficulty, and may be omitted here.

A particular case of this theorem (that in which $n=2$) occurs in the 'Disquisitiones Arithmeticæ' (see Art. 234 of that work).

3. If every determinant of the augmented matrix of a redundant system of linear congruences be divisible by the modulus, while the greatest divisor of the unaugmented matrix is prime to the modulus, the system is resoluble and admits of only one solution. For, if the modulus be represented by $P\times Q\times R\dots$, P, Q, R, ... denoting powers of unequal primes, one (at least) of the determinants of the unaugmented matrix is prime to P, one (at least) is prime to Q, &c.; whence it may be inferred that the system is resoluble for each of the modules P, Q, R, ..., and admits of only one solution for each of them; it is therefore resoluble for their product $P\times Q\times R\dots$, and admits of only one solution for that modulus.

Let $\|K\|$ denote (as in the preceding article) a given matrix of the type $n\times(n+m)$, of which k is the greatest divisor; and let it be required to find the complete solution of the symbolic equation

$$\|K\|=\|k\|\times\|A\|, \quad . \quad . \quad . \quad . \quad . \quad . \quad . \quad . \quad (4)$$

in which $\|k\|$ is a square matrix of which the determinant is k, $\|A\|$ a prime matrix of the same type as $\|K\|$, and in which the constituents of $\|A\|$ and $\|k\|$ are the unknown numbers.

We shall first obtain a particular solution of this equation, and then show how from any particular solution the complete solution may be deduced.

We may suppose that the constituents of any horizontal row of $\|K\|$ admit of no common divisor but unity; for, if δ_1, δ_2, ..., δ_n be the greatest common divisors of the constituents of the horizontal rows of $\|K\|$, we find

$$\|K\|=\|\delta_1, \delta_2, \delta_3, \dots, \delta_n\|\times\|K'\|, \quad . \quad . \quad . \quad . \quad . \quad . \quad (5)$$

$\|K'\|$ denoting a matrix the constituents of which are derived from those of $\|K\|$ by the relation

$$K'_{r,s} = \frac{1}{\delta_r} K_{r,s}; \quad \ldots \quad (6)$$

so that the solution of equation (4) depends on the solution of a similar equation for the matrix $\|K'\|$, in which the constituents of each horizontal row are relatively prime. Let then the matrix $\left\| \begin{matrix} r \times (n+m) \\ K \end{matrix} \right\|$, *i.e.* the matrix

$$\left\| \begin{matrix} K_{1,1}, & K_{1,2}, & \ldots, & K_{1,n+m} \\ K_{2,1}, & K_{2,2}, & \ldots, & K_{2,n+m} \\ \cdot & \cdot & \cdot & \cdot \\ K_{r,1}, & K_{r,2}, & \ldots, & K_{r,n+m} \end{matrix} \right\| \quad [1 \leqq r < n],$$

be a prime matrix, but let the matrix $\left\| \begin{matrix} (r+1) \times (n+m) \\ K \end{matrix} \right\|$ admit of a greatest divisor μ. Determine $\omega_1, \omega_2, \ldots, \omega_r$ by the system of congruences

$$\left. \begin{matrix} K_{1,i}\,\omega_1 + K_{2,i}\,\omega_2 + \ldots + K_{r,i}\,\omega_r \equiv K_{r+1,i}, \bmod \mu \\ i = 1, 2, 3 \ldots, n+m \end{matrix} \right\} \quad \ldots \quad (7)$$

(which, as we have just seen, is always resoluble), and in $\|K\|$ replace the constituents $K_{r+1,i}$ by the numbers

$$\frac{1}{\mu}[K_{r+1,i} - \sum_{s=1}^{s=r} \omega_s K_{s,i}];$$

we thus deduce from $\|K\|$ another matrix $\|K''\|$ connected with it by the relation $|K| = \mu \times |K''|$, and such that the matrix of its first $r+1$ horizontal rows is prime. By proceeding in this manner, we shall at last obtain a prime matrix $\|A_0\|$, which satisfies the equation $|K| = k \times |A_0|$; we may then, by the method of the last article, determine a square matrix $\|k_0\|$ satisfying the equation

$$\|K\| = \|k_0\| \times \|A_0\|, \quad \ldots \quad (8)$$

and thus obtain a particular solution of the proposed equation (4).

To deduce the general solution of that equation, let $\|k_1\|$ and $\|A_1\|$ be any two matrices satisfying it. We have therefore the equality

$$\|k_1\| \times \|A_1\| = \|k_0\| \times \|A_0\|, \quad \ldots \quad (9)$$

which evidently implies that $|A_1| = |A_0|$; . . . . (10)

whence, by the theorem of the last article,

$$\|A_1\| = \|a\| \times \|A_0\|, \quad \ldots \quad (11)$$

$\|\alpha\|$ denoting a unit-matrix. Combining (9) and (11), we find

$$\|k_1\| \times \|\alpha\| \times \|A_0\| = \|k_0\| \times \|A_0\|; \quad \text{(12)}$$

whence, by the same theorem, it follows that

$$\|k_1\| \times \|\alpha\| = \|k_0\|, \quad \text{(13)}$$

or, which is the same thing, $\|k_1\| = \|k_0\| \times \|\alpha\|^{-1}$, ... (14)

$\|\alpha\|^{-1}$ denoting the matrix reciprocal to $\|\alpha\|$. The complete solution of equation (4) is therefore contained in the formulae

$$\left.\begin{aligned} \|A\| &= \|\ \alpha\ \| \times \|A_0\| \\ \|K\| &= \|K_0\| \times \|\ \alpha\ \|^{-1} \end{aligned}\right\}, \quad \text{(15)}$$

$\|\alpha\|$ denoting an arbitrary unit-matrix of the type $n \times n$, and $\|A_0\|$, $\|k_0\|$ being any two matrices that satisfy the equation.

In this, as in the preceding article, we have, for simplicity, excluded the case in which $m = 0$ and the matrices $\|K\|$ and $\|A\|$ are squares; but it is readily seen that no exception is presented by this particular case.

4. Let

$$\left.\begin{gathered} A_{i,1} x_1 + A_{i,2} x_2 + \ldots + A_{i,n+m} x_{n+m} = 0 \\ i = 1, 2, 3, \ldots, n \end{gathered}\right\} \quad \text{(16)}$$

represent a system of indeterminate equations of which the matrix is $\|A\|$. We shall suppose that the determinants of $\|A\|$ are not all equal to zero, *i.e.* that the system is independent; so that its *index of indeterminates* (or the excess of the number of indeterminates above the number of really independent equations) is m. If we take r solutions of the system, for example the solutions

$$\left.\begin{gathered} x_{s,1}, x_{s,2}, x_{s,3}, \ldots, x_{s,n+m} \\ s = 1, 2, 3, \ldots, r \end{gathered}\right\}, \quad \text{(17)}$$

it is evident that, if $r > m$, the determinants of the matrix $\|x\|$ are all equal to zero. If $r \leqq m$, and if the determinants of the matrix $\|x\|$ be not all equal to zero, the solutions (17) are said to form a *set of* r *independent solutions*; if $r = m$, they form a *complete set of independent solutions.* A *set of relatively prime solutions* is an independent set of which the matrix is prime; a complete set of relatively prime solutions may be called, for a reason which will presently appear, a *fundamental set of solutions.* It is always possible, in an infinite number of ways, to assign complete sets of independent solutions of a system of equations of the form (16). Among the methods by which this may be accomplished we shall select one which depends on the following principle:—

If $\left\| \begin{matrix} r \times (m+r) \\ a \end{matrix} \right\|$ represent any matrix of the type $r \times (m+r)$ the determinants of which are not all equal to zero, and if $\lambda_1, \lambda_2, \ldots, \lambda_{m+r}$ be integers which satisfy the equations

$$\left. \begin{matrix} \sum_{k=1}^{k=m+r} a_{i,k} \lambda_k = 0 \\ i = 1, 2, 3, \ldots, r \end{matrix} \right\}, \qquad (18)$$

while $a_{r+1,1}, a_{r+1,2}, a_{r+1,3}, \ldots, a_{r+1,m+r}$ are integers satisfying the inequality

$$\sum_{k=1}^{k=m+r} a_{r+1,k} \lambda_k \gtrless 0, \qquad (19)$$

the determinants of the matrix

$$\left\| \begin{matrix} (r+1) \times (m+r) \\ a \end{matrix} \right\|$$

are not all equal to zero.

For, if $\sum_{k=1}^{k=m+r} a_{r+1,k} \lambda_k = \theta$, it is evident that, by combining this equation with the equation (18), we may express each of the determinants

$$\theta \times \left| \begin{matrix} r \times (m+r) \\ a \end{matrix} \right|$$

in succession as a linear function of the determinants of $\left\| \begin{matrix} (r+1) \times (m+r) \\ a \end{matrix} \right\|$. If, therefore, the former determinants do not all vanish, neither can the latter.

Let, then, $a_{m,1}, a_{m,2}, \ldots, a_{m,n+m}$ represent any particular solution (other, of course, than that in which every indeterminate is equal to zero) of the system (16); and let $A_{n+1,1}, A_{n+1,2}, \ldots, A_{n+1,n+m}$ be integral numbers satisfying the inequality

$$\sum_{k=1}^{k=n+m} A_{n+1,k} a_{m,k} \gtrless 0; \qquad (20)$$

the system

$$\left. \begin{matrix} A_{i,1} x_1 + A_{i,2} x_2 + \ldots + A_{i,n+m} x_{n+m} = 0 \\ i = 1, 2, 3, \ldots, n+1 \end{matrix} \right\}. \qquad (21)$$

which is obtained by the addition of a single equation to the system (16), is itself an independent system, as appears from the principle just enunciated; its index of indeterminateness is therefore $m-1$. Let $\left\| \begin{matrix} (m-1) \times (n+m) \\ a \end{matrix} \right\|$ represent a complete set of independent solutions of (21); it may then be inferred, from a second application of the same principle, that $\left\| \begin{matrix} m \times (n+m) \\ a \end{matrix} \right\|$

represents a complete set of independent solutions of the proposed system (16). Thus the determination of a complete set of independent solutions of a system of which the index of indeterminateness is m depends on the determination of a similar set of solutions for a system of which the index is lower by a unit. By successive reductions, therefore, we shall at last arrive at a system of which the index of indeterminateness is unity, the complete solution of which is of course immediately found by evaluating the determinants of its matrix.

The practical application of this method supposes only that we can always assign a *particular* solution of a system of the form (16) or (21). And this, it may be observed, can always be done, either by trial or by other obvious and not unsymmetrical expedients.

5. If $\|a\|$ represent the matrix of a complete set of independent solutions of the proposed system (16), and $\|b\|$ be any matrix of the same type as $\|a\|$ and connected with $\|a\|$ by the equation

$$\|b\| = \|k\| \times \|a\|, \qquad (22)$$

in which $\|k\|$ denotes a square matrix of which the determinant is not zero, it is evident that the constituents of $\|b\|$ are also a complete set of independent solutions. And, conversely, if $\|b\|$ be the matrix of a complete set of independent solutions, $\|a\|$ is also the matrix of a similar set. For, if $\|K\|$ be the matrix composed of the first minors of $\|k\|$, so that

$$\|K\| \times \|k\| = \|k, k, k, \ldots\|,$$

we have from (22),

$$\|K\| \times \|b\| = \|k, k, \ldots, k, k\| \times \|a\|;$$

from which it appears that $\|k, k, \ldots\| \times \|a\|$, and therefore $\|a\|$ itself, is the matrix of an independent set of solutions.

This observation enables us to obtain a complete set of relatively prime solutions, as soon as we have obtained an independent set. If $\|b\|$ be the matrix of the independent set, we have only to determine, by the method of Art. 3, a square matrix $\|k\|$ and an oblong prime matrix $\|a\|$, satisfying the equation

$$\|b\| = \|k\| \times \|a\|;$$

the constituents of $\|a\|$ are then the terms of a set of fundamental solutions.

Or again, if in Art. 4 we employ, instead of the inequality (19), the equation

$$\sum_{k=1}^{k=n+m} a_{r+1,k}\,\lambda_k = 1, \qquad (23)$$

it is easily shown that, if $\left\|\begin{matrix} r \times (n+m) \\ a \end{matrix}\right\|$ be a prime matrix, $\left\|\begin{matrix} (r+1) \times (n+m) \\ a \end{matrix}\right\|$ is also a prime matrix; so that, by following the method of that article, we may obtain directly a set of fundamental solutions of any proposed system. Only it will be observed that, in this mode of obtaining such a set, we suppose that we can assign particular solutions, not only of systems of the form (16), but also of equations of the form (23).

6. The importance of fundamental sets of solutions in the theory of linear indeterminate equations is evident from the following proposition:—

'If $\|a\|$ represent a set of fundamental solutions of the system (16), the complete solution of that system is contained in the formula

$$\left.\begin{array}{c} x_i = \sum_{k=1}^{k=m} \xi_k \, a_{k,\,i} \\ i = 1, 2, 3, \ldots, n+m \end{array}\right\}, \qquad (24)$$

in which $\xi_1, \xi_2, \ldots, \xi_m$ are absolutely indeterminate integral numbers.'

For it is evident that every set of numbers included in (24) satisfies (16); and, conversely, if $a_{m+1,1}, a_{m+1,2}, \ldots, a_{m+1,n+m}$ be any solution of (16), the determinants of the matrix $\left\|\begin{array}{c}(m+1)\times(n+m)\\ a\end{array}\right\|$ are all zero, while the matrix $\left\|\begin{array}{c}m\times(n+m)\\ a\end{array}\right\|$ is prime; whence, by a principle employed in Art. 2, the system

$$\begin{array}{c} a_{m+1,\,i} = \sum_{k=1}^{k=m} \xi_k \, a_{k,\,i} \\ i = 1, 2, 3, \ldots, n+m \end{array}$$

is satisfied by one, and only one, system of integral values of $\xi_1, \xi_2, \ldots, \xi_m$; or, which is the same thing, the numbers $a_{m+1,1}, a_{m+1,2}, \ldots, a_{m+1,n+m}$ are included in the formula (24).

It may be added that no fractional values of $\xi_1, \xi_2, \ldots, \xi_m$ can give integral values to $x_1, x_2, \ldots, x_{n+m}$; and that the same values of $x_1, x_2, x_3, \ldots, x_{n+m}$ cannot arise from different values of $\xi_1, \xi_2, \ldots, \xi_m$.

The converse of the proposition just established is also true; *i.e.*

'If the formula

$$x_i = \sum_{k=1}^{k=m} \xi_k \, a_{k,\,i} \qquad (24)$$

represent every solution of an indeterminate system of equations, the matrix $\|a\|$ is a prime matrix.'

For, if $\|b\|$ represent a set of fundamental solutions of the indeterminate system, we may express the constituents of $\|b\|$ as linear functions of the constituents of $\|a\|$, by means of the equations (24), so as to obtain an equation of the form

$$\|b\| = \|\xi\| \times \|a_1\|,$$

$\|\xi\|$ denoting a square matrix; whence it immediately appears that $\|a\|$ is a prime matrix, and $\|\xi\|$ a unit-matrix.

Thus, if we apply Euler's method for the resolution of indeterminate equations to the system (16), we obtain, as the final result of the process, a system of equations of the form (24); and, as it is demonstrable from the nature of the method itself that these final equations contain the complete solution of the proposed system, their matrix is a prime matrix.

If $\|a\|$ and $\|b\|$ be any two sets of fundamental solutions of the same system, we shall have the equation

$$\|b\| = \|\xi\| \times \|a\|,$$

$\|\xi\|$ denoting a unit-matrix. The matrices, therefore, of all sets of fundamental solutions are deducible, by premultiplication with unit-matrices, from the matrix of any given set of such solutions.

7. If $\|a\|$ and $\|b\|$ represent two complete sets of independent solutions of the same system, the determinants of $\|a\|$ and $\|b\|$ are evidently connected by the relation

$$\beta \times |a| = \pm \alpha \times |b|,$$

α and β denoting the greatest divisors of $\|a\|$ and $\|b\|$ respectively. A similar relation subsists between the matrix of the system and the matrix of any complete set of independent solutions of it.

Let $\|A\|$ and $\|a\|$ represent those matrices respectively, K and k their greatest divisors; the relation in question is expressed by the formula

$$k \times |A| = K \times |a|, \quad \ldots \ldots \ldots \quad (25)$$

where it is to be remembered that the types of the matrices $\|A\|$ and $\|a\|$ are complementary; so that, as has been already observed (see Art. 1), there is an ambiguity of sign in the equation (25).

To obtain the demonstration, let Q and q denote the sums of the squares of the determinants of $\|A\|$ and $\|a\|$ respectively, and consider the determinant $\begin{vmatrix} A \\ a \end{vmatrix}$. This determinant is certainly not zero, for multiplying it by itself, we find

$$\begin{vmatrix} A \\ a \end{vmatrix}^2 = Q \times q. \quad \ldots \ldots \ldots \quad (26)$$

Let, then, $\begin{vmatrix} A \\ a \end{vmatrix}$ be multiplied by any determinant of $\|A\|$; for example, by

$$\Sigma \pm A_{1,1} A_{2,2} \ldots A_{n,n}.$$

Observing that $\Sigma \pm A_{1,1} A_{2\,2} \ldots A_{n,n}$ may assume the form

$$\begin{vmatrix} A_{1,1} & , A_{2,1} & , \ldots, A_{n,1} & , 0, 0, \ldots, 0 \\ A_{1,2} & , A_{2,2} & , \ldots, A_{n,2} & , 0, 0, \ldots, 0 \\ \cdot & \cdot & \cdot & \cdot \\ A_{1,n} & , A_{2,n} & , \ldots, A_{n,n} & , 0, 0, \ldots, 0 \\ A_{1,n+1}, & A_{2,n+1}, & \ldots, A_{n,n+1}, & 1, 0, \ldots, 0 \\ A_{1,n+2}, & A_{2,n+2}, & \ldots, A_{n,n+2}, & 0, 1, \ldots, 0 \\ \cdot & \cdot & \cdot & \cdot \\ A_{1,n+m}, & A_{2,n+m}, & \ldots, A_{n,n+m}, & 0, 0, \ldots, 1 \end{vmatrix},$$

we obtain the equation

$$\begin{vmatrix} A \\ a \end{vmatrix} \times \Sigma \pm A_{1,1}\, A_{2,2} \ldots A_{n,n} = Q \times \Sigma \pm a_{1,n+1}\, a_{2,n+2} \ldots a_{m,n+m}, \quad . \quad . \quad . \quad (27)$$

in which we may permute the second set of indices in any manner consistent with the condition that $\begin{vmatrix} A \\ a \end{vmatrix}$ should not change its sign; so that we may write

$$\begin{vmatrix} A \\ a \end{vmatrix} \times |A| = Q \times |a|, \quad . \quad . \quad . \quad . \quad . \quad . \quad . \quad . \quad . \quad . \quad (28)$$

the correspondence of the determinants in $|A|$ and $|a|$ being fixed by the matrix $\begin{Vmatrix} A \\ a \end{Vmatrix}$. The equation (25) is an immediate consequence of this result; and, if in that equation we suppose the correspondence of the determinants to be still fixed by the matrix $\begin{Vmatrix} A \\ a \end{Vmatrix}$, we shall have to write

$$k \times |A| = K \times |a|$$

or

$$k \times |A| = -K \times |a|,$$

according as $\begin{vmatrix} A \\ a \end{vmatrix}$ is a positive or a negative number.

8. From the preceding principles we may deduce the solution of the following problem, which admits of important applications in other parts of arithmetic:—

'To find all the matrices of a given type, of which the determinants have given values, not all equal to zero.'

Two particular cases of this problem (those in which the matrices are of the types 2×3 and 2×4) occur in the 'Disquisitiones Arithmeticae' (see Arts. 279 and 236). In both places Gauss has suppressed the analysis of the problem, and has only given a synthetical demonstration that its conditions are satisfied by the solution he assigns. This, indeed, in Art. 279, he expressly observes. He has also suppressed his method of deducing the complete solution from any particular solution,—an omission, however, which may probably be supplied by a comparison of Art. 234 with Art. 213, I. The very general and most important case of a matrix of the type $n \times (n+1)$ has been subsequently treated by M. Hermite *.

Let $\|x\|$ denote a matrix of the type $n \times (n+m)$, of which the constituents are absolutely indeterminate quantities; writing λ for $\frac{\Pi(n+m)}{\Pi n \,.\, \Pi m}$, we shall represent its determinants by $X_1, X_2, \ldots, X_\lambda, \ldots$. If $m > 1$, these determinants are not all independent, but are connected by certain identities of the form

$$\Phi(X_1, X_2, \ldots, X_\lambda) = 0, \quad \ldots \ldots \quad (29)$$

Φ denoting a rational and integral homogeneous function with numerical coefficients. If, therefore, $C_1, C_2, \ldots, C_\lambda$ be a given set of integral numbers, which can be represented as the determinants of a matrix of the type $n \times (n+m)$, these numbers will satisfy every relation of the form (29); so that the identity

$$\Phi(X_1, X_2, \ldots, X_\lambda) = 0$$

will involve also the numerical equation

$$\Phi(C_1, C_2, \ldots, C_\lambda) = 0. \quad \ldots \ldots \quad (30)$$

To obtain a convenient notation for $C_1, C_2, \ldots, C_\lambda$, let us imagine that we have formed a square matrix of the type $(n+m) \times (n+m)$ by the addition of m horizontal rows to the matrix $\|x\|$; if, in the development of the determinant of this matrix, the coefficient of X_i be the determinant

$$|x_{n+r,\mu_s}|, \quad \begin{matrix} r = 1, 2, 3, \ldots, m \\ s = 1, 2, 3, \ldots, m \end{matrix}$$

($\mu_1, \mu_2, \ldots, \mu_m$ denoting m of the numbers $1, 2, \ldots, n+m$), we may represent X_i and C_i by the symbols $[\mu_1, \mu_2, \ldots, \mu_m]$ and $(\mu_1, \mu_2, \ldots, \mu_m)$ respectively; observing, however, that, if two of the numbers $\mu_1, \mu_2, \ldots$ are equal, the value zero is to be attributed to each of these symbols.

If r denote one of the numbers $1, 2, 3, \ldots, n$, the determinants of the matrix obtained by adding the horizontal row

$$x_{r,1}, \; x_{r,2}, \; \ldots\ldots, \; x_{r,n+m}$$

to the matrix $\left\| \begin{matrix} n \times (n+m) \\ x \end{matrix} \right\|$ are identically equal to zero. We thus obtain $\lambda \times \frac{m}{n+1}$ equations of the form

* Crelle's Journal, vol. xl. p. 264; see also Eisenstein, ibid. vol. xxviii. p. 327.

$$\sum_{i=1}^{i=m+n} [i, \mu_1, \mu_2, \ldots, \mu_{m-1}] x_{r,i} = 0, \quad . \quad . \quad . \quad . \quad . \quad . \quad . \quad . \quad (31)$$

$\mu_1, \mu_2, \ldots, \mu_{m-1}$ representing any combination of $m-1$ of the numbers

$$1, 2, 3, \ldots, m+n.$$

In connexion with these equations, consider also the similarly formed system,

$$\sum_{i=1}^{i=m+n} (i, \mu_1, \mu_2, \ldots, \mu_{m-1}) y_i = 0. \quad . \quad . \quad . \quad . \quad . \quad . \quad . \quad . \quad (32)$$

This system, which is in appearance redundant (containing $\lambda \times \frac{m}{n+1}$ equations, and only $m+n$ indeterminates), is in reality defective, and is equivalent to m independent equations. For, if $(k_1, k_2, \ldots, k_m)$ be one of the given numbers C which is not equal to zero, the partial system of m equations

$$\left.\begin{array}{c} \sum_{i=1}^{i=m+n} (i, k_1, k_2, \ldots, k_{j-1}, k_{j+1}, \ldots, k_m) y_i = 0 \\ j = 1, 2, 3, \ldots, m \end{array}\right\} \quad . \quad . \quad . \quad . \quad . \quad . \quad (33)$$

is certainly an independent system, because the determinant of the coefficients of $y_{k_1}, y_{k_2}, \ldots, y_{k_m}$ is $(k_1, k_2, \ldots, k_m)^m$, and is therefore different from zero. Again, every equation of (32) which is not already comprised in (33) may be obtained by linearly combining the equations of that partial system. To verify this assertion, let

$$\left.\begin{array}{c} \Sigma [i, \mu_1, \mu_2, \ldots, \mu_{m-1}] x_{r,i} = 0 \\ r = 1, 2, 3, \ldots, n \end{array}\right\} \quad . \quad . \quad . \quad . \quad . \quad . \quad . \quad . \quad (34)$$

be the system of n equations obtained by attributing to r the n values of which it is susceptible in any one of the equations (31). Eliminating from this system those $n-1$ determinants $[i, \mu_1, \mu_2, \ldots, \mu_{m-1}]$ in which i has a value *not* included in a set of $m+1$ numbers $\nu_1, \nu_2, \ldots, \nu_{m+1}$ arbitrarily selected from the series $1, 2, 3, \ldots, n+m$, we obtain a relation which may be expressed in the form

$$\sum_{i=1}^{i=m+1} (-1)^i [\nu_i, \mu_1, \mu_2, \ldots, \mu_{m-1}] \times [\nu_1, \nu_2, \ldots, \nu_{i-1}, \nu_{i+1}, \ldots, \nu_{m+1}] = 0, \quad . \quad . \quad (35)$$

representing $\frac{mn\lambda^2}{(m+1)(n+1)}$ equations, since the sets

$$\mu_1, \mu_2, \ldots, \mu_{m-1},$$

$$\nu_1, \nu_2, \ldots, \nu_{m+1}$$

may respectively denote any sets of $m-1$ and $m+1$ numbers taken from the series $1, 2, \ldots, m+n$. Since (35) is of the form $\Phi(X_1, X_2, \ldots, X_\lambda) = 0$, we may at once infer the corresponding relation

$$\sum_{i=1}^{i=m+1} (-1)^i (\nu_i, \mu_1, \mu_2, \ldots, \mu_{m-1}) \times (\nu_1, \nu_2, \ldots, \nu_{i-1}, \nu_{i+1}, \ldots, \nu_{m+1}) = 0, \quad . \quad . \quad (36)$$

by means of which any one of the equations (32) may be deduced from the equations of the partial system (33). Thus, if we multiply the equations of that system taken in order by the determinants $(-1)^j (k_j, h_1, h_2, \ldots, h_{m-1})$, and add the results, we obtain

$$(k_1, k_2, \ldots, k_m) \sum_{i=1}^{i=n+m} (i, h_1, h_2, \ldots, h_{m-1})\, y_i = 0,$$

i.e. since $(k_1, k_2, \ldots, k_m)$ is not zero,

$$\sum_{i=1}^{i=n+m} (i, h_1, h_2, \ldots, h_{m-1})\, y_i = 0.$$

The system (32) is therefore equivalent to a system of m independent equations.

Let $\left\| \begin{matrix} m \times (m+n) \\ \gamma \end{matrix} \right\|$ represent the matrix of (33), or of any other independent system equivalent to (32) (the determinants of all such matrices are proportional); let $\Gamma_1, \Gamma_2, \ldots, \Gamma_\lambda$ be the determinants of $\|\gamma\|$; $\|\xi\|$ and $\Xi_1, \Xi_2, \ldots, \Xi_\lambda$ the matrix and determinants of the system similarly derived from (31). By the theorem of Art. 7, we have

$$\left| \begin{matrix} \xi \\ x \end{matrix} \right| \times |\xi| = \Sigma \Xi^2 \times |x|, \quad . \quad . \quad . \quad . \quad . \quad . \quad . \quad . \quad (37)$$

or observing that $\left| \begin{matrix} \xi \\ x \end{matrix} \right| = \Sigma \Xi X$, and that (37) is an identity of the form $\Phi = 0$,

$$\Sigma \Gamma C \times |\gamma| = \Sigma \Gamma^2 \times |C|, \quad . \quad . \quad . \quad . \quad . \quad . \quad . \quad . \quad (38)$$

where $|C|$ symbolizes the numbers $C_1, C_2, \ldots, C_\lambda$ which correspond to the determinants of $\|\gamma\|$ in the same inverse order in which, in equation (37), the determinants of $\|x\|$ correspond to those of $\|\xi\|$. But, if $\left\| \begin{matrix} n \times (m \times n) \\ \theta \end{matrix} \right\|$ represent a system of fundamental solutions of (32) or (33), we have also

$$\left| \begin{matrix} \gamma \\ \theta \end{matrix} \right| \times |\gamma| = \Sigma \Gamma^2 \times |\theta|; \quad . \quad . \quad . \quad . \quad . \quad . \quad . \quad . \quad . \quad (39)$$

whence, combining (38) and (39), and representing the greatest common divisor of $C_1, C_2, \ldots, C_\lambda$ by c, we find

$$c \times |\theta| = |C|. \quad . \quad . \quad . \quad . \quad . \quad . \quad . \quad . \quad . \quad (40)$$

If, then, $\|c\|$ denote any square matrix of determinant c and of the type $n \times n$, the formula $\|c\| \times \|\theta\|$ contains the complete solution of the problem.

If γ represent the greatest divisor of $\| \gamma \|$, we infer from (38)

$$c \times | \gamma | = \gamma \times | C |; \quad \ldots \ldots \ldots \quad (41)$$

whence, if $\| \gamma' \|$ be a prime matrix of the type $m \times (m+n)$ satisfying the equation $| \gamma | = \gamma \times | \gamma' |$ (see Art. 3), we find

$$| C | = c \times | \gamma' |. \quad \ldots \ldots \ldots \quad (42)$$

The preceding analysis enables us therefore to obtain simultaneously the representation of the determinants $| C |$ as the determinants of two complementary matrices of the types $n \times (m+n)$ and $m \times (m+n)$ respectively. We have thus two distinct methods of arriving at the solution of the problem; of these one requires the determination of a set of fundamental solutions of a system of linear equations, the other the reduction (by the method of Art. 3) of a given matrix to a prime matrix. The greatest divisor of $\| \gamma \|$, which we have represented by γ, is evidently $(k_1, k_2, \ldots, k_m)^{m-1} \times c$. If, therefore, C, one of the given numbers $C_1, C_2, \ldots, C_\mu$, be a unit, we have only to take C for $(k_1, k_2, \ldots, k_m)$, and we shall immediately obtain a matrix $\| \gamma \|$ of the type $m \times (m+n)$ satisfying the equation

$$| \gamma | = | C |.$$

And similarly might a matrix of the type $n \times (m+n)$, satisfying the same equation, be written down without calculation.

9. The importance of the case in which $m=1$ is so great that we may be allowed to point out the identity of the solution obtained by the preceding method with that already given by M. Hermite. Let, then, $C_1, C_2, \ldots, C_{n+1}$ represent the determinants of a matrix of the type $n \times (n+1)$ taken in their natural order (*i.e.* so taken that if the matrix be completed by an additional row of constituents, $c_1, c_2, \ldots, c_{n+1}$, the value of its determinant would be

$$c_1 C_1 + c_2 C_2 + c_3 C_3 + \ldots + c_{n+1} C_{n+1}).$$

We have then to obtain a set of fundamental solutions of the equation

$$C_1 y_1 + C_2 y_2 + C_3 y_3 + \ldots + C_{n+1} y_{n+1} = 0. \quad \ldots \ldots \quad (43)$$

Such a set may always be obtained by the following particular method. Supposing that C_1 is not zero, consider the equations

$$\left.\begin{array}{l} 0 = C_1 y_1 + C_2 y_2 \\ 0 = C_1 y_1 + C_2 y_2 + C_3 y_3 \\ \ldots \ldots \ldots \ldots \ldots \\ 0 = C_1 y_1 + C_2 y_2 + C_3 y_3 + \ldots + C_{n+1} y_{n+1} \end{array}\right\}, \quad \ldots \ldots \quad (44)$$

and take a particular solution of each of them, assigning to the last indeterminate in each the least value (zero excepted) of which it is susceptible. If we denote by Δ_k the greatest common divisor of $C_1, C_2, \ldots, C_k$, so that $\Delta_1 = C_1$, $\Delta_{n+1} = c$, it is evident that the value of y_{k+1} in the equation

$$C_1 y_1 + C_2 y_2 + \ldots + C_{k+1} y_{k+1} = 0$$

will be $\dfrac{\Delta_k}{\Delta_{k+1}}$; and, if in the same equation we represent the values of y_1, y_2, $y_3, \ldots, y_k$ by $r_{k,1}, r_{k,2}, r_{k,3}, \ldots, r_{k,k}$, the matrix

$$\left\|\begin{matrix} r_{1,1}, \dfrac{\Delta_1}{\Delta_2}, & 0, & 0 & . & . & . & . & . & 0 \\ r_{2,1}, & r_{2,2}, \dfrac{\Delta_2}{\Delta_3}, & 0 & . & . & . & . & . & 0 \\ r_{3,1}, & r_{3,2}, & r_{3,3}, \dfrac{\Delta_3}{\Delta_4} & . & . & . & . & . & 0 \\ . & . & . & . & . & . & . & . & 0 \\ r_{n,1}, & r_{n,2}, & r_{n,3}, & r_{n,4} & . & . & . & . & \dfrac{\Delta_n}{\Delta_{n-1}} \end{matrix}\right\| \quad \ldots \ldots \quad (45)$$

will represent a set of fundamental solutions of (43). For, in the first place, it represents a set of independent solutions, because its first determinant is

$$\frac{\Delta_1}{\Delta_2} \times \frac{\Delta_2}{\Delta_3} \times \ldots \times \frac{\Delta_n}{\Delta_{n+1}}, \text{ or } \frac{\Delta_1}{\Delta_{n+1}}, \text{ or } \frac{C_1}{c};$$

therefore its determinants are proportional to $C_1, C_2, \ldots$, &c.; or, since the first of them is $\dfrac{C_1}{c}$, they are respectively equal to the numbers

$$\frac{C_1}{c}, \frac{C_2}{c}, \frac{C_3}{c}, \ldots, \frac{C_{n+1}}{c},$$

which admit of no common divisor.

To obtain a set of values for the constituents $r_{i,j}$, which occur in the matrix (45), we may form the series of equations

$$\left.\begin{array}{r} \lambda_1 C_2 + \mu_1 \Delta_1 = \Delta_2 \\ \lambda_2 C_3 + \mu_2 \Delta_2 = \Delta_3 \\ \ldots\ldots\ldots \\ \lambda_{n-1} C_n + \mu_n \Delta_n = \Delta_{n+1} \end{array}\right\} \quad \ldots \ldots \quad (46)$$

It will then be found that the equations (44) are satisfied by the values of r comprised in the formula

$$r_{i,j} = -\lambda_{j-1} \mu_j, \ldots, \mu_{i-1} \frac{C_{i+1}}{\Delta_{i+1}} \quad [j \leqq i]; \quad \ldots \ldots \quad (47)$$

and on substituting these values in the matrix (45), it will coincide, after an unimportant modification, with that occurring in M. Hermite's solution of the problem.

But, in practice, the simplest method of obtaining a solution of the problem considered in this article is to solve the equation (43) by Euler's method, and to employ, in the place of the matrix (45), the matrix of the set of fundamental solutions thus obtained (see Art. 6).

10. Another problem, closely connected with the preceding, and of no less frequent application, has also been completely solved by M. Hermite*; but as it may serve to illustrate the utility of the methods employed in this paper, we shall venture to resume and generalize it here. The problem is:—

'Given a set of $n+1$ numbers $C_1, C_2, \ldots, C_{n+1}$ without any common divisor, to assign all the matrices $\|x\|$ of the type $n \times (n+1)$ which satisfy the equation

$$\begin{vmatrix} C \\ x \end{vmatrix} = 1.\text{'}$$

Let $c_1, c_2, \ldots, c_{n+1}$ be any particular solution of the equation

$$C_1 y_1 + C_2 y_2 + \ldots + C_{n+1} y_{n+1} = 1 \quad . \quad . \quad . \quad . \quad . \quad . \quad (48)$$

(which is always possible because $C_1, C_2, \ldots, C_{n+1}$ are relatively prime); and let $\|\gamma\|$ represent a set of fundamental solutions of the equation

$$c_1 y_1 + c_2 y_2 + \ldots + c_{n+1} y_{n+1} = 0. \quad . \quad . \quad . \quad . \quad . \quad . \quad . \quad . \quad (49)$$

Then, if $\|u\|$ represent any unit-matrix of the type $n \times n$, and $\lambda_1, \lambda_2, \ldots, \lambda_n$ absolutely indeterminate integers, the complete solution of the problem is contained in the formula

$$\|u\| \times \|\gamma_{i,j} + \lambda_i C_j\| \qquad \left.\begin{matrix} i = 1, 2, 3, \ldots, n \\ j = 1, 2, 3, \ldots, n+1 \end{matrix}\right\}. \quad . \quad . \quad . \quad . \quad (50)$$

For, if $\|x\|$ be any one of the matrices contained in that formula, it is readily seen that

$$\begin{vmatrix} C \\ x \end{vmatrix} = \begin{vmatrix} C \\ \gamma \end{vmatrix} = C_1 c_1 + C_2 c_2 + \ldots + C_{n+1} c_{n+1} = 1.$$

Conversely, if $\|x\|$ be a matrix satisfying the equation $\begin{vmatrix} C \\ x \end{vmatrix} = 1$, $\|x\|$ is included in the formula (50). To show this, we observe that the complete solution of equation (48) is contained in the formula

$$y_j = c_j + \sum_{i=1}^{i=n} \theta_{i,j} \lambda_i, \quad j = 1, 2, \ldots, n+1, \quad . \quad . \quad . \quad . \quad . \quad . \quad (51)$$

---

* Liouville, vol. xiv. p. 21.

in which $\| \theta \|$ is *any* set of fundamental solutions of the equation

$$C_1 y_1 + C_2 y_2 + \ldots + C_{n+1} y_{n+1} = 0, \quad \ldots \quad (52)$$

and $\lambda_1, \lambda_2, \ldots, \lambda_n$ are indeterminate integers. The complete solution of the same equation (48) is therefore supplied by the determinants of the matrix $\| \gamma_{i,j} + \lambda_i C_j \|$. For those determinants may be represented by the formula

$$c_j + \sum_{i=1}^{i=n} [i, j] \lambda_i, \quad j = 1, 2, 3, \ldots, n+1,$$

in which $[i, j]$ symbolizes a first minor of the determinant $\begin{vmatrix} C \\ \gamma \end{vmatrix}$, so that

$$[i, j] = \frac{d \begin{vmatrix} C \\ \gamma \end{vmatrix}}{d\gamma_{i,j}}.$$

But the numbers $[i, 1], [i, 2], \ldots, [i, n+1]$ satisfy (52) for every value of i; and, since $\begin{vmatrix} C \\ \gamma \end{vmatrix} = 1$, the determinants of the matrix

$$\| [i, j] \| \quad \left. \begin{matrix} i = 1, 2, 3, \ldots, n \\ j = 1, 2, 3, \ldots, n+1 \end{matrix} \right\} \quad \ldots \quad (53)$$

are the numbers $C_1, C_2, \ldots, C_{n+1}$, and are therefore relatively prime. It follows from this that (53) represents a set of fundamental solutions of (52); *i.e.* that the complete solution of (48) is represented by the determinants of $\| \gamma_{i,j} + \lambda_i C_j \|$. If, then, $\| x \|$ be a matrix satisfying the equation $\begin{vmatrix} C \\ x \end{vmatrix} = 1$, since the determinants of $\| x \|$ evidently satisfy (48), values can be assigned to $\lambda_1, \lambda_2, \ldots, \lambda_n$ which shall verify the equation

$$| \gamma_{i,j} + \lambda_i C_j | = | x |;$$

whence it follows that

$$\| x \| = \| u \| \times \| \gamma_{i,j} + \lambda_i C_j \|,$$

$\| u \|$ denoting a unit-matrix, *i.e.* $\| x \|$ is one of the matrices included in the formula (50).

The result, incidentally obtained in the foregoing analysis, that the complete solution of an equation of the form

$$C_1 x_1 + C_2 x_2 + \ldots + C_{n+1} x_{n+1} = 1$$

can be exhibited in the determinantal form (50) is occasionally useful.

The preceding problem is a particular case of the following more general enunciation :—

'Given a prime matrix $\| C \|$ of the type $m \times (m+n)$, to find all the matrices $\| x \|$ of the type $n \times (m+n)$ which satisfy the equation

$$\begin{vmatrix} C \\ x \end{vmatrix} = 1.\text{'} \quad \ldots \quad (54)$$

Let $\|\gamma\|$ be a matrix which satisfies (54), let the numbers $\mu_{i,j}$ represent absolute indeterminates, and $\|u\|$ any unit-matrix; the complete solution of the problem is contained in the formula

$$\|x\| = \|u\| \times \|\gamma + \Sigma\mu C\|, \quad . \quad . \quad . \quad . \quad . \quad . \quad . \quad . \quad . \quad . \qquad (55)$$

where $\|\gamma + \Sigma\mu C\|$ represents the matrix,

$$\left\| \gamma_{i,j} + \sum_{\theta=1}^{\theta=m} \mu_{i,\theta} C_{\theta,j} \right\| \qquad \left. \begin{matrix} i = 1, 2, 3, \ldots, n \\ j = 1, 2, 3, \ldots, n+m \end{matrix} \right\}.$$

For if $\|x\|$ be a matrix satisfying the equation (54), we have

$$\begin{vmatrix} C \\ x \end{vmatrix} = 1 = \begin{vmatrix} C \\ \gamma \end{vmatrix};$$

and consequently
$$\begin{Vmatrix} C \\ x \end{Vmatrix} = \|v\| \times \begin{Vmatrix} C \\ \gamma \end{Vmatrix},$$

$\|v\|$ denoting a unit of the type $(m+n) \times (m+n)$. But, because the first m horizontal rows in $\begin{Vmatrix} C \\ x \end{Vmatrix}$ and $\begin{Vmatrix} C \\ \gamma \end{Vmatrix}$ are identical, it is evident that

$$v_{i,j} = 0 \qquad \left. \begin{matrix} i = 1, 2, 3, \ldots, m \\ j = 1, 2, 3, \ldots, m+n \end{matrix} \right\},$$

except when $i = j$, in which case

$$v_{1,1} = v_{2,2} = \ldots = v_{m,m} = 1.$$

The unit-matrix $\|v\|$ therefore arises from the composition of two unit-matrices of the forms

$$\begin{Vmatrix} 1 & , 0 & , 0 & , \ldots, & 0 & , 0, & 0, & \ldots, & 0 \\ 0 & , 1 & , 0 & , \ldots, & 0 & , 0, & 0, & \ldots, & 0 \\ 0 & , 0 & , 1 & , \ldots, & 0 & , 0, & 0, & \ldots, & 0 \\ \cdot & \cdot & \cdot & \cdot & \cdot & \cdot & \cdot & \cdot & \cdot \\ \lambda_{1,1}, & \lambda_{1,2}, & \lambda_{1,3}, & \ldots, & \lambda_{1,m}, & 1, & 0, & \ldots, & 0 \\ \lambda_{2,1}, & \lambda_{2,2}, & \lambda_{2,3}, & \ldots, & \lambda_{2,m}, & 0, & 1, & \ldots, & 0 \\ \cdot & \cdot & \cdot & \cdot & \cdot & \cdot & \cdot & \cdot & \cdot \\ \lambda_{n,1}, & \lambda_{n,2}, & \lambda_{n,3}, & \ldots, & \lambda_{n,m}, & 0, & 0, & \ldots, & 1 \end{Vmatrix}$$

and

$$\begin{Vmatrix} 1, & 0, & 0, & \ldots, & 0 & , 0 & , \ldots, & 0 \\ 0, & 1, & 0, & \ldots, & 0 & , 0 & , \ldots, & 0 \\ 0, & 0, & 1, & \ldots, & 0 & , 0 & , \ldots, & 0 \\ \cdot & \cdot & \cdot & \cdot & \cdot & \cdot & \cdot & \cdot \\ 0, & 0, & 0, & \ldots, & u_{1,1}, & u_{1,2}, & \ldots, & u_{1,n} \\ 0, & 0, & 0, & \ldots, & u_{2,1}, & u_{2,2}, & \ldots, & u_{2,n} \\ \cdot & \cdot & \cdot & \cdot & \cdot & \cdot & \cdot & \cdot \\ 0, & 0, & 0, & \ldots, & u_{n,1}, & u_{n,2}, & \ldots, & u_{n,n} \end{Vmatrix},$$

$\|\lambda\|$ denoting a matrix of the type $n\times m$ of which the constituents may be any numbers whatever, and $\|u\|$ a unit-matrix of the type $n\times n$. If for $\|\lambda\|$ we substitute the matrix

$$\|\mu\| = \|u\|^{-1}\times\|\lambda\|,$$

it is readily seen that we may invert the order of the factors in the expression of $\|v\|$; so that, using an abbreviated notation, the signification of which is evident, we may write either

$$\|v\| = \left\|\begin{matrix}1, & 0\\ \lambda, & 1\end{matrix}\right\| \times \left\|\begin{matrix}1, & 0\\ 0, & u\end{matrix}\right\|,$$

or

$$\|v\| = \left\|\begin{matrix}1, & 0\\ 0, & u\end{matrix}\right\| \times \left\|\begin{matrix}1, & 0\\ \mu, & 1\end{matrix}\right\|.$$

Substituting the latter expression of $\|v\|$ in the equation

$$\left\|\begin{matrix}C\\ x\end{matrix}\right\| = \|v\| \times \left\|\begin{matrix}C\\ \gamma\end{matrix}\right\|,$$

we immediately infer

$$\|x\| = \|u\|\times\|\gamma + \Sigma\mu C\|.$$

Every matrix satisfying the equation $\left|\begin{matrix}C\\ x\end{matrix}\right| = 1$ is therefore comprised in the formula (55); and since it is evident, conversely, that every matrix comprised in (55) satisfies the equation, that formula contains the complete solution of the question.

A particular solution of the problem (which may be taken for $\|\gamma\|$) can be obtained as follows:—Complete the matrix $\|C\|$ by any n horizontal rows of constituents which do not cause the determinant of the resulting matrix to vanish. From this matrix a prime (*i.e.* a unit) matrix of the same type is to be deduced by the method of Art. 3, a reduction which can always be effected without changing the prime matrix $\|C\|$.

11. The consideration of sets of fundamental solutions of linear systems is also of use in the theory of indeterminate systems containing terms not affected by any indeterminate. Let

$$\left.\begin{matrix}A_{i,0} + A_{i,1}x_1 + A_{i,2}x_2 + \dots + A_{i,n+m}x_{n+m} = 0\\ i = 1, 2, 3, \dots, n\end{matrix}\right\} \quad . \quad . \quad . \quad . \quad . \quad (56)$$

represent such a system; its general solution will assume the form

$$\left.\begin{matrix}x_k = a_k + \sum_{\theta=1}^{\theta=m} \mu_\theta\, a_{k,\theta}\\ k = 1, 2, 3, \dots, n+m\end{matrix}\right\}, \quad . \quad . \quad . \quad . \quad . \quad . \quad . \quad . \quad . \quad (57)$$

where $a_1, a_2, \dots, a_{n+m}$ is a particular solution of (56), $\mu_1, \mu_2, \dots, \mu_m$ indeterminate numbers, and $\|\alpha\|$ a set of fundamental solutions of the system

$$\left.\begin{array}{c} A_{i,1}\,x_1 + A_{i,2}\,x_2 + \ldots + A_{i,\,n+m}\,x_{n+m} = 0 \\ i = 1,\ 2,\ 3,\ \ldots,\ n \end{array}\right\} \quad . \quad . \quad . \quad . \quad . \quad . \quad (58)$$

Whenever, therefore, the proposed system is resoluble, its complete solution involves m indeterminates; but in order that it should be resoluble, a certain condition must be satisfied by its coefficients. This condition is 'that the greatest divisors of its augmented and unaugmented matrices must be equal *.' We shall call these divisors D and D_0 respectively, representing the matrices themselves by $\|A\|$ and $\|A_0\|$. That the condition is necessary may be seen by eliminating in turn every combination of $n-1$ indeterminates from (56). We thus find that every determinant of $\|A\|$ is divisible by D_0, *i.e.* that D is divisible by D_0; but evidently D divides D_0, so that $D_0 = D$. To show that the condition is sufficient, as well as necessary, consider the system

$$\left.\begin{array}{c} A_{i,0}\,x_0 + A_{i,1}\,x_1 + A_{i,2}\,x_2 + \ldots + A_{i,\,n+m}\,x_{n+m} = 0 \\ i = 1,\ 2,\ 3,\ \ldots,\ n \end{array}\right\}, \quad . \quad . \quad . \quad . \quad (59)$$

and let $\left\|\begin{array}{c}(m+1)\times(n+m+1)\\ \theta\end{array}\right\|$ represent a set of its fundamental solutions. To say that (56) is resoluble is the same thing as to say that (59) admits of solutions in which the value of x_0 is unity; and (59) will not, or will, admit of such solutions according as $\theta_{1,0}$, $\theta_{2,0}$, ..., $\theta_{m,0}$ do, or do not, admit of any common divisor beside unity. But, by the theorem of Art. 7, those determinants of $\|\theta\|$ into which the column $\theta_{1,0}$, $\theta_{2,0}$, ... enters, are equal to the determinants of $\|A_0\|$ taken in a proper order and divided by D. If $D = D_0$, the determinants of $\|A_0\|$,

---

* [This theorem has already been given by M. Ignaz Heger (Memoirs of the Vienna Academy, vol. xiv. second part, p. 111). I regret that in the abstract of the present paper, which has been inserted in the 'Proceedings of the Royal Society,' no reference was made to M. Heger's memoir, with the contents of which I was unacquainted at the time at which that abstract was prepared. M. Heger's demonstration (adapted to the terminology here employed) is, in the main, as follows. (1) If the unaugmented matrix of an indeterminate system be prime, the system is always resoluble. For, every determinate system, of which the matrix is a unit-matrix, is resoluble in integral numbers; and we may suppose the given indeterminate system to form part of such a determinate system (see Art. 10, *suprà*). (2) The equation $\|A\| = \|D\| \times \|A'\|$, in which $\|D\|$ is a square matrix, having D for its determinant, and $\|A'\|$ a prime matrix of the same type as $\|A\|$, is always resoluble (see Art. 3). We can therefore replace the given system (56) by a system of which the augmented matrix is $\|A'\|$, and which is resoluble or irresoluble at the same time with the given system. But if $D_0 = D$, the unaugmented matrix of this derived system is prime; *i.e.* if $D_0 = D$, the proposed system is resoluble. (3) That the condition is necessary as well as sufficient may be proved as in the text.—Sept. 1861, H. J. S. S.]

divided by D, are relatively prime, and consequently those determinants of $\|\theta\|$ which contain $\theta_{1,0}, \theta_{2,0}, \ldots, \theta_{m,0}$ are also relatively prime; a conclusion which implies that $\theta_{1,0}, \theta_{2,0}, \ldots, \theta_{m,0}$ are themselves relatively prime, *i.e.* that the system (56) is resoluble.

This criterion is not immediately applicable if the system (56) be not independent, *i.e.* if the determinants of its augmented matrix $\|A\|$ be all equal to zero. But it may be applied to any independent system equivalent to the proposed system and deduced linearly from it.

If we represent by D_k the greatest divisor of the matrix deduced from the matrix of (59) by omitting from it the column $A_{1,k}, A_{2,k}, \ldots, A_{n,k}$, we may enunciate the following proposition :—

'In every solution of the system (59), the value of x_k is divisible by $\frac{D_k}{D}$; and, conversely, a solution of that system can always be assigned in which x_k shall have any given value divisible by $\frac{D_k}{D}$.'

It will be seen that the solution of (56) depends, first, on the solution of (59), and, secondly, on that of the indeterminate equation

$$\theta_{1,0}\, x_1 + \theta_{2,0}\, x_2 + \ldots + \theta_{m+1,0}\, x_{m+1} = 1. \qquad (60)$$

If we represent the values of the indeterminates in this equation as the determinants of the matrix

$$\| \gamma_{i,j} + \mu_i\, \theta_{j,0} \| \qquad \left.\begin{matrix} i = 1, 2, 3, \ldots, m \\ j = 1, 2, \ \ldots, m+1 \end{matrix}\right\},$$

(see Art. 10), we may express the most general values of the indeterminates which satisfy (56) in the determinantal form

$$x_k = \begin{vmatrix} \theta_{1,k}, & \theta_{2,k}, & & \theta_{m+1,k} \\ \gamma_{1,1} + \mu_1\, \theta_{1,0}, & \gamma_{1,2} + \mu_1\, \theta_{2,0}, & \ldots, & \gamma_{1,m+1} + \mu_1\, \theta_{m+1,0} \\ \gamma_{2,1} + \mu_2\, \theta_{1,0}, & \gamma_{2,2} + \mu_2\, \theta_{2,0}, & \ldots, & \gamma_{2,m+1} + \mu_2\, \theta_{m+1,0} \\ \cdot & \cdot & \cdot & \cdot \\ \gamma_{m,1} + \mu_m\, \theta_{1,0}, & \gamma_{m,2} + \mu_m\, \theta_{2,0}, & \ldots, & \gamma_{m,m+1} + \mu_m\, \theta_{m+1,0} \end{vmatrix} \qquad (61)$$

12. We shall now indicate an important transformation of which any square matrix of integral numbers is susceptible. We begin with the following theorem :—

'If a given rectangular matrix be premultiplied by a unit-matrix, the greatest common divisor of any vertical column of minor determinants is the same in the resulting as in the given matrix.'

For it is evident that any minor, either in the given or in the resulting matrix, is an integral and linear function of the minors formed from the same vertical columns in the other matrix.

Similarly, it may be shown that :—

'When a square matrix is postmultiplied by any prime rectangular matrix, the greatest common divisor of any horizontal row of minors is the same in the resulting rectangular matrix as in the given square matrix.'

For, if
$$\left\| \begin{matrix} n \times (n+m) \\ A \end{matrix} \right\| = \left\| \begin{matrix} n \times n \\ B \end{matrix} \right\| \times \left\| \begin{matrix} n \times (n+m) \\ C \end{matrix} \right\|,$$
where $\|C\|$ is a prime matrix, it is clear that every minor of $\|A\|$ is a linear function of the minors formed from the same horizontal rows of $\|B\|$; so that, if a and b be the greatest common divisors of any corresponding horizontal rows of minors in those two matrices, a is divisible by b. But again, if θ be any one of the determinants of $\|C\|$, and s be the order of the minors under consideration, any minor of $\|B\|$, after multiplication by θ^s, may be expressed as a linear function of a certain group of the minors taken from the same horizontal rows of $\|A\|$. Consequently $\theta^s \times b$ is divisible by a; or, since θ may have any one of a series of values which are relatively prime, b is divisible by a, that is $b=a$.

By combining these results, we obtain the theorem :—

'If $\nabla_n, \nabla_{n-1}, \nabla_{n-2}, \ldots, \nabla_1$ represent the greatest common divisors of all the minors of order $n, n-1, \ldots, 1$, respectively, which can be formed out of a given square matrix, these numbers will remain unchanged, when the given matrix is premultiplied by any unit-matrix and postmultiplied by any prime matrix whatsoever.'

13. Let θ, the determinant of the square matrix $\left\| \begin{matrix} n \times n \\ a \end{matrix} \right\|$, be a positive number, different from zero. It may be shown that by postmultiplication with a properly assumed unit $\|\alpha\|$, the matrix $\|a\|$ can be reduced to the form
$$\left\| \begin{matrix} \mu_1, & r_{1,2}, & r_{1,3}, & \ldots, & r_{1,n} \\ 0\ , & \mu_2\ , & r_{2,3}, & \ldots, & r_{2,n} \\ 0\ , & 0\ , & \mu_3\ , & \ldots, & r_{3,n} \\ \cdot & \cdot & \cdot & \cdot & \cdot \\ 0\ , & 0\ , & 0\ , & \ldots, & \mu_n \end{matrix} \right\|, \quad \ldots \ldots \quad (62)$$
where $\mu_1, \mu_2, \ldots, \mu_n$ are positive numbers, such that $\mu_1 \times \mu_2 \times \ldots \times \mu_n = \theta$, and the constituents $r_{i,k}$ satisfy the inequalities
$$0 \leqq r_{i,k} < \mu_i. \quad \ldots \ldots \quad (63)$$

This was first observed by Gauss for the case $n=2$; by Seeber for $n=3$; and the general theorem has been enunciated by M. Hermite *. Its precise statement is :—

'Every matrix of the type $n \times n$ is equivalent (by postmultiplication) to one, and only one, of the *reduced* matrices included in the formula (62).'

To show this, let $v_{1,1}, v_{2,1}, \ldots, v_{n,1}$ be the integral and relatively prime numbers which satisfy the equations

$$\left.\begin{array}{c} a_{i,1}\, v_{1,1} + a_{i,2}\, v_{2,1} + \ldots + a_{i,n}\, v_{n,1} = 0 \\ i = 2, 3, \ldots, n \end{array}\right\}, \quad \ldots \ldots \quad (64)$$

and the inequality $\qquad a_{1,1}\, v_{1,1} + a_{1,2}\, v_{2,1} + \ldots + a_{1,n}\, v_{n,1} > 0.$

Then it is evident that, if $\|v\|$ be a unit-matrix of which $v_{1,1}, v_{2,1}, \ldots, v_{n,1}$ form the first column, the matrix $\|a\| \times \|v\|$ will assume the form

$$\left\|\begin{array}{l} \mu_1,\ b_{1,2},\ b_{1,3},\ \ldots,\ b_{1,n} \\ 0\ ,\ b_{2,2},\ b_{2,3},\ \ldots,\ b_{2,n} \\ 0\ ,\ b_{3,2},\ b_{3,3},\ \ldots,\ b_{3,n} \\ \cdot\quad \cdot\quad \cdot\quad \cdot\quad \cdot\quad \cdot\quad \cdot\quad \cdot \\ 0\ ,\ b_{n,2},\ b_{n,3},\ \ldots,\ b_{n,n} \end{array}\right\|, \quad \ldots \ldots \quad (65)$$

where $\qquad \mu_1 = a_{1,1}\, v_{1,1} + a_{1,2}\, v_{2,1} + \ldots + a_{1,n}\, v_{n,1}.$

If this matrix be postmultiplied by the unit

$$\left\|\begin{array}{l} 1,\ k_2,\ k_3,\ \ldots,\ k_n \\ 0,\ 1\ ,\ 0\ ,\ \ldots,\ 0 \\ 0,\ 0\ ,\ 1\ ,\ \ldots,\ 0 \\ \cdot\quad \cdot\quad \cdot\quad \cdot\quad \cdot\quad \cdot \\ 0,\ 0\ ,\ 0\ ,\ \ldots,\ 1 \end{array}\right\|,$$

the constituents $b_{1,i}$ will be changed into $b_{1,i} + \mu_1 k_i$, while all the other constituents will remain unaltered; so that, by assigning proper values to the numbers $k_2, \ldots, k_n$, we may bring the given matrix $\|a\|$ into the form

$$\left\|\begin{array}{l} \mu_1,\ r_{1,2},\ r_{1,3},\ \ldots,\ r_{1,n} \\ 0\ ,\ b_{2,2},\ b_{2,3},\ \ldots,\ b_{2,n} \\ 0\ ,\ b_{3,2},\ b_{3,3},\ \ldots,\ b_{3,n} \\ \cdot\quad \cdot\quad \cdot\quad \cdot\quad \cdot\quad \cdot\quad \cdot\quad \cdot \\ 0\ ,\ b_{n,2},\ b_{n,3},\ \ldots,\ b_{n,n} \end{array}\right\|,$$

* Gauss, Disq. Arith., Art. 213; Seeber, 'Untersuchungen ueber die Eigenschaften der positiven ternären quadratischen Formen' (Mannheim, 1831), Art. 31; M. Hermite, Crelle, vol. xli. p. 192.

where $r_{1,i}$ verifies the inequality

$$0 \leqq r_{1,i} < \mu_1.$$

From this it is easy to infer that if a matrix of the type $(n-1)\times(n-1)$ can be reduced to the form (62), the same reduction is possible for a matrix of the type $n \times n$, *i.e.* since that reduction is possible when $n=1$, $n=2$, ..., it is possible for every value of n.

To prove that $\|a\|$ is equivalent (by postmultiplication) to only one of the *reduced* matrices (62), it is sufficient to show that no two reduced matrices can be equivalent. If $\|a\|$ and $\|a'\|$ be two reduced matrices and $\|v\|$ a unit-matrix such that $\|a\| \times \|v\| = \|a'\|$, it may be inferred, by comparing the corresponding constituents of the two matrices $\|a\| \times \|v\|$ and $\|a'\|$ (beginning with the lowest horizontal rows of each and proceeding upwards), that all the constituents of $\|v\|$ which lie below its principal diameter are zero; and consequently that the constituents of the principal diameter itself are all positive units. Further, that the constituents above the principal diameter of $\|v\|$ are likewise zero may be established (for each line of constituents parallel to the diameter, beginning with that nearest to it) by means of the inequalities (63) which are satisfied by the constituents both of $\|a\|$ and $\|a'\|$. It thus appears that two reduced matrices cannot be equivalent, without being identical. It will be observed that the reducing unit is unique; *i.e.* that only one postmultiplying unit can be assigned by which a given matrix can be reduced to the form (62).

If, instead of reducing the given matrix $\|a\|$ by postmultiplication, we employ a premultiplying unit, we obtain the following theorem :—

'Every matrix of the type $n \times n$ and of determinant θ is equivalent (by premultiplication) to one, and only one, of the matrices included in the formula (62), in which $\mu_1, \mu_2, \ldots$ are positive, $\mu_1 \times \mu_2 \times \ldots \times \mu_n = \theta$, and $r_{i,k}$ satisfies the inequality

$$0 \leqq r_{i,k} < \mu_k.\text{'} \qquad (66)$$

14*. The transformation to which we have referred in Art. 12 is obtained by employing simultaneously a premultiplying and a postmultiplying unit-matrix. It is expressed by the equation

$$\|a\| = \|\alpha\| \times \left\| \frac{\nabla_n}{\nabla_{n-1}}, \frac{\nabla_{n-1}}{\nabla_{n-2}}, \ldots, \frac{\nabla_1}{\nabla_0} \right\| \times \|\beta\|, \qquad (67)$$

in which $\|a\|$ is a given square matrix of the type $n \times n$, $\|\alpha\|$ and $\|\beta\|$ are unit-matrices, and $\nabla_n, \nabla_{n-1}, \nabla_{n-2}, \ldots, \nabla_1, \nabla_0$ are the determinant and greatest

* [This article has been in great part rewritten since the paper was read. The demonstration is not essentially changed, but is presented in what seems to be a simpler form.—Sept. 1861, H. J. S. S.]

common divisors of the minor determinants of $\|a\|$, so that, in particular, ∇_n is the determinant of $\|a\|$, ∇_{n-1} the greatest common divisor of its minor determinants of order $n-1$, ∇_1 the greatest common divisor of its constituents, and $\nabla_0=1$. The units $\|\alpha\|$ and $\|\beta\|$ are not absolutely determined, but admit, when $n>1$, of an infinite number of different values. If $n=1$, it is evident that the formula (67) is verified; for we have the identical equation

$$\|a\| = \|1\| \times \left\| \frac{\nabla_1}{\nabla_0} \right\| \times \|1\|.$$

It is therefore sufficient to show that, if the transformation indicated in the formula can be effected for matrices of the type $(n-1)\times(n-1)$, it can also be effected for matrices of the type $n\times n$. The demonstration depends on an elementary principle, which it is worth while to enunciate separately.

'If

$$\left.\begin{array}{c} U_i = A_{i,1}x_1 + A_{i,2}x_2 + \ldots + A_{i,n+m}x_{n+m} \\ i=1,2,3,\ldots,n \end{array}\right\} \quad \ldots\ldots \quad (68)$$

denote a system of n linear functions of $n+m$ indeterminates, $(m \geqq 0)$, and if the constituents of the matrix $\|A\|$ do not admit of any common divisor, it is always possible to assign integral values to $x_1, x_2, \ldots, x_{n+m}$, which shall render $U_1, U_2, \ldots, U_n$ relatively prime.'

For, in the first place, we can obtain values for $U_1, U_2, \ldots, U_n$ which shall not have any common divisor with a given number M. Let $p, q, r, \ldots$ be the different prime divisors of M; one at least of the constituents of $\|A\|$, for example $A_{i,j}$, is prime to p. Attributing to x_j a value prime to p, and values divisible by p to the remaining indeterminates, we shall obtain for U_i a value which is certainly prime to p. Similarly, by subjecting the indeterminates to proper congruential conditions with respect to the modules, $q, r, \ldots$, we can render one, at least, of the functions U prime to q, one prime to r, and so on; *i.e.* since we can assign to the indeterminates values simultaneously satisfying all these congruential conditions, we can give to $U_1, U_2, \ldots, U_n$ values the greatest common divisor of which is prime to M. Let D_n be the greatest divisor of $\|A\|$, D_{n-1} the greatest common divisor of the first minors of $\|A\|$; and let $C_1, C_2, \ldots, C_n$ be a set of simultaneous values of $U_1, U_2, \ldots, U_n$, having a greatest common divisor c, which is prime to $\dfrac{D_n}{D_{n-1}}$. Since the equations

$$\left.\begin{array}{c} A_{i,1}x_1 + A_{i,2}x_2 + \ldots + A_{i,n+m}x_{n+m} = C_i \\ i=1,2,3,\ldots,n \end{array}\right\}$$

are resoluble, it will follow from the condition of resolubility (see Art. 11), that the determinants of its augmented matrix, and in particular those which contain the column $C_1, C_2, \ldots, C_n$, are divisible by D_n. Let $\theta \times c \times D_{n-1}$ be the greatest common divisor of these last determinants; then $\theta \times c \times D_{n-1}$ is divisible by D_n, *i.e.* θ is divisible by $\dfrac{D_n}{D_{n-1}}$. It appears from this, that the condition of resolubility is satisfied by the system

$$A_{i,1}x_1 + A_{i,2}x_2 + \ldots + A_{i,n+m}x_{n+m} = \frac{C_i}{c},$$
$$i = 1, 2, 3, \ldots, n,$$

that is to say, it is possible to obtain a simultaneous system of relatively prime values for $U_1, U_2, \ldots, U_n$.

To apply this principle to the transformation of the matrix $\| a \|$, let

$$[a_{i,j}] = \frac{1}{\nabla_{n-1}} \frac{d\nabla_n}{da_{i,j}}. \quad \ldots \ldots \ldots \quad (69)$$

The constituents of the matrix $\| [a] \|$ do not admit of any common divisor; consequently, in the system

$$\left. \begin{array}{c} [a_{i,1}]b_{1,1} + [a_{i,2}]b_{2,1} + \ldots + [a_{i,n}]b_{n,1} = u_{i,1} \\ i = 1, 2, 3, \ldots, n \end{array} \right\}, \quad \ldots \ldots \quad (70)$$

we can assign values to $b_{1,1}, b_{2,1}, \ldots, b_{n,1}$ which shall render $u_{1,1}, u_{2,1}, \ldots, u_{n,1}$ relatively prime. Let $\| u \|$ denote a unit-matrix of which the first column is $u_{1,1}, u_{2,1}, \ldots, u_{n,1}$; and $\| b \|$ a square matrix of which the first column is $b_{1,1}, b_{2,1}, \ldots, b_{n,1}$, and of which the remaining constituents are defined by the equations

$$\left. \begin{array}{l} b_{i,j} = a_{i,1}u_{1,j} + a_{i,2}u_{2,j} + \ldots + a_{i,n}u_{n,j} \\ i = 1, 2, 3, \ldots, n \\ j = \ 2, 3, \ldots, n \end{array} \right\} \quad \ldots \ldots \quad (71)$$

Observing that the systems (69) and (70) involve the inverse system,

$$\left. \begin{array}{c} a_{i,1}u_{1,1} + a_{i,2}u_{2,1} + \ldots + a_{i,n}u_{n,1} = \dfrac{\nabla_n}{\nabla_{n-1}} b_{i,1} \\ i = 1, 2, 3, \ldots, n \end{array} \right\}, \quad \ldots \ldots \quad (72)$$

we infer that the matrices $\| u \|$ and $\| b \|$ verify the equation

$$\| a \| \times \| u \| = \| b \| \times \left\| \frac{\nabla_n}{\nabla_{n-1}}, 1, 1, \ldots \right\|, \quad \ldots \ldots \quad (73)$$

in which $\left\| \dfrac{\nabla_n}{\nabla_{n-1}}, 1, 1, \ldots \right\|$ denotes a matrix of the type $n \times n$. It follows from (73) that ∇_{n-1} is the determinant of $\| b \|$; let that matrix be reduced by premultiplication with a unit-matrix; and let

$$\| b \| = \| v \| \times \| \nabla_{n-1} \|, \quad \ldots \ldots \ldots \quad (74)$$

where $\|v\|$ is the reducing unit, and $\|\nabla_{n-1}\|$ the reduced matrix

$$\left\| \begin{matrix} \mu_1, & r_{1,2}, & r_{1,3}, & \ldots, & r_{1,n} \\ 0, & \mu_2, & r_{2,3}, & \ldots, & r_{2,n} \\ 0, & 0, & \mu_3, & \ldots, & r_{3,n} \\ \cdot & \cdot & \cdot & \cdot & \cdot \\ 0, & 0, & 0, & \ldots, & \mu_n \end{matrix} \right\|, \quad (75)$$

so that (73) assumes the form

$$\|a\| = \|v\| \times \|\nabla_{n-1}\| \times \left\| \frac{\nabla_n}{\nabla_{n-1}}, 1, 1, \ldots \right\| \times \|u\|^{-1}. \quad (76)$$

It may be proved that in (75)

$$\mu_1 = 1,\ r_{1,2} = 0,\ r_{1,3} = 0,\ \ldots,\ r_{1,n} = 0.$$

For, since the matrix

$$\|\nabla_{n-1}\| \times \left\| \frac{\nabla_n}{\nabla_{n-1}}, 1, 1, \ldots \right\|$$

is derived from $\|a\|$ by multiplication with unit-matrices, ∇_{n-1} is the greatest common divisor of the first minors of

$$\|\nabla_{n-1}\| \times \left\| \frac{\nabla_n}{\nabla_{n-1}}, 1, 1, \ldots \right\|.$$

Therefore ∇_{n-1} divides $\mu_2 \times \mu_3 \times \ldots \times \mu_n$, which is one of those minors; but also

$$\nabla_{n-1} = \mu_1 \times \mu_2 \times \ldots \times \mu_n; \quad \textit{i.e.}\ \mu_1 = 1,\ \mu_2 \times \mu_3 \times \ldots \times \mu_n = \nabla_{n-1},$$

and the product $\|\nabla_{n-1}\| \times \left\| \frac{\nabla_n}{\nabla_{n-1}}, 1, 1, \ldots \right\|$ assumes the form

$$\left\| \begin{matrix} \frac{\nabla_n}{\nabla_{n-1}}, & r_{1,2}, & r_{1,3}, & \ldots, & r_{1\,n} \\ 0 & , \mu_2, & r_{2,3}, & \ldots, & r_{2,n} \\ 0 & , 0, & \mu_3, & \ldots, & r_{3,n} \\ \cdot & \cdot & \cdot & \cdot & \cdot \\ 0 & , 0, & 0, & \ldots, & \mu_n \end{matrix} \right\|.$$

One of the minors of this matrix is $r_{1,2} \times \mu_3 \times \ldots \times \mu_n$, which cannot be divisible by ∇_{n-1} or $\mu_2 \times \mu_3 \times \ldots \times \mu_n$, unless $r_{1,2}$ is a multiple of μ_2; but $r_{1,2} < \mu_2$, because $\|\nabla_{n-1}\|$ is reduced, therefore $r_{1,2} = 0$. Similarly, it may successively be shown that $r_{1,3} = 0, \ldots, r_{1,n} = 0$. Now if the matrix

$$\left\| \begin{matrix} \mu_2, & r_{2,3}, & \ldots, & r_{2,n} \\ 0, & \mu_3, & \ldots, & r_{3,n} \\ \cdot & \cdot & \cdot & \cdot \\ 0, & 0, & \ldots, & \mu_n \end{matrix} \right\|, \quad (77)$$

which is of the type $(n-1)\times(n-1)$, be reduced to the form

$$\|v'\| \times \left\| \frac{\nabla_{n-1}}{\nabla_{n-2}}, \frac{\nabla_{n-2}}{\nabla_{n-3}}, \ldots, \frac{\nabla_1}{\nabla_0} \right\| \times \|u'\|,$$

in which $\nabla_{n-2}, \nabla_{n-3}, \ldots$ represent the greatest common divisors of the minors of (77), we may replace $\|\nabla_{n-1}\|$ by the matrix

$$\|\nabla_{n-1}\| = \left\|\begin{matrix}1 & 0\\ 0 & v'\end{matrix}\right\| \times \left\| 1, \frac{\nabla_{n-1}}{\nabla_{n-2}}, \frac{\nabla_{n-2}}{\nabla_{n-3}}, \ldots, \frac{\nabla_1}{\nabla_0} \right\| \times \left\|\begin{matrix}1 & 0\\ 0 & u'\end{matrix}\right\|,$$

where $\left\|\begin{matrix}1 & 0\\ 0 & v'\end{matrix}\right\|$ and $\left\|\begin{matrix}1 & 0\\ 0 & u'\end{matrix}\right\|$ denote unit-matrices of the type $n\times n$, the forms of which are sufficiently indicated by the symbols themselves. Hence, observing that

$$\left\|\begin{matrix}1 & 0\\ 0 & u'\end{matrix}\right\| \times \left\| \frac{\nabla_n}{\nabla_{n-1}}, 1, 1, \ldots \right\| = \left\| \frac{\nabla_n}{\nabla_{n-1}}, 1, 1, \ldots \right\| \times \left\|\begin{matrix}1 & 0\\ 0 & u'\end{matrix}\right\|,$$

and that

$$\left\| 1, \frac{\nabla_{n-1}}{\nabla_{n-2}}, \frac{\nabla_{n-2}}{\nabla_{n-3}}, \ldots, \frac{\nabla_1}{\nabla_0} \right\| \times \left\| \frac{\nabla_n}{\nabla_{n-1}}, 1, 1, \ldots \right\| = \left\| \frac{\nabla_n}{\nabla_{n-1}}, \frac{\nabla_{n-1}}{\nabla_{n-2}}, \ldots, \frac{\nabla_1}{\nabla_0} \right\|,$$

we obtain, from (76),

$$\|a\| = \|v\| \times \left\|\begin{matrix}1 & 0\\ 0 & v'\end{matrix}\right\| \times \left\| \frac{\nabla_n}{\nabla_{n-1}}, \frac{\nabla_{n-1}}{\nabla_{n-2}}, \ldots, \frac{\nabla_1}{\nabla_0} \right\| \times \left\|\begin{matrix}1 & 0\\ 0 & u'\end{matrix}\right\| \times \|u^{-1}\|,$$

or more simply,

$$\|a\| = \|\alpha\| \times \left\| \frac{\nabla_n}{\nabla_{n-1}}, \frac{\nabla_{n-1}}{\nabla_{n-2}}, \ldots, \frac{\nabla_1}{\nabla_0} \right\| \times \|\beta\|.$$

It has, however, still to be shown that $\nabla_{n-2}, \nabla_{n-3}, \ldots$ which have been defined with reference to the matrix (77) are the greatest common divisors of the successive systems of minors of $\|a\|$. These greatest common divisors are the same for the given matrix $\|a\|$ and for the matrix

$$\left\| \frac{\nabla_n}{\nabla_{n-1}}, \frac{\nabla_{n-1}}{\nabla_{n-2}}, \ldots, \frac{\nabla_1}{\nabla_0} \right\|,$$

which is derived from it by multiplication with unit-matrices; consequently ∇_{n-1} divides every first minor of

$$\left\| \frac{\nabla_n}{\nabla_{n-1}}, \frac{\nabla_{n-1}}{\nabla_{n-2}}, \ldots, \frac{\nabla_1}{\nabla_0} \right\|,$$

and, in particular, it divides

$$\frac{\nabla_n}{\nabla_{n-1}} \times \frac{\nabla_{n-2}}{\nabla_{n-3}} \times \frac{\nabla_{n-3}}{\nabla_{n-4}} \times \ldots \times \frac{\nabla_1}{\nabla_0} = \frac{\nabla_n \times \nabla_{n-2}}{\nabla_{n-1}}; \quad i.e.\ \frac{\nabla_{n-1}}{\nabla_{n-2}} \text{ divides } \frac{\nabla_n}{\nabla_{n-1}}.$$

Again, $\nabla_{n-1}, \nabla_{n-2}, \ldots, \nabla_1, \nabla_0$, which are the determinant and greatest common divisors of the minors of (77), are also the determinant and greatest common divisors of the minors of the matrix

$$\left\| \frac{\nabla_{n-1}}{\nabla_{n-2}},\ \frac{\nabla_{n-2}}{\nabla_{n-3}},\ \ldots,\ \frac{\nabla_1}{\nabla_0} \right\|; \quad . \quad . \quad . \quad . \quad . \quad . \quad . \quad . \quad . \quad (78)$$

so that if $s \leqq n-2$, ∇_s divides every minor of order s in (78), and, consequently, the minor $\frac{\nabla_{s+1}}{\nabla_s} \times \frac{\nabla_{s-1}}{\nabla_{s-2}} \times \frac{\nabla_{s-2}}{\nabla_{s-3}} \times \ldots \times \frac{\nabla_1}{\nabla_0}$; or $\frac{\nabla_s}{\nabla_{s-1}}$ divides $\frac{\nabla_{s+1}}{\nabla_s}$.

It thus appears that in the series of numbers

$$\frac{\nabla_n}{\nabla_{n-1}},\ \frac{\nabla_{n-1}}{\nabla_{n-2}},\ \ldots,\ \frac{\nabla_2}{\nabla_1},\ \frac{\nabla_1}{\nabla_0}$$

each term is divisible by that which comes after it. Every product of s terms of that series is therefore divisible by the product

$$\frac{\nabla_s}{\nabla_{s-1}} \times \frac{\nabla_{s-1}}{\nabla_{s-2}} \times \ldots \times \frac{\nabla_1}{\nabla_0} = \nabla_s;$$

or, which is the same thing, ∇_s is the greatest common divisor of the minors of order s in the reduced matrix (78), and therefore in the given matrix $\|a\|$.

15. If the proposed matrix $\|a\|$ be not square, but of the type $n \times (n+m)$, let $\|a\| = \|\nabla_n\| \times \|a'\|$, where $\|a'\|$ is a prime matrix of the same type as $\|a\|$, and $\|\nabla_n\|$ a square matrix of which the determinant is ∇_n the greatest divisor of $\|a\|$. Then if $\|\nabla_n\|$ be expressed in the form

$$\|v\| \times \left\| \frac{\nabla_n}{\nabla_{n-1}},\ \frac{\nabla_{n-1}}{\nabla_{n-2}},\ \ldots,\ \frac{\nabla_1}{\nabla_0} \right\| \times \|u\|,$$

and if, for brevity, we write $\|V\|$ for $\|u\| \times \|a'\|$, we obtain for $\|a\|$ the expression

$$\|a\| = \|v\| \times \left\| \frac{\nabla_n}{\nabla_{n-1}},\ \frac{\nabla_{n-1}}{\nabla_{n-2}},\ \ldots,\ \frac{\nabla_1}{\nabla_0} \right\| \times \|V\|. \quad . \quad . \quad . \quad . \quad . \quad . \quad (79)$$

The numbers $\nabla_n, \nabla_{n-1}, \ldots$, which are the greatest common divisors of the minors of $\|\nabla_n\|$, are also by the theorem of Art. **12**, the greatest common divisors of the minors of $\|a\|$. We see therefore that $\frac{\nabla_s}{\nabla_{s-1}}$ is always divisible by $\frac{\nabla_{s-1}}{\nabla_{s-2}}$, in the case of an oblong as well as a square matrix.

16. To show still more clearly the nature of the quotients $\frac{\nabla_s}{\nabla_{s-1}}$, we add the following proposition :—

'If, in any rectangular matrix, we divide each minor determinant of order s by the greatest common divisor of its own first minors, the greatest common divisor of all the quotients thus obtained is $\frac{\nabla_s}{\nabla_{s-1}}$.'

By this proposition, $\dfrac{\nabla_s}{\nabla_{s-1}}$ is itself defined as a greatest common divisor, instead of being defined as the quotient of one greatest common divisor divided by another.

To establish its truth, we may first consider the quotient $\dfrac{\nabla_n}{\nabla_{n-1}}$ in any rectangular matrix $\|A\|$ of the type $n\times(m+n)$. Let ω denote the greatest common divisor of the quotients obtained by dividing each determinant of $\|A\|$ by the greatest common divisor of the first minors of that determinant: we have then to show that

$$\frac{\nabla_n}{\nabla_{n-1}}=\omega.$$

Since the greatest common divisor of any vertical column of minors in $\|A\|$ is not altered by premultiplication with a unit-matrix, it is evident that ω as well as $\dfrac{\nabla_n}{\nabla_{n-1}}$ will remain unchanged by that operation. If, therefore,

$$\|A\|=\|v\|\times\left\|\frac{\nabla_n}{\nabla_{n-1}},\ \frac{\nabla_{n-1}}{\nabla_{n-2}},\ \ldots,\ \frac{\nabla_1}{\nabla_0}\right\|\times\|V\|, \quad . \quad . \quad . \quad . \quad . \quad (79)$$

where $\|v\|$ is a unit, and $\|V\|$ a prime matrix, we may consider instead of $\|A\|$ the simpler matrix

$$\left\|\frac{\nabla_n}{\nabla_{n-1}},\ \frac{\nabla_{n-1}}{\nabla_{n-2}},\ \ldots,\ \frac{\nabla_1}{\nabla_0}\right\|\times\|V\|. \quad . \quad . \quad . \quad . \quad . \quad . \quad . \quad . \quad (80)$$

Let $\|\theta_1\|$, $\|\theta_2\|$, ..., &c. be the different square matrices of $\|V\|$; θ_1, θ_2, ... their determinants; ψ_i the greatest common divisor of those first minors in $\|\theta_i\|$ which do not contain the constituents of its uppermost row, so that $\dfrac{\theta_i}{\psi_i}$ is integral; lastly, let ω_i be the quotient obtained by dividing the determinant of

$$\left\|\frac{\nabla_n}{\nabla_{n-1}},\ \frac{\nabla_{n-1}}{\nabla_{n-2}},\ \ldots,\ \frac{\nabla_1}{\nabla_0}\right\|\times\|\theta_i\| \quad . \quad . \quad . \quad . \quad . \quad . \quad . \quad . \quad (81)$$

by the greatest common divisor of its first minors, so that ω is the greatest common divisor of ω_1, ω_2, ... Now the greatest common divisor of the first minors of (81) is evidently divisible by ∇_{n-1}, and divides $\nabla_{n-1}\times\psi_i$ (because $\nabla_{n-1}\,\psi_i$ is the greatest common divisor of one of its rows of minors). Consequently ω_i divides $\nabla_n\,\theta_i\div\nabla_{n-1}$, and is divisible by $\nabla_n\,\theta_i\div\nabla_{n-1}\,\psi_i$. Therefore $\dfrac{\nabla_n}{\nabla_{n-1}}$ is a common divisor of certain numbers respectively dividing the numbers ω_1, ω_2, ..., viz. the numbers $\dfrac{\nabla_n}{\nabla_{n-1}}\cdot\dfrac{\theta_i}{\psi_i}$; it is also (because θ_1, θ_2, ... are relatively prime) the greatest common divisor of the numbers $\dfrac{\nabla_n}{\nabla_{n-1}}\,\theta$, in which the same numbers ω_i are respectively contained; *i.e.* $\frac{\nabla_n}{\nabla_{n-1}}$ is the greatest common divisor of the numbers $\omega_1, \omega_2, \ldots$ themselves, or

$$\frac{\nabla_n}{\nabla_{n-1}} = \omega.$$

By the aid of this particular case of the theorem the general proposition itself may be proved as follows:—

If, in any rectangular matrix, of the type $n \times (m+n)$, we propose to determine Ω_s, the greatest common divisor of the quotients obtained by dividing each minor determinant of order s by the greatest common divisor of its own first minors, we may begin by selecting any s vertical columns $[s < n]$, and forming the proper quotient for each determinant of order s contained in this partial matrix of the type $n \times s$. Let λ_i denote the greatest common divisor of these quotients; then, as we have just seen, λ_i is the greatest common divisor of all the determinants of the partial matrix, divided by the greatest common divisor of all its first minors. Hence (by Art. 12) λ_i will remain unchanged when the given matrix is premultiplied by a unit-matrix. But Ω_s is the greatest common divisor of all the divisors $\lambda_1, \lambda_2, \ldots$ corresponding to every group of s vertical columns; therefore Ω_s is itself unchanged by premultiplication. Similarly, if a *square* matrix be postmultiplied by a rectangular prime matrix, it may be shown that Ω_s is the same for the given square matrix and for the resulting rectangular matrix. Hence, if as before

$$\|A\| = \|v\| \times \left\| \frac{\nabla_n}{\nabla_{n-1}}, \ldots, \frac{\nabla_1}{\nabla_0} \right\| \times \|V\|,$$

Ω_s and $\frac{\nabla_s}{\nabla_{s-1}}$ are the same for $\left\| \frac{\nabla_n}{\nabla_{n-1}}, \frac{\nabla_{n-1}}{\nabla_{n-2}}, \ldots, \frac{\nabla_1}{\nabla_0} \right\|$ and for $\|A\|$. But in the matrix $\left\| \frac{\nabla_n}{\nabla_{n-1}}, \ldots, \frac{\nabla_1}{\nabla_0} \right\|$ it is evident that $\frac{\nabla_s}{\nabla_{s-1}}$ and Ω_s coincide; therefore in any rectangular matrix

$$\frac{\nabla_s}{\nabla_{s-1}} = \Omega_s.$$

From the definition of $\frac{\nabla_s}{\nabla_{s-1}}$ as a greatest common divisor, which we have now obtained, we infer that if $\|D\|$ be any matrix containing another matrix $\|\nabla\|$, and if $D_s, D_{s-1}, \ldots, \nabla_s, \nabla_{s-1}, \ldots$ be the greatest common divisors of the corresponding minors in $\|D\|$ and $\|\nabla\|$ respectively, not only is ∇_s divisible by D_s, and ∇_{s-1} by D_{s-1}, but also $\frac{\nabla_s}{\nabla_{s-1}}$ by $\frac{D_s}{D_{s-1}}$.

It is not difficult to show that in any matrix $\frac{\nabla_s}{\nabla_{s-k}}$ is the greatest common divisor of all the quotients obtained by dividing each minor of order s by the greatest common divisor of its minors of order $s-k$. But, as this extension of the preceding result is not needed in what follows, we may omit it here.

We may add that the theorem of this article is precisely equivalent to the following, which may be demonstrated by a different method.

'If P^{I_s} be the highest power of a given prime that divides all the minors of order s in a given matrix, and if all the minors of order $s-1$ contained in one particular minor of order s are divisible by $P^{I_{s-1}+m}$, that minor is itself divisible by P^{I_s+m}.'

It should be observed that, whenever all the minors of any determinant are zero, the quotient obtained by dividing the determinant by the greatest common divisor of its minors is also zero.

17. These results admit of immediate application to the theory of systems of linear congruences. The general type of such systems is

$$\left.\begin{array}{c} A_{i,1}\,x_1 + A_{i,2}\,x_2 + \ldots + A_{i,n}\,x_n \equiv A_{i,n+1}, \text{ mod } M \\ i = 1,\ 2,\ 3,\ \ldots,\ n' \end{array}\right\} ; \quad . \quad . \quad . \quad . \quad (82)$$

and to construct a complete theory of them it is requisite, first, to assign a criterion for their resolubility or irresolubility; secondly, when they are resoluble, to investigate the number of incongruous solutions of which they are susceptible; and, lastly, to exhibit a method for obtaining all these solutions. We shall first suppose that $n'=n$; *i.e.* that the proposed system is neither defective nor redundant.

Let D_n, D_{n-1}, ..., ∇_n, ∇_{n-1}, ..., respectively, denote the greatest common divisors of the determinants and minors of the augmented and unaugmented matrices of the system (82); also let δ_n, δ_{n-1}, ..., δ_1 denote the greatest common divisors of M with $\frac{\nabla_n}{\nabla_{n-1}}$, of M with $\frac{\nabla_{n-1}}{\nabla_{n-2}}$, ...; and let d_n, d_{n-1}, ... similarly represent the greatest common divisors of M with $\frac{D_n}{D_{n-1}}$, of M with $\frac{D_{n-1}}{D_{n-2}}$, ...; then, if $d = d_n \times d_{n-1} \times \ldots \times d_1$, $\delta = \delta_n \times \delta_{n-1} \times \ldots \times \delta_1$, we have the two following theorems:—

(i) 'The necessary and sufficient condition for the resolubility of the system (81) is $d=\delta$.'

(ii) 'When this condition is satisfied, the number of its incongruous solutions is d.'

To demonstrate the first of these theorems, we revert to the principle of Art. **11**, from which it appears that the necessary and sufficient condition for the resolubility of the system (82) is that the greatest divisors of the two matrices

$$\left\|\begin{matrix} M, & 0, & 0, & . & 0, & A_{1,1}, & \ldots, & A_{1,n} \\ 0, & M, & 0, & . & 0, & A_{2,1}, & \ldots, & A_{2,n} \\ 0, & 0, & M, & . & 0, & A_{3,1}, & \ldots, & A_{3,n} \\ . & . & . & . & . & . & . & . \\ 0, & 0, & 0, & . & M, & A_{n,1}, & \ldots, & A_{n,n} \end{matrix}\right\| \quad . \quad . \quad . \quad . \quad . \quad . \quad (83)$$

and

$$\left\|\begin{matrix} M, & 0, & 0, & . & 0, & A_{1,1}, & \ldots, & A_{1,n+1} \\ 0, & M, & 0, & . & 0, & A_{2,1}, & \ldots, & A_{2,n+1} \\ 0, & 0, & M, & . & 0, & A_{3,1}, & \ldots, & A_{3,n+1} \\ . & . & . & . & . & . & . & . \\ 0, & 0, & 0, & . & M, & A_{n,1}, & \ldots, & A_{n,n+1} \end{matrix}\right\| \quad . \quad . \quad . \quad . \quad . \quad . \quad (84)$$

are to be equal to one another. Now the first of these greatest common divisors is evidently the greatest common divisor of

$$M^n, M^{n-1}\nabla_1, M^{n-2}\nabla_2, \ldots, M\nabla_{n-1}, \nabla_n;$$

which, for brevity, we shall represent by the symbol

$$[M^n, M^{n-1}\nabla_1, M^{n-2}\nabla_2, \ldots, M\nabla_{n-1}, \nabla_n]. \quad . \quad . \quad . \quad . \quad . \quad . \quad (85)$$

Let $M = P \times Q \times R \ldots$, $P, Q, R, \ldots$ denoting powers of different primes; we may then, in (85), replace M by $P, Q, R, \ldots$ successively, since

$$[M^n, M^{n-1}\nabla_1, \ldots, M\nabla_{n-1}, \nabla_n]$$
$$= [P^n, P^{n-1}\nabla_1, \ldots, P\nabla_{n-1}, \nabla_n] \times [Q^n, Q^{n-1}\nabla, \ldots, \nabla_n] \times \ldots.$$

If P divide any one of the numbers $\dfrac{\nabla_s}{\nabla_{s-1}}, \ldots$, let $\dfrac{\nabla_s}{\nabla_{s-1}}$ be the least of them that it divides; also let $P_i = \left[P, \dfrac{\nabla_i}{\nabla_{i-1}}\right]$; so that $P_i = P$, if $i \overline{\gtrless} s$. Then

$$[P^n, P^{n-1}\nabla_1, \ldots, P\nabla_{n-1}, \nabla_n] = P_1 \times \left[\frac{P^n}{P_1}, P^{n-1}\frac{\nabla_1}{P_1}, P^{n-2}\frac{\nabla_2}{P_1}, \ldots, \frac{\nabla_n}{P_1}\right]$$
$$= P_1 \times \left[P^{n-1}, P^{n-2}\frac{\nabla_2}{\nabla_1}, P^{n-3}\frac{\nabla_3}{\nabla_1}, \ldots, \frac{\nabla_n}{\nabla_1}\right],$$

observing that $\dfrac{\nabla_1}{P_1}$ is prime to P [if $s > 1$], and that we may therefore divide the last n numbers by $\dfrac{\nabla_1}{P_1}$, and may then omit $\dfrac{P_n}{P_1}$ which is divisible by P^{n-1}.

Continuing this process, we find

$$[P^n, P^{n-1}\nabla_1, \ldots, P\nabla_{n-1}, \nabla_n]$$
$$= P_1 \times P_2 \times \ldots \times P_{s-1}\left[P^{n-s+1}, P^{n-s}\frac{\nabla_s}{\nabla_{s-1}}, P^{n-s-1}\frac{\nabla_{s+1}}{\nabla_{s-1}}, \ldots, \frac{\nabla_n}{\nabla_{s-1}}\right];$$

or, since $\frac{\nabla_s}{\nabla_{s-1}}$ is divisible by P, and $\frac{\nabla_{s+k}}{\nabla_{s-1}} = \frac{\nabla_{s+k}}{\nabla_{s+k-1}} \times \frac{\nabla_{s+k-1}}{\nabla_{s+k-2}} \times \ldots \times \frac{\nabla_s}{\nabla_{s-1}}$ by P^{k+1},

$$[P^n, P^{n-1}\nabla_1, P^{n-2}\nabla_2, \ldots, P\nabla_{n-1}, \nabla_n] = P_1 \times P_2 \times P_3 \times \ldots \times P_{s-1} \times P^{n-s+1} = \prod_{i=1}^{i=n} . P_i.$$

But $\delta_i = P_i \times Q_i \times R_i \times \ldots$; and consequently the greatest common divisor of the determinants of (83) is $\delta_1 \times \delta_2 \times \ldots \times \delta_n$ or δ. Similarly, the greatest divisor of (84) is $d_1 \times d_2 \times \ldots \times d_n$ or d. The necessary and sufficient condition for the resolubility of the proposed system of congruences is therefore contained in the formula

$$d = \delta.$$

It should, however, be observed that, since $\frac{D_s}{D_{s-1}}$ divides $\frac{\nabla_s}{\nabla_{s-1}}$ (Art. 16), d_s divides δ_s, and therefore the equation

$$d = \delta$$

involves the coexistence of the n equations

$$d_1 = \delta_1,\ d_2 = \delta_2,\ \ldots,\ d_n = \delta_n. \qquad (86)$$

To investigate the number of solutions of the systems (82), supposed to be resoluble, let $\|\alpha\|$ and $\|\beta\|$ be two unit-matrices satisfying the equation

$$\|\alpha\| \times \|A\| \times \|\beta\| = \left\| \frac{\nabla_n}{\nabla_{n-1}}, \frac{\nabla_{n-1}}{\nabla_{n-2}}, \ldots, \frac{\nabla_1}{\nabla_0} \right\|; \qquad (87)$$

also let

$$\left.\begin{array}{c} x_i = \beta_{i,1} v_1 + \beta_{i,2} v_2 + \ldots + \beta_{i,n} v_n \\ i = 1, 2, 3, \ldots, n \end{array}\right\};$$

and

$$\left.\begin{array}{c} c_i = \alpha_{i,1} A_{1,n+1} + \alpha_{i,2} A_{2,n+1} + \ldots + \alpha_{i,n} A_{n,n+1} \\ i = 1, 2, 3, \ldots, n \end{array}\right\}.$$

Then it is evident that the proposed system of congruences is precisely equivalent to the system

$$\left.\begin{array}{c} \frac{\nabla_{n-i+1}}{\nabla_{n-i}} v_i \equiv c_i, \text{ mod } M \\ i = 1, 2, 3, \ldots, n \end{array}\right\}, \qquad (88)$$

in such a manner that the two systems are simultaneously resoluble or irresoluble;

and that from any number of incongruous solutions of the one an equal number of incongruous solutions of the other is deducible. But the whole number of incongruous solutions of (88) is $\delta_1 \times \delta_2 \times \ldots \times \delta_n = \delta$; *i.e.* the number of solutions of the proposed system is δ.

By the use of the unit-matrices $\|\alpha\|$ and $\|\beta\|$, the actual resolution of the proposed system is made to depend on the resolution of the n congruences contained in (88). But this method of solving a system of linear congruences, though very symmetrical, is perhaps too tedious for the purposes of computation.

18*. Let the proposed system of congruences be the *defective* system

$$\left.\begin{array}{c} A_{i,1}x_1 + A_{i,2}x_2 + \ldots + A_{i,n+m}x_{n+m} \equiv A_{i,n+m+1}, \bmod M \\ i = 1, 2, 3, \ldots, n \end{array}\right\}, \quad . \quad . \quad (89)$$

and let the notation of the last article be retained. It is easily seen that the condition of resolubility of the system (89) is, as before,

$$\delta = d.$$

But the number of its incongruous solutions, when that condition is satisfied, is not δ, but $\delta \times M^m$. For we have seen that we can find a unit-matrix $\|\alpha\|$, and a prime matrix $\|A'\|$ of the type $n \times (n+m)$, satisfying the equation

$$\|\alpha\| \times \|A\| = \left\| \frac{\nabla_n}{\nabla_{n-1}}, \frac{\nabla_{n-1}}{\nabla_{n-2}}, \ldots, \frac{\nabla_2}{\nabla_1}, \frac{\nabla_1}{\nabla_0} \right\| \times \|A'\|;$$

we may therefore replace the system (89) by a system of the form

$$\frac{\nabla_{n-i+1}}{\nabla_{n-i}} U_i \equiv C_i, \bmod M, \quad . \quad . \quad . \quad . \quad . \quad . \quad . \quad . \quad . \quad (90)$$

in which $\quad U_i = A'_{i,1}x_1 + A'_{i,2}x_2 + \ldots + A'_{i,n+m}x_{n+m},$

and $\quad C_i = \alpha_{i,1}A_{1,n+m+1} + \alpha_{i,2}A_{2,n+m+1} + \ldots + \alpha_{i,n}A_{n,n+m+1}.$

If the system (89) is resoluble, the system (90) will be so too, and will give d or δ different systems of values for $U_1, U_2, \ldots, U_n$, any one of which may be represented by the formula

$$\left.\begin{array}{c} U_i \equiv u_i, \bmod M \\ i = 1, 2, 3, \ldots, n \end{array}\right\}. \quad . \quad . \quad . \quad . \quad . \quad . \quad . \quad . \quad . \quad (91)$$

Let us replace the modulus M by P, the highest power of one of its prime divisors. Since $\|A'\|$ is a prime matrix, one at least of its determinants, for example, the determinant $\Sigma \pm A'_{1,1}A'_{2,2} \ldots A'_{n,n}$, is prime to P. It will follow

* [This article has been added since the paper was read. The theorems contained in it are supplementary to that of the preceding article.—September 1861. H. J. S. S.]

from this that, whatever values we attribute to $x_{n+1}, x_{n+2}, \ldots, x_{n+m}$, each of the δ systems represented by (91) is resoluble for the modulus P, and gives, for any assumed values of $x_{n+1}, x_{n+2}, \ldots, x_{n+m}$, only one set of values of $x_1, x_2, \ldots, x_n$. Each of those δ systems admits, therefore, of P^m solutions for the modulus P, *i.e.* of M^m for the modulus M. The system (89) will consequently admit of $\delta \times M^m$ solutions.

Let us also consider the *redundant* system of congruences,

$$\left.\begin{array}{c} A_{i,1}x_1 + A_{i,2}x_2 + \ldots + A_{i,n}x_n \equiv A_{i,n+1}, \text{ mod } M \\ i = 1, 2, 3, \ldots, n+m \end{array}\right\}, \quad . \quad . \quad . \quad (92)$$

and let D_{n+1} denote the greatest divisor of its augmented matrix. Let p represent a prime divisor of M, and let p^θ, p^{I_s}, p^{i_s} be the highest powers of p, which divide M, D_s, ∇_s respectively. The condition of resolubility of Art. 11, applied to the system (92), considered with respect to the modulus p^θ, becomes, after division by $p^{(m-1)\theta}$,

$$[p^{I_{n+1}}, p^{I_n+\theta}, p^{I_{n-1}+2\theta}, \ldots, p^{(n+1)\theta}] = [p^{i_n+\theta}, p^{i_{n-1}+2\theta}, \ldots, p^{(n+1)\theta}]. \quad (93)$$

And this equation is impossible if $\theta > I_{n+1} - I_n$. For $I_{s+1} - I_s \geqq I_s - I_{s-1}$, because $\dfrac{D_{s+1}}{D_s}$ is divisible by $\dfrac{D_s}{D_{s-1}}$; the inequality $I_{n+1} < I_n + \theta$ involves, therefore, the inequalities

$$\left.\begin{array}{c} I_{n+1} < I_{n-s+1} + s\theta \\ s = 1, 2, 3, \ldots, n+1 \end{array}\right\}, \quad . \quad . \quad . \quad . \quad . \quad . \quad . \quad . \quad . \quad (94)$$

and these, again, imply the corresponding inequalities

$$\left.\begin{array}{c} I_{n+1} < i_{n-s+1} + s\theta \\ s = 1, 2, 3, \ldots, n+1 \end{array}\right\}, \quad . \quad . \quad . \quad . \quad . \quad . \quad . \quad . \quad . \quad (95)$$

because $I_{n-s+1} \leqq i_{n-s+1}$. From (94) it appears that the value of

$$[p^{I_{n+1}}, p^{I_n+\theta}, \ldots, p^{(n+1)\theta}] \text{ is } p^{I_{n+1}},$$

and from (95) that the value of

$$[p^{i_n+\theta}, p^{i_{n-1}+2\theta}, \ldots, p^{(n+1)\theta}]$$

is a power of p superior to $p^{I_{n+1}}$; *i.e.* the equation (93) is impossible. We thus obtain, as a first condition for the resolubility of the proposed system (92), the congruence

$$\frac{D_{n+1}}{D_n} \equiv 0, \text{ mod } M. \quad . \quad . \quad . \quad . \quad . \quad . \quad . \quad . \quad . \quad (96)$$

When this condition is satisfied, we obtain from (93), omitting the term $p^{I_{n+1}}$

and dividing by p^θ, the equation of condition,

$$[p^{I_n}, p^{I_{n-1}+\theta}, p^{I_{n-2}+2\theta}, \ldots, p^{n\theta}] = [p^{i_n}, p^{i_{n-1}+\theta}, p^{i_{n-2}+2\theta}, \ldots, p^{n\theta}],$$

which leads us (as in the preceding article) to the simple formula

$$d = \delta.$$

This equation, therefore, and the congruence (96), express the necessary and sufficient conditions for the resolubility of the proposed redundant system.

When these two conditions are simultaneously satisfied, the number of incongruous solutions is δ. For, if we again consider the proposed system of congruences with respect to the modulus p^θ, and select from it a partial system of n congruences such that the determinants of its augmented matrix, which are necessarily divisible by p^{I_n}, are not divisible by any higher power of p, it is readily seen that every set of values of the indeterminates $x_1, x_2, \ldots, x_n$, which satisfies the partial system, will also (by virtue of the inequality $\theta \leqq I_{n+1} - I_n$) satisfy the remaining congruences of the proposed system. The number of solutions of the proposed system is therefore the same as that of the partial system. And because p^{I_n}, the highest power of p which divides every determinant of order n in the augmented matrix of the proposed system, is also the highest power of p which divides the augmented matrix of the partial system, it follows from the last theorem of Art. 16, that $p^{I_{n-1}}, p^{I_{n-2}}, \ldots$ are the highest powers of p which divide the corresponding orders of determinants in the latter, as well as in the former matrix. The number of solutions of the partial system (and consequently of the proposed system), considered with respect to the modulus p^θ, is therefore expressed by the formula

$$[p^{I_n}, p^{I_{n-1}+\theta}, \ldots, p^{n\theta}];$$

or, finally, the number of solutions of the proposed system, considered with respect to M as modulus, is d or δ.

19. We shall terminate this paper with an elementary theorem, relating to linear systems of equations, which admits of frequent application in other parts of the theory of numbers.

Resuming the notation of Art. 11, we may see from the theorem of that article, that if the system (56) be resoluble for any given values of the numbers $A_{1,0}, A_{2,0}, \ldots, A_{n,0}$, it is also resoluble for any other values of those numbers, respectively congruous, for the modulus D, to the given values; so that the resolubility or irresolubility of the system depends exclusively on the residues of the numbers $A_{i,0}$, mod D. There are D^n possible combinations of these residues, and we shall now show that for D^{n-1} of them the system is resoluble, while for the remaining $D^{n-1}(D-1)$ it is irresoluble. For this purpose let

$$\|a\| \times \|A\| = \left\| \frac{D_n}{D_{n-1}}, \frac{D_{n-1}}{D_{n-2}}, \ldots, \frac{D_1}{D_0} \right\| \times \|A'\|, \quad . \quad . \quad . \quad . \quad . \quad (97)$$

$\|a\|$ denoting a unit-matrix, and $\|A'\|$ a prime matrix of the same type as $\|A\|$, while $D_n, D_{n-1}, \ldots, D_0$ are of course the greatest common divisors of the determinants and minors of $\|A\|$. Let also

$$-C_i = A_{1,0}\, a_{i,1} + A_{2,0}\, a_{i,2} + \ldots + A_{n,0}\, a_{i,n}.$$

The given system is then exactly equivalent to the system

$$\left. \begin{array}{c} \dfrac{D_{n-i+1}}{D_{n-i}} [A'_{i,1} x_1 + A'_{i,2} x_2 + \ldots + A'_{i,n+m} x_{n+m}] = C_i \\ i = 1, 2, 3, \ldots, n \end{array} \right\} . \quad . \quad . \quad . \quad (98)$$

For the resolubility of this system it is requisite that C_i should be divisible by $\dfrac{D_{n-i+1}}{D_{n-i}}$; and this condition is sufficient as well as necessary, because $\|A'\|$ is a prime matrix. Now, of the D or D_n values, incongruous mod D, which may be attributed to C_i, $\dfrac{D_n \times D_{n-i}}{D_{n-i+1}}$ are divisible by $\dfrac{D_{n-i+1}}{D_{n-i}}$; whence it is evident that of the D^n systems of values which may be attributed to $C_1, C_2, \ldots, C_n$, $D^n \div \left[\dfrac{D_1}{D_0} \cdot \dfrac{D_2}{D_1} \cdot \dfrac{D_3}{D_2} \cdots \dfrac{D_n}{D_{n-1}}\right]$, *i.e.* D^{n-1} render the system (98) resoluble. Consequently the given system is also resoluble for D^{n-1}, and no more, of the systems of values that can be attributed (mod D) to $A_{1,0}, A_{2,0}, \ldots, A_{n,0}$.

20. The methods employed in the present paper are without exception such as to be immediately applicable to any species of complex numbers which admit of resolution into actual or ideal prime factors. And the greater part of the results at which we have arrived may be transferred, *mutatis mutandis*, to the theories of such numbers. For example, if in the equations (56) we suppose the constituents of $\|A\|$ to represent complex numbers, it will be found that the criterion for the resolubility or irresolubility of the system, which we have demonstrated in the case of ordinary integers, applies equally in the case of complex numbers; and again, the condition of resolubility of a system of congruences of which the modulus as well as the coefficients are complex numbers, is precisely the same as in the case of common whole numbers; while the expression for the number of the solutions (when the condition of resolubility is satisfied) is simply the *norm* of M.

But without entering into the developments which this extension of the subject of this paper would require, we shall confine ourselves to an application of the result of the preceding article to a demonstration of the fundamental principle in the arithmetical theory of complex numbers, that the number of incongruous residues for any complex modulus is represented by the norm of the modulus.

Let a be one of the roots $a_1, a_2, \ldots, a_n$ of the equation $F_n(x)=0$, which is supposed to be of n dimensions, to be irreducible, and to have all its coefficients integral, that of its first term being unity. Let also $\phi_{n-1}(a)$ be the complex modulus under consideration; its norm, which we shall symbolize by N, is defined by the equation

$$N = N.\phi_{n-1}(a) = \prod_{i=1}^{i=n} \phi_{n-1}(a_i).$$

Consider the N^{2n-1} residues (incongruous, mod N) which are included in the formula

$$R_{2n-2}(a), \qquad (99)$$

where R_{2n-2} denotes an integer function of $2n-2$ dimensions; it is evident that every complex number is congruous, for the modulus $\phi_{n-1}(a)$, to one at least of these N^{2n-1} residues. If R and R' be any two (the same or different) of the same residues, it is also plain that the congruence

$$R \equiv R', \text{ mod } \phi_{n-1}(a)$$

will, or will not, be satisfied, according as it is, or is not, possible to assign two functions of x, $F_{n-1}(x)$ and $\phi_{n-2}(x)$ having integer coefficients, and satisfying the equation

$$F_n(x)\,\phi_{n-2}(x) + F_{n-1}(x)\,\phi_{n-1}(x) = R(x) - R'(x). \qquad (100)$$

This equation is equivalent to a system of $2n-1$ linear equations, in which the unknown quantities are the $2n-1$ coefficients of $\phi_{n-2}(x)$ and $F_{n-1}(x)$, and of which the determinant is the dialytic resultant of $F_n(x)$ and $\phi_{n-1}(x)$, *i.e.* the norm of $\phi_{n-1}(a)$, or N. If then we suppose $R(a)$ to represent any given residue included in the formula (99), it will appear from the theorem of the preceding article that the equation (100) is resoluble for N^{2n-2} different values of $R'(x)$, *i.e.* that every complex number is congruous, for the modulus $\phi_{n-1}(a)$, to precisely N^{2n-2} of the N^{2n-1} residues contained in the formula (99), or that the number of residues, incongruous mod $\phi_{n-1}(a)$, is precisely N.

It is, however, proper to observe that a complete demonstration of this important theorem has already been given by Professor Sylvester (see a paper signed 'Lanavicensis,' in the 'Quarterly Journal of Pure and Applied Mathematics,' vol. iv. pp. 94 and 124).

[The following abstract of the preceding paper was published in the Proceedings of the Royal Society, vol. xi. pp. 87–89.]

The present communication relates to the theory of the solution, in positive and negative integral numbers, of systems of linear indeterminate equations, having integral coefficients. In connexion with this theory, a solution is also given of certain problems relating to rectangular matrices, composed of integral numbers, which are of frequent use in the higher arithmetic. Of this kind are the two following:—

1. 'Given (in integral numbers) the values of the determinants of any rectangular matrix of given dimensions, to find all the matrices, the constituents of which are integers, and the determinants of which have those given values.

2. 'Given any rectangular matrix, the determinants of which have a given number D for their greatest common divisor, to find all the supplementary matrices, which, with the given matrix, form square matrices, of which the determinant is D.'

A solution of particular, but still very important cases of these two problems, has been already given by M. Hermite. The method by which in this paper their general solution has been obtained, depends on an elementary, but apparently fertile, principle in the theory of indeterminate linear systems; viz., that if m be the *index of indeterminateness* of such a system (*i.e.* the excess of the number of indeterminates above the number of really independent equations), it is always possible to assign a set of m solutions, such that the determinants of the matrix formed by them shall admit of no common divisor but unity.

Such a set of solutions is termed a *fundamental* set, and possesses the characteristic property that every other solution of the system can be integrally expressed by means of the solutions contained in it. A set of *independent* solutions is one in which the determinants of the matrix have a finite common divisor, *i.e.* are not all zero. The theory of independent and fundamental sets of solutions in some respects resembles that of independent and fundamental systems of units in Lejeune Dirichlet's celebrated generalisation of the solution of the Pellian equation.

By the aid of the same principle of fundamental sets, the following criterion is obtained for the resolubility or irresolubility of indeterminate linear systems* :—

'A linear system is, or is not, resoluble in integral numbers, according as the greatest common divisor of the determinants of the matrix of the system is, or is not, equal to the corresponding greatest common divisor of its *augmented* matrix.'

* [See note on p. 387. Ed.]

[The matrix of a linear system of equations is, of course, the rectangular matrix formed by the coefficients of the indeterminates; the *augmented* matrix is the matrix derived from that matrix, by adding to it a vertical column composed of the absolute terms of the equations.]

A system of linear congruences may, of course, be regarded as a system of linear indeterminate equations of a particular form; and the criterion for its resolubility or irresolubility is implicitly contained in that just given for any indeterminate system. But this criterion may be expressed in a form in which its relation to the modulus is very clearly seen.

Let $A_{i,1}x_1 + A_{i,2}x_2 + \ldots + A_{i,n}x_n \equiv A_{i,n+1}$, mod M, $i = 1, 2, 3, \ldots, n$

represent a system of congruences; let us denote by $\nabla_n, \nabla_{n-1}, \ldots, \nabla_1, \nabla_0$, the greatest common divisors of the determinant, first minors, &c., of the matrix of the system [so that, in fact, ∇_n is the determinant itself, ∇_1 the greatest common divisor of the coefficients $A_{i,j}$, and $\nabla_0 = 1$]; by $D_n, D_{n-1}, \ldots, D_1, D_0$ the corresponding members for the *augmented* matrix; let also δ_i and d_i respectively represent the greatest common divisors of M with $\dfrac{\nabla_i}{\nabla_{i-1}}$, and of M with $\dfrac{D_i}{D_{i-1}}$,

and put
$$m = d_n \times d_{n-1} \times \ldots \times d_1,$$
$$\mu = \delta_n \times \delta_{n-1} \times \ldots \times \delta_1. \;\ldots.$$

Then the necessary and sufficient condition for the resolubility of the system is
$$m = \mu;$$
and when this condition is satisfied, the number of solutions is precisely m.

The demonstration of this result (which seems to exhaust the theory of these systems) is obtained by means of the following theorem :—

'If $\|A\|$ represent any square matrix in integral numbers, ∇_n its determinant, $\nabla_{n-1}, \nabla_{n-2}, \ldots, \nabla_1, \nabla_0$ the greatest common divisors of its successive orders of minors, it is always possible to assign two unit-matrices $\|\alpha\|$ and $\|\beta\|$, of the same dimensions as $\|A\|$, and satisfying the equation

$$\|A\| = \|\alpha\| \times \left\| \begin{matrix} \dfrac{\nabla_n}{\nabla_{n-1}}, & 0\ , & 0\ , & \ldots, & 0 \\ 0\ , & \dfrac{\nabla_{n-1}}{\nabla_{n-2}}, & 0\ , & \ldots, & 0 \\ 0\ , & 0\ , & \dfrac{\nabla_{n-2}}{\nabla_{n-3}}, & \ldots, & 0 \\ \cdot & \cdot & \cdot & \cdot & \cdot \\ 0\ , & 0\ , & \ldots\ldots\ldots, & & \dfrac{\nabla_1}{\nabla_0} \end{matrix} \right\| \times \|\beta\|.\text{'}$$

The following result (among many which may be deduced from this transformation of a square matrix) admits of frequent applications :—

'If D be the greatest common divisor of the determinants of the matrix of any system of n independent linear equations ; of the D^n sets of values (incongruous mod D) that may be attributed to the absolute terms of the equations, the system is resoluble for D^{n-1}, and irresoluble for $D^{n-1}(D-1)$.'

As an example of the use that may be made of this result, it is shown in conclusion, that it supplies an immediate demonstration of a fundamental principle in the general theory of complex integral numbers, composed of the root of any irreducible equation having its first coefficient unity and all its coefficients integral; viz. that the number of incongruous residues, for any modulus, is always represented by the norm of the modulus. A demonstration of this principle has, however, already been given in the 'Quarterly Journal of Pure and Applied Mathematics,' in a paper signed *Lanavicensis* * ; to whom, therefore, the honour of priority in this inquiry is due.

---

* [The author of this paper was Professor Sylvester.]

XIII.

ON THE CRITERION OF RESOLUBILITY IN INTEGRAL NUMBERS OF THE INDETERMINATE EQUATION

$$f = ax^2 + a'x'^2 + a''x''^2 + 2bx'x'' + 2b'xx'' + 2b''x'x = 0.$$

[Proceedings of the Royal Society, vol. xiii. pp. 110, 111. Received January 20; Read January 28, 1864.]

IT is sufficient to consider the case in which f is an indefinite form of a determinant different from zero. We may also suppose that f is primitive, *i.e.* that the six numbers a, a', a'', b, b', b'' do not admit of any common divisor. We represent by Ω the greatest common divisor of the minors of the matrix of f, by $\Delta\Omega^2$ the determinant of f, and by ΩF the contravariant of f, *i.e.* the form

$$(b^2 - a'a'')\,x^2 + \dots;$$

$\Omega\Delta^2$ will then be the determinant of F, and Δf its contravariant. By $\overline{\Omega}$, $\overline{\Delta}$, and $\overline{\Omega\Delta}$ we denote the quotients obtained by dividing Ω, Δ, and $\Omega\Delta$ by the greatest squares contained in them respectively; ω is any uneven prime dividing $\overline{\Omega}$, but not $\overline{\Delta}$; δ is any uneven prime dividing $\overline{\Delta}$, but not $\overline{\Omega}$; and θ is any uneven prime dividing both $\overline{\Omega}$ and $\overline{\Delta}$, and consequently not dividing $\overline{\Omega\Delta}$. We may then enunciate the theorem:—

'The equation $f = 0$ will, or will not, be resoluble in integral numbers different from zero according as the equations included in the formulae

$$\left(\frac{\overline{\Omega}}{\delta}\right) = \left(\frac{F}{\delta}\right), \quad \left(\frac{\overline{\Delta}}{\omega}\right) = \left(\frac{f}{\omega}\right), \quad \left(\frac{-\overline{\Omega\Delta}}{\theta}\right) = \left(\frac{f}{\theta}\right)\left(\frac{F}{\theta}\right)$$

are, or are not, satisfied.'

The symbols $\left(\frac{\overline{\Omega}}{\delta}\right)$, $\left(\frac{\overline{\Delta}}{\omega}\right)$, and $\left(\frac{-\overline{\Omega\Delta}}{\theta}\right)$ are the quadratic symbols of Legendre; the symbols $\left(\frac{F}{\delta}\right)$, $\left(\frac{F}{\theta}\right)$, $\left(\frac{f}{\omega}\right)$, $\left(\frac{f}{\theta}\right)$ are generic characters of f (see the memoir of Eisenstein, 'Neue Theoreme der höheren Arithmetik,' in his 'Mathematische Abhandlungen,' p. 185, or in Crelle's Journal, vol. XXXV. p. 125).

The theorem includes those of Legendre and Gauss on the resolubility of equations of the form $ax^2+a'x'^2+a''x''^2=0$ (Legendre, Théorie des Nombres, vol. i. p. 47; Gauss, Disq. Arith., Arts. 294, 295, & 298). It is equally applicable whether the coefficients and indeterminates of f are real integers or complex integers of the type $p+qi$.

It will be observed that if $f, f', f'', \ldots$ are forms contained in the same genus, the equations $f=0$, $f'=0$, $f''=0$, &c. are either all resoluble or all irresoluble.

XIV.

ON THE ORDERS AND GENERA OF QUADRATIC FORMS CONTAINING MORE THAN THREE INDETERMINATES.

[Proceedings of the Royal Society, vol. xiii. pp. 199–203. Received March 22; Read April 21, 1864.]

LET us represent by f_1 a homogeneous form or quantic of any order containing n indeterminates; by $(a^{(1)})$, a square matrix of order n; by $(a^{(i)})$ its ith derived matrix, *i.e.* the matrix of order $\dfrac{\underline{|n}}{\underline{|i}\;\underline{|n-i}} = I$, the constituents of which are the minor determinants of order i of the matrix $(a^{(1)})$; and lastly, by f_i, a form of any order containing I indeterminates, the coefficients of which depend on the coefficients of f_1. When f_1 is transformed by $(a^{(1)})$, let f_i be transformed by $(a^{(i)})$; if, after division or multiplication by a power of the modulus of transformation, the metamorphic of f_i depends on the metamorphic of f_1, in the same way in which f_i depends on f_1, f_i is said to be a concomitant of the ith species of f_1. Thus, a concomitant of the first species is a covariant; a concomitant of the $(n-1)$th species is a contravariant; if $n=2$ there are only covariants; if $n=3$ there are only covariants and contravariants; but if $n>3$, there will exist in general concomitants of the intermediate species.

There is an obvious difference between covariants and contravariants on the one hand, and the intermediate concomitants on the other. The number of indeterminates in a covariant or contravariant is the same as in its primitive; in an intermediate concomitant, the number of indeterminates is always greater than in its primitive. Again, to every metamorphic of a covariant or contravariant, there corresponds a metamorphic of its primitive; whereas, in the case of a concomitant of the intermediate order i, a metamorphic of the primitive will correspond, not to every metamorphic of the concomitant, but only to such metamorphics as result from transformations the matrices of which are the ith derived matrices of matrices of order n.

It is also obvious that, besides the $n-1$ species of concomitance here defined, there are, when $n > 3$, an infinite number of other species of concomitance of the same general nature. For, from any derived matrix we may form another derived matrix, and so on continually; and to every such process of derivation a distinct species of concomitance will correspond.

The notion of intermediate concomitance appears likely to be of use in many researches; in what follows, it is employed to obtain a definition of the ordinal and generic characters of quadratic forms containing more than three indeterminates. (The case of quadratic forms containing three indeterminates has been considered by Eisenstein in his memoir, 'Neue Theoreme der höheren Arithmetik,' Crelle, vol. xxxv. pp. 121 and 125.) Let

$$f_1 = \sum_{p=1}^{p=n} \sum_{q=1}^{q=n} A^{(1)}_{p,q}\, x_p\, x_q$$

represent a quadratic form of n indeterminates; let $(A^{(1)})$ be the symmetrical matrix of this form, and $(A^{(i)})$ the ith derived matrix of $(A^{(1)})$; $(A^{(i)})$ will also be a symmetrical matrix, and the quadratic form

$$f_i = \sum_{p=1}^{p=I} \sum_{q=1}^{q=I} A^{(i)}_{p,q}\, X_p\, X_q \quad \ldots\ldots\ldots \quad \text{(A)}$$

will be a concomitant of the ith species of f_1. It is immaterial what principle of arrangement is adopted in writing the quadratic matrix $(A^{(i)})$, and the transforming matrix $(a^{(i)})$; provided only that the arrangement be the same in the two matrices, and that in each matrix it be the same in height and in breadth.

For example, if

$$f_1 = a_1 x_1^2 + a_2 x_2^2 + a_3 x_3^2 + a_4 x_4^2 + 2b_1 x_1 x_2 + 2b_2 x_1 x_3 + 2b_3 x_1 x_4 + 2b_4 x_2 x_3 + 2b_5 x_2 x_4 + 2b_6 x_3 x_4$$

be a quadratic form containing four indeterminates, the form f_2 defined by the equation

$$\begin{aligned}
f_2 = {} & (b_1^2 - a_1 a_2) X_1^2 + (b_2^2 - a_1 a_3) X_2^2 + (b_3^2 - a_1 a_4) X_3^2 \\
& + (b_4^2 - a_2 a_3) X_4^2 + (b_5^2 - a_2 a_4) X_5^2 + (b_6^2 - a_3 a_4) X_6^2 \\
& + 2(b_1 b_2 - a_1 b_4) X_1 X_2 + 2(b_1 b_3 - a_1 b_5) X_1 X_3 \\
& - 2(b_1 b_4 - a_2 b_2) X_1 X_4 - 2(b_1 b_5 - a_2 b_3) X_1 X_5 \\
& - 2(b_2 b_5 - b_3 b_4) X_1 X_6 + 2(b_2 b_3 - a_1 b_6) X_2 X_3 \\
& + 2(b_2 b_4 - a_3 b_1) X_2 X_4 - 2(b_1 b_6 - b_3 b_4) X_2 X_5 \\
& - 2(b_2 b_6 - a_3 b_3) X_2 X_6 - 2(b_1 b_6 - b_2 b_5) X_3 X_4 \\
& + 2(b_3 b_5 - a_4 b_1) X_3 X_5 + 2(b_3 b_6 - a_4 b_2) X_3 X_6 \\
& + 2(b_4 b_5 - a_2 b_6) X_4 X_5 - 2(b_4 b_6 - a_3 b_5) X_4 X_6 \\
& + 2(b_5 b_6 - a_4 b_4) X_5 X_6
\end{aligned}$$

is the concomitant of the second species of f_1.

The $n-1$ forms defined by the formula (A), of which the first is the form f_1 itself, and the last the contravariant of f_1, we shall term *the fundamental concomitants of* f_1; in contradistinction to those other quadratic concomitants (infinite in number) of which the matrices are the symmetrical matrices that may be derived, by multiplicate derivation, from $(A^{(1)})$. Passing to the arithmetical theory of quadratic forms, *i.e.* supposing that the constituents of $(A^{(1)})$ are integral numbers, we shall designate by $\nabla_1, \nabla_2, \ldots, \nabla_n$ the greatest common divisors (taken positively) of the minors of different orders of the matrix $(A^{(1)})$, so that, in particular, ∇_1 is the greatest common divisor of its constituents, and ∇_n is the absolute value of its determinant, here supposed to be different from zero. By the primary divisor of a quadratic form we shall understand the greatest common divisor of the coefficients of the squares and double rectangles in the quadratic form; by the secondary divisor we shall understand the greatest common divisor of the coefficients of the squares and of the rectangles; so that the primary divisor is equal to, or is half of, the secondary divisor, according as the quadratic form (to use the phraseology of Gauss) is derived from a form properly or improperly primitive. It will be seen that $\nabla_1, \nabla_2, \ldots, \nabla_{n-1}$ are the primary divisors of the forms $f_1, f_2, \ldots, f_{n-1}$ respectively.

We now consider the totality of arithmetical quadratic forms, containing n indeterminates, and having a given *index of inertia* *, and a given deter-

* If a quadratic form be reduced to a sum of squares by any linear transformation, the number of positive and of negative squares is the same, whatever be the real transformation by which the

minant. The distribution of these forms into orders depends on the following principle :—

'Two forms belong to the same order when the primary and secondary divisors of their corresponding concomitants are identical.'

Since, as has been just pointed out, there are, besides the fundamental concomitants, an infinite number of other concomitants, it is important to know whether, in order to obtain the distribution into orders, it is, or is not, necessary to consider those other concomitants. With regard to the primary divisors, it can be shown that it is unnecessary to consider any concomitants other than the fundamental ones; *i.e.* it can be shown that the equality of the primary divisors of the corresponding fundamental concomitants of two quadratic forms implies the equality of the primary divisors of all their corresponding concomitants. And it is probable (but it seems difficult to prove) that the same thing is true for the secondary divisors also.

Confining our attention (in the next place) to the forms contained in any given order, we proceed to indicate the principle from which the subdivision of that order into genera is deducible.

If F_1 be any quadratic form containing r indeterminates, and F_2 be its concomitant of the second species, we have the identical equation

$$F_1(x_1, x_2, \ldots, x_r) \times F_1(y_1, y_2, \ldots, y_r) - \tfrac{1}{4}\left[\sum_{k=1}^{k=r} y_k \frac{dF_1}{dx_k}\right]^2 = F_2\begin{pmatrix} x_1, x_2, \ldots, x_r \\ y_1, y_2, \ldots, y_r \end{pmatrix}, \quad \text{(E)}$$

in which the symbol $F_2\begin{pmatrix} x_1, x_2, \ldots, x_r \\ y_1, y_2, \ldots, y_r \end{pmatrix}$ indicates that the determinants $\begin{pmatrix} x_1, x_2, \ldots, x_r \\ y_1, y_2, \ldots, y_r \end{pmatrix}$ are to be taken for the indeterminates of F_2, the order in which they are taken being the same as the order in which the determinants of any two horizontal rows of the matrix of F_1 are taken in forming the matrix of F_2. Let $\theta_i = \frac{1}{\nabla_i} f_i$ for every value of i from 1 to $n-1$; it will be found that, if we form the concomitant of the second species of θ_i, its primary divisor is the

---

reduction is effected. For the index of inertia we may take the number of the positive squares; it is equal to the number of continuations of sign in a series of ascending principal minors of the matrix of the quadratic form; the series commencing with unity, *i.e.* with a minor of order 0, and each minor being so taken as to contain that which precedes it in the series (see Professor Sylvester 'On formulae connected with Sturm's theorem,' Phil. Trans. vol. cxliii. p. 481).

quotient $\frac{\nabla_{i+1}}{\nabla_i} \div \frac{\nabla_i}{\nabla_{i-1}}$, which, as has been shown elsewhere (see Phil. Trans. vol. cli. p. 317*), is always an integral number. Let δ_i be any uneven prime dividing $\frac{\nabla_{i+1}}{\nabla_i} \div \frac{\nabla_i}{\nabla_{i-1}}$; we infer from the identity (B) that the numbers prime to δ_i, which can be represented by θ_i, are either all quadratic residues of δ_i, or all non-quadratic residues of δ_i. In the former case, we attribute to f_i the *particular character* $\left(\frac{\theta_i}{\delta_i}\right) = +1$; in the latter, the particular character $\left(\frac{\theta_i}{\delta_i}\right) = -1$. If $\nabla_1 = 1$, *i.e.* if the form f_1 itself do not admit of any primary divisor beside unity (which is the only important case), the product

$$\left(\frac{\nabla_n}{\nabla_{n-1}} \div \frac{\nabla_{n-1}}{\nabla_{n-2}}\right) \times \left(\frac{\nabla_{n-1}}{\nabla_{n-2}} \div \frac{\nabla_{n-2}}{\nabla_{n-3}}\right) \times \dots$$

is equal to $\frac{\nabla_n}{\nabla_{n-1}}$; whence, inasmuch as every prime that divides ∇_n also divides $\frac{\nabla_n}{\nabla_{n-1}}$, it appears that a primitive quadratic form will always have one particular character at least, with respect to every uneven prime dividing its determinant, and will have more than one if the uneven prime divide more than one of the quotients $\frac{\nabla_{i+1}}{\nabla_i} \div \frac{\nabla_i}{\nabla_{i-1}}$.

The subdivision of an order into genera can now be effected by assigning to the same genus all those forms whose particular characters coincide. But it remains to consider whether the above enumeration of particular characters is complete. It is evident that we might apply the theorem (B) to other concomitants besides those included in the fundamental system; and it might appear as if in this manner we could obtain other particular characters besides those which we have given. But it can be shown that such other particular characters are implicitly contained in ours; *i.e.* it can be shown that two quadratic forms which coincide in respect of the particular characters deducible from their fundamental concomitants will also coincide in respect of the particular characters deducible from any other concomitant. Again, it will be found that if the determinant be uneven, there are no particular characters with regard to 4 or 8. For this case, therefore, our enumeration is complete. But when the determinant is even, besides the particular characters arising from its uneven prime

* [See p. 395 of this volume.]

divisors, there may also be particular characters with regard to 4 or 8. There is no difficulty in enumerating these particular characters; nevertheless we suppress the enumeration here, not only because it would require a detailed distinction of cases, but also because there appears to be some difficulty in showing that the characters with regard to 4 or 8, which may arise from the excluded concomitants, are virtually included in those which arise from the concomitants of the fundamental set.

XV.

ON COMPLEX BINARY QUADRATIC FORMS.

[Proceedings of the Royal Society, vol. xiii. pp. 278–298. Received May 18; Read June 16, 1864.]

THE purpose of this note is to extend to complex quadratic forms some important investigations of Gauss relating to real quadratic forms. We shall consider in order (I.) the definition of the genera, (II.) the theory of composition, (III.) the determination of the number of ambiguous classes, (IV.) the representation of forms of the principal genus by ternary quadratic forms of determinant 1. For the comparison of the numbers of classes of different orders, we may refer to a paper by M. Lipschitz (Crelle's Journal, vol. liv. p. 193); and, for the principles of the theory of complex numbers and complex quadratic forms, to Lejeune Dirichlet's memoir, 'Recherches sur les formes quadratiques à coefficients et à indéterminées complexes' (Crelle, vol. xxiv. p. 291).

I. *The Definition of the Genera.*

Let $f=(a, b, c)$ be an uneven* primitive form of determinant D, and $m=ax^2+2bxy+cy^2$, $m'=ax'^2+2bx'y'+cy'^2$ two numbers represented by f. The generic characters of f are deducible from the equation

$$(ax^2+2bxy+cy^2)(ax'^2+2bx'y'+cy'^2)=(axx'+b[xy'+x'y]+cyy')^2-D(xy'-x'y)^2,$$

or, as we shall write it, $$mm'=P^2-DQ^2.$$

---

* A primitive form (a, b, c) is uneven, semi-even, or even, according as the greatest common divisor of a, $2b$, c is 1, $1+i$, or $(1+i)^2$; *i.e.* in Lejeune Dirichlet's nomenclature, according as (a, b, c) is of the first, second, or third species. In this paper, when we speak of an uneven, semi-even, or even form or class, we shall always suppose the form or class to be primitive. A semi-even number is a number divisible by $1+i$, but not by $(1+i)^2$.

Thus, supposing that p is an uneven prime dividing D, and that m and m' are prime to p, the numbers prime to p which are represented by f are either all quadratic residues of p, or else all non-quadratic residues of p; in the former case, we attribute to f the character $\left[\frac{f}{p}\right] = +1$, in the latter, the character $\left[\frac{f}{p}\right] = -1$.

Again, to investigate the supplementary characters relating to powers of the even prime $1+i$, let $m = \mu + \mu' i$ be an uneven number, μ and μ' representing real numbers, and, for brevity, let

$$(-1)^{\frac{1}{4}(Nm-1)} = \alpha,$$
$$(-1)^{\frac{1}{8}[(\mu+\gamma\mu')^2-1]} = \beta,$$
$$(-1)^{\mu'} = \gamma.$$

The values of the units, or *characters*, α, β, γ depend on the residue of m for the modulus $(1+i)^5$, as is shown in the following table.

TABLE I.

$m \equiv$	$\alpha =$	$\beta =$	$\gamma =$
± 1	$+1$	$+1$	$+1$
$\pm i$	$+1$	$+1$	-1
± 3	$+1$	-1	$+1$
$\pm 3i$	$+1$	-1	-1
$\pm(1-2i)$	-1	$+1$	$+1$
$\pm(2+i)$	-1	$+1$	-1
$\pm(1+2i)$	-1	-1	$+1$
$\pm(2-i)$	-1	-1	-1

An inspection of the table shows that, of the sixteen uneven residues of $(1+i)^5$, eight have the character $\omega = +1$, and eight the character $\omega = -1$, ω representing any one of the seven characters α, β, γ, $\beta\gamma$, $\alpha\gamma$, $\alpha\beta$, $\alpha\beta\gamma$. It will also be seen that any character of a product of two uneven factors is found by multiplying together the corresponding characters of the factors; so that, conversely, according as any character of a product of two uneven factors is $+1$ or -1, the two factors agree or differ in respect of that character.

The next table assigns the supplementary characters proper to any given determinant; they depend on the residue of the determinant for the modulus $(1+i)^5$.

TABLE II.

$D \equiv$	Characters	$D \equiv$	Characters
$\pm(\ +i)$	β	± 1	γ
$\pm(1-i)$	$\alpha\beta$	$\pm i$	α
$\pm(3+i)$	$\alpha\beta\gamma$	± 3	γ
$\pm(3-i)$	$\beta\gamma$	$\pm 3i$	α
± 2	$\alpha,\ \gamma$	$\pm(1-2i)$	γ
$\pm 2i$	γ	$\pm(2+i)$	$\alpha\gamma$
$2(1+i)$	$\alpha\beta,\ \gamma$	$\pm(1+2i)$	γ
$2(1-i)$	$\beta,\ \gamma$	$\pm(2-i)$	$\alpha\gamma$
4	$\alpha,\ \gamma$		
0	$\alpha,\ \beta,\ \gamma$		

Of the eighteen propositions contained in this table, it will suffice to enunciate and demonstrate one :—

'If $D \equiv \pm(3+i)$, mod $(1+i)^5$, and f is an uneven form of determinant D, the uneven numbers represented by f, all have the character $\alpha\beta\gamma = +1$, or else all have the character $\alpha\beta\gamma = -1$.'

In the equation $P^2 - DQ^2 = mm'$, let us suppose that m and m' are uneven; then P is uneven because D is semi-even; also $Q^2 \equiv \pm 1$, $\pm 2i$, 4, or 0, mod $(1+i)^5$, according as the index of the highest power of $1+i$ dividing Q is 0, 1, 2, or > 2.

If Q is uneven, $\quad mm' \equiv \pm 3i$ or $\equiv \pm(2+i)$, mod $(1+i)^5$;
if Q is semi-even, $\quad mm' \equiv \pm(1+2i)$, mod $(1+i)^5$;
if Q is even, $\quad mm' \equiv \pm 1$, mod $(1+i)^5$;

i.e. in all three cases, mm' has the character $\alpha\beta\gamma = 1$, and m and m' both have the character $\alpha\beta\gamma = +1$, or else both have the character $\alpha\beta\gamma = -1$.

We add a third table for the purpose of distinguishing between the possible and impossible genera. In this table, S^2 is the greatest square dividing D, P is uneven and primary *, I is the index of the highest power of $1+i$ dividing S, ϖ represents an uneven prime dividing P, σ an uneven prime dividing S but not P. For brevity, the symbols ϖ and σ are written instead of $\left[\frac{f}{\varpi}\right]$ and $\left[\frac{f}{\sigma}\right]$.

TABLE III.

(i) $D=PS^2$, $P\equiv 1$, mod 4.

$I=0, 1$	ϖ	σ, γ
$I=2$	ϖ	σ, γ, a
$I>2$	ϖ	σ, γ, a, β.

(ii) $D=PS^2$, $P\equiv 1+2i$, mod 4.

$I=0, 1$	ϖ, γ	σ
$I=2$	ϖ, γ	σ, a
$I>2$	ϖ, γ	σ, a, β.

(iii) $D=iPS^2$, $P\equiv 1$, mod 4.

$I=0$	ϖ, a	σ
$I=1, 2$	ϖ, a	σ, γ
$I>2$	ϖ, a	σ, γ, β.

(iv) $D=iPS^2$, $P\equiv 1+2i$, mod 4.

$I=0$	$\varpi, a\gamma$	σ
$I=1, 2$	ϖ, a, γ	σ
$I>2$	ϖ, a, γ	σ, β.

(v) $D=(1+i)PS^2$, $P\equiv 1$, mod 4.

$I=0$	ϖ, β	σ
$I=1$	ϖ, β	σ, γ
$I>1$	ϖ, β	σ, γ, a.

(vi) $D=(1+i)PS^2$, $P\equiv 1+2i$, mod 4.

$I=0$	$\varpi, \beta\gamma$	σ
$I=1$	ϖ, β, γ	σ
$I>1$	ϖ, β, γ	σ, a.

* By a primary uneven number we understand (with Lejeune Dirichlet) an uneven number $\mu+\mu' i$ satisfying the congruences $\mu\equiv 1$, mod 4, $\mu'\equiv 0$, mod 2.

(vii) $D = i(1+i)PS^2$, $P \equiv 1$, mod 4.

$I=0$	ϖ, $\alpha\beta$	σ
$I=1$	ϖ, $\alpha\beta$	σ, γ
$I>1$	ϖ α, β	σ, γ.

(viii) $D = i(1+i)PS^2$, $P \equiv 1+2i$, mod 4.

$I=0$	ϖ, $\alpha\beta\gamma$	σ
$I=1$	ϖ, $\alpha\beta$, γ	σ
$I>1$	ϖ, α, β, γ	σ.

The characters preceding the vertical line by which the table is divided are not independent, but are subject to the condition (arising from the laws of quadratic residues) that their product must be a positive unit. To show that this is so, let $D = i^{\alpha'}(1+i)^{\beta'}PS^2$, where α' and β' are each either 0 or 1; also let $\gamma' = 0$, or 1, according as $P \equiv 1$, or $\equiv 1+2i$, mod 4. If m is a number prime to $(1+i)D$ and capable of primitive representation * by f, the congruence $\omega^2 \equiv D$, mod m, is resoluble; and its resolubility implies the condition

$$\left[\frac{D}{m}\right] = \left[\frac{i^{\alpha'}}{m}\right] \times \left[\frac{(1+i)^{\beta'}}{m}\right] \times \left[\frac{P}{m}\right] = 1.$$

But, by the laws of quadratic residues,

$$\left[\frac{i}{m}\right] = \alpha, \qquad \left[\frac{1+i}{m}\right] = \beta, \qquad \left[\frac{P}{m}\right] = \gamma^{\gamma'}\left[\frac{m}{P}\right];$$

and the condition just written becomes

$$\alpha^{\alpha'}\beta^{\beta'}\gamma^{\gamma'}\left[\frac{m}{P}\right] = 1,$$

which is coincident with that indicated in the table. Thus (as in the real theory) one half of the whole number of assignable generic characters are impossible †; we shall presently obtain a different proof of this result, and shall also show that the remaining half correspond to actually existing genera.

For the characters of a semi-even form f, it is convenient to take the characters of the numbers represented by $\frac{f}{1+i}$; and, for the characters of an

---

* If $m = ax^2 + 2bxy + cy^2$, the representation of m by (a, b, c) is said to be primitive when the values of the indeterminates are relatively prime.

† The determinant is supposed not to be a square.

even form, the characters of the numbers represented by $\frac{f}{2i}$. The following table will serve to form the complete generic character in each case.

For a semi-even form.

(i) $D \equiv PS^2, \ P \equiv 1$, mod 4.

$$I = 0 \mid \varpi \quad \mid \sigma.$$

(ii) $D \equiv PS^2, \ P \equiv 1 + 2i$, mod 4.

$$I = 0 \mid \varpi, \gamma \mid \sigma.$$

For an even form.

$$I = 0 \mid \varpi \quad \mid \sigma.$$

II. *The Theory of Composition.*

The theory of composition given in the 'Disquisitiones Arithmeticae' is immediately applicable to complex quadratic forms. There are, however, a few points to which we must direct attention.

(1) If m_1, m_2, m_3 are the greatest common divisors of

$$a, \ 2b, \ c; \quad a, \ (1+i)\,b, \ c; \quad a, \ b, \ c,$$

we have

(i) $m_1 = m_2 = m_3$,

(ii) $m_1 = m_2 = (1+i)\,m_3$,

(iii) $m_1 = (1+i)\,m_2 = (1+i)^2\,m_3$,

according as (a, b, c) either is or is derived from (i) an uneven, (ii) a semi-even, (iii) an even primitive. Hence the order of a form is given when m_1 and m_3 are given. Thus, if F is compounded of f and f', and if $M_1\, M_2\, M_3$, $m_1\, m_2\, m_3$, $m'_1\, m'_2\, m'_3$ refer to F, f, f' respectively, the order of F is completely determined by the two theorems:—'M_1 is the product of m_1 and m'_1.' '$\frac{M_1}{M_3}$ is the least common multiple of $\frac{m_1}{m_3}$ and $\frac{m'_1}{m'_3}$.' (Gauss's 5th and 6th conclusions, Disq. Arith., Art. 235.)

It will be found that Gauss's proof of these theorems can be transferred to the complex theory; only, when f and f' are both semi-even or derived from semi-even primitives, the proof of the sixth conclusion is incomplete and, while showing that F cannot be derived from an uneven primitive, fails to show whether it is derived from a semi-even or from an even primitive. But, in the same way in which Gauss has shown that M_1 is divisible by $m_1 \times m'_1$, it can also be shown that M_2 is divisible by $m_2 \times m'_2$*; *i.e.* in the case which we are considering, M_2 is divisible by M_1, because $m_2 = m_1$, $m'_2 = m'_1$, and $m_1 m'_1 = M_1$. Therefore $M_2 = M_1$, and F is derived from a semi-even primitive in accordance with our enunciation of Gauss's sixth conclusion.

(2) In the real theory, when two or more forms are compounded, each form may be taken either directly or inversely; but, however the forms are taken, the determinant of the resulting form is the same. In the complex theory, not only may each of the forms to be compounded be taken in either of two different ways, but also the determinant of the resulting form may receive either of two values, differing, however, only in sign; and it is important to attend to the ambiguities which thus arise. If a complex rational number n be written in the form $i^\lambda (1+i)^\mu \frac{P}{Q}$, where λ is 0, 1, 2, or 3, μ is any positive or negative integer, and P, Q are primary uneven complex integers, we may term i^λ the sign of n. Let F, of which the determinant is D, be transformed into the product $f_1 \times f_2 \times \ldots \times f_h$, by a substitution $[X, Y]$ linear and homogeneous in respect of h binary sets; we have, as in the real theory, h equations of the type

$$\left(\frac{dX}{dx_k}\frac{dY}{dy_k} - \frac{dX}{dy_k}\frac{dY}{dx_k}\right)^2 = \frac{d_k}{D} \times \frac{\Pi . f^2}{f_k^2},$$

d_k representing the determinant of f_k. Let

$$n_k = \left(\frac{dX}{dx_k}\frac{dY}{dy_k} - \frac{dX}{dy_k}\frac{dY}{dx_k}\right) \div \frac{\Pi . f}{f_k},$$

so that $n_k^2 = \frac{d_k}{D}$; if i^{λ_k} is the sign of n_k, we shall say that f_k is taken with the sign i^{λ_k}. We can thus enunciate the theorem :—'Forms compounded of the same forms, taken with the same signs, are equivalent.' If $f_1, f_2, \ldots, f_h$ are given forms which it is required to compound, the signs of $d_1, d_2, \ldots, d_h$ must be all real or else all unreal; and the sign of D will be real or unreal accordingly. The value of D (irrespective of its sign) is ascertained as in the real theory; but it may receive at our option, in the one case, either of the two real signs, and in the other case, either of the two unreal signs. And, whichever sign we give to D, the form f_k may be taken with either of the two real signs, if the sign of

* Disq. Arith., Art. 235. The proof that $2(bb' + \Delta)$ and $2(bb' - \Delta)$ are divisible by $m_1 \times m'_1$, may be employed (*mutatis mutandis*) to show that $(1+i)(bb'+\Delta)$ and $(1+i)(bb'-\Delta)$ are divisible by $m_2 \times m'_2$.

$\frac{d_k}{D}$ is $+1$, and with either of the two unreal signs, if the sign of $\frac{d_k}{D}$ is -1. In the important case in which $d_1, d_2, \ldots$ all have the same sign, we shall always suppose D to have that sign, and $f_1, f_2, \ldots$ to be all taken with the sign $+1$. Adopting this convention, we see that the class compounded of given classes of the same determinant, or of different determinants having the same sign, is defined without ambiguity.

(3) By the general formulae of M. Arndt (Crelle, vol. lvi. p. 69), which on account of their great utility we transcribe here, we can always obtain a form (A, B, C), compounded in any given manner of two forms, (a, b, c) and (a', b', c'), of which the determinants d and d' are to one another as two squares.

$$A = \frac{aa'}{\mu^2},$$

$$\left.\begin{aligned} \frac{an'}{\mu} B &\equiv \frac{ab'}{\mu} \\ \frac{a'n}{\mu} B &\equiv \frac{a'b}{\mu} \\ \frac{bn' + b'n}{\mu} B &\equiv \frac{bb' + Dnn'}{\mu} \end{aligned}\right\}, \text{ mod } A,$$

$$C = \frac{B^2 - D}{A}.$$

In these formulae, D is the greatest common divisor of dm'^2 and $d'm^2$, m and m' representing the greatest common divisors of a, $2b$, c, and a', $2b'$, c'; n and n' are the square roots of $\frac{d}{D}$ and $\frac{d'}{D}$; μ is the greatest common divisor of an', $a'n$, and $bn' + b'n$. The signs of D, n, and n' are given, because the manner of the composition is supposed to be given; to μ we may attribute any sign we please, because the forms (A, B, C) and $(-A, B, -C)$ are equivalent.

(4) If $F = (A, B, C)$ is compounded of two primitive forms f and f', and if M is the highest power of $1+i$ dividing A, B, C, so that M is 1, or $1+i$, or $(1+i)^2$, the complete character of the primitive form $\frac{1}{M}F$ is obtained by the following rule:—

'If ω is any character common to f and f', $\frac{1}{M}F$ will have the character $\omega = +1$ or $\omega = -1$, according as f and f' agree or differ in respect of that character.'

In comparing the characters of f and f', it is to be observed that if ω and ω' are two supplementary characters of f, and $\omega\times\omega'$ a supplementary character of f', $\omega\times\omega'$ is to be regarded as a character common to f and f'.

(5) Let us represent by (1), (σ), and (Σ)*, respectively, the principal uneven, semi-even, and even classes of determinant D; *i.e.* the classes containing the forms

$$(1,\,0,\,-D),\qquad \left(1+i,\,1,\,-\frac{D-1}{1+i}\right),\qquad\text{and}\qquad \left(2i,\,i^k,\,-\frac{D-i^{2k}}{2i}\right),$$

the existence of the last two classes implying the congruences

$$D\equiv 1,\ \text{mod}\ 2,\qquad D\equiv i^{2k},\ \text{mod}\ 4,$$

respectively. Employing the formulae of M. Arndt, we find

$$(f)\times(1)=(f),$$

if (f) is any class of determinant D;

$$(f)\times(\sigma)=(1+i)\,(f),$$

if f is derived from a semi-even or even primitive;

$$(f)\times(\Sigma)=2i\,(f),$$

if f is derived from an even primitive; and, in particular,

$$(1)\times(1)=(1),\quad (\sigma)\times(\sigma)=(1+i)\,(\sigma),\quad (\Sigma)\times(\Sigma)=2i\,(\Sigma).$$

Also, if (f) and (f^{-1}) are two opposite primitive classes,

$$(f)\times(f)^{-1}=(1),\quad\text{or}\quad=(1+i)\,(\sigma),\quad\text{or}\quad=2i\,(\Sigma),$$

according as f and f^{-1} are uneven, semi-even, or even. Hence the three equations

$$(f_1)\times(\phi)=(f_2),\quad (f_1)\times(\phi)=(1+i)\,(f_2),\quad (f_1)\times(\phi)=2i\,(f_2),$$

in which (f_1) and (f_2) are given primitive classes, uneven in the first, semi-even in the second, and even in the third, are respectively satisfied by the uneven, semi-even, and even classes

$$(\phi)=(f_2)\times(f_1)^{-1},\qquad (\phi)=\frac{(f_2)\times(f_1)^{-1}}{1+i},\qquad (\phi)=\frac{(f_2)\times(f_1)^{-1}}{2i},$$

but by no other classes whatever. Again, let $D=\Delta m^2$, and let the forms

$$(mp,\,mq,\,mr),\quad ([1+i]\,mp,\,mq,\,[1+i]\,mr),\quad (2imp,\,mq,\,2imr)$$

represent classes derived by the multiplier m from uneven, semi-even, and even primitives of determinant Δ; in all three forms we suppose p prime to $2D$;

---

* It is often convenient to symbolize a class by placing within brackets a symbol representing a form contained in the class; thus (f) may be used to symbolize the class containing the form f.

in the second and third, we suppose q uneven and $\Delta \equiv 1$, mod 2; in the third, we suppose $\Delta \equiv i^{2k}$, mod 4. The formulae of M. Arndt will then establish the six equations :—

$$(m,\ 0,\ -\Delta m) \times (p,\ mq,\ m^2 r) = (mp,\ mq,\ mr),$$

$$\left([1+i]m,\ m,\ -m\frac{\Delta-1}{1+i}\right) \times (p,\ mq,\ 2im^2 r) = ([1+i]mp,\ mq,\ [1+i]mr),$$

$$\left(2im,\ i^k m,\ -m\frac{\Delta-i^{2k}}{2i}\right) \times (p,\ mq,\ -4m^2 r) = (2imp,\ mq,\ 2imr),$$

$$\left([1+i]m,\ m,\ -m\frac{\Delta-1}{1+i}\right) \times ([1+i]p,\ mq,\ [1+i]m^2 r)$$
$$= (1+i)\,([1+i]mp,\ mq,\ [1+i]mr),$$

$$\left(2im,\ i^k m,\ -m\frac{\Delta-i^{2k}}{2i}\right) \times ([1+i]p,\ mq,\ 2i[1+i]m^2 r) = (1+i)\,(2imp,\ mq,\ 2imr),$$

$$\left(2im,\ i^k m,\ -m\frac{\Delta-i^{2k}}{2i}\right) \times (2ip,\ mq,\ 2im^2 r) = 2i\,(2imp,\ mq,\ 2imr).$$

From these equations, which contain a solution (for complex numbers) of the problem solved for real numbers in Art. 250 of the 'Disquisitiones Arithmeticæ,' we may infer the following theorems (Disq. Arith., Arts. 251 and 253) :—

'The number ω of classes of any order Ω is a divisor of the number n of uneven classes of the same determinant D; and, given any two classes of order Ω, there are always $\frac{n}{\omega}$ uneven classes which compounded with one of them produce the other.'

'If $D \equiv 1$, mod 2, and if the classes of Ω are derived from semi-even or even primitives, ω is a divisor of the number n' of semi-even classes of determinant D; and, given any two classes of order Ω, there are always $\frac{n'}{\omega}$ semi-even classes which compounded with one of them produce $1+i$ times the other.'

'If $D \equiv \pm 1$, mod 4, and if the classes of Ω are derived from even primitives, ω is a divisor of the number n'' of even classes of determinant D; and, given any two classes of order Ω, there are always $\frac{n''}{\omega}$ even classes which compounded with one of them produce $2i$ times the other.'

III. *Determination of the number of Ambiguous Classes.*

Any form (A, B, C), in which $2B \equiv 0$, mod A, is called by Gauss an ambiguous form; but in the investigation which follows we shall, for brevity, understand by an ambiguous form an uneven form of one of the four types

(i) $(A, 0, C)$,

(ii) $([1+i]B, B, C)$,

(iii) $(2B, B, C)$,

(iv) $(2iB, B, C)$.

To determine the number of uneven ambiguous classes of any determinant D supposed not to be a square, we shall determine, first, the number of ambiguous forms of determinant D, and secondly, the number of ambiguous forms in each ambiguous class.

(1) Let μ be the number of different uneven primes dividing D. The number of ambiguous forms of the type (i) is $4 \times 2^\mu$, or $8 \times 2^\mu$, according as D is, or is not, uneven. For we may resolve $-D$ into any two relatively prime factors, and may take one of them (with any sign we please) for A, and the other for C.

There are no ambiguous forms of the type (ii), unless $D \equiv i$, mod 2, or $\equiv 0$, mod $(1+i)^3$. For, in the equation $D = B[B-(1+i)C]$, if B is uneven, we have $D \equiv i$, mod 2, because C must be uneven; if B is semi-even or even, we have $D \equiv 0$, mod $(1+i)^3$. If $D \equiv i$, mod 2, we resolve D into any two relatively prime factors X and Y, and, writing $B = X$, $B-(1+i)C = Y$, we find $C = \dfrac{X-Y}{1+i}$, which is integral because X and Y are uneven, and uneven because X is not $\equiv Y$, mod 2. Thus, if $D \equiv i$, mod 2, there are $4 \times 2^\mu$ ambiguous forms of the type (ii). Again, if $D \equiv 0$, mod $(1+i)^3$, we may resolve D in any way we please into two factors having $1+i$ for their greatest common divisor; we find in this way $8 \times 2^\mu$ ambiguous forms of the type (ii).

There are no ambiguous forms of the types (iii) or (iv), unless $D \equiv 1$, mod 2, or $\equiv 2$, mod 4, or $\equiv 0$, mod $(1+i)^5$. For, if in the equation $D = B(B-2C)$ we suppose B uneven, we find $D \equiv 1$, mod 2; if B is semi-even, $B^2 \equiv 2i$, and $2BC \equiv 2(1+i)$, mod 4, whence $D \equiv 2$, mod 4; lastly, if B is even, $D \equiv 0$, mod $(1+i)^5$. The same reasoning applies to the equation $D = B(B-2iC)$. If $D \equiv 1$, mod 2, we resolve D in every possible way into the product of two factors relatively prime; let $D = X \times Y$ be such a resolution, then $D = iX \times -iY$ is another; and it will be seen that according as the last coefficient in the two forms

$$\left[2X,\ X,\ \frac{X-Y}{2}\right],\quad \left[2iX,\ X,\ \frac{X-Y}{2i}\right]$$

is uneven, or is not uneven, so the last coefficient in the two forms

$$\left[2iX,\ iX,\ \frac{iX+iY}{2}\right],\quad \left[-2X,\ iX,\ \frac{X+Y}{2}\right],$$

is not, or is, uneven; *i.e.* there are $2\times 2^\mu$ ambiguous forms of each of the types (iii) and (iv). If $D\equiv 2$, mod 4, we resolve D in every possible way into two factors, of which $1+i$ is the greatest common divisor; we thus find $4\times 2^\mu$ uneven forms of each of the types (iii) and (iv). Lastly, if $D\equiv 0$, mod $(1+i)^5$, we resolve D in every possible way into two factors of which $1+i$ is the greatest common divisor, and we obtain $8\times 2^\mu$ forms of each of the types (iii) and (iv).

The result of this enumeration is that, if D be uneven, or semi-even, or $\equiv 2i$, mod 4, there are $8\times 2^\mu$ ambiguous forms; if $D\equiv 2$, mod 4, or $\equiv 0$, mod $(1+i)^3$, but not $\equiv 0$, mod $(1+i)^5$, there are $16\times 2^\mu$; and if $D\equiv 0$, mod $(1+i)^5$, there are $32\times 2^\mu$. On comparing this result with table III., it will be seen that in every case there are four times as many ambiguous forms as there are assignable generic characters.

(2) Let $f=(a,\ b,\ c)$ be any form of an ambiguous class; if $(I)=\begin{vmatrix}\mu, & -\lambda\\ \nu, & -\mu\end{vmatrix}$ is an improper automorphic of f, then $\lambda,\ \mu,\ \nu$ satisfy the equations

$$\mu^2-\lambda\nu \qquad =1, \quad \ldots \ldots \quad (1)$$

$$\lambda a+2\mu b+\nu c=0; \quad \ldots \ldots \quad (2)$$

and, conversely, if $\lambda,\ \mu,\ \nu$ satisfy the equations (1) and (2), $(I)=\begin{vmatrix}\mu, & -\lambda\\ \nu, & -\mu\end{vmatrix}$ is an improper automorphic of f. Let $\alpha,\ \gamma,\ p,\ q$ (of which α and γ are relatively prime) be a system of integral numbers satisfying the equations

$$\left.\begin{array}{ll} p\alpha=\lambda & p\gamma=\mu-1\\ q\alpha=\mu+1 & q\gamma=\nu\end{array}\right\}; \quad \ldots \ldots \quad (3)$$

and let $\theta=0,\ 1-i,\ 1$, or $-i$, according as $0,\ 1-i,\ 1$, or $-i$ satisfies the congruences

$$p+\theta\alpha\equiv 0,\ \text{mod } 2,$$
$$q+\theta\gamma\equiv 0,\ \text{mod } 2,$$

which are simultaneously resoluble, and admit of only one solution, because α and γ are relatively prime, while $q\alpha-p\gamma=2$. Then it will be found that, by the proper transformation

$$(J)=\begin{vmatrix}\alpha, & \tfrac{1}{2}(p+\theta\alpha)\\ \gamma, & \tfrac{1}{2}(q+\theta\gamma)\end{vmatrix},$$

f is transformed into an ambiguous form ϕ, which will be of the type (i), (ii), (iii), or (iv), according as $\theta = 0$, $1-i$, 1, or $-i$. It will also be seen that, subject to the condition that α and γ are relatively prime, there are always four, and only four, solutions of the system (3), represented by the formula

$$i^k\alpha, \quad i^k\gamma, \quad i^{-k}p, \quad i^{-k}q.$$

There are thus four transformations included in the formula (J), two of them transforming f into the same ambiguous form ϕ, and the other two transforming f into the same form taken negatively. The four transformations (J), and the two ambiguous forms ϕ and $-\phi$, we shall term respectively the transformations and the ambiguous forms appertaining to the improper automorphic (I). If we now form the transformations appertaining to every improper automorphic of f, it can be proved (A) that these transformations will all be different, and (B) that they will include every proper transformation of f into an ambiguous form.

(A) As the four transformations appertaining to the same improper automorphic are evidently different, it will be sufficient to show that, if (J) and (J') appertain to the improper automorphics (I) and (I'), the supposition $(J) = (J')$ implies $(I) = (I')$. From the equations

$$\alpha = \alpha', \quad \gamma = \gamma', \quad p + \theta\alpha = p' + \theta'\alpha', \quad q + \theta\gamma = q' + \theta'\gamma',$$

which are equivalent to the symbolic equation $(J) = (J')$, combined with the system (3) and with a similar system containing the accented letters, we find

$$(\theta - \theta')\,\alpha^2 = \lambda' - \lambda, \quad (\theta - \theta')\,\alpha\gamma = \mu' - \mu, \quad (\theta - \theta')\,\gamma^2 = \nu' - \nu;$$

whence again $(\theta - \theta')(a\alpha^2 + 2b\alpha\gamma + c\gamma^2) = 0$ by virtue of equation (2). The coefficient of $\theta - \theta'$ is not zero, for $D = b^2 - ac$ is not a square; therefore $\theta - \theta' = 0$; *i.e.* $\lambda = \lambda'$, $\mu = \mu'$, $\nu = \nu'$, or $(I) = (I')$.

(B) Let $\begin{vmatrix} \alpha, & \beta \\ \gamma, & \delta \end{vmatrix}$ be a proper transformation of f into an ambiguous form ϕ; according as ϕ is of the type (i), (ii), (iii), or (iv), let $\theta = 0$, $1-i$, 1, or $-i$: let also

$$\lambda = 2\alpha\beta - \theta\alpha^2, \quad \mu = \alpha\delta + \beta\gamma - \theta\alpha\gamma, \quad \nu = 2\gamma\delta - \theta\gamma^2;$$

then $\begin{vmatrix} \mu, & -\lambda \\ \nu, & -\mu \end{vmatrix} = (I)$ is an improper automorphic of f; for

$$\mu^2 - \lambda\nu = (\alpha\delta - \beta\gamma)^2 = 1, \quad \text{and} \quad \lambda a + 2\mu b + \nu c = 0,$$

because of the ambiguity of the form into which f is transformed by $\begin{vmatrix} \alpha, & \beta \\ \gamma, & \delta \end{vmatrix}$. Also $\begin{vmatrix} \alpha, & \beta \\ \gamma, & \delta \end{vmatrix}$ appertains to (I); for, writing p and q instead of $2\beta - \theta\alpha$ and $2\delta - \theta\gamma$, we have

$$\begin{vmatrix} \alpha, & \beta \\ \gamma, & \delta \end{vmatrix} = \begin{vmatrix} \alpha, & \frac{1}{2}(p + \theta\alpha) \\ \gamma, & \frac{1}{2}(q + \theta\gamma) \end{vmatrix},$$

α, γ, p, q (of which α and γ are relatively prime) being four numbers which satisfy the system (3); *i.e.* $\begin{vmatrix} \alpha, \beta \\ \gamma, \delta \end{vmatrix}$ appertains to (I), an improper automorphic of f.

It follows from (B) that, if we calculate the ambiguous forms ϕ and $-\phi$ appertaining to every improper automorphic of f, we shall obtain all the ambiguous forms to which f is equivalent; it remains to see how many of these ambiguous forms are different from one another. If

$$(I) = \begin{vmatrix} \mu, & -\lambda \\ \nu, & -\mu \end{vmatrix}$$

is any given improper automorphic of f, all its similar automorphics are contained in the four formulae

$$(T)^{2k} \times (I), \qquad (T)^{2k+1} \times (I),$$

$$(T)^{2k} \times \begin{vmatrix} -1, & 0 \\ 0, & -1 \end{vmatrix} \times (I), \qquad (T)^{2k+1} \times \begin{vmatrix} -1, & 0 \\ 0, & -1 \end{vmatrix} \times (I),$$

where k is any positive or negative number, and

$$(T) = \begin{vmatrix} t_1 - u_1 b, & -u_1 c \\ u_1 a, & t_1 + u_1 b \end{vmatrix},$$

$[t_1, u_1]$ representing a fundamental solution of the equation $t^2 - Du^2 = 1$. Similarly, if (J) represent the four transformations, appertaining to (I), by which f passes into ϕ or $-\phi$, all the proper transformations of f into ϕ or $-\phi$ are included in the formula $(T)^k \times (J)$. We shall now show that the four transformations included in the formula $(T)^k \times (J)$ appertain to the improper automorphic $(T)^{2k} \times (I)$. Writing

$$\alpha_k = (t_k - bu_k)\,\alpha - cu_k\,\gamma, \qquad p_k = (t_k - bu_k)\,p - cu_k\,q,$$

$$\gamma_k = au_k\,\alpha + (t_k + bu_k)\,\gamma, \qquad q_k = au_k\,p + (t_k + bu_k)\,q,$$

$$\lambda_{2k} = (t_{2k} - bu_{2k})\,\lambda - cu_{2k}\,\mu,$$

$$\mu_{2k} = (t_{2k} - bu_{2k})\,\mu - cu_{2k}\,\nu = au_{2k}\,\lambda + (t_{2k} + bu_{2k})\,\mu,$$

$$\nu_{2k} = au_{2k}\,\mu + (t_{2k} + bu_{2k})\,\nu,$$

we find immediately

$$(T)^k \times (J) = \begin{vmatrix} \alpha_k, & \tfrac{1}{2}(p_k + \theta\alpha_k) \\ \gamma_k, & \tfrac{1}{2}(q_k + \theta\gamma_k) \end{vmatrix}, \quad (T)^{2k} \times (I) = \begin{vmatrix} \mu_{2k}, & -\lambda_{2k} \\ \nu_{2k}, & -\mu_{2k} \end{vmatrix}.$$

Also, attending to the equations (2) and (3) and to the relations

$$t_{2k} = t_k^2 - Du_k^2, \quad u_{2k} = 2\,t_k\,u_k,$$

we obtain, after substitution and reduction,

$$p_k \alpha_k = \lambda_{2k}, \qquad p_k \gamma_k = \mu_{2k} - 1,$$

$$q_k \alpha_k = \mu_{2k} + 1, \quad q_k \gamma_k = \nu_{2k},$$

i.e. $(T)^k \times (J)$ appertains to $(T)^{2k} \times (I)$, if (J) appertains to (I).

It follows from this result that the ambiguous forms appertaining to (I) and to $(T) \times (I)$ are the same; for f is transformed into the same forms by (J) and $(T) \times (J)$; and conversely, if the ambiguous forms appertaining to two different automorphics (I) and (I') are identical, an equation of the form $(I') = (T)^{2k} \times (I)$ will subsist; for, if (J) and (J') are the transformations appertaining to (I) and (I'), since by hypothesis (J) and (J') transform f into the same form, we must have an equation of the form $(J') = (T)^k \times (J)$; but (J') appertains to (I'), and $(T)^k \times (J)$ to $(T)^{2k} \times (I)$; therefore $(I') = (T)^{2k} \times (I)$, by what has been shown above in (A).

If then we calculate the eight ambiguous forms appertaining to the four improper automorphics

$$(I), \quad \begin{vmatrix} -1, & 0 \\ 0, & -1 \end{vmatrix} \times (I), \quad (T) \times (I), \quad \begin{vmatrix} -1, & 0 \\ 0, & -1 \end{vmatrix} \times (T) \times (I),$$

these eight forms will be the only ambiguous forms equivalent to f. Thus every uneven ambiguous class contains eight ambiguous forms.

Combining this result with the preceding we obtain the theorem :—

'The number of uneven ambiguous classes is one half of the whole number of assignable generic characters.'

The number of semi-even and even ambiguous classes is determined by the two following theorems :—

'When $D \equiv \pm 1$, mod 4, there are as many even as semi-even ambiguous classes.'

'When $D \equiv 1$, mod 2, there are as many semi-even as uneven ambiguous classes, or only half as many, according as there are altogether as many semi-even as uneven classes, or only half as many.'

To prove the first of these theorems, let $D \equiv i^{2k}$, mod 4, and let

$$\Sigma = \left(2i, \ i^k, \ \frac{i^{2k} - D}{2i}\right);$$

it is evident from the principles of the composition of forms that if (ϕ) is a given semi-even ambiguous class, the equation

$$(\Sigma) \times (\phi) = (1 + i)\,(f)$$

is satisfied by one, and only one, even ambiguous class (f); in addition to this we shall now show that, if (f) is a given even ambiguous class, the same equation is satisfied by one, and only one, semi-even ambiguous class (ϕ); from which two things the truth of the theorem is manifest. First, let the whole number of even classes be equal to the whole number of semi-even classes *; then the equation

$$(\Sigma) \times (\phi) = (1+i)\,(f)$$

is satisfied by only one semi-even class (ϕ); and this class is ambiguous, for the equation is satisfied by the opposite of (ϕ) as well as by (ϕ) itself; therefore (ϕ) and its opposite are the same class, or (ϕ) is an ambiguous class. Secondly, let the number of semi-even classes be three times the number of even classes; then the equation

$$(\Sigma) \times (\phi) = (1+i)\,(f)$$

---

* That, if $D \equiv \pm 1$, mod 4, there are either as many semi-even as even classes, or else three times as many, is a theorem of M. Lipschitz (Crelle, vol. liv. p. 196), of which it is worth while to give a proof here. The number of even classes is to the number of semi-even classes as unity to the number of semi-even classes satisfying the equation

$$(\Sigma) \times (\phi) = (1+i)\,(f),$$

f representing any given even form. To investigate the semi-even classes satisfying this equation, apply to f a complete system of transformations for the modulus $1+i$, for example, the transformations

$$\begin{vmatrix} 1, & 0 \\ 0, & 1+i \end{vmatrix}, \qquad \begin{vmatrix} 1+i, & 0 \\ 0, & 1 \end{vmatrix}, \qquad \begin{vmatrix} 1+i, & 1 \\ 0, & 1 \end{vmatrix},$$

and divide the resulting forms by $1+i$; of the quotients, one or three will be semi-even, according as

$$D \equiv \pm 1 \quad \text{or} \quad \equiv \pm 5, \text{ mod } (1+i)^5.$$

It will be found that each of these semi-even forms satisfies the equation

$$\Sigma \times \phi = (1+i)\, f;$$

and, conversely, every semi-even form ϕ satisfying that equation is equivalent to one of these forms; for, from any transformation of $(1+i)f$ into $\Sigma \times \phi$, we may (by attributing to the indeterminates of Σ the values 1, 0) deduce a transformation of modulus $1+i$ by which f passes into $(1+i)\phi$; *i.e.* ϕ is equivalent to one of the forms obtained by the preceding process. It only remains to show that, when there are three of these forms, they constitute either one or three classes, but never two. For this purpose it is sufficient to consider the three semi-even forms

$$\sigma_0 = \left(1+i,\ 1,\ -\frac{D-1}{1+i}\right), \quad \sigma_1, \quad \text{and} \quad \sigma_2,$$

obtained by the preceding process from the form Σ. These forms satisfy the equations

$$\sigma_0 \times \sigma_0 = (1+i)\,\sigma_0, \quad \sigma_1 \times \sigma_1 = (1+i)\,\sigma_2, \quad \sigma_2 \times \sigma_2 = (1+i)\,\sigma_1, \quad \sigma_1 \times \sigma_2 = (1+i)\,\sigma_0;$$

from which it follows that any one of the suppositions $\sigma_1 = \sigma_2$, $\sigma_2 = \sigma_0$, $\sigma_0 = \sigma_1$ involves the other two.

is satisfied by three, and only three, different classes (ϕ); but it is also satisfied by the opposites of these classes; therefore one of them is necessarily an ambiguous class. Let that class be (ϕ_0); the other two are defined by the equations

$$(1+i)(\phi_1) = (\sigma_1) \times (\phi_0), \qquad (1+i)(\phi_2) = (\sigma_2) \times (\phi_0),$$

and cannot be ambiguous classes, for by duplication we find

$$(\phi_1) \times (\phi_1) = (1+i)(\sigma_2), \qquad (\phi_2) \times (\phi_2) = (1+i)(\sigma_1);$$

whereas every semi-even ambiguous class produces $(1+i)(\sigma_0)$ by its duplication *.

The second theorem may be proved as follows. Let

$$f = ([1+i]p,\ q,\ [1+i]r)$$

be a semi-even form of determinant D; and let

$$\sigma_0 = \left((1+i),\ 1,\ -\frac{D-1}{1+i}\right):$$

we suppose that p is uneven. The equation $(\sigma_0) \times (\phi) = (f)$ is satisfied by one uneven class (ϕ_0), or by two (ϕ_0) and (ϕ_1), according as the forms

$$\phi_0 = (p,\ q,\ 2ir) \quad \text{and} \quad \phi_1 = (2ip,\ q,\ r),$$

if r is uneven, or the forms

$$\phi_0 = (p,\ q,\ 2ir) \quad \text{and} \quad \phi_1 = (2ip,\ [1+i]p+q,\ p+[1-i]q+r),$$

if r is even, are, or are not, equivalent †. If any one of the forms f, ϕ_0, ϕ_1 is ambiguous, the others are so too; the same thing is therefore true for the classes (f), (ϕ_0), (ϕ_1). Thus the number of semi-even ambiguous classes is equal to, or is one half of, the number of uneven ambiguous classes, according as the classes (ϕ_0) and (ϕ_1) are, or are not, identical; *i.e.* according as the whole number of semi-even classes is equal to, or is one half of, the whole number of even classes.

The demonstration in the 'Disquisitiones Arithmeticae,' that the number of genera of uneven forms of any determinant cannot exceed the number of uneven ambiguous classes of the same determinant, may be transferred

* For the definition of the classes (σ_0), (σ_1), (σ_2) see the preceding note.

† The forms ϕ_0 and ϕ_1 are obtained by applying to f a complete set of transformations of modulus $1+i$, dividing the resulting forms by $1+i$, and retaining only those quotients which are uneven forms.

without change to the complex theory. We thus obtain a proof (independent of the law of quadratic reciprocity and of the theorems which determine the quadratic characters of i and of $1+i$) of the impossibility of one half of the whole number of assignable generic characters; and from that impossibility, as we shall now show, the quadratic theorems are themselves deducible.

(1) If p is an uneven prime $\equiv 1$, mod 2, there are two genera of uneven forms of determinant p: of these one is the principal genus and has the complete characters $\left[\frac{f}{p}\right]=1$, $\gamma=1$; the other, containing the form $(i, 0, ip)$, has the particular character $\gamma=-1$; whence it follows that every uneven form of determinant p which has the character $\gamma=1$ is a form of the principal genus and has the character $\left[\frac{f}{p}\right]=1$. Again, if $p\equiv 1$, mod 4, the form $\left(2i, i, -\frac{p+1}{2i}\right)$ is an uneven form of determinant p; this form has the particular character $\gamma=-1$, because $-\frac{p+1}{2i}\equiv i$, mod 2; it is therefore not a form of the principal genus; but it has the character $\left[\frac{f}{p}\right]=1$, because $2i$ is a square; therefore, if $p\equiv 1$, mod 4, every uneven form of determinant p has the character $\left[\frac{f}{p}\right]=1$.

(2) There is but one genus of forms of determinant i, and its complete character is $\alpha=1$; there is also but one genus of forms of determinant $1+i$, and its complete character is $\beta=1$.

(3) Let p and q be uneven primes of which the imaginary parts are even; to prove the law of reciprocity it will suffice to show that, if $\left[\frac{p}{q}\right]=1$, then $\left[\frac{q}{p}\right]=1$. The equation $\left[\frac{p}{q}\right]=1$ implies the existence of a congruence of the type $\omega^2-p\equiv 0$, mod q, and consequently of an uneven form of determinant p and of the type $\left(q, \omega, \frac{\omega^2-p}{q}\right)$. This form has the character $\gamma=1$, because $q\equiv 1$, mod 2; it therefore has the character $\left[\frac{f}{p}\right]=1$; *i.e.* $\left[\frac{q}{p}\right]=1$.

(4) To prove the equation

$$\left[\frac{i}{p}\right]=(-1)^{\frac{1}{4}(Np-1)},$$

in which we may suppose that the uneven prime p is primary, it will suffice to show that,

(i) if $\left[\frac{i}{p}\right]=1$, then $(-1)^{\frac{1}{4}(Np-1)}=1$;

(ii) if $(-1)^{\frac{1}{4}(Np-1)}=1$, then $\left[\frac{i}{p}\right]=1$.

(i) Let $\left[\frac{i}{p}\right]=1$; then, if $\omega^2-i\equiv 0$, mod p, $\left(p, \omega, \frac{\omega^2-i}{p}\right)$ is a form of determinant i; it therefore has the character $\alpha=1$, *i.e.* $(-1)^{\frac{1}{4}(Np-1)}=1$.

(ii) Let $(-1)^{\frac{1}{4}(Np-1)}=1$; then $p\equiv 1$, mod 4, and the form $(i, 0, ip)$ is an uneven form of determinant p; it therefore has the character $\left[\frac{f}{p}\right]=1$; whence $\left[\frac{i}{p}\right]=1$.

(5) Similarly, if $p=p_0+ip_1$ is an uneven and primary prime, to prove the equation

$$\left[\frac{1+i}{p}\right]=(-1)^{\frac{1}{8}[(p_0+p_1)^2-1]}$$

we shall show that,

(i) if $\left[\frac{1+i}{p}\right]=1$, then $(-1)^{\frac{1}{8}[(p_0+p_1)^2-1]}=1$;

(ii) if $(-1)^{\frac{1}{8}[(p_0+p_1)^2-1]}=1$, then $\left[\frac{1+i}{p}\right]=1$.

(i) Let $\left[\frac{1+i}{p}\right]=1$; then there is a form of determinant $1+i$ and of the type $\left(p, \omega, \frac{\omega^2-1-i}{p}\right)$; this form has the character $\beta=1$; therefore $(-1)^{\frac{1}{8}[(p_0+p_1)^2-1]}=1$.

(ii) Let $(-1)^{\frac{1}{8}[(p_0+p_1)^2-1]}=1$; then p is either $\equiv 1-2i$ or $\equiv 1$, mod $(1+i)^5$. If $p=(1+i)^5k+1-2i$, $([1+i]^3, i, 1-2ki)$ is an uneven form of determinant p; this form has the character $\gamma=1$, and consequently it also has the character $\left[\frac{f}{p}\right]=1$; therefore $\left[\frac{1+i}{p}\right]=\left[\frac{(1+i)^3}{p}\right]=1$. If $p=(1+i)^5k+1$, one or other of the forms $([1+i]^5, 1, -k)$ and $([1+i]^5, 1+[1+i]^3, 1-k)$ is an uneven form of determinant p, having the character $\left[\frac{f}{p}\right]=1$; therefore, in this case also, $\left[\frac{1+i}{p}\right]=\left[\frac{(1+i)^5}{p}\right]=1$.

IV. *The representation of Binary Forms of the Principal Genus by Ternary Forms of determinant* 1.

The solution of the general problem, 'To find the representations (if any) of a given binary by a given ternary quadratic form,' depends, in the case of complex as of real numbers, on the solution of the problem of equivalence for ternary forms. Extending the methods of Gauss to the complex theory, we find the necessary and sufficient condition for the primitive * representation of a binary form f of determinant D by a ternary form of determinant 1 to be, that f should be a form of the principal genus; or, if $D \equiv \pm 1$, mod 4, that f should be a form either of the principal genus, or else of that genus which differs from the principal genus only in having the character $\gamma = -1$, instead of $\gamma = +1$. Again, because the reduction of Lagrange is applicable to complex binary forms, the reduction of Gauss † is applicable to complex ternary forms. It is thus found

---

* If a matrix of the type
$$\begin{vmatrix} \alpha, & \beta \\ \alpha', & \beta' \\ \alpha'', & \beta'' \end{vmatrix}$$
transforms a ternary into a binary quadratic form, the representation of the binary by the ternary form is said to be primitive when the three determinants of the matrix are relatively prime.

† If
$$F = ax^2 + a'y^2 + a''z^2 + 2byz + 2b'xz + 2b''xy$$
is a ternary form of determinant Δ, and
$$Ax^2 + A'y^2 + A''z^2 + 2Byz + 2B'xz + 2B''xy$$
its contravariant, by applying the reduction of Lagrange to the form $ax^2 + 2b''xy + a'y^2$, we can render $N.a \leqq 2\sqrt{N.A''}$ (Dirichlet in Crelle's Journal, vol. xxiv. p. 348); and by applying the same reduction to the form $A'y^2 + 2Byz + A''z^2$, we can render $N.A'' \leqq 2\sqrt{N.a\Delta}$. The reduction of Gauss consists in the alternate application of these two reductions until we arrive at a form in which we have simultaneously
$$N.a \leqq 2\sqrt{N.A''}, \quad N.A'' \leqq 2\sqrt{N.a\Delta};$$
and consequently
$$N.a \leqq 4\sqrt[3]{N.\Delta}, \quad N.A'' \leqq 4\sqrt[3]{N.\Delta^2}.$$
If $\Delta = 1$, we have $N.a \leqq 4$, $N.A'' \leqq 4$; whence a and A'' can only have the values
$$0, \ \pm 1, \ \pm i, \ \pm(1+i), \ \pm(1-i) \ \pm 2, \ \pm 2i;$$
and it will be found, on an examination of the different cases that can arise, that the reduction can always be continued until a and A'' are either both units, or both zero. In the former case, by applying a further transformation of the type
$$\begin{vmatrix} 1, & \mu'', & \mu' \\ 0, & 1, & \mu \\ 0, & 0, & 1 \end{vmatrix},$$
the coefficients b, b', b'' may be made to disappear, and we obtain a form equivalent to F and of the

that the number of classes of such forms of a given determinant is finite; and, in particular, that every form of determinant 1 is equivalent to one or other of the forms $-x^2-y^2-z^2$ and $x^2+iy^2+iz^2$, of which the former cannot represent numbers $\equiv i$ or $\equiv 1+i$, mod 2; and the latter cannot primitively represent numbers $\equiv 2$ or $\equiv 2(1+i)$, mod 4. The method of reduction itself supplies a transformation of any given form of determinant 1 into one or other of those two forms.

If $D\equiv i$ or $\equiv 1+i$, mod 2, no binary form of determinant D can be represented by $-x^2-y^2-z^2$, because D cannot be represented by the contravariant of that form, *i.e.* by the form $-x^2-y^2-z^2$ itself. Consequently, if $D\equiv i$ or $\equiv 1+i$, mod 2, the binary forms of its principal genus are certainly capable of primitive representation by $x^2+iy^2+iz^2$.

If $D\equiv 1$, mod 2, no form of the principal genus can be primitively represented by $x^2+iy^2+iz^2$. Let $f=(a, b, c)$ be such a form, and let us suppose, as we may do, that b is even, so that $ac\equiv 1$, mod 2, and $a\equiv c\equiv 1$, mod 2 (the supposition $a\equiv c\equiv i$ is inadmissible, because f is of the principal genus); if possible, let the prime matrix

$$\begin{vmatrix} \alpha, & \beta \\ \alpha', & \beta' \\ \alpha'', & \beta'' \end{vmatrix}$$

---

type $\epsilon x^2+\epsilon' y^2+\epsilon'' z^2$, ϵ, ϵ', ϵ'' representing units of which the product is -1. In the latter case, the form obtained by applying the reduction of Gauss is of the type

$$a'y^2+a''z^2+2byz+2b'xz;$$

whence $a'b'^2=1$, so that b' is a unit which we shall call ϵ; and the form

$$\epsilon^2y^2+a''z^2+2byz+2\epsilon xz,$$

by a transformation of the type

$$\begin{vmatrix} 1, & 0, & \mu' \\ 0, & 1, & \mu \\ 0, & 0, & 1 \end{vmatrix},$$

is changed into one of the four forms

$$\epsilon^2y^2+2\epsilon xz,\quad \epsilon^2y^2+z^2+2\epsilon xz,\quad \epsilon^2y^2+iz^2+2\epsilon xz,\quad \epsilon^2y^2+(1+i)z^2+2\epsilon xz;$$

of which the first two, by the transformations

$$\begin{vmatrix} \epsilon^{-1}i, & 0, & \epsilon^{-1} \\ \epsilon i, & \epsilon i, & \epsilon \\ 0, & -i, & -1 \end{vmatrix},\quad \begin{vmatrix} 0, & 0, & -\epsilon \\ \epsilon^{-1}i, & 0, & 0 \\ 0, & i, & \epsilon^2 \end{vmatrix},$$

are changed into the form $-x^2-y^2-z^2$, and the last two, by the transformations

$$\begin{vmatrix} 0, & -\epsilon, & 0 \\ -\epsilon^{-1}, & 0, & 0 \\ 0, & -\epsilon^2 i, & -1 \end{vmatrix},\quad \begin{vmatrix} \epsilon^{-1}, & \epsilon^{-1}, & \epsilon^{-1}(1-i) \\ -\epsilon, & -\epsilon, & \epsilon i \\ 0, & -1, & i \end{vmatrix},$$

are changed into $x^2+iy^2+iz^2$. (See Disq. Arith., Arts. 272–274.)

(of which A, B, C are the determinants) transform $x^2+iy^2+iz^2$ into f; we have the equations

$$a=\alpha^2+i\alpha'^2+i\alpha''^2, \quad c=\beta^2+i\beta'^2+i\beta''^2, \quad D=A^2-iB^2-iC^2,$$

from which, and from the congruences $D\equiv a\equiv c\equiv 1$, mod 2, we infer the incompatible conditions

$$\alpha'+i\alpha''\equiv\beta'+i\beta''\equiv 0, \text{ mod } (1+i), \quad A\equiv 1, \text{ mod } (1+i);$$

i.e. f is incapable of primitive representation by $x^2+iy^2+iz^2$. If, therefore, $D\equiv 1$, mod 2, the forms of its principal genus are capable of primitive representation by $-x^2-y^2-z^2$. We may add that, when $D\equiv\pm 1$, mod 4, the forms of that genus which differs from the principal genus only in having the character $\gamma=-1$, instead of $\gamma=+1$, are capable of primitive representation by $x^2+iy^2+iz^2$, but not by $-x^2-y^2-z^2$.

Lastly, let $D\equiv 0$, mod 2. If $D\equiv 2$ or $\equiv 2(1+i)$, mod 4, D cannot be primitively represented by $x^2-iy^2-iz^2$, the contravariant of $x^2+iy^2+iz^2$; *i.e.* no form of determinant D can be primitively represented by $x^2+iy^2+iz^2$; so that forms of the principal genus are certainly capable of primitive representation by $-x^2-y^2-z^2$. But, if $D\equiv 2i$ or $\equiv 0$, mod 4, the forms of the principal genus are capable of primitive representation by both the ternary forms $-x^2-y^2-z^2$ and $x^2+iy^2+iz^2$. For, if $f=(a, b, c)$ be a form of the principal genus of any even determinant, f can only represent numbers $\equiv 0$ or $\equiv 1$, mod 2; so that a ternary form of determinant 1 and of the type

$$f+p''z^2+2qyz+2q'xz$$

will be equivalent to $-x^2-y^2-z^2$, or to $x^2+iy^2+iz^2$, according as $p''\equiv 0$ or $\equiv 1$, mod 2, on the one hand, or $p''\equiv i$ or $\equiv 1+i$, on the other hand. Again, if (k, k') is a value of the expression $\sqrt{(a, -b, c)}$, mod D, (in which we now suppose a uneven and b semi-even or even), $\left(k+\frac{D}{1+i}, k'\right)$ is another value of the same expression; and it can be shown * that when $D\equiv 2i$ or $\equiv 0$, mod 4, one of

* If $f+p''z^2+2qyz+2q'xz$ is a ternary form of determinant 1, derived from the value (k, k') of the expression $\sqrt{(a, -b, c)}$, mod D, k is the coefficient of yz in the contravariant form. Hence

$$a=k^2-D(q'^2-ap''), \quad \text{or} \quad ap''=q'^2+\frac{a-k^2}{D}.$$

Observing that $a\equiv 1$, mod 2, and $q'^2\equiv 0$ or $\equiv 1$, mod 2, we see that $p''\equiv 0$ or $\equiv 1$, mod 2, on the one hand, or $\equiv i$ or $\equiv 1+i$, mod 2, on the other hand, according as $\frac{a-k^2}{D}\equiv 0$ or $\equiv 1$, mod 2, on the one the two forms of determinant 1, and of the type

$$f + p''z^2 + 2qyz + 2q'xz,$$

which are deducible by the method of Gauss from those two values, satisfies the condition $p'' \equiv 0$ or $\equiv 1$, mod 2, while the other satisfies the condition $p'' \equiv i$ or $\equiv 1+i$, mod 2; that is, f is capable of primitive representation by both the forms $-x^2-y^2-z^2$ and $x^2+iy^2+iz^2$.

The preceding theory supplies a solution of the problem, 'Given a form of the principal genus of forms of determinant D, to investigate a form from the duplication of which it arises.' Let $f=(a, b, c)$ be the given form, and let us suppose (as we may do) that a and c are uneven. When $D \equiv i$ or $\equiv 1+i$, mod 2, let

$$\begin{vmatrix} \alpha, & \beta \\ \alpha', & \beta' \\ \alpha'', & \beta'' \end{vmatrix}$$

be a prime matrix (of which the determinants are A, B, C) transforming $x^2+iy^2+iz^2$ into $(a, -b, c)$; and let ϕ represent the binary form $(C-iB, A, iC-B)$; then the matrix

$$\begin{vmatrix} \beta'+i\beta'', & \beta, & \beta, & -i(\beta'-i\beta'') \\ \alpha'+i\alpha'', & \alpha, & \alpha, & -i(\alpha'-i\alpha'') \end{vmatrix} \quad . \quad . \quad . \quad . \quad . \quad . \quad . \quad . \quad \text{(Z)}$$

transforms f into $\phi \times \phi$ *; and is a prime matrix, for its determinants $C-iB$,

---

hand, or $\equiv i$ or $\equiv 1+i$, mod 2, on the other hand. But

$$\frac{a-k^2}{D} - \frac{a-\left(k+\frac{D}{1+i}\right)^2}{D} = (1-i)k + \frac{D}{2i},$$

which $\equiv 1+i$, mod 2, if $D \equiv 0$, mod 4, and $\equiv i$, mod 2, if $D \equiv 2i$, mod 4, since k is evidently uneven in either case. From this it appears that, if $\frac{a-k^2}{D} \equiv 0$ or $\equiv 1$, mod 2, then

$$\frac{a-\left(k+\frac{D}{1+i}\right)^2}{D} \equiv i \text{ or } \equiv 1+i, \text{ mod } 2;$$

that is, in one of the two forms $f+p''z^2+2qyz+2q'xz$, $p'' \equiv 0$ or $\equiv 1$, mod 2, and in the other, $p'' \equiv i$ or $\equiv 1+i$, mod 2.

* This assertion may be verified by means of the identity

$$\begin{aligned}
&(q_1q_2-q_0q_3)(p_0xx'+p_1xy'+p_2x'y+p_3x'y')^2 \\
&\quad + (p_0q_3+p_3q_0-p_1q_2-p_2q_1)(p_0xx'+p_1xy'+p_2x'y+p_3x'y')(q_0xx'+q_1xy'+q_2x'y+q_3x'y') \\
&\qquad\qquad + (p_1p_2-p_0p_3)(q_0xx'+q_1xy'+q_2x'y+q_3x'y')^2 \\
&= [(p_0q_2-p_2q_0)x'^2+(p_0q_3-p_3q_0+p_1q_2-p_2q_1)x'y'+(p_1q_3-p_3q_1)y'^2] \\
&\qquad \times [(p_0q_1-p_1q_0)x^2+(p_0q_3-p_3q_0+p_2q_1-p_1q_2)xy+(p_2q_3-p_3q_2)y^2],
\end{aligned}$$

in which we have to replace the quantities $\begin{matrix} p_0 & p_1 & p_2 & p_3 \\ q_0 & q_1 & q_2 & q_3 \end{matrix}$ by the elements of the matrix (Z).

$2A$, and $iC-B$ are not simultaneously divisible by any uneven prime (because A, B, and C are relatively prime), and are not simultaneously divisible by $1+i$, because (Z) is congruous, for the modulus $1+i$, to the first, or to the second of the matrices

$$\begin{vmatrix} 0, 1, 1, 0 \\ 1, 0, 0, 1 \end{vmatrix} \text{ and } \begin{vmatrix} 1, 0, 0, 1 \\ 0, 1, 1, 0 \end{vmatrix}, \quad . \quad . \quad . \quad . \quad . \quad . \quad . \quad . \quad (Z')$$

according as $a \equiv i$ and $c \equiv 1$, or $a \equiv 1$ and $c \equiv i$, mod 2. Consequently ϕ is a form the duplication of which produces f. When $D \equiv 1$ or $\equiv 0$, mod 2, let the prime matrix

$$\begin{vmatrix} \alpha \,, & \beta \\ \alpha' , & \beta' \\ \alpha'', & \beta'' \end{vmatrix}$$

transform $-x^2-y^2-z^2$ into $(a, -b, c)$. As we cannot have simultaneously $\alpha \equiv \beta$, $\alpha' \equiv \beta'$, $\alpha'' \equiv \beta''$, mod $(1+i)$, we may suppose that α and β are incongruous, mod $(1+i)$. If $\phi = (B+iC, iA, B-iC)$, the matrix

$$\begin{vmatrix} \beta'+i\beta'', & i\beta, & i\beta, & \beta'-i\beta'' \\ \alpha'+i\alpha'', & i\alpha, & i\alpha, & \alpha'-i\alpha'' \end{vmatrix} \quad . \quad . \quad . \quad . \quad . \quad . \quad . \quad . \quad (Z)$$

transforms f into $\phi \times \phi$, and is a prime matrix, being congruous to one or other of the matrices (Z′), for the modulus $1+i$, in consequence of the two suppositions that a and c are uneven, and that α and β are incongruous, mod $(1+i)$: so that f arises from the duplication of ϕ.

From the resolubility of this problem we can infer (precisely as Gauss has done in the real theory) that that half of the assignable generic characters which is not impossible corresponds to actually existing genera. We can also deduce a demonstration of the theorem 'Any form of determinant D can be transformed into any other form of the same genus by a transformation of which the coefficients are rational fractions having denominators prime to $2D$.' For, every form which arises from the duplication of an uneven primitive form (that is, every form of the principal genus) represents square numbers prime to $2D$, and is therefore equivalent to a form of the type $\left(\lambda^2, \mu, \dfrac{\mu^2-D}{\lambda^2}\right)$. But $(1, 0, -D)$ is transformed into $\left(\lambda^2, \mu, \dfrac{\mu^2-D}{\lambda^2}\right)$ by the matrix $\begin{vmatrix} \lambda, & \dfrac{\mu}{\lambda} \\ 0, & \dfrac{1}{\lambda} \end{vmatrix}$; *i.e.* any two forms of the principal genus can be transformed into one another by transformations of the kind indicated. Again, if f_1, f_2 be two forms of any other genus, a form ϕ of the principal genus exists satisfying the equation $f_2 = \phi \times f_1$. But, since ϕ can be transformed into the principal form, we can assign to the indeterminates of φ rational values, having denominators prime to $2D$, which shall cause ϕ to acquire the value $+1$; and thus, from the transformation of f_2 into $f_1 \times \phi$, we deduce a rational transformation of f_2 into f_1, the coefficients of which have denominators prime to $2D$. The truth of the converse proposition, 'Two forms which are transformable into one another by rational transformations having denominators prime to $2D$ belong to the same genus,' is evident from the definition of the generic characters themselves. The proposition itself is of some importance, as it furnishes a verification of the completeness of the enumeration of generic characters contained in table III.

XVI.

ON A FORMULA FOR THE MULTIPLICATION OF FOUR THETA FUNCTIONS.

[Proceedings of the London Mathematical Society, vol. i. No 8, pp. 3–14. Read May 21, 1866.]

1. In a letter addressed to M. Hermite, and dated Berlin, Aug. 6, 1845 *, Jacobi says :—

'Dans mes Leçons universitaires de Königsberg, moi aussi j'ai eu coutûme de partir des fonctions θ. Dans ces Leçons, en multipliant quatre séries $\sum_{i\,-\infty}^{+\infty} \epsilon^{-(ax+bi)^2}$ pour différentes valeurs de x, et en transformant les exposants par la formule

$$i^2+i'^2+i''^2+i'''^2=\left(\frac{i+i'+i''+i'''}{2}\right)^2+\left(\frac{i+i'-i''-i'''}{2}\right)^2+\dots$$

j'ai obtenu tout de suite une formule de laquelle découlent, comme cas particuliers et sans le moindre calcul, les expressions fractionnaires des fonctions elliptiques, les théorèmes sur l'addition des trois espèces, et plusieurs centaines de formules intéressantes auxquelles on ne saurait arriver que par un calcul algébrique fatiguant.'

The formula to which this passage refers has not (it would seem) been given by any writer on elliptic functions. The object of the present paper is to enunciate and demonstrate it; and to justify what Jacobi says of it by showing that many of the fundamental formulae of the theory of elliptic functions are either particular cases of it, or corollaries from it. For

* Jacobi, Mathematische Werke, vol. i. p. 358. [The 'Leçons universitaires' here referred to have since been printed from a MS. of Borchardt's. They occupy pp. 497–538 of Jacobi's Gesammelte Werke, vol. i. Berlin, 1881. Ed.]

the sake of symmetry, however, it is convenient to employ the arithmetical equality

$$i^2 + i'^2 + i''^2 + i'''^2 = \left(\frac{-i + i' + i'' + i'''}{2}\right)^2 + \left(\frac{i - i' + i'' + i'''}{2}\right)^2$$
$$+ \left(\frac{i + i' - i'' + i'''}{2}\right)^2 + \left(\frac{i + i' + i'' - i'''}{2}\right)^2,$$

instead of that indicated by Jacobi.

2. The Theta functions being defined by the equation

$$\theta_{\mu, \mu'}(x) = \sum_{n=-\infty}^{n=+\infty} (-1)^{n\mu'} \epsilon^{i\pi \left[\frac{1}{4}(2n+\mu)^2 \omega + (2n+\mu)\frac{x}{a}\right]}, \quad . \quad . \quad . \quad . \quad (1)$$

(in which μ and μ' are given integral numbers, a any constant, ω an imaginary constant having for the coefficient of i* in its imaginary part a quantity different from zero and positive) or, if $q = \epsilon^{i\pi\omega}$, $v = \epsilon^{\frac{i\pi}{a}}$, by the equation

$$\theta_{\mu, \mu'}(x) = \sum_{n=-\infty}^{n=+\infty} (-1)^{n\mu'} q^{\frac{1}{4}(2n+\mu)^2} v^{(2n+\mu)x}, \quad . \quad . \quad . \quad . \quad . \quad . \quad (2)$$

the formula for the multiplication of four Theta functions is

$$\begin{aligned}
&2\,\theta_{\mu_1, \mu'_1}(x_1)\, \theta_{\mu_2, \mu'_2}(x_2)\, \theta_{\mu_3, \mu'_3}(x_3)\, \theta_{\mu_4, \mu'_4}(x_4) \\
&= \theta_{\sigma-\mu_1, \sigma'-\mu'_1}(s - x_1)\, \theta_{\sigma-\mu_2, \sigma'-\mu'_2}(s - x_2)\, \theta_{\sigma-\mu_3, \sigma'-\mu'_3}(s - x_3)\, \theta_{\sigma-\mu_4, \sigma'-\mu'_4}(s - x_4) \\
&+ \theta_{\sigma-\mu_1, \sigma'-\mu'_1+1}(s - x_1)\, \theta_{\sigma-\mu_2, \sigma'-\mu'_2+1}(s - x_2)\, \theta_{\sigma-\mu_3, \sigma'-\mu'_3+1}(s - x_3) \\
&\qquad \times \theta_{\sigma-\mu_4, \sigma'-\mu'_4+1}(s - x_4) \\
&+ (-1)^{\sigma'}\, \theta_{\sigma-\mu_1+1, \sigma'-\mu'_1}(s - x_1)\, \theta_{\sigma-\mu_2+1, \sigma'-\mu'_2}(s - x_2)\, \theta_{\sigma-\mu_3+1, \sigma'-\mu'_3}(s - x_3) \\
&\qquad \times \theta_{\sigma-\mu_4+1, \sigma'-\mu'_4}(s - x_4) \\
&+ (-1)^{\sigma'+1}\, \theta_{\sigma-\mu_1+1, \sigma'-\mu'_1+1}(s - x_1)\, \theta_{\sigma-\mu_2+1, \sigma'-\mu'_2+1}(s - x_2) \\
&\qquad \times \theta_{\sigma-\mu_3+1, \sigma'-\mu'_3+1}(s - x_3)\, \theta_{\sigma-\mu_4+1, \sigma'-\mu'_4+1}(s - x_4), \quad (3)
\end{aligned}$$

where

$$2s = x_1 + x_2 + x_3 + x_4, \quad 2\sigma = \mu_1 + \mu_2 + \mu_3 + \mu_4, \quad 2\sigma' = \mu'_1 + \mu'_2 + \mu'_3 + \mu'_4,$$

the sums $\mu_1 + \mu_2 + \mu_3 + \mu_4$ and $\mu'_1 + \mu'_2 + \mu'_3 + \mu'_4$ being supposed even.

Its proof depends on the two identities

$$\mu_1^2 + \mu_2^2 + \mu_3^2 + \mu_4^2 = (\sigma - \mu_1)^2 + (\sigma - \mu_2)^2 + (\sigma - \mu_3)^2 + (\sigma - \mu_4)^2, \quad . \quad . \quad . \quad (4)$$

---

* Here, and in the rest of this paper, i is used for $\sqrt{(-1)}$. The effect of the condition stated in the text is to render the analytical modulus of q inferior to unity, and thus ensure the convergence of the Theta series for all values of x real or imaginary.

$$\mu_1 \mu_1' + \mu_2 \mu_2' + \mu_3 \mu_3' + \mu_4 \mu_4' = (\sigma - \mu_1)(\sigma' - \mu_1') + (\sigma - \mu_2)(\sigma' - \mu_2') + (\sigma - \mu_3)(\sigma' - \mu_3') + (\sigma - \mu_4)(\sigma' - \mu_4'), \quad (5)$$

both of which admit of immediate verification.

Representing the product

$$\theta_{\mu_1, \mu'_1}(x_1)\, \theta_{\mu_2, \mu'_2}(x_2)\, \theta_{\mu_3, \mu'_3}(x_3)\, \theta_{\mu_4, \mu'_4}(x_4)$$

by Σ, we have, evidently,

$$2\Sigma = 2\Sigma(-1)^{\mu'_1 n_1 + \mu'_2 n_2 + \mu'_3 n_3 + \mu'_4 n_4} \times q^{\frac{1}{4}[(2n_1+\mu_1)^2 + (2n_2+\mu_2)^2 + (2n_3+\mu_3)^2 + (2n_4+\mu_4)^2]} \times v^{(2n_1+\mu_1)x_1 + (2n_2+\mu_2)x_2 + (2n_3+\mu_3)x_3 + (2n_4+\mu_4)x_4}, \quad (6)$$

the sign of summation Σ extending to all integral values of n_1, n_2, n_3, n_4, from $-\infty$ to $+\infty$.

Let

$$\left.\begin{aligned} N_1 &= -n_1 + n_2 + n_3 + n_4, \\ N_2 &= \quad n_1 - n_2 + n_3 + n_4, \\ N_3 &= \quad n_1 + n_2 - n_3 + n_4, \\ N_4 &= \quad n_1 + n_2 + n_3 - n_4\,; \end{aligned}\right\} \quad \dots\dots\dots \quad (7)$$

so that, conversely,

$$\left.\begin{aligned} 4n_1 &= -N_1 + N_2 + N_3 + N_4, \\ 4n_2 &= \quad N_1 - N_2 + N_3 + N_4, \\ 4n_3 &= \quad N_1 + N_2 - N_3 + N_4, \\ 4n_4 &= \quad N_1 + N_2 + N_3 - N_4. \end{aligned}\right\} \quad \dots\dots\dots \quad (8)$$

Transforming in (6) the index of q by the formula (4), and the indices of -1 and v by the formula (5), we find

$$2\Sigma = 2\Sigma(-1)^{\frac{1}{2}[N_1(\sigma' - \mu'_1) + N_2(\sigma' - \mu'_2) + N_3(\sigma' - \mu'_3) + N_4(\sigma' - \mu'_4)]} \times q^{\frac{1}{4}[(N_1 + \sigma - \mu_1)^2 + (N_2 + \sigma - \mu_2)^2 + (N_3 + \sigma - \mu_3)^2 + (N_4 + \sigma - \mu_4)^2]} \times v^{(N_1 + \sigma - \mu_1)(s - x_1) + (N_2 + \sigma - \mu_2)(s - x_2) + (N_3 + \sigma - \mu_3)(s - x_3) + (N_4 + \sigma - \mu_4)(s - x_4)}, \quad (9)$$

the sign of summation Σ extending to all values of N_1, N_2, N_3, N_4, defined by the equations (7); *i.e.*, to all integral values of N_1, N_2, N_3, N_4, from $-\infty$ to $+\infty$, which, substituted in the equations (8), give integral values to n_1, n_2, n_3, n_4. We have, therefore, to ascertain what values of N_1, N_2, N_3, N_4 satisfy this condition. In the first place, we see, from the equations (7), that the difference between any two of the four numbers N_1, N_2, N_3, N_4 is even; *i.e.* that N_1, N_2, N_3, N_4 are all even or all uneven. (α) Let them be all even, and let

$$N_1 = 2\nu_1, \quad N_2 = 2\nu_2, \quad N_3 = 2\nu_3, \quad N_4 = 2\nu_4\,;$$

substituting these values in the equations (8), we find that, in order to render

n_1, n_2, n_3, n_4 integral, it is necessary and sufficient that $\nu_1+\nu_2+\nu_3+\nu_4$ should be even. (β) Let N_1, N_2, N_3, N_4 be all uneven, and let

$$N_1=2\nu_1+1, \quad N_2=2\nu_2+1, \quad N_3=2\nu_3+1, \quad N_4=2\nu_4+1\,;$$

substituting as before, we find that, in order to render n_1, n_2, n_3, n_4 integral, it is necessary and sufficient that $\nu_1+\nu_2+\nu_3+\nu_4$ should be uneven. Separating the terms in (9) in which N_1, N_2, N_3, N_4 are all even from those in which N_1, N_2, N_3, N_4 are all uneven, we obtain

$$\begin{aligned}2\Sigma = 2\Sigma'(-1)^{\nu_1(\sigma'-\mu'_1)+\nu_2(\sigma'-\mu'_2)+\nu_3(\sigma'-\mu'_3)+\nu_4(\sigma'-\mu'_4)}\\ \times q^{\frac{1}{4}[(2\nu_1+\sigma-\mu_1)^2+(2\nu_2+\sigma-\mu_2)^2+(2\nu_3+\sigma-\mu_3)^2+(2\nu_4+\sigma-\mu_4)^2]}\\ \times v^{(2\nu_1+\sigma-\mu_1)(s-x_1)+(2\nu_2+\sigma-\mu_2)(s-x_2)+(2\nu_3+\sigma-\mu_3)(s-x_3)+(2\nu_4+\sigma-\mu_4)(s-x_4)}\\ +2(-1)^{\sigma'}\Sigma''(-1)^{\nu_1(\sigma'-\mu'_1)+\nu_2(\sigma'-\mu'_2)+\nu_3(\sigma'-\mu'_3)+\nu_4(\sigma'-\mu'_4)}\\ \times q^{\frac{1}{4}[(2\nu_1+\sigma-\mu_1+1)^2+(2\nu_2+\sigma-\mu_2+1)^2+(2\nu_3+\sigma-\mu_3+1)^2+(2\nu_4+\sigma-\mu_4+1)^2]}\\ \times v^{(2\nu_1+\sigma-\mu_1+1)(s-x_1)+(2\nu_2+\sigma-\mu_2+1)(s-x_2)+(2\nu_3+\sigma-\mu_3+1)(s-x_3)+(2\nu_4+\sigma-\mu_4+1)(s-x_4)}, \quad (10)\end{aligned}$$

the signs of summation Σ' and Σ'' referring to all values of $\nu_1, \nu_2, \nu_3, \nu_4$, from $-\infty$ to $+\infty$, for which the sum $\nu_1+\nu_2+\nu_3+\nu_4$ is even and uneven respectively. If, for a moment, we represent the general terms of the series Σ' and Σ'' by P and Q, we have evidently,

$$2\Sigma'.\,P=\Sigma\,.\,P+\Sigma(-1)^{\nu_1+\nu_2+\nu_3+\nu_4}P, \quad 2\Sigma''.\,Q=\Sigma\,.\,Q-\Sigma(-1)^{\nu_1+\nu_2+\nu_3+\nu_4}Q,$$

the signs of summation Σ extending to all values of $\nu_1, \nu_2, \nu_3, \nu_4$ from $-\infty$ to $+\infty$ without any limitation. Substituting these values in (10), we obtain, finally,

$$2\Sigma=\Sigma\,.\,P+\Sigma(-1)^{\nu_1+\nu_2+\nu_3+\nu_4}P+(-1)^{\sigma'}\Sigma\,.\,Q+(-1)^{\sigma'+1}\Sigma(-1)^{\nu_1+\nu_2+\nu_3+\nu_4}Q, \quad (11)$$

an equation which, on replacing P and Q by the expressions which they represent, will be found to coincide with (3).

3. We shall now develope some of the particular results included in the formula (3): in doing so, we shall have occasion to employ the equations

$$\theta_{\mu+2,\mu'}(x)=(-1)^{\mu'}\theta_{\mu,\mu'}(x)\,; \quad \theta_{\mu,\mu'+2}(x)=\theta_{\mu,\mu'}(x), \quad \ldots \quad (12)$$

$$\theta_{\mu,\mu'}(-x)=(-1)^{\mu\mu'}\theta_{\mu,\mu'}(x), \quad \ldots \quad (13)$$

which are immediate consequences of the equation of definition (1) or (2). From (12) we infer that there are only four distinct Theta functions,

$$\theta_{0,0}(x), \quad \theta_{0,1}(x), \quad \theta_{1,0}(x), \quad \theta_{1,1}(x)\,;$$

from (13) we infer that the first three of these are even functions, the last an uneven function; so that $\theta_{1,1}(0)=0$, and each of the three derived functions

$$\theta'_{0,0}(0), \quad \theta'_{0,1}(0), \quad \theta'_{1,0}(0)$$

is equal to zero.

We successively attribute to the symbols

$$\left|\begin{matrix} x_1, & x_2, & x_3, & x_4 \\ \mu_1, & \mu_2, & \mu_3, & \mu_4 \\ \mu_1', & \mu_2', & \mu_3', & \mu_4' \end{matrix}\right|$$

the systems of values

$$\left|\begin{matrix} 0, & 0, & 0, & 0 \\ 0, & 0, & 0, & 0 \\ 0, & 0, & 0, & 0 \end{matrix}\right|, \quad \left|\begin{matrix} x, & x, & 0, & 0 \\ 0, & 0, & 0, & 0 \\ 0, & 0, & 0, & 0 \end{matrix}\right|, \quad \left|\begin{matrix} x, & x, & 0, & 0 \\ 0, & 0, & 0, & 0 \\ 1, & 1, & 0, & 0 \end{matrix}\right|, \quad \left|\begin{matrix} x, & x, & 0, & 0 \\ 1, & 1, & 0, & 0 \\ 0, & 0, & 0, & 0 \end{matrix}\right|, \quad \left|\begin{matrix} x, & x, & 0, & 0 \\ 1, & 1, & 0, & 0 \\ 1, & 1, & 0, & 0 \end{matrix}\right|,$$

and we obtain the equations

$$\theta_{0,0}^4(0) = \theta_{0,1}^4(0) + \theta_{1,0}^4(0), \qquad \text{(14)}$$

$$\left.\begin{aligned} \theta_{0,0}^2(x)\,\theta_{0,0}^2(0) &= \theta_{0,1}^2(x)\,\theta_{0,1}^2(0) + \theta_{1,0}^2(x)\,\theta_{1,0}^2(0), \\ \theta_{0,1}^2(x)\,\theta_{0,0}^2(0) &= \theta_{0,0}^2(x)\,\theta_{0,1}^2(0) - \theta_{1,1}^2(x)\,\theta_{1,0}^2(0), \\ \theta_{1,0}^2(x)\,\theta_{0,0}^2(0) &= \theta_{1,1}^2(x)\,\theta_{0,1}^2(0) + \theta_{0,0}^2(x)\,\theta_{1,0}^2(0), \\ \theta_{1,1}^2(x)\,\theta_{0,0}^2(0) &= \theta_{1,0}^2(x)\,\theta_{0,1}^2(0) - \theta_{0,1}^2(x)\,\theta_{1,0}^2(0), \end{aligned}\right\} \qquad \text{(15)}$$

of which (14) arises from the first of (15) by putting 0 for x; and the four equations (15) are equivalent to two independent equations.

4. Again, attributing to the elements x_1, x_2, x_3, x_4 the values $x-y, x+y$, 0, 0, we may obtain formulae expressing the product of the Theta functions of the sum and difference of two quantities in terms of the Theta functions of the two quantities themselves. Thus, to obtain the simplest possible expressions for the six products

$$\begin{matrix} \theta_{0,0}(x-y)\,\theta_{0,1}(x+y), & \theta_{0,0}(x-y)\,\theta_{1,0}(x+y), & \theta_{0,0}(x-y)\,\theta_{1,1}(x+y), \\ \theta_{0,1}(x-y)\,\theta_{1,0}(x+y), & \theta_{0,1}(x-y)\,\theta_{1,1}(x+y), & \theta_{1,0}(x-y)\,\theta_{1,1}(x+y), \end{matrix}$$

(in which the two Theta functions are different) we attribute to the elements

$$\left|\begin{matrix} \mu_1, & \mu_2, & \mu_3, & \mu_4 \\ \mu_1', & \mu_2', & \mu_3', & \mu_4' \end{matrix}\right|$$

the systems of values

$$\left|\begin{matrix} 0, & 0, & 0, & 0 \\ 0, & 1, & 0, & 1 \end{matrix}\right|, \quad \left|\begin{matrix} 0, & 1, & 0, & 1 \\ 0, & 0, & 0, & 0 \end{matrix}\right|, \quad \left|\begin{matrix} 0, & 1, & 0, & 1 \\ 0, & 1, & 1, & 0 \end{matrix}\right|,$$

$$\left|\begin{matrix} 0, & 1, & 0, & 1 \\ 1, & 0, & 1, & 0 \end{matrix}\right|, \quad \left|\begin{matrix} 0, & 1, & 0, & 1 \\ 1, & 1, & 0, & 0 \end{matrix}\right|, \quad \left|\begin{matrix} 1, & 1, & 0, & 0 \\ 0, & 1, & 0, & 1 \end{matrix}\right|,$$

and we obtain the results

$$\begin{aligned} &\theta_{0,0}(x-y)\,\theta_{0,1}(x+y)\,\theta_{0,0}(0)\,\theta_{0,1}(0) \\ &\qquad = \theta_{0,1}(y)\,\theta_{0,0}(y)\,\theta_{0,1}(x)\,\theta_{0,0}(x) - \theta_{1,1}(y)\,\theta_{1,0}(y)\,\theta_{1,1}(x)\,\theta_{1,0}(x), \qquad \text{(16)} \end{aligned}$$

$$\theta_{0,0}(x-y)\,\theta_{1,0}(x+y)\,\theta_{0,0}(0)\,\theta_{1,0}(0)$$
$$=\theta_{1,0}(y)\,\theta_{0,0}(y)\,\theta_{1,0}(x)\,\theta_{0,0}(x)+\theta_{1,1}(y)\,\theta_{0,1}(y)\,\theta_{1,1}(x)\,\theta_{0,1}(x), \quad (17)$$

$$\theta_{0,0}(x-y)\,\theta_{1,1}(x+y)\,\theta_{0,1}(0)\,\theta_{1,0}(0)$$
$$=\theta_{1,1}(y)\,\theta_{0,0}(y)\,\theta_{1,0}(x)\,\theta_{0,1}(x)+\theta_{1,0}(y)\,\theta_{0,1}(y)\,\theta_{1,1}(x)\,\theta_{0,0}(x), \quad (18)$$

$$\theta_{0,1}(x-y)\,\theta_{1,0}(x+y)\,\theta_{0,1}(0)\,\theta_{1,0}(0)$$
$$=\theta_{1,0}(y)\,\theta_{0,1}(y)\,\theta_{1,0}(x)\,\theta_{0,1}(x)+\theta_{1,1}(y)\,\theta_{0,0}(y)\,\theta_{1,1}(x)\,\theta_{0,0}(x), \quad (19)$$

$$\theta_{0,1}(x-y)\,\theta_{1,1}(x+y)\,\theta_{0,0}(0)\,\theta_{1,0}(0)$$
$$=\theta_{1,0}(y)\,\theta_{0,0}(y)\,\theta_{1,1}(x)\,\theta_{0,1}(x)+\theta_{1,1}(y)\,\theta_{0,1}(y)\,\theta_{1,0}(x)\,\theta_{0,0}(x), \quad (20)$$

$$\theta_{1,0}(x-y)\,\theta_{1,1}(x+y)\,\theta_{0,0}(0)\,\theta_{0,1}(0)$$
$$=\theta_{0,1}(y)\,\theta_{0,0}(y)\,\theta_{1,1}(x)\,\theta_{1,0}(x)+\theta_{1,1}(y)\,\theta_{1,0}(y)\,\theta_{0,1}(x)\,\theta_{0,0}(x), \quad (21)$$

of which the number may be doubled by changing $+y$ into $-y$.

It will be observed that the values attributed to $\begin{vmatrix}\mu_1, \mu_2\\ \mu_1', \mu_2'\end{vmatrix}$ are determined in each of these equations by the indices of the Theta functions in the given product $\theta_{\mu_1,\mu'_1}(x-y)\,\theta_{\mu_2,\mu'_2}(x+y)$: the values of $\begin{vmatrix}\mu_3, \mu_4\\ \mu_3', \mu_4'\end{vmatrix}$ are then determined by the conditions that 2σ and $2\sigma'$ must be even, and that, since $\theta_{1,1}(0)=0$, the combination $\begin{vmatrix}1\\1\end{vmatrix}$ is inadmissible.

5. If we differentiate the formulae (16)–(21) with respect to y, and then put $y=0$, we arrive at the following equations, which serve to express the differential coefficient of a quotient of two Theta functions, in terms of the Theta functions themselves:—

$$\theta_{0,1}(x)\,\theta'_{0,0}(x)-\theta_{0,0}(x)\,\theta'_{0,1}(x)=\frac{\theta'_{1,1}(0)\,\theta_{1,0}(0)}{\theta_{0,0}(0)\,\theta_{0,1}(0)}\,\theta_{1,1}(x)\,\theta_{1,0}(x), \quad . \; . \; (22)$$

$$\theta_{1,0}(x)\,\theta'_{0,0}(x)-\theta_{0,0}(x)\,\theta'_{1,0}(x)=-\frac{\theta'_{1,1}(0)\,\theta_{0,1}(0)}{\theta_{0,0}(0)\,\theta_{1,0}(0)}\,\theta_{1,1}(x)\,\theta_{0,1}(x), \quad . \; . \; (23)$$

$$\theta_{1,1}(x)\,\theta'_{0,0}(x)-\theta_{0,0}(x)\,\theta'_{1,1}(x)=-\frac{\theta'_{1,1}(0)\,\theta_{0,0}(0)}{\theta_{0,1}(0)\,\theta_{1,0}(0)}\,\theta_{1,0}(x)\,\theta_{0,1}(x), \quad . \; . \; (24)$$

$$\theta_{1,0}(x)\,\theta'_{0,1}(x)-\theta_{0,1}(x)\,\theta'_{1,0}(x)=-\frac{\theta'_{1,1}(0)\,\theta_{0,0}(0)}{\theta_{0,1}(0)\,\theta_{1,0}(0)}\,\theta_{1,1}(x)\,\theta_{0,0}(x), \quad . \; . \; (25)$$

$$\theta_{1,1}(x)\,\theta'_{0,1}(x)-\theta_{0,1}(x)\,\theta'_{1,1}(x)=-\frac{\theta'_{1,1}(0)\,\theta_{0,1}(0)}{\theta_{0,0}(0)\,\theta_{1,0}(0)}\,\theta_{1,0}(x)\,\theta_{0,0}(x), \quad . \; . \; (26)$$

$$\theta_{1,1}(x)\,\theta'_{1,0}(x)-\theta_{1,0}(x)\,\theta'_{1,1}(x)=-\frac{\theta'_{1,1}(0)\,\theta_{1,0}(0)}{\theta_{0,0}(0)\,\theta_{0,1}(0)}\,\theta_{0,1}(x)\,\theta_{0,0}(x). \quad . \; . \; (27)$$

6. Each of the four products

$$\theta_{0\,0}(x-y)\,\theta_{0,0}(x+y),\quad \theta_{0,1}(x-y)\,\theta_{0,1}(x+y),$$
$$\theta_{1,0}(x-y)\,\theta_{1,0}(x+y),\quad \theta_{1\,1}(x-y)\,\theta_{1,1}(x+y),$$

(in which the two Theta functions are the same) can be expressed in six different ways in terms of the squares of the Theta functions of x and y: thus

$$\begin{aligned}\theta_{\mu\,\mu'}(x-y)\,\theta_{\mu,\mu'}(x+y)\,\theta^2_{0\,0}(0)\\ &=\theta^2_{0,1}(y)\,\theta^2_{\mu,\mu'+1}(x)+(-1)^{\mu'}\theta^2_{1,0}(y)\,\theta^2_{\mu+1,\mu'}(x)\\ &=\theta^2_{0,0}(y)\,\theta^2_{\mu,\mu'}(x)+(-1)^{\mu'}\theta^2_{1,1}(y)\,\theta^2_{\mu+1,\mu'+1}(x),\end{aligned}\quad\text{. . (28)}$$

$$\begin{aligned}\theta_{\mu\,\mu'}(x-y)\,\theta_{\mu,\mu'}(x+y)\,\theta^2_{0,1}(0)\\ &=\theta^2_{0,0}(y)\,\theta^2_{\mu,\mu'+1}(x)+(-1)^{\mu'}\theta^2_{1,0}(y)\,\theta^2_{\mu+1,\mu'+1}(x)\\ &=\theta^2_{0,1}(y)\,\theta^2_{\mu,\mu'}(x)+(-1)^{\mu'}\theta^2_{1,1}(y)\,\theta^2_{\mu+1,\mu'}(x),\end{aligned}\quad\text{. . (29)}$$

$$\begin{aligned}\theta_{\mu\,\mu'}(x-y)\,\theta_{\mu,\mu'}(x+y)\,\theta^2_{1,0}(0)\\ &=(-1)^{\mu'}\theta^2_{0,0}(y)\,\theta^2_{\mu+1\,\mu'}(x)-(-1)^{\mu'}\theta^2_{0,1}(y)\,\theta^2_{\mu+1,\mu'+1}(x)\\ &=\theta^2_{1\,0}(y)\,\theta^2_{\mu\,\mu'}(x)-\theta^2_{1,1}(y)\,\theta^2_{\mu,\,\mu'+1}(x).\end{aligned}\quad (30)$$

To form these equations, we represent by (A), (A′), (B), (B′), (C), (C′), the formulae obtained by attributing to the indices in (3) the values

$$\begin{vmatrix}\mu, & \mu, & 0, & 0\\ \mu', & \mu', & 0, & 0\end{vmatrix},\ \text{(A)}\qquad \begin{vmatrix}\mu+1, & \mu+1, & 1, & 1\\ \mu'+1, & \mu'+1, & 1, & 1\end{vmatrix},\ \text{(A}'\text{)}$$

$$\begin{vmatrix}\mu, & \mu, & 0, & 0\\ \mu', & \mu', & 1, & 1\end{vmatrix},\ \text{(B)}\qquad \begin{vmatrix}\mu+1, & \mu+1, & 1, & 1\\ \mu', & \mu', & 1, & 1\end{vmatrix},\ \text{(B}'\text{)}$$

$$\begin{vmatrix}\mu, & \mu, & 1, & 1\\ \mu', & \mu', & 0, & 0\end{vmatrix},\ \text{(C)}\qquad \begin{vmatrix}\mu, & \mu, & 1, & 1\\ \mu'+1, & \mu'+1, & 1, & 1\end{vmatrix};\ \text{(C}'\text{)}$$

the indeterminates x_1, x_2, x_3, x_4 receiving the values $x-y, x+y, 0, 0$: the left-hand members of the three formulae (A′), (B′), (C′) contain the factor $\theta^2_{1,1}(0)$, and are therefore zero; and the equations (A) ± (A′), (B) ± (B′), (C) ± (C′) will be found to coincide respectively with (28), (29), (30)*.

---

* All the equations (16)–(21), and some of the equations (28)–(30), will be found in an excellent memoir of M. Betti, 'La Teorica delle funzioni ellittiche e sue applicazioni' ('Tortolini, Vol. III; New Series, p. 126). M. Betti derives these equations from the formula

$$\begin{aligned}2\theta_{\mu,\nu}(x+y)\,\theta_{\mu',\nu'}(x-y)\,\theta_{\alpha,0}(0)\,\theta_{0\,\beta}(0)\\ &=\theta_{\mu,\nu}(x)\,\theta_{\mu',\nu'}(x)\,\theta_{\alpha,0}(y)\,\theta_{0,\beta}(y)+(-1)^{\nu}\theta_{\mu+1,\nu}(x)\,\theta_{\mu'+1,\nu'}(x)\,\theta_{\alpha+1,0}(y)\,\theta_{1,\beta}(y)\\ &\quad+(-1)^{\nu}\theta_{\mu+1,\nu+1}(x)\,\theta_{\mu'+1,\nu'+1}(x)\,\theta_{\alpha+1\,1}(y)\,\theta_{1,\beta+1}(y)+\theta_{\mu,\nu+1}(x)\,\theta_{\mu',\nu'+1}(x)\,\theta_{\alpha,1}(y)\,\theta_{0,\beta+1}(y),\end{aligned}$$

One of the formulae (29), by putting $\mu=0$, $\mu'=1$, becomes

$$\theta_{0,1}(x-y)\,\theta_{0,1}(x+y)\,\theta^2_{0,1}(0)=\theta^2_{0,1}(y)\,\theta^2_{0,1}(x)-\theta^2_{1,1}(y)\,\theta^2_{1,1}(x), \quad . \quad . \quad . \quad (31)$$

an equation which is particularly important in the theory of elliptic functions.

7. Differentiating the equations (28)–(30) twice with respect to y, and putting $y=0$, we obtain a system of equations, of which the following, derived from (31), is one:—

$$\frac{d}{dx}\left[\frac{\theta'_{0,1}(x)}{\theta_{0,1}(x)}\right]=\frac{\theta''_{0,1}(0)}{\theta_{0,1}(0)}-\frac{\theta'^2_{1,1}(0)\,\theta^2_{1,1}(x)}{\theta^2_{0,1}(0)\,\theta^2_{0,1}(x)}. \quad . \quad . \quad . \quad . \quad . \quad (32)$$

If, in certain of the equations of this system*, we put $x=0$; or, more simply, if we differentiate the equations (15) twice with respect to x, and then put $x=0$, we find

$$\left.\begin{aligned}
\theta^3_{0,0}\,\theta''_{0,0}&=\theta^3_{0,1}\,\theta''_{0,1}+\theta^3_{1,0}\,\theta''_{1,0},\\
\theta^2_{0,0}\,\theta'^2_{1,1}&=\theta^2_{0,1}\,\theta_{1,0}\,\theta''_{1,0}-\theta^2_{1\,0}\,\theta_{0,1}\,\theta''_{0,1},\\
\theta^2_{0,1}\,\theta'^2_{1,1}&=\theta^2_{0,0}\,\theta_{1,0}\,\theta''_{1,0}-\theta^2_{1,0}\,\theta_{0,0}\,\theta''_{0,0},\\
\theta^2_{1,0}\,\theta'^2_{1,1}&=\theta^2_{0,1}\,\theta_{0,0}\,\theta''_{0,0}-\theta^2_{0,0}\,\theta_{0,1}\,\theta''_{0,1},
\end{aligned}\right\} \quad . \quad . \quad . \quad . \quad . \quad . \quad . \quad . \quad (33)$$

where $\theta_{0,0}$, $\theta''_{0,0}$, &c. are written for $\theta_{0,0}(0)$, $\theta''_{0,0}(0)$, &c. This system is equivalent to two independent relations between $\theta''_{0,0}$, $\theta''_{0,1}$, and $\theta''_{1,0}$.

8. Lastly, in the equation (3), let

$$x_1+x_2+x_3=0, \quad \mu_1=\mu_2=\mu_3=\mu_4=\mu, \quad \mu'_1=\mu'_2=\mu'_3=\mu'_4=\nu\,;$$

differentiate the equation with respect to x_4, and afterwards put $x_4=0$: we find, on combining with one another (by addition or subtraction) the three equations answering to the three combinations

$$\mu=0,\ \nu=0\,; \quad \mu=0,\ \nu=1\,; \ \mu=1,\ \nu=0,$$

$$\begin{aligned}\theta_{1,1}(x_1)\,\theta_{1,1}(x_2)\,\theta_{1,1}(x_3)\,\theta'_{1,1}(0)=(-1)^{\mu+\nu}\,\theta_{\mu,\nu}(0)\,[&\theta_{\mu,\nu}(x_1)\,\theta_{\mu,\nu}(x_2)\,\theta'_{\mu,\nu}(x_3)\\
&+\theta_{\mu,\nu}(x_1)\,\theta'_{\mu,\nu}(x_2)\,\theta_{\mu,\nu}(x_3)+\theta'_{\mu,\nu}(x_1)\,\theta_{\mu,\nu}(x_2)\,\theta_{\mu,\nu}(x_3)],\end{aligned} \quad (34)$$

where, on the right-hand side, μ and ν may have any one of the three sets of values $\mu=0,\ \nu=0\,;\ \ \mu=0,\ \nu=1\,;\ \ \mu=1,\ \nu=0.$

9. It remains to apply these formulae to the demonstration of the fundamental properties of elliptic functions.

---

where $\alpha=\mu-\mu'$, $\beta=\nu'-\nu$; which is a particular case of the formula (3), and which has also been given by M. Hermite ('Liouville,' New Series, Vol. III, p. 27). The notation of M. Betti is different from that employed here and by M. Hermite in the memoir just cited.

* Viz., in (28) let $\mu=\mu'=0$, and in (28), (29), (30) let $\mu=\mu'=1$.

If $\sqrt{k}$, $\sqrt{k'}$ represent the fractions $\frac{\theta_{1,0}(0)}{\theta_{0,0}(0)}$, $\frac{\theta_{0,1}(0)}{\theta_{0,0}(0)}$, the three elliptic functions of the first species are defined by the equations

$$\sin \operatorname{am} x = \frac{1}{\sqrt{k}} \cdot \frac{1}{i} \cdot \frac{\theta_{1,1}(x)}{\theta_{0,1}(x)}, \quad \ldots \ldots \quad (35)$$

$$\cos \operatorname{am} x = \frac{\sqrt{k'}}{\sqrt{k}} \cdot \frac{\theta_{1,0}(x)}{\theta_{0,1}(x)}, \quad \ldots \ldots \quad (36)$$

$$\Delta \operatorname{am} x = \sqrt{k'} \cdot \frac{\theta_{0,0}(x)}{\theta_{0,1}(x)}, \quad \ldots \ldots \quad (37)$$

in which the constant a of equation (1) is to receive a certain value which will presently be assigned. Introducing these functions into the equations (14) and (15), we find

$$k^2 + k'^2 = 1, \quad \ldots \ldots \quad (38)$$

$$\cos^2 \operatorname{am} x + \sin^2 \operatorname{am} x = 1, \quad \ldots \ldots \quad (39)$$

$$\Delta^2 \operatorname{am} x + k^2 \sin^2 \operatorname{am} x = 1. \quad \ldots \ldots \quad (40)$$

Again, the differential equation (26) assumes the form

$$\frac{d \,.\, \sin \operatorname{am} x}{dx} = \frac{1}{i} \frac{\theta_{0,0}(0)\, \theta'_{1,1}(0)}{\theta_{1,0}(0)\, \theta_{0,1}(0)} \cos \operatorname{am} x \, \Delta \operatorname{am} x.$$

To identify the function $\sin \operatorname{am} x$ with the function u which satisfies the differential equation

$$\left(\frac{du}{dx}\right)^2 = (1 - u^2)(1 - k^2 u^2)$$

and the initial conditions $u = 0$ and $\frac{du}{dx} = +1$ when $x = 0$, it is necessary and sufficient that the coefficient $\frac{1}{i} \frac{\theta_{0,0}(0)\, \theta'_{1,1}(0)}{\theta_{1,0}(0)\, \theta_{0,1}(0)}$ should be equal to unity. This condition is satisfied by assigning to the constant a, which has remained undetermined in the equations (1) and (2), the value

$$\pi \times \frac{\sum\limits_{-\infty}^{+} q^{n^2} \times \sum\limits_{-\infty}^{+\infty} (-1)^n (2n+1)\, q^{\frac{1}{4}(2n+1)^2}}{\sum\limits_{-\infty}^{+\infty} (-1)^n q^{n^2} \times \sum\limits_{-\infty}^{+\infty} q^{\frac{1}{4}(2n+1)^2}}.$$

Adopting this value of the constant a, we obtain from (26), (25), and (22),

$$\frac{d \,.\, \sin \operatorname{am} x}{dx} = \cos \operatorname{am} x \, \Delta \operatorname{am} x, \quad \ldots \ldots \quad (41)$$

$$\frac{d.\cos \operatorname{am} x}{dx} = -\sin \operatorname{am} x \,\Delta \operatorname{am} x, \quad . \quad . \quad . \quad . \quad . \quad . \quad . \quad . \quad (42)$$

$$\frac{d.\Delta \operatorname{am} x}{dx} = -k^2 \sin \operatorname{am} x \cos \operatorname{am} x. \quad . \quad . \quad . \quad . \quad . \quad . \quad . \quad (43)$$

Next, dividing the equations (29), (19), and (16) by equation (31), and changing y into $-y$ in (16), we find,

$$\frac{\theta_{0,0}(0)\,\theta_{1,0}(0)}{\theta_{0,1}^2(0)}\frac{\theta_{1,1}(x+y)}{\theta_{0,1}(x+y)} = \frac{\theta_{1,0}(y)\,\theta_{0,0}(y)\,\theta_{1,1}(x)\,\theta_{0,1}(x) + \theta_{1,1}(y)\,\theta_{0,1}(y)\,\theta_{1,0}(x)\,\theta_{0,0}(x)}{\theta_{0,1}^2(y)\,\theta_{0,1}^2(x) - \theta_{1,1}^2(y)\,\theta_{1,1}^2(x)},$$

$$\frac{\theta_{1,0}(0)}{\theta_{0,1}(0)}\frac{\theta_{1,0}(x+y)}{\theta_{0,1}(x+y)} = \frac{\theta_{1,0}(y)\,\theta_{0,1}(y)\,\theta_{1,0}(x)\,\theta_{0,1}(x) + \theta_{1,1}(y)\,\theta_{0,0}(y)\,\theta_{1,1}(x)\,\theta_{0,0}(x)}{\theta_{0,1}^2(y)\,\theta_{0,1}^2(x) - \theta_{1,1}^2(y)\,\theta_{1,1}^2(x)},$$

$$\frac{\theta_{0,0}(0)}{\theta_{0,1}(0)}\frac{\theta_{0,0}(x+y)}{\theta_{0,1}(x+y)} = \frac{\theta_{0,1}(y)\,\theta_{0,0}(y)\,\theta_{0,1}(x)\,\theta_{0,0}(x) + \theta_{1,1}(y)\,\theta_{1,0}(y)\,\theta_{1,1}(x)\,\theta_{1,0}(x)}{\theta_{0,1}^2(y)\,\theta_{0,1}^2(x) - \theta_{1,1}^2(y)\,\theta_{1,1}^2(x)};$$

or, introducing the elliptic functions,

$$\sin \operatorname{am}(x+y) = \frac{\cos \operatorname{am} y \,\Delta \operatorname{am} y \sin \operatorname{am} x + \cos \operatorname{am} x \,\Delta \operatorname{am} x \sin \operatorname{am} y}{1 - k^2 \sin^2 \operatorname{am} x \sin^2 \operatorname{am} y}, \quad . \quad . \quad (44)$$

$$\cos \operatorname{am}(x+y) = \frac{\cos \operatorname{am} x \cos \operatorname{am} y - \Delta \operatorname{am} y \sin \operatorname{am} y \,\Delta \operatorname{am} x \sin \operatorname{am} x}{1 - k^2 \sin^2 \operatorname{am} x \sin^2 \operatorname{am} y}, \quad . \quad . \quad (45)$$

$$\Delta \operatorname{am}(x+y) = \frac{\Delta \operatorname{am} x \,\Delta \operatorname{am} y - k^2 \cos \operatorname{am} y \cos \operatorname{am} x \sin \operatorname{am} y \sin \operatorname{am} x}{1 - k^2 \sin^2 \operatorname{am} x \sin^2 \operatorname{am} y}. \quad (46)$$

These formulae of addition may be obtained in various other well-known forms, by using other combinations of the equations (16)–(21) with the equations (28)–(30).

10. The elliptic function of the second species is defined by the equation

$$Z(x) = \int_0^x k^2 \sin^2 \operatorname{am} x \, dx. \quad . \quad . \quad . \quad . \quad . \quad . \quad . \quad . \quad . \quad (47)$$

Observing that
$$\sqrt{k} = \frac{\theta_{1,0}(0)}{\theta_{0,0}(0)} = \frac{1}{i}\frac{\theta'_{1,1}(0)}{\theta_{0,1}(0)},$$

we may write the equation (32) in the form

$$k^2 \sin^2 \operatorname{am} x = \frac{\theta''_{0,1}(0)}{\theta_{0,1}(0)} - \frac{d}{dx}\left[\frac{\theta'_{0,1}(x)}{\theta_{0,1}(x)}\right];$$

whence, integrating, we obtain

$$Z(x) = x\frac{\theta''_{0,1}(0)}{\theta_{0,1}(0)} - \frac{\theta'_{0,1}(x)}{\theta_{0,1}(x)}, \quad . \quad . \quad . \quad . \quad . \quad . \quad . \quad . \quad . \quad (48)$$

which is Jacobi's expression for the elliptic function of the second species.

If x_1, x_2, x_3 are three arguments of which the sum is zero, we have

$$Z(x_1)+Z(x_2)+Z(x_3)=-\frac{\theta'_{0,1}(x_1)}{\theta_{0,1}(x_1)}-\frac{\theta'_{0,1}(x_2)}{\theta_{0,1}(x_2)}-\frac{\theta'_{0,1}(x_3)}{\theta_{0,1}(x_3)}.$$

The right-hand member of this equation may be transformed, by means of the formula (34), so as to contain only elliptic functions of the first species. We thus obtain

$$\begin{aligned} Z(x_1)+Z(x_2)+Z(x_3) &= \frac{\theta'_{1,1}(0)\,\theta_{1,1}(x_1)\,\theta_{1,1}(x_2)\,\theta_{1,1}(x_3)}{\theta_{0,1}(0)\,\theta_{0,1}(x_1)\,\theta_{0,1}(x_2)\,\theta_{0,1}(x_3)} \\ &= k^2 \sin\operatorname{am} x_1 \sin\operatorname{am} x_2 \sin\operatorname{am} x_3, \quad . \; . \; . \; . \; . \quad (49) \end{aligned}$$

which is the theorem of addition for elliptic functions of the second species.

11. The elliptic function of the third species is defined by the equation

$$\Pi(x,\, y)=\int_0^x \frac{k^2 \sin\operatorname{am} y \cos\operatorname{am} y\,\Delta\operatorname{am} y \sin^2\operatorname{am} x\,dx}{1-k^2\sin^2\operatorname{am} y\sin^2\operatorname{am} x}, \quad . \; . \; . \quad (50)$$

and is a function of two quantities, the argument x and the parameter y.

Dividing the equation (31) by $\theta^2_{0,1}(x)\,\theta^2_{0,1}(y)$, we find

$$\begin{aligned} 1-\frac{\theta^2_{1,1}(y)\,\theta^2_{1,1}(x)}{\theta^2_{0,1}(y)\,\theta^2_{0,1}(x)} &= 1-k^2\sin^2\operatorname{am} y\sin^2\operatorname{am} x \\ &= \frac{\theta_{0,1}(x+y)\,\theta_{0,1}(x-y)\,\theta^2_{0,1}(0)}{\theta^2_{0,1}(y)\,\theta^2_{0,1}(x)}. \end{aligned}$$

Take the logarithm of each side, differentiate with respect to y, and then integrate with respect to x from 0 to x; we obtain

$$\Pi(x,\, y)=x\frac{\theta'_{0,1}(y)}{\theta_{0,1}(y)}+\tfrac{1}{2}\log\frac{\theta_{0,1}(x-y)}{\theta_{0,1}(x+y)}, \quad . \; . \; . \; . \; . \; . \; . \quad (51)$$

the important equation by which Jacobi has expressed the function $\Pi(x, y)$ containing two indeterminates by means of functions of one indeterminate. If x_1, x_2, x_3 are three arguments of which the sum is zero, we have

$$\Pi(x_1,\, y)+\Pi(x_2,\, y)+\Pi(x_3,\, y)=\tfrac{1}{2}\log\frac{\theta_{0,1}(x_1-y)\,\theta_{0,1}(x_2-y)\,\theta_{0,1}(x_3-y)}{\theta_{0,1}(x_1+y)\,\theta_{0,1}(x_2+y)\,\theta_{0,1}(x_3+y)}. \quad (52)$$

The quantity after the logarithmic sign can be transformed (and that in various ways) by means of the formula (3), so as to contain only elliptic functions of the first species. Thus, attributing to the indeterminates and indices in the formula (3) the values

$$\left|\begin{matrix} -y, & -x_1+y, & -x_2+y, & -x_3+y \\ 0\;, & 0\;, & 0\;, & 0 \\ 1\;, & 1\;, & 1\;, & 1 \end{matrix}\right|,$$

we get

$$2\,\theta_{0,1}(y)\,\theta_{0,1}(x_1-y)\,\theta_{0,1}(x_2-y)\,\theta_{0,1}(x_3-y)$$
$$=\theta_{0,1}(2y)\,\theta_{0,1}(x_1)\,\theta_{0,1}(x_2)\,\theta_{0,1}(x_3)+\theta_{0,0}(2y)\,\theta_{0,0}(x_1)\,\theta_{0,0}(x_2)\,\theta_{0,0}(x_3)$$
$$+\theta_{1,1}(2y)\,\theta_{1,1}(x_1)\,\theta_{1,1}(x_2)\,\theta_{1,1}(x_3)-\theta_{1,0}(2y)\,\theta_{1,0}(x_1)\,\theta_{1,0}(x_2)\,\theta_{1,0}(x_3). \quad (53)$$

In this equation change y into $-y$, and divide the first result by the second: the formula (52) becomes

$$\Pi(x_1,y)+\Pi(x_2,y)+\Pi(x_3,y)$$
$$=\tfrac{1}{2}\log\frac{1+\Delta\,\mathrm{am}\,2y[\Delta\,\mathrm{am}\,x]-\cos\mathrm{am}\,2y[\cos\mathrm{am}\,x]+\sin\mathrm{am}\,2y[\sin\mathrm{am}\,x]}{1+\Delta\,\mathrm{am}\,2y[\Delta\,\mathrm{am}\,x]-\cos\mathrm{am}\,2y[\cos\mathrm{am}\,x]-\sin\mathrm{am}\,2y[\sin\mathrm{am}\,x]}, \quad (54)$$

where $[\sin\mathrm{am}\,x]$, $[\Delta\,\mathrm{am}\,x]$, $[\cos\mathrm{am}\,x]$ are written as abbreviations for

$$k^2\sin\mathrm{am}\,x_1\sin\mathrm{am}\,x_2\sin\mathrm{am}\,x_3,\qquad \frac{1}{k'^2}\Delta\,\mathrm{am}\,x_1\,\Delta\,\mathrm{am}\,x_2\,\Delta\,\mathrm{am}\,x_3,$$
$$\frac{k^2}{k'^2}\cos\mathrm{am}\,x_1\cos\mathrm{am}\,x_2\cos\mathrm{am}\,x_3.$$

This is a very symmetrical form of the theorem of addition of functions of the third species: the equation of Legendre

$$\Pi(x_1,y)+\Pi(x_2,y)+\Pi(x_3,y)$$
$$=\tfrac{1}{2}\log\frac{1+k^2\sin\mathrm{am}\,y\sin\mathrm{am}\,x_1\sin\mathrm{am}\,x_2\sin\mathrm{am}\,(x_3+y)}{1-k^2\sin\mathrm{am}\,y\sin\mathrm{am}\,x_1\sin\mathrm{am}\,x_2\sin\mathrm{am}\,(x_3-y)}\quad. \quad (55)$$

is, however, more convenient for computation. It may be obtained by attributing to the indeterminates and indices in the formula (3) the values

$$\begin{vmatrix} x_1-y, & x_2-y, & x_3, & 0\\ 0, & 0, & 0, & 0\\ 1, & 1, & 1, & 1\end{vmatrix},\qquad \begin{vmatrix} x_1-y, & x_2-y, & x_3, & 0\\ 1, & 1, & 1, & 1\\ 1, & 1, & 1, & 1\end{vmatrix},$$

and adding the results: we thus find

$$2\,\theta_{0,1}(x_1-y)\,\theta_{0,1}(x_2-y)\,\theta_{0,1}(x_3)\,\theta_{0,1}(0)$$
$$=\theta_{0,1}(x_1)\,\theta_{0,1}(x_2)\,\theta_{0,1}(x_3+y)\,\theta_{0,1}(y)+\theta_{1,1}(x_1)\,\theta_{1,1}(x_2)\,\theta_{1,1}(x_3+y)\,\theta_{1,1}(y). \quad (56)$$

Changing y into $-y$, dividing the first result by the second, and substituting the resulting expression of $\dfrac{\theta_{0,1}(x_1-y)\,\theta_{0,1}(x_2-y)}{\theta_{0,1}(x_1+y)\,\theta_{0,1}(x_2+y)}$ in the equation (52), we arrive at the formula (55).

Jacobi has given, in the 'Fundamenta Nova,' two other expressions of this addition theorem, both of which are easily inferred from (52) by using appropriate particularisations of the formula (3).

XVII.

ON THE ORDERS AND GENERA OF TERNARY QUADRATIC FORMS.

[Received February 21; Read February 27, 1867.]

---

1. EISENSTEIN, in a memoir entitled 'Neue Theoreme der höheren Arithmetik'*, has defined the ordinal and generic characters of ternary quadratic forms of an uneven determinant; and, in the case of definite forms, has assigned the weight of any given order or genus. But he has not considered forms of an even determinant, neither has he given any demonstrations of his results. To supply these omissions, and so far to complete the work of Eisenstein, is the object of the present memoir.

2. We represent by f the ternary quadratic form

$$ax^2 + a'y^2 + a''z^2 + 2byz + 2b'xz + 2b''xy; \quad . \quad . \quad . \quad . \quad . \quad . \qquad (1)$$

we suppose that f is *primitive* (*i.e.* that the six integral numbers a, a', a'', b, b', b'' admit of no common divisor other than unity), and that its discriminant is different from zero; this discriminant, or the determinant of the matrix

$$\begin{vmatrix} a, & b'', & b' \\ b'', & a', & b \\ b', & b, & a'' \end{vmatrix}, \quad . \quad . \quad . \quad . \quad . \quad . \quad . \quad . \quad . \quad . \qquad (2)$$

we represent by D; by Ω we denote the greatest common divisor of the minor determinants of the matrix (2); by ΩF the contravariant of f, or the form

$$(a'a'' - b^2)x^2 + (a''a - b'^2)y^2 + (aa' - b''^2)z^2$$
$$+ 2(b'b'' - ab)yz + 2(b''b - a'b')zx + 2(bb' - a''b'')xy; \quad . \quad . \quad . \quad . \qquad (3)$$

we shall term F the *primitive contravariant* of f, and we shall write

$$F = Ax^2 + A'y^2 + A''z^2 + 2Byz + 2B'xz + 2B''xy. \quad . \quad . \quad . \quad . \qquad (4)$$

---

* Crelle's Journal, vol. xxxv. p. 117.

If $D=\Delta\Omega^2$, Δ is an integral number, and the discriminant, contravariant, and primitive contravariant of F are respectively $\Omega\Delta^2$, Δf, and f. The numbers Ω and Δ are arithmetical invariants of f; *i.e.* they remain unaltered when f is transformed by any substitution of which the determinant is unity and the coefficients integral numbers. We shall accordingly describe the primitive form f, and the class of forms containing f, as a form, and class, of the invariants $[\Omega, \Delta]$. Similarly, F is a form of the invariants $[\Delta, \Omega]$, and the class containing F is a class of those invariants. The relation between the forms f and F is reciprocal; and this reciprocity extends throughout the whole theory, the contravariants f and F, and the invariants Ω and Δ, being everywhere simultaneously interchangeable.

Of definite forms we shall consider only those which are positive; and in the case of such forms we shall suppose Ω, as well as Δ, to be positive, in order that F as well as f may be positive. In the case of indefinite forms we shall always attribute opposite signs to Ω and Δ; so that, in this case, the discriminants of f and F will be of opposite signs. Thus the definiteness, or indefiniteness, of a form is indicated by the signs of its invariants; if, for example, p and q are positive numbers, the forms

$$x^2+py^2+pqz^2, \quad x^2-py^2-pqz^2, \quad -x^2+py^2+pqz^2$$

are respectively of the invariants

$$[p, q], \quad [-p, q], \quad [p, -q];$$

and their primitive contravariants

$$pqx^2+qy^2+z^2, \quad -pqx^2+qy^2+z^2, \quad pqx^2-qy^2-z^2,$$

are respectively of the invariants

$$[q, p], \quad [q, -p], \quad [-q, p].$$

3. A primitive form f is properly primitive when one at least of its three *principal* coefficients a, a', a'' is uneven; it is improperly primitive when those coefficients are all even. In an improperly primitive form, one at least of the three coefficients b, b', b'' is uneven (or the form would not be primitive); if, therefore, f is improperly primitive, Ω is uneven and F properly primitive; and, reciprocally, if F is improperly primitive, Δ is uneven and f properly primitive. Again, the discriminant of an improperly primitive form is always even. Whenever, therefore, Ω and Δ are both even, or both uneven, neither f nor F is improperly primitive. Primitive forms of the same invariants $[\Omega, \Delta]$ are said to belong to the same order when they and their primitive contravariants are alike properly or alike improperly primitive. An order of properly primitive

forms of the invariants $[\Omega, \Delta]$ always exists, for the form

$$x^2+\Omega y^2+\Omega\Delta z^2$$

is a form of that order. And we shall show hereafter that, when Ω is uneven and Δ even, there is always an improperly primitive order of forms of the invariants $[\Omega, \Delta]$, in which f is improperly and F properly primitive, except when Ω is an uneven square and $\frac{1}{2}\Delta$ an even or uneven square. And, reciprocally, when Δ is uneven and Ω even, there is always an improperly primitive order of forms of the invariants $[\Omega, \Delta]$, in which f is properly and F improperly primitive, except when Δ is an uneven square and $\frac{1}{2}\Omega$ an even or uneven square. These exceptions cannot occur if the forms are indefinite.

For example, there are two orders of forms of the invariants [1, 12]. The properly primitive order contains three classes, represented by the forms

$$\begin{pmatrix}1, & 1, & 12\\ 0, & 0, & 0\end{pmatrix}, \quad \begin{pmatrix}1, & 3, & 4\\ 0, & 0, & 0\end{pmatrix}, \quad \begin{pmatrix}2, & 3, & 3\\ 1, & 1, & 1\end{pmatrix}.$$

The improperly primitive order, in which the forms are improperly primitive but their contravariants properly primitive, contains two classes, represented by the forms

$$\begin{pmatrix}2, & 2, & 4\\ -1, & -1, & 0\end{pmatrix}, \quad \begin{pmatrix}2, & 2, & 4\\ 0, & 0, & -1\end{pmatrix}.$$

4. From the identical equations

$$f(x_1, y_1, z_1) \times f(x_2, y_2, z_2) - \tfrac{1}{4}\left[x_1\frac{df}{dx_2}+y_1\frac{df}{dy_2}+z_1\frac{df}{dz_2}\right]^2$$
$$= \Omega F(y_1z_2-z_1y_2,\ z_1x_2-x_1z_2,\ x_1y_2-y_1x_2), \quad . \quad . \quad (5)$$

$$F(x_1, y_1, z_1) \times F(x_2, y_2, z_2) - \tfrac{1}{4}\left[x_1\frac{dF}{dx_2}+y_1\frac{dF}{dy_2}+z_1\frac{dF}{dz_2}\right]^2$$
$$= \Delta f(y_1z_2-z_1y_2,\ z_1x_2-x_1z_2,\ x_1y_2-y_1x_2), \quad . \quad . \quad (6)$$

we obtain the subdivision of the orders into genera. If ω represent any uneven prime dividing Ω, δ any uneven prime dividing Δ, these equations imply the theorems—

I. 'The numbers, prime to ω, which are represented by f, are either all quadratic residues of ω, or all non-quadratic residues of ω.' In the first case, we attribute to f the *particular generic character* $\left(\frac{f}{\omega}\right)=+1$; in the second, we attribute to f the particular generic character $\left(\frac{f}{\omega}\right)=-1$.

II. 'The numbers, prime to δ, which are represented by F, are either all quadratic residues of δ, or all non-quadratic residues of δ.' We attribute to F

the particular generic character $\left(\frac{F}{\delta}\right) = +1$ in the first case, $\left(\frac{F}{\delta}\right) = -1$ in the second.

If Ω and Δ are both divisible by any uneven prime, f and F will both have particular characters with respect to that prime. These theorems are due to Eisenstein.

Besides its particular characters with respect to uneven primes dividing Ω, f, if properly primitive, will have, in certain cases, particular characters (which we shall call *supplementary*) with respect to 4 and 8. If the uneven numbers represented by f are all $\equiv 1$, mod 4, we attribute to f the particular character $(-1)^{\frac{1}{2}(f-1)} = +1$; if they are all $\equiv 3$, mod 4, we attribute to f the particular character $(-1)^{\frac{1}{2}(f-1)} = -1$. If they are all either $\equiv 1$ or $\equiv 7$, mod 8, we attribute to f the particular character $(-1)^{\frac{1}{8}(f^2-1)} = +1$; if they are all either $\equiv 3$ or $\equiv 5$, mod 8, we attribute to f the particular character $(-1)^{\frac{1}{8}(f^2-1)} = -1$. Lastly, if they are all either $\equiv 1$ or $\equiv 3$, mod 8, f has the character

$$(-1)^{\frac{1}{2}(f-1)+\frac{1}{8}(f^2-1)} = +1;$$

if they are all either $\equiv 5$ or $\equiv 7$, mod 8, it has the character

$$(-1)^{\frac{1}{2}(f-1)+\frac{1}{8}(f^2-1)} = -1.$$

Similarly, if F is properly primitive, it will, in certain cases, acquire the characters $(-1)^{\frac{1}{2}(F-1)} = +1$, or $= -1$; $(-1)^{\frac{1}{8}(F^2-1)} = +1$, or $= -1$;

$$(-1)^{\frac{1}{2}(F-1)+\frac{1}{8}(F^2-1)} = +1, \text{ or } = -1.$$

The following table is useful for ascertaining the supplementary characters of any proposed form.

TABLE I.

A: f and F properly primitive.

	$\Omega \equiv 1$, mod 2.	$\Omega \equiv 2$, mod 4.	$\Omega \equiv 4$, mod 8.	$\Omega \equiv 0$, mod 8.
$\Delta \equiv 1$, mod 2.	Ψ	$(-1)^{\frac{1}{8}(f^2-1)}\,\Psi$	$(-1)^{\frac{1}{2}(f-1)}$, $*(-1)^{\frac{1}{2}(F-1)}$	$(-1)^{\frac{1}{2}(f-1)}$, $*(-1)^{\frac{1}{2}(F-1)}$ $(-1)^{\frac{1}{8}(f^2-1)}$
$\Delta \equiv 2$, mod 4.	$(-1)^{\frac{1}{8}(F^2-1)}\,\Psi$	$(-1)^{\frac{1}{8}(f^2-1)+\frac{1}{8}(F^2-1)}\,\Psi$	$(-1)^{\frac{1}{2}(f-1)}$, †$(-1)^{\frac{1}{8}(F^2-1)}$ $*(-1)^{\frac{1}{2}(F-1)+\frac{1}{8}(F^2-1)}$	$(-1)^{\frac{1}{2}(f-1)}$, $*(-1)^{\frac{1}{2}(F-1)+}$ $(-1)^{\frac{1}{8}(f^2-1)}$, †$(-1)^{\frac{1}{8}(F^2-1)}$
$\Delta \equiv 4$, mod 8.	$*(-1)^{\frac{1}{2}(f-1)}$, $(-1)^{\frac{1}{2}(F-1)}$	†$(-1)^{\frac{1}{8}(f^2-1)}$, $(-1)^{\frac{1}{2}(F-1)}$ $*(-1)^{\frac{1}{2}(f-1)+\frac{1}{8}(f^2-1)}$	$(-1)^{\frac{1}{2}(f-1)}$, $(-1)^{\frac{1}{2}(F-1)}$	$(-1)^{\frac{1}{2}(f-1)}$, $(-1)^{\frac{1}{2}(F-1)}$ $(-1)^{\frac{1}{8}(f^2-1)}$
$\Delta \equiv 0$, mod 8.	$*(-1)^{\frac{1}{2}(f-1)}$, $(-1)^{\frac{1}{2}(F-1)}$ $(-1)^{\frac{1}{8}(F^2-1)}$	$*(-1)^{\frac{1}{2}(f-1)+\frac{1}{8}(f^2-1)}$, $(-1)^{\frac{1}{2}(F-1)}$ †$(-1)^{\frac{1}{8}(f^2-1)}$, $(-1)^{\frac{1}{8}(F^2-1)}$	$(-1)^{\frac{1}{2}(f-1)}$, $(-1)^{\frac{1}{2}(F-1)}$ $(-1)^{\frac{1}{8}(F^2-1)}$	$(-1)^{\frac{1}{2}(f-1)}$, $(-1)^{\frac{1}{2}(F-1)}$ $(-1)^{\frac{1}{8}(f^2-1)}$, $(-1)^{\frac{1}{8}(F^2-1)}$

B : f improperly, F properly primitive.

$\Omega \equiv 1$, mod 2 ; $(-1)^{\frac{1}{2}(F-1)} = -(-1)^{\frac{1}{2}(\Omega-1)}$.

$\Delta \equiv 2$, mod 4.	$(-1)^{\frac{1}{2}(F-1)}$
$\Delta \equiv 0$, mod 4.	$(-1)^{\frac{1}{2}(F-1)}$, $(-1)^{\frac{1}{8}(F^2-1)}$

C : f properly, F improperly primitive.

$\Delta \equiv 1$, mod 2 ; $(-1)^{\frac{1}{2}(f-1)} = -(-1)^{\frac{1}{2}(\Delta-1)}$.

$\Omega \equiv 2$, mod 4.	$(-1)^{\frac{1}{2}(f-1)}$
$\Omega \equiv 0$, mod 4.	$(-1)^{\frac{1}{2}(f-1)}$, $(-1)^{\frac{1}{8}(f^2-1)}$

In this table the asterisk, prefixed to a supplementary character of f, indicates that that character is attributed to f only when $(-1)^{\frac{1}{2}(F-1)} = (-1)^{\frac{1}{2}(\Omega'-1)}$; prefixed to a supplementary character of F, it indicates that that character is attributed to F only when $(-1)^{\frac{1}{2}(f-1)} = (-1)^{\frac{1}{2}(\Delta'-1)}$, Ω' and Δ' denoting the greatest uneven divisors of Ω and Δ, taken with the same signs as Ω and Δ. Similarly, the obelisk prefixed to a character of f or F indicates that that character is attributable to f or F only when $(-1)^{\frac{1}{2}(F-1)} = -(-1)^{\frac{1}{2}(\Omega'-1)}$ in the first case, and when $(-1)^{\frac{1}{2}(f-1)} = -(-1)^{\frac{1}{2}(\Delta'-1)}$ in the second.

The use of the table is most easily explained by an example. Let the proposed form be

$$f = 2x^2 + 7y^2 + 7z^2 - 2yz ;$$

its invariants are [2, 24], and its primitive contravariant is

$$F = 24x^2 + 7y^2 + 7z^2 + 2yz.$$

Since $\Omega \equiv 2$, mod 4, $\Delta \equiv 0$, mod 8, F has the supplementary characters $(-1)^{\frac{1}{2}(F-1)}$ and $(-1)^{\frac{1}{8}(F^2-1)}$; the values of these characters are found by an inspection of the coefficients, and are -1 and $+1$ respectively. Again, since $\Omega' = 1$,

$$(-1)^{\frac{1}{2}(F-1)} = -(-1)^{\frac{1}{2}(\Omega'-1)} ;$$

the character $(-1)^{\frac{1}{8}(f^2-1)}$ is therefore attributable to f, and an inspection of its coefficients shows that $(-1)^{\frac{1}{8}(f^2-1)} = +1$.

The demonstration of the assertions implied in the table (so far as they relate to supplementary characters) is obtained without difficulty from the equations (5) and (6). It will suffice to consider one case as an example of the rest. Let f and F be both properly primitive, and let $\Omega=2\Omega'\equiv 2$, mod 4; $\Delta\equiv 0$, mod 8. If $M_1=F(x_1, y_1, z_1)$, $M_2=F(x_2, y_2, z_2)$ are two uneven numbers represented by F, we infer from equation (6) that

$$\tfrac{1}{2}\left(x_1\frac{dF}{dx_2}+y_1\frac{dF}{dy_2}+z_1\frac{dF}{dz_2}\right)$$

is an uneven number, and consequently that $M_1\times M_2\equiv 1$, mod 8; M_1 and M_2 are therefore congruous to one another, mod 8; *i.e.* all uneven numbers represented by F are congruous, mod 8, or F has the characters $(-1)^{\frac{1}{2}(F-1)}$ and $(-1)^{\frac{1}{8}(F^2-1)}$. To prove that f has the supplementary character attributed to it in the table, we observe, first of all, that F cannot represent unevenly even numbers; for, if possible, let $F(x_1, y_1, z_1)$ be unevenly even, and let $F(x_2, y_2, z_2)$ be any uneven number represented by F; then in the equation (6) we have a square congruous, mod 8, to an unevenly even number, which is impossible. Now let

$$m_1=f(x_1, y_1, z_1), \quad m_2=f(x_2, y_2, z_2)$$

be any two uneven numbers represented by f; the number

$$\tfrac{1}{2}\left(x_1\frac{df}{dx_2}+y_1\frac{df}{dy_2}+z_1\frac{df}{dz_2}\right)$$

is uneven in equation (5); and, considering that equation as a congruence for the modulus 8, we find $m_1\times m_2\equiv 1$ or $\equiv 1+2(-1)^{\frac{1}{2}(\Omega'-1)+\frac{1}{2}(F-1)}$, according as

$$F(y_1z_2-y_2z_1, \quad z_1x_2-z_2x_1, \quad x_1y_2-x_2y_1)$$

is evenly even or uneven. If, then, $(-1)^{\frac{1}{2}(F-1)}=(-1)^{\frac{1}{2}(\Omega'-1)}$, $m_1\times m_2\equiv 1$ or $\equiv 3$, mod 8; *i.e.* the uneven numbers represented by f are either all of one or other of the linear forms $8k+1$, $8k+3$, or else all of one or other of the linear forms $8k+5$, $8k+7$; so that f has the supplementary character $(-1)^{\frac{1}{2}(f-1)+\frac{1}{8}(f^2-1)}$. But, if $(-1)^{\frac{1}{2}(F-1)}=-(-1)^{\frac{1}{2}(\Omega'-1)}$, $m_1\times m_2\equiv +1$ or $\equiv -1$, mod 8, and the uneven numbers represented by f are either all of the linear forms $8k\pm 1$, or else all of the linear forms $8k\pm 3$, so that f has the character $(-1)^{\frac{1}{8}(f^2-1)}$.

The signification of the symbols

$$\Psi, \quad (-1)^{\frac{1}{8}(f^2-1)}\Psi, \quad (-1)^{\frac{1}{8}(F^2-1)}\Psi, \quad (-1)^{\frac{1}{8}(f^2-1)+\frac{1}{8}(F^2-1)}\Psi,$$

which occur in the table, is explained in Arts. 6 and 7. In the next article we shall establish an auxiliary proposition which is frequently useful.

5. 'There exist pairs of forms ϕ and Φ, equivalent to f and F, and satisfying the congruences

$$\left.\begin{aligned}\phi&\equiv\alpha x^2+\beta\Omega y^2+\gamma\Omega\Delta z^2,\\ \Phi&\equiv\beta\gamma\Omega\Delta x^2+\alpha\gamma\Delta y^2+\alpha\beta z^2,\\ \alpha\beta\gamma&\equiv 1,\end{aligned}\right\} \quad \ldots\ldots\ldots \quad (7)$$

for any proposed modulus ∇; but this modulus must be uneven, if either f or F is improperly primitive.'

In the proof of this proposition we shall employ two lemmas of a very elementary character.

(i) A properly primitive form f represents numbers prime to any given number ∇; and an improperly primitive form f represents the doubles of numbers prime to any given number ∇.

Let p be any prime divisor of ∇, and, if f is improperly primitive, let p be an uneven prime. If one of the numbers a, a', a'' is prime to p, let a be prime to p; then, if x is prime to p and y and z are divisible by p, f will acquire a value prime to p. If a, a', a'' are all divisible by p, one of the three numbers b, b', b'' must be prime to p; let b be prime to p; then, if x is divisible by p and y and z are prime to p, f will acquire a value prime to p.

If f is improperly primitive and $p=2$, we may consider $\frac{1}{2}f$ instead of f, and $\frac{1}{2}a$, $\frac{1}{2}a'$, $\frac{1}{2}a''$ instead of a, a', a''; and we may prove in the same way that $\frac{1}{2}f$ represents uneven numbers.

Thus, among the p^3 systems of values which can be attributed to x, y, z for the modulus p, there are always some which render f (or $\frac{1}{2}f$) prime to p; there are, therefore, among the ∇^3 systems of values which can be attributed to x, y, z for the modulus ∇, some which render f (or $\frac{1}{2}f$) simultaneously prime to every prime dividing ∇.

(ii) If $\Omega\Delta$ is uneven, f represents numbers of both the linear forms $4k+1$ and $4k+3$.

One at least of the principal coefficients of f is uneven, because its discriminant is uneven: let then a be uneven, and let $a' \equiv \lambda$, mod 2, $a'' \equiv \mu$, mod 2; the substitution $x = x + \lambda y + \mu z$ will transform f into a form f_1, in which a_1, a'_1, a''_1 are all uneven, and in which, because the discriminant is uneven, either only one, or else all three, of the coefficients b_1, b'_1, b''_1 are even. The four numbers $a_1, a'_1, a''_1, a_1+a'_1+a''_1+2b_1+2b'_1+2b''_1$ are then all uneven; they are all represented by f_1, that is by f; but they are not all congruous to one another for the modulus 4; therefore f represents numbers of both the linear forms $4k+1$ and $4k+3$.

It follows from these lemmas (i) that, if f is an improperly primitive form, we can find a form equivalent to f and having one of its principal coefficients unevenly even and prime to any uneven number; (ii) that, if f is properly primitive, we can find a form equivalent to f and having one of its principal coefficients prime to any given number; (iii) that, if $\Omega\Delta$ is uneven, we may suppose this principal coefficient of either of the two linear forms $4k+1$ or $4k+3$, at our option.

We shall first suppose that the forms f and F, which it is proposed to transform into forms ϕ and Φ satisfying the congruences (7), are properly primitive. Let $\nabla' = \nabla\Omega^2\Delta$, and let us assume that, in the form F, A'' is prime to ∇', and also that $A'' \equiv \Omega$, mod 4, if $\Omega\Delta$ is uneven. Let $\gamma \equiv \frac{1}{A''}$, mod ∇ ; the redundant system of congruences

$$\left.\begin{aligned} a\,x + b''y + b' &\equiv 0, \\ b''x + a'y + b &\equiv 0, \\ b'x + b\,y + a'' &\equiv \gamma\Omega\Delta, \end{aligned}\right\} \text{mod } \nabla',$$

is resoluble, admitting Ω incongruous solutions*. Let

$$x \equiv \lambda, \text{ mod } \nabla',$$
$$y \equiv \mu, \text{ mod } \nabla'$$

be any one of these solutions, and let us transform f by the substitution

$$x = x + \lambda z,$$
$$y = y + \mu z,$$

into an equivalent form f_1. The coefficients a_1, b_1'', a_1' are the same as a, b'', a'; the coefficients a_1'', b_1, b_1' are respectively congruous for the modulus ∇' to $\gamma\Omega\Delta$, 0, 0 ; so that f_1 satisfies the congruence

$$f_1 \equiv ax^2 + 2b''xy + a'y^2 + \gamma\Omega\Delta z^2, \text{ mod } \nabla'.$$

The binary form (a, b'', a') is primitive; for if d is a prime dividing a, b'', a', it divides $-\Omega A''$, the determinant of (a, b'', a'), and $\Omega^2\Delta$, the discriminant of f; it therefore divides Ω (because A'' and Δ are relatively prime), and is a common divisor of the coefficients of the primitive form f_1, *i.e.* $d=1$. Again, (a, b'', a') is not improperly primitive; if $\Omega\Delta$ is even, this is manifest, for f_1 is not improperly primitive; if $\Omega\Delta$ is uneven, $\Omega A''$ is by hypothesis of the form $4k+1$, and there are no improperly primitive binary forms of the determinant $-\Omega A''$. We may now suppose that, in the properly primitive binary form (a, b'', a'), a is uneven and prime to ∇'; let $\beta \equiv \frac{A''}{a}$, mod ∇'; then the congruences

$$a\,x + b'' \equiv 0, \text{ mod } \nabla',$$
$$b''x + a' \equiv \beta\Omega, \text{ mod } \nabla'$$

are resoluble and admit of but one solution. Let $x \equiv \lambda$, mod ∇', be that solu-

* Philosophical Transactions, vol. cli. p. 323.

tion; if f_1 be transformed by the substitution $x = x + \lambda y$, the resulting form ϕ will satisfy the congruence

$$\phi \equiv ax^2 + \beta\Omega y^2 + \gamma\Omega\Delta z^2, \text{ mod } \nabla',$$

and the forms ϕ and Φ will satisfy the congruences (7) for the modulus ∇.

Attributing to ∇' the value $\nabla\Omega'^2\Delta'$, we may apply the same demonstration to the case in which either f or F is improperly primitive, the modulus ∇ being supposed uneven.

6. When neither f nor F is improperly primitive and neither Ω nor Δ are evenly even, f and F, considered separately, have no particular characters, properly so called, with respect to 4 or 8. But, considered jointly, they have a certain character with respect to 4 or 8, which we shall term their *simultaneous generic character*, and which it is important to consider here.

If $m = f(x, y, z)$, $M = F(X, Y, Z)$, the representations of m by f and of M by F are said to be *simultaneous* when x, y, z, X, Y, Z satisfy the equation

$$xX + yY + zZ = 0. \quad . \quad . \quad . \quad . \quad . \quad . \quad . \quad . \quad . \quad . \quad (8)$$

This definition of simultaneous representation suffices for our immediate purpose; we add, however, that if the two representations are *primitive* * as well as simultaneous, the equations

$$\left\| \begin{matrix} x, & y, & z \\ x', & y', & z' \end{matrix} \right\| = X, Y, Z, \quad . \quad . \quad . \quad . \quad . \quad . \quad . \quad . \quad . \quad (9)$$

$$\left\| \begin{matrix} X, & Y, & Z \\ X', & Y', & Z' \end{matrix} \right\| = x, y, z, \quad . \quad . \quad . \quad . \quad . \quad . \quad . \quad . \quad . \quad (10)$$

are resoluble in integral numbers x', y', z', X', Y', Z'. For, considering the first of these equations, we observe that $[x, y, z]$ is a given solution, in relatively prime numbers, of the equation (8); let $[x', y', z']$ be a solution of the same equation which with $[x, y, z]$ forms a fundamental set; the equation (9) is then satisfied by virtue of the characteristic property of the fundamental set (Philosophical Transactions, vol. cli. p. 297). Thus, if the representations of m by f and of M by F, are primitive and simultaneous, m is primitively represented by a binary form of determinant $-\Omega M$, which is itself primitively represented by f†; and, reciprocally, M is primitively represented by a binary form of determinant $-\Delta m$, which is itself primitively represented by F.

In the four cases in which neither Ω nor Δ is evenly even, the simultaneous

---

* The representation of a number by a form is said to be *primitive* when the values of the indeterminates do not admit of any common divisor besides unity.

† See Art. 10.

character of f and F is given in the table of Art. 4. The symbol Ψ in that table represents the unit $(-1)^{\frac{1}{2}(\Delta' f+1)\cdot\frac{1}{2}(\Omega' F+1)}$, *i.e.* the unit $(-1)^{\frac{1}{2}(\Delta' m+1)\cdot\frac{1}{2}(\Omega' M+1)}$, Ω' and Δ' denoting the same numbers as in Art. 4, and m, M being any two uneven numbers simultaneously represented by f and F. Thus, if $\Omega\equiv\Delta\equiv 1$, mod 2, the simultaneous character Ψ is attributed in the table to f and F; *i.e.* the uneven numbers simultaneously represented by f and F either all satisfy the equation $\Psi=1$, or else all satisfy the equation $\Psi=-1$.

To demonstrate these simultaneous characters we consider the four cases separately, and, for the forms f and F, we substitute forms ϕ and Φ equivalent to them and satisfying certain congruences for the modulus 4 or 8. The existence of the equivalent forms thus assumed results, in each case, from the theorem of the preceding article.

Case (i). Let $\Omega\equiv\Delta\equiv 1$, and let ϕ and Φ satisfy the congruences

$$\left.\begin{array}{rl} \Delta\phi & \equiv \alpha x^2+\beta y^2+\gamma z^2, \\ \Omega\Phi & \equiv \beta\gamma X^2+\alpha\beta Y^2+\alpha\beta Z^2, \\ \text{or}\qquad \Omega\Phi & \equiv \alpha X^2+\beta Y^2+\gamma Z^2, \\ \alpha\beta\gamma & \equiv 1, \end{array}\right\}\ \text{mod } 4.$$

Attributing in succession to the indeterminates

$$\begin{array}{ccc} x, & y, & z \\ X, & Y, & Z \end{array}$$

all systems of values, mod 2, which satisfy the congruence

$$xX+yY+zZ\equiv 0,\ \text{mod } 2,$$

and which render m and M simultaneously uneven, we find that in every case Δm is congruous, for the modulus 4, to one of the three numbers α, β, γ, and ΩM to one of the remaining two. Thus $\frac{1}{2}(\Delta m+1)\times\frac{1}{2}(\Omega M+1)$ is necessarily congruous, for the modulus 2, to one of the three numbers

$$\tfrac{1}{4}(\beta+1)(\gamma+1),\quad \tfrac{1}{4}(\gamma+1)(\alpha+1),\quad \tfrac{1}{4}(\alpha+1)(\beta+1).$$

But these numbers are all congruous to one another for the modulus 2, because the congruence $\alpha\beta\gamma\equiv 1$, mod 4, implies the congruence $\alpha+\beta+\gamma+1\equiv 0$, mod 4. Therefore the unit Ψ has always the same value for every pair of uneven numbers simultaneously represented by f and F.

It will be seen that $\Psi=-1$, or $\Psi=+1$, according as the congruences $\alpha\equiv\beta\equiv\gamma\equiv 1$, mod 4, are, or are not, satisfied.

Case (ii). Let $\Omega\equiv 2$, mod 4, $\Delta\equiv 1$, mod 2, and let

$$\begin{aligned} \Delta\phi &\equiv \alpha x^2+2\beta y^2+2\gamma z^2,\ \text{mod } 8, \\ \Omega'\Phi &\equiv 2\alpha X^2+\beta Y^2+\gamma Z^2,\ \text{mod } 4, \\ \alpha\beta\gamma &\equiv 1,\ \text{mod } 4. \end{aligned}$$

The admissible combinations of the values of x, y, z, X, Y, Z, mod 2, give rise to eight cases,

$$\begin{array}{lll} \Delta m \equiv \alpha, & \text{mod } 8; & \Omega' M \equiv \beta, \text{ or } \equiv \gamma, \text{ mod } 4, \\ \Delta m \equiv \alpha + 2\beta, & \text{mod } 8; & \Omega' M \equiv -\beta, \text{ or } \equiv \gamma, \text{ mod } 4, \\ \Delta m \equiv \alpha + 2\gamma, & \text{mod } 8; & \Omega' M \equiv \beta, \text{ or } \equiv -\gamma, \text{ mod } 4, \\ \Delta m \equiv \alpha + 2\beta + 2\gamma, & \text{mod } 8; & \Omega' M \equiv -\beta, \text{ or } \equiv -\gamma, \text{ mod } 4, \end{array}$$

and, in all of them, the value of the unit $(-1)^{\frac{1}{8}(\Delta^2 m^2 - 1)}\Psi$, and therefore of the unit $(-1)^{\frac{1}{8}(f^2-1)}\Psi$, is the same, because, by virtue of the congruence

$$\alpha + \beta + \gamma + 1 \equiv 0, \text{ mod } 4,$$

the four numbers

$$\tfrac{1}{8}(\alpha+1)(\alpha+2\beta+1), \quad \tfrac{1}{8}(\alpha+1)(\alpha+2\gamma+1),$$
$$\tfrac{1}{8}(\alpha+2\beta+2\gamma+1)(\alpha+2\beta+1), \quad \tfrac{1}{8}(\alpha+2\beta+2\gamma+1)(\alpha+2\gamma+1)$$

are all congruous to one another for the modulus 2.

Case (iii). $\Omega \equiv 1$, mod 2, $\Delta \equiv 2$, mod 4. In this case the simultaneous character of the forms f and F may be demonstrated as in case (ii), or may be inferred by reciprocation from the result in that case.

Case (iv). $\Omega \equiv \Delta \equiv 2$, mod 4. Let

$$\Delta' \phi \equiv \alpha x^2 + 2\beta y^2 + 4\gamma z^2, \text{ mod } 8,$$
$$\Omega' \Phi \equiv 4\alpha X^2 + 2\beta Y^2 + \gamma Z^2, \text{ mod } 8,$$
$$\alpha\beta\gamma \equiv 1, \text{ mod } 4.$$

Here again there are eight cases,

$$\begin{array}{lll} \Delta' m \equiv \alpha; & \Omega' M \equiv \gamma, & \text{or } \gamma + 2\beta, \quad \text{mod } 8, \\ \Delta' m \equiv \alpha + 2\beta; & \Omega' M \equiv \gamma, & \text{or } \gamma + 2\beta + 4, \text{ mod } 8, \\ \Delta' m \equiv \alpha + 4; & \Omega' M \equiv \gamma + 4, & \text{or } \gamma + 2\beta + 4, \text{ mod } 8, \\ \Delta' m \equiv \alpha + 2\beta + 4; & \Omega' M \equiv \gamma + 4, & \text{or } \gamma + 2\beta, \quad \text{mod } 8, \end{array}$$

and in all of them the value of the unit $(-1)^{\frac{1}{8}(\Delta'^2 m^2 - 1) + \frac{1}{8}(\Omega'^2 M^2 - 1)}\Psi$, and therefore of the unit $(-1)^{\frac{1}{8}(f^2-1) + \frac{1}{8}(F^2-1)}\Psi$, is the same; because, by virtue of the congruence $\alpha + \beta + \gamma + 1 \equiv 0$, mod 4, the two numbers

$$\tfrac{1}{8}(\alpha+\gamma)(\alpha+\gamma+2), \quad \tfrac{1}{8}(\alpha+\gamma+2\beta)(\alpha+\gamma+2\beta+2)$$

are congruous to one another for the modulus 2.

7. The following observations will serve to show more clearly the import of the simultaneous character in each of the four cases.

Case (i). Let $\Psi = -1$; then, if m and M are any two uneven numbers simultaneously represented by f and F, $m \equiv \Delta$, mod 4, and $M \equiv \Omega$, mod 4. Also, f cannot represent numbers congruous to 7Δ, mod 8, nor F numbers congruous to 7Ω, mod 8; for the congruences

$$\tfrac{1}{4}(\beta+1)(\gamma+1) \equiv \tfrac{1}{4}(\gamma+1)(\alpha+1) \equiv \tfrac{1}{4}(\alpha+1)(\beta+1) \equiv 1, \text{ mod } 2,$$

imply that $\alpha \equiv \beta \equiv \gamma \equiv 1$, mod 4; *i.e.* that ϕ, or, which is the same thing, f can only represent uneven numbers congruous to Δ, 3Δ, 5Δ. And similarly of uneven numbers F can only represent those which are congruous to Ω, 3Ω, 5Ω. Numbers congruous to 3Δ are represented by f, and numbers congruous to 3Ω are represented by F; but these representations are not simultaneous with the representation of any uneven number by F in the first case, and by f in the second.

Let $\Psi = +1$; then if m and M are uneven numbers simultaneously represented by f and F, one at least of the two congruences $m \equiv -\Delta$, mod 4, $M \equiv -\Omega$, mod 4, must be satisfied. Subject to this restriction, m and M may have any of the four linear forms $8k+1$, $8k+3$, $8k+5$, $8k+7$.

Case (ii). The restrictions imposed on the numbers m and M by the simultaneous characters are exhibited in the annexed table.

If	$(-1)^{\frac{1}{8}(f^2-1)}\Psi = (-1)^{\frac{1}{8}(\Delta^2-1)}$	$(-1)^{\frac{1}{8}(f^2-1)}\Psi = -(-1)^{\frac{1}{8}(\Delta^2-1)}$
$M \equiv \Omega'$, mod 4	$m \equiv 5\Delta$, 7Δ, mod 8	$m \equiv \Delta$, 3Δ, mod 8
$M \equiv 3\Omega'$, mod 4	$m \equiv \Delta$, 7Δ, mod 8	$m \equiv 3\Delta$, 5Δ, mod 8

Except when Ω and Δ are both uneven, it will be found that, in the case of any two properly primitive forms f and F, every representation of an uneven number by either of the two is simultaneous with the representation of uneven numbers by the other. If, therefore, $(-1)^{\frac{1}{8}(f^2-1)}\Psi = (-1)^{\frac{1}{8}(\Delta^2-1)}$, f cannot represent numbers congruous to 3Δ, mod 8, because it cannot represent them simultaneously with uneven numbers, and, if $(-1)^{\frac{1}{8}(f^2-1)}\Psi = -(-1)^{\frac{1}{8}(\Delta^2-1)}$, f cannot represent numbers congruous to 7Δ, mod 8.

Case (iii). In this case, which is the reciprocal of the last, we have the table,

If	$(-1)^{\frac{1}{8}(F^2-1)}\Psi = (-1)^{\frac{1}{8}(\Omega^2-1)}$	$(-1)^{\frac{1}{8}(F^2-1)}\Psi = -(-1)^{\frac{1}{8}(\Omega^2-1)}$
$m \equiv \Delta'$, mod 4	$M \equiv 5\Omega$, 7Ω, mod 8	$M \equiv \Omega$, 3Ω, mod 8
$m \equiv 3\Delta'$, mod 4	$M \equiv \Omega$, 7Ω, mod 8	$M \equiv 3\Omega$, 5Ω, mod 8

And F cannot represent numbers congruous to 3Ω, or cannot represent numbers congruous to 7Ω, according as $(-1)^{\frac{1}{8}(F^2-1)}\Psi = (-1)^{\frac{1}{8}(\Omega^2-1)}$, or $= -(-1)^{\frac{1}{8}(\Omega^2-1)}$.

Case (iv). In this case both f and F represent numbers of all the four linear forms $8k+1$, $8k+3$, $8k+5$, $8k+7$. The table, in which the modulus is everywhere 8, exhibits the restrictions imposed by the simultaneous character.

If	$(-1)^{\frac{1}{8}(f^2-1)+\frac{1}{8}(F^2-1)}\Psi=(-1)^{\frac{1}{8}(\Delta'^2-1)+\frac{1}{8}(\Omega'^2-1)}$	$(-1)^{\frac{1}{8}(f^2-1)+\frac{1}{8}(F^2-1)}\Psi=-(-1)^{\frac{1}{8}(\Delta'^2-1)+\frac{1}{8}(\Omega'^2-1)}$
$m\equiv \Delta'$	$M\equiv 5\Omega',\ 7\Omega'$	$M\equiv \Omega',\ 3\Omega'$
$m\equiv 3\Delta'$	$M\equiv 3\Omega',\ 5\Omega'$	$M\equiv \Omega',\ 7\Omega'$
$m\equiv 5\Delta'$	$M\equiv \Omega',\ 3\Omega'$	$M\equiv 5\Omega',\ 7\Omega'$
$m\equiv 7\Delta'$	$M\equiv \Omega',\ 7\Omega'$	$M\equiv 3\Omega',\ 5\Omega'$

8. The complete generic character of a form or class is the complex of all the particular characters attributable to the form or class and to its primitive contravariant, including their simultaneous character, if any. And two forms or classes which have the same complete generic character are said to belong to the same genus. But not every complete generic character that can be assigned *à priori* is the character of any really existing genus of forms. The table on the next two pages will serve, in the case of any given order, to distinguish those complete generic characters which are possible, *i.e.* to which actually existing genera correspond, from those which are impossible.

In this table Ω_2^2 and Δ_2^2 are the greatest squares dividing Ω and Δ; the quotients $\Omega\div\Omega_2^2$, $\Delta\div\Delta_2^2$ are respectively represented, if uneven, by Ω_1 and Δ_1, if even, by $2\Omega_1$ and $2\Delta_1$, so that Ω_1 and Δ_1 are always uneven and not divisible by any square; ω_1 and δ_1 are any primes dividing Ω_1 and Δ_1, ω_2 and δ_2 are any uneven primes dividing Ω_2 and Δ_2, but ω_2 must not divide Ω_1 nor must δ_2 divide Δ_1; lastly, Ψ is still the unit $(-1)^{\frac{1}{2}(\Omega'F+1)\cdot\frac{1}{2}(\Delta'f+1)}$, or, which is the same thing, the unit $(-1)^{\frac{1}{2}(\Omega_1F+1)\cdot\frac{1}{2}(\Delta_1f+1)}$, f and F in the exponents of these units denoting uneven numbers simultaneously represented by the forms f and F.

The table A of properly primitive generic characters contains twenty-five compartments corresponding to the twenty-five cases indicated in its margins; the tables B and C of improperly primitive genera contain three such compartments each. In each compartment are inscribed all the particular characters which make up the complete generic character of a form coming under the

TABLE II. OF COMPLETE

A : f and F properly primitive.

	$\Omega = \Omega_1 \Omega_2^2$. $\Omega_2 \equiv 1$, mod 2.		$\Omega = \Omega_1 \Omega_2^2$. $\Omega_2 \equiv 2$, mod 4.		$\Omega = \Omega_1 \Omega_2^2$. $\Omega_2 \equiv 0$, mod 4.	
$\Delta = \Delta_1 \Delta_2^2$. $\Delta_2 \equiv 1$, mod 2.	Ψ $\left(\frac{f}{\omega_1}\right), \left(\frac{F}{\delta_1}\right)$ S	$\left(\frac{f}{\omega_2}\right), \left(\frac{F}{\delta_2}\right)$ $\Gamma = 2^\gamma$	Ψ $\left(\frac{f}{\omega_1}\right), \left(\frac{F}{\delta_1}\right)$ Q	$(-1)^{\frac{1}{2}(f-1)}$ $\left(\frac{f}{\omega_2}\right), \left(\frac{F}{\delta_2}\right)$ $\Gamma = 3 \times 2^{\gamma-1}$	Ψ $\left(\frac{f}{\omega_1}\right), \left(\frac{F}{\delta_1}\right)$ Q	$(-1)^{\frac{1}{2}(f-1)}$ $(-1)^{\frac{1}{8}(f^2-1)}$ $\left(\frac{f}{\omega_2}\right), \left(\frac{F}{\delta_2}\right)$ $\Gamma = 3 \times 2^\gamma$
$\Delta = \Delta_1 \Delta_2^2$. $\Delta_2 \equiv 2$, mod 4.	Ψ $\left(\frac{f}{\omega_1}\right), \left(\frac{F}{\delta_1}\right)$ Q	$(-1)^{\frac{1}{2}(F-1)}$ $\left(\frac{f}{\omega_2}\right), \left(\frac{F}{\delta_2}\right)$ $\Gamma = 3 \times 2^{\gamma-1}$	Ψ $\left(\frac{f}{\omega_1}\right), \left(\frac{F}{\delta_1}\right)$ P	$(-1)^{\frac{1}{2}(f-1)}$ $(-1)^{\frac{1}{2}(F-1)}$ $\left(\frac{f}{\omega_2}\right), \left(\frac{F}{\delta_2}\right)$ $\Gamma = 2^{\gamma+1}$	Ψ $\left(\frac{f}{\omega_1}\right), \left(\frac{F}{\delta_1}\right)$ P	$(-1)^{\frac{1}{2}(f-1)}$ $(-1)^{\frac{1}{8}(f^2-1)}$ $(-1)^{\frac{1}{2}(F-1)}$ $\left(\frac{f}{\omega_2}\right), \left(\frac{F}{\delta_2}\right)$ $\Gamma = 2^{\gamma+2}$
$\Delta = \Delta_1 \Delta_2^2$. $\Delta_2 \equiv 0$, mod 4.	Ψ $\left(\frac{f}{\omega_1}\right), \left(\frac{F}{\delta_1}\right)$ Q	$(-1)^{\frac{1}{2}(F-1)}$ $(-1)^{\frac{1}{8}(F^2-1)}$ $\left(\frac{f}{\omega_2}\right), \left(\frac{F}{\delta_2}\right)$ $\Gamma = 3 \times 2^\gamma$	Ψ $\left(\frac{f}{\omega_1}\right), \left(\frac{F}{\delta_1}\right)$ P	$(-1)^{\frac{1}{2}(f-1)}$ $(-1)^{\frac{1}{2}(F-1)}$ $(-1)^{\frac{1}{8}(F^2-1)}$ $\left(\frac{f}{\omega_2}\right), \left(\frac{F}{\delta_2}\right)$ $\Gamma = 2^{\gamma+2}$	Ψ $\left(\frac{f}{\omega_1}\right), \left(\frac{F}{\delta_1}\right)$ P	$(-1)^{\frac{1}{2}(f-1)}$ $(-1)^{\frac{1}{8}(f^2-1)}$ $(-1)^{\frac{1}{2}(F-1)}$ $(-1)^{\frac{1}{8}(F^2-1)}$ $\left(\frac{f}{\omega_2}\right), \left(\frac{F}{\delta_2}\right)$ $\Gamma = 2^{\gamma+3}$
$\Delta = 2\Delta_1 \Delta_2^2$. $\Delta_2 \equiv 1$, mod 2.	$(-1)^{\frac{1}{8}(F^2-1)}\Psi$ $\left(\frac{f}{\omega_1}\right), \left(\frac{F}{\delta_1}\right)$ S	$\left(\frac{f}{\omega_2}\right), \left(\frac{F}{\delta_2}\right)$ $\Gamma = 2^\gamma$	$(-1)^{\frac{1}{8}(F^2-1)}\Psi$ $\left(\frac{f}{\omega_1}\right), \left(\frac{F}{\delta_1}\right)$ R	$(-1)^{\frac{1}{2}(f-1)}$ $\left(\frac{f}{\omega_2}\right), \left(\frac{F}{\delta_2}\right)$ $\Gamma = 2^{\gamma+1}$	$(-1)^{\frac{1}{8}(F^2-1)}\Psi$ $\left(\frac{f}{\omega_1}\right), \left(\frac{F}{\delta_1}\right)$ R	$(-1)^{\frac{1}{2}(f-1)}$ $(-1)^{\frac{1}{8}(f^2-1)}$ $\left(\frac{f}{\omega_2}\right), \left(\frac{F}{\delta_2}\right)$ $\Gamma = 2^{\gamma+2}$
$\Delta = 2\Delta_1 \Delta_2^2$. $\Delta_2 \equiv 0$, mod 2.	Ψ $(-1)^{\frac{1}{8}(F^2-1)}$ $\left(\frac{f}{\omega_1}\right), \left(\frac{F}{\delta_1}\right)$ Q	$(-1)^{\frac{1}{2}(F-1)}$ $\left(\frac{f}{\omega_2}\right), \left(\frac{F}{\delta_2}\right)$ $\Gamma = 3 \times 2^\gamma$	Ψ $(-1)^{\frac{1}{8}(F^2-1)}$ $\left(\frac{f}{\omega_1}\right), \left(\frac{F}{\delta_1}\right)$ P	$(-1)^{\frac{1}{2}(f-1)}$ $(-1)^{\frac{1}{2}(F-1)}$ $\left(\frac{f}{\omega_2}\right), \left(\frac{F}{\delta_2}\right)$ $\Gamma = 2^{\gamma+2}$	Ψ $(-1)^{\frac{1}{8}(F^2-1)}$ $\left(\frac{f}{\omega_1}\right), \left(\frac{F}{\delta_1}\right)$ P	$(-1)^{\frac{1}{2}(f-1)}$ $(-1)^{\frac{1}{8}(f^2-1)}$ $(-1)^{\frac{1}{2}(F-1)}$ $\left(\frac{f}{\omega_2}\right), \left(\frac{F}{\delta_2}\right)$ $\Gamma = 2^{\gamma+3}$

GENERIC CHARACTERS.

A: f and F properly primitive.

$\Omega = 2\,\Omega_1\,\Omega_2^2.$ $\Omega_2 \equiv 1, \bmod 2.$		$\Omega = 2\,\Omega_1\,\Omega_2^2.$ $\Omega_2 \equiv 0, \bmod 2.$	
$(-1)^{\frac{1}{8}(f^2-1)}\Psi$ $\left(\frac{f}{\omega_1}\right), \left(\frac{F}{\delta_1}\right)$ S	$\left(\frac{f}{\omega_2}\right), \left(\frac{F}{\delta_2}\right)$ $\Gamma = 2^\gamma$	Ψ $(-1)^{\frac{1}{8}(f^2-1)}$ $\left(\frac{f}{\omega_1}\right), \left(\frac{F}{\delta_1}\right)$ Q	$(-1)^{\frac{1}{2}(f-1)}$ $\left(\frac{f}{\omega_2}\right), \left(\frac{F}{\delta_2}\right)$ $\Gamma = 3 \times 2^\gamma$
$(-1)^{\frac{1}{8}(f^2-1)}\Psi$ $\left(\frac{f}{\omega_1}\right), \left(\frac{F}{\delta_1}\right)$ R	$(-1)^{\frac{1}{2}(F-1)}$ $\left(\frac{f}{\omega_2}\right), \left(\frac{F}{\delta_2}\right)$ $\Gamma = 2^{\gamma+1}$	Ψ $(-1)^{\frac{1}{8}(f^2-1)}$ $\left(\frac{f}{\omega_1}\right), \left(\frac{F}{\delta_1}\right)$ P	$(-1)^{\frac{1}{2}(f-1)}$ $(-1)^{\frac{1}{2}(F-1)}$ $\left(\frac{f}{\omega_2}\right), \left(\frac{F}{\delta_2}\right)$ $\Gamma = 2^{\gamma+2}$
$(-1)^{\frac{1}{8}(f^2-1)}\Psi$ $\left(\frac{f}{\omega_1}\right), \left(\frac{F}{\delta_1}\right)$ R	$(-1)^{\frac{1}{2}(F-1)}$ $(-1)^{\frac{1}{8}(F^2-1)}$ $\left(\frac{f}{\omega_2}\right), \left(\frac{F}{\delta_2}\right)$ $\Gamma = 2^{\gamma+2}$	Ψ $(-1)^{\frac{1}{8}(f^2-1)}$ $\left(\frac{f}{\omega_1}\right), \left(\frac{F}{\delta_1}\right)$ P	$(-1)^{\frac{1}{2}(f-1)}$ $(-1)^{\frac{1}{2}(F-1)}$ $(-1)^{\frac{1}{8}(F^2-1)}$ $\left(\frac{f}{\omega_2}\right), \left(\frac{F}{\delta_2}\right)$ $\Gamma = 2^{\gamma+3}$
$(-1)^{\frac{1}{8}(f^2-1)+\frac{1}{8}(F^2-1)}\Psi$ $\left(\frac{f}{\omega_1}\right), \left(\frac{F}{\delta_1}\right)$ S	$\left(\frac{f}{\omega_2}\right), \left(\frac{F}{\delta_2}\right)$ $\Gamma = 2^\gamma$	$(-1)^{\frac{1}{8}(F^2-1)}\Psi$ $(-1)^{\frac{1}{8}(f^2-1)}$ $\left(\frac{f}{\omega_1}\right), \left(\frac{F}{\delta_1}\right)$ R	$(-1)^{\frac{1}{2}(f-1)}$ $\left(\frac{f}{\omega_2}\right), \left(\frac{F}{\delta_2}\right)$ $\Gamma = 2^{\gamma+2}$
$(-1)^{\frac{1}{8}(f^2-1)}\Psi$ $(-1)^{\frac{1}{8}(F^2-1)}$ $\left(\frac{f}{\omega_1}\right), \left(\frac{F}{\delta_1}\right)$ R	$(-1)^{\frac{1}{2}(F-1)}$ $\left(\frac{f}{\omega_2}\right), \left(\frac{F}{\delta_2}\right)$ $\Gamma = 2^{\gamma+1}$	Ψ $(-1)^{\frac{1}{8}(f^2-1)}$ $(-1)^{\frac{1}{8}(F^2-1)}$ $\left(\frac{f}{\omega_1}\right), \left(\frac{F}{\delta_1}\right)$ P	$(-1)^{\frac{1}{2}(f-1)}$ $(-1)^{\frac{1}{2}(F-1)}$ $\left(\frac{f}{\omega_2}\right), \left(\frac{F}{\delta_2}\right)$ $\Gamma = 2^{\gamma+3}$

B: f improperly, F properly primitive.

$(-1)^{\frac{1}{2}(F-1)} = -(-1)^{\frac{1}{2}(\Omega-1)}$; $\Omega \equiv 1, \bmod 2.$

	$\Omega = \Omega_1\,\Omega_2^2.$ $\Omega_2 \equiv 1, \bmod 2.$	
$\Delta = \Delta_1\,\Delta_2^2.$ $\Delta_2 \equiv 0, \bmod 2.$	$(-1)^{\frac{1}{8}(F^2-1)}$ $\left(\frac{f}{\omega_1}\right), \left(\frac{F}{\delta_1}\right)$	$\left(\frac{f}{\omega_2}\right), \left(\frac{F}{\delta_2}\right)$ $\Gamma = 2^\gamma$
$\Delta = 2\,\Delta_1\,\Delta_2^2.$ $\Delta_2 \equiv 1, \bmod 2.$	$\left(\frac{f}{\omega_1}\right), \left(\frac{F}{\delta_1}\right)$	$\left(\frac{f}{\omega_2}\right), \left(\frac{F}{\delta_2}\right)$ $\Gamma = 2^{\gamma-1}$
$\Delta = 2\,\Delta_1\,\Delta_2^2.$ $\Delta_2 \equiv 0, \bmod 2.$	$\left(\frac{f}{\omega_1}\right), \left(\frac{F}{\delta_1}\right)$	$(-1)^{\frac{1}{8}(F^2-1)}$ $\left(\frac{f}{\omega_2}\right), \left(\frac{F}{\delta_2}\right)$ $\Gamma = 2^\gamma$

C: f properly, F improperly primitive.

$(-1)^{\frac{1}{2}(f-1)} = -(-1)^{\frac{1}{2}(\Delta-1)}$; $\Delta \equiv 1, \bmod 2.$

	$\Delta = \Delta_1\,\Delta_2^2.$ $\Delta_2 \equiv 1, \bmod 2.$	
$\Omega = \Omega_1\,\Omega_2^2.$ $\Omega_2 \equiv 0, \bmod 2.$	$(-1)^{\frac{1}{8}(f^2-1)}$ $\left(\frac{f}{\omega_1}\right), \left(\frac{F}{\delta_1}\right)$	$\left(\frac{f}{\omega_2}\right), \left(\frac{F}{\delta_2}\right)$ $\Gamma = 2^\gamma$
$\Omega = 2\,\Omega_1\,\Omega_2^2.$ $\Omega_1 \equiv 1, \bmod 2.$	$\left(\frac{f}{\omega_1}\right), \left(\frac{F}{\delta_1}\right)$	$\left(\frac{f}{\omega_2}\right), \left(\frac{F}{\delta_2}\right)$ $\Gamma = 2^{\gamma-1}$
$\Omega = 2\,\Omega_1\,\Omega_2^2.$ $\Omega_1 \equiv 0, \bmod 2.$	$\left(\frac{f}{\omega_1}\right), \left(\frac{F}{\delta_1}\right)$	$(-1)^{\frac{1}{8}(f^2-1)}$ $\left(\frac{f}{\omega_2}\right), \left(\frac{F}{\delta_2}\right)$ $\Gamma = 2^\gamma$

case to which the compartment corresponds; the symbols

$$\left(\frac{f}{\omega_1}\right),\ \left(\frac{f}{\omega_2}\right),\ \left(\frac{F}{\delta_1}\right),\ \left(\frac{F}{\delta_2}\right)$$

implying that f has a particular character with respect to every prime ω_1 or ω_2, F a particular character with respect to every prime δ_1 or δ_2. Each compartment is divided into two parts by a vertical line, and the particular characters (one of which in table A either is or contains Ψ) placed to the left of this line are subject to the condition that their product is equal, in table A, to the unit $(-1)^{\frac{1}{2}(\Omega_1+1)\cdot\frac{1}{2}(\Delta_1+1)}$, in table B, to the unit $(-1)^{\frac{1}{8}(\Omega_1^2-1)}\times(-1)^{\frac{1}{2}(\Omega_1+1)\cdot\frac{1}{2}(\Delta_1+1)}$, in table C, to the unit $(-1)^{\frac{1}{8}(\Delta_1^2-1)}\times(-1)^{\frac{1}{2}(\Omega_1+1)\cdot\frac{1}{2}(\Delta_1+1)}$. If $\alpha=+1$ or $=-1$, according as Ω is of the form $\Omega_1\Omega_2^2$ or $2\Omega_1\Omega_2^2$, and if, similarly, $\beta=+1$ or $=-1$, according as Δ is of the form $\Delta_1\Delta_2^2$ or $2\Delta_1\Delta_2^2$, we may express this condition, in table A, by the equation

$$\Psi\alpha^{\frac{1}{8}(f^2-1)}\beta^{\frac{1}{8}(F^2-1)}\left(\frac{f}{\Omega_1}\right)\left(\frac{F}{\Delta_1}\right)=(-1)^{\frac{1}{2}(\Omega_1+1)\cdot\frac{1}{2}(\Delta_1+1)}, \quad . \quad . \quad . \quad . \quad (11)$$

and in tables B and C respectively, by the equations

$$(-\beta)^{\frac{1}{8}(F^2-1)}\left(\frac{f}{\Omega_1}\right)\left(\frac{F}{\Delta_1}\right)=(-1)^{\frac{1}{8}(\Omega_1^2-1)+\frac{1}{2}(\Omega_1+1)\cdot\frac{1}{2}(\Delta_1+1)}, \quad . \quad . \quad . \quad . \quad (12)$$

$$(-\alpha)^{\frac{1}{8}(f^2-1)}\left(\frac{f}{\Omega_1}\right)\left(\frac{F}{\Delta_1}\right)=(-1)^{\frac{1}{8}(\Delta_1^2-1)+\frac{1}{2}(\Omega_1+1)\cdot\frac{1}{2}(\Delta_1+1)}. \quad . \quad . \quad . \quad . \quad (13)$$

The condition distinguishes the possible and impossible genera, every generic character which satisfies it being the character of an actually existing genus, and every generic character which does not satisfy it belonging to no forms whatever. The demonstration of this important theorem will occupy the next articles; it is, however, requisite to show in the first place that the enumeration of the supplementary characters in table II. is in accordance with the table I. of Art. 4. For the tables B and C this is evident; in table A it is necessary to attend to the signification of the symbol Ψ, which serves to represent the simultaneous character of f and F (as has been already explained in Arts. 6 and 7) in those cases (marked S in the table) in which neither $(-1)^{\frac{1}{2}(f-1)}$ nor $(-1)^{\frac{1}{2}(F-1)}$ is a character, but which also appears in every compartment of the table without exception.

(1) When $(-1)^{\frac{1}{2}(f-1)}$ and $(-1)^{\frac{1}{2}(F-1)}$ are both characters (cases P in the table), Ψ is not an independent character, because its value is determined by the values of $(-1)^{\frac{1}{2}(f-1)}$ and $(-1)^{\frac{1}{2}(F-1)}$. It is retained in the table only because it serves to express the criterion of possibility.

(2) When $(-1)^{\frac{1}{2}(f-1)}$ and Ψ, but not $(-1)^{\frac{1}{2}(F-1)}$, are inscribed as characters, Ψ represents the character $(-1)^{\frac{1}{2}(F-1)}$, if $(-1)^{\frac{1}{2}(f-1)}=(-1)^{\frac{1}{2}(\Delta_1-1)}$, and is simply $+1$ (*i.e.* not a character at all) if $(-1)^{\frac{1}{2}(f-1)}=-(-1)^{\frac{1}{2}(\Delta_1-1)}$. This is in accordance with table I., according to which, in the cases under consideration, $(-1)^{\frac{1}{2}(F-1)}$ is, or is not, a character, according as $(-1)^{\frac{1}{2}(f-1)}=(-1)^{\frac{1}{2}(\Delta_1-1)}$, or $=-(-1)^{\frac{1}{2}(\Delta_1-1)}$. Similarly, if $(-1)^{\frac{1}{2}(F-1)}$ and Ψ, but not $(-1)^{\frac{1}{2}(f-1)}$, are inscribed as characters, Ψ represents the character $(-1)^{\frac{1}{2}(f-1)}$, or is not a character at all, according as $(-1)^{\frac{1}{2}(F-1)}=(-1)^{\frac{1}{2}(\Omega_1-1)}$, or $=-(-1)^{\frac{1}{2}(\Omega_1-1)}$; which again agrees with table I.

In these cases, marked Q in the table, the symbol Ψ supersedes the asterisks and obelisks of table I., and also serves to express the criterion of possibility.

(3) When $(-1)^{\frac{1}{2}(f-1)}$ and $(-1)^{\frac{1}{8}(F^2-1)}\times\Psi$, but not $(-1)^{\frac{1}{2}(F-1)}$, are characters in the table, $(-1)^{\frac{1}{8}(F^2-1)}\times\Psi$ represents the character $(-1)^{\frac{1}{2}(F-1)+\frac{1}{8}(F^2-1)}$, or $(-1)^{\frac{1}{8}(F^2-1)}$, according as $(-1)^{\frac{1}{2}(f-1)}=(-1)^{\frac{1}{2}(\Delta_1-1)}$, or $-(-1)^{\frac{1}{2}(\Delta_1-1)}$. And again, when $(-1)^{\frac{1}{2}(F-1)}$ and $(-1)^{\frac{1}{8}(f^2-1)}\times\Psi$, but not $(-1)^{\frac{1}{2}(f-1)}$, are characters in the table, $(-1)^{\frac{1}{8}(f-1)}\times\Psi$ represents the character $(-1)^{\frac{1}{2}(f-1)+\frac{1}{8}(f^2-1)}$, or $(-1)^{\frac{1}{8}(f^2-1)}$, according as $(-1)^{\frac{1}{2}(F-1)}=(-1)^{\frac{1}{2}(\Omega_1-1)}$, or $=-(-1)^{\frac{1}{2}(\Omega_1-1)}$.

The result in these cases (marked R in the table) is again in accordance with table I.; and the use of the symbol Ψ is the same as in the cases Q.

Thus the units Ψ, $(-1)^{\frac{1}{8}(f^2-1)}\times\Psi$, $(-1)^{\frac{1}{8}(F^2-1)}\times\Psi$, which properly represent simultaneous characters of the forms f and F, are employed, in the cases Q and R of the table, to represent supplementary characters. This use of these symbols is admissible, because, when $\Omega\Delta$ is even (as it is in the cases Q and R), every representation of an uneven number by f or F is simultaneous with the representation of uneven numbers by F or f.

In the lower right-hand corner of each compartment in the table, the number of possible genera contained in the order to which the compartment refers is represented by Γ; γ is the number of uneven primes dividing Ω, together with the number of uneven primes dividing Δ, so that, if the same prime divide both Ω and Δ, it is to be counted twice. But it is to be observed that, when Ω and Δ are both perfect squares (a case which can only arise when the forms are definite), the number of possible genera is two thirds of the number stated in the table in the cases Q, and one half in the cases P. And again (as has been already stated in Art. 3), in table B, when Ω is an uneven square and Δ the double of a square, there are no possible genera; and when Δ is an uneven square and Ω the double of an uneven square, there are none in table C.

9. It results from the theorem of Art. 5 that, if f and F are properly primitive, they simultaneously and primitively represent uneven numbers prime to $\Omega\Delta$. We may, therefore, suppose that in f and F a and A'' are uneven and prime to $\Omega\Delta$; we may also suppose that these numbers are prime to one another, because A'' being prime to $\Omega\Delta$, and a being uneven, the binary form (a, b'', a') is properly primitive (Art. 5), and so represents numbers prime to its determinant. Lastly, we may assume that a and A'' are positive. If the forms f and F are definite, a and A'' are certainly positive; if they are indefinite, Δ and Ω are of opposite signs; supposing, for example, that Δ is positive and Ω negative, let m be any positive number primitively represented by f, and M any number simultaneously and primitively represented by F, then M is positive as well as m; otherwise mMf, which is of the type $MX^2+\Omega Y^2+m\Omega\Delta Z^2$, would be a definite form. Positive numbers are therefore simultaneously and primitively represented by f and F; *i.e.* we may suppose a and A'' simultaneously positive. The complete generic character of f is then determined by the characters of a and A''. But

$$aa'-b^2=\Omega A'', \quad A'A''-B^2=\Delta a,$$

whence it follows that

$$\left(\frac{-\Omega}{a}\right)\left(\frac{A''}{a}\right)=1, \qquad \left(\frac{-\Delta}{A''}\right)\left(\frac{a}{A''}\right)=1;$$

multiplying these equations together and observing that, by the laws of quadratic residues,

$$\left(\frac{A''}{a}\right)\left(\frac{a}{A''}\right)=(-1)^{\frac{1}{2}(a-1)\cdot\frac{1}{2}(A''-1)},$$

we find

$$(-1)^{\frac{1}{2}(a-1)\cdot\frac{1}{2}(A''-1)}\left(\frac{-\Omega}{a}\right)\left(\frac{-\Delta}{A''}\right)=1. \quad \ldots \ldots \quad (14)$$

Let α and β retain the significations assigned to them in equation (11), Art. 8; transforming $\left(\frac{-\Omega}{a}\right)$ and $\left(\frac{-\Delta}{A''}\right)$ by the law of quadratic reciprocity, we find

$$\left(\frac{-\Omega}{a}\right)=(-1)^{\frac{1}{2}(a-1)\cdot\frac{1}{2}(\Omega_1+1)}\,\alpha^{\frac{1}{8}(a^2-1)}\left(\frac{a}{\Omega_1}\right),$$

$$\left(\frac{-\Delta}{A''}\right)=(-1)^{\frac{1}{2}(A''-1)\cdot\frac{1}{2}(\Delta_1+1)}\,\beta^{\frac{1}{8}(A''^2-1)}\left(\frac{A''}{\Delta_1}\right),$$

and equation (14) becomes

$$(-1)^{\frac{1}{2}(\Omega_1+1)\cdot\frac{1}{2}(\Delta_1+1)+\frac{1}{2}(\Omega_1+A'')\cdot\frac{1}{2}(\Delta_1+a)}\,\alpha^{\frac{1}{8}(a^2-1)}\,\beta^{\frac{1}{8}(A''^2-1)}\left(\frac{a}{\Omega_1}\right)\left(\frac{A''}{\Delta_1}\right)=1;$$

or observing that

$$(-1)^{\frac{1}{2}(\Omega_1+A'')\cdot\frac{1}{2}(\Delta_1+a)}=(-1)^{\frac{1}{2}(\Omega_1 A''+1)\cdot\frac{1}{2}(\Delta_1 a+1)}=\Psi,$$

and writing f and F for a and A'',

$$\Psi\, \alpha^{\frac{1}{8}(f^2-1)}\, \beta^{\frac{1}{8}(F^2-1)} \left(\frac{f}{\Omega_1}\right) \left(\frac{F}{\Delta_1}\right) = (-1)^{\frac{1}{2}(\Omega_1+1)\cdot\frac{1}{2}(\Delta_1+1)},$$

i.e. the generic character of f satisfies the condition of possibility (11).

Again, if f is improperly and F properly primitive, let A'' be prime to $2\Omega\Delta$; then the binary form (a, b'', a') is primitive because A'' is prime to $\Delta\Omega$, and improperly primitive because f is improperly primitive. We may, therefore, suppose that $\frac{1}{2}a$ is uneven and prime to $\Omega A''$, and, as before, that a and A'' are positive. Multiplying together the equations

$$\left(\frac{-\Omega}{\frac{1}{2}a}\right)\left(\frac{A''}{\frac{1}{2}a}\right) = 1, \quad \left(\frac{-2\Delta}{A''}\right)\left(\frac{\frac{1}{2}a}{A''}\right) = 1,$$

and transforming the result by the law of reciprocity, we find

$$(-\beta)^{\frac{1}{8}(A''^2-1)} \left(\frac{a}{\Omega_1}\right)\left(\frac{A''}{\Delta_1}\right) = (-1)^{\frac{1}{8}(\Omega_1^2-1)+\frac{1}{2}(\Omega_1+1)\cdot\frac{1}{2}(\Delta_1+1)},$$

i.e. the condition (12) is satisfied by the generic character of f.

The case in which f is properly and F improperly primitive is the reciprocal of the preceding.

To show that the conditions (11), (12), (13) are sufficient as well as necessary, other principles are required. These principles relate to the representation of binary by ternary quadratic forms, and will be found in the 'Disquisitiones Arithmeticae,' Arts. 282–284; it will, however, be convenient briefly to restate them here in a form suited for our present purpose.

10. A binary quadratic form (p, q'', p') or ϕ is said to be represented by a ternary form f when f is transformed into ϕ by a substitution of the type

$$x = \alpha_1 x + \beta_1 y,$$
$$y = \alpha_2 x + \beta_2 y,$$
$$z = \alpha_3 x + \beta_3 y.$$

The representation is said to be *primitive* when the determinants of the matrix

$$\begin{vmatrix} \alpha_1, & \beta_1 \\ \alpha_2, & \beta_2 \\ \alpha_3, & \beta_3 \end{vmatrix}$$

are relatively prime. If ϕ is primitively represented by f, f is equivalent to a form containing ϕ as a part, *i.e* to a form f' of the type

$$f' = px^2 + p'y^2 + p''z^2 + 2qyz + 2q'xz + 2q''xy,$$

for f is transformed into such a form by a substitution of which the matrix is

$$\begin{vmatrix} \alpha_1, & \beta_1, & \gamma_1 \\ \alpha_2, & \beta_2, & \gamma_2 \\ \alpha_3, & \beta_3, & \gamma_3 \end{vmatrix},$$

$\gamma_1, \gamma_2, \gamma_3$ denoting any three numbers which render the determinant of that matrix equal to $+1$.

Let $$F' = Px^2 + P'y^2 + P''z^2 + 2Qyz + 2Q'xz + 2Q''xy$$
be the primitive contravariant of f', so that, in particular,

$$\Omega P'' = q''^2 - pp' \; ; \quad \ldots \ldots \ldots \quad (15)$$

multiplying the equations

$$\left.\begin{aligned} P'P'' - \quad Q^2 &= \Delta p, \\ QQ' - P''Q'' &= \Delta q'', \\ PP'' - \quad Q'^2 &= \Delta p', \end{aligned}\right\} \quad \ldots \ldots \ldots \quad (16)$$

(which result from the contravariance of f' and F') by x^2, $2xy$, y^2 respectively, we obtain

$$-\Delta\,(px^2 + 2q''xy + p'y^2) = (Q^2 - P'P'')\,x^2 - 2\,(QQ' - P''Q'')\,xy + (Q'^2 - PP'')\,y^2 \, ;$$

and this equation, considered as a congruence for the modulus P'', becomes

$$\Delta\phi + (Qx - Q'y)^2 \equiv 0, \text{ mod } P'', \quad \ldots \ldots \quad (17)$$

the coefficients of x^2, $2xy$, y^2 in the left-hand member being all divisible by P''. If, therefore, ϕ is a binary quadratic form of determinant $-\Omega P''$, admitting of primitive representation by a ternary form of order $[\Omega, \Delta]$, $-\Delta\phi$ is a quadratic residue of P''. And we shall now show that if ϕ is a primitive (and not negative) binary form of determinant $-\Omega P''$, P'' being of the same sign as Δ and prime to Δ, ϕ admits of primitive representation by ternary forms of the invariants $[\Omega, \Delta]$ whenever $-\Delta\phi$ is a quadratic residue of P''.

Because $-\Delta\phi$ is a quadratic residue of P'', the congruence (17) admits of solution in integral numbers Q, Q'. Any solution of this congruence supplies a system of five numbers, P, P', Q, Q', Q'', satisfying the equations (16). The greatest common divisor of these five numbers divides Δ because p, q'', p' are relatively prime; but P'' is prime to Δ; therefore the six numbers P, P', P'', Q, Q', Q'' are relatively prime. Let q and q' be determined by the equations

$$\left.\begin{aligned} qq'' - q'p' &= \quad \Omega Q', \\ qp \; - q'q'' &= -\Omega Q, \end{aligned}\right\} \quad \ldots \ldots \ldots \quad (18)$$

which are always resoluble because their determinant $q''^2 - pp' = \Omega P''$ is different from zero. Also let p'' be determined by the equation

$$q'Q' + qQ + p''P'' = \Omega\Delta. \quad \ldots \ldots \ldots \quad (19)$$

The values of q, q', p'' are rational; and, if they are fractions, their denominators, when they are expressed in their lowest terms, are divisors of P''. Substituting in (19) for P'', Q', Q their values derived from the equations (15) and (18), we find that $\Omega^2\Delta$ is the determinant of the matrix

$$\begin{vmatrix} p, & q'', & q' \\ q'', & p', & q \\ q', & q, & p'' \end{vmatrix} . \qquad (20)$$

Let

$$\begin{vmatrix} \Omega[p], & \Omega[q''], & \Omega[q'] \\ \Omega[q''], & \Omega[p'], & \Omega[q] \\ \Omega[q'], & \Omega[q], & \Omega[p''] \end{vmatrix} \qquad (21)$$

be the matrix reciprocal to the matrix (20); we know, from the equations (15) and (18), that $[p'']=P''$, $[q']=Q'$, $[q]=Q$. Again, in the reciprocal matrices (20) and (21), we must have

$$\begin{aligned} &[p'][p'']-[q]^2 = \Delta p, \\ &[q][q']-[p''][q''] = \Delta q'', \\ &[p][p'']-[q'^2] = \Delta p', \end{aligned}$$

or, substituting for $[p'']$, $[q]$, $[q']$ their values,

$$\begin{aligned} &[p']P''-Q^2 = \Delta p, \\ &QQ' - P''[q''] = \Delta q'', \\ &[p]P''-Q'^2 = \Delta p'. \end{aligned}$$

Comparing these equations with the equations (16), and observing that P'' is not zero, we infer that

$$[p']=P', \quad [q'']=Q'', \quad [p]=P.$$

The matrix reciprocal to the matrix (20) is therefore

$$\begin{vmatrix} \Omega P, & \Omega Q'', & \Omega Q' \\ \Omega Q'', & \Omega P', & \Omega Q \\ \Omega Q', & \Omega Q, & \Omega P'' \end{vmatrix}, \qquad (22)$$

and, consequently,

$$\begin{aligned} \Delta q &= Q'Q''-PQ, \\ \Delta q' &= QQ''-P'Q', \\ \Delta p'' &= PP'-Q''^2. \end{aligned}$$

These equations prove that the denominators of q, q', p'' are divisors of Δ; *i.e.* that q, q', p'' are integral numbers, because P'' is prime to Δ. The coefficients of the ternary form

$$f' = px^2+p'y^2+p''z^2+2qyz+2q'xz+2q''xy$$

are therefore integral; this form is primitive, and represents primitively the form (p, q'', p'); it is also a form of the given invariants $[\Omega, \Delta]$; for its discriminant is $\Delta\Omega^2$ and the greatest common divisor of the first minors of its matrix is Ω; hence its second invariant is Δ and its first invariant either $+\Omega$ or $-\Omega$. But, when the given invariants Ω and Δ are both positive, ϕ is a positive binary form of the negative determinant $-\Omega P''$, and such a form cannot be represented by an indefinite ternary form of a positive discriminant; f' is therefore definite, and its first invariant is $+\Omega$. When the given invariants Ω and Δ are of opposite signs, ϕ is a binary form of the positive determinant $-\Omega P''$; such a form cannot be represented by a definite ternary form; f' is therefore indefinite, and, as its invariants must be of opposite signs, in this case also its first invariant is $+\Omega$.

Also, if ϕ is properly primitive and P'' uneven, the forms f' and F' are both properly primitive, one of the principal coefficients of each being uneven. In this case, therefore, ϕ is represented by a form of the properly primitive order $[\Omega, \Delta]$. If ϕ is improperly primitive (a supposition which implies that $\Omega P'' \equiv 3$, mod 4), and if Δ is even, f' is improperly primitive. For no properly primitive ternary form of even discriminant can represent primitively an improperly primitive binary form, the supposition that (p, q'', p') is improperly primitive and p'' uneven implying that the discriminant is uneven. And the same thing follows from the preceding analysis; for, considering the equations (18) as congruences for the modulus 2, we find, on the supposition that ϕ is improperly primitive, $q \equiv Q'$, mod 2, $q' \equiv Q$, mod 2, and, substituting in (19), $p'' \equiv 0$, mod 2, so that f' is improperly primitive.

11*. We can now assign a properly primitive form of any given invariants $[\Omega, \Delta]$, and of any given generic character satisfying the condition of possibility. Let M be any number prime to 2Δ, of the same sign as Δ, and possessing all the particular characters (except the simultaneous character, if any) which are

---

* [At the end of a reprint of his paper 'On the Orders and Genera of Quadratic Forms containing more than three Indeterminates,' Professor Smith adds the note :—

'I avail myself of the opportunity afforded by this reprint to append a rectification to Art. 11 of my memoir on the Orders and Genera of Ternary Quadratic Forms (Phil. Trans. vol. clvii. pp. 272 *sqq.*). The numbers M, considered in that article, are necessarily prime to 2Δ, but are not necessarily prime to Ω, whether the given genus be properly or improperly primitive. But, if μ is any uneven prime dividing both M and Ω, μ must satisfy the equation $\left(\frac{-\Delta}{\mu}\right) = \left(\frac{f}{\mu}\right)$. To this limitation, which is indeed implied by the equations (23) and (24), but which ought to have been expressly stated, the theorems in the concluding paragraph of the article are subject.—H. J. S. S. April 22, 1868.']

attributed to F in the given generic character; also, if Ω is uneven and Δ uneven or unevenly even, we shall suppose that $M \equiv \Omega$, mod 4. Let ϕ be any properly primitive, and not negative, binary quadratic form of determinant $-\Omega M$; and let m be any number prime to $2\Omega M$ which is represented by ϕ. By the theory of binary quadratic forms, the generic characters which are attributable to ϕ are (i) its characters with respect to primes dividing M, (ii) its characters with respect to primes dividing Ω, (iii) its supplementary characters. These last are exhibited in the following table,

If $-\Omega M \equiv$	Supplementary characters
1, mod 4.	None
3, mod 4.	$(-1)^{\frac{1}{2}(\phi-1)}$
2, mod 8.	$(-1)^{\frac{1}{8}(\phi^2-1)}$
6, mod 8.	$(-1)^{\frac{1}{2}(\phi-1)+\frac{1}{8}(\phi^2-1)}$
4, mod 8.	$(-1)^{\frac{1}{2}(\phi-1)}$
0, mod 8.	$(-1)^{\frac{1}{2}(\phi-1)}$, $(-1)^{\frac{1}{8}(\phi^2-1)}$

Let μ be any prime divisor of M, and let us determine the first set of characters by the equations

$$\left(\frac{\phi}{\mu}\right) = \left(\frac{-\Delta}{\mu}\right), \quad \quad (23)$$

the second set by the equations

$$\left(\frac{\phi}{\omega}\right) = \left(\frac{f}{\omega}\right), \quad \quad (24)$$

$\left(\frac{f}{\omega}\right)$ being a particular character of f, of which the value is assigned in the proposed generic character. With respect to the supplementary characters of ϕ, it will be found on a comparison of the above table with table II. A, that, when the proposed generic character includes no simultaneous character, the supplementary characters attributable to ϕ are the same as those attributable to f; we then assign to the supplementary characters of ϕ the same values which are assigned to the supplementary characters of f in the proposed generic character.

But, when the proposed generic character includes a simultaneous character, there is always a supplementary character (and only one) attributable to ϕ and not to f; this character of ϕ we determine so that the value of the simultaneous character of f and F and the value of the unit similarly formed with m and M may be coincident. This determination is always possible, as will be seen on a comparison of the cases S of table II. A with the above table of supplementary characters of binary forms. As we have now assigned a value to every particular character attributable to ϕ, it is necessary to inquire whether a form ϕ, possessing such a complete character, actually exists; *i.e.* whether the character which we have assigned to ϕ satisfies the condition of possibility for binary forms of determinant $-\Omega M$.

If, as in Art. 8, $\alpha = +1$ or $= -1$, according as Ω is of the form $\Omega_1 \Omega_2^2$ or $2\Omega_1 \Omega_2^2$, that condition is

$$(-1)^{\frac{1}{2}(\Omega_1 M+1)\cdot\frac{1}{2}(\phi-1)}\, \alpha^{\frac{1}{8}(\phi^2-1)} \left(\frac{\phi}{\Omega_1 M}\right) = 1, \quad \ldots\ldots\ldots \quad (25)$$

or, since $(-1)^{\frac{1}{2}(\Delta_1+1)+\frac{1}{2}(\Delta_1\phi+1)} = (-1)^{\frac{1}{2}(\phi-1)}$, the condition is

$$(-1)^{\frac{1}{2}(\Omega_1 M+1)\cdot\frac{1}{2}(\Delta_1+1)+\frac{1}{2}(\Omega_1 M+1)\cdot\frac{1}{2}(\Delta_1\phi+1)}\, \alpha^{\frac{1}{8}(\phi^2-1)} \left(\frac{\phi}{\Omega_1 M}\right) = 1. \quad . \quad (26)$$

But $\left(\frac{\phi}{\Omega_1}\right) = \left(\frac{f}{\Omega_1}\right)$, by the equations (24), and, if (again as in Art. 8) $\beta = +1$ or $= -1$, according as Δ is of the form $\Delta_1 \Delta_2^2$ or $2\Delta_1 \Delta_2^2$,

$$\left(\frac{\phi}{M}\right) = \left(\frac{-\Delta}{M}\right) = (-1)^{\frac{1}{2}(M-1)\cdot\frac{1}{2}(\Delta_1+1)}\, \beta^{\frac{1}{8}(M^2-1)} \left(\frac{M}{\Delta_1}\right).$$

Substituting these values in (26), and observing that in every case

$$(-1)^{\frac{1}{2}(\Omega_1 M+1)\cdot\frac{1}{2}(\Delta_1\phi+1)}\, \alpha^{\frac{1}{8}(\phi^2-1)}\, \beta^{\frac{1}{8}(M^2-1)} = \Psi \times \alpha^{\frac{1}{8}(f^2-1)}\, \beta^{\frac{1}{8}(F^2-1)},$$

we obtain

$$\Psi \times \alpha^{\frac{1}{8}(f^2-1)}\, \beta^{\frac{1}{8}(F^2-1)} \left(\frac{f}{\Omega_1}\right) \left(\frac{F}{\Delta_1}\right) = (-1)^{\frac{1}{2}(\Omega_1+1)\cdot\frac{1}{2}(\Delta_1+1)}.$$

But this equation is the equation (11) of Art. 8, which is by hypothesis satisfied by the proposed generic character; therefore the equation (25) is also satisfied; *i.e.* a properly primitive binary form ϕ exists, of determinant $-\Omega M$, possessing the generic character which we have assigned to it. This form, multiplied by $-\Delta$, is a quadratic residue of M; for the equation

$$\left(\frac{-\Delta\phi}{\mu}\right) = 1$$

is satisfied for every prime dividing M, by virtue of the equations (23). Let, then, a ternary form f, of the properly primitive order and of the invariants

$[\Omega, \Delta]$, be determined, representing ϕ primitively. The generic character of this form is completely determined by the numbers m and M, which are uneven numbers simultaneously represented by f and F; it is therefore a form of the proposed generic character.

Of the two improperly primitive orders, it will suffice to consider that in which f is improperly and F properly primitive; so that Ω is uneven and Δ even. Let M be a number prime to $2\Omega\Delta$, of the same sign as Δ, and satisfying the generic characters of F, including the congruence $M \equiv -\Omega$, mod 4; also let ϕ be an improperly primitive binary form of determinant $-\Omega M$; the generic characters attributable to ϕ are (i) its characters $\left(\frac{\phi}{\mu}\right)$, (ii) its characters $\left(\frac{\phi}{\omega}\right)$. These characters we determine, as before, by the equations (23) and (24). The complete generic character thus assigned to ϕ is possible; for the condition that it should be possible is

$$\left(\frac{\phi}{\Omega_1 M}\right) = (-1)^{\frac{1}{8}(\Omega_1^2-1)+\frac{1}{8}(M^2-1)},$$

or

$$\left(\frac{f}{\Omega_1}\right)\left(\frac{-2\Delta}{M}\right) = (-1)^{\frac{1}{8}(\Omega_1^2-1)}.$$

Transforming $\left(\frac{-2\Delta}{M}\right)$ by the law of reciprocity, we find

$$(-\beta)^{\frac{1}{8}(F^2-1)}\left(\frac{f}{\Omega_1}\right)\left(\frac{F}{\Delta_1}\right) = (-1)^{\frac{1}{2}(\Omega_1+1)\cdot\frac{1}{2}(\Delta_1+1)+\frac{1}{8}(\Omega_1^2-1)},$$

an equation which the proposed generic character satisfies by hypothesis (equation (12) Art. 8). An improperly primitive form ϕ of determinant $-\Omega M$, therefore, actually exists, having the generic character which we have assigned to it; *i.e.* ternary forms exist having the proposed generic character.

It is evident from the demonstration that if M is of the same sign as Δ, prime to 2Δ, and also (when Ω is uneven and Δ uneven or unevenly even) congruous to Ω, mod 4, there is always one genus of properly primitive binary forms, of determinant $-\Omega M$, capable of primitive representation by a given genus of ternary forms of the properly primitive order $[\Omega, \Delta]$, of which the contravariant characters coincide with the characters of M. And, similarly, if Δ is even, Ω uneven, M prime to Δ and $\equiv -\Omega$, mod 4, there is always one genus of improperly primitive binary forms of determinant $-\Omega M$ capable of primitive representation by a given genus of ternary forms of the improperly primitive order $[\Omega, \Delta]$, of which the contravariant characters coincide with the characters of M. And, in both cases, no other primitive form (if M is prime to Ω, no other form, primitive or derived) of determinant $-\Omega M$ is capable of such representation.

12. By a rational substitution we shall understand in this article a substitution of which the determinant is unity, and of which the coefficients are rational. If the common denominator of the coefficients is prime to any number m, we shall say that the substitution is prime to m.

If f_1 and f_2 are ternary forms, having integral coefficients, of which f_1 is a form of the invariants (Ω, Δ), and is transformed, by a rational substitution prime to $2\Omega\Delta$, into f_2, f_2 is a form of the same invariants, of the same order, and of the same genus as f_1. This may be proved nearly in the same way in which it is proved that equivalent forms have the same invariants and are of the same order and genus; it is only necessary to observe that F_1 and F_2, as well as f_1 and f_2, are transformable into one another by rational substitutions prime to $2\Omega\Delta$.

The converse proposition, 'if f_1 and f_2 are two forms of the same invariants (Ω, Δ), of the same order, and of the same genus, they are transformable, each into the other, by rational substitutions prime to $2\Omega\Delta$,' is also true, and is of importance in the present theory, because it establishes the completeness of the enumeration of the generic characters of ternary forms. To avoid the introduction, in this place, of principles relating to quaternary quadratic forms, we shall give an indirect demonstration of it, depending on the following lemma which relates to binary quadratic forms.

'If ϕ_1, ϕ_2 are two primitive binary quadratic forms of the same determinant and of the same genus, the resolubility of the equation $\phi_1(x, y) = M$ implies the resolubility of the equation $\phi_2(x, y) = Mz^2$; and, in the solution of this equation, the value of z may be supposed prime to any given number k.'

Because ϕ_1 and ϕ_2 are of the same genus, ϕ_2 is transformable, by a bipartite linear substitution, into the product $\chi \times \phi_1$, χ representing a properly primitive form of the principal genus (Disq. Arith., Art. 251). But χ is transformable, by a quadratic substitution, into the square of a properly primitive form ψ (ibid., Art. 287). Therefore, by a mixed quadratic and linear substitution, ϕ_2 is transformed into the product $\psi^2 \times \phi_1$. Attributing, in this mixed substitution, to the indeterminates of ϕ_1 the values which satisfy the equation $\phi_1 = M$, and to the indeterminates of ψ any values whatever for which ψ acquires a value z prime to k, we obtain a solution of the equation $\phi_2 = Mz^2$.

Let us first suppose that the given ternary forms f_1 and f_2 belong to the properly primitive order of the invariants (Ω, Δ); let M_1, M_2 be two numbers of the same sign as Δ, prime to $2\Omega\Delta$ and primitively represented by F_1, F_2 respectively; we may suppose that $M_1 \equiv M_2$, mod 8; and that the representations of M_1 and M_2 are simultaneous with the representations of uneven numbers by f_1 and f_2. Let ϕ_1, ϕ_2 be two binary quadratic forms, of the determinants $-\Omega M_1$, $-\Omega M_2$ respectively, represented by f_1 and f_2 simultaneously with the representations of M_1 and M_2 by F_1 and F_2*. Then ϕ_1 and ϕ_2 are properly primitive; their generic characters with respect to uneven primes dividing Ω will coincide, because

$$\left(\frac{\phi}{\omega}\right)=\left(\frac{f_1}{\omega}\right)=\left(\frac{f_2}{\omega}\right)=\left(\frac{\phi_2}{\omega}\right);$$

their supplementary characters will also coincide, for the same supplementary characters are attributable to ϕ_1 and ϕ_2, and these supplementary characters are determined for ϕ_1 in accordance with the supplementary characters of f_1, or the simultaneous character of f_1 and F_1, and for ϕ_2 in accordance with characters which are the same with these; lastly, if μ is any prime dividing both M_1 and M_2, the characters of ϕ_1 and ϕ_2 with respect to μ will also coincide; for

$$\left(\frac{\phi_1}{\mu}\right)=\left(\frac{-\Delta}{\mu}\right)=\left(\frac{\phi_2}{\mu}\right).$$

The remaining characters of ϕ_1 and ϕ_2 (*i.e.* their characters with respect to primes dividing only one of the two numbers M_1 and M_2), being characters with respect to different primes, cannot be incompatible. The complete generic characters of ϕ_1 and ϕ_2 are therefore compatible, and are satisfied by the numbers contained in certain arithmetical progressions. Each of these progressions contains (by the theorem of Lejeune Dirichlet) an infinite number of positive and negative primes. Let p be one of these primes of the same sign as Ω, and not dividing $2\Omega\Delta$; p will satisfy the generic characters both of ϕ_1 and ϕ_2, and will be represented by some form of determinant $-\Omega M_1$ and of the same genus as ϕ_1, and also by some form of determinant $-\Omega M_2$ and of the same genus as ϕ_2. Therefore, by the lemma of this article, $p\theta_1^2$ will be primitively represented by ϕ_1, and $p\theta_2^2$ by ϕ_2, θ_1 and θ_2 denoting numbers prime to $2\Omega\Delta$. Let Φ_1, Φ_2 be two properly primitive binary forms represented by F_1, F_2, simultaneously with the representations of $p\theta_1^2$, $p\theta_2^2$, by f_1, f_2. The determinants of Φ_1, Φ_2, are $-\Delta p\theta_1^2$,

---

* If $$M = F(\alpha'\beta''-\alpha''\beta',\ \alpha''\beta-\alpha\beta'',\ \alpha\beta'-\alpha'\beta),$$
and if f is transformed into a binary form ϕ by the substitution

$$x=\alpha\ x+\beta\ y,$$
$$y=\alpha'\ x+\beta'\ y,$$
$$z=\alpha''x+\beta''y,$$

the representations of M by F and of ϕ by f are said to be simultaneous, or to appertain to one another (Gauss, Disq. Arith., Art. 280).

$-\Delta p\theta_2^2$; and it will be found (as in the case of the forms ϕ_1, ϕ_2) that the generic characters of Φ_1, Φ_2 are compatible; and that a prime P of the same sign as Δ, and not dividing $2\Omega\Delta$, is assignable, such that $P\Theta_1^2$, $P\Theta_2^2$ are primitively represented by Φ_1, Φ_2 respectively, Θ_1 and Θ_2 denoting numbers prime to $2\Omega\Delta$. Thus the numbers $p\theta_1^2$, $P\Theta_1^2$ are simultaneously and primitively represented by f_1 and F_1; the numbers $p\theta_2^2$, $P\Theta_2^2$ are simultaneously and primitively represented by f_2 and F_2. We may therefore suppose that ψ_1 is a form equivalent to f_1, in which $a_1 = p\theta_1^2$, $A_1'' = P\Theta_1^2$, and that ψ_2 is a form equivalent to f_2, in which $a_2 = p\theta_2^2$, $A_2'' = P\Theta_2^2$. The fractional form

$$\frac{1}{Pp}[PX^2 + \Omega Y^2 + p\Omega\Delta Z^2]$$

is then transformed into ψ_1 by the substitution

$$\begin{vmatrix} \dfrac{a_1}{\theta_1}, & \dfrac{b_1''}{\theta_1}, & \dfrac{b_1'}{\theta_1} \\ 0, & \dfrac{A_1''}{\Theta_1\theta_1}, & -\dfrac{B_1}{\Theta_1\theta_1} \\ 0, & 0, & \dfrac{1}{\Theta_1} \end{vmatrix},$$

of which the determinant is Pp, and into ψ_2 by a similar substitution of the same determinant. Either of the two forms ψ_1, ψ_2 (and consequently either of the two f_1, f_2) is, therefore, transformable into the other by a rational substitution prime to $2\Omega\Delta$. It will be found that, if the signs of Θ_1, Θ_2, θ_1, θ_2 are properly determined, the primes P, p will not appear in the denominators of these substitutions.

If f_1 and f_2 belong to an improperly primitive order, the preceding proof requires very little modification. It will suffice to consider the case in which f_1 and f_2 are improperly, F_1 and F_2 properly primitive. We take

$$M_1 \equiv M_2 \equiv -\Omega, \text{ mod } 4;$$

ϕ_1 and ϕ_2 are then improperly primitive and have compatible generic characters; let $2p\theta_1^2$ be represented by ϕ_1, and $2p\theta_2^2$ by ϕ_2; Φ_1 and Φ_2 are properly primitive and of the determinants $-2\Delta p\theta_1^2$, $-2\Delta p\theta_2^2$; these forms have compatible generic characters (their supplementary characters, in particular, being determined by those of F_1 and F_2); let, then, $P\Theta_1^2$ be represented by Φ_1 and $P\Theta_2^2$ by Φ_2, and let us suppose that ψ_1, ψ_2 are forms equivalent to f_1, f_2, in which

$$a_1 = 2p\theta_1^2, \quad A_1'' = P\Theta_1^2, \quad a_2 = 2p\theta_2^2, \quad A_2'' = P\Theta_2^2;$$

the fractional form

$$\frac{1}{Pp}\left[\tfrac{1}{2}(P+\Omega)X^2+(\Omega-P)XY+\tfrac{1}{2}(P+\Omega)Y^2+p\Omega\Delta Z^2\right]$$

is transformed into ψ_1 by the substitution

$$\left|\begin{array}{ccc} \tfrac{1}{2}\dfrac{a_1}{\theta_1}, & \tfrac{1}{2}\dfrac{b''\Theta_1+A_1''}{\Theta_1\theta_1}, & \tfrac{1}{2}\dfrac{b_1'\Theta_1-B_1}{\Theta_1\theta_1} \\ -\tfrac{1}{2}\dfrac{a_1}{\theta_1}, & -\tfrac{1}{2}\dfrac{b''\Theta_1-A_1''}{\Theta_1\theta_1}, & -\tfrac{1}{2}\dfrac{b_1'\Theta_1+B_1}{\Theta_1\theta_1} \\ 0\ , & 0\ , & \dfrac{1}{\Theta_1} \end{array}\right|,$$

and into ψ_2 by a similar substitution. The determinant of each of these substitutions is Pp, and the denominators of their coefficients do not contain the prime 2, because b_1'', b_2'', A_1'', A_2'', Θ_1, Θ_2 are all uneven, and because $B_1\equiv b_1'$, mod 2, $B_2\equiv b_2'$, mod 2. Each of the forms f_1, f_2 is therefore transformable into the other by a rational substitution prime to $2\Omega\Delta$.

13. We have hitherto considered ternary forms of a negative determinant, definite or indefinite; we shall now confine our attention to definite forms. By a binary form we shall henceforward understand a positive form of negative determinant, by a ternary form a positive and definite form; and we shall occupy ourselves in the remainder of this memoir with the determination of the weight of a given genus or order of such ternary forms.

A ternary form has always 1, 2, 4, 6, 8, 12 or 24 *positive* automorphics, *i.e.* automorphics of which the determinant is a positive unit. The weight of a form is the reciprocal of the number of its positive automorphics; so that a form and its contravariant have the same weight; the weight of a class is the weight of any form contained in the class; the weight of a genus or of an order is the sum of the weights of the non-equivalent classes contained in the genus or order. When a number is represented by a ternary form, the weight of the representation is the weight of the ternary form. The weight of a binary form, or class, is also the reciprocal of the number of its positive automorphics; thus the weight of a binary form is always $\frac{1}{2}$, except when the form either is, or is derived from, a form of determinant -1, or an improperly primitive form of determinant -3; in these excepted cases the weight of the binary form is $\frac{1}{4}$ and $\frac{1}{6}$ respectively. When a binary form is represented by a ternary form, the weight of the representation is the product of the weights of the two forms.

To determine the weight of a given genus of ternary forms we avail ourselves of the principles introduced into arithmetic by Gauss and Dirichlet, and employed by them to determine the number of binary forms of any given determinant. Let (f, F) represent a given genus of ternary forms of the invariants $[\Omega, \Delta]$, and either of the properly primitive order or of that improperly primitive order in which f is improperly and F properly primitive. Let $f_1, f_2, \ldots$ or (f) denote a system of forms representing the classes of the given genus; $F_1, F_2, \ldots$ or (F), the primitive contravariants of those forms. Let M represent any positive number prime to $2\Omega\Delta$ and satisfying the generic characters of F; when (f, F) is a properly primitive genus, Ω being uneven and Δ uneven or unevenly even, we shall also suppose that M satisfies the congruence $\Omega M \equiv 1$, mod 4: the numbers designated by M will be subject to the restrictions here stated throughout the whole investigation. Lastly, let L be a positive quantity which we shall afterwards suppose to increase without limit; and let T be the sum of the weights of the representations by the forms (F) of all the numbers M which do not surpass L. The quotient $T \div L^{\frac{3}{2}}$ approximates to a finite limit, when L is increased without limit. Of this limit, we shall obtain two distinct expressions, the one containing as a factor the weight W of the genus (f, F), the other not containing that factor, and depending on the arithmetical relation which subsists between the sum of the weights of the representations of a given number M by the forms (F) and the sum of the weights of the properly or improperly primitive binary classes of determinant $-\Omega M$. A comparison of the two expressions will then give the required weight of the genus (f, F).

14. The first determination of the limit of the quotient $T \div L^{\frac{3}{2}}$ depends on the following auxiliary propositions, in which F represents any form of the system (F).

(1) If δ is an uneven prime dividing Δ, F acquires a value prime to δ for $\delta^2(\delta-1)$ systems of values of x, y, z, mod δ.

As, instead of f and F, we may consider any forms equivalent to f and F, we may suppose that f and F satisfy, for any assigned powers of the uneven primes dividing $\Omega\Delta$, the congruences of Art. 5,

$$f \equiv \alpha x^2 + \beta\Omega y^2 + \gamma\Omega\Delta z^2,$$
$$F \equiv \beta\gamma\Omega\Delta x^2 + \alpha\gamma\Delta y^2 + \alpha\beta z^2,$$
$$\alpha\beta\gamma \equiv 1.$$

The congruence $F \equiv 0$, mod δ, is then satisfied by δ^2 systems of values of x, y, z, mod δ; for z must be divisible by δ, but x and y may have any values, mod δ; F is therefore prime to δ for the remaining $\delta^2(\delta-1)$ systems of values of x, y, z, mod δ.

(2) If ω is an uneven prime dividing Ω but not Δ, F is prime to Ω for $\omega(\omega-1)\left[\omega-\left(\frac{-\Delta f}{\omega}\right)\right]$ systems of values of x, y, z, mod ω.

For, if $F\equiv 0$, mod ω, x may have any value, mod ω, but y and z must have values satisfying the congruence $\gamma\Delta y^2+\beta z^2\equiv 0$, mod ω. If $\left(\frac{-\Delta\beta\gamma}{\omega}\right)=-1$, the only values of y and z that satisfy this congruence are $y\equiv 0$, $z\equiv 0$, mod ω; and the congruence $F\equiv 0$, mod ω, is satisfied by ω systems of values of x, y, z, mod ω. If $\left(\frac{-\Delta\beta\gamma}{\omega}\right)=+1$, the congruence $\gamma\Delta y^2+\beta z^2\equiv 0$, mod ω, is satisfied by $2\omega-1$ systems of values of y and z; in this case, therefore, the congruence $F\equiv 0$, mod ω, admits of $\omega(2\omega-1)$ solutions. And, observing that

$$\left(\frac{-\Delta\beta\gamma}{\omega}\right)=\left(\frac{-\Delta\alpha}{\omega}\right)=\left(\frac{-\Delta f}{\omega}\right),$$

we find in both cases alike that F is prime to ω for $\omega(\omega-1)\left[\omega-\left(\frac{-\Delta f}{\omega}\right)\right]$ systems of values of x, y, z, mod ω.

(3) It is evident from the congruence

$$F\equiv Ax^2+A'y^2+A''z^2, \text{ mod } 2,$$

in which one at least of the numbers A, A', A'' is uneven, that F acquires an uneven value for four out of the eight systems of values, mod 2, which can be attributed to x, y, z.

(4) If $\Omega\Delta$ is uneven, the number of solutions of the congruence $\Omega F\equiv 1$, mod 4, is $8(2-\Psi)$.

For this congruence may be written in the form (Art. 6)

$$\alpha x^2+\beta y^2+\gamma z^2\equiv 1, \text{ mod } 4,$$

α, β, γ representing uneven numbers which satisfy the congruence

$$\alpha+\beta+\gamma+1\equiv 0, \text{ mod } 4.$$

Of the three numbers x, y, z one must be uneven, the other two even. The number of solutions in which x is uneven, y and z even, is 8 or 0, according as $\alpha\equiv +1$ or $\equiv -1$, mod 4. The whole number of solutions is therefore

$$12+4\left[(-1)^{\frac{1}{2}(\alpha-1)}+(-1)^{\frac{1}{2}(\beta-1)}+(-1)^{\frac{1}{2}(\gamma-1)}\right],$$

i.e. 24, or 8, according as the congruences $\alpha\equiv\beta\equiv\gamma\equiv 1$, mod 4, are, or are not, satisfied; or again (Art. 6), according as $\Psi=-1$, or $\Psi=+1$. The congruence $\Omega F\equiv 1$, mod 4, admits therefore of $8(2-\Psi)$ solutions.

(5) If Ω is uneven and Δ unevenly even, f as well as F being properly primitive, there are 16 solutions of the congruence $\Omega F \equiv 1$, mod 4; for this congruence may be written in the form (Art. 6)

$$2\alpha x^2 + 2\beta y^2 + \gamma z^2 \equiv 1, \text{ mod } 4.$$

For clearness, we shall henceforward represent by r any uneven prime dividing both Ω and Δ, by δ any uneven prime dividing Δ but not Ω, by ω any uneven prime dividing Ω but not Δ. Let $\theta = 2 - \Psi$, when $\Omega \equiv \Delta \equiv 1$, mod 2; let $\theta = 2$, when f and F being properly primitive, Ω is uneven and Δ unevenly even; and let $\theta = 4$ in every other case; also let

$$\nabla = 4\,\Pi r \times \Pi\omega \times \Pi\delta,$$

$$\psi(\nabla) = \frac{\theta}{8}\nabla^3 \Pi\left[1 - \frac{1}{r}\right]\Pi\left[1 - \frac{1}{\omega}\right]\Pi\left[1 - \frac{1}{\delta}\right]\Pi\left[1 - \left(\frac{-\Delta f}{\omega}\right)\frac{1}{\omega}\right].$$

Combining the lemmas (1) ... (5), we obtain the theorem :—

'The form F represents numbers of the series M for $\psi(\nabla)$ of the ∇^3 systems of values, mod ∇, that can be attributed to x, y, z.'

Let x_i, y_i, z_i represent one of these $\psi(\nabla)$ systems of values; it is evident that F represents a number of the series M for every system of values of x, y, z included in the formulae

$$\left.\begin{aligned} x &= \nabla X + x_i, \\ y &= \nabla Y + y_i, \\ z &= \nabla Z + z_i, \end{aligned}\right\} \qquad (27)$$

in which X, Y, Z represent any integral numbers whatever. It is also evident that there are as many systems of values of x, y, z included in the formulae (27), for which F acquires a value not surpassing L, as there are points having their rectangular coordinates of the form

$$x = \frac{\nabla X + x_i}{\sqrt{L}},$$

$$y = \frac{\nabla Y + y_i}{\sqrt{L}},$$

$$z = \frac{\nabla Z + z_i}{\sqrt{L}},$$

and lying inside, or on the surface of, the ellipsoid

$$F(x, y, z) = 1. \qquad (28)$$

Let ν_i be the number of these points, and let L be increased without limit;

the limit of the fraction $\frac{\nabla^3 \nu_i}{L^{\frac{3}{2}}}$ is the volume of the ellipsoid (28), or $\frac{4}{3}\frac{\pi}{\Delta\sqrt{\Omega}}$. Extending this result to all the $\psi(\nabla)$ values of i, we find

$$\lim \frac{\Sigma \nu_i}{L^{\frac{3}{2}}} = \frac{4}{3}\frac{\psi(\nabla)}{\nabla^3}\cdot\frac{\pi}{\Delta\sqrt{\Omega}}. \quad (29)$$

Let τ be the sum of the weights of the representations of the numbers M which do not surpass L by the form F, and let w be the weight of f or F, so that $\tau = w\Sigma\nu_i$; the equation (29) becomes

$$\lim \frac{\tau}{L^{\frac{3}{2}}} = \frac{4}{3}\frac{\psi(\nabla)}{\nabla^3}\cdot\frac{\pi}{\Delta\sqrt{\Omega}}\cdot w; \quad (30)$$

or, considering in succession all the forms of (F), and observing that $T = \Sigma\tau$, $W = \Sigma w$,

$$\lim \frac{T}{L^{\frac{3}{2}}} = \frac{\theta W}{6}\,\Pi\left(1-\frac{1}{r}\right)\Pi\left(1-\frac{1}{\delta}\right)\Pi\left(1-\frac{1}{\omega}\right)\,\Pi\left[1-\left(\frac{-\Delta f}{\omega}\right)\frac{1}{\omega}\right], \quad (31)$$

which is the first determination of the limit of the quotient $\frac{T}{L^{\frac{3}{2}}}$.

15. The second determination of the limit of the quotient $T \div L^{\frac{3}{2}}$ depends on the following theorem :—

'The sum of the weights of the primitive representations by the forms (F) of a given number M divisible by μ unequal primes is 2^μ times the weight of a genus of binary forms, of determinant $-\Omega M$, and properly or improperly primitive according as the forms (f) are properly or improperly primitive.'

The principles which give the demonstration of this theorem are contained in Arts. 280–284 of the 'Disquisitiones Arithmeticae,' and have been in part already employed in Art. 10 of this memoir. We have shown in Art. 11 that one genus, and only one, of binary forms of determinant $-\Omega M$ admits of primitive representation by the forms (f) of the ternary genus (f, F). Let $\phi_1, \phi_2, \ldots$ or (ϕ) be a system of forms representing the classes of that binary genus; these forms are properly or improperly primitive, according as the forms (f) are properly or improperly primitive: let n be their number and ν the sum of their weights; as their weights are all equal, the weight of each of them is $\frac{\nu}{n}$; so that each has $\frac{n}{\nu}$ positive automorphics, and is transformed into any equivalent form by $\frac{n}{\nu}$ positive substitutions. We shall first show that the sum of the weights of the primitive representations of the forms (ϕ) by the forms (f) is equal to

$2^\mu \times \nu$; and, secondly, that the sum of the weights of the primitive representations of the numbers M by the forms (F) is equal to the sum of the weights of the primitive representations of the forms (ϕ) by the forms (f).

(i) Each of the n congruences

$$-\Delta\phi \equiv (Qx - Q'y)^2, \text{ mod } M, \quad \ldots \quad (32)$$

in which Q, Q' are the numbers to be determined, is resoluble, and admits of 2^μ incongruous solutions. From each such solution we deduce, by the method of Gauss employed in Art. 10, a ternary form f' of the given genus, containing one of the forms (ϕ) as a part, and having Q, Q', M for the coefficients of $2yz$, $2xz$, z^2 in its primitive contravariant. There are $2^\mu \times n$ of these forms (f'); none of them is the same as any other, and none of them can be transformed into any other by a substitution of the type

$$\begin{vmatrix} 1, & 0, & \kappa' \\ 0 & 1, & \kappa \\ 0, & 0, & 1 \end{vmatrix}; \quad \ldots \quad (33)$$

for, if one of them could be so transformed into another, these two would contain as a part the same form ϕ, and the values of Q, Q' in the primitive contravariant of the one would be congruous, for the modulus M, to the values of Q, Q' in the primitive contravariant of the other; the two forms would thus be derived from the same solution of the same congruence (32). Again, the primitive representations of the forms (ϕ) by the forms (f) are equal in number to the positive transformations of the forms (f) into the forms (f'). For, every positive transformation of a form of (f) into a form of (f') supplies a primitive representation of some form of (ϕ) by that form of (f); and these representations are all different because the same form f cannot be transformed into two of the forms (f'), or twice into one of them, by positive substitutions of which the first two columns are the same; otherwise, one of the forms (f') could be transformed into another by a substitution of the type (33), or else one of those forms would have an automorphic of that type, whereas no substitution of the type (33), in which κ and κ' are different from zero, can be an automorphic of any ternary form. There are, therefore, at least as many different primitive representations of the forms (ϕ) by the forms (f) as there are positive transformations of the forms (f) into the forms (f'). And there are no more; for, if

$$\begin{vmatrix} \alpha, & \beta \\ \alpha', & \beta' \\ \alpha'', & \beta'' \end{vmatrix}$$

is a given primitive representation of ϕ by f, let $\gamma, \gamma', \gamma''$ be numbers which render the determinant of the substitution

$$\begin{vmatrix} \alpha, & \beta, & \gamma \\ \alpha', & \beta', & \gamma' \\ \alpha'', & \beta'', & \gamma'' \end{vmatrix} \quad . \quad . \quad . \quad . \quad . \quad . \quad . \quad . \quad . \quad . \quad . \quad . \quad (34)$$

equal to $+1$; and let f_1 be the form, containing ϕ as a part, into which f is transformed by the substitution (34). The coefficient of z^2 in the primitive contravariant of f_1 is M; and, if the coefficients of $2yz$, $2xz$ in that contravariant are Q_1, Q'_1, these numbers supply a solution of the congruence (32). Let f' be that form of (f') which is deduced from this solution; then f_1 is equivalent to f', and is transformed into it by a substitution of the type (33), in which

$$\kappa = \frac{Q_1 - Q}{M}, \qquad \kappa' = \frac{Q_1' - Q'}{M}.$$

Therefore f is transformed into f' by the substitution

$$\begin{vmatrix} \alpha, & \beta, & \gamma + \kappa'\alpha + \kappa\beta \\ \alpha', & \beta', & \gamma' + \kappa'\alpha' + \kappa\beta' \\ \alpha'', & \beta'', & \gamma'' + \kappa'\alpha'' + \kappa\beta'' \end{vmatrix},$$

i.e. the given primitive representation of ϕ by f is included among those supplied by the positive transformations of the forms (f) into the forms (f'). Thus the number of the primitive representations of the forms (ϕ) by the forms (f) is equal to the number of the positive transformations of the forms (f) into the forms (f'). To obtain the sum of the weights of these representations we consider, in particular, f one of the forms of (f); let d be the number of its positive automorphics, so that $\frac{1}{d}$ is its weight, and let s be the number of the forms (f') which are equivalent to it. Then there are $d \times s$ primitive representations of the forms (ϕ) by f; but the weight of each of these representations is $\frac{1}{d} \times \frac{\nu}{n}$; the sum of the weights of the primitive representations of the forms (ϕ) by f is therefore $s \times \frac{\nu}{n}$. Extending this conclusion to all the forms of (f), and observing that Σs is equal to the number of the forms (f'), *i.e.* to $2^\mu \times n$, we find that the sum of the weights of the primitive representations of the forms (ϕ) by the forms (f) is $2^\mu \times \nu$.

(ii) Let $M = F(\Gamma, \Gamma', \Gamma'')$ be a given primitive representation of M by F; and let

$$\begin{vmatrix} \alpha, & \beta \\ \alpha', & \beta' \\ \alpha'', & \beta'' \end{vmatrix} \quad . \quad . \quad . \quad . \quad . \quad . \quad . \quad . \quad . \quad . \quad . \quad . \quad (35)$$

be a matrix of which the constituents satisfy the equations

$$\alpha'\beta'' - \alpha''\beta' = \Gamma, \quad \alpha''\beta - \alpha\beta'' = \Gamma', \quad \alpha\beta' - \alpha'\beta = \Gamma''. \quad . \quad . \quad . \quad (36)$$

All the matrices, of which the constituents satisfy these equations, are then included in the formula

$$\begin{vmatrix} \alpha, & \beta \\ \alpha', & \beta' \\ \alpha'', & \beta'' \end{vmatrix} \times |v|, \quad . \quad . \quad . \quad . \quad . \quad . \quad . \quad . \quad . \quad . \quad (37)$$

in which $|v|$ is a square binary matrix of which the determinant is $+1$. Thus the binary forms which are represented by f simultaneously with the given representation of M by F are all equivalent to one another and to some form of (ϕ); let ϕ be that form of (ϕ) to which they are equivalent, and let us suppose (as we may do) that f is transformed into ϕ by the substitution (35). Substituting successively for $|v|$ in the formula (37) the $\frac{n}{\nu}$ positive automorphics of ϕ, we obtain $\frac{n}{\nu}$ representations of ϕ by f: these representations are all different, and they include every representation of ϕ by f which is simultaneous with the given representation of M by F: the weight of each of them is $\frac{1}{d} \times \frac{\nu}{n}$; the sum of their weights is therefore equal to $\frac{1}{d}$, or to the weight of the given representation of M by F. Hence the sum of the weights of all the primitive representations of M by the forms (F) is equal to the sum of the weights of the simultaneous representations of the forms (ϕ) by the forms (f), or, which is the same thing, to the sum of the weights of all the primitive representations of the forms (ϕ) by the forms (f); because every primitive representation of a form (ϕ) by a form (f) is simultaneous with one, and only one, primitive representation of M by a form (F).

Combining the conclusions (i) and (ii), we obtain the result enunciated at the beginning of this article.

16. Let σ represent the number of uneven primes dividing Ω, counting those which also divide Δ; let $\sigma' = -1$ when $\Omega M \equiv -1$, mod 4*; let $\sigma' = +1$ when $\Omega \equiv 0$, mod 8; and let $\sigma' = 0$ in all other cases. Let also $h(\Omega M)$ and $h'(\Omega M)$ be the weights of the properly and improperly primitive orders of binary forms of determinant $-\Omega M$; then

$$2^{\mu} \times \nu = \frac{h(\Omega M)}{2^{\sigma+\sigma'}}, \quad \text{or} \quad = \frac{h'(\Omega M)}{2^{\sigma+\sigma'}},$$

* If this congruence is satisfied by any one number of the series M, it is satisfied by every number of that series.

according as the forms (f) are properly or improperly primitive. If λ^2 is any square divisor of M, the sum of the weights of those representations of M by the forms (F) which are derived from primitive representations of $\frac{M}{\lambda^2}$ by the same forms, is

$$\frac{h\left(\frac{\Omega M}{\lambda^2}\right)}{2^{\sigma+\sigma'}}, \quad \text{or} \quad \frac{h'\left(\frac{\Omega M}{\lambda^2}\right)}{2^{\sigma+\sigma'}}.$$

Therefore the sum of the weights of all the representations of M by the forms (F) is

$$\frac{\Sigma . h\left(\frac{\Omega M}{\lambda^2}\right)}{2^{\sigma+\sigma'}}, \quad \text{or} \quad \frac{\Sigma . h'\left(\frac{\Omega M}{\lambda^2}\right)}{2^{\sigma+\sigma'}},$$

the signs of summation extending to every square divisor of M. Or, if we represent by $H(\Omega M)$ the sum of the weights of those uneven binary classes of determinant $-\Omega M$ which are prime to Ω, and by $H'(\Omega M)$ the sum of the weights of those even classes of determinant $-\Omega M$ which are prime to Ω, the sum of the weights of all the representations of M by the forms (F) is

$$\frac{H(\Omega M)}{2^{\sigma+\sigma'}}, \quad \text{or} \quad \frac{H'(\Omega M)}{2^{\sigma+\sigma'}},$$

according as the forms (f) are properly or improperly primitive.

17. We now consider the sums

$$\Sigma\,[xz - y^2 = \Omega M], \quad \ldots \ldots \quad (38)$$

$$\Sigma'\,[xz - y^2 = \Omega M]. \quad \ldots \ldots \quad (39)$$

In both the sign of summation extends to every solution in integral numbers of the equation $xz - y^2 = \Omega M$, in which the greatest common divisor of x, y, z is prime to Ω, and in which x, y, z satisfy the inequalities

$$\left.\begin{array}{lll} x > 0, & y \geqq 0, & z > 0, \\ x \geqq 2y \leqq z, & & x \leqq z. \end{array}\right\} \quad \ldots \ldots \quad (40)$$

But, in the first sum, one at least of the two numbers x and z is uneven; in the second, x and z are even, and y is uneven. The symbol $[xz - y^2 = \Omega M]$ has the value 1, $\frac{1}{2}$, $\frac{1}{4}$, or $\frac{1}{6}$, according as the inequalities (40) are satisfied, excluding all signs of equality, or admitting one, two, or three such signs. Again, representing by $(2y)$ the absolute value of $2y$, we observe that a *reduced* binary form is a form (x, y, z) of which the coefficients satisfy the inequalities

$$\left.\begin{array}{lll} \text{(i)} & x > 0, & z > 0, \\ & x \geqq (2y) \leqq z, & x \leqq z, \\ \text{(ii)} & \text{if } x = (2y), & y > 0, \\ & \text{if } x = z, & y \geqq 0, \end{array}\right\} \quad \ldots \ldots \quad (41)$$

and that, by a fundamental proposition in the theory of binary forms, every class contains one, and only one, reduced form. Attending only to those uneven classes of determinant $-\Omega M$ which are prime to Ω, and comparing the inequalities (40) and (41), we find that the sum (38) contains (i) an unit corresponding to every pair of reduced forms (x, y, z), $(x, -y, z)$ of which the coefficients satisfy none of the equalities $y=0$, $x=2y$, $x=z$; (ii) one half of an unit corresponding to every reduced form of which the coefficients satisfy one of them; and (iii) one fourth of an unit corresponding to a reduced form (if there be such a form of determinant $-\Omega M$ prime to Ω) of which the coefficients satisfy the two equalities, $y=0$, $x=z$, and of which the weight is consequently $\frac{1}{4}$. We thus obtain the equation

$$H(\Omega M)=\Sigma[xz-y^2=\Omega M].$$

Again, attending only to those even classes of the uneven determinant $-\Omega M$ which are prime to Ω, we find that the sum (39) contains units corresponding to pairs of reduced forms and half units corresponding to single reduced forms; it also contains one sixth of an unit corresponding to a reduced form (if there be such a reduced form of determinant $-\Omega M$ prime to Ω) of which the coefficients satisfy the three equalities $x=2y$, $2y=z$, $x=z$, and of which the weight is consequently $\frac{1}{6}$. We therefore have the equation

$$H'(\Omega M)=\Sigma'[xz-y^2=\Omega M].$$

18. According as the forms (f) are properly or improperly primitive, let

$$\Upsilon=\Sigma\,.\,\Sigma[xz-y^2=\Omega M],$$

or

$$\Upsilon=\Sigma\,.\,\Sigma'[xz-y^2=\Omega M],$$

the first sign of summation extending to all values of M not surpassing L; so that, in both cases alike,

$$T=\frac{\Upsilon}{2^{\sigma+\sigma'}}.$$

To determine the limit of the quotient $\frac{\Upsilon}{L^{\frac{3}{2}}}$, when L is increased without limit, we shall again employ the geometric method of Gauss. For its application here the following preliminary lemmas relating to the arithmetical properties of the function $xz-y^2$ are requisite.

I. If p is any uneven prime and m any given number, the congruence

$$xz-y^2\equiv m, \bmod p, \quad . \quad . \quad . \quad . \quad . \quad . \quad . \quad . \quad . \quad . \quad (42)$$

admits of $p\left[p+\left(\frac{-m}{p}\right)\right]$ solutions.

For, if $\left(\frac{-m}{p}\right) = +1$, y^2+m is prime to p for $p-2$ values of y, and is divisible by p for 2 values of y. When y^2+m is prime to p, we may assign to z any value prime to p, determining x by the congruence $xz \equiv y^2+m$; we thus obtain $(p-1)(p-2)$ solutions of (42). When y^2+m is divisible by p, the congruence $xz \equiv 0$, mod p, admits of $2p-1$ solutions; we thus obtain in all $(p-1)(p-2)+2(2p-1) = p(p+1)$ solutions of (42).

If $\left(\frac{-m}{p}\right) = -1$, y^2+m is prime to p for every value of y; there are thus $p(p-1)$ solutions of (42).

Lastly, if $\left(\frac{-m}{p}\right) = 0$, *i.e.* if $m \equiv 0$, mod p, y^2+m is prime to p for $p-1$ values of y and divisible by p for one value of y; there are thus $(p-1)^2+2p-1 = p^2$ solutions of (42).

We shall have to use the following corollary of this lemma.

If m is prime to p, and if we successively attribute to x, y, z the p^3 systems of values, mod p, of which they are susceptible, $xz-y^2$ will have the same quadratic character as m for $\frac{1}{2}p(p-1)\left[p+\left(\frac{-m}{p}\right)\right]$ of these systems.

II. The congruences $xz-y^2 \equiv 1, \equiv 3, \equiv 5, \equiv 7$, mod 8, each admit of 48 solutions in which x and z are not simultaneously even; of the congruences, $xz-y^2 \equiv 3, \equiv 7$, mod 8, the first admits of 16, the second of 48 solutions in which x and z are simultaneously even.

For example, let the proposed congruence be $xz-y^2 \equiv 3$, mod 8. If y has one of its four even values, mod 8, we may give to z any one of its four uneven values, mod 8, and determine the value of x in the resulting congruence; we thus obtain 4×4 solutions in which x and z are uneven. If y has one of its four uneven values, mod 8, the congruence becomes $xz \equiv 4$, mod 8, which admits of 8 solutions in which x and z are not simultaneously even, and 4 in which they are simultaneously even. There are thus $4 \times 4 + 4 \times 8 = 48$ solutions of the congruence $xz-y^2 \equiv 3$, mod 8, in which x and z are not simultaneously even, and $4 \times 4 = 16$ in which x and z are simultaneously even.

III. If p is any prime, even or uneven, i and i' integral exponents, of which $i > 0$, $i' \geqq 0$, and m any given number prime to p or divisible by any power of p, the congruence

$$xz - y^2 \equiv mp^i, \text{ mod } p^{i+i'} \quad . \quad . \quad . \quad . \quad . \quad . \quad . \quad (43)$$

admits of $p^{2i+2i'}\left(1-\frac{1}{p^2}\right)$ *primitive* solutions, *i.e.* solutions in which x, y, z, or, which is the same thing, x, z are not simultaneously divisible by p.

(i) If the assertion is true for i, i', and if $j \leqq i'$, it is true for $i+j,\ i'-j$. For, on writing mp^j for m in (43), it becomes

$$xz-y^2 \equiv mp^{i+j}, \text{ mod } p^{(i+j)+(i'-j)};$$

if, therefore, the former congruence admits of

$$p^{2i+2i'}\left(1-\frac{1}{p^2}\right)=p^{2(i+j)+2(i'-j)}\left(1-\frac{1}{p^2}\right)$$

primitive solutions, the latter does so too.

(ii) If the assertion is true for i, 0, it is also true for i, i', where $i' \leqq i$.

For, if x, y, z is a given primitive solution of

$$xz-y^2 \equiv mp^i, \text{ mod } p^i, \qquad (44)$$

$Xp^i+x,\ Yp^i+y,\ Zp^i+z$ is a primitive solution of (43), whenever X, Y, Z satisfy the congruence

$$Xz-2Yy+Zx+\frac{xz-y^2}{p^i} \equiv m, \text{ mod } p^{i'}.$$

This congruence admits of $p^{2i'}$ solutions; for the given numbers x and z are not simultaneously divisible by p. Thus, from each primitive solution of (44) we obtain $p^{2i'}$ primitive solutions of (43). These solutions are all different, and they exhaust all the solutions of (43); if, therefore, (44) admits of $p^{2i}\left(1-\frac{1}{p^2}\right)$ solutions, (43) admits of $p^{2i+2i'}\left(1-\frac{1}{p^2}\right)$ solutions.

(iii) The assertion is true if $i=1,\ i'=0$. For (lemma I.) there are p^2 solutions of the congruence $xz-y^2 \equiv 0$, mod p, and of these one is not primitive.

The proposition is, therefore, true universally. We shall have to employ the following corollaries from it.

1. The function $xz-y^2$ is divisible by p^i, but not divisible by p^{i+1}, for $p^{2i}(p-1)^2(p+1)$ systems of values of x, y, z, mod p^{i+1}; the values of x, y, z not being simultaneously divisible by p.

2. If p is an uneven prime, one half of the quotients obtained by dividing these $p^{2i}(p-1)^2(p+1)$ values of $xz-y^2$ by p^i are quadratic residues, and one half are non-quadratic residues of p.

3. If $p=2$, the function $xz-y^2$ is divisible by 2^i, but not by 2^{i+1}, for $3 \times 2^{i+6}$ systems of values of x, y, z, mod 2^{i+3}, the values of x, y, z not being simultaneously even. And, if these $3 \times 2^{i+6}$ values of $xz-y^2$ be divided by 2^i, one fourth part of the quotients is contained in each of the linear forms $8k+1,\ 8k+3,\ 8k+5,\ 8k+7$.

19. Let $\nabla = 8\Omega \times \Pi r \times \Pi \omega \times \Pi \delta$, and let us successively attribute to x, y, z in the function $xz - y^2$ the ∇^3 systems of values, mod ∇, of which they are susceptible; let $\phi(\nabla)$ represent the number of those systems in which the greatest common divisor of x, y, z is prime to ∇, and which give to $xz - y^2$ a value divisible by Ω and such that the quotient $\frac{xz - y^2}{\Omega}$ is a number of the series M. If the forms (f) are properly primitive, x and z are not to be simultaneously even; if those forms are improperly primitive, x and z are to be simultaneously even. We shall now show that $\phi(\nabla)$ is determined by the equation

$$\phi(\nabla) = \tfrac{3}{8}\eta\,\nabla^3 \frac{1}{\Omega}\,\Pi\left(1 - \frac{1}{r}\right)\Pi\left(1 - \frac{1}{\delta}\right)\Pi\left(1 - \frac{1}{\omega}\right)$$
$$\times\,\Pi\tfrac{1}{2}\left(1 - \frac{1}{r^2}\right)\Pi\left(1 - \frac{1}{\omega^2}\right)\Pi\tfrac{1}{2}\left[1 + \left(\frac{-\Omega F}{\delta}\right)\frac{1}{\delta}\right], \quad . \quad . \quad (45)$$

η being a coefficient of which the value is 1, $\frac{1}{2}$, $\frac{1}{3}$, $\frac{1}{4}$ or $\frac{1}{12}$*, as shown in the following table.

(i): (f) properly primitive.

	$\Omega \equiv 1$, mod 2.	$\Omega \equiv 2$, mod 4.	$\Omega \equiv 4$, mod 8.	$\Omega \equiv 0$, mod 8.
$\Delta \equiv 1$, mod 2.	$\frac{1}{2}$	1	$\frac{1}{4}[3+(-1)^{\frac{1}{2}(\Delta f+1)}]$	$\frac{1}{4}[3+(-1)^{\frac{1}{2}(\Delta f+1)}]$
$\Delta \equiv 2$, mod 4.	$\frac{1}{2}$	1	$\frac{1}{2}$	$\frac{1}{2}$
$\Delta \equiv 4$, mod 8.	$\frac{1}{2}$	$\frac{1}{2}$	$\frac{1}{2}$	$\frac{1}{2}$
$\Delta \equiv 0$, mod 8.	$\frac{1}{4}$	$\frac{1}{4}$	$\frac{1}{4}$	$\frac{1}{4}$

(ii): (f) improperly primitive.

$\Delta \equiv 2$, mod 4.	$\frac{1}{3}$
$\Delta \equiv 0$, mod 4.	$\frac{1}{12}[2+(-1)^{\frac{1}{8}(\Omega^2-1)+\frac{1}{8}(F^2-1)}]$

* It will be seen that 4η in the table (i) is, in every case, the number of the linear forms $8k+1$, $8k+3$, $8k+5$, $8k+7$, in which the numbers M are contained.

To establish the equation (45) we consider separately the different primes dividing ∇. And first let us take an uneven prime δ, dividing Δ but not Ω. Of the δ^3 systems of values of x, y, z, mod δ,

$$\delta^3 \times \left(1 - \frac{1}{\delta}\right) \times \tfrac{1}{2}\left[1 + \left(\frac{-\Omega F}{\delta}\right)\frac{1}{\delta}\right]$$

give to $xz - y^2$ a value prime to δ, and satisfying the equation (lemma I. Cor.)

$$\left(\frac{xz - y^2}{\delta}\right) = \left(\frac{\Omega F}{\delta}\right).$$

Secondly, let us consider an uneven prime ω dividing Ω but not Δ; and let ω^i be the highest power of ω dividing Ω. Of the ω^{3i+3} systems of values of x, y, z, mod ω^{i+1},

$$\omega^{3i+3} \times \frac{1}{\omega^i} \times \left(1 - \frac{1}{\omega}\right) \times \left(1 - \frac{1}{\omega^2}\right),$$

in which x, y, z are not simultaneously divisible by ω, render $xz - y^2$ divisible by ω^i, and also render the quotient $\frac{xz - y^2}{\omega^i}$ prime to ω (lemma III. Cor. 1).

Thirdly, let us consider an uneven prime r dividing both Δ and Ω, and let r^i be the highest power of r dividing Ω. Of the r^{3i+3} systems of values of x, y, z, mod r^{i+1},

$$r^{3i+3} \times \frac{1}{r} \times \left(1 - \frac{1}{r}\right) \times \tfrac{1}{2}\left(1 - \frac{1}{r^2}\right),$$

in which x, y, z are not simultaneously divisible by r, render $xz - y^2$ divisible by r^i but not by r^{i+1}, and also render the quotients $\frac{xz - y^2}{r^i}$ all quadratic residues or all non-quadratic residues of r (lemma III. Cor. 2).

Lastly, let us consider the even prime 2, and let 2^i be the highest power of 2 dividing Ω. Considering separately the eighteen cases of the tables (i) and (ii), we find that, of the 2^{3i+9} systems of values of x, y, z, mod 2^{i+3},

$$\tfrac{3}{8} \times \eta \times 2^{3i+9} \times \frac{1}{2^i}$$

(in which x, y, z are not simultaneously even, but x and z are or are not simultaneously even, according as the forms (f) are improperly or properly primitive) give to $\frac{xz - y^2}{2^i}$ an integral and uneven value satisfying the supplementary character (if any) of $\frac{\Omega}{2^i} F$, and, if the forms (f) are properly primitive, satisfying the congruence $xz-y^2\equiv 1$, mod 4, when Ω is uneven and Δ uneven or unevenly even.

For example, let $i\geqq 1$, $\Delta\equiv 0$, mod 8. Here F, or $\frac{\Omega}{2^i}F$, has two supplementary characters, and, of the 2^{3i+9} systems of values of x, y, z, mod 2^{i+3},

$$\tfrac{3}{8}\times\tfrac{1}{4}\times 2^{3i+9}\times\frac{1}{2^i},$$

in which x and z are not simultaneously even, give to $\frac{xz-y^2}{2^i}$ an integral and uneven value satisfying the supplementary characters of $\frac{\Omega}{2^i}F$ (Lemma III. Cor. 3).

Again, let $i\geqq 2$, $\Delta\equiv 1$, mod 2. Here F has, or has not, a supplementary character, according as $(-1)^{\frac{1}{4}(\Delta f+1)}=-1$, or $=+1$. In the former case, of the 2^{3i+9} systems of values of x, y, z, mod 2^{i+3},

$$\tfrac{3}{8}\times\tfrac{1}{2}\times 2^{3i+9}\times\frac{1}{2^i},$$

in which x and z are not simultaneously even, give to $\frac{xz-y^2}{2^i}$ an integral and uneven value satisfying the supplementary character of $\frac{\Omega}{2^i}F$. In the latter case, of the same 2^{3i+9} systems of values,

$$\tfrac{3}{8}\times 1\times 2^{3i+9}\times\frac{1}{2^i},$$

in which x and z are not simultaneously even, give to $\frac{xz-y^2}{2^i}$ an integral and uneven value. Both results are comprised in the formula

$$\tfrac{3}{8}\times\tfrac{1}{4}\left[3+(-1)^{\frac{1}{2}(\Delta f+1)}\right]\times 2^{3i+9}\times\frac{1}{2^i}.$$

As a third example, let $i=0$, $\Delta\equiv 0$, mod 4, and let the forms considered be of an improperly primitive order. Then $\Omega F\equiv 3$, mod 4; and either $\Omega F\equiv 3$, mod 8, or $\Omega F\equiv 7$, mod 8. The congruence $xz-y^2\equiv\Omega F$, mod 8, in which only even values of x and z are to be admitted, is satisfied in the former case by 16, in the latter by 48 systems; *i.e.* in either case the number of systems is

$$\tfrac{3}{8}\times\tfrac{1}{12}\left[2+(-1)^{\frac{1}{8}(\Omega^2-1)+\frac{1}{8}(F^2-1)}\right]\times 2^9.$$

The formula (45) results immediately from the combination of these determinations relative to the primes δ, ω, r, and 2.

20. Let x_i, y_i, z_i be one of the $\phi(\nabla)$ systems of values of x, y, z, mod ∇, defined in the last article; and let us decompose the sum Υ of Art. 18 into

$\phi(\nabla)$ partial sums $\Upsilon_1, \Upsilon_2, \ldots$, comprising in the sum Υ_i all those terms of Υ in which x, y, z are of the linear forms

$$x = \nabla X + x_i,$$
$$y = \nabla Y + y_i,$$
$$z = \nabla Z + z_i,$$

X, Y, Z denoting any integral numbers whatever. The sum Υ_i is equal to the number of points having their positive rectangular coordinates of the forms

$$x = \frac{\nabla X + x_i}{\sqrt{\Omega L}},$$
$$y = \frac{\nabla Y + y_i}{\sqrt{\Omega L}},$$
$$z = \frac{\nabla Z + z_i}{\sqrt{\Omega L}},$$

and lying within the hyperboloidal cuneus, bounded by the planes $y = 0$, $x = z$, $x = 2y$, and the hyperboloid $xz - y^2 = 1$; points lying on the hyperboloidal boundary are counted as lying within the cuneus; points lying on its plane boundaries are counted as $\frac{1}{2}$ each, and points lying on the intersection of $y = 0$ with $x = z$, and with $x = 2y$ respectively as $\frac{1}{4}$ and $\frac{1}{6}$. Let V be the volume of the cuneus, and let L be increased without limit; we have

$$\lim \frac{\Upsilon_i}{L^{\frac{3}{2}}} = \Omega^{\frac{3}{2}} \times \frac{V}{\nabla^3};$$

and, since this limit is thus ascertained to be the same for all the partial sums $\Upsilon_1, \Upsilon_2, \ldots$,

$$\lim \frac{\Upsilon}{L^{\frac{3}{2}}} = \Omega^{\frac{3}{2}} \times \frac{\phi(\nabla)}{\nabla^3} \times V,$$

or, which is the same thing,

$$\lim \frac{\mathrm{T}}{L^{\frac{3}{2}}} = \frac{1}{2^{\sigma+\sigma'}} \times \Omega^{\frac{3}{2}} \times \frac{\phi(\nabla)}{\nabla^3} \times V.$$

The value of V may be determined by dividing the cuneus into laminae parallel to the plane of xz; if A be the area of a section at a distance y from that plane, we find

$$A = (1 + y^2)\left[\tfrac{1}{2}\log(1 + y^2) - \log 2y\right] - \tfrac{1}{2}(1 - 3y^2);$$

whence $$V = \int_0^{\sqrt{\frac{1}{3}}} A\,dy = \tfrac{1}{9}\pi. \qquad (46)$$

Substituting for $\frac{\phi(\nabla)}{\nabla^3}$ and for V their values given by the equations (45) and (46), we find

$$\lim \frac{T}{L^{\frac{3}{2}}} = \tfrac{1}{24}\pi \frac{\eta}{2^{\sigma'}} \sqrt{\Omega} \times \Pi\left(1-\frac{1}{r}\right) \Pi\left(1-\frac{1}{\delta}\right) \Pi\left(1-\frac{1}{\omega}\right)$$

$$\times \Pi \tfrac{1}{4}\left(1-\frac{1}{r^2}\right) \cdot \Pi \tfrac{1}{2}\left(1-\frac{1}{\omega^2}\right) \Pi \tfrac{1}{2}\left[1+\left(\frac{-\Omega F}{\delta}\right)\frac{1}{\delta}\right], \quad . \quad (47)$$

which is the second determination of the limit of the quotient $\dfrac{T}{L^{\frac{3}{2}}}$.

Finally, equating the two values of this limit, and denoting the coefficient $\dfrac{1}{2^{\sigma'}} \times \dfrac{\eta}{\theta}$ by $\frac{1}{2}\zeta$, we obtain the following determination of the weight of the proposed genus,

$$W = \tfrac{1}{8}\Delta\Omega\,\zeta \times \Pi \tfrac{1}{4}\left(1-\frac{1}{r^2}\right) \Pi \tfrac{1}{2}\left[1+\left(\frac{-\Delta f}{\omega}\right)\frac{1}{\omega}\right] \Pi \tfrac{1}{2}\left[1+\left(\frac{-\Omega F}{\delta}\right)\frac{1}{\delta}\right], \quad . \quad (48)$$

the values of ζ (which are computed from those of σ', η, θ) being as follows :—

A : (f) and (F) properly primitive.

	$\Omega \equiv 1$, mod 2.	$\Omega \equiv 2$, mod 4.	$\Omega \equiv 4$, mod 8.	$\Omega \equiv 0$, mod 8.
$\Delta \equiv 1$, mod 2.	$\frac{1}{3}[2+\Psi]$	$\frac{1}{2}$	$\frac{1}{8}[3+(-1)^{\frac{1}{2}(\Delta f+1)}]$	$\frac{1}{16}[3+(-1)^{\frac{1}{2}(\Delta f+1)}]$
$\Delta \equiv 2$, mod 4.	$\frac{1}{2}$	$\frac{1}{2}$	$\frac{1}{4}$	$\frac{1}{8}$
$\Delta \equiv 4$, mod 8.	$\frac{1}{8}[3+(-1)^{\frac{1}{2}(\Omega F+1)}]$	$\frac{1}{4}$	$\frac{1}{4}$	$\frac{1}{8}$
$\Delta \equiv 0$, mod 8.	$\frac{1}{16}[3+(-1)^{\frac{1}{2}(\Omega F+1)}]$	$\frac{1}{8}$	$\frac{1}{8}$	$\frac{1}{16}$

B: (f) improperly, (F) properly primitive.

$\Omega \equiv 1$, mod 2 ; $\Omega F \equiv 3$, mod 4.

$\Delta \equiv 2$, mod 4.	$\frac{1}{3}$
$\Delta \equiv 0$, mod 4.	$\frac{1}{12}[2+(-1)^{\frac{1}{8}(\Omega^2 F^2-1)}]$

C: (f) properly, (F) improperly primitive.

$\Delta \equiv 1$, mod 2 ; $\Delta f \equiv 3$, mod 4.

$\Omega \equiv 2$, mod 4.	$\frac{1}{3}$
$\Omega \equiv 0$, mod 4.	$\frac{1}{12}[2+(-1)^{\frac{1}{8}(\Delta^2 f^2-1)}]$

The last of these tables is obtained by reciprocation from the second.

The result in the case $\Omega \equiv \Delta \equiv 1$, mod 2, is given in the memoir of Eisenstein (Crelle, vol. xxxv. p. 128).

21. The equation (47) may also be deduced from the theorem of Art. 15 by another method. We consider first and principally the case in which the forms (f) and (F) are both properly primitive.

From Art. 16 we obtain the equation

$$T = \frac{1}{2^{\sigma+\sigma'}} \sum_{M \geqq 1}^{M \leqq L} \Sigma h \left(\frac{\Omega M}{\lambda^2}\right),$$

the second sign of summation extending to every square divisor of M. Inverting the order of the summations, and designating by m any number prime to $2\Omega\Delta$, we may write this equation in the form

$$T = \frac{1}{2^{\sigma+\sigma'}} \sum_{m=1}^{m \leqq \sqrt{L}} \sum_{M \geqq 1}^{M \leqq \frac{L}{m^2}} h(\Omega M).$$

But, by a theorem of Lejeune Dirichlet,

$$h(\Omega M) = \frac{1}{\pi} \sqrt{\Omega M} \,.\, \Sigma \left(\frac{-\Omega M}{n}\right) \frac{1}{n},$$

the sign of summation extending to all uneven numbers prime to ΩM. The limit of $\frac{T}{L^{\frac{3}{2}}}$ is therefore the limit of the expression

$$\frac{1}{2^{\sigma+\sigma'}} \frac{\sqrt{\Omega}}{\pi} \frac{1}{L^{\frac{3}{2}}} \sum_{m=1}^{m \leqq \sqrt{L}} \sum_{M \geqq 1}^{M \leqq \frac{L}{m^2}} \sqrt{M} . \Sigma \left(\frac{-\Omega M}{n}\right) \frac{1}{n},$$

or, leaving the summation with respect to n to be effected last, of the expression

$$\frac{1}{2^{\sigma+\sigma'}} \frac{\sqrt{\Omega}}{\pi} \sum_{n=1}^{n=\infty} \frac{1}{n} \sum_{m=1}^{m \leqq \sqrt{L}} \frac{1}{L^{\frac{3}{2}}} \sum_{M \geqq 1}^{M \leqq \frac{L}{m^2}} \left(\frac{-\Omega M}{n}\right) \sqrt{M}. \quad \ldots \quad (49)$$

In this expression n is uneven and prime to Ω; but n is not necessarily prime to Δ. Let $n = n_1^2 n_2$, n_1^2 denoting the greatest square dividing n, so that n_2 is a product of unequal primes; also let ν represent any prime dividing n, other than one of the primes δ; and let η represent $\frac{1}{4}$, $\frac{1}{2}$, or 1, according as the numbers M are contained in one, two, or all four of the linear forms $8k+1$, $8k+3$, $8k+5$, $8k+7$; so that η has the same value as in Art. 19. The limit of the sum

$$\frac{1}{L^{\frac{3}{2}}} \sum_{M \geqq 1}^{M \leqq \frac{L}{m^2}} \left(\frac{-\Omega M}{n}\right) \sqrt{M} \quad \ldots \ldots \ldots \quad (50)$$

is zero, or

$$\tfrac{1}{3}\eta\left(\frac{-\Omega F}{n_2}\right)\Pi\left(1-\frac{1}{\omega}\right)\Pi\tfrac{1}{2}\left(1-\frac{1}{r}\right)\Pi\tfrac{1}{2}\left(1-\frac{1}{\delta}\right)\Pi\left(1-\frac{1}{\nu}\right)\frac{1}{m^3}, \quad . \quad . \quad . \qquad (51)$$

according as n_2 does or does not contain any primes other than the primes δ. For, in the sum (50), it is only necessary to consider those numbers M which are prime to n, because $\left(\frac{-\Omega M}{n}\right)=0$ when M is not prime to n; and if

$$\nabla = 8\,\Pi\omega\;\Pi r\;\Pi\delta\;\Pi\nu,$$

$$\chi(\nabla)=\nabla\times\tfrac{1}{2}\eta\;\Pi\left(1-\frac{1}{\omega}\right)\Pi\tfrac{1}{2}\left(1-\frac{1}{r}\right)\Pi\tfrac{1}{2}\left(1-\frac{1}{\delta}\right)\Pi\left(1-\frac{1}{\nu}\right),$$

the sum (50) contains $\chi(\nabla)$ numbers M inferior to ∇; let these be represented by $x_1, x_2, \ldots, x_i$; then all the numbers M, which enter into that sum, are contained in the $\chi(\nabla)$ linear forms $x\nabla + x_i$; and the sum (50) may be decomposed into $\chi(\nabla)$ partial sums, of which the sum

$$\left(\frac{-\Omega x_i}{n_2}\right)\frac{1}{L^{\frac{3}{2}}}\sum_{x=0}^{x\nabla+x_i \leqq \frac{L}{m^2}}\sqrt{x\nabla+x_i}$$

is one. The limit of this sum is

$$\tfrac{2}{3}\,\frac{1}{\nabla}\,\frac{1}{m^3}\left(\frac{-\Omega x_i}{n_2}\right),$$

so that the limit of the sum (50) is

$$\tfrac{2}{3}\,\frac{1}{\nabla}\,\frac{1}{m^3}\,\Sigma_i\left(\frac{-\Omega x_i}{n_2}\right).$$

If n_2 is divisible by any prime other than the primes δ, one half of the symbols $\left(\frac{-\Omega x_i}{n_2}\right)$ are equal to $+1$, and the other half are equal to -1; in this case, therefore, the limit of the sum (50) is zero. But, if n_2 contain no prime other than the primes δ, the symbols $\left(\frac{-\Omega x_i}{n_2}\right)$ are all equal to one another and to $\left(\frac{-\Omega F}{n_2}\right)$; and the limit of the sum (50) is

$$\tfrac{2}{3}\,\frac{\chi(\nabla)}{\nabla}\,\frac{1}{m^3}\left(\frac{-\Omega F}{n_2}\right),$$

in accordance with the formula (51). Substituting in the expression (49) for the sum (50) its limiting value, we find

$$\lim\frac{T}{L^{\frac{3}{2}}}=\frac{1}{2^{\sigma+\sigma'}}\,\frac{\sqrt{\Omega}}{\pi}\,\tfrac{1}{3}\eta\;\Pi\left(1-\frac{1}{\omega}\right)\Pi\tfrac{1}{2}\left(1-\frac{1}{r}\right)\Pi\tfrac{1}{2}\left(1-\frac{1}{\delta}\right)$$

$$\times\sum_{m=1}^{m=\infty}\frac{1}{m^3}\;\Sigma\Sigma\;\frac{\left(\frac{-\Omega F}{n_2}\right)\Pi\left(1-\frac{1}{\nu}\right)}{n_1^2\,n_2}. \qquad (52)$$

In the sum $\Sigma\Sigma \dfrac{\left(\frac{-\Omega F}{n_2}\right)\Pi\left(1-\frac{1}{\nu}\right)}{n_1^2 n_2}$ the summations extend to all values of n_2 composed of unequal primes δ, and to all values of n_1 prime to 2Ω; ν is any prime divisor of n_1, other than one of the primes δ. Thus the two summations are independent, and

$$\Sigma\Sigma\left(\frac{-\Omega F}{n_2}\right)\frac{\Pi\left(1-\frac{1}{\nu}\right)}{n_1^2 n_2} = \Sigma\left(\frac{-\Omega F}{n_2}\right)\frac{1}{n_2}\times\Sigma\frac{\Pi\left(1-\frac{1}{\nu}\right)}{n_1^2}.$$

But

$$\Sigma\left(\frac{-\Omega F}{n_2}\right)\frac{1}{n_2} = \Pi\left[1+\left(\frac{-\Omega F}{\delta}\right)\frac{1}{\delta}\right]$$

and

$$\Sigma\frac{\Pi\left(1-\frac{1}{\nu}\right)}{n_1^2} = \Pi\frac{1}{1-\frac{1}{\delta^2}}\,\Pi\left[1+\frac{1-\frac{1}{\nu}}{\nu^2}+\frac{1-\frac{1}{\nu}}{\nu^4}+\frac{1-\frac{1}{\nu}}{\nu^6}+\ldots\right] = \Pi\frac{1}{1-\frac{1}{\delta^2}}\,\Pi\frac{1+\frac{1}{\nu}+\frac{1}{\nu^2}}{1+\frac{1}{\nu}},$$

the last sign of multiplication extending to all primes ν which do not divide $2\Omega\Delta$. Also

$$\sum_{m=1}^{m=\infty}\frac{1}{m^3} = \Pi\frac{1}{1-\frac{1}{\nu^3}} = \Pi\frac{1}{1-\frac{1}{\nu}}\,\Pi\frac{1}{1+\frac{1}{\nu}+\frac{1}{\nu^2}},$$

so that the product

$$\Sigma\frac{1}{m^3}\,\Sigma\Sigma\left(\frac{-\Omega F}{n_2}\right)\frac{\Pi\left(1-\frac{1}{\nu}\right)}{n_1^2 n_2}$$

is equal to
$$\Pi\left[1+\left(\frac{-\Omega F}{\delta}\right)\frac{1}{\delta}\right]\Pi\frac{1}{1-\frac{1}{\delta^2}}\,\Pi\frac{1}{1-\frac{1}{\nu^2}},$$

or to
$$\tfrac{1}{8}\pi^2\,\Pi\left[1-\frac{1}{\omega^2}\right]\Pi\left[1-\frac{1}{r^2}\right]\Pi\left[1+\left(\frac{-\Omega F}{\delta}\right)\frac{1}{\delta}\right], \quad . \quad . \quad . \quad . \quad (53)$$

because
$$\Pi\frac{1}{1-\frac{1}{\omega^2}}\,\Pi\frac{1}{1-\frac{1}{r^2}}\,\Pi\frac{1}{1-\frac{1}{\delta^2}}\,\Pi\frac{1}{1-\frac{1}{\nu^2}}$$

is equal to the sum of the squares of the reciprocals of the uneven numbers, that is to $\frac{1}{8}\pi^2$. Substituting for the sum in equation (52) its equivalent (53), we obtain the formula (47).

If the forms (f) are improperly primitive, we have to employ the equation

$$h'(\Omega M) = \tfrac{1}{3}\left[2 + (-1)^{\frac{1}{8}(\Omega^2-1)+\frac{1}{8}(M^2-1)}\right] \frac{\sqrt{\Omega M}}{\pi} \Sigma \left(\frac{-\Omega M}{n}\right)\frac{1}{n};$$

and the proof is the same as in the former case. Only, if $\Delta \equiv 2$, mod 4, it is convenient on account of the factor $2 + (-1)^{\frac{1}{8}(\Omega^2-1)+\frac{1}{8}(M^2-1)}$, separately to determine the limit $T \div L^{\frac{3}{2}}$ for the numbers M which satisfy the congruences $M \equiv 3\,\Omega$, $M \equiv 7\,\Omega$, mod 8; and then to add the results.

22. The weight of an order (Art. 13) is the sum of the weights of the genera contained in the order. The determination of this sum may in every case be effected by means of the formulae

$$R = \Sigma \left\{ \Pi \tfrac{1}{4}\left[1 - \frac{1}{r^2}\right] \Pi \tfrac{1}{2}\left[1 + \left(\frac{-\Delta f}{\omega}\right)\frac{1}{\omega}\right] \Pi \tfrac{1}{2}\left[1 + \left(\frac{-\Omega F}{\delta}\right)\frac{1}{\delta}\right]\right\}$$
$$= \Pi\left(1 - \frac{1}{r^2}\right),$$

$$R' = \Sigma \left\{ \left(\frac{f}{\Omega_1}\right)\left(\frac{F}{\Delta_1}\right) \Pi \tfrac{1}{4}\left[1 - \frac{1}{r^2}\right] \Pi \left[1 + \left(\frac{-\Delta f}{\omega}\right)\frac{1}{\omega}\right] \Pi \tfrac{1}{2}\left[1 + \left(\frac{-\Omega F}{\delta}\right)\frac{1}{\delta}\right]\right\}$$
$$= 0, \text{ or } = -\frac{(-1)^{\frac{1}{2}(\Omega_1+1)\cdot\frac{1}{2}(\Delta_1+1)}}{\Omega_1 \Delta_1} \alpha^{\frac{1}{8}(\Delta_1^2-1)} \beta^{\frac{1}{8}(\Omega_1^2-1)} \Pi\left(1 - \frac{1}{r^2}\right),$$

according as $\Omega_1 \Delta_1$ is or is not divisible by any of the primes r; *i.e.* according as $\Omega_1 \Delta_1$ is not or is prime to the greatest common divisor of Ω and Δ. In the expressions for R and R' the signs of summation extend to every combination of the equations

$$\left(\frac{f}{r}\right) = +1 \text{ or } = -1, \quad \left(\frac{F}{r}\right) = +1 \text{ or } = -1,$$
$$\left(\frac{f}{\omega}\right) = +1 \text{ or } = -1, \quad \left(\frac{F}{\delta}\right) = +1 \text{ or } = -1;$$

i.e. the value of the continued product is to be determined on each of these suppositions, and the sum of these values is to be taken. From this definition, it is evident that in the sum R we may substitute for any factor of the form

$$\tfrac{1}{2}\left[1 + \left(\frac{-\Delta f}{\omega}\right)\frac{1}{\omega}\right],$$

or

$$\tfrac{1}{2}\left[1 + \left(\frac{-\Omega F}{\delta}\right)\frac{1}{\delta}\right],$$

a factor of the form

$$\tfrac{1}{2}\left\{\left[1+\left(\frac{-\Delta}{\omega}\right)\frac{1}{\omega}\right]+\left[1-\left(\frac{-\Delta}{\omega}\right)\frac{1}{\omega}\right]\right\}, \quad \dots \dots \quad (54)$$

or

$$\tfrac{1}{2}\left\{\left[1+\left(\frac{-\Omega}{\delta}\right)\frac{1}{\delta}\right]+\left[1-\left(\frac{-\Omega}{\delta}\right)\frac{1}{\delta}\right]\right\}. \quad \dots \dots \quad (55)$$

outside the sign of summation. And, similarly, for any factor $\tfrac{1}{4}\left(1-\frac{1}{r^2}\right)$ we may substitute the factor

$$1-\frac{1}{r^2}$$

outside the sign of summation. Observing that the factors (54) and (55) are all positive units, we obtain immediately

$$R = \Pi\left(1-\frac{1}{r^2}\right).$$

Again, if a prime r divide Ω_1 or Δ_1, the sum R' vanishes, being composed of pairs of terms equal in absolute magnitude and opposite in sign; if, for example, r divide Ω_1, the two terms, in one of which $\left(\frac{f}{r}\right)$ contained in $\left(\frac{f}{\Omega_1}\right)$, is $+1$ and in the other -1, but which are in other respects identical, will destroy one another. But, if none of the primes r divide Ω_1 or Δ_1, we replace those factors of the general term of R', which contain primes not dividing Ω_1 and Δ_1, by factors placed outside the sign of summation; we thus find

$$R' = \Pi\left(1-\frac{1}{r^2}\right)\Sigma\left\{\Pi\,\tfrac{1}{2}\left[\left(\frac{f}{\omega_1}\right)+\left(\frac{-\Delta}{\omega_1}\right)\frac{1}{\omega_1}\right]\Pi\,\tfrac{1}{2}\left[\left(\frac{F}{\delta_1}\right)+\left(\frac{-\Omega}{\delta_1}\right)\frac{1}{\delta_1}\right]\right\},$$

where only primes ω_1 which divide Ω_1 and primes δ_1 which divide Δ_1 occur after the sign of summation. We then substitute for each factor containing ω_1 or δ_1 a factor of the form

$$\tfrac{1}{2}\left\{\left[1+\left(\frac{-\Delta}{\omega_1}\right)\frac{1}{\omega_1}\right]+\left[-1+\left(\frac{-\Delta}{\omega_1}\right)\frac{1}{\omega_1}\right]\right\}=\left(\frac{-\Delta}{\omega_1}\right)\frac{1}{\omega_1},$$

or

$$\tfrac{1}{2}\left\{\left[1+\left(\frac{-\Omega}{\delta_1}\right)\frac{1}{\delta_1}\right]+\left[-1+\left(\frac{-\Omega}{\delta_1}\right)\frac{1}{\delta_1}\right]\right\}=\left(\frac{-\Omega}{\delta_1}\right)\frac{1}{\delta_1},$$

outside the sign of summation; and, observing that by the law of reciprocity

$$\left(\frac{-\Delta}{\Omega_1}\right)\left(\frac{-\Omega}{\Delta_1}\right) = -(-1)^{\frac{1}{2}(\Omega_1+1)\frac{1}{2}(\Delta_1+1)}\,\alpha^{\frac{1}{8}(\Delta_1^2-1)}\,\beta^{\frac{1}{8}(\Omega_1^2-1)},$$

we find

$$R' = -\frac{(-1)^{\frac{1}{2}(\Omega_1+1)\cdot\frac{1}{2}(\Delta_1+1)}}{\Omega_1\Delta_1}\,\alpha^{\frac{1}{8}(\Delta_1^2-1)}\,\beta^{\frac{1}{8}(\Omega_1^2-1)}\,\Pi\left(1-\frac{1}{r^2}\right).$$

As an example of the application of these formulae, let us consider the properly primitive order in the case in which $\Delta\equiv 1$, mod 2, $\Omega\equiv 4$, mod 8. We may determine separately the weights of those genera for which $(-1)^{\frac{1}{2}(\Delta f+1)} = -1$ and of those for which $(-1)^{\frac{1}{2}(\Delta f+1)} = +1$. In a genus of the former kind the characters

$$\left(\frac{f}{r}\right), \left(\frac{F}{r}\right), \left(\frac{f}{\omega}\right), \left(\frac{F}{\delta}\right) \quad . \quad . \quad . \quad . \quad . \quad . \quad . \quad . \quad (56)$$

may have any assigned values because the condition of possibility is

$$(-1)^{\frac{1}{2}(\Omega_1 F+1)} \left(\frac{f}{\Omega_1}\right) \left(\frac{F}{\Delta_1}\right) = (-1)^{\frac{1}{2}(\Omega_1+1)\cdot\frac{1}{2}(\Delta_1+1)}.$$

Therefore the sum of the weights of these genera is $\frac{1}{8}\Omega\Delta\,\zeta\,R$, or $\frac{1}{32}\Omega\Delta\,R$, because $\zeta = \frac{1}{4}$. But, in a genus of the latter kind, the characters (56), or some of them, are subject to the conditions

$$\left(\frac{f}{\Delta_1}\right) \left(\frac{F}{\Delta_1}\right) = (-1)^{\frac{1}{2}(\Omega_1+1)\cdot\frac{1}{2}(\Delta_1+1)}; \quad . \quad . \quad . \quad . \quad . \quad . \quad . \quad . \quad (57)$$

we have therefore to consider a sum of which the general term is the same as that of R, but into which only those terms are admitted which are formed with values of $\left(\frac{f}{\omega_1}\right)$ and $\left(\frac{F}{\delta_1}\right)$ satisfying the condition (57). This sum is expressed by the formula

$$\tfrac{1}{2}\left[R + (-1)^{\frac{1}{2}(\Omega_1+1)\cdot\frac{1}{2}(\Delta_1+1)} R'\right];$$

so that the sum of the weights of the genera of the latter kind is

$$\tfrac{1}{16}\Omega\Delta \times \tfrac{1}{2}\left[R + (-1)^{\frac{1}{2}(\Omega_1+1)\cdot\frac{1}{2}(\Delta_1+1)} R'\right].$$

Adding the two sums together, and substituting for R and R' their values, we find, for the weight of the proposed order, the expression

$$\tfrac{1}{16}\Omega\Delta\,\Pi\left(1-\frac{1}{r^2}\right), \quad \text{or} \quad \tfrac{1}{32}\Omega\Delta\left(2-\frac{1}{\Omega_1\Delta_1}\right)\Pi\left(1-\frac{1}{r^2}\right),$$

according as $\Omega_1\Delta_1$ is not or is prime to the greatest common divisor of Ω and Δ.

If, in general, we represent the weight of any proposed order of the invariants $[\Omega, \Delta]$ by the expression

$$\tfrac{1}{8}\Omega\Delta\,Z\,\Pi\left(1-\frac{1}{r^2}\right),$$

the following table (with which we shall conclude this memoir) will assign the value of the coefficient Z, and will thus serve to ascertain the weight of the order *. The determinations contained in it have been obtained by the method

---

* For the case of uneven invariants, the result has been given by Eisenstein (Crelle, vol. xxxv. p. 123); there is, however, a slight discrepancy. According to Eisenstein, λ is not zero when the greatest common divisor of Δ and Ω is a square; according to the definition in the text, λ is always zero, except when the exponent of every uneven prime common to Δ and Ω is even both in Δ and Ω.

just described; λ is $\frac{1}{\Omega_1 \Delta_1}$ or 0, according as $\Omega_1 \Delta_1$ is or is not prime to the greatest common divisor of Ω and Δ; I_1, I_2 are the exponents of the highest powers of 2 dividing Ω and Δ respectively.

A: (f) and (F) properly primitive.

	$I_1 = 0$.	I_1 even.	I_1 uneven.
$I_2 = 0$.	$\frac{1}{3}(2-\lambda)$	$\frac{1}{4}(2-\lambda)$	$\frac{1}{2}$
I_2 even.	$\frac{1}{4}(2-\lambda)$	$\frac{1}{4}(2-\lambda)$	$\frac{1}{2}$
I_2 uneven.	$\frac{1}{2}$	$\frac{1}{2}$	$\frac{1}{2}$

B: (f) improperly, (F) properly primitive.

	$I_1 = 0,\ I_2 > 0$.
I_2 even.	$\frac{1}{12}(2-\lambda)$
I_2 uneven.	$\frac{1}{6}(1-\lambda)$

C: (f) properly, (F) improperly primitive.

	$I_2 = 0,\ I_1 > 0$.
I_1 even.	$\frac{1}{12}(2-\lambda)$
I_1 uneven.	$\frac{1}{6}(1-\lambda)$

For the invariants (p^2, p^3) the weight assigned by the formula of Eisenstein is $\frac{1}{24}p^5\left(2-\frac{1}{p}\right)\left(1-\frac{1}{p^2}\right)$, p denoting an uneven prime; a result which can hardly be right, because the weight of each genus separately is congruous to 0, mod p^3.

[The following abstract of the preceding paper was published in the Proceedings of the Royal Society, vol. xv. pp. 387–389.]

The object of this paper is to supply demonstrations of the undemonstrated results, relating to ternary quadratic forms, which are contained in an important memoir of Eisenstein's ('Neue Theoreme der höheren Arithmetik,' Crelle's Journal, vol. xxxv. p. 117),—and, at the same time, to extend those results to the cases not considered by him in that memoir. The following are the principal points in which the theory of Eisenstein has been thus further developed:—

1. In Eisenstein's memoir forms of an uneven discriminant only are considered. Such forms, and their contravariants, are always properly primitive; they have particular generic characters with respect to uneven primes dividing the discriminant, but have no supplementary characters (*i.e.* characters with respect to 4 or 8). The case of forms of an even discriminant is more complicated. Besides the properly primitive order, there may exist, in this case, an improperly primitive order in which the forms themselves are improperly primitive and their contravariants properly primitive,—or, again, an improperly primitive order in which the forms themselves are properly primitive and their contravariants improperly primitive. Further, forms of an even discriminant may have characters with respect to 4 or 8; and a complete enumeration of these supplementary characters requires a careful distinction of cases. To facilitate this enumeration, a table is given in the paper for finding the supplementary characters of any proposed form.

2. A table is also given for forming the complete generic character of any proposed form. This table is intended to serve the same purposes, in the theory of ternary quadratic forms, for which the table of Lejeune Dirichlet is available in the binary theory (Crelle, vol. xix. p. 338). The table, like that of Lejeune Dirichlet, distinguishes between the possible and impossible generic characters; and the paper contains a complete demonstration of the criterion by which they are distinguished.

3. Besides the particular characters relating to uneven primes dividing the discriminant, it is convenient, in those cases in which there is no supplementary character, to consider a certain particular generic character which does not appear to have been regarded as such by Eisenstein. This character is termed in the paper the simultaneous character of the form and its contravariant: its existence is demonstrated; and its introduction as an element of the complete generic character is justified by its use in the distinction of possible and impossible genera.

4. It has been proposed to define a genus of forms as consisting of all those forms which can be transformed into one another by substitutions of which the coefficients are rational and the determinant a unit. It is desirable (in the case of quadratic forms) to add to this definition the limitation that the denominators of the fractional coefficients are to be uneven and prime to the discriminant. And it is shown, in this paper, that two ternary quadratic forms are or are not transformable into one another by such substitutions, according as their complete generic characters do or do not coincide.

5. The preceding observations apply equally to the cases of definite and indefinite forms. These two cases are included in the same analysis by means of a convention as to the signs of the two numbers defined by Eisenstein, and termed in this paper the arithmetical invariants of the ternary form. The first invariant of a form is the greatest common divisor of the first minors of the matrix of the form; the second invariant is the quotient obtained by dividing the discriminant by the square of the first invariant. According to the convention adopted in the paper, the second invariant has the same sign as the discriminant; and the first invariant has or has not the same sign as the second, according as the form is definite or indefinite.

6. The latter part of the paper is occupied exclusively with the theory of definite and positive forms. In the case of these forms, the weight (or, as Eisenstein has termed it, the density) of a class is the reciprocal of the number of automorphics (of determinant $+1$) of any form of the class; the weight of a representation of a number by a form is the weight of the form, *i.e.* the weight of the class containing the form; the weight of a genus or order is the sum of the weights of the classes comprised in the genus or order. In his memoir, Eisenstein has given (but without demonstration) the formulae which assign the weight of a given genus or order of forms of an uneven discriminant. These formulae are demonstrated in the present paper, and, with them, the corresponding formulae relating to the cases in which the discriminant is even. The demonstration is obtained by a method similar to that employed by Gauss and Dirichlet for the determination of the number of binary classes of a given determinant. The sum of the weights of the representations, by a system of forms representing the classes of any proposed genus, of all the numbers contained in certain arithmetical progressions, and not surpassing a given number, is in a finite ratio to the sesquiplicate power of the given number when that number is supposed to increase without limit. Of this limiting ratio, two distinct determinations are obtained; of which the first contains, as a factor, the weight of the

proposed genus; and an expression for that weight is obtained by a comparison of the two determinations. Of these determinations, the first is obtained immediately by an elementary application of the integral calculus; the second depends on an arithmetical theorem, which is deduced in the paper from the analysis employed by Gauss in Arts. 279–284 of the 'Disquisitiones Arithmeticae,' and which may be expressed as follows:—

'The sum of the weights of the representations of a given number (contained in one of certain Arithmetical Progressions) by a system of forms representing the classes of a ternary genus is equal to the weight of a genus of binary forms of which the determinant is the product, taken negatively, of the given number by the second invariant of the ternary forms.'

By this proposition the determination of the limiting ratio is made to depend on an approximate determination of the weight (or, which is here the same thing, the number) of the binary classes of certain series of negative determinants. Two methods are given in the paper for effecting this approximate determination. The first method presupposes Lagrange's definition of a reduced form, and depends ultimately on the evaluation of the definite integral

$$\iiint (xz - y^2)\, dx\, dy\, dz = \frac{\pi}{g},$$

of which the limits are given by the inequalities

$$x \geqq 0, \quad y \geqq 0, \quad z \geqq 0, \quad x \leqq z, \quad 2y \leqq x.$$

The second method employs the expression obtained by Lejeune Dirichlet in the form of an infinite series for the number of binary classes of a given determinant, and is thus independent of the definition of a reduced form. The same result is obtained by both methods; but the second is more easily extended to the case of quadratic forms containing more than three indeterminates.

XVIII.

ON THE ORDERS AND GENERA OF QUADRATIC FORMS CONTAINING MORE THAN THREE INDETERMINATES.

[Proceedings of the Royal Society, vol. xvi. pp. 197–208. Received October 30; Read December 5, 1867.]

THE principles upon which quadratic forms are distributed into orders and genera have been indicated in a former notice (Proceedings of the Royal Society, vol. xiii. p. 199) *. Some further results relating to the same subject are contained in the present communication.

I. *The Definition of the Orders and Genera.*

Retaining, with some exceptions to which we shall now direct attention, the notation and nomenclature of the former notice, we represent by f_1 a primitive quadratic form containing n indeterminates, of which the matrix is $\left\| \begin{matrix} n \times n \\ A_{i,j} \end{matrix} \right\|$; by $f_2, f_3, \ldots, f_{n-1}$, the fundamental concomitants of f_1, of which the last is the contravariant. The matrices of these concomitants are the matrices derived from the matrix of f_1, so that the first coefficients of $f_2, f_3, \ldots, f_{n-1}$, are respectively the determinants $\left| \begin{matrix} 2 \times 2 \\ A_{i,j} \end{matrix} \right|$, $\left| \begin{matrix} 3 \times 3 \\ A_{i,j} \end{matrix} \right|$, $\ldots$, $\left| \begin{matrix} n-1 \times n-1 \\ A_{i,j} \end{matrix} \right|$, taken with their proper signs. The discriminant of f_1, *i.e.* the determinant of the matrix $\left| \begin{matrix} n \times n \\ A_{i,j} \end{matrix} \right|$, which is supposed to be different from zero, and which is to be taken with its proper sign, is represented by ∇_n. The greatest common divisors of the minors of the orders $n-1, n-2, \ldots, 2, 1$ in the same matrix are denoted by $\nabla_{n-1}, \nabla_{n-2}, \ldots,$

* No. XIV, p. 412 of this volume.

∇_2, ∇_1, of which the last is a unit; we shall presently attribute signs to each of these greatest common divisors. The quotients

$$\frac{\nabla_n}{\nabla_{n-1}} \div \frac{\nabla_{n-1}}{\nabla_{n-2}}, \quad \frac{\nabla_{n-1}}{\nabla_{n-2}} \div \frac{\nabla_{n-2}}{\nabla_{n-3}}, \quad \ldots, \quad \frac{\nabla_2}{\nabla_1} \div \frac{\nabla_1}{1},$$

which are always integral, we represent by $I_{n-1}, I_{n-2}, \ldots, I_1$; so that

$$I_k = \frac{\nabla_{k+1}}{\nabla_k} \div \frac{\nabla_k}{\nabla_{k-1}}.$$

The numbers $I_1, I_2, \ldots, I_{n-1}$ are the first, second, ..., last invariants of the form f_1, and remain unchanged when f_1 is transformed by any substitution of which the determinant is unity and the coefficients integral numbers. Forms which have the same invariants have of course the same discriminant; but (if the number of indeterminates is greater than two) forms which have the same discriminant do not necessarily have the same invariants; for example, the quaternary forms

$$x_1^2 + x_2^2 + 2x_3^2 + 6x_4^2, \quad x_1^2 + x_2^2 + x_3^2 + 12x_4^2$$

have the same discriminant 12, but their invariants I_1, I_2, I_3 are respectively 1, 2, 3, and 1, 1, 12. As forms which have the same discriminant but different invariants do not necessarily have any close relation to one another, we shall not employ the discriminant in the classification of quadratic forms; but we shall regard the infinite number of forms which have the same invariants as corresponding, in the general theory, to the infinite number of forms which have the same determinant, in the theory of binary quadratic forms.

If the index of inertia of the form f is k, *i.e.* if f_1 can be transformed by a substitution of which the coefficients are real into a sum of k positive and $n-k$ negative squares, we attribute to the invariant I_k the sign $-$, and to every other invariant the sign $+$. Thus the numbers $\nabla_1, \nabla_2, \ldots, \nabla_k$ are all positive; ∇_{k+1}, $\nabla_{k+2}, \ldots, \nabla_n$ are alternately negative and positive, so that the discriminant ∇_n is of the same sign as $(-1)^{n-k}$, as it ought to be. This convention with respect to the signs of the invariants will enable us to comprehend in the same formulae the theory of the generic characters of forms of any index of inertia. We shall, however, suppose that the index of inertia is at least 1, *i.e.* we shall exclude negative definite forms. The invariants of a positive definite form are all positive; and the index of inertia of any indefinite form, of which the invariants are given, is always indicated by the ordinal index of its negative invariant. We shall represent by D the product $-I_1 \times -I_3 \times -I_5 \times \ldots$, the last factor being $-I_{n-1}$ or $-I_{n-2}$, according as n is even or uneven.

If $\theta_i = \frac{1}{\nabla_1} f_i$, the forms $\theta_1, \theta_2, \theta_3, \ldots, \theta_{n-1}$ are the primitive concomitants.

and the last the primitive contravariant, of f_1 or θ_1; each one of them is either uneven, *i.e.* properly primitive, or even, *i.e.* improperly primitive. Two forms which have the same invariants are said to belong to the same order when the corresponding primitive concomitants of the two forms are alike uneven or alike even. When the invariants are all uneven, and the number of the indeterminates is also uneven, there is but one order, none of the primitive concomitants being in this case even. Again, when the invariants are all uneven, and the number n of the indeterminates is even, there is either one order or two, according as $D \equiv -1$ or $\equiv +1$, mod 4; for, in both cases, there is an order in which all the primitive concomitants are uneven, and in the latter case, besides this uneven order, there is an even order in which these forms are alternately even and uneven, the two extreme forms θ_1 and θ_{n-1} being even. In the general case, when the invariants have any values even or uneven, if I_i is even, θ_i cannot be even; again, if I_i is one of a sequence of an even number of uneven invariants, preceded and followed by even invariants, θ_i cannot be even. But, if there be a sequence of an uneven number of uneven invariants $I_i, I_{i+1}, \ldots, I_{i+2j}$ preceded and followed by even invariants, the sequence of primitive concomitants $\theta_i, \theta_{i+1}, \ldots, \theta_{i+2j}$ are all uneven if θ_i is uneven, and are alternately even and uneven if θ_i is even; a *sequence* of forms or invariants may consist of a single form or invariant. We attribute the value 0 to the symbols I_0 and I_n, the value 1 to the symbols θ_0 and θ_n; thus the invariant I_1 is always to be regarded as preceded by an even invariant, and I_{n-1} as followed by an even invariant; similarly the forms θ_1 and θ_{n-1} are to be regarded as respectively preceded and followed by uneven forms. Two even forms cannot be consecutive in the series $\theta_1, \ldots, \theta_{n-1}$.

The preceding observations enable us to assign all the orders which may exist for any given invariants; if the series of invariants $I_0, I_1, \ldots, I_{n-1}, I_n$ present ω different sequences each consisting of an uneven number of uneven invariants, preceded and followed by even invariants, there are 2^ω assignable orders. These orders, in general, all exist; there are, however, the following exceptions to this statement:—

(1) If, the number of indeterminates being even and equal to 2ν, D is uneven, there is an assignable order in which the concomitants $\theta_1, \theta_2, \ldots, \theta_{n-1}$ are alternately even and uneven. But, as has been already said, this order does not exist if $D \equiv -1$, mod 4; and, if the invariants are all squares, it does not exist, even if $D \equiv 1$, mod 4, unless the equation

$$(-1)^{\frac{1}{2}\nu(\nu-1)} = (-1)^{\frac{1}{2}k(k-1)},$$

in which k is the index of inertia, is also satisfied.

(2) If, the number of indeterminates being uneven and equal to $2\nu+1$, D is uneven and $I_{2\nu}$ even, there is again an assignable order in which the concomitants $\theta_1, \ldots, \theta_{2\nu}$ are alternately even and uneven. But, when $I_{2\nu}$ is the double of a square and the other invariants are squares, this order does not exist unless the equation $(-1)^{\frac{1}{8}(n^2-1)} = (-1)^{\frac{1}{2}k(n-k)}$, in which k is still the index of inertia, is satisfied.

The reciprocal case (that obtained by changing I_s and θ_s into I_{n-s}, and θ_{n-s} for every value of s from 0 to n) presents a similar exception which it is not necessary to enunciate separately.

The generic characters of the form θ_1, or more properly of the system of concomitant forms $\theta_1, \theta_2, \ldots, \theta_{n-1}$, so far as they depend on uneven primes dividing the invariants, have been already defined in the former notice, and the definition need not be repeated here. These characters we shall term the *principal* generic characters of the system. When the invariants and primitive concomitants are all uneven, the principal characters are the only generic characters, with the exception of a certain character which we shall define hereafter and of which the value is not independent of the principal characters. In other cases, the forms of the concomitant system may acquire generic characters with respect to 4 or 8: these we shall term *supplementary*. What supplementary characters exist in any given case may always be ascertained by applying the following rules. In their enunciation we represent by I_i' the greatest uneven divisor of I_i taken with the same sign as I_i, by μ_i the exponent of the highest power of 2 contained in I_i, increased by 1 if one of the two forms $\theta_{i-1}, \theta_{i+1}$ is even, and by 2 if both those forms are even; we suppose $0 < i < n$.

I. If $\mu_i \geqq 2$, θ_i has the character $(-1)^{\frac{1}{2}(\theta_i-1)}$.

II. If $\mu_i \geqq 3$, θ_i, in addition to the character $(-1)^{\frac{1}{2}(\theta_i-1)}$, has also the character $(-1)^{\frac{1}{8}(\theta_i^2-1)}$.

III. If $\mu_i = 1$, and also $\mu_{i-1} \geqq 2$, $\mu_{i+1} \geqq 2$, θ_i (which, as well as θ_{i-1} and θ_{i+1}, is necessarily uneven) has the character $(-1)^{\frac{1}{8}(\theta_i^2-1)}$, or $(-1)^{\frac{1}{2}(\theta_i-1)+\frac{1}{8}(\theta_i^2-1)}$ according as $(-1)^{\frac{1}{2}(\theta_{i-1}-1)+\frac{1}{2}(\theta_{i+1}-1)} = (-1)^{\frac{1}{2}(I_i'+1)}$, or $= (-1)^{\frac{1}{2}(I_i'-1)}$.

It will be observed that, by I., the forms θ_{i-1} and θ_{i+1} have the characters $(-1)^{\frac{1}{2}(\theta_{i-1}-1)}$ and $(-1)^{\frac{1}{2}(\theta_{i+1}-1)}$.

IV. If $\mu_i = 0$ and also $\mu_{i-1} \geqq 2$, $\mu_{i+1} \geqq 2$, θ_i, if uneven, has the character $(-1)^{\frac{1}{2}(\theta_i-1)}$, or no character at all, according as $(-1)^{\frac{1}{2}(\theta_{i-1}-1)+\frac{1}{2}(\theta_{i+1}-1)} = (-1)^{\frac{1}{2}(I_i-1)}$ or $= (-1)^{\frac{1}{2}(I_i+1)}$.

No even concomitant has any supplementary character. But, if θ_i is an even concomitant, the uneven forms preceding and following it have, by I., the characters $(-1)^{\frac{1}{2}(\theta_{i-1}-1)}$ and $(-1)^{\frac{1}{2}(\theta_{i+1}-1)}$. These characters are not independent but are connected by the equation

$$(-1)^{\frac{1}{2}(\theta_{i-1}-1)+\frac{1}{2}(\theta_{i+1}-1)}=(-1)^{\frac{1}{2}(I_i+1)}.$$

Thus, if $I_i, I_{i+1}, \ldots, I_{i+2j}$ is a sequence of an uneven number of uneven invariants preceded and followed by even invariants, and corresponding to a sequence of alternately even and uneven concomitants $\theta_i, \theta_{i+1}, \ldots, \theta_{i+2j}$, the character, mod 4, of every uneven form of this sequence and of the next following form θ_{i+2j+1} is determined by the character of the form θ_{i-1}. We have, in fact, if $s=1, 2, \ldots, j+1$,

$$(-1)^{\frac{1}{2}(\theta_{i+2s-1}-1)}=(-1)^s\times(-1)^{\frac{1}{2}[I_i\times I_{i+2}\times\ldots\times I_{i+2s-2}-1]}\times(-1)^{\frac{1}{2}(\theta_{i-1}-1)}.$$

Besides these supplementary characters which, no less than the principal characters, are attributable to individual forms of the concomitant system, there exist, or may exist, other characters which we shall term simultaneous, attributable to certain sequences of those forms considered conjointly. Such a character is attributable to every sequence of uneven forms of which none possesses any supplementary character but which are immediately preceded and followed by forms having such characters. The following definition is requisite, in order to explain the nature of these simultaneous characters.

If $\left\| \begin{matrix} n-1\times n \\ a_{i,j} \end{matrix} \right\|$ is a matrix of the type $n-1\times n$, of which the determinants are not all zero, and if m_k represents the value acquired by θ_k, when we attribute to the indeterminates of that form the values of the determinants

$$\left\| \begin{matrix} k\times n \\ a_{i,j} \end{matrix} \right\| \qquad \begin{matrix} i=1, 2, \ldots, k, \\ j=1, 2, \ldots, n, \end{matrix}$$

taken in the same order in which the determinants of any k horizontal rows of the matrix $\left\| \begin{matrix} n\times n \\ A_{i,j} \end{matrix} \right\|$ are taken in forming the matrix of θ_k, then the numbers $m_1, m_2, \ldots, m_{n-1}$ are said to be simultaneously represented by the forms $\theta_1, \theta_2, \ldots, \theta_{n-1}$.

Let $\theta_{i+1}, \ldots, \theta_{i+i'}$ be a sequence of i' uneven concomitants, $\mu_{i+1}, \mu_{i+2}, \ldots, \mu_{i+i'}$ being 0 or 1, but μ_i and $\mu_{i+i'+1}$ being greater than 1; the uneven numbers simultaneously represented by $\theta_{i+1}, \theta_{i+2}, \ldots, \theta_{i+i'}$ are all such as to render the unit

$$(-1)^{\sum_{s=i}^{s=i+i'}\frac{1}{4}(\theta_s-1)(\theta_{s+1}-1)}\times(-1)^{\sum_{s=i+1}^{s=i+i'}\frac{1}{4}(I_s'+1)(\theta_s-1)}\times(-1)^{\sum_{s=i+1}^{s=i+i'}\mu_s\frac{1}{8}(\theta_s^2-1)},$$

which we shall symbolize by $\psi(i, i')$, equal to $+1$, or else are all such as to render that unit equal to -1. We therefore attribute to the sequence of forms $\theta_{i+1}, \ldots, \theta_{i+i'}$, the simultaneous character $\psi(i, i') = +1$ or $\psi(i, i') = -1$, according as the former or latter of those equations is satisfied. If $i' = 1$, the sequence consists of but one form, so that the character $\psi(i, i')$ ceases to be a simultaneous character; in fact, if $\mu_{i+1} = 1$, it coincides with the supplementary character attributable to θ_{i+1} by III.; if $\mu_{i+1} = 0$, it either becomes nugatory (*i.e.* identically equal to $+1$, irrespective of the value of m_{i+1}) or it coincides with the supplementary character of θ_{i+1}, according as that form (by IV.) has not or has a supplementary character.

The complex of all the particular characters (principal, supplementary, and simultaneous) constitutes the complete character of the system of concomitants $\theta_1, \theta_2, \ldots, \theta_{n-1}$. Not every complete generic character, assignable *à priori*, corresponds to actually existing forms, but only such characters as satisfy a certain condition of possibility. This condition is expressed by the equation

$$\psi(0, n-1) \times \prod_{s=1}^{s=n-1} \left(\frac{\theta_s}{I'_s}\right) = +1, \quad . \quad . \quad . \quad . \quad . \quad . \quad . \quad (\mathrm{A})$$

in which, if θ_s is an even form, we understand by the symbol $\left(\frac{\theta_s}{I_s}\right)$ the quadratic character with respect to I_s of the *half* of any number, prime to I_s, which is represented by θ_s. The unit $\psi(0, n-1)$ is formed in the same way as the unit $\psi(i, i')$: we may omit, however, from the exponent of -1 in its expression every term into which an even form enters; if, for example, θ_s is an even form, that exponent contains the terms

$$[\tfrac{1}{2}(\theta_{s-1} - 1) + \tfrac{1}{2}(\theta_{s+1} - 1) + \tfrac{1}{2}(I_s + 1)] \times \tfrac{1}{2}(\theta_s - 1) + \mu_s \tfrac{1}{8}(\theta_s^2 - 1),$$

and no other term into which θ_s enters; but $\mu_s = 0$, and the coefficient of $\frac{1}{2}(\theta_s - 1)$ is even; so that θ_s disappears from the expression of the unit $\psi(0, n-1)$. It will thus be seen that the equation (A) involves only generic characters (principal, supplementary, or simultaneous) of the concomitant system: that equation therefore expresses a relation which the complete character must satisfy.

In using these formulae, we must attend to the significations which we have assigned to the symbols I_0, I_n, θ_0, and θ_n. Thus

$$(-1)^{\frac{1}{2}(\theta_0 - 1)} = 1 = (-1)^{\frac{1}{8}(\theta_0^2 - 1)}, \quad \mu_0 > 3, \text{ etc.}$$

We shall conclude this part of our subject with the two theorems:—

(i) 'Every genus, of which the character satisfies the condition of possibility, actually exists.'

(ii) 'Two forms of the same invariants of the same order and of the same genus are transformable, each into the other, by rational linear substitutions of which the determinants are units and in which the denominators of the coefficients are prime to any given number.'

The first of these theorems shows that the condition of possibility is sufficient as well as necessary; the second establishes the completeness of the enumeration of ordinal and generic characters.

II. *Determination of the Weight of a given Genus of Definite Forms.*

It has been shown by Gauss, in the digression on ternary forms in the fifth section of the 'Disquisitiones Arithmeticae,' that the solution of the problems 'to obtain all the representations of a given binary form, or of a given number, by a given ternary form,' depends on the solution of the problem 'to determine whether two given ternary forms are equivalent, and, if they are, to obtain all the transformations of either of them into the other.' Similarly the solution of the problem 'to obtain all the representations of a given quadratic form of i indeterminates ($i=1, 2, \dots, n-1$) by a given form of n indeterminates' depends on the solution of the problem of equivalence for quadratic forms of n indeterminates. The following proposition is here of primary importance:—

'If the form ϕ_1 of $n-1$ indeterminates and of the invariants $I_1, I_2, \dots, I_{n-3}, MI_{n-2}$ is capable of primitive representation by the form θ_1 of n indeterminates and of the invariants $I_1, I_2, \dots, I_{n-3}, I_{n-2}, I_{n-1}$, then $-I_{n-1}\times\phi_{n-2}$ (where ϕ_{n-2} is the primitive contravariant of ϕ_1) is a quadratic residue of M.'

The converse is true, subject to certain limitations:—

'If M is prime to I_{n-1} and not negative except when I_{n-1} is negative, and if $-I_{n-1}\times\phi_{n-2}$ is a quadratic residue of M, ϕ_1 is capable of primitive representation by f_1.'

'If, in addition, M is prime to I_{n-2}, there is always either one or two genera of forms of the invariants $I_1, I_2, \dots, MI_{n-2}$ capable of primitive representation by forms of a given genus of the invariants $I_1, I_2, \dots, I_{n-2}, I_{n-1}$; and, if there are two genera capable of such representation, they are of different orders.'

These theorems are especially useful in the theory of definite forms, to which, for the remainder of this paper, we shall confine our attention. In the case of such forms we understand by the weight of a form the reciprocal of the number of its positive automorphics, and by the weight of a class the weight of any form representing the class; the weight of a genus or order is the sum of the weights of the classes contained in the genus or order; the weight of a representation of a number by a form is the weight of the representing form; the weight of a representation of a form by a form is the product of the weights of the representing and represented forms.

Let Γ denote a system of forms, representatives of a given genus of the invariants $I_1, I_2, \ldots, I_{n-1}$; let M be a number divisible by μ different uneven primes, none of which divide any of the invariants; and let M be uneven or unevenly even, according as the contravariants of the forms Γ are uneven or even: we then have the theorem:—

'The sum of the weights of the representations of M by the contravariants of the forms Γ, is 2^μ times the weight of the single genus, or the two genera, of invariants $I_1, I_2, \ldots, MI_{n-2}$ which admit of representation by the forms Γ.'

The method which this theorem may serve to indicate supplies a solution of the problem 'to determine the weight of a given genus of definite forms of n indeterminates and of the invariants $I_1, I_2, \ldots, I_{n-1}$.' We shall represent the weight of the given genus by the formula

$$W = \zeta_{2\nu+1} \times \Pi\, \chi(\delta) \times B_{2\nu+1} \times \prod_{s=1}^{s=2\nu} I_s^{\frac{1}{2}s(n-s)}$$

when n is uneven and equal to $2\nu+1$, and by the formula

$$W = \zeta_{2\nu} \times \Pi\, \chi(\delta) \times B_{2\nu} \times \prod_{s=1}^{s=2\nu-1} I_s^{\frac{1}{2}s(n-s)} \times \frac{1}{\pi^\nu} \sum_1^\infty \left(\frac{D}{m}\right) \frac{1}{m^\nu}$$

when n is even and equal to 2ν; and we shall consider separately the factors of which these formulae are composed.

(i) In the infinite series $\frac{1}{\pi^\nu} \sum_1^\infty \left(\frac{D}{m}\right) \frac{1}{m^\nu}$, which enters into the expression of W only when the number of indeterminates is even, D still represents the product $(-1)^\nu I_1 \times I_3 \times \ldots \times I_{2\nu-1}$, and the summation extends to all uneven values of m, which are prime to D, from 1 to ∞. The sum of this infinite series can in every case be obtained in a finite form by the methods employed by Dirichlet (in the 21st volume of Crelle's Journal) and by Cauchy (in the 17th volume of the Mémoires de l'Académie des Sciences, p. 679). As the result of the summation does not seem to have been given, we shall present it here in one of many various forms which it may assume. Let D_1 represent the quotient obtained by dividing D by its greatest square divisor; let q be any uneven prime dividing D but not D_1, and let $V = \frac{1}{\pi^\nu} \sum_1^\infty \left(\frac{D_1}{m}\right) \frac{1}{m^\nu}$, the sign of summation extending to all values of m prime to $2D_1$; we then have the equation

$$\frac{1}{\pi^\nu}\sum_1^\infty \left(\frac{D}{m}\right)\frac{1}{m^\nu} = \Pi\left[1-\left(\frac{D_1}{q}\right)\frac{1}{q^2}\right]\times V.$$

To obtain the value of V, let Δ represent the positive value of D_1, so that $\Delta = D_1$ when ν is even, and $\Delta = -D_1$ when ν is uneven. Also let

$$F_{2\sigma}(x) = \frac{x^{2\sigma+1}}{2\sigma+1} - \tfrac{1}{2}x^{2\sigma} + \frac{\Pi(2\sigma)}{\Pi(2\sigma-1)\,\Pi(2)}\beta_1 x^{2\sigma-1}$$
$$-\frac{\Pi(2\sigma)}{\Pi(2\sigma-3)\,\Pi(4)}\beta_3 x^{2\sigma-3} + \dots + (-1)^{\sigma-1}\frac{\Pi(2\sigma)}{\Pi(1)\,\Pi(2\sigma)}\beta_{2\sigma-1}x,$$

$$F_{2\sigma-1}(x) = \frac{x^{2\sigma}}{2\sigma} - \tfrac{1}{2}x^{2\sigma-1} + \frac{\Pi(2\sigma-1)}{\Pi(2\sigma-2)\,\Pi(2)}\beta_1 x^{2\sigma-2}$$
$$-\frac{\Pi(2\sigma-1)}{\Pi(2\sigma-4)\,\Pi(4)}\beta_3 x^{2\sigma-4} + \dots + (-1)^{\sigma}\frac{\Pi(2\sigma-1)}{\Pi(2)\,\Pi(2\sigma-2)}\beta_{2\sigma-3}x^2,$$

where $\beta_1, \beta_3, \dots$ are the fractions of Bernoulli, so that $F_k(x)$ is the function which, when x is an integral number, is equivalent to the sum $\sum\limits_{s=1}^{s=x-1} s^k$. Then, putting $\epsilon = (-1)^{\frac{1}{2}(\nu+2)}$ or $= (-1)^{\frac{1}{2}(\nu+1)}$, according as ν is even or uneven, we have

(1) when $D_1 \equiv 1$, mod 4,

$$\epsilon V = \frac{2^{\nu-1}}{\Pi(\nu-1)}\times\left[1-\left(\frac{2}{\Delta}\right)\frac{1}{2^\nu}\right]\times\frac{1}{\sqrt{\Delta}}\times\sum_{s=1}^{s=\Delta}\left(\frac{s}{\Delta}\right)F_{\nu-1}\left(\frac{s}{\Delta}\right),$$

(2) in every other case,

$$\epsilon V = \frac{2^{\nu-1}}{\Pi(\nu-1)}\times\frac{1}{2\sqrt{\Delta}}\times\sum_{s=1}^{s=4\Delta}\left(\frac{D_1}{s}\right)F_{\nu-1}\left(\frac{s}{4\Delta}\right),$$

the summation $\sum\limits_1^\Delta$ extending to every integral value of s inferior to Δ and prime to Δ, the summation $\sum\limits_1^{4\Delta}$ extending to every integral value of s inferior to 4Δ and prime to 4Δ. The formula (1) is inapplicable when $\Delta = D_1 = 1$; but, in this case, ν is even and the sum of the series $\frac{1}{\pi^\nu}\sum\frac{1}{m^\nu}$ is known.

(ii) The factor $\prod\limits_{s=1}^{s=n-1} I_s^{\frac{1}{2}s(n-s)}$ requires no explanation; it is rational when n is uneven, and is a multiple of $\sqrt{\Delta}$ when n is even.

(iii) The factor B_n is determined by the equations

$$B_{2\nu} = \tfrac{1}{2}\beta_1\times\tfrac{1}{2}\beta_3\times\tfrac{1}{2}\beta_5\times\dots\times\tfrac{1}{2}\beta_{2\nu-3},$$
$$B_{2\nu+1} = \tfrac{1}{2}\beta_1\times\tfrac{1}{2}\beta_3\times\tfrac{1}{2}\beta_5\times\dots\times\tfrac{1}{2}\beta_{2\nu-1}\times\frac{1}{\Pi(\nu)},$$

where $\beta_1, \beta_3, \dots$ are again the fractions of Bernoulli, so that $\beta_1 = \frac{1}{6}$, $\beta_3 = \frac{1}{30}$, etc.

(iv) The factors (i) and (ii) depend only on the invariants and on the number of the indeterminates, the factor (iii) only on the number of indeterminates. These factors are therefore the same for all genera of the invariants $I_1, I_2, \ldots, I_{n-1}$. But the two remaining factors involve, or may involve, certain of the generic characters, and are, therefore, not always the same for all genera. In the factor $\Pi\,\chi(\delta)$ the sign of multiplication extends to every uneven prime δ dividing any one or more of the invariants $I_1, I_2, \ldots, I_{n-1}$: it will suffice, therefore, to define the function $\chi(\delta)$, which depends on only one of those primes. Let $i_1, i_2, \ldots$ be the indices of all the invariants which are divisible by δ; let these indices be arranged in order of magnitude, beginning with 0 and ending with n (because I_0 and I_n may be considered as divisible by δ). The positive differences $i_{s+1} - i_s$ we shall term *intervals*. By the *moiety* of any whole number a we understand $\frac{1}{2}a$ when a is even, $\frac{1}{2}(a-1)$ when a is uneven. Let κ_s be the moiety of the interval $i_{s+1} - i_s$; when that interval is even, let the barred symbol $\bar{\kappa}_s$ represent the product $(-1)^{\kappa_s} I_{1+i_s} \times I_{3+i_s} \times \ldots \times I_{-1+i_{s+1}}$; and let $\Gamma(\bar{\kappa}_s) = 1 + \left(\frac{\bar{\kappa}_s \times \theta_{i_s} \times \theta_{i_{s+1}}}{\delta}\right)\frac{1}{\delta^{\kappa_s}}$. Lastly, let $\Omega(h)$ represent the product $\prod_{s=1}^{s=h}\left(1 - \frac{1}{\delta^{2s}}\right)$; let σ be the moiety of $n-1$, and μ the number of the invariants $I_1, I_2, \ldots, I_{n-1}$ which are divisible by δ. Then $\chi(\delta)$ is the integral function of $\frac{1}{\delta}$ defined by the equation

$$\chi(\delta) = \frac{1}{2^{\mu}} \times \frac{\Omega(\sigma)}{\Pi\,\Omega(\kappa_s)} \times \Pi\,\Gamma(\bar{\kappa}_s)$$

when n is uneven, and by the equation

$$\chi(\delta) = \frac{1}{2_{\mu}} \times \frac{\Omega(\sigma)}{\Pi\,\Omega(\kappa_s)} \times \Pi\,\Gamma(\bar{\kappa}_s) \times \left[1 - \left(\frac{D}{\delta}\right)\frac{1}{\delta^{\frac{1}{2}n}}\right]$$

when n is even. If D is divisible by δ, the symbol $\left(\frac{D}{\delta}\right)$ is zero. In both formulae the sign of multiplication Π extends to every value of κ_s or $\bar{\kappa}_s$; the value $+1$ is, as before, to be attributed to the symbols θ_0 and θ_n.

(v) Each factor $\chi(\delta)$ of the product $\Pi\,\chi(\delta)$ thus depends on an uneven prime δ dividing the invariants, on the indices of the invariants divisible by δ, on the principal generic characters with respect to δ, and on the quadratic characters with respect to δ of the invariants not divisible by δ. The remaining factor ζ_n may be said to depend on the relation of the concomitants and invariants to the prime 2 and its powers. The determination of this factor presents no theoretical difficulty; but, on account of the multiplicity of the cases to be considered, we shall confine ourselves in this place to the two cases in which the invariants are all uneven.

(A) When the invariants are all uneven and the given genus is of an uneven order, let Σ_n represent the unit $(-1)^h \psi(0, n-1)$, where $\psi(0, n-1)$ is the simultaneous character of the given genus, and h is determined by the equation

$$4h = (I_1 - 1)(I_2 + 1) + (I_2 - 1)(I_1 I_3 + 1) + (I_1 I_3 - 1)(I_2 I_4 + 1) + (I_2 I_4 - 1)(I_1 I_3 I_5 + 1) + \dots + (\dots I_{n-4} I_{n-2} - 1)(\dots I_{n-3} I_{n-1} + 1).$$

The value of ζ_n then is as follows:—

(1) if $n = 4\lambda$,

$$\zeta_n = \frac{1}{2^{2\lambda-1}}[2^{2\lambda-1} + (-1)^\lambda \Sigma_n], \text{ or } = 1,$$

according as $D \equiv 1$, or $\equiv -1$, mod 4;

(2) if $n = 4\lambda + 2$,

$$\zeta_n = \frac{1}{2^{2\lambda}}[2^{2\lambda} + (-1)^\lambda \Sigma_n], \text{ or } = 1,$$

according as $D \equiv 1$, or $\equiv -1$, mod 4;

(3) if $n = 4\lambda + 1$,

$$\zeta_n = \frac{1}{2^{2\lambda}}[2^{2\lambda} + (-1)^\lambda \Sigma_n];$$

(4) if $n = 4\lambda + 3$,

$$\zeta_n = \frac{1}{2^{2\lambda+1}}[2^{2\lambda+1} + (-1)^{\lambda + \frac{1}{2}(D-1)} \Sigma_n].$$

(B) When the invariants are all uneven and the given genus of an even order, so that $n = 2\nu$ is even, the value of ζ_n is

$$\frac{1}{2^{n-2}} \times \frac{1}{1 - \left(\frac{2}{D}\right)\frac{1}{2^\nu}}.$$

It is easy to apply these general formulae to particular examples; but our imperfect knowledge of quadratic forms containing many indeterminates renders it practically impossible to test the results by any independent process. The demonstrations are simple in principle, but require attention to a great number of details with respect to which it is very easy to fall into error. As soon as they can be put into a convenient form, they shall be submitted to the Royal Society.

Eisenstein has observed that, when the number of indeterminates does not surpass eight, there is but one class of quadratic forms of the discriminant 1, but that, when the number of indeterminates surpasses eight, there is always more than one such class. This observation is in accordance with our general formulae, except that they imply the existence of an improperly primitive class of eight indeterminates and of the discriminant 1.

The theorems which have been given by Jacobi, Eisenstein, and recently in great profusion by M. Liouville, relating to the representation of numbers by four squares and other simple quadratic forms, appear to be deducible by a uniform method from the principles indicated in this paper. So also are the theorems relating to the representation of numbers by six and eight squares, which are implicitly contained in the developments given by Jacobi in the 'Fundamenta Nova.' As the series of theorems relating to the representation of numbers by sums of squares ceases, for the reason assigned by Eisenstein, when the number of squares supasses eight, it is of some importance to complete it. The only cases which have not been fully considered are those of five and seven squares. The principal theorems relating to the case of five squares have indeed been given by Eisenstein (Crelle's Journal, vol. xxxv. p. 368); but he has considered only those numbers which are not divisible by any square. We shall here complete his enunciation of those theorems, and shall add the corresponding theorems for the case of seven squares. We attend only to primitive representations.

Let Δ represent a number not divisible by any square, Ω^2 an uneven square, α any exponent. By $\Phi_5(4^\alpha\Omega^2\Delta)$, $\Phi_7(4^\alpha\Omega^2\Delta)$, we denote the number of representations of $4^\alpha\Omega^2\Delta$ by five and seven squares respectively; by $Q_5(4^\alpha\Omega^2\Delta)$, $Q_7(4^\alpha\Omega^2\Delta)$, we represent the products

$$5\times 2^{3\alpha}\times\Omega^3\times\Pi\left[1-\left(\frac{\Delta}{q}\right)\frac{1}{q^2}\right]\times\frac{1}{\Delta},$$

$$7\times 2^{5\alpha}\times\Omega^5\times\Pi\left[1-\left(\frac{-\Delta}{q}\right)\frac{1}{q^3}\right]\times\frac{1}{\Delta},$$

the sign of multiplication Π extending to every prime dividing Ω but not dividing Δ; we then have the formulae :—

(A) For five squares.

(1) If $\Delta\equiv 1$, mod 4,

$$\Phi_5(4^\alpha\Omega^2\Delta)=Q_5(4^\alpha\Omega^2\Delta)\times\eta\times\sum_1^\Delta\left(\frac{s}{\Delta}\right)s(s-\Delta),$$

where, if $\Delta \equiv 1$, mod 8, $\eta = 12$; if $\Delta \equiv 5$, mod 8, $\eta = 28$ or $= 20$, according as $a = 0$ or > 0. If, however, $\Delta = 1$, we are to replace $\eta \times \Sigma$ by 2.

(2) In every other case,

$$\Phi_5(4^a \Omega^2 \Delta) = Q_5(4^a \Omega^2 \Delta) \times \eta \times \sum_1^{4\Delta} \left(\frac{\Delta}{s}\right) s(s - 4D),$$

where $\eta = 1$ or $= \frac{1}{2}$, according as $a = 0$ or > 0.

(B) For seven squares.

(1) If $\Delta \equiv 3$, mod 4,

$$\Phi_7(4^a \Omega^2 \Delta) = Q_7(4^a \Omega^2 \Delta) \times \eta \times \sum_1^{\Delta} \left(\frac{s}{\Delta}\right) s(s - \Delta)(2s - \Delta),$$

where $\eta = 30$, if $a = 0$, $\Delta \equiv 3$, mod 8; $\eta = \frac{2}{3} \times 37$, if $a = 0$, $\Delta \equiv 7$, mod 8; $\eta = \frac{1}{3} \times 140$, if $a > 0$.

(2) In every other case,

$$\Phi_7(4^a \Omega^2 \Delta) = Q_7(4^a \Omega^2 \Delta) \times \eta \times \sum_1^{4\Delta} \left(\frac{-\Delta}{s}\right) s(s - 2\Delta)(s - 4\Delta),$$

where $\eta = \frac{1}{3}$ or $= \frac{5}{12}$, according as $a = 0$ or > 0.

The sums $\sum_1^{\Delta}$, and $\sum_1^{4\Delta}$ in these formulae are easily reduced (by distinguishing different linear forms of the number Δ) to others more readily calculated (see the note of Eisenstein, to which we have already referred); but, in the present notice, we have preferred to retain them in the form in which they first present themselves.

We shall conclude this paper by calling attention to a class of theorems which have a certain resemblance to the important results established by M. Kronecker for binary quadratic forms.

Let $\frac{1}{4} \frac{F_4(M)}{\Pi(4)}$ represent the weight of the quaternary classes of the invariants $[1, 1, M]$, $\frac{1}{4} \frac{F_6(M)}{\Pi(6)}$ the weight of the senary classes of the invariants $[1, 1, 1, 1, M]$, then

$$F_4(M) + 2F_4(M - 1^2) + 2F_4(M - 2^2) + \ldots = \Sigma(-1)^{\frac{1}{2}(d+1)} d^3,$$
$$F_6(2M) + 2F_6(2M - 1^2) + 2F_6(2M - 2^2) + \ldots = \Sigma d^3.$$

In the first of these formulae M is any unevenly even number, or any number $\equiv 3$, mod 4; in the second M is any uneven number: the series in both are to be continued so long as the numbers $M - s^2$ or $2M - s^2$ are positive;

d is any uneven divisor of M. The origin of these formulae (which may serve as examples of many others) is exactly analogous to that which M. Kronecker has pointed out as characteristic of the more elementary of the two classes into which his formulae are naturally divided. Whether, for forms of four and six indeterminates, similar formulae exist comparable to the less elementary formulae of M. Kronecker, and whether, for forms containing more than six indeterminates, such formulae exist at all, are questions well worthy of the attention of arithmeticians.

XIX.

ON SOME GEOMETRICAL CONSTRUCTIONS.

[Proceedings of the London Mathematical Society, vol. ii. pp. 85–100. Read May 28, 1868.]

---

1. A CONIC A is said to *circumscribe harmonically* a conic B, when A circumscribes a triangle which is self-conjugate with regard to B. Similarly, A is said to be *inscribed harmonically in* B, when A is inscribed in a triangle which is self-conjugate with regard to B. Though this mode of expression is not very accurate, it has the advantage of brevity, and it may serve to fix in the memory the well known theorems :—

I. 'If A circumscribe B harmonically, B is harmonically inscribed in A.'

II. 'If A circumscribe B harmonically, the conic corresponding to A in any correlative figure is harmonically inscribed in the conic corresponding to B.'

Of these theorems we shall have to make frequent use, as also of the two following and their correlatives :—

III. 'If A circumscribe B harmonically, A circumscribes an infinite number of triangles self-conjugate with regard to B; viz. if x_1 be any point of A, and $x_2 x_3$ the chord intercepted by A on the polar of x_1 with regard to B, $x_1 x_2 x_3$ is a self-conjugate triangle with regard to B. Or, which is the same thing, the *harmonic envelope of A and B* (*i.e.* the conic enveloped by lines cutting A and B harmonically) coincides with the polar reciprocal of A with regard to B.'

IV. 'If A circumscribe B harmonically, the centre of homology of any triangle inscribed in A and its polar triangle with regard to B will lie on A.' (Dr. Salmon's Conic Sections, p. 326.)

The pairs of points, conjugate with regard to a conic A, which lie upon a line L form a system in involution. Similarly, the pairs of lines, conjugate with respect to A, which intersect at a point P form a pencil in involution. These involutions we shall term *the involutions of A upon the line L, and at the point P*, respectively. When we say that a conic is given, we shall understand that the polar system of the conic is given; *i.e.* that the involution of the conic upon any line, and at any point in its plane, is given, or can be determined linearly. If any single element (point or tangent) of a given conic is given, we can determine linearly as many elements of it as we please. But, if no single element of a given conic is given, the determination of any single element will require a quadratic construction, and the conic itself may be imaginary.

When there are two involutions I_1 and I_2 upon the same line L, the involution of which the double points are the extremities of the segment common to I_1 and I_2, is said to be *compounded of* I_1 *and* I_2. To obtain the involution compounded of two given involutions I_1 and I_2, let x be any point of L, x_1, x_2 the conjugates of x in I_1 and I_2 respectively, y_1 the conjugate of x_2 in I_1, y_2 the conjugate of x_1 in I_2; the harmonic conjugate of x with regard to $y_1 y_2$ is also the conjugate of x in the involution composed of I_1 and I_2. This construction may be demonstrated by projecting the involutions I_1 and I_2 upon a conic in the usual manner, and applying Pascal's theorem to the pentagon formed by the projections of the points x, x_1, y_2, y_1, x_2, x.

All the constructions which we shall employ in this paper are linear, except when the contrary is expressly stated. We shall, for the most part, leave the correlative of each proposition to be supplied by the reader.

2. *Problem* 1.—'To determine the conic σ which passes through three given points a, b, c and circumscribes harmonically two given conics S_1 and S_2.'

Solution.—Let $a_1 b_1 c_1$, $a_2 b_2 c_2$ be the polar triangles of abc with regard to S_1 and S_2 respectively; and let aa_1, bb_1, cc_1 meet in x_1; aa_2, bb_2, cc_2 in x_2; then x_1, x_2 are points of σ (Theorem IV, Art. 1), which is thus completely determined by the five points a, b, c, x_1, x_2.

Problem 2.—'To determine the conic σ which passes through two given points a, b and circumscribes harmonically three given conics S_1, S_2, S_3.'

Solution.—Let c_1, c_2, c_3 be the poles of ab with regard to S_1, S_2, S_3 respectively. Through a draw any line aP, not passing through any one of the points c_1, c_2, c_3; let p_1, p_2, p_3 be the poles of aP with regard to S_1, S_2, S_3; and let P be the unknown point in which σ meets aP for the second time. Considering the triangle abP, inscribed in σ, with regard to each of the conics S_1, S_2, S_3 in succession, we see that the three intersections (Pc_1, bp_1), (Pc_2, bp_2), (Pc_3, bp_3), as well as a, b, P, lie upon σ. We have, therefore, the anharmonic equation

$$P.[a, c_1, c_2, c_3] = b.[a, p_1, p_2, p_3],$$

which implies that the conic passing through a, c_1, c_2, c_3, and satisfying the anharmonic equation

$$[a, c_1, c_2, c_3]=b\,.\,[a, p_1, p_2, p_3],$$

also passes through P. Thus P is determined linearly, and with it σ, on which we now have six points. (The actual construction of P is as follows:—Let aP cut c_2c_3 in a'; determine on c_2c_3 a point c', satisfying the anharmonic equation

$$[a', c', c_2, c_3]=b\,.\,[a, p_1, p_2, p_3];$$

the point P is the intersection of aP and $c'c_1$.)

Problem 3.—'To determine the conic σ which passes through a given point a and circumscribes harmonically four given conics S_1, S_2, S_3, S_4.'

Solution.—Let aP, aQ be any two straight lines passing through a, but not conjugate with regard to any one of the given conics. Let p_1, p_2, p_3, p_4 and q_1, q_2, q_3, q_4 be the poles of aP and of aQ respectively with regard to the conics S_1, S_2, S_3, S_4; and let P, Q be the unknown points in which aP, aQ meet σ for the second time. Considering the triangle aPQ, inscribed in σ, with regard to the conics S_1, S_2, S_3, S_4 in succession, we obtain the anharmonic equation

$$P\,.\,[a, q_1, q_2, q_3, q_4]=Q\,.\,[a, p_1, p_2, p_3, p_4],$$

which suffices for the linear determination of P and Q, by a kind of double position. Assume any point x on aP as the true position of P, and determine on aQ the corresponding position y of Q, first by the equation

$$y_3\,.\,[a, p_1, p_2, p_3]=x\,.\,[a, q_1, q_2, q_3],$$

and then by the equation

$$y_4\,.\,[a, p_1, p_2, p_4]=x\,.\,[a, q_1, q_2, q_4].$$

The two positions of y thus obtained will form, when x varies, two homographic ranges y_3 and y_4; of the double points, one is at the intersection of aQ and p_1p_2, the other can be determined linearly, and is the true position of Q. The details of the construction are as follows:—Denote by Q the intersection (aP, q_1q_2), and by p the intersection (aQ, p_1p_2); let x be a point varying its position on aP, and let xq_3, xq_4 intersect q_1q_2 in q_3', q_4'. On the line p_1p_2 determine the points p_3', p_4', which satisfy the equation

$$[p_3', p_4', p, p_1, p_2]=[q_3', q_4', q, q_1, q_2],$$

and let p_3p_3', p_4p_4' cut aQ in y_3, y_4 respectively. The points y_3, y_4 will form two homographic ranges on aQ; for each of these ranges is homographic with the range x. One of the double points of the two ranges is at p; for, if we imagine x to coincide with q, q_3' and q_4' will coincide with one another and with q;

whence p_3' and p_4', and with them y_3 and y_4, will coincide with one another and with p. The other double point can therefore be obtained linearly; it will be the point Q, and the corresponding position of the point x will be the point P. We shall thus have seven points on σ; viz., the three points a, P, Q, and the four intersections (Pq_r, Qp_r), where $r=1, 2, 3, 4$.

Problem 4.—'To determine the conic σ which circumscribes harmonically five given conics S_1, S_2, S_3, S_4, S_5.' Only the polar system of σ can be determined linearly, and σ itself may be imaginary.

Solution.—(1) If each of two conics A and B harmonically circumscribe a third S, every conic C which passes through the intersections of A and B also circumscribes S harmonically.* To prove this geometrically, let x be any one of the points of intersection of A and B, and let the polar of x with respect to S cut A, B, C in the points $a_1 a_2$, $b_1 b_2$, $c_1 c_2$ respectively. Then $a_1 a_2$ and $b_1 b_2$ are pairs of conjugate points with respect to S; therefore also $c_1 c_2$, which is in involution with $a_1 a_2$ and $b_1 b_2$, is a pair of conjugate points with respect to S; *i.e.* C circumscribes a triangle $xc_1 c_2$ which is self-conjugate with respect to S. From this it appears that the conics which harmonically circumscribe four given conics all pass through four fixed points; for, if A, B are two conics circumscribing S_1, S_2, S_3, S_4, the conic C which passes through a given point c and circumscribes harmonically those four conics is no other than the conic of the system (A, B) which passes through the point c.

(2) Let L be any line in the plane of the five given conics: to obtain the involution of σ upon L, we have first to determine the intersections of L by two of the conics which harmonically circumscribe the four conics S_2, S_3, S_4, S_5; and to do this, we have only, in Problem 3, to take successively for a two different points on L. Let I_1 represent the involution determined by the two pairs of intersections; similarly, let I_2 represent the involution determined on L by the system of conics which circumscribe harmonically the four conics S_1, S_3, S_4, S_5. The involution of σ upon L is the involution compounded of I_1 and I_2.

3. In the preceding problems, any number of the given conics may degenerate into pairs of points, but the solutions will remain applicable if we observe that the pole of a straight line L, with regard to a system of two points, is the harmonic conjugate, with regard to the two points, of the intersection of L by

---

* The theorem of M. Hesse, 'If two pairs of opposite vertices of a quadrilateral are conjugate pairs with regard to a conic, the third pair of opposite vertices is also a conjugate pair,' is a particular case of the correlative theorem.

the line joining the two points. Since a conic circumscribing a conic which has degenerated into a pair of points is a conic with regard to which the two points are conjugate, our last problem includes that of M. de Jonquières ('Annales de Mathématiques,' par MM. Terquem et Gerono, vol. xiv. p. 435), 'To determine the conic which divides harmonically five given segments.' Again, the two points of a degenerate conic may become coincident, in which case a conic circumscribing harmonically the degenerate conic is simply a conic passing through the point which represents that conic. Thus the problem 4 includes the problems 1, 2, and 3. We might have made the solution of these problems depend on the corresponding cases of M. de Jonquières' problem. For example, if $x_1 y_1$, $x_2 y_2$, $x_3 y_3$, $x_4 y_4$ are the polar chords of the point a with regard to the conics S_1, S_2, S_3, S_4, the conic passing through a, and dividing harmonically the four segments $x_1 y_1$, $x_2 y_2$, $x_3 y_3$, $x_4 y_4$, harmonically circumscribes the conics S_1, S_2, S_3, S_4. But the solutions which M. de Jonquières has given of the various cases of his problem are, perhaps, less direct than those which we have deduced from the theorem of Dr. Salmon (Theorem 4, Art. 1).

We may add that it follows from the solution of the problem of M. de Jonquières that the polar system of a conic is given when five pairs of conjugate points with regard to the conic are given (but these five pairs must be independent; see Art. 4).

4. Some remarks, which may not be without interest, are suggested by the analysis corresponding to the problems 1–4.

Let $$\alpha_i x^2 + \beta_i y^2 + \gamma_i z^2 + 2\alpha'_i yz + 2\beta'_i xz + 2\gamma'_i xy = 0 \quad . \quad . \quad . \quad . \quad . \quad (1)$$
be the equation in point coordinates of the conic σ_i, and let
$$A_i \xi^2 + B_i \eta^2 + C_i \zeta^2 + 2A'_i \eta\zeta + 2B'_i \xi\zeta + 2C'_i \xi\eta = 0 \quad . \quad . \quad . \quad . \quad (2)$$
be the equation in line coordinates of the conic S_i, the two sets of coordinates being connected by the relation
$$\xi x + \eta y + \zeta z = 0.$$
The equation which expresses that σ harmonically circumscribes S_i is
$$A_i \alpha + B_i \beta + C_i \gamma + 2A'_i \alpha' + 2B'_i \beta' + 2C'_i \gamma' = 0\,; \quad . \quad . \quad . \quad . \quad (3)$$
and the problem 4 is the geometrical equivalent of the analytical problem, 'To determine the ratios of $\alpha\beta\gamma\ \alpha'\beta'\gamma'$ from five linear and independent equations of the type (3).' In these equations the coefficients may have any values whatever; whereas, in the problem of M. de Jonquières, the five equations are subject to the condition that in each of them the symmetrical determinant formed with the six coefficients must be equal to zero; and in the problem, 'to determine the

conic passing through five given points,' there is the still further limitation that the first minors of those determinants must also be equal to zero.

We shall denote the systems of conics represented by the equations

$$\text{(i)}\quad \lambda_1\sigma_1+\lambda_2\sigma_2=0,$$

$$\text{(ii)}\quad \lambda_1\sigma_1+\lambda_2\sigma_2+\lambda_3\sigma_3=0,$$

$$\text{(iii)}\quad \lambda_1\sigma_1+\lambda_2\sigma_2+\lambda_3\sigma_3+\lambda_4\sigma_4=0,$$

$$\text{(iv)}\quad \lambda_1\sigma_1+\lambda_2\sigma_2+\lambda_3\sigma_3+\lambda_4\sigma_4+\lambda_5\sigma_5=0$$

by the symbols (σ_1, σ_2), $(\sigma_1, \sigma_2, \sigma_3)$, $(\sigma_1, \sigma_2, \sigma_3, \sigma_4)$, and $(\sigma_1, \sigma_2, \sigma_3, \sigma_4, \sigma_5)$; and we shall describe them as *systems of the orders* 1, 2, 3, 4 respectively. The coefficients λ are absolutely indeterminate, and it is understood that the conics σ are *independent, i.e.* that σ_3 does not belong to the system (σ_1, σ_2), nor σ_4 to the system $(\sigma_1, \sigma_2, \sigma_3)$, nor σ_5 to the system $(\sigma_1, \sigma_2, \sigma_3, \sigma_4)$. If the equations (i), (ii), (iii), (iv) are in line-coordinates, we shall describe the corresponding systems of conics as *tangential systems of the orders* 1, 2, 3, and 4. In the enunciations and solutions of the problems 1–4, we have tacitly supposed that the data are such as to render them determinate; the necessary and sufficient condition for this determinateness is that the five conics S_1, S_2, S_3, S_4, S_5 (or the pairs of points, real, imaginary, or coincident, by which any, or all of them, are replaced) should be independent, or should form a tangential system of order 4. A system of order 1 is the 'faisceau,' a system of order 2 is the 'réseau,' of French geometers. A single conic may be regarded as forming a system of order 0.

It is evident that a system of order k is determined by $k+1$ independent conics, and that, if $k+1$ independent conics of a system of order k harmonically circumscribe a given conic, every conic of the system harmonically circumscribes that conic. (The geometrical proof for the case $k=1$, which has been given above, might easily be extended to the other cases.) We have also the important theorem:—

'All the conics of a given system of order k circumscribe harmonically all the conics of a certain tangential system of order $4-k$; and, conversely, all the conics which circumscribe harmonically the conics of a given tangential system of order $4-k$ form a system of order k.'

The tangential system of order $4-k$, which thus corresponds to a given system of order k, we shall call *the system contravariant to the given system.* The relations between the two systems may be inferred from the known properties of indeterminate systems of linear equations. Thus, given $k+1$ independent conics σ $(k \gtreqless 0, k \leqq 4)$, we have a system of $k+1$ indeterminate equations of the type,

$$\alpha_i A + \beta_i B + \gamma_i C + 2\alpha_i' A' + 2\beta_i' B' + 2\gamma_i' C' = 0, \quad [i = 1, 2, \ldots, k+1,] \quad . \quad (4)$$

in which α_i, β_i, γ_i, α_i', β_i', γ_i' are the given coefficients, and A, B, C, A', B', C', the indeterminates. The order of indeterminateness of the system is $5-k$; and, if

$$A_j,\ B_j,\ C_j,\ A_j',\ B_j',\ C_j' \quad [j = 1, 2, \ldots, 5-k,] \quad . \quad . \quad . \quad . \quad . \quad (5)$$

represent $5-k$ independent solutions, every solution is included in the formulae

$$\Sigma . \lambda_j A_j, \quad \Sigma . \lambda_j B_j, \quad \Sigma . \lambda_j C_j, \quad \Sigma . \lambda_j A_j', \quad \Sigma . \lambda_j B_j', \quad \Sigma . \lambda_j C_j',$$

the $5-k$ coefficients λ being absolutely indeterminate. This establishes the first part of the theorem; the second is the correlative, as well as the converse, of the first. Again, considering the matrix of the system (4) and the matrix of the system (5) of independent solutions as two complementary matrices, we know (Phil. Trans., vol. cli. p. 301)* that the determinants of the one are proportional to the complementary determinants of the other; so that, in particular, if any determinant of either matrix is zero, the complementary determinant of the other matrix is also zero. As, perhaps, no geometrical application has hitherto been given of this analytical property, we shall refer to a few of its consequences here, though the results are such as might be obtained by simple geometrical reasoning.

(*a*) Let $k=4$; the system is determined by five independent conics, and the contravariant system is of order 0, *i.e.* it is the single conic S harmonically inscribed in these five conics. The tangents of this conic are the conics of the given system which degenerate into pairs of coincident straight lines; for, if the line ($\eta = 0$, $\zeta = 0$) be a tangent of S, we must have $A = 0$; and, therefore, the complementary determinant in the matrix of the given system is also zero; *i.e.* $x^2 = 0$ is one of the conics contained in the given system. Similarly, we may show that those pairs of straight lines which are degenerate conics of the given system are no other than the pairs of straight lines which are harmonically conjugate with respect to S; viz. if the lines ($\xi = 0$, $\eta = 0$), ($\xi = 0$, $\zeta = 0$) are conjugate with respect to S, we must have $A' = 0$, a condition which implies that $yz = 0$ is one of the conics of the given system.

(*b*) Let $k=3$; the given system is determined by four independent conics, the contravariant system (S_1, S_2) being a system of conics inscribed in the same quadrilateral. As in the former case, we may show that the only conics of the given system which resolve themselves into two straight lines are represented

* No. XII p. 376 of this volume.

by the four common tangents of the system (S_1, S_2); and that the conics which degenerate into two straight lines are the pairs of lines harmonically conjugate with respect to the same system. Further, the three pairs of opposite vertices of the quadrilateral circumscribing the system (S_1, S_2) are conics of that system, and therefore are conics harmonically inscribed in the conics of the given system; *i.e.* the three diagonals of that quadrilateral cut the given system in involutions of which the double points upon each diagonal are the opposite vertices of the quadrilateral on that diagonal. (Mathematical questions from the Educational Times, vol. iv. p. 110; M. Cremona, in Crelle's Journal, vol. lxi. p. 110.)

(*c*) Let $k=2$, so that the given system is a 'réseau,' and the contravariant system a tangential 'réseau.' As it is necessary to consider two contravariant 'réseaux' in the theory of cubic curves, we shall place here the solution of two elementary problems relating to them.

Problem 5.—'Given three independent conics of the system $(\sigma_1, \sigma_2, \sigma_3)$; to determine the contravariant system.'

We may determine as many elements as we please of a conic touching two given lines, and harmonically inscribed in $\sigma_1, \sigma_2, \sigma_3$ (Problem 2). This conic is one of the conics of the contravariant system.

Problem 6.—'To determine the conic of the system $(\sigma_1, \sigma_2, \sigma_3)$ which harmonically circumscribes two given conics.'

Determine three independent conics of the contravariant system; the conic circumscribing them and the two given conics is the conic required; but, if the three contravariant conics and the two given conics are not independent, when considered tangentially, the problem is indeterminate.

Two particular cases of the problem are of frequent occurrence:—

(1) 'To determine the conic of the given system which passes through two given points.'

(2) 'To determine the conic of the given system with regard to which a given point and line are pole and polar.'

In the latter case (as indeed in the general case), only the polar system of the required conic can be obtained linearly, and the conic itself may be imaginary.

The 'double points' of the given system $(\sigma_1, \sigma_2, \sigma_3)$ lie on a cubic curve, the Jacobian locus, or Hessian of the system. Each point of a pair of points self-conjugate with respect to the system is a double point of the system; and the three vertices of any triangle self-conjugate with respect to two conics of the system are three double points, of which the conjugates lie in the axis of homology of the triangle and of its polar triangle with regard to any third independent conic of the system. The contravariant system possesses the correlative properties; and its Jacobian envelope is the Pippian, or Cayleyan, of the given system. Every common chord of two conics of the given system is a tangent of the Cayleyan; for, if L be a common chord of σ_2 and σ_3, and if $\omega_1\,\omega_2$ be the double points of the involution determined on L by the system $(\sigma_1, \sigma_2, \sigma_3)$, $\omega_1\,\omega_2$ is a conic of the contravariant system, so that L is a 'double line' of that system, and consequently a tangent of the Cayleyan. Thus the Cayleyan is the involution-envelope of the given system; and, reciprocally, the Hessian is the involution-locus of the contravariant system. These well-known properties are introduced here to show the importance of considering explicitly the contravariant system, for the relation between the two systems of conics may be said to be the source of the contravariant relation of the Hessian and the Cayleyan.

5. *Problem* 7.—'Given two conics σ_1 and σ_2; to determine the conic of the system (σ_1, σ_2) which harmonically circumscribes a given conic.'

Solution.—Determine four independent conics harmonically inscribed in σ_1 and σ_2; the conic harmonically circumscribing these conics and the given conic is the conic required. For the four auxiliary conics it will be convenient to take four pairs of points reciprocal with regard to the system (σ_1, σ_2).

Problem 8.—'Given two conics σ_1, σ_2 and two straight lines L_1, L_2; to determine the conics of the system (σ_1, σ_2) with regard to which L_1, L_2 are a pair of conjugate lines.' The problem is quadratic.

Solution.—Let λ_1 be the conic reciprocal to L_1* with regard to the system (σ_1, σ_2), and let λ_1 cut L_2 in the points a, b. The conics required are the two conics A and B with regard to which the poles of L_1 are a and b. When these points have been determined by a quadratic construction, the polar systems of the two conics will be known. But this quadratic determination we shall not require; and the following construction, which is linear, will suffice. Let X be any point on L_1, x the point reciprocal to X. The polars of X with regard to the conics (σ_1, σ_2) form a pencil of lines at x, of which the rays correspond anharmonically to the conics themselves; in this pencil xa, xb are the rays

* The point reciprocal to a given point P with regard to a system of conics (σ_1, σ_2) is the point in which the polars of P with regard to that system intersect. The conic reciprocal to a line is the locus of points reciprocal to the points of the line. Every such reciprocal conic passes through the three vertices of the *harmonic triangle* of the system; *i.e.* of the triangle self-conjugate with regard to all the conics of the system.

corresponding to A and B. And, since the involution of λ_1 upon L_2 (of which a, b are the double points) may be obtained linearly, we can determine linearly a pencil in involution at x of which the double rays are the rays corresponding to the conics A and B.

Problem 9.—'Given three conics σ_1, σ_2, S; to determine the conics of the system (σ_1, σ_2) which are harmonically inscribed in S.' This problem depends upon the preceding, which is a particular case of it; it is, of course, quadratic.

Solution.—If any single element of S be given, let p, q be any two points of S, L_1 the line joining them, P, Q the points reciprocal to p, q with regard to the system (σ_1, σ_2). Let Σ be the conic reciprocal to $P\,Q$; Σ will pass through p, q, and L_1 will be one of the chords of intersection of S and Σ; the opposite chord of intersection can then be determined linearly; let it be L_2; the two conics of the system (σ_1, σ_2) with regard to which L_1, L_2 are conjugate lines (Problem 8), are harmonically inscribed in S. For, if A be either of those conics, A is harmonically circumscribed by Σ, because Σ circumscribes the harmonic triangle of the system (σ_1, σ_2); but A is also harmonically circumscribed by the degenerate conic $L_1\,L_2$; therefore A is harmonically circumscribed by S, which is a conic of the system $(\Sigma_1, L_1\,L_2)$.

If only the polar system of S be given, let I be the involution of S upon any line L_1, and let λ_1 be the conic reciprocal to L_1 with regard to (σ_1, σ_2). To the involution I upon L_1 there will correspond an involution of points upon the reciprocal conic λ_1; let PQ be the polar line of the involution upon λ_1, and let Σ be the conic reciprocal to PQ; then, as before, L_1 will be one of the common chords of S and Σ, and the opposite common chord can be determined linearly.

Problem 10.—'Given three conics σ_1, σ_2, S; to find the fourth point common to the conics which circumscribe the harmonic triangle of (σ_1, σ_2) and which also harmonically circumscribe S.' The harmonic triangle of (σ_1, σ_2) must not also be a self-conjugate angle of S.

Solution.—Let P be any point in the plane of the conics, L the polar of P with regard to S, p the point reciprocal to P, and λ the conic reciprocal to L with regard to the system (σ_1, σ_2). To the involution of S upon L there corresponds an involution upon the conic λ; let q be the pole of this involution; the conic reciprocal to pq will circumscribe S harmonically, for it will pass through P and will cut L in two points which will be reciprocal to a pair of points of the involution upon λ, and which will therefore be a pair of points of the involution of S upon L. Let $p'q'$ be a second line of which the reciprocal conic harmonically circumscribes S; let θ be the point of intersection of pq, $p'q'$; the point reciprocal to θ is the fourth point of intersection required.

This fourth point is evidently (Theorem IV, Art. 1) the centre of homology of the harmonic triangle of (σ_1, σ_2) and of its polar triangle with regard to S. We may obtain the axis of homology either by the correlative construction, or by observing that it is the polar, with regard to S, of the centre of homology.

The solution of the problem 'To determine the conic which circumscribes the harmonic triangle of (σ_1, σ_2), passes through a given point and harmonically circumscribes a given conic' is explicitly contained in what precedes. To determine the conic which circumscribes the harmonic triangle of (σ_1, σ_2) and also harmonically circumscribes two given conics S_1, S_2, we should have to substitute successively S_1 and S_2 for S in the preceding solution, and to determine the two corresponding positions θ_1, θ_2 of the point θ; the conic reciprocal to $\theta_1\theta_2$ would then be the conic required. Though no single element of this conic is given, yet an *uneven* number of its points (the three vertices of the harmonic triangle) are given symmetrically. This explains why the conic is necessarily real, and why we can determine points on it linearly.

6. The polar conics of a cubic curve form a system of conics of order 2. Conversely, every system of conics of order 2 may be regarded as the polar system of a certain cubic curve which we shall call the *fundamental cubic* of the system. Let $\Delta_i = x_i\frac{d}{dx} + y_i\frac{d}{dy} + z_i\frac{d}{dz}$ $[i = 1, 2, 3]$; the analytical determination of the equation $C = 0$ of the cubic curve of which three given independent conics σ_1, σ_2, σ_3 are polar conics requires the determination of the coordinates x_i, y_i, z_i of the poles of the three given conics. Nine equations, determining the ratios of these nine coordinates, are obtained by equating to zero the coefficients of x, y, z in the expressions

$$\Delta_2\sigma_3 - \Delta_3\sigma_2, \quad \Delta_3\sigma_1 - \Delta_1\sigma_3, \quad \Delta_1\sigma_2 - \Delta_2\sigma_1.$$

The matrix of these nine equations is skew-symmetrical; the determinant is therefore zero, and the equations can be satisfied by at least one system of ratios of the nine coordinates; and, except in special cases, by only one such system. When values have been assigned to the nine coordinates, the coefficients of C may be ascertained from the equations,

$$\Delta_1 C = \sigma_1, \quad \Delta_2 C = \sigma_2, \quad \Delta_3 C = \sigma_3.$$

There are thus two curves of the third order, and two curves of the third class, which we have to consider in connection with a given system of conics of order 2 :—(1), the fundamental cubic; (2), the Hessian; (3), the fundamental contravariant, *i.e.* the curve of the third class related to the contravariant system of conics precisely as the fundamental cubic is related to the given system; (4), the Cayleyan, which, as we have seen, is related to the contravariant system as the Hessian is to the given system. If the equation of the fundamental cubic is

$$x^3 + y^3 + z^3 + 6m\,xyz = 0,$$

the equation of the Hessian is

$$m^2(x^3 + y^3 + z^3) - (1 + 2m^3)\,xyz = 0\,;$$

the equation of the fundamental contravariant in line coordinates is

$$m(\xi^3 + \eta^3 + \zeta^3) - 3\xi\eta\zeta = 0,$$

and the equation of the Cayleyan in line coordinates is

$$-m(\xi^3 + \eta^3 + \zeta^3) + (-1 + 4m^3)\,\xi\eta\zeta = 0.$$

The fundamental contravariant is mentioned by Professor Cayley (Phil. Trans., vol. cxlvii. p. 427) as a curve of the third class of which the Cayleyan is the Hessian envelope. It is also the curve designated as K_3 by M. Cremona ('Introduzione ad una Teoria Geometrica delle Curve Piane,' p. 117). It may be described as the evectant of $\frac{S^3}{T^2}$, S and T being the invariants of M. Aronhold. The harmonic relation between the polar conics of the fundamental cubic and the polar conics of the fundamental contravariant may be immediately verified by means of their equations.

7. *Problem* 11.—'Given three independent conics of the system $(\sigma_1, \sigma_2, \sigma_3)$; to determine the polar systems of the fundamental cubic and of the fundamental contravariant.' It will be convenient to exclude the exceptional case in which the three given conics have a pole and polar in common.

Solution.—The conics of the system (σ_2, σ_3) are, of course, all conics of the system $(\sigma_1, \sigma_2, \sigma_3)$, and their poles all lie on a right line L_1. Let abc be the harmonic triangle of (σ_2, σ_3), $a'b'c'$ the polar triangle of abc with regard to σ_1; A, B, C the pairs of common chords of σ_1 and σ_2, which intersect at a, b, c respectively. By a known property of cubic curves, the polar conic of one of two points which are self-conjugate with regard to every conic of the polar system is the degenerate conic of that system which consists of two straight lines intersecting at the other of the two points. Thus the poles of the conics A, B, C are respectively the intersections $(bc, b'c')$, $(ac, a'c')$, $(ab, a'b')$. Hence L_1, the locus of the poles of the system (σ_2, σ_3) which includes the conics

A, B, C, is the axis of homology of the triangles abc, $a'b'c'$, and can be determined linearly (Problem 10). Similarly, let L_2, L_3 be the loci of the poles of the conics (σ_3, σ_1), (σ_1, σ_2); the vertices P_1, P_2, P_3 of the triangle L_1, L_2, L_3 will be respectively the poles of the conics σ_1, σ_2, σ_3.

Let P be a given point in the plane, and let σ be the polar conic of P. If p_1 be the polar line of P with regard to σ_1, p_1 is also the polar line of P_1 with regard to σ. Thus the polar system of σ is determined linearly (Problem 6). Again, if σ be given and its pole P be required, let p_1 be the polar of P_1 with regard to σ; the pole of p_1 with regard to σ_1 is P. Thus, in the preceding construction, it will suffice to determine the two axes of homology L_2 and L_3, since, when the pole of one conic σ_1 is known, the pole of every other conic of the system is known also.

To obtain the polar system of the fundamental contravariant, we have only to determine three conics of the contravariant system (Problem 5), and to apply to them the correlative of the preceding construction.

8. *Problem* 12.—'Given three independent conics of the system $(\sigma_1, \sigma_2, \sigma_3)$; to determine the polar system of the Hessian.'

The solution of this problem depends on the following propositions:—

(1) 'The locus Σ of the poles of those conics of the system $(\sigma_1, \sigma_2, \sigma_3)$ which are harmonically inscribed in a given conic S is a conic.'

For consider any straight line L; let (λ_1, λ_2) be the system of conics (contained in the given system) of the poles of which L is the locus; then two points of the locus Σ, and only two, lie upon L; viz. the poles of the two conics which belong to the system (λ_1, λ_2) and are harmonically inscribed in S (Problem 9). Or, we may prove the same thing analytically; for, if $(x_i y_i z_i)$ is the pole of σ_i, a polar conic of the fundamental cubic C, the coefficients of σ_2 are linear in $x_i y_i z_i$; and the condition which expresses that σ_i is harmonically inscribed in S will be quadratic in $x_i y_i z_i$; *i.e.* the locus of $(x_i y_i z_i)$ is a conic section.

(2) 'Given three independent conics of the system $(\sigma_1, \sigma_2, \sigma_3)$ and the conic S; to determine Σ.'

Let K be any straight line, (κ_1, κ_2) the system of conics, contained in the system $(\sigma_1, \sigma_2, \sigma_3)$, of which the poles lie on K. The poles of the conics $(\sigma_1, \sigma_2, \sigma_3)$ correspond correlatively to the polars of a fixed point X with regard to the conics themselves; for these points and lines are poles and polars with regard to the polar conic of X. If, in the construction of Problem 9, we draw the arbitrary straight line L_1 through the point X, we shall obtain a pencil in involution at the point x reciprocal to X with regard to the system (κ_1, κ_2), of which the double lines are precisely the lines corresponding to the two conics of the system (κ_1, κ_2) which are harmonically inscribed in S. The correlative involution will be an involution of points on the line K, for K is the polar of x with regard to the polar conic of X; and this involution will be no other than the involution of Σ upon K, because its double points will be the poles of the two conics (κ_1, κ_2) which are harmonically inscribed in S.

(3) 'If σ be any conic of the system $(\sigma_1, \sigma_2, \sigma_3)$ and P its pole, the locus of the poles of the conics of the system which are harmonically inscribed in σ is Σ, the Hessian polar conic of P.'

This theorem is easily verified analytically. For the same equation which expresses that (xyz) is a point of the polar conic of $(x'y'z')$ with regard to the Hessian also expresses that the polar conic of (xyz) with regard to the fundamental cubic is harmonically inscribed in the polar conic of $(x'y'z')$ with regard to the same curve. But the theorem may also be inferred geometrically from a known property of curves of the third order. For the six points R, in which Σ cuts the Hessian, are the points conjugate (upon the Hessian) to the six points r in which σ cuts the Hessian. (M. Cremona, loc cit. p. 108.) The six degenerate polar conics which are composed of pairs of straight lines intersecting at the points r are to be considered as conics harmonically inscribed in σ; their poles are the six points R. Therefore the locus of the poles of the conics of the system $(\sigma_1, \sigma_2, \sigma_3)$ which are harmonically inscribed in σ passes through the six points R; *i.e.* that locus coincides with the Hessian polar conic of R.

(4) 'If σ be any conic of the system $(\sigma_1, \sigma_2, \sigma_3)$ and P its pole, the locus of the poles of those conics of the system which harmonically circumscribe σ is the Hessian polar line of P.'

This theorem may be established by the same analysis as the last. Or it may be deduced from it geometrically; for, if σ' is any conic of the system which harmonically circumscribes σ, and if P' is the pole of σ', P lies on the Hessian polar conic of P', because σ is harmonically inscribed in σ'; *i.e.* P' lies on the Hessian polar line of P.'

It is evident that these propositions determine linearly the Hessian polar system. The polar system of the Cayleyan may be obtained by a correlative construction.

9. If C be a cubic and Γ its Hessian, the cubics $C+\lambda\Gamma$ are termed the *syzygetic cubics* of C. If P be any point, σ the polar conic of P and Σ its Hessian polar conic; its polar conic with regard to the syzygetic $C+\lambda\Gamma$ is $\sigma+\lambda\Sigma$. Thus the polar conics of a fixed point correspond anharmonically to the syzygetic

cubics. Let P', σ', Σ' represent a second given point, and its two polar conics; the reciprocal point of P', with regard to the system (σ, Σ), will be the same as the reciprocal point of P with regard to the system (σ', Σ'). Let this point be Ω, and let Ωx be the mixed derivative of P, P', with regard to the syzygetic $C+\lambda\Gamma$; the lines Ωx will correspond anharmonically to the polar conics of P, or of P', and therefore to the syzygetic cubics. If we suppose that the polar systems of C and Γ are both given, the polar system of the syzygetic $C+\lambda\Gamma$, corresponding to any given line Ωx, is also given. For $\sigma+\lambda\Sigma$, the polar conic of P with regard to $C+\lambda\Gamma$, is given, since it is the conic of the system (σ, Σ) with regard to which P' and Ωx are pole and polar. And, similarly, if Q be any point in the plane, the polar conic of Q with regard to $C+\lambda\Gamma$ is given; for it belongs to a given system of order 1, and the polar line of P with regard to it is known, being the same as the polar line of Q with regard to $\sigma+\lambda\Sigma$.

There are three cubics of which any given cubic is the Hessian, and these three cubics are syzygetic with the given cubic. We proceed to determine the polar systems of these three cubics.

Problem 13.—'To determine the polar systems of the cubics of which a given cubic C is the Hessian.' By a given cubic, we understand a cubic of which the polar system is given. The problem is, of course, cubical.

Solution.—Determine the polar system of Γ, the Hessian of C. Let P, P' be any two points of which the second lies on the polar conic of the first; and, as before, let σ, σ' be the polar conics of P, P', Σ, Σ' the Hessian polar conics, Ωx the mixed derivative of P, P' with regard to $C+\lambda\Gamma$. Consider the lines Ωx as corresponding anharmonically to the conics $\sigma'+\lambda\Sigma'$ and $\sigma+\lambda\Sigma$. For any given conic $\sigma+\lambda\Sigma$, determine the pencil in involution having its centre at the point Ω, of which the double rays Ωy_1, Ωy_2 correspond to the two conics of the system (σ', Σ'), which are inscribed harmonically in $\sigma+\lambda\Sigma$: we may obtain this determination by taking for L, in the construction of Problem 9, a straight line passing through P which is the point reciprocal to Ω with regard to the system (σ', Σ'). We thus have, at the point Ω, a pencil of lines Ωx, and a pencil of pairs of lines Ωy_1, Ωy_2. These two pencils correspond anharmonically to one another; for to each ray Ωx there corresponds but one pair Ωy_1, Ωy_2; and to each ray Ωy there corresponds but one ray Ωx, because there is in the system (σ, Σ) but one conic which circumscribes harmonically a given conic in the system (σ', Σ'). The ray Ωx will in three different directions coincide with one of the corresponding rays Ωy. Let Ωx_1, Ωx_2, Ωx_3 be the three directions of coincidence, which are to be determined by the cubic construction of

M. Chasles.* The three syzygetic cubics corresponding to the three rays Ωx_1, Ωx_2, Ωx_3 are the three cubics of which C is the Hessian; because, for each of those three syzygetics, the Hessian polar conic of P passes through P', and therefore coincides with σ; so that the Hessian itself coincides with C.

The problem, 'To determine the two cubics which have the same Hessian as a given cubic C,' may be solved in the same manner, but is only quadratic. In fact, when one of the three directions of coincidence Ωx_1, Ωx_2, Ωx_3 is known *à priori*, a pencil in involution of which the other two are the double lines may be determined linearly.

10. *Problem* 14.—'Given the polar system of a cubic C; to determine its nine points of inflexion.' The problem requires one biquadratic and two cubic constructions.

Solution.—The syzygetic cubics comprise four triangles; of which one is real and one consists of one real line and a pair of conjugate imaginary lines; the two others are two imaginary triangles. The first two of these triangles will serve to determine the three real and the six imaginary points of inflexion. Each of the four triangles, considered as a syzygetic cubic, is characterized by the property that it is its own Hessian. Retaining the notation of the last article, let Ωx be the ray corresponding to any given syzygetic $C+\lambda\Gamma$; and similarly let Ωh correspond to the Hessian of $C+\lambda\Gamma$, Ωx_1, Ωx_2 to those two syzygetics, other than $C+\lambda\Gamma$, which have its Hessian for their Hessians. We shall thus have, at the point Ω, a pencil of lines Ωh and a pencil of triplets of lines Ωx, Ωx_1, Ωx_2. And, since to every Hessian only one triplet of fundamental cubics corresponds, and to every cubic only one Hessian, the rays of the one pencil will correspond anharmonically to the triplets of the other. The ray Ωh will in four different directions coincide with one of the rays of its corresponding triplet; and the syzygetic cubics corresponding to the four directions of coincidence are precisely the four syzygetic triangles. Let K be any conic passing through Ω, and let the point in which any ray Ωx meets K for the second time be designated by x; also let ω, ω' be two given points, of which the first is, and the second is not, a point of K. The rays Ωh, and the conics $(\omega, \omega', x, x_1, x_2)$, will correspond to one another anharmonically; and the locus of the intersections of the corresponding rays and conics will be a cubic curve, which will cut K in the two points Ω, ω, and in four other points h_1, h_2, h_3, h_4, which may be determined by a biquadratic

* Comptes Rendus, vol. xli. p. 681.

construction indicated by M. Chasles.* The four rays Ωh_1, Ωh_2, Ωh_3, Ωh_4 (of which two, and only two, are real) are the four directions of coincidence. Let Ωh be either of the two real rays. The polar system of the syzygetic triangle corresponding to Ωh is real, and its polar conics all pass through the three vertices of the triangle. Determine two of these polar conics which pass through one and the same point taken arbitrarily in the plane; the three remaining intersections of the two conics are to be obtained by a cubic construction, and are the three vertices of the syzygetic triangle.

* Comptes Rendus, vol. xli. p. 1193.

XX.

OBSERVATIO GEOMETRICA.

[Annali di Matematica, Ser. II. vol. ii. pp. 318–321 : 1869.]

VIR clarissimus, H. Siebeck, in commentatione eximia mense Augusto edita *, solutionem dedit problematis non inelegantis :—

'Datis quinis punctis in una quaque quatuor sectionum conicarum Σ_1, Σ_2, Σ_3, Σ_4, invenire nonum punctum Ω omnibus curvis cubicis commune, quae transeunt per octo puncta (Σ_1, Σ_2) et (Σ_3, Σ_4).'

Cujus problematis solutionem et nos paullo ante inveneramus, et cum Societate Britannica eodem mense Augusto communicavimus †; nec aegre ferimus disquisitionem viri clarissimi nostrae antevertisse. Illud vero nobis proprium fortasse fuit, quod docuimus problematis solutionem revera linearem esse, cum determinatione cubica trianguli chordalis omnino supersederi possit. Cujus rei rationem paucis hic explicare liceat.

Quoties duarum conicarum altera A alterius B triangulo polari circumscripta est, hanc illi harmonice inscriptam, illam huic harmonice circumscriptam, brevitatis causa dicere consuevimus : conicae autem harmonice circumscriptae ordinem, conicae harmonice inscriptae classem, praecipue respicere oportet. Sint S_1, S_2 duae conicae quatuor conicis datis harmonice inscriptae ; omnis conica systematis

$$ {}_1 S_1 + l_2 S_2 \quad . \quad . \quad . \quad . \quad . \quad . \quad . \quad . \quad . \quad . \quad (S) $$

(quod est simplex secundae classis) omni conicae systematis

$$ \lambda_1 \Sigma_1 + \lambda_2 \Sigma_2 + \lambda_3 \Sigma_3 + \lambda_4 \Sigma_4 \quad . \quad . \quad . \quad . \quad . \quad . \quad . \quad . \quad . \quad (\Sigma) $$

---

* Hujus Diarii vol. ii. p. 65.

† British Association for the Advancement of Science, Norwich Meeting, Aug. 22, 1868.

(quod est triplex secundi ordinis) harmonice inscripta erit. Atque in universum, cuivis systemati lineari curvarum secundi ordinis respondebit per inscriptionem harmonicam systema lineare curvarum secundae classis, triplici simplex, duplici duplex, simplici triplex: quin etiam (si rem ulterius persequi placet) curvae unicae secundi ordinis systema quadruplex secundae classis, respondebit, et similiter systemati quadruplici secundi ordinis curva unica secundae classis. Cujusmodi bina systemata non sine caussa *contravariantia* appellavimus; eorumque proprietates in commentatione peculiari aliquatenus explicavimus*; in qua etiam tractavimus linearem solutionem problematis:—

'Datis quatuor conicis ex systemate triplici Σ, invenire quotlibet elementa conicae pertinentis ad systema S et rectam datam tangentis.'

Quo problemate hic utimur ad determinationem conicarum S_1 et S_2; ipsam vero solutionem, cum sit maxime elementaris, hoc loco iterare supervacaneum putamus.

Systema S continet tres conicas evanescentes; quae, quum conicis Σ nihilo secius inscriptae esse debeant, fiunt ipsarum poli conjugati. Itaque triangulus chordalis systematis Σ idem est atque triangulus polaris systematis S; datur autem implicite, quinis punctis inventis in utrâque conicarum S_1 et S_2. Praeterea apparet universum systema conicarum triangulo chordali $\alpha\beta\gamma$ circumscriptarum contineri in systemate Σ; unde sequitur tria puncta $\alpha\beta\gamma$ et quaterna puncta intersectionis binarum quarumvis conicarum ex systemate Σ ad unam eandemque conicam pertinere. Systema enim duplex et simplex, quae ambo in eodem systemate triplici continentur, unam conicam semper communem habent. Cum autem compertum sit punctum quaesitum Ω esse quartum punctum intersectionis conicarum $(\alpha, \beta, \gamma, \Sigma_1, \Sigma_2)$ et $(\alpha, \beta, \gamma, \Sigma_3, \Sigma_4)$, in eo versatur cardo quaestionis, ut demonstretur quotlibet harum elementa lineari ratione determinari posse. Quarum in neutra ne unum quidem elementum datur separatim; dantur enim implicite tum tria puncta $\alpha\beta\gamma$ utrique communia, tum in singulis quaterna (Σ_1, Σ_2) et (Σ_3, Σ_4).

Sint $a_1 a_2$, $p_1 p_2$ duo paria polorum conjugatorum in systemate (Σ_1, Σ_2). Puncta $a_1 a_2$ erunt etiam poli conjugati unius ex conicis (Σ_3, Σ_4); cujus conicae ponemus (quod licet) puncta $p_1 p_2$ nequaquam esse polos conjugatos; ita ut omnis conica ex systemate Σ, quae utramque rectam $a_1 a_2$, $p_1 p_2$, harmonice secet, in systemate (Σ_1, Σ_2) necessario comprehendatur. Sit (S_1, S_2) systema curvarum secundi ordinis, determinatum per conicas S_1 et S_2, quarum utramque (ut supra

* Lecta est Societati Mathematicae Londinensi, die 28[mo] Maii, 1868. [XIX. ante p. 524.]

diximus) quinis punctis definiri intelligimus. Hujusmodi systematis poli conjugati ita se habent, ut punctis in recta sitis reciproce respondeant puncta in conica triangulo $\alpha\beta\gamma$ circumscripta. Sint igitur A, P conicae rectis $a_1 a_2$, $p_1 p_2$, reciprocae; sit praeterea, in conica A, a polus rectae $a_1 a_2$; et similiter p polus rectae $p_1 p_2$ in conica P. Conica $[ap]$, rectae ap reciproca, erit ipsa conica quaesita $(\alpha, \beta, \gamma, \Sigma_1, \Sigma_2)$. Nam, quod primum est, conica $[ap]$ circumscribitur triangulo $\alpha\beta\gamma$, utpote rectae reciproca; ideoque continetur in systemate Σ. Deinde, quod secundum est, secat harmonice utramque rectam $a_1 a_2$, $p_1 p_2$; rectae enim $a_1 a_2$, ap conicam A quatuor punctis harmonice secant; punctis autem harmonicis in circumferentia conicae A respondent reciproce puncta in recta $a_1 a_2$ harmonica; hoc est, conica $[ap]$ rectam $a_1 a_2$ secat harmonice; quod idem de recta $p_1 p_2$ similiter probatur. Sit denique bq recta conicae $(\alpha, \beta, \gamma, \Sigma_3, \Sigma_4)$ eodem modo respondens quo recta ap conicae $(\alpha, \beta, \gamma, \Sigma_1, \Sigma_2)$: punctum quaesitum Ω erit punctum conjugatum puncto intersectionis rectarum ap, bq: eamque determinationem linearem esse patet.

Etiam solutionem problematis :—

'Determinare puctum μ quatuor punctis (Σ_3, Σ_4) oppositum in cubica quae transit per octo puncta (Σ_1, Σ_2), (Σ_3, Σ_4), et per punctum m separatim datum' ita tractari posse demonstravimus, ut ad constructionem regula tantum opus sit. Sint r_1, r_2, r_3, r_4 puncta conicis Σ_1 et Σ_2 communia; constat ex notissimo theoremate conicam $(\Omega, r_1, r_2, r_3, r_4)$, quae est ipsa illa conica $(\alpha, \beta, \gamma, \Sigma_1, \Sigma_2)$ rectae ap reciproca, satisfacere aequationi anharmonicae

$$[r_1, r_2, r_3, r_4] = (\Sigma_3, \Sigma_4) . [r_1, r_2, r_3, r_4].$$

Itaque res eo redit ut in ejusdem conicae circumferentia inveniatur punctum M, quod satisfaciat aequationi

$$[r_1, r_2, r_3, r_4, M] = (\Sigma_3, \Sigma_4) . [r_1, r_2, r_3, r_4, m].$$

Designante litera σ conicam (Σ_3, Σ_4, m), systema conicarum

$$\lambda\sigma + \lambda_1 \Sigma_1 + \lambda_2 \Sigma_2$$

cum systemate conicarum triangulo chordali circumscriptarum systema simplex Θ commune habet; quia utrumque systema duplex est, et in uno eodemque systemate triplici continetur. Cujus systematis Θ omnes conicae per puncta $\alpha\beta\gamma$ manifesto transeunt; habent autem quartum punctum intersectionis X, quod erit in circumferentia conicae $(\alpha, \beta, \gamma, \Sigma_1, \Sigma_2)$, quippe quae et ipsa pertineat ad systema Θ. Jam apparet punctis X in circumferentia conicae $(\alpha, \beta, \gamma, \Sigma_1, \Sigma_2)$ respondere conicas σ in systemate simplici (Σ_3, Σ_4); respondent autem singulis singulae; hoc est, respondent anharmonice. Praeterea conicae $(\Sigma_3, \Sigma_4, r_1)$, quam per literam σ_i designabimus, manifesto respondet punctum r_i; quia id punctum omnibus conicis

$$\lambda\sigma_i+\lambda_1\Sigma_1+\lambda_2\Sigma_2,$$

atque adeo omnibus conicis Θ, commune est. Unde sequitur punctum X ipsum esse punctum quaesitum M, cum satisfaciat aequationi

$$[r_1, r_2, r_3, r_4, X]=(\Sigma_3, \Sigma_4).[r_1, r_2, r_3, r_4, m].$$

Hinc nanciscimur sequentem determinationem puncti M respondentis datae conicae $\sigma=(\Sigma_3, \Sigma_4, m)$. Designante litera ρ quamvis conicam ex systemate (Σ_1, Σ_2), conica $(a, \beta, \gamma, \sigma, \rho)$ pertinebit ad systema Θ. Capiatur recta R huic conicae reciproce respondens, quod quomodo fieri possit jam supra explicavimus; punctum M erit punctum reciproce respondens puncto intersectionis rectarum R et ap. Simplicissima autem erit constructionis ratio, si incipimus a determinatione punctorum M_3 et M_4 conicis Σ_3 et Σ_4 respondentium; quo facto, habebimus tria puncta Ω, M_3, M_4, tribus conicis $(\Omega, r_1, r_2, r_3, r_4)$, Σ_3, Σ_4 respondentia; quartum vero punctum quartae cuivis conicae respondens ex aequalitate rationis anharmonicae facillime determinabitur.

XXI.

ON THE FOCAL PROPERTIES OF HOMOGRAPHIC FIGURES.

[Proceedings of the London Mathematical Society, vol. ii. pp. 196–248. Read April 8, 1869.]

A.—Focal Properties of Two Homographic Plane Figures.

1. *Two Plane Figures in Perspective.*

We consider two plane figures Ω and ω in perspective with one another; we denote the centre of the perspective by S, and the axis of the perspective (or the line of intersection of the two planes) by $\Omega_1\,\omega_1$; we exclude the cases in which the straight lines at an infinite distance in the two planes are corresponding lines; *i.e.* we suppose that the centre of perspective is not at an infinite distance, and that the planes are not parallel. Let OY, $o'y$ be the vanishing lines of the planes Ω and ω, or the straight lines which in the planes Ω and ω correspond to the straight lines at an infinite distance in the planes ω and Ω; the plane Ω is divided by OY into two regions (Ω_1) and (Ω_2); similarly, $o'y$ divides ω into two corresponding regions (ω_1) and (ω_2). Let (Ω_1) be that region of Ω in which $\Omega_1\,\omega_1$ is situated; then $\Omega_1\,\omega_1$ is also situated in (ω_1); and it will be seen that, if P, p are corresponding points in the regions (Ω_1), (ω_1), the radii vectores SP, Sp are of the same sign; but, if P, p are corresponding points in (Ω_2), (ω_2), the radii vectores SP, Sp are of opposite signs; or, in the language of some writers on perspective, (Ω_1) and (ω_1) are projections of one another, but (Ω_2) and (ω_2) are transprojections of one another.

2. *The Correspondence of Directions.*

If the positive and negative directions on any straight line in either of the planes Ω and ω are regarded as determined, the corresponding directions on the corresponding line are also determined; viz., if a point move in the positive direction on a straight line in either plane, its image in the other plane moves in the positive direction on the corresponding straight line. Hence, if P, Q are two points in the same region of Ω, and p, q their images, which are of course in the corresponding region of ω, the direction from P to Q along the finite segment PQ is of the same sign as the direction from p to q along the finite segment pq; but if P, Q are in opposite regions of Ω, so that the finite segment PQ is divided internally by OY in the point A, the direction $p \infty q$ will correspond to the direction PAQ, and the directions of the finite segments PQ, pq will be of opposite signs. We may add that if A is any point whatever on OY, to the directions PA, QA there will correspond similar or dissimilar directions on the parallel straight lines which are the images of the lines PA, QA, according as P and Q are in the same region or in different regions of Ω. And in particular, if in the plane Ω there be drawn any parallel to the vanishing lines, the corresponding line in the plane ω will also be parallel to the vanishing lines, but the corresponding directions on the two parallels will be similar or dissimilar according as they lie in the regions (Ω_1), (ω_1) or in the regions (Ω_2), (ω_2).

Again, if in the plane Ω we consider one of the two directions of rotation round any point as positive, (say, for example, that direction of rotation which viewed from S appears right-handed,) the signs of the directions of rotation will thereby be fixed for each point of the plane ω; but for all points in the region (ω_1) that direction of rotation which viewed from S is right-handed will be positive, whereas for all points of the region (ω_2) the same direction of rotation must be considered negative; it being inconsistent with the perspective relation to regard one and the same direction of rotation as being positive for all points of the one plane, and also to regard one and the same direction of rotation as being positive for all points of the other plane.

3. *The Equiangular Points, or Foci.*

Through S draw two lines perpendicular to the planes which bisect the dihedral angle formed by the intersecting planes Ω and ω. Let these perpendiculars meet the plane Ω in F_1, F_2, and the plane ω in $f_1 . f_2$; let also SF_2f_2 be perpendicular to that bisecting plane which lies in the same angle with S: then SF_1, Sf_1 are of the same sign, and F_1, f_1 lie in the regions (Ω_1), (ω_1) respectively; but SF_2, Sf_2 are of opposite signs, and F_2, f_2 lie respectively in the regions (Ω_2), (ω_2).

Since each of the lines F_1f_1, F_2f_2 is at right angles to the line of intersection of Ω and ω, and is besides equally inclined to those two planes, any dihedral angle of which the axis is either F_1f_1 or F_2f_2 is intersected in two equal rectilineal angles by the planes Ω and ω. We thus obtain the theorem :—

'Angles in the plane Ω at the points F_1, F_2 are projected into equal angles in the plane ω at the points f_1, f_2.'

Or, more precisely :—

'The angle contained by given directions on two straight lines lying in the plane Ω, and intersecting at F_1 or F_2, is equal to the angle formed in the plane ω by the corresponding directions on the corresponding straight lines which intersect at f_1 or f_2.'

For brevity, we shall call F_1, F_2, f_1, f_2, the *foci* of the perspective in the planes Ω and ω respectively. And, since the directions of rotation as viewed from S are the same round F_1 and f_1, but are opposite round F_2 and f_2, we shall call F_1, f_1 the *similar foci*, and F_2, f_2 the *dissimilar foci*.

In the annexed diagram the plane of the paper is the plane of symmetry;

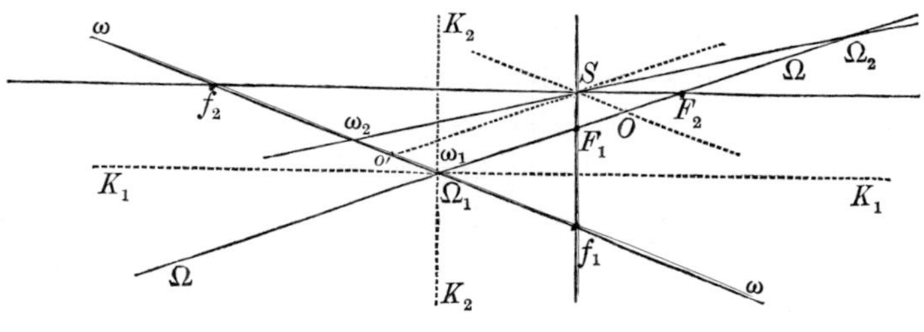

i.e. it is the plane passing through S and cutting Ω, ω at right angles; $O\Omega_1$, $o'\omega_1$ are the traces of the given planes Ω and ω, K_1, K_2 are the traces of the bisecting planes; F_1F_2, f_1f_2 are the foci, which lie of course in the plane of symmetry; the axis of the perspective $\Omega_1\omega_1$ is perpendicular to the plane of the paper at Ω_1 or ω_1, and the vanishing lines OY, $o'y$ are perpendicular to the same plane at the points O, o', which we shall term the *centres* of the two planes. If Y, y are the points at infinity on the vanishing lines, Y and y are corresponding points; but it will be observed that the vanishing lines are not corresponding lines, nor the centres corresponding points. We may term the lines F_1OF_2, $f_1o'f_2$, which are corresponding lines at right angles to the vanishing lines, the *focal axes* of the two planes; so that the centre of each plane corresponds to the point at an infinite distance on the focal axis of the other. Since SO, So' are parallel to the traces of the planes Ω, ω, and SF_1f_1, F_2Sf_2 to the traces of the bisecting planes, the figure $SO\Omega_1o'$ is a parallelogram, the triangles SOF_1, SOF_2,

$So'f_1$, $So'f_2$ are isosceles, F_1F_2 and f_1f_2 are bisected at the centres O and o', SO is equal to OF_1 or OF_2, So' to $o'f_1$ or $o'f_2$; whence we find

$$\Omega_1 F_1 = \quad \omega_1 f_1 = \tfrac{1}{2}(f_2 f_1 + F_2 F_1),$$
$$\Omega_1 F_2 = -\omega_1 f_2 = \tfrac{1}{2}(f_2 f_1 - F_2 F_1).$$

These equations, as all others in this paper, are to be interpreted, with regard to sign as well as magnitude; we shall avoid the use of the sign = when we have to speak of equality irrespective of sign.

4. *The Equi-Segmental Axes, or Cyclic Lines.*

Any point of the line $\Omega_1\omega_1$ in which the planes Ω and ω intersect, considered as a point in either plane, has itself for its corresponding point in the other plane. If we do not attend to the coincidence of the corresponding points, we may express this by saying that $\Omega_1\omega_1$ is an *equi-segmental* line in either plane; *i.e.* that to any segment of $\Omega_1\omega_1$, considered as a line in either plane, an equal segment corresponds in the other plane. But besides this coincident pair of equi-segmental lines, there is another pair of corresponding equi-segmental lines which are not coincident. Through S extend a plane parallel to the plane containing the vanishing lines, and let it meet the focal axes of the two planes in Ω_2 and ω_2. The lines O_2Y, $\omega_2 y$, parallel to the vanishing lines, are equi-segmental lines. For, if P, p are corresponding points on those lines, the radii vectores SP, Sp, $S\Omega_2$, $S\omega_2$, are respectively equal and of opposite signs, and the angles $\Omega_1 SP$, $\omega_1 Sp$ are equal; so that P, p lie at equal distances from the plane of symmetry but on opposite sides of that plane; *i.e.* the lines $\Omega_2 Y$, $\omega_2 y$ are equi-segmental. We may term the coincident lines $\Omega_1 y_1$, $\omega_1 y$, the *similar axes*, and the lines $\Omega_2 Y$, $\omega_2 y$ the *dissimilar axes*. Since $\Omega_1\Omega_2$ is double of $\Omega_1 O$, *i.e.* of $o'S$ or $o'f_1$, and similarly $\omega_1\omega_2$ is double of $\omega_1 o'$, *i.e.* of OS or OF_1, we see that in either plane the equi-segmental axes are situated symmetrically with respect to the centre of that plane, and that the semi-distance between the foci is equal to the semi-distance between the equi-segmental axes in the other plane. The semi-distance between the foci in either plane may conveniently be called the *parameter* of that plane. Designating the parameters of the planes Ω and ω by C and c, we shall find that, if $C=c$, the foci of each plane lie on its equi-segmental axes, and the centre of perspective lies on one of the two bisecting planes. But, if C and c are unequal, for example, if $C<c$, the foci lie between the equi-segmental axes in Ω, and outside them in ω, and the distances $c-C$ and $c+C$, between a focus and the nearer and further equi-segmental axis, are the same for both planes.

If we cause one of the two planes, for example the plane ω, to revolve round the axis of intersection of the two planes, the two figures, as is well known, will continue in perspective; and the locus of the centre of perspective will be a circle lying in the plane of symmetry and described on $F_1 F_2$ as diameter. At one of the coincidences of the two planes which take place during a complete revolution the similar foci come to coincide with one another and with the centre of perspective; and, in like manner, at the other coincidence the dissimilar foci coincide with one another and with the centre of perspective; the similar foci continuing similar and the dissimilar foci continuing dissimilar during the whole revolution. If, however, we bring together corresponding points in the dissimilar axes, (which we may conceive done by causing either plane to rotate through an angle of 180° round an axis perpendicular to the plane $\Omega_2 \omega_2$ at S,) the two figures will again be in perspective, but the foci, which were before similar, will become dissimilar, and *vice versâ*. Thus the two foci, and the two equi-segmental axes, in either plane, are not absolutely distinguished as similar to, or dissimilar from, their corresponding foci or axes; these denominations being, in fact, relative to one or other of the two ways in which the planes can be placed in perspective with one another, and changing when we pass from one of those ways to the other. In every case, if a pair of corresponding foci or axes be regarded as similar, the nearer axes or foci are also to be regarded as similar.

The equi-segmental axes may, perhaps, be called the *cyclic lines*. This denomination is suggested by an anology which will come before us presently.

5. *Any two Homographic Plane Figures.*

Since any two homographic plane figures, such that the lines at an infinite distance in the two figures are not corresponding lines, (this limitation is to be always understood in what follows when we speak of two homographic plane figures,) can be placed in perspective with one another, it appears that in any two such homographic systems there exist two pairs of corresponding foci, and two pairs of equi-segmental axes. This we shall now show independently of all considerations of perspective. Let P_1, P_2, q_1, q_2 be the imaginary circular points at an infinite distance in the planes Ω and ω respectively; and to $P_1 P_2$, $q_1 q_2$ let $p_1 p_2$, $Q_1 Q_2$ correspond in the planes ω and Ω respectively. The lines $P_1 P_2$, $p_1 p_2$ and $Q_1 Q_2$, $q_1 q_2$ will be pairs of corresponding lines; $P_1 P_2$, $q_1 q_2$ being the lines at an infinite distance in the two planes, and $p_1 p_2$, $Q_1 Q_2$ the vanishing lines. Further, the three diagonal points of the quadrangle $P_1 P_2 Q_1 Q_2$ (which are all real) will correspond to the three diagonal points of the quadrangle $p_1 p_2 q_1 q_2$; of these three pairs of corresponding points, one pair are the points Y, y at an infinite distance on the vanishing lines; the two other pairs are the two pairs of foci. For, if F, f be corresponding diagonal points (other than Y, y) of the two imaginary quadrangles, the homographic pencils at F, f will be equiangular, because the imaginary circular asymptotes FP_1, FP_2 correspond in the pencil at F to the imaginary lines fp_1, fp_2, or fq_1, fq_2, *i.e.* to the imaginary circular asymptotes in the pencil at f. To determine the two pairs of foci in two given homographic planes Ω and ω, we consider a pair of rectangular points at an infinite distance in each plane; let A_1A_2, b_1b_2 be these pairs of points; a_1a_2, B_1B_2 the pairs of points corresponding to them. The lines B_1B_2, a_1a_2 are the vanishing lines of the two planes; the centre of either plane is the point corresponding to the point at an infinite distance in the direction perpendicular to that of the vanishing line in the other plane; the focal axes are the lines perpendicular to the vanishing lines of the two planes at their respective centres; lastly, the foci are the points of intersection of the focal axes by the circles described on B_1B_2, a_1a_2 as diameters, and are situated in each plane symmetrically with regard to its vanishing line. If, assuming that we view each plane from a determinate region in space, we regard the rotations round F_1, f_1 as similar, it is evident that the rotations round F_2, f_2 must be dissimilar, and *vice versâ*; otherwise the two homographic figures would be similar, and the lines at an infinite distance would be corresponding lines contrary to the hypothesis. We might prove the same thing, by imagining the planes of the two figures to coincide. The circular asymptotes at F_1, f_1, and again at F_2, f_2, will then be corresponding lines. But the correspondence in one case will be direct, and in the other inverse (*i.e.* in the one case those asymptotes which run to the same imaginary circular point at an infinite distance will be corresponding lines, in the other case asymptotes running to opposite circular points will correspond). And, since the locus of the intersections of corresponding rays in two equiangular pencils is a circle or an equilateral hyperbola, according as the rotations of the two pencils are in the same or in opposite directions, we infer that two equiangular pencils in the same plane have the same direction of rotation or opposite directions, according as the circular asymptotes of the two pencils correspond directly or inversely.

As we have obtained the foci in each plane by a quadratic construction, (which seems inevitable,) we have still to determine their correspondence and the corresponding directions of rotation round each. To do this, we have only to observe that each plane is divided by its focal and vanishing axes into four regions which correspond to one another in a manner which is readily ascertained because, when we pass in either plane from one region A into another region B across one of their common boundaries, we must simultaneously pass in the other plane from the region corresponding to A into the region corresponding to B, and must traverse the corresponding boundary. (The line at an infinite distance is a common boundary, it will be observed, of two diametrically opposite regions.) Thus we have only to ascertain in either plane the region which corresponds to a given region in the other; the correspondence of the remaining regions is then known, and with it the correspondence of the foci. Lastly, if O', o be the points at an infinite distance on the focal axes of the two planes, the direction of rotation from FO' to FY corresponds to the direction of rotation from fo to fy, the rotating radii vectores being supposed to move in corresponding regions in the two planes.

If A, a are corresponding points in the regions (Ω_2), (ω_2) respectively, the angles F_2F_1A, f_2f_1a are both acute; they are, therefore, equal to one another, since, by virtue of the equiangularity of the pencils at F_1, f_1, they must be either supplementary or equal. Observing that the directions F_1F_2, F_1A are opposite in sign to the directions f_1f_2, f_1a, we see that angles between corresponding directions on corresponding radii vectores are equal, in which form we have already stated the equiangular property of the foci.

The determination of the foci requires (as we have seen) the construction of three points in each plane corresponding to three points at an infinite distance in the other plane. And thus the determination of the foci, though very elementary in theory, is in actual practice somewhat troublesome. But, when the foci have once been determined, the homographic representation of either plane upon the other can be carried out very rapidly; since, if A be any given point of Ω, F_1 the further, and F_2 the nearer focus, we have only to make the angle f_2f_1a equal to the angle F_2F_1A, and the angle f_1f_2a supplementary to the angle F_1F_2A; the directions in which these angles are to be measured being at once indicated by the correspondence of the regions in which A and a are situated.

The anharmonic equation $(O, X, F_1, \infty) = (\infty, x, f_1, o')$ or $OX.o'x = -C.c$, in which X, x denote corresponding points on the focal axes, suffices to prove that the parallels to the vanishing lines at Ω_1, ω_1 and again at Ω_2, ω_2, are corresponding lines. And that these lines are equi-segmental will then follow from the equiangular property of the foci, since $F_1\Omega_1 = f_1\omega_1$, $F_1\Omega_2 = -f_1\omega_2$. The image in either plane of any given indefinite line in the other is most easily found by making the intercepts on the equi-segmental axes in the first plane equal in sign and in magnitude to the corresponding intercepts in the second plane;

so that, if the two intercepts are drawn in the same direction from the focal axis in one plane, they are drawn in opposite directions in the other plane.

6. *Homographic Plane Figures placed Homologically.*

It will be observed that any two homographic plane figures can be made homological, or put in plane perspective with one another, in four different ways. For we can take either pair of equi-segmental axes for the axis of homology, and either pair of foci for the centre of homology. And it is sometimes of importance, in the theory of homological figures, to consider the non-coincident foci and equi-segmental axes, as well as the two foci which are united in the centre of homology, and the two equi-segmental axes which are united in the axis of homology.

For example, if we regard a conic section as homological with itself, any point in the plane of the conic being the centre of homology, and its polar the axis of homology, the foot of the perpendicular let fall from the pole upon the polar will represent the second pair of foci (which in this case are coincident because the parameters are equal); and in like manner the second pair of equi-segmental axes will be represented by the parallel to the polar through the pole. Thus we have the elementary properties of a conic section, 'angles subtended at the foot of the perpendicular by chords passing through the pole are bisected by the polar,' 'the pole is the point of bisection of intercepts on the parallel to the polar made by tangents at the extremities of chords passing through the pole,' &c.

Again, let two conics be regarded as homological, a point of intersection of their common tangents being the centre of homology, and the axis of homology being either of the two common chords which pass through the intersection of the polars of the centre of homology with regard to the two conics; then it will be found that there is a second pair of foci, situated on the perpendicular let fall from the centre of homology upon the axis of homology, and that corresponding points of the two conics subtend equiangular pencils (with opposite rotations) at these two points. And, in like manner, corresponding lines in the two figures determine equal intercepts (but measured in opposite directions) upon two axes situated at distances from the second pair of foci equal to the distance of the axis of homology from the centre of homology.

7. *Homographic Plane Figures placed Symmetrically.*

The equality of the parameters of two plane homographic figures is the necessary and sufficient condition that they should be capable of being so placed in the same plane that each point shall have but one corresponding point. For, if the two figures can be so placed as to have this symmetrical relation to one another, the imaginary points corresponding to the imaginary circular points at an infinite distance must coincide; and hence the imaginary chords $Q_1 Q_2$, $p_1 p_2$, and with them the real parameters, must be equal. Conversely, we can always render the two figures capable of a symmetric position by altering the linear dimensions of either of them in the ratio of its parameter to the parameter of the other; since after this alteration the imaginary chords $Q_1 Q_2$, $p_1 p_2$ will be equal and can be made to coincide. There are then two positions of symmetry, viz., the two positions of homology in which the vanishing lines coincide. It is sometimes convenient to imagine the scale of one of the figures altered in the parametric ratio; we shall express this by saying that the figures are reduced to the same scale.

8. *Metrical Properties of the Focal Radii Vectores.*

The following elementary properties of the foci of two homographic plane figures are frequently useful:—

(α) The focal radii vectores of two corresponding points P and p satisfy the equation

$$\frac{F_1 P}{F_2 P} + \frac{f_1 p}{f_2 p} = 0.$$

The truth of this equation, so far as absolute magnitude is concerned, appears immediately from a comparison of the triangles $F_1 P F_2$, $f_1 p f_2$; the two ratios are of opposite signs, because the radii vectores drawn from the nearer foci are of the same sign and the radii vectores drawn from the further foci are of opposite signs.

(β) The rectangle contained by the sum of the radii vectores of P and the difference of the radii vectores of p is equal to the rectangle contained by the difference of the radii vectores of P and the sum of the radii vectores of p; and either of these rectangles is equal to four times the rectangle of the parameters. This relation is expressed by the single equation

$$(F_1 P + F_2 P)(f_1 p + f_2 p) = F_1 F_2 \cdot f_1 f_2,$$

if we observe that in it the signs of $F_1 P$ and $F_2 P$ are arbitrary, and that the signs of $f_1 p$ and $f_2 p$ depend on the signs of $F_1 P$ and $F_2 P$ respectively. The truth of the equation may be inferred immediately from the elementary theorem that, if a straight line bisecting the angle of a triangle either internally or externally be produced to meet the base, the square of the bisecting line is equal to the rectangle contained by the sum of either side and the segment of the base adjacent to it and the difference between the other side and the segment adjacent to it.

(γ) Let N, n be the feet of the perpendiculars let fall from P and p upon the vanishing lines. The distances NP, np are of opposite sign, and their rectangle (see Art. 5) is equal to the rectangle of the parameters. We then have the relations

$$\frac{F_1P}{NP}=\frac{f_1p}{o'f_1},\quad \frac{F_2P}{NP}=\frac{f_2p}{o'f_2},\quad \frac{f_1p}{np}=\frac{F_1P}{OF_1},\quad \frac{f_2p}{np}=\frac{F_2P}{OF_2},$$

any one of which, combined with the equation $NP.np=OF_1.o'f_1$ and with the equation (α), gives the other three. To prove them, we have only to observe that the quadrilaterals NPF_1O, $o'f_1pn$ are not only equiangular, but also similar, because $NP.np=OF_1.o'f_1$.

If, in the construction of the point p corresponding to a given point P, we wish to avoid the use of points lying on the further side of the vanishing line, we may either determine the ratio of f_1p to F_1P by one of the formulae (γ); or, preferably, we may make the angle $f_1o'p$ equal to the angle OF_1N, the point p being then the intersection of $o'p$ and f_1p.

(δ) If the two planes are in perspective, we have for the similar foci,

$$\frac{F_1P}{f_1p}=\frac{SP}{Sp}$$

and for the dissimilar foci,

$$\frac{F_2P}{f_2p}=-\frac{SP}{Sp}.$$

9. *Circles changed into Circles.*

The circles of the system of which F_1, F_2 are the limiting points and the vanishing line is the radical axis are transformed into circles of the system of which f_1, f_2 are the limiting points and the vanishing line is the radical axis; and these are the only circles in either figure which are changed into circles in the other figure. For it is evident that to the conics (P_1, P_2, Q_1, Q_2) there will correspond the conics (p_1, p_2, q_1, q_2); or again, if the locus of A be a circle of the system of which F_1, F_2 are the limiting points, the ratio $\dfrac{F_1A}{F_2A}$ is constant; therefore the ratio $\dfrac{f_1a}{f_2a}$ is also constant, Art. 8, (α); *i.e.* the locus of a is a circle of the system of which f_1, f_2 are the limiting points. We shall, for brevity, call these two systems of circles the *focal circles* of the two planes.

It will be found that the radii of corresponding circles are to one another as the parameters; and that, if the figures be reduced to the same scale and superposed so that their foci coincide, the corresponding circles will coincide (but not in respect of their corresponding points).

10. *The Homographic Modulus of Corresponding Pencils.*

In any two homographic pencils (A) and (a) there exists a pair of corresponding right angles (Steiner, 'Systematische Entwickelung,' p. 31); and, if the pencils are not equiangular, there is only one such pair. We shall term these corresponding right angles the *right angles of the pencils* (A) and (a). Let A and a be any two corresponding points in the planes Ω and ω; the lines bisecting the angles F_1AF_2, f_1af_2, internally and externally, are the lines containing the right angles of the pencils at A and a. For the double rays of the pencil $A.[P_1P_2,\ Q_1Q_2,\ F_1F_2]$, (which is a pencil in involution, because P_1P_2, Q_1Q_2, F_1F_2 are the vertices of a quadrangle,) correspond to the double rays of the corresponding pencil $a.[p_1p_2,\ q_1q_2,\ f_1f_2]$. We might prove the same thing, without using imaginary points, by considering the corresponding circles which pass through A and a. And, since the rectangle of the central abscissas of corresponding points is equal to the rectangle of the parameters, we see that the external bisector at either of the two points A or a corresponds to the internal bisector at the other.

If, in any two homographic pencils, Φ and ϕ are corresponding angles measured from either pair of the corresponding rectangular lines, the ratio $\tan\Phi : \tan\phi$ is constant. This constant ratio we may term the *homographic modulus* of the two pencils. We observe (1) that the definition is applicable to homographic pencils in involution; (2) that the homographic modulus is positive or negative according as corresponding directions of rotation in the two pencils are regarded as having the same sign or opposite signs; (3) that the homographic modulus of two equiangular pencils is $+1$ or -1; (4) that the definition is relative to a given pair of the corresponding rectangular lines, and that, if for this pair we substitute the other pair, the homographic modulus changes into its reciprocal.

The homographic modulus of the pencils at A and a, taken relatively to the external bisector at A and the internal bisector at a, is evidently

$$\begin{aligned}\cot\tfrac{1}{2}A\ \cot\tfrac{1}{2}a &= \tan\tfrac{1}{2}(F_1+F_2)\ \tan\tfrac{1}{2}(f_1+f_2) = \tan\tfrac{1}{2}(F_1+F_2) : \tan\tfrac{1}{2}(F_1-F_2)\\ &= R_2+R_1 : R_2-R_1 = r_2-r_1 : r_2+r_1;\end{aligned}$$

the letters A, F_1, F_2, a, f_1, f_2 denoting the internal angles of the triangles F_1AF_2, f_1af_2, and R_1, R_2, r_1, r_2 representing the absolute values of the focal radii vectores of A and a. We have supposed that F_1 is the nearer, F_2 the further focus, so that $F_2=f_2$, $F_1+f_1=\pi$.

11. *Angles changed into Equal or Supplementary Angles.*

In any two homographic pencils (A) and (a) there exists an infinite number of equal corresponding angles, and again an infinite number of supplementary corresponding angles (M. Chasles, 'Géométrie Supérieure,' Art. 147). We may add that the angles in either pencil (A) which are equal to their corresponding angles in the other pencil form a pencil in involution of which the right angle is the right angle of the pencil (A), and of which the modulus is the homographic modulus of the two pencils taken positively; and, similarly, the angles in the pencil (A) which are supplementary to their corresponding angles form a pencil in involution which has the same right angle as the pencil (A) and the same modulus taken negatively. The former involution always has real double lines, the latter never.

To find the involutions of equal angles at the corresponding points A and a, let F_1, f_1 be the nearer foci, let any circle passing through A and F cut the vanishing line in M_1, M_2, and let m_1, m_2 be the corresponding points at an infinite distance in the plane ω. The angle M_1AM_2 will be transformed into an equal angle m_1am_2; for the angles M_1AM_2, $M_1F_1M_2$ are equal (not supplementary, since the chords M_1M_2, AF_1 do not intersect); *i.e.* the angles M_1AM_2 and $m_1f_1m_2$, or finally M_1AM_2 and m_1am_2, are equal. It will be observed that the directions AM_1, am_1, and again AM_2, am_2, are corresponding directions.

To find the involutions of supplementary angles at the points A and a, we have only, in the preceding construction, to substitute the further focus F_2 for the nearer focus F_1. The angles M_1AM_2, $M_1F_2M_2$ will be supplementary (and not equal); so that M_1AM_2 will be transformed into a supplementary angle m_1am_2.

12. *Segments changed into Equal Segments.*

If H and h' are the *centres* of two homographically divided lines, (*i.e.* the points which on each line correspond to the points at an infinite distance upon the other line,) and if A, a are any two corresponding points on the two lines, the rectangle $HA \times h'a$, which we may term the *rectangle of the homography*, is constant.

In any two homographically divided lines there is an infinite number of segments equal to their corresponding segments and having the same sign, and again an infinite number of segments equal in absolute magnitude to their corresponding segments but having opposite signs. Upon either line either set of segments form an involution of which the centre is the homographic centre of the line, and of which the rectangle is the rectangle of the homography taken positively or negatively according as the segments considered are equal to their corresponding segments with the same sign or with opposite signs.

Thus, if $2D$, $2d$ are the intercepts made by the cyclic axes on any two corresponding lines meeting the vanishing lines in H and h', the segments X_1X_2 of the involution determined by the equation $HX_1 \times HX_2 = D \times d$ are equal to the corresponding segments x_1x_2 of the involution $h'x_1 \times h'x_2 = D \times d$; and, again, the corresponding segments of the involutions determined by $HX_1 \times HX_2 = -D \times d$, and by $h'x_1 \times h'x_2 = -D \times d$ are equal but have opposite signs.

And, in general, the segments of the involution $HX_1 \times HX_2 = \mu \times D \times d$, where μ is any multiplier, are μ times the corresponding segments of the involution $\mu \times h'x_1 \times h'x_2 = D \times d$. The segments of the two involutions are divided externally or internally by the vanishing lines according as μ is positive or negative.

It thus appears that there are an infinite number of triangles in either plane, similar to their corresponding triangles and having a given ratio to them. For take any point A in the plane Ω, any straight line passing through it, and any positive ratio; there is always one triangle, (and only one, if A is not a focus and the given line not a cyclic axis,) having a vertex at A and a side in the given straight line, which is transformed into an equiangular triangle of which the sides are to the sides of the triangle in the given positive ratio. The two corresponding triangles are not intersected by the vanishing lines, so that points interior to either triangle correspond to points interior to the other. If we do not attend to the sign of the given ratio, there are in all four triangles, each having a vertex at the point A and a side upon the given line, which are equiangular to their corresponding triangles, and which have the given ratio for their ratio of similarity. For, if we do not attend to signs, there are two segments of the given line having a common extremity at A which are in the given ratio to their corresponding segments, and there are two lines passing through A which make angles with the given line equal to the corresponding angles. Of these triangles, that just considered is one; the other three are all intersected by the vanishing line, so that two of the sides of any one of them are to their corresponding lines in a negative ratio.

13. *The Confocal Conics.*

Every conic which has a focus at F_1 or F_2 is transformed into a conic having a focus at f_1 or f_2. For right angles at F are transformed into right angles at f, so that if the involution determined by the given conic at the point F be rectangular, the involution determined by the corresponding conic at the point f is also rectangular. And conversely, if a point in either figure and its image in the other be both foci of corresponding conics, the point and its image are corresponding foci in the two figures.

Two cases of this property are of special interest :—

(i) A circle having its centre at F is transformed into a conic of which f is the focus and the vanishing line the directrix. This follows, independently of the general property, from the equations (γ), which also show that the eccentricity of the conic is equal to the radius of the circle divided by the parameter: the eccentricity thus varies directly as the radius of the circle.

(ii) Conics in the plane Ω of which F_1, F_2 are the foci are transformed into conics of which f_1, f_2 are the foci, the ellipses into hyperbolas, and the hyperbolas into ellipses. We shall term these conics the *confocal conics* of the two homographic figures. An independent proof of the theorem is supplied by the formula (β); and conversely, the theorem may be used to establish that formula, since the rectangle contained by the major semi-axes of two corresponding confocal conics is evidently equal to the rectangle of the parameters.

The eccentricity of any confocal in either figure is the reciprocal of the eccentricity of the corresponding confocal, the asymptotes of the hyperbola containing the same angle as the focal radii vectores of the extremities of the minor axis of the ellipse. If the figures be reduced to the same scale and the foci be superposed, corresponding confocals will intersect on the cyclic axes and will thus have the same latus rectum.

If Σ and σ are any two corresponding confocals, the normals of Σ and the normals of σ are corresponding lines (Art. 10). Hence also the centre of curvature at any point of Σ corresponds to the centre of curvature at the corresponding point of σ, and the evolute of either is transformed into the evolute of the other. Again, any two *similar* arcs of Σ (*i.e.* arcs of which the difference is geometrically rectifiable) have for their corresponding arcs two similar arcs of σ; and the rectifying tangents in either figure (*i. e.* the tangents of which the difference is equal to the difference of the arcs) are images of the rectifying tangents in the other figure. In the same way, the polygon of a given number of sides and of minimum perimeter, inscribed in any arc of a confocal conic of either figure, corresponds to the polygon of the same number of sides and of minimum perimeter, inscribed in the corresponding confocal arc. It is hardly necessary to observe that the elliptic integrals which express the lengths of corresponding arcs of corresponding confocal curves are not themselves equal to one another, and are not transformed into one another by the homographic transformation.

14. *The Indicatrix, or Strain Ellipse.*

The *indicatrix* at any point A of the plane Ω is the evanescent ellipse which is the image of an evanescent circle, having its centre at the corresponding point a. The indicatrix is, in fact, the 'strain ellipse,' if we regard any part of the plane Ω as a deformation of the corresponding part of the plane ω, produced by a mechanical strain. It is readily seen that, if we consider the radius of the evanescent circle at a as an infinitesimal of the first order, the distance of the centre of the strain ellipse from A will be an infinitesimal of the second order. For the determination in species of the strain ellipse at the point A, we have the theorem, 'The strain ellipse is similar and similarly situated to the ellipse of which the principal axes are normal at the point A to the confocal ellipse and hyperbola intersecting at that point, and are respectively equal to the major axes of those curves.'

This auxiliary ellipse is no other than the ellipse employed by M. Chasles in his solution of the problem, 'To determine the principal axes of an ellipse of which one pair of conjugate diameters are given in magnitude and position' ('Aperçu historique des Méthodes en Géométrie,' Note 25). M. Chasles has shown that a reciprocal relation subsists between the auxiliary ellipse and the ellipse of the confocal system which passes through its centre. Thus the centre of either ellipse lies on the circumference of the other; the major axis of either is normal to the other; the asymptotes of either pass through the imaginary foci of the other; the major and minor axis of either are respectively equal to the sum and difference of the focal radii vectores of its centre considered as a point on the circumference of the other; lastly, the distance between the real foci of either is equal to that diameter of the other which is conjugate to the diameter passing through the two centres.

To prove that the strain ellipse is similar and similarly situated to the auxiliary ellipse, it is sufficient to observe that the asymptotes of the evanescent circle at a are aq_1, aq_2; and that, consequently, the strain ellipse touches the imaginary lines AQ_1, AQ_2 at the points Q_1, Q_2. But the strain ellipse is infinitesimal; the imaginary lines AQ_1, AQ_2 are therefore its asymptotes, *i.e.* it is similar and similarly situated to the auxiliary ellipse.

To determine, then, the strain ellipse in species, we have only to draw the focal radii vectores of the point A, and to bisect the angle contained by them internally and externally; the major and minor axes of the strain ellipse are respectively in the directions of the bisecting lines, and are proportional to the sum and difference of the radii vectores. It will be seen that the confocal hyperbolas are lines of greatest elongation or least compression, and that the confocal ellipses are lines of least elongation or greatest compression. The focal circles are lines of *similar distortion*, because for all points on any one of them the ratio of the two radii vectores is constant, and therefore the ratio of their sum and difference is constant; *i.e.* all points on the same focal circle have similar indicatrices.

It remains to find the absolute dimensions of the indicatrix. Let $d\Sigma_1$, $d\Sigma_2$ represent the elements of the arcs of the confocal hyperbola and ellipse which intersect at A; let $d\sigma_1$, $d\sigma_2$ represent the corresponding elements in the plane ω; and let Λ_1, Λ_2, λ_1, λ_2 be the major semi-axes of the curves Σ_1, Σ_2, σ_1, σ_2; so that, supposing F_1 the nearer focus, we have

$$\Lambda_2 = \tfrac{1}{2}(R_2 + R_1), \qquad \lambda_2 = \tfrac{1}{2}(r_2 - r_1),$$

$$\Lambda_1 = \tfrac{1}{2}(R_2 - R_1), \qquad \lambda_1 = \tfrac{1}{2}(r_2 + r_1),$$

$$\Lambda_1\lambda_1 = \Lambda_2\lambda_2 = Cc.$$

Considering two points on Σ_1 and Σ_2 indefinitely near to A, and denoting, as in Art. 10, the angles F_1AF_2 and f_1af_2 by A and a, we find

$$d\Sigma_1 = \frac{d\Lambda_2}{\cos\frac{1}{2}A}, \qquad d\sigma_1 = -\frac{d\lambda_2}{\sin\frac{1}{2}a},$$

$$d\Sigma_2 = -\frac{d\Lambda_1}{\sin\frac{1}{2}A}, \qquad d\sigma_2 = \frac{d\lambda_1}{\cos\frac{1}{2}a}.$$

But
$$\frac{d\Lambda_1}{\Lambda_1} + \frac{d\lambda_1}{\lambda_1} = 0, \qquad \frac{d\Lambda_2}{\Lambda_2} + \frac{d\lambda_2}{\lambda_2} = 0;$$

so that
$$d\Sigma_1 = \frac{\sin\frac{1}{2}a}{\cos\frac{1}{2}A} \cdot \frac{1}{\lambda_2} \cdot \Lambda_2\, d\sigma_1,$$

$$d\Sigma_2 = \frac{\cos\frac{1}{2}a}{\sin\frac{1}{2}A} \cdot \frac{1}{\lambda_1} \cdot \Lambda_1\, d\sigma_2.$$

Again, substituting for $\frac{1}{2}a$ its value $\frac{1}{2}(F_1 - F_2)$, we find, from the triangles $F_1 A F_2$, $f_1 a f_2$,

$$\frac{\sin \frac{1}{2} a}{\cos \frac{1}{2} A} = \frac{\Lambda_1}{C}, \qquad \frac{\cos \frac{1}{2} a}{\sin \frac{1}{2} A} = \frac{\Lambda_2}{C};$$

whence, if i be the radius of the evanescent circle at a, the major and minor semi-axes of the strain ellipse at A are respectively

$$\frac{\Lambda_1 \Lambda_2}{Cc} \cdot \frac{\Lambda_2}{C} i, \qquad \frac{\Lambda_1 \Lambda_2}{Cc} \cdot \frac{\Lambda_1}{C} i. \quad \ldots \ldots \ldots \quad \text{(A)}$$

From these expressions it follows that, given in magnitude the parameters of the two planes, and given in position a single pair of corresponding points A and a, given also in position and magnitude the indicatrix at one of these points, for example at A, then the homography of the two planes is determined. For the coefficients,

$$\frac{\Lambda_1 \Lambda_2}{Cc} \times \frac{\Lambda_2}{C}, \qquad \frac{\Lambda_1 \Lambda_2}{Cc} \times \frac{\Lambda_1}{C},$$

being given, the values of Λ_1 and Λ_2 may be found (by the extraction of a cube root): thus the auxiliary ellipse at A is completely determined. The vanishing line of the plane Ω is one of the four tangents to the auxiliary ellipse which are parallel to a diameter equal to $2C$, and the homographic centre is the point of contact. Similarly the vanishing line and centre in the plane ω may be determined, since the indicatrix at a can be found when that at A is given.

15. *The Canonical and Elliptic Equations of a Plane Homography.*

If X, Y, x, y are the coordinates of two corresponding points, the focal axis and vanishing line in each plane being taken as the axes of coordinates, we have

$$Xx = Cc,$$

$$\frac{Y}{X+C} = \frac{y}{x+c},$$

the former equation being equivalent to the anharmonic equation of Art. 5, the latter expressing the property of the corresponding foci of which the abscissae are $-C$ and $-c$. These equations, written in either of the forms

$$\left. \begin{array}{l} Xx = Cc, \\ Yx = Cy, \end{array} \right\} \quad \text{or} \quad \left. \begin{array}{l} Xx = Cc, \\ Xy = cY, \end{array} \right\} \quad \ldots \ldots \ldots \quad \text{(B)}$$

may be termed the *canonical equations* of the homography (M. Chasles, 'Géométrie Supérieure,' Art. 533), and may be employed to verify analytically the preceding results. It will be remembered that the axes of y and Y are not corresponding lines, neither are the origins corresponding points. Thus the abscissae of corresponding points are not corresponding lines, and indeed are not measured in corresponding directions; but the ordinates of corresponding points (considered as lines drawn from the extremities of the abscissae parallel to the vanishing lines) are corresponding lines.

The elliptic coordinates Λ_1, Λ_2 and λ_1, λ_2 of two corresponding points (*i.e.* the major semi-axes of the confocal conics passing through the points) are, as we have seen, connected by the relations

$$\Lambda_1 \lambda_1 - \Lambda_2 \lambda_2 = Cc. \quad \ldots\ldots\ldots \quad (B')$$

Thus every homographic transformation of a plane figure in which the line at an infinite distance is transformed into a line at a finite distance is equivalent to an inverse transformation of the elliptic coordinates of the points of the plane. In this way the expressions already given for the semi-axes of the indicatrix may be immediately deduced from the elementary elliptic formulae

$$d\Sigma_1 = \sqrt{\left(\frac{\Lambda_2^2 - \Lambda_1^2}{\Lambda_2^2 - C^2}\right)}\, d\Lambda_2, \qquad d\Sigma_2 = -\sqrt{\left(\frac{\Lambda_2^2 - \Lambda_1^2}{C^2 - \Lambda_1^2}\right)}\, d\Lambda_1,$$

combined with the corresponding formulae for the plane ω. Again, using the formulae $\Lambda_1 \Lambda_2 = CX$, $\lambda_1 \lambda_2 = cx$, we may write those expressions in either of the forms

$$\frac{X}{\lambda_2} i, \quad \frac{X}{\lambda_1} i; \quad \text{or} \quad \frac{\Lambda_2}{x} i, \quad \frac{\Lambda_1}{x} i. \quad \ldots\ldots \quad (A')$$

Thus the ratio of corresponding elementary areas at A and a is that of X^2 to $\lambda_1 \lambda_2$, or of $\Lambda_1 \Lambda_2$ to x^2, or of $C^{\frac{1}{2}} X^{\frac{3}{2}}$ to $c^{\frac{1}{2}} x^{\frac{3}{2}}$; *i.e.* it varies in the sesquiplicate ratio of the distances of the two areas from the vanishing lines. The lines $X^3 = \pm Cc^2$, $x^3 = \pm C^2 c$, (of which two in each plane are real and four imaginary,) are lines at which corresponding elementary areas are equal. More generally, the lines $kX = \pm KC$, $Kx = \pm kc$ are the real lines, in the planes Ω and ω, at which corresponding evanescent areas are to one another in the ratio of $K^3 C^2$ to $k^3 c^2$.

16. *Theorems relating to Curvature.*

Since evanescent segments at the same point and upon the same straight line are altered in one and the same ratio in any homographic transformation, the curvature of all curves which touch one another at a given point is altered in one and the same ratio. Thus, if a curve touch a focal circle of the plane Ω, its radius of curvature at the point of contact is altered in the transformation in the ratio of C to c. Again, it will be found that the radius of curvature of a curve at a point at which its tangent is parallel to the vanishing line is altered in the same parametric ratio. Hence, if we consider in the plane Ω any conic which passes through the imaginary points Q_1, Q_2, (and which, consequently, is transformed into a circle,) it has the same curvature at the two points where it is touched by focal circles and at the two points where it is touched by parallels to the vanishing line; for the radius of curvature at any one of these four points is to the radius of the corresponding circle in the ratio of the parameters. We thus obtain incidentally a solution of the problem, 'Given a system of circles, and a conic, having the same radical axis, to determine the two circles of the system which touch the conic'; for the points of contact are at the extremities of the diameter equal to the diameter conjugate to the radical axis. In particular, the radius of curvature of the indicatrix, at the points where its tangent is parallel to the vanishing line or to the tangent to the focal circle passing through its centre, is to the radius of the corresponding evanescent circle in the ratio of the parameters: thus, if R is the radius of curvature of the auxiliary ellipse at the point O, and Ai, Bi the principal semi-axes of the indicatrix, we have the equations

$$\frac{Ai}{\Lambda_2} R = \frac{C}{c} i, \qquad \frac{Bi}{\Lambda_1} R = \frac{C}{c} i,$$

which are in accordance with the expressions (A), since $R = \dfrac{C^3}{\Lambda_1 \Lambda_2}$. More generally, if Di is any semi-diameter of the indicatrix at the point A, the radius of curvature of any curve touching that semi-diameter at the point A is altered in the ratio of D^3 to AB (since $\dfrac{D^3}{AB} i$ is the radius of curvature of the indicatrix at the extremities of the diameter conjugate to Di). It will be seen that, of all curves passing through the point A, those which touch the confocal hyperbola at A experience in the transformation the greatest augmentation (or the least diminution) of curvature, and those which touch the confocal ellipse experience the greatest diminution (or the least augmentation) of curvature; so that the confocal conics may be said to be loci of greatest and least augmentation (or diminution) of curvature. The ratio of the radius of curvature of any curve passing through the point A to the radius of curvature of the corresponding curve is thus intermediate between the ratios $\Lambda_2^3 : C^2 c$ and $\Lambda_1^3 : C^2 c$. Let $K^3 : C^2 c$ be any ratio intermediate between these two; there are evidently two equal semi-diameters of the indicatrix at the point A such that the radii of curvature of curves touching either of them are altered in the ratio $K^3 : C^2 c$. If Φ be the angle made by either of these semi-diameters with the major axis of the indicatrix, the equation $\frac{D^3 i}{AB} = \frac{K^3}{C^2 c} i$ becomes, on substituting for A and B their values given by the expressions (A),

$$\frac{\cos^2 \Phi}{\Lambda_2^2} + \frac{\sin^2 \Phi}{\Lambda_1^2} = \frac{1}{K^2}, \quad . \ . \ . \ . \ . \ . \ . \ . \ . \quad \text{(C)}$$

or

$$\lambda_2^2 \cos^2 \Phi + \lambda_1^2 \sin^2 \Phi = \frac{C^2 c^2}{K^2}. \quad . \ . \ . \ . \ . \ . \ . \ . \ . \quad \text{(C')}$$

From the first of these equations we learn that the two semi-diameters coincide in direction with those semi-diameters of the auxiliary ellipse which are equal to K; the second implies that the angles made by the two semi-diameters with the major axis of the indicatrix at A are equal to the angles made in the plane ω with the major axis of the indicatrix at a by the tangents drawn from the point a to that confocal conic which, in the plane ω, corresponds to the confocal conic of major semi-axis K in the plane Ω.

17. *Curves of Constant Alteration of Curvature.*

We may also regard the equation (C) as equivalent to the differential equation of a system of curves such that at any point on any one of them the radius of curvature is altered in the constant ratio of K^3 to $C^2 c$. Substituting for $\tan^2 \Phi$ its value,

$$-\frac{\Lambda_2^2 - C^2}{\Lambda_1^2 - C^2} \cdot \frac{d\,\Lambda_1^2}{d\,\Lambda_2^2},$$

we find for this differential equation the expression

$$\frac{d\,\Lambda_2}{\Lambda_2} \sqrt{\left(\frac{\Lambda_2^2 - K^2}{\Lambda_2^2 - C^2}\right)} = \pm \frac{d\,\Lambda_1}{\Lambda_1} \sqrt{\left(\frac{K^2 - \Lambda_1^2}{C^2 - \Lambda_1^2}\right)}.$$

The integral of this equation is easily obtained in a finite form; it seems, however, too cumbrous for discussion. It contains an algebraical function raised to the power $\frac{K}{C}$, but no other transcendental function. If, therefore, $\frac{K}{C}$ is rational, *i.e.* if the given ratio $K^3 : C^2 c$ is a multiple of the parametric ratio by the cube of a rational number, the curves of constant alteration of curvature are algebraic and of finite dimensions; in every other case they are transcendental. They lie entirely outside the confocal conic (K), and seem to meet it in cusps at which the tangents are normal to it. If $K=C$, *i.e.* if the ratio is the parametric ratio, the curves of constant alteration of curvature are given by the equations

$$\frac{d\Lambda_2}{\Lambda_2} = \pm\frac{d\Lambda_1}{\Lambda_1},$$

or

$$\frac{\Lambda_1}{\Lambda_2} = \text{constant}, \qquad \Lambda_1\Lambda_2 = \text{constant};$$

they are thus the focal circles and the parallels to the vanishing line, as we have already seen.

The orthogonal trajectories of the curves of constant alteration of curvature are always algebraic; they have for their differential equation

$$\frac{\Lambda_2 d\Lambda_2}{\sqrt{[(\Lambda_2^2 - C^2)(\Lambda_2^2 - K^2)]}} = \frac{\Lambda_1 d\Lambda_1}{\sqrt{[(C^2 - \Lambda_1^2)(K^2 - \Lambda_1^2)]}}.$$

18. *Curves of Constant Elongation.*

Let $\theta C : c$ represent any given ratio; there are in general two equal semi-diameters of the indicatrix at any point A which are in that ratio to the radius i of the corresponding evanescent circle. If, however, $\theta\frac{C}{c}i$ be greater than the major axis or less than the minor axis of the indicatrix, the two diameters are imaginary. The equations

$$\Lambda_2^2\Lambda_1 = \theta C^3, \quad . \quad . \quad . \quad . \quad . \quad . \quad . \quad . \quad . \quad . \quad . \quad . \quad (\alpha)$$

$$\Lambda_1^2\Lambda_2 = \theta C^3, \quad . \quad . \quad . \quad . \quad . \quad . \quad . \quad . \quad . \quad . \quad . \quad . \quad (\beta)$$

represent two loci which separate the parts of the plane Ω in which the two diameters are real from those parts in which they are imaginary; at points on the locus (α) the two diameters coincide with the major axis of the indicatrix; at points on the locus (β) they coincide with its minor axis. The two loci are included in the same Cartesian equation of the sixth order

$$\theta^2 C^2 X^2 Y^2 = (X^2 - \theta^2 C^2)(X^2 - \theta C^2)(X^2 + \theta C^2),$$

which represents a curve symmetrical with respect to the vanishing line and focal axis, and having a quadruple point at Y (the point at an infinite distance on the vanishing line). The two branches, one on each side of the vanishing axis, which touch it and one another at the point Y form the locus (α); the locus (β) consists of the two branches which have the line at an infinite distance for their common tangent at the point Y. In the space included between the two branches of the locus (α) the major semi-axis of the indicatrix is less than $\theta \frac{C}{c} i$; in the spaces intermediate between the loci (α) and (β) the two diameters are real; and they are again imaginary in the spaces interior to the locus (β). To determine the angle Φ which either of the two diameters makes with the major axis of the indicatrix, we observe that these lines coincide in direction with the two diameters of the auxiliary ellipse which are equal to $2\theta \frac{C^3}{\Lambda_1 \Lambda_2}$; we thus obtain the equation

$$\Lambda_1^2 \cos^2 \Phi + \Lambda_2^2 \sin^2 \Phi = \frac{\Lambda_1^4 \Lambda_2^4}{\theta^2 C^6} = \frac{X^4}{\theta^2 C^2}, \quad . \quad . \quad . \quad . \quad . \quad . \quad \text{(D)}$$

which implies that the two diameters coincide in direction with the tangents drawn from the point A to the confocal conic of which the major semi-axis is $\frac{X^2}{\theta C}$. Substituting for $\tan^2 \Phi$ its value, we find

$$\frac{\Lambda_2^4 \Lambda_1^2 - \theta^2 C^6}{\Lambda_2^2 - C^2} \cdot \frac{d\,\Lambda_2^2}{\Lambda_2^2} = \frac{\Lambda_1^4 \Lambda_2^2 - \theta^2 C^6}{\Lambda_1^2 - C^2} \cdot \frac{d\,\Lambda_1^2}{\Lambda_1^2},$$

which is the differential equation of the curves of constant elongation, and seems not to admit of integration in any finite form. Its equivalent in Cartesian coordinates is

$$\frac{dX^2 + dY^2}{C^2} = \theta^2 \frac{dx^2 + dy^2}{c^2},$$

or

$$\frac{X^4 (dX^2 + dY^2)}{\theta^2 C^2} = C^2 dX^2 + (YdX - XdY)^2.$$

It will be observed that $\Lambda_1 \Lambda_2 = \theta C^2$, or $X = \theta C$, is a particular integral of the equation.

The following is an important property of the curves of constant elongation:—

'The intercept in the plane Ω on any tangent to one of these curves between the point of contact and the vanishing line is in the given ratio to the similarly defined segment on the corresponding tangent in the plane ω.'

To establish this property, we have only to observe that the point of contact is one of the double points of that involution upon the tangent of which the segments are to their corresponding segments in the given ratio. Or we may infer it from the equation (D) with the help of the easily demonstrated theorem:—

'If Ψ and ψ are the angles made with the vanishing line by any two corresponding tangents to the confocals (Γ) and (γ), the ratio $\frac{\sin\Psi}{\sin\psi}$ is constant and equal to $\frac{\Gamma}{C}$ or $\frac{c}{\gamma}$.'

19. *Curves of Equal Tangential Deflexion.*

Through any point A of the plane Ω there pass two curves such that the angle between any two tangents to either of them is equal to the angle between the two corresponding tangents to the corresponding curve. We may term these curves the *curves of equal tangential deflexion.* If Di be the length of the semi-diameter of the indicatrix which touches one of these curves at the point A, and if $d\Theta = d\theta$ be the angle contained between this semi-diameter and a consecutive tangent to the curve, we find, since all areas at the point A are altered in the same ratio, $D^2 i^2 d\Theta : i^2 d\theta :: ABi^2 : i^2$, or simply

$$D^2 = AB.$$

Denoting by Φ the angle made by the semi-diameter Di with the major axis of the indicatrix, and substituting for A, B, and D their values, we have for the differential equation of the curves of equal tangential deflexion

$$\frac{\cos^2\Phi}{\Lambda_2} = \frac{\sin^2\Phi}{\Lambda_1}, \quad . \quad . \quad . \quad . \quad . \quad . \quad . \quad . \quad . \quad . \quad . \quad (E)$$

or

$$\frac{d\Lambda_1}{\sqrt{\Lambda_1}\sqrt{(C^2-\Lambda_1^2)}} = \pm\frac{d\Lambda_2}{\sqrt{\Lambda_2}\sqrt{(\Lambda_2^2-C^2)}}.$$

These curves are always algebraic; for, putting $\Lambda_1 = Cu_1^2$, $\Lambda_2 = \frac{C}{u_2^2}$, we have

$$\frac{du_1}{\sqrt{(1-u_1^4)}} = \pm\frac{du_2}{\sqrt{(1-u_2^4)}},$$

of which the integral is algebraic.

The equation (E) may also be obtained by observing that the curves of equal tangential deflexion which pass through the point A must touch at that point the double lines of the pencil in involution which is equiangular to the corresponding pencil, and that the equation (E) is the equation determining these double lines (see Arts. 10 and 11). It appears from this that the angle at which the two curves intersect is always supplementary to the corresponding angle.

20. *Curves Similar and Similarly situated to their Images.*

As an additional example of the use of the formulae (B), let us propose to determine the conic sections which in either figure are transformed into conics similar and similarly situated with regard to the vanishing line and focal axis of the other figure. If

$$aX^2 + a'Y^2 + a''C^2 + 2bCY + 2b'CX + 2b''XY = 0$$

is the equation of any conic in the plane Ω, the equation of the corresponding conic is

$$a''x^2 + a'y^2 + ac^2 + 2b''cy + 2b'cx + 2bxy = 0.$$

And if these two conics are similar and similarly situated, we must have $a = a''$, $b = \pm b''$; *i.e.* every conic for which one of the two foci is the pole of the line parallel to the vanishing axis and passing through the other focus (or, which is the same thing, every conic for which one of the pairs of lines joining its points at an infinite distance to its points on the vanishing axis intersect at a focus) is transformed into a similar and similarly situated conic, the ratio of similarity being that of the parameters. There are thus two sets of conics (each forming a triply indeterminate linear system) which satisfy the conditions of the problem; but the conics of only one set at a time can be regarded as similarly situated to the corresponding conics, because in determining the two sets different directions on the vanishing lines are taken to determine the similarity of position.

Again, the corresponding conics will have their areas in the ratio of the squares of the parameters, if $(aa' - b''^2)^3 = (a'a'' - b^2)^3$; *i.e.* the areas of all conics with regard to which the lines parallel to the vanishing line and passing through the foci are self-conjugate lines are to the areas of their corresponding conics in the duplicate ratio of the parameters. The only real conics of which the area is changed in this ratio are those defined by this geometrical condition; they form a quadruply indeterminate linear tangential system. But the analytical condition is also satisfied by the imaginary conics in the plane Ω, with regard to which the imaginary lines $X = \pm\rho C$, or $X = \pm\rho^2 C$, are harmonically conjugate, ρ denoting an imaginary cube root of unity. More generally, it will be found that the conics of which the area is changed in any given ratio are those which have for a pair of conjugate lines the two straight lines at which elementary areas are changed in the given ratio (Art. 15). If the corresponding conics are hyperbolas, we may substitute for the area in this result the triangle contained by the asymptotes and any tangent.

Lastly, the geometrical condition that a conic in either plane should be similar to its corresponding conic is that the pairs of points in which it intersects the vanishing line and the line at an infinite distance should subtend equal angles at a focus. But the quadruply indeterminate system determined by this condition is not a linear one.

Theorems of a similar kind to the preceding, but relating to curves of a higher order, may be obtained by observing that symmetrical functions of X, x; Y, y; or again of Λ_1, λ_1; Λ_2, λ_2 are unchanged by the transformation. Thus any curve represented by

$$F\left(X+\frac{Cc}{X},\ \frac{cY^2}{X},\ Y+\frac{cY}{X}\right)=0,$$

or again by

$$f\left(\Lambda_1+\frac{Cc}{\Lambda_1},\ \Lambda_2+\frac{Cc}{\Lambda_2}\right)=0,$$

is transformed into a curve similar and similarly situated with regard to the focal axis.

B.—Focal Properties of Two Homographic Point-Figures*.

21. *The Imaginary Cones corresponding to Evanescent Spheres.*

By a point-figure we shall here understand a system of straight lines and planes passing through a point which is termed the centre of the point-figure. Let S, s be the centres of two point-figures, homographically related to one another; and let P, q represent the evanescent spheres (here to be regarded as imaginary cones), which have their centres at S, s. Excluding altogether from consideration the very particular case in which these two imaginary cones correspond to one another homographically, and in which, consequently, the two figures admit of exact coincidence with one another, let us represent by p, Q the imaginary cones, which in the figures s, S correspond to the cones P, q. We observe that if either p or Q is a cone of revolution, the other is so too; for if the cones P, Q have double contact, so also have the corresponding cones p, q. We shall hereafter (Art. 38) return for a moment to this particular case, but for the present we shall suppose that neither p nor Q is a cone of revolution.

---

* The focal properties of homographic point-figures might be obtained by simple considerations of perspective (see Arts. 21 and 37). We have, however, preferred to deduce these properties from their genuine source—the properties of the imaginary circle in which all spheres intersect one another at an infinite distance. In the case of homographic plane figures, we have ventured to employ both methods successively (Arts. 1–4, and Art. 5). This has been done at some risk of repetition; but it seemed desirable to exhibit this part of the theory in its most elementary and practical, as well as in its most abstract form, in the hope that some of the simpler results may be found of use in the actual practice of perspective.

22. *The Principal Axes.*

On this supposition there exists in each pencil one, and only one, system of straight lines at right angles to one another, such that their corresponding lines are also at right angles to one another. These lines are the principal axes of the cones Q and p. For the principal axes of Q are the system of self-conjugate axes common to the cones P and Q; these principal axes, therefore, correspond to the system of axes self-conjugate with regard to p and q; *i.e.* to the principal axes of p. We shall call these two sets of rectangular axes the principal axes of the two figures, and we shall distinguish them as the axes of XYZ, xyz.

23. *The Focal Lines and Cyclic Planes.*

To the four imaginary lines of intersection of P and Q, and to the four imaginary tangent planes common to those two cones, there correspond the four lines of intersection, and the four common tangent planes, of p and q. Hence to C_1 and C_2, the two real cyclic planes of Q, and to F_1, F_2, the two real focal lines of Q, there correspond c_1, c_2, the cyclic planes, and f_1, f_2, the focal lines of p. It will be observed that the real focal lines of an imaginary cone (differing in this respect from the focal lines of a real cone) lie in that principal plane of the cone to which the cyclic planes are perpendicular. We shall call the axis, in which the cyclic planes intersect, and which is perpendicular to the focal plane (*i.e.* to the plane containing the focal lines), the mean axis; of the two axes in the focal plane, we shall term that the major axis which makes with either cyclic plane, and with either focal line, acute angles together less than a right angle.

24. *The Reciprocity of the Imaginary Cones.*

The imaginary cones Q and p are reciprocal. Let Y, y be their mean axes; XZ, xz, their focal planes; P_1P_2, Q_1Q_2, and p_1p_2, q_1q_2, the imaginary lines in which these planes meet the cones P, Q and p, q respectively. From the anharmonic equation

$$[P_1, P_2, Q_1, Q_2, X, Z] = [p_1, p\ , q_1, q_2, x, z] \quad . \quad . \quad . \quad . \quad . \quad (1)$$

we infer the equation

$$[P_1, P_2, Q_1, Q_2] = [q_1, q_2, p_1, p_2],$$

which implies that the imaginary angles Q_1SQ_2, p_1sp_2 are equal, because P_1, P_2 and q_1, q_2 are pairs of lines representing evanescent circles. Again, X, Z and x, z are harmonic conjugates of the pairs P_1P_2, Q_1Q_2, and p_1p_2, q_1q_2 respectively.

Hence we must have either the equation

$$[P_1, P_2, Q_1, Q_2, X, Z] = [q_1, q_2, p_1, p_2, x, z],$$

or else the equation

$$[P_1, P_2, Q_1, Q_2, X, Z] = [q_1, q_2, p_1, p_2, z, x].$$

But the former equation is inadmissible; for, on combining it with (1), we obtain

$$[p_1, q_1, x, z] = [q_1, p_1, x, z],$$

which is untrue, since p_2, and not q_1, is the harmonic conjugate of p_1 with regard to xz. It is, therefore, the latter equation which subsists; it implies that Q_1 or Q_2 makes the same angles with X, that p_1 or p_2 makes with z; *i.e.*, that the angle Q_1SX is the complement of p_1sx. Similarly, the angle which the axis of X makes with either of the lines of Q which lie in the plane XY, is the complement of the corresponding angle in the plane xy; that is to say, the two cones are reciprocal.

It is evident that the mean axis of Q corresponds to the mean axis of p. Let C, c be the acute angles F_1SX, f_1sx; then the acute angle contained by YX and either cyclic plane of Q is the complement of c, and the acute angle contained by yx and either cyclic plane of p is the complement of C. Hence X is the major or minor axis of Q, according as $C < c$, or $C > c$; and to the major axis of Q the minor axis of p corresponds, and *vice versâ*. Neither of the angles C, c can be zero, nor a right angle, nor can they be equal to one another.

The reciprocity of the cones Q and p gives rise to a reciprocal relation between the two homographic figures, which may be thus stated. Conceive the two figures placed with their corresponding principal axes coincident. Let A, a be any two corresponding planes in the figures S and s; let b be the normal to A at the common centre of the figures; and B the normal to a at the same point; then B and b are corresponding lines in the figures S and s. If, therefore, we consider any two corresponding systems of planes and lines in S and s, the *reciprocal* systems of lines and planes will also be corresponding systems in s and S. Thus all the properties (metrical as well as descriptive) of two homographic point-figures are double, and we have an uniform method for passing from any property to its correlative.

25. *The Correspondence of Directions.*

The angles contained by planes intersecting in a focal line of S are equal to the corresponding angles contained by planes intersecting in a focal line of s; and, correlatively, the angles contained by lines intersecting at S in one of the cyclic planes of S, are equal to the corresponding angles in a cyclic plane of s. These theorems are evident, because the imaginary tangent planes of P, which intersect in F_1, correspond to the imaginary tangent planes of q, which intersect in f_1; and similarly, the lines in which P is intersected by either cyclic plane of Q correspond to the lines in which q is intersected by either cyclic plane of p.

To fix the correspondence of the directions of rotation round either pair of corresponding focal lines, or in either pair of corresponding cyclic planes, we consider the intersections of the planes and lines of S and s by the surfaces of two spheres of radius unity having their centres at S and s. Let A, B, C be three points on the sphere S, forming a spherical triangle; it will be remembered that three points, not in the same great circle, always form one, and only one, spherical triangle, if by a spherical triangle we understand (as is usually done), a triangle formed by arcs of great circles, each of which is less than two right angles. As corresponding point to any point A on the sphere S, we might take either of two diametrically opposite points a, a' on the sphere s. But for one of these points (for example a) the corresponding directions of rotation round A and a are similar (*i.e.* both right handed or both left handed, when viewed from the centres of the spheres); while for the other point a' the corresponding directions of rotation are dissimilar. Let then a, b, c be the three points which on the sphere s correspond with similar rotations to the points A, B, C. These three points are thus determined without any ambiguity, and we shall now show that to points in the interior of the triangle ABC there correspond, with similar rotations, points in the interior of abc. The proof of this important theorem depends on the two principles: (i.), that if a point move continuously on either sphere, and traverse any curve on that sphere, its corresponding point on the other sphere simultaneously traverses the corresponding curve; (ii.), that if A and a are corresponding points with similar rotations, and if, while A moves continuously to B, a moves continuously to b, then a and b are also corresponding points with similar rotations. The first of these principles may be considered as evident; to establish the second, it will suffice to consider A and B as consecutive positions of A, so that while A describes the element AB, a describes the element ab. Let E be any great circle not intersecting AB, then the corresponding great circle e does not intersect ab, and if these two great circles be described by corresponding points V and v, the vector arcs AV, av will by hypothesis revolve in similar directions. But the arcs AV, BV evidently revolve in similar directions, and so do the arcs av, bv; *i.e.* the corresponding rotations round B and b are similar. Let us now suppose that a point sets out from B, and describes the side BC of the triangle ABC; the corresponding point will at the same time describe the side bc of the triangle abc; for as it must not traverse either of the great circles ab, ac, it cannot describe an arc greater than a semicircle. Thus, to the points of any side of ABC there correspond, with similar rotations, the points of the corresponding side of abc. Let V be any point internal to ABC, let AV cut BC in A_1, and let a_1 on bc correspond to A_1 on BC; then AA_1B, aa_1c are corresponding spherical triangles, with similar rotations at their corresponding points; therefore the points of aa_1 correspond, with similar rotations, to the points of AA_1; *i.e.*, the point v, which corresponds with similar rotation to V, lies on aa_1 in the interior of the triangle abc.

The great circles which form the triangles ABC, abc, divide the spheres S and s each into eight spherical triangles, which correspond to one another one by one, with similar rotations at their corresponding vertices, just as the triangles ABC, abc. Thus each sphere is divided into eight regions, corresponding to the eight regions of the other sphere, in such a manner, that, if any point be taken on either sphere, the point which corresponds to it with similar rotation lies in the corresponding region of the other sphere.

We shall now take for ABC one of the eight octantal triangles XYZ, and for abc the corresponding octant xyz; we shall denote by F_1, f_1; $Y\Omega_1$, $y\omega_1$, the foci and cyclic arcs which lie in the octants XYZ, xyz; and by F_2, f_2; $Y\Omega_2$, $y\omega_2$; the foci and cyclic arcs which lie in the octants $XY\bar{Z}$, $xy\bar{z}$; so that the directions of rotation round the corresponding foci F_1, f_1 and F_2, f_2, will be similar, and the directions $Y\Omega_1$, $y\omega_1$; $Y\Omega_2$, $y\omega_2$, will be corresponding directions on the cyclic arcs. It will be convenient to consider only the hemispheres of which the points X, x are the spheric centres, and the planes YZ, yz the bases. Thus, to any given point on the hemisphere S (not lying on the base circle itself), there corresponds one point, and only one, on the hemisphere s; and again any two great circles upon either hemisphere (neither of which is the base circle) intersect one another only in one point. To find the point a of the hemisphere s, which corresponds to a given point A of the hemisphere S, we draw the vector arcs F_1A, F_2A, and make the angles f_2f_1a, f_1f_2a equal in sign and magnitude to the angles F_2F_1A, F_1F_2A; the point of intersection of the arcs f_1a, f_2a is the point a required. Similarly, to find the great circle a of the hemisphere s, which corresponds to a given great circle A of the hemisphere S, we find the points D_1, D_2, in which A intersects the cyclic arcs of S, and we make the arcs ω_1d_1, ω_2d_2, equal in sign and magnitude to the arcs Ω_1D_1, Ω_2D_2; the arc d_1d_2 is the arc required.

26. *The Confocal Spherical Conics.*

The spherical conics of which F_1, F_2 are the foci are transformed into the spherical conics of which f_1, f_2 are the foci. This is evident from the equiangular property of the foci; or, again, if R_1, R_2, r_1, r_2 are the focal radii vectores of the corresponding points Δ and δ, the spherical triangles $F_1 \Delta F_2$, $f_1 \delta f_2$ give the equations

$$\frac{\tan \frac{1}{2}(R_1+R_2)}{\tan \frac{1}{2}(r_1+r_2)} = \frac{\tan \frac{1}{2}(R_1-R_2)}{\tan \frac{1}{2}(r_1-r_2)} = \frac{\tan C}{\tan c},$$

which imply that if $R_1 \pm R_2$ is constant, $r_1 \pm r_2$ is also constant. Thus the ellipses are transformed into ellipses, and the hyperbolas into hyperbolas, these denominations being relative to the two foci lying on each of the hemispheres S and s. If $\Lambda = \frac{1}{2}(R_1+R_2)$, or $=\frac{1}{2}(R_1-R_2)$, is the focal semi-axis of one of the confocal conics of S, the quotient $\dfrac{\tan \Lambda}{\tan C}$, which is one of the spherical eccentricities of the conic, remains unchanged in the transformation; for, if λ be the semi-axis of the corresponding conic, we have the equation

$$\frac{\tan \Lambda}{\tan \lambda} = \frac{\tan C}{\tan c},$$

which results from the homography of corresponding points of the great circles XZ, xz. It is also evident that if we consider any two corresponding conics of the two confocal systems, there will correspond to one another in the two figures:—the normal arcs of the two curves, their spherical centres of curvature, their evolutes, their *similar* arcs, as also the spherical polygons of minimum perimeter circumscribing corresponding arcs, and the spherical polygons of maximum perimeter inscribed in corresponding arcs.

27. *The Concyclic Spherical Conics.*

Correlatively, the system of concyclic conics of which $Y\Omega_1$, $Y\Omega_2$ are the cyclic arcs are transformed into concyclic conics of which $y\omega_1$, $y\omega_2$ are the cyclic arcs, the ellipses into ellipses, and the conics of the third species into conics of the third species; these denominotions being again relative to the hemispheres which we are considering. (See M. Chasles' 'Sur les propriétés générales des coniques sphériques,' Arts. 1–4.) If $D_1 D_2$, $d_1 d_2$ are corresponding arcs, cutting the cyclic arcs in D_1, D_2, d_1, d_2, the spherical triangles $D_1 Y D_2$, $d_1 y d_2$, in which

$D_1 Y = d_1 y$, $D_2 Y = d_2 y$, supply the equations

$$\frac{\tan \frac{1}{2}(D_1 + D_2)}{\tan \frac{1}{2}(d_1 + d_2)} = \frac{\tan \frac{1}{2}(D_1 - D_2)}{\tan \frac{1}{2}(d_1 - d_2)} = \frac{\tan c}{\tan C}.$$

Let E, e be the areas of the spherical quadrilaterals $\Omega_1 D_1 D_2 \Omega_2$, $\omega_1 d_1 d_2 \omega_2$, we find $E = \pi - D_1 - D_2$, $e = \pi - d_1 - d_2$, whence

$$\frac{\tan \frac{1}{2} E}{\tan \frac{1}{2} e} = \frac{\tan C}{\tan c},$$

a formula which expresses a remarkable property of the cyclic arcs.

To chords of any conic (Φ) of the concyclic system of S which cut off equal spherical areas from that conic, there will correspond chords cutting off from the corresponding conic (ϕ) areas equal to one another. To a spherical polygon of maximum area inscribed in any arc of (Φ), or to a polygon of minimum area circumscribing any arc of (Φ), there will correspond polygons possessing a similar maximum or minimum property with regard to the corresponding arc of (ϕ). These results follow from the known properties of concyclic spherical conics; or they may be deduced by reciprocation from the properties of the confocal conics of the two homographic systems.

28. *Arcs and Angles changed into equal Arcs and Angles.*

On any great circle A of S there are two points at right angles to one another, such that their corresponding points, on the corresponding great circle a, are also at right angles. These points are the external and internal points of bisection of the intercept made on the great circles by the cyclic arcs; they are also the points at which the great circles are touched by conics of the concyclic systems. Let $2D$, $2d$ be the intercepts; the homographic modulus of the two great circles (relative to the internal points of bisection) is $\dfrac{\tan D}{\tan d}$; the arcs of the involution

$$\tan H_1 \tan H_2 = \frac{\tan D}{\tan d}$$

are equal to the corresponding arcs of the involution

$$\tan h_1 \tan h_2 = \frac{\tan d}{\tan D};$$

and the arcs of the involution

$$\tan H_1 \tan H_2 = -\frac{\tan D}{\tan d}$$

are equal to the supplements of the corresponding arcs of the involution

$$\tan h_1 \tan h_2 = -\frac{\tan d}{\tan D}.$$

The determination of the angles which at any point Δ are transformed into equal or supplementary angles at the point δ is correlative to the preceding. The external and internal bisectors of the angles between the radii vectores at Δ and δ are the right angles of the homographic pencils at A and a, and if $F_1 \Delta F_2 = 2\Delta$, $f_1 \delta f_2 = 2\delta$, the homographic modulus of the pencils, relative to the internal bisectors, is $\dfrac{\tan\Delta}{\tan\delta}$. The equiangular and supplementary involutions are respectively

$$\tan H_1 \tan H_2 = \frac{\tan\Delta}{\tan\delta}, \qquad \tan h_1 \tan h_2 = \frac{\tan\delta}{\tan\Delta},$$

and

$$\tan H_1 \tan H_2 = -\frac{\tan\Delta}{\tan\delta}, \qquad \tan h_1 \tan h_2 = -\frac{\tan\delta}{\tan\Delta}.$$

Combining the results relating to equal arcs and to equal angles, we see that, given any arc of a great circle in either figure, and a point upon it, there is always a spherical triangle having a vertex at the given point, and a second vertex upon the given arc, which is transformed into an equal and superposable spherical triangle.

The homographic modulus of the pencils at Δ and δ may be also expressed in terms of the radii vectores of the points Δ and δ, since from the triangles $F_1 \Delta F_2$, $f_1 \delta f_2$ we find

$$\frac{\tan\Delta}{\tan\delta} = \frac{\sin\frac{1}{2}(R_1 - R_2)}{\sin\frac{1}{2}(R_1 + R_2)} : \frac{\sin\frac{1}{2}(r_1 - r_2)}{\sin\frac{1}{2}(r_1 + r_2)}$$
$$= \frac{\cos\frac{1}{2}(R_1 - R_2)}{\cos\frac{1}{2}(R_1 + R_2)} : \frac{\cos\frac{1}{2}(r_1 - r_2)}{\cos\frac{1}{2}(r_1 + r_2)}.$$

29. *The Equations of the Homography in Spherical Coordinates.*

The equation of the cone Q, referred to its principal axes, is

$$\frac{\sin^2 C}{\sin^2 c} X^2 + Y^2 + \frac{\cos^2 C}{\cos^2 c} Z^2 = 0; \quad \ldots \ldots \ldots \quad (Q)$$

the equations of its cyclic planes, and of its focal lines, are respectively

$$Y^2 + Z^2 \quad \sec^2 c = 0,$$
$$Z^2 - X^2 \quad \cot^2 c = 0,$$
$$X^2 + Y^2 \quad \sin^2 c = 0;$$

and

$$Y^2 + Z^2 \quad \cos^2 C = 0, \qquad X = 0,$$
$$Z^2 - X^2 \quad \tan^2 C = 0, \qquad Y = 0,$$
$$X^2 + Y^2 \operatorname{cosec}^2 C = 0, \qquad Z = 0.$$

The equations of the cone p, and of its cyclic planes and focal lines, are obtained by interchanging C and c.

Let Δ, δ be corresponding points on the two spheres, and let the arcs $X\Delta$, $Y\Delta$, $Z\Delta$, $x\delta$, $y\delta$, $z\delta$ meet the arcs YZ, ZX, XY, yz, zx, xy in the points A, B, C, a, b, c respectively. If we take the ratios of the cosines

$$X = \cos \Delta X, \quad Y = \cos \Delta Y, \quad Z = \cos \Delta Z,$$

$$x = \cos \delta x, \quad y = \cos \delta y, \quad z = \cos \delta z,$$

as the spherical coordinates of the points Ω and ω respectively, the homographic relation of the two figures is expressed by the equations

$$\frac{\sin C \cos X}{\sin c \cos x} = \frac{\cos Y}{\cos y} = \frac{\cos C \cos Z}{\cos c \cos z}.$$

Or, again, if we take one of the following systems of tangents as the coordinates of the points Δ and δ,

$$Y = \tan XB, \quad Z = \tan XC; \quad y = \tan xb, \quad z = \tan xc; \quad . \quad . \quad . \quad . \quad (1)$$

$$Z = \tan YC, \quad X = \tan YA; \quad z = \tan yc, \quad x = \tan ya; \quad . \quad . \quad . \quad . \quad (2)$$

$$X = \tan ZA, \quad Y = \tan ZB; \quad x = \tan za, \quad y = \tan zb; \quad . \quad . \quad . \quad . \quad (3)$$

the homographic relation is expressed by the equations

$$Y = y\frac{\tan C}{\tan c}, \quad Z = z\frac{\sin C}{\sin c}; \quad . \quad . \quad . \quad . \quad . \quad . \quad . \quad . \quad . \quad (1)$$

$$Z = z\frac{\sin c}{\sin C}, \quad X = x\frac{\cos c}{\cos C}; \quad . \quad . \quad . \quad . \quad . \quad . \quad . \quad . \quad . \quad (2)$$

$$X = x\frac{\cos C}{\cos c}, \quad Y = y\frac{\tan c}{\tan C}. \quad . \quad . \quad . \quad . \quad . \quad . \quad . \quad . \quad . \quad (3)$$

30. *The Parameters of the Confocal and Concyclic Cones.*

Instead of the equation (Q), it will be convenient to employ the equation

$$\frac{X^2}{\mathrm{A}^2} + \frac{Y^2}{\mathrm{B}^2} + \frac{Z^2}{\Gamma^2} = 0$$

to represent the cone Q; so that

$$\mathrm{A} : \mathrm{B} : \Gamma :: \frac{\sin c}{\sin C} : 1 : \frac{\cos c}{\cos C},$$

$$\tan C = \sqrt{\left(\frac{\mathrm{B}^2 - \Gamma^2}{\mathrm{A}^2 - \mathrm{B}^2}\right)}, \quad \tan c = \frac{\mathrm{A}}{\Gamma} \tan C = \frac{\mathrm{A}}{\Gamma} \sqrt{\left(\frac{\mathrm{B}^2 - \Gamma^2}{\mathrm{A}^2 - \mathrm{B}^2}\right)}.$$

We suppose A, B, Γ all positive, and $A > B > \Gamma$, *i.e.* $c > C$. The figure S is then transformed into s by the equations

$$X = \mathrm{A}x, \quad Y = \mathrm{B}y, \quad Z = \Gamma z, \quad . \quad . \quad . \quad . \quad . \quad . \quad . \quad . \quad (1)$$

or, writing $\alpha = \frac{1}{\mathrm{A}}$, $\beta = \frac{1}{\mathrm{B}}$, $\gamma = \frac{1}{\Gamma}$, by the equations

$$X = \frac{x}{\alpha}, \quad Y = \frac{y}{\beta}, \quad Z = \frac{z}{\gamma}.$$

We shall term the quantities Ψ and Φ the *parameters* of the confocal cone

$$\frac{X^2}{\mathrm{A}^2 - \Psi^2} + \frac{Y^2}{\mathrm{B}^2 - \Psi^2} + \frac{Z^2}{\Gamma^2 - \Psi^2} = 0, \quad . \quad . \quad . \quad . \quad . \quad . \quad . \quad . \quad (2)$$

and of the concyclic cone

$$X^2\left(\frac{1}{\mathrm{A}^2} - \frac{1}{\Phi^2}\right) + Y^2\left(\frac{1}{\mathrm{B}^2} - \frac{1}{\Phi^2}\right) + Z^2\left(\frac{1}{\Gamma^2} - \frac{1}{\Phi^2}\right) = 0, \quad . \quad . \quad . \quad . \quad (3)$$

respectively. These quantities are of frequent use in the theory, as will appear from the following observations :—

(α) If ψ and ϕ are the parameters of the cones corresponding to (Ψ) and (Φ), we have $\psi = \frac{1}{\Psi}$, $\phi = \frac{1}{\Phi}$; for the cones (2) and (3) are transformed by the equations (1) into the cones

$$\frac{x^2}{\alpha^2 - \psi^2} + \frac{y^2}{\beta^2 - \psi^2} + \frac{z^2}{\gamma^2 - \psi^2} = 0,$$

$$x^2\left(\frac{1}{\alpha^2} - \frac{1}{\phi^2}\right) + y^2\left(\frac{1}{\beta^2} - \frac{1}{\phi^2}\right) + z^2\left(\frac{1}{\gamma^2} - \frac{1}{\phi^2}\right) = 0,$$

of which the parameters are respectively ψ and ϕ.

(β) If we imagine the principal axes of the two pencils coincident, the cone reciprocal to that confocal cone in S of which the parameter is Ψ is the concyclic cone in s of which the parameter is $\frac{1}{\Psi}$.

(γ) If Ψ_1, Ψ_2 are the parameters of the two confocal spherical conics which pass through a given point, the parameter of the concyclic conic passing through that point is $\Phi = \frac{\mathrm{AB}\Gamma}{\Psi_1\Psi_2}$; and, reciprocally, if Φ_1, Φ_2 are the parameters of the concyclic conics touching a given arc, the parameter of the confocal conic touching that arc is $\Psi = \frac{\mathrm{AB}\Gamma}{\Phi_1 \Phi}$.

(δ) Let Λ, λ represent the focal semi-axes of the corresponding confocal conics (Ψ) and (ψ); we have

$$\sin^2\Lambda = \frac{\Psi^2 - \Gamma^2}{\mathrm{A}^2 - \Gamma^2}, \quad \cos^2\Lambda = \frac{\mathrm{A}^2 - \Psi^2}{\mathrm{A}^2 - \Gamma^2}, \quad \tan^2\Lambda = \frac{\Psi^2 - \Gamma^2}{\mathrm{A}^2 - \Psi^2},$$

with similar values for $\sin^2\lambda$, $\cos^2\lambda$, $\tan^2\lambda$; and hence

$$\sin\Lambda = \frac{\Psi}{\mathrm{A}}\sin\lambda, \quad \cos\Lambda = \frac{\Psi}{\Gamma}\cos\lambda, \quad \tan\Lambda = \frac{\Gamma}{\mathrm{A}}\tan\lambda.$$

(ϵ) Thus, for the homographic modulus of the pencils at the corresponding points Δ, δ, we have the expression (see Art. 28)

$$\frac{\tan\Delta}{\tan\delta} = \frac{\sin\Lambda_2}{\sin\lambda_2} : \frac{\sin\Lambda_1}{\sin\lambda_1} = \frac{\Psi_2}{\Psi_1},$$

the angles being measured from the tangents to the confocals (Ψ_2), (ψ_2). And correlatively for the modulus of the homography on any corresponding arcs D, d, we have

$$\frac{\tan D}{\tan d} = \frac{\Phi_2}{\Phi_1},$$

the arcs D, d being measured from the points of contact of the concyclic conics (Φ_2) and (ϕ_2).

(ζ) Lastly, if (Ψ_1), (Ψ_2) are the two confocals intersecting at Δ, and (Φ) the concyclic conic passing through Δ, we have

$$\frac{\sin(\Lambda_1 + \Lambda_2)}{\sin(\lambda_1 + \lambda_2)} = \frac{\Psi_1\Psi_2}{\mathrm{A}\Gamma}, \quad \text{or} \quad \frac{\sin R_1}{\sin r_1} = \frac{\sin R_2}{\sin r_2} = \frac{\mathrm{B}}{\Phi},$$

an equation which corresponds to the equations (γ) of Art. 8.

31. *The Indicatrix on the Sphere.*

Let $d\Sigma_1$, $d\Sigma_2$, $d\sigma_1$, $d\sigma_2$ be corresponding elements of the spherical ellipses and spherical hyperbolas which pass through the corresponding points Δ and δ; let also the arcs Λ_1, Λ_2, λ_1, λ_2 be the focal semi-diameters of these conics; and let $2\Delta = F_1\Delta F_2$, $2\delta = f_1\delta f_2$. Considering two consecutive corresponding points on the two ellipses, and again on the two hyperbolas, we find

$$d\Sigma_1 = \frac{d\Lambda_2}{\sin\Delta}, \quad d\Sigma_2 = \frac{d\Lambda_1}{\cos\Delta},$$

$$d\sigma_1 = \frac{d\lambda_2}{\sin\delta}, \quad d\sigma_2 = \frac{d\lambda_1}{\cos\delta}.$$

But, differentiating the equations

$$\frac{\tan \Lambda_1}{\tan \lambda_1} = \frac{\tan \Lambda}{\tan \lambda_2} = \frac{\tan C}{\tan c},$$

we have

$$\frac{d\Lambda_1}{\sin 2\Lambda_1} = \frac{d\lambda_1}{\sin 2\lambda_1}, \quad \frac{d\Lambda_2}{\sin 2\Lambda_2} = \frac{d\lambda_2}{\sin 2\lambda_2};$$

and from the triangles F_1PF_2, f_1pf_2,

$$\frac{\sin \Delta}{\sin \delta} = \frac{\cos C}{\cos \Lambda_1} : \frac{\cos c}{\cos \lambda_1} = \frac{\sin C}{\sin \Lambda_1} : \frac{\sin c}{\sin \lambda_1},$$

$$\frac{\cos \Delta}{\cos \delta} = \frac{\cos C}{\cos \Lambda_2} : \frac{\cos c}{\cos \lambda_2} = \frac{\sin C}{\sin \Lambda_2} : \frac{\sin c}{\sin \lambda_2};$$

whence

$$\frac{d\Sigma_1}{d\sigma_1} = \frac{\sin c}{\sin C} \times \frac{\sin \Lambda_1}{\sin \lambda_1} \times \frac{\sin 2\Lambda_2}{\sin 2\lambda_2},$$

$$\frac{d\Sigma_2}{d\sigma_2} = \frac{\sin c}{\sin C} \times \frac{\sin \Lambda_2}{\sin \lambda_2} \times \frac{\sin 2\Lambda_1}{\sin 2\lambda_1};$$

or, substituting from the equations (δ) and (γ), Art. 30,

$$\frac{d\Sigma_1}{d\sigma_1} = \frac{\Psi_1 \Psi_2^2}{\mathrm{AB\Gamma}} = \frac{\Psi_2}{\Phi},$$

$$\frac{d\Sigma_2}{d\sigma_2} = \frac{\Psi_2 \Psi_1^2}{\mathrm{AB\Gamma}} = \frac{\Psi_1}{\Phi}.$$

If in these formulae we put $d\sigma_1 = d\sigma_2 = i$, the corresponding values of $d\Sigma_1$ and $d\Sigma_2$ are the principal semi-axes of the evanescent ellipse corresponding to the circle of which the centre is δ and i is the infinitesimal radius.

32. *Curves of Equal Tangential Deflexion and of Constant Elongation.*

Since $d\Sigma_1$ is the circular measure of the infinitesimal angle contained between the two lines in which (Ψ_1) is cut by (Ψ_2) and $(\Psi_2 + d\Psi_2)$, we have

$$d\Sigma_1^2 = \frac{(\Psi_1^2 - \Psi_2^2)\, \Psi_2^2\, d\Psi_2^2}{(\mathrm{A}^2 - \Psi_2^2)(\mathrm{B}^2 - \Psi_2^2)(\Psi_2^2 - \Gamma^2)},$$

$$d\Sigma_2^2 = \frac{(\Psi_1^2 - \Psi_2^2)\, \Psi_1^2\, d\Psi_1^2}{(\mathrm{A}^2 - \Psi_1^2)(\Psi_1^2 - \mathrm{B}^2)(\Psi_1^2 - \Gamma^2)},$$

which may also be deduced from the ordinary formulae of elliptic coordinates in space. We may use these expressions to obtain the differential equations of certain loci analogous to those considered in Arts. 17, 18, and 19. Thus, observing that the homographic modulus of the pencil at Δ is $\frac{\Psi_2}{\Psi_1}$, we have, for the curves of equal tangential deflexion, the differential equation

$$\frac{\sqrt{\Psi_2}\,.\,d\Psi_2}{\sqrt{[(A^2-\Psi_2^2)(B^2-\Psi_2^2)(\Psi_2^2-\Gamma^2)]}}=\frac{\sqrt{\Psi_1}\,.\,d\Psi_1}{\sqrt{[(A^2-\Psi_1^2)(\Psi_1^2-B^2)(\Psi_1^2-\Gamma^2)]}}. \quad (1)$$

The curves of 'constant elongation' are defined by the equation

$$d\Sigma_1^2+d\Sigma_2^2=K^2(d\sigma_1^2+d\sigma_2^2),$$

or

$$\left(1-\frac{K^2\Psi_1^2\Psi_2^4}{A^2B^2\Gamma^2}\right)d\Sigma_1^2+\left(1-\frac{K^2\Psi_2^2\Psi_1^4}{A^2B^2\Gamma^2}\right)d\Sigma_2^2=0,$$

in which the variables are not separated. If, however, we attend only to the curves of no elongation, and consider any tangent to one of them as determined by the parameters Φ_1 and Φ_2 of the two concyclic conics which it touches, its differential equation, in this system of tangential coordinates, is obtained by writing Φ_1 and Φ_2 for Ψ_1 and Ψ_2 in the equation (1). For, substituting $\frac{1}{\alpha}, \frac{1}{\beta}, \frac{1}{\gamma}, \frac{1}{\psi_1}, \frac{1}{\psi_2}$ for A, B, Γ, Ψ_1, Ψ_2, in that equation, we have an equation between ψ_1 and ψ_2 of the same form as (1), which represents a curve of no elongation on the hemisphere s; ψ_1 and ψ_2 being the parameters of the concyclic conics which touch any tangent of that curve.

It is evident that an infinitesimal spherical area at any point of the concyclic conic (Φ) is altered in the ratio of $AB\Gamma:\Phi^3$; and, in particular, that this is the ratio of the area contained between (Φ) and ($\Phi+d\Phi$) on S to the area contained between the corresponding curves (ϕ) and ($\phi+d\phi$) on s. The concyclic conics, as curves of constant alteration of adjacent infinitesimal areas, resemble the parallels to the vanishing line in the theory of two homographic plane figures.

33. *Circles of which a Focus is the Centre.*

Since, in general, spherical conics of which F_1 or F_2 is a focus are transformed into spherical conics of which f_1 or f_2 is a focus, and the director arcs of the corresponding curves are corresponding arcs, it follows that circles of which F_2 is the spherical centre are transformed into conics of which f_2 is a focus and of which the director arc is an arc $y\theta$ perpendicular to xz at a distance θ from x defined by the equation

$$\tan\theta=\frac{\tan c}{\tan^2 C}.$$

If R be the spherical radius of one of the given circles, and if r and δ denote the spherical distances from the focus and from the director arc of any point on the corresponding conic, we shall have the equation

$$\frac{\sin r}{\sin \delta} = \frac{\sin c}{\sin \theta} \times \frac{\tan R}{\tan C}.$$

Similarly, spherical conics of which $Y\Omega_1$ or $Y\Omega_2$ is a cyclic arc are transformed into spherical conics of which $y\omega_1$ or $y\omega_2$ is a cyclic arc, and the cyclic poles of corresponding curves are corresponding points. In particular, circles parallel to $Y\Omega_2$ are transformed into conics of which $y\omega_2$ is a cyclic arc, and of which the cyclic pole is a point (ϕ) on xz at a distance ϕ from x defined by the equation

$$\tan \phi = \frac{\tan^2 c}{\tan C}.$$

If R be the radius of a circle parallel to $Y\Omega_2$, p the spherical perpendicular let fall from the cyclic pole on any tangent arc to the corresponding conic, and ρ the angle contained between the tangent arc and the cyclic arc, we shall have the equation

$$\frac{\sin p}{\sin \rho} = \frac{\sin \phi}{\cos C} \times \frac{\tan R}{\tan c}.$$

The two arcs $y\theta$ may be termed the *director arcs*, and the two points (φ) the *cyclic poles*, of the figure s. It is evident that we shall have the relation

$$\tan \theta \tan \Phi = \tan \phi \tan \Theta = 1, \quad \text{or} \quad \Phi + \theta = \tfrac{1}{2}\pi = \phi + \Theta.$$

34. *Circles changed into Circles.*

To determine the small circles of the sphere S which are transformed into small circles of the sphere s, we make use of the principle that a small circle of a sphere is a spherical conic having double contact with the imaginary asymptotic circle; the chord (or arc) of contact being the parallel great circle. Hence the circles required are the spherical conics which have double contact with both P and Q. Of these circles there are three series corresponding to the three pairs of chords of intersection of P and Q. For the chords of contact of any one of the circles with P and Q are a pair of harmonic conjugates of one of the pairs of chords of intersection of P and Q; and, conversely, any such pair of harmonic conjugates may be taken for the chords of contact of a circle with P and Q, or again with Q and P. But the circles of only one of these series are real; their chords of contact being harmonic conjugates of the cyclic arcs, and their centres being on the great circle of the foci. Let R be the radius of one of these circles, Φ the distance of its centre (Φ) from X. The harmonic conjugate of the great circle of which (Φ) is the spherical pole with regard to the cyclic arcs must have the same pole with regard to the imaginary conic Q and with regard to the circle. This condition supplies the equation

$$\tan^2 R = -\frac{\sin(\Phi - C)\sin(\Phi + C)}{\sin(\Phi - c)\sin(\Phi + c)},$$

which determines the radius of the circle when the position of its centre is given, and shows that the circle is real only when Φ is intermediate between C and c.

If r be the radius of the corresponding circle, and ϕ' the distance of its centre from x, we shall have the equations

$$\frac{\tan R}{\tan r} = \frac{\sin 2C}{\sin 2c},$$

$$\tan \Phi \tan \phi' = \tan C \tan c.$$

The corresponding formulae for the circles of the imaginary systems whose centres lie on XY, xy are

$$\tan^2 R = -\frac{\tan^2 C + \sec^2 C \tan^2 \Phi}{\tan^2 c + \sec^2 c \tan^2 \Phi},$$

$$\tan \Phi \tan \phi' = -\sin C \sin c,$$

$$\frac{\tan^2 R}{\tan^2 r} = \frac{\cos^2 c}{\cos^2 C} \cdot \frac{\tan^2 C}{\tan^2 c},$$

where Φ and ϕ' are the distances of the centres of the corresponding circles from X and x. Changing in these formulae c and C into their complements, we have the formulae for the corresponding imaginary circles of which the centres lie on ZY, zy.

35. *Theorems relating to Curvature.*

If two curves on either sphere touch one another at any point, the ratio of the tangents of their spherical radii of curvature remains unchanged in the transformation. This is evident from the corresponding theorem relating to plane homographical figures, because the two planes touching the spheres at two corresponding points are homographic. Thus the ellipses and hyperbolas on either sphere are lines of greatest or least alteration of curvature, as well as lines of greatest or least elongation or contraction. The circles which are transformed into circles are of course loci of points at which the tangent of the radius of curvature is altered in a constant ratio.

36. *Connexion with the Plane Theory.*

If we suppose the arcs C and c to become infinitely small, retaining a finite ratio to one another, the parts of the two spherical figures which lie infinitely near to X and x will ultimately become two plane similar figures. But we can also regard two dissimilar homographic plane figures as a limiting case of two homographic figures upon a sphere. The points in the two hemispheres, which we have hitherto considered, correspond to one another throughout the whole of each surface with similar directions of rotation. But if, in the hemisphere S, we substitute for the quadrant containing F_1 the opposite quadrant, so as to consider the hemisphere of which Z is the spheric centre and the great circle XY the base, we shall obtain a figure of which one quadrant (F_2) answers with similar rotation to the corresponding quadrant (f_2), and the other quadrant (F_1) answers with dissimilar rotation to the corresponding quadrant (f_1). If, for example, in the formulae of Arts. 26–28, we change C, D, $\frac{1}{2}D_1$ or $\frac{1}{2}D_2$, H_1, H_2, Δ, $\frac{1}{2}R_1$ or $\frac{1}{2}R_2$, into their complements, we shall have the equations which express the metrical relations of the two figures considered in this particular manner. In these new formulae, $2c$ and $2C$ are the angles contained between the foci and between the cyclic planes of s; or, again, they are the angles contained between the cyclic planes and the foci of S. The new arcs D, H_1, H_2 are not measured from the points corresponding to the original points of the arcs d, h_1, h_2, but from points distant by a quadrant from the points corresponding to those original points; they are also measured backward—*i.e.* in the direction opposite to that which corresponds to the direction in which the arcs d, h_1, h_2 are measured. And a correlative statement is true for the angles Δ, H_1, H_2. It will be observed that R_1 or R_2 is changed into its supplement according as the points considered lie in the regions of similar or of dissimilar rotation. Again, it is immaterial whether we change D_2 or D_1 into its supplement; in the former case, we consider (in the figure S) the triangle D_1YD_2, in the latter the triangle $D_1\overline{Y}D_2$.

If we now suppose the arcs c and C to become evanescent, the parts of the two figures adjacent to x and Z respectively will become two dissimilar homographic plane figures, and we may pass from the spherical formulae to the corresponding formulae of the plane theory.

37. *Point-Figures in Perspective.*

When two homographic point-figures are in a perspective position, (*i.e.* when the corresponding planes and lines of the two figures intersect upon the same plane,) one of the focal lines of each pencil is, evidently, the line joining the centres S and s of the two pencils. To find the other focal lines, let Ss meet the plane of intersection in O_1, let O be the point harmonically conjugate to O_1 with regard to Ss, and O_2 the orthogonal projection of O on the plane of intersection; SO_2 and sO_2 are the focal lines required.

To place two given homographic point-figures in a perspective position, we first of all place a pair of corresponding focal lines in the same straight line, the vertices of the two pencils not coinciding, but corresponding vectorial planes coinciding. Let O_2 be the point of intersection of the two remaining focal lines, and let V_1 and V_2 be the planes which bisect the angle SO_2s externally and internally; then, according as the corresponding directions of rotation round SO_2 and so_2 are similar or dissimilar, V_1 or V_2 is the plane of intersection of the two homographic figures. It is evident that the two figures will continue in perspective if their centres be moved nearer to or further from one another in the coincident focal lines; or, again, if either of them be rotated through an angle of 180° round these coincident lines. Of the cyclic planes, one pair are parallel to the plane of intersection, the other pair intersect in that plane, and in the plane bisecting Ss at right-angles.

38. *Case in which the Homography is Spheroidal.*

The theory of the particular case in which the transformation is *spheroidal* —*i.e.* in which the imaginary cones Q and p, corresponding to the evanescent sphere-cones q and P, are cones of revolution—presents no difficulty whatever. If X and x are the centres of the imaginary small circles Q and p, the *azimuths* of any two corresponding points A and a are equal, and their *zenith-distances* are connected by the relation $\dfrac{\tan XA}{\tan xa} = \text{constant}$.

This constant ratio we may term the *modulus* of the transformation.

C.—Focal Properties of Two Homographic Spaces.

39. *The Imaginary Conics and the Parameters.*

We proceed, in the last place, to consider two spaces S and s homographically related to one another. Let Ω and σ be the imaginary circles at an infinite distance in which all spheres in the two spaces intersect one another; ω and Σ the imaginary conics corresponding to them. As we shall suppose that the planes at an infinite distance in the two spaces are not corresponding planes, the imaginary conics Ω and Σ, ω and σ, are certainly different. If either ω or Σ is an imaginary circle, the other is so too; for if Ω and Σ have a common chord, ω and σ must also have a common chord, and *vice versâ*. We shall, however, for the present, exclude this important particular case, and shall suppose that neither ω nor Σ is an imaginary circle. Let O, o' be the centres of Σ and ω respectively (these conics have no real tangents, and therefore are not parabolas); let X, Y, x', y' be the points at an infinite distance on their principal axes, Z, z' the points at an infinite distance on the normals to their planes, and let Σ_1, Σ_2, ω_1, ω_2 be the asymptotic points of Σ, ω, lying on the lines XY, $x'y'$, which are the lines at an infinite distance in the planes of the two conics, and which we shall suppose to meet the imaginary circles Ω and σ in the points $\Omega_1\,\Omega_2$ and $\sigma_1\,\sigma_2$. The lines XY, $x'y'$ are evidently corresponding lines; and because the poles of XY with regard to Ω and Σ correspond to the poles of $x'y'$ with regard to ω and σ, the points Z, o' and the points O, z' are corresponding points. The anharmonic equation $[\Sigma_1, \Sigma_2, \Omega_1, \Omega_2]=[\sigma_1, \sigma_2, \omega_1, \omega_2]$, which is implied by the homographic relation of the figures, may also be written $[\Omega_1, \Omega_2, \Sigma_1, \Sigma_2]=[\sigma_1, \sigma_2, \omega_1, \omega_2]$; and in this form it expresses that the imaginary angles $\Sigma_1' O \Sigma_2$, $\omega_1 \sigma' \omega_2$ are superposable; *i.e.* that the imaginary conics Σ and ω are similar. Again, because X, Y are harmonic conjugates of $\Omega_1\,\Omega_2$ and $\Sigma_1\,\Sigma_2$, while x, y are harmonic conjugates of $\omega_1\,\omega_2$ and $\sigma_1\,\sigma_2$, x, y correspond to X, Y; and we may suppose the correspondence fixed by the equation

$$[\Omega_1, \Omega_2, \Sigma_1, \Sigma_2, X, Y]=[\omega_1, \omega_2, \sigma_1, \sigma_2, x, y].$$

This equation implies one or other of the equations

$$[\Omega_1, \Omega_2, \Sigma_1, \Sigma_2, X, Y]=[\sigma_1, \sigma_2, \omega_1, \omega_2, x, y]$$

or

$$[\Omega_1, \Omega_2, \Sigma_1, \Sigma_2, X, Y]=[\sigma_1, \sigma_2, \omega_1, \omega_2, y, x].$$

Of these, the former is inadmissible, as it would imply that

$$[\sigma_1, \omega_1, x, y]=[\omega_1, \sigma_1, x, y],$$

which is impossible, since ω_2, and not σ_1, is the harmonic conjugate of ω_1 with regard to xy. We infer, therefore, that the point at infinity on the major axis of Σ corresponds to the point at infinity on the minor axis of ω, and *vice versâ*. Let $A\sqrt{(-1)}$, $B\sqrt{(-1)}$, $a\sqrt{(-1)}$, $b\sqrt{(-1)}$ be the principal semi-axes of Σ and ω; A, B are the parameters of S, and a, b of s; they are connected by the equation $Aa=Bb$, which results from the similarity of Σ and ω.

40. *The Correspondence of Directions—the Principal Axes.*

From the homographic relation of the two figures it follows that to each direction on any straight line in either figure there corresponds a definite direction on the corresponding line. And again, to each direction of rotation round any line there corresponds a definite direction of rotation round the corresponding line. It is easily shown (by considering in each figure two infinitesimally near positions of a straight line in relation to a line at a finite distance) that the two figures are either similar in respect to all rotations, or dissimilar in respect to all rotations; *i.e.* that corresponding rotations round corresponding directions are either always similar, or else always dissimilar. For clearness we may suppose that corresponding rotations in the two figures are similar. We shall call the lines OX, OY, OZ, $o'x$, $o'y$, $o'z$ the *principal axes* of the two figures, and the planes OYZ, OXZ, OXY, $o'yz$, $o'xz$, $o'xy$ the *principal planes*; the axes OZ, $o'z$, which alone are corresponding lines, we shall call the *focal axes*, and the planes of XY, xy the *vanishing planes.*

Each space is divided by its three principal planes into eight octants, corresponding respectively to the eight octants of the other space. Considering these octants as tetrahedra, of which the plane at an infinite distance is one boundary, and observing that in either space the plane at an infinite distance corresponds to the vanishing plane of the other space, we find that to adjacent octants on the same side of the vanishing plane in either space there correspond in the other space adjacent octants on the same side of the vanishing plane, but that adjacent octants on opposite sides of the vanishing plane in either space correspond to octants diametrically opposite in the other space; so that, if the correspondence of two octants is given, that of the remaining octants is immediately ascertained. Again, if P, Q are any two points on the same side of the vanishing plane of S, and if p, q are the points corresponding to P, Q, the directions PQ, pq are corresponding directions in the two spaces; and similarly the corresponding directions of rotation round any two corresponding lines may be ascertained. We may add that if V be any closed figure in S, which lies wholly on one side of the vanishing plane, points in the interior of V will correspond to points in the interior of a closed figure v corresponding in the space s to V.

41. *Determination of the Principal Axes and Parameters.*

The geometrical construction for the determination of the principal axes in each figure and of the parameters A, B, a, b, is as follows. We first obtain the vanishing plane of each figure; *i.e.* we determine in each figure three points corresponding to three points at an infinite distance in the other figure; the points at an infinite distance in the directions normal to the vanishing planes are the points Z and z', and the points corresponding to these are the centres o' and O of the imaginary conics ω and Σ; thus the focal axes OZ and $o'z'$ are known. At the point O in the vanishing plane of S take two pairs of lines corresponding to two pairs of rectangular lines intersecting at z' in the plane at an infinite distance in s. The axes OX and OY are the pair of lines at right angles to one another in the involution determined by the two pairs so constructed; x and y, which determine $o'x$ and $o'y$, are the points corresponding to X and Y. Lastly, to find the parameters, we observe that, if in any two corresponding planes the chords intercepted by Σ and ω are $2D\sqrt{(-1)}$ and $2d\sqrt{(-1)}$ respectively, the parameters of the two homographic plane figures are D and d; their homographic centres are the points of bisection of the chords, and their focal axes are the perpendiculars to the chords at their points of bisection. Hence we obtain the four parameters A, B, a, b by constructing the homographic foci of the principal planes XZ, YZ, xz, yz.

42. *The Confocal Quadrics.*

The imaginary conic Σ, in which we may suppose $A > B$, determines a system of confocal quadrics of which it is the imaginary focal conic. One of the two real focal conics is an ellipse in the plane of YZ, of which the foci, in the axis of Z, are the homographic foci of the plane YZ, and of which the vertices, in the same axis, are the homographic foci of the plane XZ. The other is an hyperbola in the plane of XZ, having of course the vertices of the ellipse for foci and its foci for vertices. The system of confocal quadrics of which ω is the imaginary focal conic corresponds homographically to the confocal quadrics of the system S. For, since the conics σ and ω correspond to the conics Σ and Ω, the imaginary developable circumscribing the two former conics corresponds to the imaginary developable circumscribing the two latter conics, and therefore the quadrics inscribed in these corresponding developables are themselves corresponding surfaces. In particular, to the focal ellipse of S there corresponds the focal hyperbola of s, and *vice versâ*; the extremities of the focal axes of the ellipses being transformed into the extremities of the focal axes of the hyperbolas, and the extremities of the minor axes of the ellipses into the asymptotic points of the hyperbolas. Again, the ellipsoids of either confocal system are changed into the hyperboloids of two sheets of the other system; and the hyperboloids of one sheet into the hyperboloids of one sheet. And by considering the two pairs of homographic planes XZ, xz, YZ, yz, we see immediately that the eccentricities of the sections of corresponding confocals made by corresponding principal planes are reciprocal, and that the rectangle of their major semi-axes is equal to the rectangle of the parameters $A \times a$ or $B \times b$. Again, to the normals of any confocal there correspond the normals of the corresponding confocal; the lines of curvature of the two surfaces, their umbilics, the two systems of orthogonal developables formed by the normals of each of them, their centres of curvature, and the surfaces which are the loci of those centres, all correspond homographically; the cuspidal lines of the normal developables are corresponding geodesic lines upon the surfaces of centres, and the lines of contact of two corresponding developables with those sheets of the surfaces of centres upon which their cuspidal lines do not lie are in like manner corresponding lines. Further, since the normals of corresponding confocals are corresponding lines, the geodesics of either surface correspond to the geodesics of the other; again, the confocals enveloped by the developables of two corresponding geodesics are corresponding confocals, and the lines of contact are corresponding lines. To the various modes of description of the lines of curvature of either system of confocals by means of a thread stretched upon surfaces of the system there will correspond similar modes of description of the lines of curvature of the other system of confocals. For an example, we may take the general theorem of M. Chasles:—

'If an inextensible thread, of which the extremities are fastened to two fixed points upon one of two confocal surfaces of different kinds, be strained by the point of a pencil which moves upon the second surface, so that the thread consists (in general) of six portions, two of which are geodesics of the first surface, two are geodesics of the second surface, while the other two are the portions of common tangents to the two surfaces included between the points of contact, the point of the pencil will describe a line of curvature of the second surface.'

While an inextensible thread moves in either space in the manner described in this enunciation, an inextensible thread will move in the same manner in the other space; and the six portions of the first thread will correspond homographically to the six portions of the second. But it is to be observed that the constant lengths of the two threads will be related transcendentally to one another; as also will the lengths of the corresponding curvilinear portions of the two threads.

We may add that to two geodesic arcs of which the difference is rectifiable, there will correspond two geodesic arcs of which the difference is rectifiable.

Again, when the difference of two arcs of a line of curvature can be expressed by geodesic lines in either figure, the corresponding difference can be similarly expressed in the other figure.

43. *The Point-Figures at Corresponding Points—their Focal Lines.*

We shall next consider any two corresponding points P and p in the two spaces. At these two points we have two homographic point-figures, of which the relations to one another are readily ascertained. To the cones which from P envelope the conics Ω and Σ there will correspond the cones which from p envelope ω and σ. Thus the principal axes of the point-figures at P and p are the normals to the surfaces of the confocal system which pass through P and p; and the focal lines of the figures are the generators of the hyperboloids of a single sheet which pass through P and p. We thus have the theorem :—

'Any two corresponding generators of two hyperboloids of the two confocal systems are the axes of pencils of planes of which the correspondence is equiangular.'

If the points P and p be taken on corresponding focal conics, the two generators will coincide. Thus, 'the focal conics are the loci of points at which the correspondence of the homographic point-figures is spheroidal.'

It is evident that, given in one of the two spaces a point and three generators (of the same or different hyperboloids) and the corresponding point and lines in the other space, we can immediately, by means of the equiangular pencils of planes, determine the point p in either space which corresponds to a given point P in the other. We might take for the three generators in each space any three tangents to a focal conic; the simplest construction being perhaps that in which the tangents at the vertices of the focal conics are employed as the axes of equiangular pencils.

We thus obtain the following rule, which is well adapted to the methods of descriptive geometry :—'Project the given point P orthogonally on the planes of XZ, YZ, and using the focal radii vectores of the projections, as in Art. 5, determine the points corresponding to them in the planes of xz and of yz : these points are the orthogonal projections of the point p.'

44. *The Strain Ellipsoid—its Cyclic Planes and Focal Asymptotes.*

The position of the cyclic planes of the homographic figure at P may be ascertained by means of the focal lines of the figure at p. But these cyclic planes are also the cyclic planes of the 'strain ellipsoid' at P; *i.e.* of the evanescent ellipsoid which has its centre at P and corresponds to an evanescent sphere having its centre at p. For this evanescent ellipsoid has for its asymptotic cone the imaginary cone which from P envelopes Σ, and is thus concentric, similar, and similarly situated with the auxiliary ellipsoid of M. Chasles, *i.e.* with the ellipsoid of which the principal axes are equal to the major axes of the three confocal surfaces passing through P, and are normal to those three surfaces respectively (Aperçu historique des Méthodes en Géométrie, Note 25). It appears at the same time that the asymptotes of the focal conic of the auxiliary ellipsoid, or of the strain ellipsoid, coincide with the focal lines of the point P.

The cyclic planes at P and p are the 'planes of no distortion' at those corresponding points; *i.e.* (1) evanescent lines passing through P and lying in either cyclic plane are altered in a constant ratio; (2) angles in a cyclic plane at P are transformed into equal angles in the corresponding cyclic plane; so that P, p are homographic foci of either pair of cyclic planes. The second property is analogous to the property that the focal lines are the axes of equal homographic pencils of planes. If we observe that the focal asymptotes of a quadric are the axes of its circumscribing right cylinders, we may enunciate a property of the focal lines analogous to the first property of the cyclic planes:—

'Planes parallel to either focal line, and infinitely near to P, are transformed into planes, which may ultimately be regarded as parallel to the corresponding focal line, and of which the distances from p are in a constant ratio to the distances of the first planes from P.'

We may express this by saying that a generating line of a confocal hyperboloid is, at any point of it, a line of equal transverse elongation. And since the right cylinder of which the focal line at P is the axis and which circumscribes the strain ellipsoid at P is transformed into a right cone of which the vertex lies on the vanishing plane of s, we see that, if the point P vary its position on a given hyperbolic generator, the ratio of transverse elongation varies inversely as the distance of p from the vanishing plane of s, or directly as the distance of P from the vanishing plane of S.

45. *The Canonical and Elliptic Equations.*

If we represent by X, Y, Z, x, y, z the coordinates of corresponding points in the two spaces referred to their principal axes, the canonical equations of the homography will be

$$Zz = Aa = Bb,$$

$$\left.\begin{matrix} Xz = Ax \\ Yz = By \end{matrix}\right\}, \text{ or } \left.\begin{matrix} xZ = aX \\ yZ = bY \end{matrix}\right\} . \qquad \text{(A)}$$

If, again, we denote the elliptic coordinates of corresponding points in either space (referred to the corresponding confocal systems) by $\Lambda_1, \Lambda_2, \Lambda_3, \lambda_1, \lambda_2, \lambda_3$, the homographic equations are

$$\Lambda_1 \lambda_1 = \Lambda_2 \lambda_2 = \Lambda_3 \lambda_3 = Aa = Bb\,; \quad . \; . \; . \; . \; . \; . \; . \quad \text{(B)}$$

so that every general homographic transformation may be represented as a transformation of the elliptic coordinates of a point into their reciprocals.

46. *Determination of the Strain Ellipsoid.*

Either of these sets of formulae will serve to determine the ratios of the axes of the strain ellipsoid at P to the radius of the evanescent sphere at p. The rectangular formulae show that the ratio of an evanescent volume at P to the corresponding volume at p is that of Z^4 to $Aa \times ab$; whence, if $\theta\Lambda_1, \theta\Lambda_2, \theta\Lambda_3$ are the semi-axes of the strain ellipsoid at P, and i the radius of the evanescent sphere at p,

$$\theta^3 \frac{\Lambda_1 \Lambda_2 \Lambda_3}{i^3} = \frac{Z^4}{Aa \times ab},$$

or, since $\Lambda_1 \Lambda_2 \Lambda_3 = ABZ$, and $Zz = Aa = Bb$, $\theta = \dfrac{i}{z}$.

Or again, transforming by the equations (B) the elliptic formula

$$d\Sigma_1 = \sqrt{\frac{(\Lambda_1^2 - \Lambda_2^2)(\Lambda_1^2 - \Lambda_3^2)}{(\Lambda_1^2 - A^2)(\Lambda_1^2 - B^2)}}\, d\Lambda_1,$$

we find

$$d\sigma_1 = \frac{Aa \times AB}{\Lambda_1 \Lambda_2 \Lambda_3} \frac{d\Sigma_1}{\Lambda_1},$$

or

$$\frac{d\Sigma_1}{\Lambda_1} = \frac{d\sigma_1}{z},$$

which agrees with the preceding determination of θ, the symbols $d\Sigma_1$ and $d\sigma_1$ representing corresponding elementary arcs normal to (Λ_1) and (λ_1).

Our limits prevent us from applying these formulae to the determination of the loci corresponding to those considered in Arts. 17–19. For the same reason, we omit the elementary theorems relating to the curvature and torsion of curve lines, and the curvature of curve surfaces.

47. *The Parameters of the Confocal and Concyclic Cones at any Point.*

The equation of the imaginary cone which from the point P envelopes Σ is

$$\frac{X^2}{\Lambda_1^2} + \frac{Y^2}{\Lambda_2^2} + \frac{Z^2}{\Lambda_3^2} = 0,$$

and the cone which from the same point envelopes the confocal surface, of which Ψ is the semi-axis major, is

$$\frac{X^2}{\Lambda_1^2-\Psi^2}+\frac{Y^2}{\Lambda_2^2-\Psi^2}+\frac{Z^2}{\Lambda_3^2-\Psi^2}=0;$$

so that the coefficients (designated by A, B, Γ, α, β, γ in Art. 30), which determine the homography of the point-figures at P and p, are, in fact, the elliptic coordinates of those points; and the parameters of the confocal cones of the point-figures are the same as the parameters of the confocal quadrics which they envelope. Thus, the formulae of Arts. 30, 31 are immediately applicable to the figures at P and p. And, if Φ is the parameter of a concyclic cone at P, so that $\Phi=\frac{\Lambda_1\Lambda_2\Lambda_3}{\Psi_1\Psi_2}$, where Ψ_1, Ψ_2 are the parameters of two confocal quadrics touching any line of (Φ), the elongation at P in the direction of any line of (Φ) is given by any one of the formulae

$$\frac{T}{\tau}=\frac{\Lambda_1\Lambda_2\Lambda_3}{Aa\times AB}\Phi=\frac{\Lambda_1^2\Lambda_2^2\Lambda_3^2}{Aa\times AB}\times\frac{1}{\Psi_1\Psi_2}=\frac{Z\Phi}{Aa}=\frac{B}{a}\times\frac{Z^2}{\Psi_1\Psi_2}=\frac{\Phi}{z},$$

T and τ representing corresponding elements at P and p. It will be observed that, at equal distances from the vanishing plane, the elongation is the same on all lines touching the same two confocal quadrics.

48. *Lines Tangent to two Confocal Quadrics.*

Let L_1 and L_2 be two straight lines in the space S, each of which touches the two confocals (Ψ_1) and (Ψ_2); let also l_1, l_2, (ψ_1), (ψ_2) be the corresponding lines and confocal quadrics in the space s. The tangent planes $L_1\Psi_1$, $L_1\Psi_2$ (*i.e.* the tangent planes to (Ψ_1), (Ψ_2) at their points of contact with L_1) are at right-angles to one another, and are transformed into two planes which are at right-angles to one another. Again, the pair of planes, tangent to any third confocal surface (Ψ_3), which intersect in L_1, make the same angles with the bisecting planes $L_1\Psi_1$, $L_1\Psi_2$ that the pair of planes, tangent to the same confocal surface (Ψ_3) and intersecting in L_2, make with the bisecting planes $L_2\Psi_1$, $L_2\Psi_2$. For the involutions of pairs of planes determined by the confocal system at the lines L_1 and L_2 are necessarily equiangular in respect of all their corresponding pairs, because they are equiangular in respect of the coincident pairs of planes determined by (Ψ_1) and (Ψ_2), and of the imaginary pair of cyclic planes determined by the imaginary circle at an infinite distance. From this theorem (of which M. Chasles has given a different demonstration; see Liouville, Vol. XI, First

Series, p. 109) we infer that equal dihedral angles, similarly placed in the pencils of planes at L_1 and L_2, are transformed into dihedral angles equal to one another, and placed similarly to one another, in the pencils of planes at l_1 and l_2. Or again, if I and i are the angles made with $L_1 \Psi_1$ and $l_1 \psi_1$ by corresponding planes passing through L_1 and l_1, and if we denote the major semi-axes of the surfaces (Ψ_1), (Ψ_2), (ψ_1), (ψ_2) by Ψ_1, Ψ_2, ψ_1, ψ_2, we shall have the equation

$$\frac{\tan I}{\tan i} = \frac{\Psi_1}{\Psi_2} = \frac{\psi_2}{\psi_1},$$

which results immediately from a formula given by M. Chasles (*loc. cit.* p. 106), combined with the equations of transformation (B); and which shows, in conformity with our theorem, that the ratio of $\tan I$ to $\tan i$ is the same, whatever common tangent of (Ψ_1) and (Ψ_2) we consider. We have, in fact, the still more general theorem:

'All pencils of planes, of which the axes are touched by two confocals having their major semi-axes in a given ratio, have that ratio for their modulus of transformation; and in all such pencils, the involutions which are transformed into equiangular involutions are equiangular with one another,' which is an immediate consequence from Arts. 47, and 30, (ϵ.).

Since a generating line of a confocal hyperboloid may be regarded as a line of which the two tangent confocals coincide, this enunciation includes, as a particular case, the equiangular property of the generating lines.

We have seen that the focal conics are the loci of points at which the transformation is spheroidal. We may now add, that at any one of these points the modulus of transformation (Art. 38) is $\frac{Q}{P}$, if Q is the semi-axis major (A or B) of the focal conic on which the point is taken, and P is the semi-axis major of the confocal quadric which passes through the point.

49. *Ivory's Theorem.*

If on two confocal surfaces of the same kind in the space S we consider two points which *correspond* to one another in the sense in which that term is employed in Ivory's theorem, these two points will be transformed into two others in the space s, which will also correspond to one another in the same sense. This principle, which is immediately verified by means of the equations (A), may serve to transform some geometrical, and even physical, propositions. For example, we see that to every focal generation of a quadric according to Jacobi's method, there corresponds homographically a similar focal generation of another quadric.

50. *Equi-Segmental Axes and Planes.*

The equi-segmental axes of all planes in the space S, which cut the vanishing plane in straight lines parallel to a given straight line, (or, which is the same thing, of all planes which pass through a given point at an infinite distance on the vanishing plane,) lie on two planes at equal distances from the vanishing plane and parallel to it.

For, in the first place, parallel planes in the space S have their equi-segmental axes at one and the same distance from the vanishing plane, since to parallel planes in the space S there correspond planes in the space s, which intersect the vanishing plane of that space in the same straight line, and of which the foci are consequently at a constant distance from one another; this constant distance being equal to the distance of the equi-segmental axes of the planes in the space S. Again, planes in the space S, which intersect the vanishing plane in the same straight line, have their equal axes situated at equal distances from the central plane. For to these planes correspond parallel planes in the space s; and, by what has just been proved, the equi-segmental axes of these planes lie in two planes parallel to the vanishing plane; therefore the equi-segmental axes of the planes in S lie in two corresponding planes, *i.e.* in two planes parallel to the vanishing plane. The theorem itself results from the combination of these two particular cases of it.

It may be worth while to verify the theorem analytically. If

$$\frac{X}{p} + \frac{Y}{q} + \frac{Z}{r} = 1$$

is the equation of any plane of S, the equation of the corresponding plane of s is

$$\frac{Ax}{pz} + \frac{By}{qz} + \frac{Aa}{rz} = 1,$$

which meets the vanishing plane of s in the line

$$z = 0, \quad \frac{Ax}{p} + \frac{By}{q} = -\frac{Aa}{r},$$

or

$$z = 0, \quad \frac{x}{a\frac{p}{r}} + \frac{y}{b\frac{q}{r}} = -1.$$

The square of the semi-chord determined on this line by the imaginary conic ω, or

$$\frac{x^2}{a^2} + \frac{y^2}{b^2} + 1 = 0,$$

is

$$-\frac{(a^2p^2+b^2q^2)(q^2r^2+p^2r^2+p^2q^2)}{r^2(p^2+q^2)^2};$$

and this square, multiplied by the square of the sine of the angle between the given plane and the vanishing plane of S, becomes

$$-\frac{a^2 p^2 + b^2 q^2}{p^2 + q^2},$$

an expression of which the value depends only on the ratio of p to q.

It follows from this theorem, that to ascertain the position in the space S of the equi-segmental axes of any plane whatever, it will suffice to attend to the principal equi-segmental axes, *i.e.* to the equi-segmental axes of planes which pass through the focal axis. Let P be such a plane, and let $D\sqrt{(-1)}$ be the semi-diameter of the imaginary focal conic lying in that plane, $d\sqrt{(-1)}$ the semi-diameter of the imaginary focal of s, determined by the corresponding plane. It will be found that $Dd = Aa$; so that we have for the semi-distance d of the equi-segmental axes of the plane P, the expression $d = \frac{Aa}{D}$. Thus all the planes, loci of real equi-segmental lines, or, as we shall term them, all the *equi-segmental planes* of S, are comprised between two planes, at distances a and b on the positive side of the vanishing plane, and between two planes symmetrically situated on the negative side of the same plane. Again, since the semi-diameters of S, which are equally inclined to its principal axes, are equal to one another, each equi-segmental plane contains two distinct series of equi-segmental parallels, the two series being equally inclined to the plane of ZX or ZY; in the two extreme pairs of equi-segmental planes these two series coincide with one another, and their common direction is that of one of the principal axes OX or OY.

If we consider two planes intersecting in the axis of Z, and inclined at an angle I to the plane of ZX, we have for the square of the semi-distance of their equi-segmental lines the expression,

$$d^2 = a^2 \cos^2 I + b^2 \sin^2 I.$$

The corresponding inclination i is given by the equation

$$\tan I = \frac{B}{A} \tan i = \frac{a}{b} \tan i,$$

and the corresponding value of D^2 is

$$D^2 = A^2 \cos^2 i + B^2 \sin^2 i.$$

These equations show that if we imagine the spaces S and s so placed that their focal axes coincide, while the axes of X and Y lie in the axes of y and x

respectively, the principal equi-segmental axes of either space will be those generating lines of hyperboloids of the other space which lie in planes parallel to the central plane.

The homographic relation of any two corresponding equi-segmental planes is very simple. If we conceive of the points of each plane as referred to its principal equi-segmental axes, the corresponding coordinates of corresponding points will be equal, and only the angle between the axes will be different in the two planes. These angles are never equal to one another (except in the excluded case $A = B$, $a = b$); they are, however, supplementary to one another in the *principal* equi-segmental planes, *i.e.* in the planes defined by the equations

$$Z = \pm\sqrt{(ab)}, \quad z = \pm\sqrt{(AB)},$$

since in these planes we have

$$\tan^2 I = \frac{B}{A}, \quad \tan^2 i = \frac{b}{a}, \quad \tan I \tan i = 1.$$

To obtain the straight line in s, which corresponds to any given straight line in S, we may either determine its projections on the two focal planes, by means of the equi-segmental axes in those planes; or we may, instead, consider the intersections of the given straight line with any pair of equi-segmental planes of S, and obtain the corresponding points in the corresponding equi-segmental planes of s. We have, however, in every case to measure the equal corresponding segments in corresponding directions; and these can always be ascertained by inspection, if we have first fixed the correspondence of the eight octants of each space to the eight octants of the other.

51. *Properties of the Hyperbolic Generators.*

The generating lines of the confocal hyperboloids possess a metrical property with regard to the equi-segmental planes which may be very variously expressed, according to the equi-segmental planes considered. Thus:

'The intercept made in the space S upon any generator of a confocal hyperboloid by the tangent planes to that hyperboloid, which are parallel to the vanishing plane, is to the corresponding intercept in the space s in the constant ratio of $\sqrt{(AB)}$ to $\sqrt{(ab)}$.'

Or, again:

'The intercept made on any hyperbolic generator of S by the two equi-segmental planes $Z = \pm\sqrt{(ab)}$ is to the corresponding intercept in the space s in the inverse ratio of the major axes of the hyperboloids to which the two generators belong.'

In connexion with this property we may mention the following, which, however, does not depend on the general homographic transformation we are considering:

'If one of two confocal hyperboloids be transformed into the other by the transformation of Ivory, segments on any generator of the one are transformed into equal segments on the generator of the other.'

52. *Homographic Spaces placed Symmetrically.*

It is in general impossible to place two homographic spaces S and s in the same space, so that any given point of that space shall have the same corresponding point, to whichever of the two spaces it is considered to belong. The conditions that this reciprocal relation of the two spaces should be possible are that corresponding rotations in the two figures should be similar and that

$$A = b, \quad B = a, \qquad \text{(C)}$$

either of these equations, of course, implying the other. For, if these equations be satisfied, and if corresponding rotations be similar, we may place the axes of OZ, OX, OY upon the axes of oz, $\pm oy$, $\pm ox$, inasmuch as the positive directions of OZ and oz are not corresponding directions. Writing, as we may then do, $\pm X$ for Y and $\pm Y$ for X in the equations (A), we find

$$zZ = ab, \quad zX = \pm by, \quad zY = \pm ax,$$

and these equations are not altered by interchanging simultaneously X, x; Y, y; Z, z. The points which coincide with their conjugates are the points of the lines

$$z = +\surd(ab), \quad y = \pm\surd\left(\frac{a}{b}\right) x,$$

$$z = -\surd(ab), \quad y = \mp\surd\left(\frac{a}{b}\right) x;$$

i.e. the principal equi-segmental planes of S coincide with their corresponding planes, and in each of these planes the points of one of the principal equi-segmental axes coincide with their corresponding points. Every plane which passes through either of these lines corresponds to itself, and so does every line which meets both of them. Again, we may also place the axes of OZ, OX, OY upon the axes of $-oz$, $\pm oy$, $\mp ox$; in this case, the equations (A) become

$$zZ = -ab, \quad zX = \mp by, \quad zY = \pm ax,$$

which are still symmetrical, but which give imaginary loci of coincident points. Either the upper signs, or else the lower signs, may be taken in each case; so that the two spaces admit of four different symmetrical positions.

We may arrive at the preceding results without using the equations (A); for it is readily seen that the necessary and sufficient conditions for the reciprocity of the two homographic systems are that the imaginary conics Σ and ω should coincide, and that those points on the two conics should be coincident, which correspond to the same points of the imaginary circle at an infinite distance. The equations (C) are the conditions that the two conics should be equal in all respects; if these equations are satisfied, the two conics can be brought into coincidence in four different ways, and in each of these four ways the points which ought to coincide will coincide, if corresponding rotations in the two spaces are similar.

53. *Case of a Spheroidal Homography.*

It is hardly necessary to do more than mention the case of a spheroidal homography, in which $A = B$, $a = b$. All meridian planes of the space S have the same foci at a distance $\pm A$ from the equatorial (or vanishing plane), and their equi-segmental axes lie in the same two parallel planes at a distance $\pm a$ from the equatorial plane.

The angle contained by any two meridian planes is unchanged in the transformation; and the homographic relation is the same for all pairs of corresponding meridian planes. Thus, all angles between planes and lines intersecting at either focus remain unchanged in the transformation, and the pencils in space at corresponding foci are superposable. Similarly, each equi-segmental plane is superposable upon the plane corresponding to it. The two spaces may, in fact, be conceived as generated by the equiangular rotation of two homographic planes round their focal axes. The condition that they should be capable of occupying a reciprocal position is that the distances between the foci in each space should be equal.

54. *Historical Note.*

The existence of two pairs of parallel equi-segmental axes in any two homographic plane figures was established by M. Moebius in 1827. ('Barycentrische Calcul,' p. 320, sect. 230.) M. Moebius also showed that, if the corresponding points of two corresponding equi-segmental axes coincide in the line of intersection of two homographic planes, the two planes are in perspective. Magnus ('Sammlung von Aufgaben und Lehrsätzen aus der Analytischen Geometrie,'

Berlin, 1833, p. 41, sect. 12) proved that in two homographic plane figures there exists a pair of corresponding points at which the corresponding pencils are equi-angular; and that, if the figures be placed in the same plane with these 'centres of collineation' coincident, and either of them rotate in its own plane round the centre of collineation, it will become homological with the other in two diametrically opposite positions, in one of which positions one pair of equi-segmental axes will coincide, while the other pair will coincide in the other position. Magnus expressly says that 'of two collinearly-related systems [*i.e.*, two homographic plane figures in which the straight lines at an infinite distance are not corresponding lines] 'each has, in general, only one centre of collineation.' As Magnus tacitly supposes that the figures are not in any position whatever with regard to one another, but are already placed in the same plane, this statement is not untrue; but it is only part of the truth, and the analysis by which Magnus obtains one centre of collineation in each figure, will also supply a second pair, if we change the sign of the constant p in the equations (1) of p. 42 *loc. cit.* It is of course quite true that, if the two figures are once placed in the same plane, there is only one point in each which can be regarded as a centre of collineation; and this, which Magnus has proved analytically, Dr. Salmon has also shown geometrically ('Higher Plane Curves,' Art. 230, p. 246). But it is to be remembered that two planes can be made to coincide in two different ways according as they are placed face to face, or both facing the same way, and, in one of these positions of coincidence one of the pairs of foci are the centres of collineation, and the other pair in the other position. It is worth while to add that though, as Dr. Salmon has observed, the position of the imaginary circular points at an infinite distance is unaffected by any motion of translation or rotation of a plane figure in its own plane, those two imaginary points are interchanged with one another if the figure be rotated through an angle of 180° round any axis in its own plane. And the change of the centre of collineation, which takes place when one of two homographic figures, of which the planes are coincident, is thus rotated, is a necessary consequence of the interchange of the imaginary cyclic points in the rotated figure.

In the 'Traité de Géométrie Supérieure,' only one pair of equi-segmental axes and one pair of foci are expressly mentioned. But the omission is only accidental, as the methods by which one pair of foci and one pair of equi-segmental axes are obtained would equally supply the other pair. The theorem, that 'if two planes are in perspective, the foci are the points in which they are intersected by the perpendiculars let fall from the centre of perspective on the planes bisecting the angles contained by the two planes,' is an immediate inference from a principle first given by M. Chasles ('Aperçu de l'Histoire des Méthodes en Géométrie,' note iv.) and subsequently employed by Mr. Mulcahy ('Principles of Modern Geometry,' cap. VIII., Art. 115).

Subsequently to the communication of this memoir to the London Mathematical Society, but (it is unnecessary to say) quite independently of it, three papers have appeared, relating in part to the same subject. (1) In the May number of the 'Nouvelles Annales de Mathématiques,' M. Abel Transon obtains the theorem of the two pairs of foci by the application of a very general analytical method; he accurately describes the similarity and dissimilarity of the foci, and speaks of the theorem itself as 'une propriété de l'homographie qui n'avait peut-être pas encore été remarquée.' (2) M. Richelot, of Königsberg, in a paper dated Oct. 29, 1868, and published in the second part of the 70th volume of Crelle's Journal, has considered the analytical theory of homographic figures in space, and has been led to the consideration of their focal properties. It would seem, however, that M. Richelot supposes the tangents of the focal conics to be the only axes of equiangular pencils of planes; whereas, as we have shown, this property is possessed by every generating line of any confocal hyperboloid. The cause of the oversight (if it is one) appears in the words: 'Es muss, in der That, eine Axe im obigen Sinne [*i.e.* if we understand M. Richelot correctly, a line which is the axis of a pencil of planes equiangular with its corresponding pencil] die Eigenschaft besitzen, dass unter den unendlich vielen auf ihr senkrechten Ebenen eine existirt, deren entsprechende Ebene auf der der Axe entsprechenden Geraden senkrecht steht' (p. 141). This property, however, is not possessed by every axis of a pencil of planes equiangular with its corresponding pencil, but only by those which lie in one of the principal planes. M. Richelot speaks of a forthcoming work of a pupil of his own, M. Maegis, as intended to contain a complete analytical theory of homography in space. (3) In the November number of the Nouvelles Annales de Mathématiques, M. Housel enunciates the theorem: 'En deplaçant sans déformation deux figures homographiques dans l'espace, on peut les rendre homologiques.' This theorem is not in accordance with Art. 50 of the present paper, because in that article we have in effect shown that corresponding equi-segmental planes are never superposable except in the case of a spheroidal homography. But the analysis of M. Housel seems insufficient to establish his conclusion, since it is not shown that the values ultimately obtained of the ten unknown quantities of Art. XIII. of M. Housel's memoir actually satisfy the twelve equations of that article. [The values of the unknown quantities are not obtained in an explicit form, and there are only ten of them, and not eleven, because p depends on X, Y, Z.] And, considered in itself, the conclusion is inadmissible; for any homological transformation of space must change the imaginary circle, in which all spheres intersect, into a circle, whereas in general that circle is changed into an imaginary ellipse by a homographic transformation. Again, the homographic relation depends on fifteen constants, the homological relation on seven, and the six constants of displacement can only reduce the fifteen constants to nine. Thus it would seem *à priori* that two conditions must be satisfied in order that two homographic spaces should be capable of a homological position. And the equation $A = B$ (or $a = b$) of Art. 50 is equivalent to two independent relations connecting the fifteen constants of the homography, since that equation is equivalent to the two conditions that a certain conic should be a circle.

XXII.

ON THE FOCAL PROPERTIES OF CORRELATIVE FIGURES.

[Proceedings of the London Mathematical Society, vol. iii p. 12. Read December 9, 1869.]

THIS paper was an Appendix to a former paper 'On the Focal Properties of Homographic Figures.' By the term 'Focal Properties' are intended those properties which arise from considering the imaginary circular points at an infinite distance in either figure, and the points corresponding to them in the other figure. These properties appear to be much less varied in their character in the case of two correlative figures than in the case of two homographic figures; and the two following theorems (of which the first is well known) will suffice to give an idea of the general nature of the results.

I. In two correlative figures in space, there are always two corresponding tetrahedra, such that three adjacent edges of each are rectangular; the three edges opposite to these being at an infinite distance, and the edges at a finite distance in either figure corresponding to the edges at an infinite distance in the other.

II. If we consider any point in either figure, and its correlative plane in the other, we have two definite planes passing through the point, and two corresponding points upon the plane which may be called respectively the cyclic planes of the point, and the foci of the plane. If we take any third point in the plane, the angles which its focal radii vectores make with the line joining the foci are equal to the angles which the traces of the corresponding planes upon the cyclic planes make with the line of intersection of those two planes.

These theorems suppose only that in the two correlative figures the plane at an infinite distance in either figure answers to a point at a finite distance in the other.

END OF VOL. I.